Code of Federal Regulations

CODE OF FEDERAL REGULATIONS

T0199930

Title 7
Agriculture

Parts 900 to 999

Revised as of January 1, 2020

Containing a codification of documents
of general applicability and future effect

As of January 1, 2020

Published by the Office of the Federal Register
National Archives and Records Administration
as a Special Edition of the Federal Register

Table of Contents

Cite this Code: CFR

To cite the regulations in this volume use title, part and section number. Thus, 7 CFR 900.1 *refers to title 7, part 900, section 1.*

Explanation

The Code of Federal Regulations is a codification of the general and permanent rules published in the Federal Register by the Executive departments and agencies of the Federal Government. The Code is divided into 50 titles which represent broad areas subject to Federal regulation. Each title is divided into chapters which usually bear the name of the issuing agency. Each chapter is further subdivided into parts covering specific regulatory areas.

Each volume of the Code is revised at least once each calendar year and issued on a quarterly basis approximately as follows:

Title 1 through Title 16...as of January 1
Title 17 through Title 27 ....as of April 1
Title 28 through Title 41 ....as of July 1
Title 42 through Title 50 ....as of October 1

The appropriate revision date is printed on the cover of each volume.

LEGAL STATUS

The contents of the Federal Register are required to be judicially noticed (44 U.S.C. 1507). The Code of Federal Regulations is prima facie evidence of the text of the original documents (44 U.S.C. 1510).

HOW TO USE THE CODE OF FEDERAL REGULATIONS

The Code of Federal Regulations is kept up to date by the individual issues of the Federal Register. These two publications must be used together to determine the latest version of any given rule.

To determine whether a Code volume has been amended since its revision date (in this case, January 1, 2020), consult the "List of CFR Sections Affected (LSA)," which is issued monthly, and the "Cumulative List of Parts Affected," which appears in the Reader Aids section of the daily Federal Register. These two lists will identify the Federal Register page number of the latest amendment of any given rule.

EFFECTIVE AND EXPIRATION DATES

Each volume of the Code contains amendments published in the Federal Register since the last revision of that volume of the Code. Source citations for the regulations are referred to by volume number and page number of the Federal Register and date of publication. Publication dates and effective dates are usually not the same and care must be exercised by the user in determining the actual effective date. In instances where the effective date is beyond the cut-off date for the Code a note has been inserted to reflect the future effective date. In those instances where a regulation published in the Federal Register states a date certain for expiration, an appropriate note will be inserted following the text.

OMB CONTROL NUMBERS

The Paperwork Reduction Act of 1980 (Pub. L. 96–511) requires Federal agencies to display an OMB control number with their information collection request.

Many agencies have begun publishing numerous OMB control numbers as amendments to existing regulations in the CFR. These OMB numbers are placed as close as possible to the applicable recordkeeping or reporting requirements.

PAST PROVISIONS OF THE CODE

Provisions of the Code that are no longer in force and effect as of the revision date stated on the cover of each volume are not carried. Code users may find the text of provisions in effect on any given date in the past by using the appropriate List of CFR Sections Affected (LSA). For the convenience of the reader, a "List of CFR Sections Affected" is published at the end of each CFR volume. For changes to the Code prior to the LSA listings at the end of the volume, consult previous annual editions of the LSA. For changes to the Code prior to 2001, consult the List of CFR Sections Affected compilations, published for 1949-1963, 1964-1972, 1973-1985, and 1986-2000.

"[RESERVED]" TERMINOLOGY

The term "[Reserved]" is used as a place holder within the Code of Federal Regulations. An agency may add regulatory information at a "[Reserved]" location at any time. Occasionally "[Reserved]" is used editorially to indicate that a portion of the CFR was left vacant and not dropped in error.

INCORPORATION BY REFERENCE

What is incorporation by reference? Incorporation by reference was established by statute and allows Federal agencies to meet the requirement to publish regulations in the Federal Register by referring to materials already published elsewhere. For an incorporation to be valid, the Director of the Federal Register must approve it. The legal effect of incorporation by reference is that the material is treated as if it were published in full in the Federal Register (5 U.S.C. 552(a)). This material, like any other properly issued regulation, has the force of law.

What is a proper incorporation by reference? The Director of the Federal Register will approve an incorporation by reference only when the requirements of 1 CFR part 51 are met. Some of the elements on which approval is based are:

(a) The incorporation will substantially reduce the volume of material published in the Federal Register.

(b) The matter incorporated is in fact available to the extent necessary to afford fairness and uniformity in the administrative process.

(c) The incorporating document is drafted and submitted for publication in accordance with 1 CFR part 51.

What if the material incorporated by reference cannot be found? If you have any problem locating or obtaining a copy of material listed as an approved incorporation by reference, please contact the agency that issued the regulation containing that incorporation. If, after contacting the agency, you find the material is not available, please notify the Director of the Federal Register, National Archives and Records Administration, 8601 Adelphi Road, College Park, MD 20740-6001, or call 202-741-6010.

CFR INDEXES AND TABULAR GUIDES

A subject index to the Code of Federal Regulations is contained in a separate volume, revised annually as of January 1, entitled CFR INDEX AND FINDING AIDS. This volume contains the Parallel Table of Authorities and Rules. A list of CFR titles, chapters, subchapters, and parts and an alphabetical list of agencies publishing in the CFR are also included in this volume.

An index to the text of "Title 3—The President" is carried within that volume.

The Federal Register Index is issued monthly in cumulative form. This index is based on a consolidation of the "Contents" entries in the daily Federal Register.

A List of CFR Sections Affected (LSA) is published monthly, keyed to the revision dates of the 50 CFR titles.

REPUBLICATION OF MATERIAL

There are no restrictions on the republication of material appearing in the Code of Federal Regulations.

INQUIRIES

For a legal interpretation or explanation of any regulation in this volume, contact the issuing agency. The issuing agency's name appears at the top of odd-numbered pages.

For inquiries concerning CFR reference assistance, call 202–741–6000 or write to the Director, Office of the Federal Register, National Archives and Records Administration, 8601 Adelphi Road, College Park, MD 20740-6001 or e-mail *fedreg.info@nara.gov.*

THIS TITLE

Title 7—AGRICULTURE is composed of fifteen volumes. The parts in these volumes are arranged in the following order: Parts 1–26, 27–52, 53–209, 210–299, 300–399, 400–699, 700–899, 900–999, 1000–1199, 1200–1599, 1600–1759, 1760–1939, 1940–1949, 1950–1999, and part 2000 to end. The contents of these volumes represent all current regulations codified under this title of the CFR as of January 1, 2020.

The Food and Nutrition Service current regulations in the volume containing parts 210–299, include the Child Nutrition Programs and the Food Stamp Program. The regulations of the Federal Crop Insurance Corporation are found in the volume containing parts 400–699.

All marketing agreements and orders for fruits, vegetables and nuts appear in the one volume containing parts 900–999. All marketing agreements and orders for milk appear in the volume containing parts 1000–1199.

For this volume, Robert J. Sheehan, III was Chief Editor. The Code of Federal Regulations publication program is under the direction of John Hyrum Martinez, assisted by Stephen J. Frattini.

Title 7—Agriculture

(This book contains parts 900 to 999)

SUBTITLE B—REGULATIONS OF THE DEPARTMENT OF AGRICULTURE (CONTINUED)

Part

Subtitle B—Regulations of the Department of Agriculture (Continued)

CHAPTER IX—AGRICULTURAL MARKETING SERVICE (MARKETING AGREEMENTS AND ORDERS; FRUITS, VEGETABLES, NUTS), DEPARTMENT OF AGRICULTURE

PART 900—GENERAL REGULATIONS

Subpart A—Procedural Requirements Governing Proceedings to Formulate Marketing Agreements and Marketing Orders

Subpart B—Supplemental Procedural Requirements Governing Proceedings to Amend Federal Milk Marketing Agreements and Marketing Orders

Subpart C—Supplemental Procedural Requirements Governing Proceedings to Amend Fruit, Vegetable and Nut Marketing Agreements and Marketing Orders

Subpart D—Procedural Requirements Governing Proceedings on Petitions To Modify or To Be Exempted From Marketing Orders

Subpart E—Supplemental Procedural Requirements for Marketing Orders, Marketing Agreements, and Requirements Covering Fruits, Vegetables, and Nuts

AUTHORITY: 7 U.S.C. 601–674; 7 U.S.C. 7401; 5 U.S.C. 301, 552; and 44 U.S.C. Ch. 35.

SOURCE: 25 FR 5907, June 28, 1960, unless otherwise noted.

Subpart A—Procedural Requirements Governing Proceedings to Formulate Marketing Agreements and Marketing Orders

AUTHORITY: 7 U.S.C. 610.

§ 900.1 Words in the singular form.

Words in this subpart in the singular form shall be deemed to import the plural, and vice versa, as the case may demand.

§900.2 Definitions.

As used in this subpart, the terms as defined in the act shall apply with equal force and effect. In addition, unless the context otherwise requires:

(a) The term *Act* means Public Act No. 10, 73 Congress (48 Stat. 31), as amended and as reenacted and amended by the Agricultural Marketing Agreement Act of 1937 (50 Stat. 246), as amended.

(b) The term *Department* means the United States Department of Agriculture.

(c) The term *Secretary* means the Secretary of Agriculture of the United States, or any officer or employee of the Department to whom authority has heretofore been delegated, or to whom authority may hereafter be delegated, to act for the Secretary.

(d) The term *judge* means any administrative law judge appointed pursuant to 5 U.S.C. 3105 or any presiding official appointed by the Secretary, and assigned to conduct the proceeding.

(e) The term *Administrator* means the Administrator of the Agricultural Marketing Service or any officer or employee of the Department to whom authority has been delegated or may hereafter be delegated to act for the Administrator.

(f) [Reserved]

(g) The term FEDERAL REGISTER means the publication provided for by the act of July 26, 1935 (49 Stat. 500), and acts supplementary thereto and amendatory thereof.

(h) The term *hearing* means that part of the proceeding which involves the submission of evidence.

(i) The term *marketing agreement* means any marketing agreement or any amendment thereto which may be entered into pursuant to section 8b of the act.

(j) The term *marketing order* means any order or any amendment thereto which may be issued pursuant to section 8c of the act, and after notice and hearing as required by said section.

(k) The term *proceeding* means a proceeding upon the basis of which a marketing agreement may be entered into or a marketing order may be issued.

(l) The term *hearing clerk* means the hearing clerk, United States Department of Agriculture, Washington, DC.

[25 FR 5907, June 28, 1960, as amended at 26 FR 7796, Aug. 22, 1961; 28 FR 579, Jan. 23, 1963; 37 FR 8059, Apr. 25, 1972; 38 FR 29798, Oct. 29, 1973; 67 FR 10829, Mar. 11, 2002; 82 FR 58098, Dec. 11, 2017]

§900.3 Proposals.

(a) A marketing agreement or a marketing order may be proposed by the Secretary or by any other person. If any person other than the Secretary proposes a marketing agreement or marketing order, he shall file with the Administrator a written application, together with at least four copies of the proposal, requesting the Secretary to hold a hearing upon the proposal. Upon receipt of such proposal, the Administrator shall cause such investigation to be made and such consideration thereof to be given as, in his opinion, are warranted. If the investigation and consideration lead the Administrator to conclude that the proposed marketing agreement or marketing order will not tend to effectuate the declared policy of the act, or that for other proper reasons a hearing should not be held on the proposal, he shall deny the application, and promptly notify the applicant of such denial, which notice shall be accompanied by a brief statement of the grounds for the denial.

(b) If the investigation and consideration lead the Administrator to conclude that the proposed marketing agreement or marketing order will tend to effectuate the declared policy of the act, or if the Secretary desires to propose a marketing agreement or marketing order, he shall sign and cause to be served a notice of hearing, as provided in this subpart.

§900.4 Institution of proceeding.

(a) *Filing and contents of the notice of hearing.* The proceeding shall be instituted by filing the notice of hearing with the hearing clerk. The notice of hearing shall contain a reference to the authority under which the marketing agreement or marketing order is proposed; shall define the scope of the hearing as specifically as may be practicable; shall contain either the terms or substance of the proposed marketing

9

agreement or marketing order or a description of the subjects and issues involved and shall state the industry, area, and class of persons to be regulated, the time and place of such hearing, and the place where copies of such proposed marketing agreement or marketing order may be obtained or examined. The time of the hearing shall not be less than 15 days after the date of publication of the notice in the FEDERAL REGISTER, as provided in this subpart, unless the Administrator shall determine that an emergency exists which requires a shorter period of notice, in which case the period of notice shall be that which the Administrator may determine to be reasonable in the circumstances: *Provided,* That, in the case of hearings on amendments to marketing agreements or marketing orders, the time of the hearing may be less than 15 days but shall not be less than 3 days after the date of publication of the notice in the FEDERAL REGISTER.

(b) *Giving notice of hearing and supplemental publicity.* (1) The Administrator shall give or cause to be given notice of hearing in the following manner:

(i) By publication of the notice of hearing in the FEDERAL REGISTER;

(ii) By mailing a true copy of the notice of hearing, using a postal or other delivery service or electronic communication, to each of the persons known to the Administrator to be interested therein;

(iii) By issuing a press release containing the complete text or a summary of the contents of the notice of hearing and making the same available to such newspapers in the area proposed to be subjected to regulation as reasonably will tend to bring the ntoice to the attention of the persons interested therein;

(iv) By forwarding copies of the notice of hearing addressed to the governors of such of the several States of the United States and to executive heads of such of the Territories and possessions of the United States as the Administrator, having due regard for the subject matter of the proposal and the public interest, shall determine, should be notified.

(2) Legal notice of the hearing shall be deemed to be given if notice is given

in the manner provided by paragraph (b)(1)(i) of this section; and failure to give notice in the manner provided in paragraph (b)(1)(ii), (iii), and (iv) of this section shall not affect the legality of the notice.

(c) *Record of notice and supplemental publicity.* There shall be filed with the hearing clerk or submitted to the judge at the hearing an affidavit or certificate of the person giving the notice provided in paragraph (b)(1) (iii) and (iv) of this section. In regard to the provisions relating to mailing in paragraph (b)(1)(ii) of this section, a determination by the Administrator that such provisions have been complied with shall be filed with the hearing clerk or submitted to the judge at the hearing. In the alternative, if notice is not given in the manner provided in paragraph (b)(1)(ii), (iii), and (iv) of this section there shall be filed with the hearing clerk or submitted to the judge at the hearing a determination by the Administrator that such notice is impracticable, unnecessary, or contrary to the public interest with a brief statement of the reasons for such determination. Determinations by the Administrator as herein provided shall be final.

[25 FR 5907, June 28, 1960, as amended at 83 FR 52944, Oct. 19, 2018]

§ 900.5 Docket number.

Each proceeding, immediately following its institution, shall be assigned a docket number by the hearing clerk and thereafter the proceeding may be referred to by such number.

§ 900.6 Judges.

(a) *Assignment.* No judge who has any pecuniary interest in the outcome of a proceeding shall serve as judge in such proceeding.

(b) *Powers of judges.* Subject to review by the Secretary, as provided elsewhere in this subpart, the judge, in any proceeding, shall have power to:

(1) Rule upon motions and requests;

(2) Change the time and place of hearing, and adjourn the hearing from time to time or from place to place;

(3) Administer oaths and affirmations and take affidavits;

(4) Examine and cross-examine witnesses and receive evidence;

(5) Admit or exclude evidence;

(6) Hear oral argument on facts or law;

(7) Do all acts and take all measures necessary for the maintenance of order at the hearing and the efficient conduct of the proceeding.

(c) *Who may act in absence of judge.* In case of the absence of the judge or his inability to act, the powers and duties to be performed by him under this part in connection with a proceeding may, without abatement of the proceeding unless otherwise ordered by the Secretary, be assigned to any other judge.

(d) *Disqualification of judge.* The judge may at any time withdraw as judge in a proceeding if he deems himself to be disqualified. Upon the filing by an interested person in good faith of a timely and sufficient affidavit of personal bias or disqualification of a judge, the Secretary shall determine the matter as a part of the record and decision in the proceeding, after making such investigation or holding such hearings, or both, as he may deem appropriate in the circumstances.

§900.7   Motions and requests.

(a) *General.* All motions and requests shall be filed with the hearing clerk, except that those made during the course of the hearing may be filed with the judge or may be stated orally and made a part of the transcript. Except as provided in §900.15(b) such motions and requests shall be addressed to, and ruled on by, the presiding officer if made prior to his certification of the transcript pursuant to §900.10 or by the Secretary if made thereafter.

(b) *Certification to Secretary.* The judge may in his discretion submit or certify to the Secretary for decision any motion, request, objection, or other question addressed to the judge.

§900.8   Conduct of the hearing.

(a) *Time and place.* The hearing shall be held at the time and place fixed in the notice of hearing, unless the judge shall have changed the time or place, in which event the judge shall file with the hearing clerk a notice of such change, which notice shall be given in the same manner as provided in §900.4 (relating to the giving of notice of the hearing): *Provided,* That, if the change

in time or place of hearing is made less than 5 days prior to the date previously fixed for the hearing, the judge, either in addition to or in lieu of causing the notice of the change to be given, shall announce, or cause to be announced, the change at the time and place previously fixed for the hearing.

(b) *Appearances*—(1) *Right to appear.* At the hearing, any interested person shall be given an opportunity to appear, either in person or through his authorized counsel or representative, and to be heard with respect to matters relevant and material to the proceeding. Any interested person who desires to be heard in person at any hearing under these rules shall, before proceeding to testify, state his name, address, and occupation. If any such person is appearing through a counsel or representative, such person or such counsel or representative shall, before proceeding to testify or otherwise to participate in the hearing, state for the record the authority to act as such counsel or representative, and the names and addresses and occupations of such person and such counsel or representative. Any such person or such counsel or representative shall give such other information respecting his appearance as the judge may request.

(2) *Debarment of counsel or representative.* Wherever, while a proceeding is pending before him, the judge finds that a person, acting as counsel or representative for any person participating in the proceeding, is guilty of unethical or unprofessional conduct, the judge may order that such person be precluded from further acting as counsel or representative in such proceeding. An appeal to the Secretary may be taken from any such order, but the proceeding shall not be delayed or suspended pending disposition of the appeal: *Provided,* That the judge may suspend the proceeding for a reasonable time for the purpose of enabling the client to obtain other counsel or other representative. In case the judge has ordered that a person be precluded from further acting as counsel or representative in the proceeding, the presiding officer, within a reasonable time thereafter shall submit to the Secretary a report of the facts and circumstances surrounding such order and

shall recommend what action the Secretary should take respecting the appearance of such person as counsel or representative in other proceedings before the Secretary. Thereafter the Secretary may, after notice and an opportunity for hearing, issue such order, respecting the appearance of such person as counsel or representative in proceedings before the Secretary, as the Secretary finds to be appropriate.

(3) *Failure to appear.* If any interested person fails to appear at the hearing, he shall be deemed to have waived the right to be heard in the proceeding.

(c) *Order of procedure.* (1) The judge shall, at the opening of the hearing prior to the taking of testimony, have noted as part of the record the notice of hearing as filed with the Office of the Federal Register and the affidavit or certificate of the giving of notice or the determination provided for in § 900.4(c).

(2) Evidence shall then be received with respect to the matters specified in the notice of the hearing in such order as the judge shall announce.

(d) *Evidence*—(1) *In general.* The hearing shall be publicly conducted, and the testimony given at the hearing shall be reported verbatim.

(i) Every witness shall, before proceeding to testify, be sworn or make affirmation. Cross-examination shall be permitted to the extent required for a full and true disclosure of the facts.

(ii) When necessary, in order to prevent undue prolongation of the hearing, the judge may limit the number of times any witness may testify to the same matter or the amount of corroborative or cumulative evidence.

(iii) The judge shall, insofar as practicable, exclude evidence which is immaterial, irrelevant, or unduly repetitious, or which is not of the sort upon which responsible persons are accustomed to rely.

(2) *Objections.* If a party objects to the admission or rejection of any evidence or to any other ruling of the judge during the hearing, he shall state briefly the grounds of such objection, whereupon an automatic exception will follow if the objection is overruled by the judge. The transcript shall not include argument or debate thereon except as ordered by the judge. The rul-

ing of the judge on any objection shall be a part of the transcript. Only objections made before the judge may subsequently be relied upon in the proceeding.

(3) *Proof and authentication of official records or documents.* An official record or document, when admissible for any purpose, shall be admissible as evidence without the production of the person who made or prepared the same. Such record or document shall, in the discretion of the judge, be evidenced by an official publication thereof or by a copy attested by the person having legal custody thereof and accompanied by a certificate that such person has the custody.

(4) *Exhibits.* All written statements, charts, tabulations, or similar data offered in evidence at the hearing shall, after identification by the proponent and upon satisfactory showing of the authenticity, relevancy, and materiality of the contents thereof, be numbered as exhibits and received in evidence and made a part of the record. Such exhibits shall be submitted in quadruplicate and in documentary form. In case the required number of copies is not made available, the judge shall exercise his discretion as to whether said exhibits shall, when practicable, be read in evidence or whether additional copies shall be required to be submitted within a time to be specified by the judge. If the testimony of a witness refers to a statute, or to a report or document (including the record of any previous hearing) the judge, after inquiry relating to the identification of such statute, report, or document, shall determine whether the same shall be produced at the hearing and physically be made a part of the evidence as an exhibit, or whether it shall be incorporated into the evidence by reference. If relevant and material matter offered in evidence is embraced in a report or document (including the record of any previous hearing) containing immaterial or irrelevant matter, such immaterial or irrelevant matter shall be excluded and shall be segregated insofar as practicable, subject to the direction of the presiding officer.

12

(5) *Official notice.* Official notice may be taken of such matters as are judicially noticed by the courts of the United States and of any other matter of technical, scientific or commercial fact of established character: *Provided,* That interested persons shall be given adequate notice, at the hearing or subsequent thereto, of matters so noticed and shall be given adequate opportunity to show that such facts are inaccurate or are erroneously noticed.

(6) *Offer of proof.* Whenever evidence is excluded from the record, the party offering such evidence may make an offer of proof, which shall be included in the transcript. The offer of proof shall consist of a brief statement describing the evidence to be offered. If the evidence consists of a brief oral statement or of an exhibit, it shall be inserted into the transcript in toto. In such event, it shall be considered a part of the transcript if the Secretary decides that the judge's ruling in excluding the evidence was erroneous. The judge shall not allow the insertion of such evidence in toto if the taking of such evidence will consume a considerable length of time at the hearing. In the latter event, if the Secretary decides that the judge erred in excluding the evidence, and that such error was substantial, the hearing shall be reopened to permit the taking of such evidence.

[25 FR 5907, June 28, 1960, as amended at 37 FR 1103, Jan. 25, 1972]

§900.9 Oral and written arguments.

(a) *Oral argument before judge.* Oral argument before the judge shall be in the discretion of the judge. Such argument, when permitted, may be limited by the judge to any extent that he finds necessary for the expeditious disposition of the proceeding and shall be reduced to writing and made part of the transcript.

(b) *Briefs, proposed findings and conclusions.* The judge shall announce at the hearing a reasonable period of time within which interested persons may file with the hearing clerk proposed findings and conclusions, and written arguments or briefs, based upon the evidence received at the hearing, citing, where practicable, the page or pages of the transcript of the testi-

mony where such evidence appears. Factual material other than that adduced at the hearing or subject to official notice shall not be alluded to therein, and, in any case, shall not be considered in the formulation of the marketing agreement or marketing order. If the person filing a brief desires the Secretary to consider any objection made by such person to a ruling of the judge, as provided in §900.8(d), he shall include in the brief a concise statement concerning each such objection, referring where practicable, to the pertinent pages of the transcript.

§900.10 Certification of the transcript.

The judge shall notify the hearing clerk of the close of a hearing as soon as possible thereafter and of the time for filing written arguments, briefs, proposed findings and proposed conclusions, and shall furnish the hearing clerk with such other information as may be necessary. As soon as possible after the hearing, the judge shall transmit to the hearing clerk an original and three copies of the transcript of the testimony and the original and all copies of the exhibits not already on file in the office of the hearing clerk. He shall attach to the original transcript of testimony his certificate stating that, to the best of his knowledge and belief, the transcript is a true transcript of the testimony given at the hearing except in such particulars as he shall specify; and that the exhibits transmitted are all the exhibits as introduced at the hearing with such exceptions as he shall specify. A copy of such certificate shall be attached to each of the copies of the transcript of testimony. In accordance with such certificate the hearing clerk shall note upon the official record copy, and cause to be noted on other copies, of the transcript each correction detailed therein by adding or crossing out (but without obscuring the text as originally transcribed) at the appropriate place any words necessary to make the same conform to the correct meaning, as certified by the judge. The hearing clerk shall obtain and file certifications to the effect that such corrections have been effected in copies other than the official record copy.

§ 900.11 Copies of the transcript.

(a) During the period in which the proceeding has an active status in the Department, a copy of the transcript and exhibits shall be kept on file in the office of the hearing clerk, where it shall be available for examination during official hours of business. Thereafter said transcript and exhibits shall be made available by the hearing clerk for examination during official hours of business after prior request and reasonable notice to the hearing clerk.

(b) Transcripts of hearings shall be made available to any person at actual cost of duplication.

[25 FR 5907, June 28, 1960, as amended at 67 FR 10829, Mar. 11, 2002]

§ 900.12 Administrator's recommended decision.

(a) *Preparation.* As soon as practicable following the termination of the period allowed for the filing of written arguments or briefs and proposed findings and conclusions the Administrator shall file with the hearing clerk a recommended decision.

(b) *Contents.* The Administrator's recommended decision shall include: (1) A preliminary statement containing a description of the history of the proceedings, a brief explanation of the material issues of fact, law, or discretion presented on the record, and proposed findings and conclusions with respect to such issues as well as the reasons or basis therefor; (2) a ruling upon each proposed finding or conclusion submitted by interested persons, and (3) an appropriate proposed marketing agreement or marketing order effectuating his recommendations.

(c) *Exceptions to recommended decision.* Immediately following the filing of his recommended decision, the Administrator shall give notice thereof, and opportunity to file exceptions thereto by publication in the FEDERAL REGISTER. Within a period of time specified in such notice any interested person may file with the hearing clerk exceptions to the Administrator's proposed marketing agreement or marketing order, or both, as the case may be, and a brief in support of such exceptions. Such exceptions shall be in writing, shall refer, where practicable, to the related pages of the transcript and may suggest appropriate changes in the proposed marketing agreement or marketing order.

(d) *Omission of recommended decision.* The procedure provided in this section may be omitted only if the Secretary finds on the basis of the record that due and timely execution of his functions imperatively and unavoidably requires such omission.

§ 900.13 Submission to Secretary.

Upon the expiration of the period allowed for filing exceptions or upon request of the Secretary, the hearing clerk shall transmit to the Secretary the record of the proceeding. Such record shall include: All motions and requests filed with the hearing clerk and rulings thereon; the certified transcript; any proposed findings or conclusions or written arguments or briefs that may have been filed; the Administrator's recommended decision, if any, and such exceptions as may have been filed.

§ 900.13a Decision by Secretary.

After due consideration of the record, the Secretary shall render a decision. Such decision shall become a part of the record and shall include: (a) A statement of his findings and conclusions, as well as the reasons or basis therefor, upon all the material issues of fact, law or discretion presented on the record, (b) a ruling upon each proposed finding and proposed conclusion not previously ruled upon in the record, (c) a ruling upon each exception filed by interested persons and (d) either (1) a denial of the proposal to issue a marketing agreement or marketing order or (2) a marketing agreement and, if the findings upon the record so warrant, a marketing order, the provisions of which shall be set forth directly or by reference, regulating the handling of the commodity or product in the same manner and to the same extent as such marketing agreement, which order shall be complete except for its effective date and any determinations to be made under § 900.14(b) or § 900.14(c): *Provided,* That such marketing order shall not be executed, issued, or made effective until and unless the Secretary determines that the requirements of §900.14(b) or §900.14(c) have been met.

§900.14 Execution and issuance of marketing agreements and marketing orders.

(a) *Execution and issuance of marketing agreement.* If the Secretary has approved a marketing agreement, as provided in §900.13a, the Administrator shall cause copies thereof to be distributed for execution by the handlers eligible to become parties thereto. If and when such number of the handlers as the Secretary shall deem sufficient shall have executed the agreement, the Secretary shall execute the agreement. After execution of a marketing agreement, such agreement shall be filed with the hearing clerk, and notice thereof, together with notice of the effective date, shall be given by publication in the FEDERAL REGISTER. The marketing agreement shall not become effective less than 30 days after its publication in the FEDERAL REGISTER, unless the Secretary, upon good cause found and published with the agreement, fixes an earlier effective date therefor: *Provided,* That no marketing agreement shall become effective as to any person signatory thereto before either (1) it has been filed with the Office of the Federal Register, or (2) such person has received actual notice that the Secretary has executed the agreement and the effective date of the marketing agreement.

(b) *Issuance of marketing order with marketing agreement.* Whenever, as provided in paragraph (a) of this section, the Secretary executes a marketing agreement, and handlers also have executed the same as provided in section 8c(8) of the Act, he shall, if he finds that it will tend to effectuate the purposes of the Act, issue and make effective the marketing order, if any, which was filed as a part of his decision pursuant to §900.13a: *Provided,* That the issuance of such order shall have been approved or favored by producers as required by section 8c(8) of the act.

(c) *Issuance of marketing order without marketing agreement.* If, despite the failure or refusal of handlers to sign the marketing agreement, as provided in section 8c(8) of the Act, the Secretary makes the determinations required

under section 8c(9) of the Act, the Secretary shall issue and make effective the marketing order, if any, which was filed as a part of his decision pursuant to §900.13a.

(d) *Effective date of marketing order.* No marketing order shall become effective less than 30 days after its publication in the FEDERAL REGISTER, unless the Secretary, upon good cause found and published with the order, fixes an earlier effective date therefor: *Provided,* That no marketing order shall become effective as to any person sought to be charged thereunder before either (1) it has been filed with the Office of the Federal Register, or (2) such person has received actual notice of the issuance and terms of the marketing order.

(e) *Notice of issuance.* After issuance of a marketing order, such order shall be filed with the hearing clerk, and notice therof, together with notice of the effective date, shall be given by publication in the FEDERAL REGISTER. (7 U.S.C. 610(c).)

[25 FR 5907, June 28, 1960, as amended at 53 FR 15659, May 3, 1988]

§900.15 Filing; extensions of time; effective date of filing; and computation of time.

(a) *Filing, number of copies.* Except as is provided otherwise in this subpart, all documents or papers required or authorized by the foregoing provisions of this subpart to be filed with the hearing clerk shall be filed in quadruplicate. Any document or paper, so required or authorized to be filed with the hearing clerk, shall, during the course of an oral hearing, be filed with the presiding officer. The provisions of this subpart concerning filing with the hearing clerk of hearing notices, recommended and final decisions, marketing agreements and orders, and all documents described in §900.17 shall be met by filing a true copy thereof with the hearing clerk.

(b) *Extensions of time.* The time for the filing of any document or paper required or authorized by the foregoing provisions of this subpart to be filed may be extended by the judge before the record is certified by the judge or by the Administrator (after the record is so certified by the judge but before it is transmitted to the Secretary), or by the Secretary (after the record is transmitted to the Secretary) upon request filed, and if, in the judgment of the judge, Administrator, or the Secretary, as the case may be, there is good reason for the extension. All rulings made pursuant to this paragraph shall be filed with the hearing clerk.

(c) *Effective date of filing.* Any document or paper required or authorized in this subpart to be filed shall be deemed to be filed at the time it is received by the Hearing Clerk.

(d) *Computation of time.* Each day, including Saturdays, Sundays, and legal public holidays, shall be included in computing the time allowed for filing any document or paper: *Provided,* That when the time for filing a document or paper expires on a Saturday, Sunday, or legal public holiday, the time allowed for filing the document or paper shall be extended to include the following business day.

[25 FR 5907, June 28, 1960, as amended at 30 FR 254, Jan. 9, 1965; 67 FR 10829, Mar. 11, 2002]

§ 900.16   Ex parte communications.

(a) At no stage of the proceeding following the issuance of a notice of hearing and prior to the issuance of the Secretary's decision therein shall an employee of the Department who is or may reasonably be expected to be involved in the decisional process of the proceeding discuss ex parte the merits of the proceeding with any person having an interest in the proceeding or with any representative of such person: *Provided,* That procedural matters and status reports shall not be included within this limitation; and *Provided further,* That an employee of the Department who is or may reasonably be expected to be involved in the decisional process of the proceeding may discuss the merits of the proceeding with such a person if all parties known to be interested in the proceeding have been given notice and an opportunity to participate. A memorandum of any such discussion shall be included in the record of the proceeding.

(b) No person interested in the proceeding shall make or knowingly cause to be made to an employee of the Department who is or may reasonably be expected to be involved in the decisional process of the proceeding an ex parte communication relevant to the merits of the proceeding except as provided in paragraph (a) of this section.

(c) If an employee of the Department who is or may reasonably be expected to be involved in the decisional process of the proceeding receives or makes a communication prohibited by this section, the Department shall place on the public record of the proceeding:

(1) All such written communications;

(2) Memoranda stating the substance of all such oral communications; and

(3) All written responses, and memoranda stating the substance of all oral responses thereto.

(d) Upon receipt of a communication knowingly made or knowingly caused to be made by a party in violation of this section, the Department may, to the extent consistent with the interest of justice and the policy of the underlying statute, take whatever steps are deemed necessary to nullify the effect of such communication.

(e) For the purposes of this section, *ex parte communication* means an oral or written communication not on the public record with respect to which reasonable prior notice to all interested parties is not given, but which shall not include requests for status reports (including requests on procedural matters) on any proceeding.

[42 FR 10833, Feb. 24, 1977]

§ 900.17   Additional documents to be filed with hearing clerk.

In addition to the documents or papers required or authorized by the foregoing provisions of this subpart to be filed with the hearing clerk, the hearing clerk shall receive for filing and shall have custody of all papers, reports, records, orders, and other documents which relate to the administration of any marketing agreement or marketing order and which the Secretary is required to issue or to approve.

§ 900.18   Hearing before Secretary.

The Secretary may act in the place and stead of a judge in any proceeding under this subpart. When he so acts the hearing clerk shall transmit the record to the Secretary at the expiration of the period provided for the filing of proposed findings of fact, conclusions and orders, and the Secretary shall thereupon, after due consideration of the record, issue his final decision in the proceeding: *Provided,* That he may issue a tentative decision in which event the parties shall be afforded an opportunity to file exceptions before the issuance of the final decision.

Subpart B—Supplemental Procedural Requirements Governing Proceedings to Amend Federal Milk Marketing Agreements and Marketing Orders

AUTHORITY: 7 U.S.C. 608c(17) and 610.

SOURCE: 73 FR 49088, Aug. 20, 2008, unless otherwise noted.

§ 900.20 Words in the singular form.

Words in this subpart in the singular form shall be deemed to import the plural, and vice versa, as the case may demand.

§ 900.21 Definitions.

As used in this subpart, the terms as defined in the Act and in § 900.2 of this part shall apply.

§ 900.22 Proposal submission requirements.

When a person other than the Secretary makes a proposal to amend a Federal milk marketing agreement or order under § 900.3 of this part, the proposal shall address the following, to the extent applicable:

(a) Explain the proposal. What is the disorderly marketing condition that the proposal is intended to address?

(b) What is the purpose of the proposal?

(c) Describe the current Federal order requirements or industry practices relative to the proposal.

(d) Describe the expected impact on the industry, including on producers and handlers, and on consumers. Explain/Quantify.

(e) What are the expected effects on small businesses as defined by the Regulatory Flexibility Act (5 U.S.C. 601–612)? Explain/Quantify.

(f) How would adoption of the proposal increase or decrease costs to producers, handlers, others in the marketing chain, consumers, the Market Administrator offices and/or the Secretary? Explain/Quantify.

(g) Would a pre-hearing information session be helpful to explain the proposal?

§ 900.23 Procedures following receipt of a proposal.

Within 30 days of receipt of a proposal to amend a Federal milk marketing agreement order under § 900.3 of this part, USDA shall either: Issue a notice providing an action plan and expected timeframes for the different steps in the formal rulemaking process for completion of the hearing not more than 120 days after the date of the issuance of the notice; request additional information from the person submitting the proposal to be used in deciding whether a hearing will be held. If the information requested is not received within a specified timeframe, the request shall be denied; or deny the request. Notice of the action plan will be made on the Dairy Programs, AMS Web site and through program releases to interested persons.

§ 900.24 Pre-hearing information sessions.

A pre-hearing information session may be held by the Secretary in response to any proposals received under § 900.3 of this part. Any person proposing an amendment to a Federal milk marketing agreement or order may request that a pre-hearing information session be held. A person submitting a proposal shall have up to 3 calendar days to modify or withdraw his or her proposal prior to the publication of a notice of hearing.

§ 900.25 Advance submission of testimony.

Any person proposing an amendment to a Federal milk marketing agreement or order under § 900.3 of this part, when participating as a witness, shall make copies of his or her testimony, if prepared as an exhibit, and any other exhibits available to USDA officials before the start of the hearing on the person's day of appearance. Individual dairy farmers shall not be subject to this requirement.

§ 900.26 Requesting USDA data for use at an amendatory hearing.

Requests for preparation of USDA data to be used at a Federal milk marketing agreement or order amendatory proceeding must be received by USDA at least 10 days before the beginning of the hearing. If an amendatory hearing is announced with less than 10 days before the start of the hearing, requests for data must be submitted within 2 days following publication of the notice of hearing in the FEDERAL REGISTER.

§ 900.27 Deadline for filing post-hearing briefs and corrections to transcript.

(a) Under § 900.10 of this part, the period of time for interested persons to file corrections to the transcript of testimony at a Federal milk marketing agreement or order amendatory proceeding shall be no more than 30 days after the hearing record is available.

(b) Under § 900.9(b) of this part, the period of time after the completion of a Federal milk marketing agreement or order amendatory hearing for interested persons to file proposed findings and conclusions, and written arguments or briefs, shall be no more than 60 days after completion of the amendatory hearing.

§ 900.28 Deadline for issuance of recommended decisions or tentative final decisions.

In a Federal milk marketing agreement or order amendatory proceeding, USDA shall issue a recommended decision under § 900.12 or, when applicable, a tentative final decision, not later than 90 days after the deadline for submission of proposed findings and conclusions, and written arguments or briefs.

§ 900.29 Deadline for filing exceptions to recommended decisions.

In a Federal milk marketing agreement or order amendatory proceeding, exceptions to a recommended decision under § 900.12 shall be filed with the hearing clerk not later than 60 days after publication of the recommended

decision in the FEDERAL REGISTER, unless otherwise specified in that decision.

§ 900.30 Deadline for issuance of Secretary's (final) decisions.

A Secretary's (final) decision under § 900.13a to a proposed amendment on marketing agreement or order shall be issued not later than 60 days after the deadline for submission of exceptions to the recommended decision.

§ 900.31 Electronic submission of hearing documents.

To the extent practicable, all documents filed with the hearing clerk in a proceeding to amend a Federal milk marketing agreement or order shall also be submitted electronically to the Dairy Programs, Agricultural Marketing Service, USDA. All documents should reference the docket number of the proceeding. Instructions for electronic filing will be provided in the notice of action plan referred to in § 900.23 of this subpart, at the amendatory hearing, and in each FEDERAL REGISTER publication regarding the amendatory proceeding.

§ 900.32 Informal rulemaking.

USDA may elect to use informal rulemaking procedures under 553 of Title 5, United States Code, to amend Federal milk marketing agreements and orders, other than provisions that directly affect milk prices. In making this determination, consideration shall be given to:

(a) The nature and complexity of the proposal;

(b) The potential regulatory and economic impacts on affected entities; and

(c) Any other relevant matters.

§ 900.33 Industry assessments.

If the Secretary determines it is necessary to improve or expedite an amendatory formal rulemaking proceeding to amend a Federal milk marketing agreement or order, USDA may impose an assessment on pooled milk to supplement appropriated funds for the procurement of such services, including but not limited to, court reporters, hearing examiners, legal counsel, hearing venue and associated travel for USDA officials. Only the milk pooled in the particular marketing area that stands to be affected by proposals heard at the amendatory proceeding may be assessed. The assessments shall be subject to the provisions of §1000.85 (7 CFR 1000.85) concerning assessments for order administration, including the provision that assessments shall not exceed $.005 per hundredweight of milk for any given month.

Subpart C—Supplemental Procedural Requirements Governing Proceedings to Amend Fruit, Vegetable and Nut Marketing Agreements and Marketing Orders

AUTHORITY: 7 U.S.C. 608c(17) and 610.

SOURCE: 73 FR 49310, Aug. 21, 2008, unless otherwise noted.

§900.36 Words in the singular form.

Words in this subpart in the singular form shall be deemed to import the plural, and vice versa, as the case may demand.

§900.37 Definitions.

As used in this subpart, the terms as defined in the Act and in §900.2 of this part shall apply.

§900.38 Pre-hearing information sessions.

A pre-hearing information session concerning a proposal to amend a fruit, vegetable or nut marketing agreement or order may be held either prior or subsequent to submission of a proposal under §900.3 of this part. Such sessions may be held by a marketing agreement or order committee or board or by the Secretary.

§900.39 Proposal submission requirements.

When a person other than the Secretary makes a proposal to amend a fruit, vegetable or nut marketing agreement or order under §900.3 of this part, the proposal shall address the following, to the extent applicable:

(a) The purpose of the proposal;

(b) The problem the proposal is designed to address with explanation and quantification;

(c) The current requirements or industry practices relative to the proposal;

(d) The expected impact on the industry, including producers, handlers, and on consumers;

(e) In the case of marketing orders, an explanation, including supporting information and data, of how the proposal would tend to improve returns to producers, and in the case of marketing agreements, how the proposal impacts the signatories to the agreement;

(f) The expected effects on small businesses as defined by the Regulatory Flexibility Act (5 U.S.C. 601–612);

(g) A description and quantification of whether the proposal would increase or decrease costs to producers, handlers, or others in the marketing chain, and to consumers, marketing order committees and boards and/or the Secretary;

(h) A description of how the proposal would be implemented; and

(i) A description, including quantification, of how compliance with the proposal would be effected.

§900.40 Written testimony and USDA data request requirements.

In addition to the provisions of §900.8(b)(4), witnesses at an amendatory fruit, vegetable or nut formal rulemaking hearing shall make, to the extent practicable, at least 8 copies of their testimony, if prepared as an exhibit, and any other exhibits available to USDA before testimony is given on the day of appearance at the hearing. Industry requests for preparation of USDA data for a rulemaking hearing should be made at least 10 days prior to the beginning of the hearing.

§900.41 Electronic document submission standards.

To the extent practicable, all documents filed with the hearing clerk in a proceeding to amend a fruit, vegetable or nut marketing agreement or order shall also be submitted electronically to the Agricultural Marketing Service, Fruit and Vegetable Programs, USDA. All documents should reference the docket number of the proceeding. Instructions for electronic filing shall be provided at the amendatory formal rulemaking hearing and in each FED-ERAL REGISTER publication regarding the amendatory proceeding.

§ 900.42   Industry assessments.

If the Secretary determines it is necessary to improve or expedite an amendatory fruit, vegetable or nut formal rulemaking proceeding, costs associated with improving or expediting the proceeding may be charged to the committees or boards. Such costs shall be paid with assessments from the handlers regulated under the marketing order to be amended or on signatories to the marketing agreement subject to amendment. Such assessments may supplement funds for costs associated with, but not limited to, court reporters, hearing examiners, legal counsel, hearing venue and associated travel for USDA officials.

§ 900.43   Use of informal rulemaking.

(a) Notwithstanding the provisions of §§ 900.1 through 900.18, and 900.36 through 900.42 of this part, the Secretary may determine that informal rulemaking procedures under § 553 of Title 5, United States Code be used to amend fruit, vegetable or nut marketing agreements and marketing orders. In making this determination, consideration shall be given to:

(1) The nature and complexity of the proposal;

(2) The potential regulatory and economic impacts on affected entities; and

(3) Any other relevant matters.

Subpart D—Procedural Requirements Governing Proceedings on Petitions To Modify or To Be Exempted From Marketing Orders

AUTHORITY: 7 U.S.C. 608c.

§ 900.50   Words in the singular form.

Words in this subpart in the singular form shall be deemed to import the plural, and vice versa, as the case may demand.

§ 900.51   Definitions.

As used in this subpart, the terms as defined in the act shall apply with equal force and effect. In addition, unless the context otherwise requires:

(a) The term *act* means Public Act No. 10, 73d Congress, as amended and as reenacted and amended by the Agricultural Marketing Agreement Act of 1937, as amended (7 U.S.C. and Sup. 601);

(b) The term *Department* means the United States Department of Agriculture;

(c) The term *Secretary* means the Secretary of Agriculture of the United States, or any officer or employee of the Department to whom authority has heretofore been delegated, or to whom authority may hereafter be delegated, to act for the Secretary;

(d) The term *judge* means any administrative law judge appointed pursuant to 5 U.S.C. 3105 or any presiding official appointed by the Secretary, and assigned to conduct the proceeding.

(e) The term *Administrator* means the Administrator of the Agricultural Marketing Service or any officer or employee of the Department to whom authority has been delegated or may hereafter be delegated to act for the Administrator.

(f) [Reserved]

(g) The term FEDERAL REGISTER means the publication provided for by the act of July 26, 1935 (49 Stat. 500), and acts supplementary thereto and amendatory thereof;

(h) The term *marketing order* means any order or any amendment thereto which may be issued pursuant to section 8c of the act;

(i) The term *handler* means any person who, by the terms of a marketing order, is subject thereto, or to whom a marketing order is sought to be made applicable;

(j) The term *proceeding* means a proceeding before the Secretary arising under section 8c(15)(A) of the Act.

(k) The term *hearing* means that part of the proceeding which involves the submission of evidence;

(l) The term *party* includes the Department;

(m) The term *hearing clerk* means the hearing clerk, United States Department of Agriculture, Washington, D.C.;

(n) [Reserved]

(o) The term *decision* means the judge's initial decision in proceedings subject to 5 U.S.C. 556 and 557, and includes the judge's (1) findings of fact and conclusions with respect to all material issues of fact, law or discretion as well as the reasons or basis thereof, (2) order, and (3) rulings on findings, conclusions and orders submitted by the parties;

(p) The term *petition* includes an amended petition.

[25 FR 5907, June 28, 1960, as amended at 26 FR 7796, Aug. 22, 1961; 28 FR 579, Jan. 23, 1963; 37 FR 8059, Apr. 25, 1972; 38 FR 29798, Oct. 29, 1973; 67 FR 10829, Mar. 11, 2002; 82 FR 58098, Dec. 11, 2017]

§900.52 Institution of proceeding.

(a) *Filing and service of petition.* Any handler desiring to complain that any marketing order or any provision of any such order or any obligation imposed in connection therewith is not in accordance with law, shall file with the hearing clerk, in quadruplicate, a petition in writing addressed to the Secretary. Promptly upon receipt of the petition, the hearing clerk shall transmit a true copy thereof to the Administrator and the General Counsel, respectively.

(b) *Contents of petition.* A petition shall contain:

(1) The correct name, address, and principal place of business of the petitioner. If petitioner is a corporation, such fact shall be stated, together with the name of the State of incorporation, the date of incorporation, and the names, addresses, and respective positions held by its officers; if an unincorporated association, the names and addresses of its officers, and the respective positions held by them; if a partnership, the name and address of each partner;

(2) Reference to the specific terms or provisions of the marketing order, or the interpretation or application thereof, which are complained of;

(3) A full statement of the facts (avoiding a mere repetition of detailed evidence) upon which the petition is based, and which it is desired that the Secretary consider, setting forth clearly and concisely the nature of the petitioner's business and the manner in which petitioner claims to be affected by the terms or provisions of the marketing order, or the interpretation or

application thereof, which are complained of;

(4) A statement of the grounds on which the terms or provisions of the marketing order, or the interpretation or application thereof, which are complained of, are challenged as not in accordance with law;

(5) Prayers for the specific relief which the petitioner desires the Secretary to grant;

(6) An affidavit by the petitioner, or, if the petitioner is not an individual, by an officer of the petitioner having knowledge of the facts stated in the petition, verifying the petition and stating that it is filed in good faith and not for purposes of delay.

(c) *Motion to dismiss petition—*(1) *Filing, contents, and responses thereto.* If the Administrator is of the opinion that the petition, or any portion thereof, does not substantially comply, in form or content, with the act or with the requirements of paragraph (b) of this section, or is not filed in good faith, or is filed for purposes of delay, the Administrator may, within thirty days after the service of the petition, file with the Hearing Clerk a motion to dismiss the petition, or any portion thereof, on one or more of the grounds stated in this paragraph. Such motion shall specify the grounds of objection to the petition and if based, in whole or in part, on an allegation of fact not appearing on the face of the petition, shall be accompanied by appropriate affidavits or documentary evidence substantiating such allegations of fact. The motion may be accompanied by a memorandum of law. Upon receipt of such motion, the Hearing Clerk shall cause a copy thereof to be served upon the petitioner, together with a notice stating that all papers to be submitted in opposition to such motion including any memorandum of law, must be filed by the petitioner with the hearing clerk not later than 20 days after the service of such notice upon the petitioner. Upon the expiration of the time specified in such notice, or upon receipt of such papers from the petitioner, the hearing clerk shall transmit all papers which have been filed in connection with the motion to the Judge for consideration.

(2) *Decision by the Judge.* The Judge, after due consideration, shall render a decision upon the motion stating the reasons for his action. Such decision shall be in the form of an order and shall be filed with the hearing clerk who shall cause a copy thereof to be served upon the petitioner and a copy thereof to be transmitted to the Administrator. Any such order shall be final unless appealed pursuant to § 900.65: *Provided,* That within 20 days following the service upon the petitioner of a copy of the order of the Judge dismissing the petition, or any portion thereof, on the ground that it does not substantially comply in form and content with the act or with paragraph (b) of this section, the petitioner shall be permitted to file an amended petition.

(3) *Oral argument.* Unless a written application for oral argument is filed by a party with the hearing clerk not later than the time fixed for filing papers in opposition to the motion, it shall be considered that the party does not desire oral argument. The granting of a request to make oral argument shall rest in the discretion of the Judge.

[25 FR 5907, June 28, 1960, as amended at 38 FR 29798, Oct. 29, 1973; 67 FR 10829, Mar. 11, 2002]

§ 900.52a   Answer to petition.

(a) *Time of filing.* Within 30 days after the filing of the petition, the Administrator shall file an answer thereto: *Provided,* That if a motion to dismiss the petition, in whole or in part, is made pursuant to § 900.52(c), the answer shall be filed within 15 days after the service of an order of the judge denying the motion or granting the motion with respect to only a portion of the petition. The answer shall be filed with the hearing clerk who shall cause a copy thereof to be served promptly upon the petitioner.

(b) *Contents.* The answer shall specify which of the material allegations of fact or of law in the petition are controverted and which are not controverted. The answer also may contain affirmative allegations of fact constituting separate defenses and statements of objections to the suffi-

ciency of the whole or any part of the petition.

[25 FR 5907, June 28, 1960, as amended at 38 FR 29798, Oct. 29, 1973; 67 FR 10829, Mar. 11, 2002]

§ 900.52b   Amended pleadings.

At any time before the close of the hearing the petition or answer may be amended, but the hearing shall, at the request of the adverse party, be adjourned or recessed for such reasonable time as the judge may determine to be necessary to protect the interests of the parties. Amendments subsequent to the first amendment or subsequent to the filing of an answer may be made only with leave of the judge or with the written consent of the adverse party.

§ 900.53   Withdrawal of petition.

If, at any time after the petition is filed, the petitioner desires to withdraw the same, he shall file with the hearing clerk (or, if filed during the course of a hearing, with the judge) a written request for permission to withdraw. The judge may, in his discretion, thereupon dismiss the petition without further procedure: *Provided,* That, if the request to withdraw is filed after a hearing has been opened, permission to withdraw shall be granted only in exceptional circumstances.

§ 900.54   Docket number.

Each proceeding, immediately following its institution, shall be assigned a docket number by the hearing clerk and thereafter the proceeding may be referred to by such number.

§ 900.55   Judges.

(a) *Assignment.* No judge who has any pecuniary interest in the outcome of the proceeding, or who has participated in any investigation preceding the institution of the proceeding, shall serve as judge in such proceeding.

(b) *Conduct.* The judge shall conduct the proceeding in a fair and impartial manner and shall not discuss ex parte the merits of the proceeding with any person who is or who has been connected in any manner with the proceeding in an advocative or investigative capacity.

(c) *Powers of judges.* Subject to review by the Secretary, as provided elsewhere in this subpart, the judge shall have power to:

(1) Rule upon motions and requests;

(2) Adjourn the hearing from time to time, and change the time and place of hearing;

(3) Administer oaths and affirmations and take affidavits;

(4) Issue subpenas, under the facsimile signature of the Secretary, requiring the attendance and testimony of witnesses and the production of books, records, contracts, papers, and other documentary evidence;

(5) Examine witnesses and receive evidence;

(6) Take or order, under the facsimile signature of the Secretary, the taking of depositions;

(7) Admit or exclude evidence;

(8) Hear oral argument on facts or law;

(9) Consolidate hearings upon two or more petitions pertaining to the same order;

(10) Do all acts and take all measures necessary for the maintenance of order at the hearing and the efficient conduct of the proceeding.

(d) *Who may act in absence of judge.* In case of the absence of the judge or his inability to act, the powers and duties to be performed by him under these rules of practice in connection with a proceeding may, without abatement of the proceeding unless otherwise ordered by the Secretary, be assigned to any other judge.

(e) *Disqualification of judge.* The judge may at any time withdraw as judge in a proceeding if he deems himself to be disqualified. Upon the filing by an interested person in good faith of a timely and sufficient affidavit of personal bias or disqualification of a judge, the Secretary shall determine the matter as a part of the record and decision in the proceeding, after making such investigation or holding such hearings, or both, as he may deem appropriate in the circumstances.

§900.56 Consolidated hearings.

At the discretion of the judge, hearings upon two or more petitions pertaining to the same order may be consolidated, and the evidence taken at such consolidated hearing may be embodied in a single record.

§900.57 Intervention.

Intervention in proceedings subject to this subpart shall not be allowed, except that, in the discretion of the Secretary or the judge, any person (other than the petitioner) showing a substantial interest in the outcome of a proceeding shall be permitted to participate in the oral argument and to file a brief.

§900.58 Prehearing conferences.

In any proceeding in which it appears that such procedure will expedite the proceeding, the judge, at any time prior to the commencement of or during the course of the hearing, may request the parties or their counsel to appear at a conference before him to consider (a) the simplification of issues; (b) the possibility of obtaining stipulations of fact and of documents which will avoid unnecessary proof; (c) the limitation of the number of expert or other witnesses; and (d) such other matters as may expedite and aid in the disposition of the proceeding. No transcript of such conference shall be made, but the judge shall prepare and file for the record a written summary of the action taken at the conference, which shall incorporate any written stipulations or agreements made by the parties at the conference or as a result of the conference. If the circumstances are such that a conference is impracticable, the judge may request the parties to correspond with him for the purpose of accomplishing any of the objects set forth in this section. The judge shall forward copies of letters and documents to the parties as the circumstances require. Correspondence in such negotiations shall not be a part of the record, but the judge shall submit a written summary for the record if any action is taken.

§900.59 Motions and requests.

(a) *General.* (1) All motions and requests shall be filed with the hearing clerk, except that those made during the course of an oral hearing may be filed with the judge or may be stated orally and made a part of the transcript.

(2) The judge is authorized to rule upon all motions and requests filed or made prior to the transmittal by the

hearing clerk to the Secretary of the record as provided in this subpart. The Secretary shall rule upon all motions and requests filed after that time.

(b) *Certification of motions.* The submission or certification of any motion, request, objection, or other question to the Secretary, as provided in this subpart, shall be in the discretion of the judge.

[25 FR 5907, June 28, 1960, as amended at 38 FR 29798, Oct. 29, 1973]

§ 900.60   Oral hearings before judge.

(a) *Time and place.* The judge shall set a time and place for hearing and shall file with the hearing clerk a notice stating the time and place of hearing. If any change in the time or place of hearing becomes necessary, it shall be made by the judge, who, in such event, shall file with the hearing clerk a notice of the change. Such notice shall be served upon the parties, unless it is made during the course of an oral hearing and made a part of the transcript.

(b) *Appearances*—(1) *Representation.* In any proceeding under the act, the parties may appear in person or by counsel or other representative. The Department, if represented by counsel, shall be represented by an attorney assigned by the General Counsel of the Department, and such attorney shall present or supervise the presentation of the position of the Department.

(2) *Debarment of counsel or representative.* Whenever, while a proceeding is pending before him, the judge finds that a person acting as counsel or representative for any party to the proceeding is guilty of unethical or unprofessional conduct, the judge may order that such person be precluded from further acting as counsel or representative in such proceeding. An appeal to the Secretary may be taken from any such order, but the proceeding shall not be delayed or suspended pending disposition of the appeal: *Provided,* That the judge may suspend the proceeding for a reasonable time for the purpose of enabling the client to obtain other counsel or representative. In case the judge has issued an order precluding a person from further acting as counsel or representative in the proceeding, the judge, within a reasonable time thereafter, shall submit to the

Secretary a report of the facts and circumstances surrounding the issuance of the order and shall recommend what action the Secretary should take respecting the appearance of such person as counsel or representative in other proceedings before the Secretary. Thereafter, the Secretary may, after notice and an opportunity for hearing, issue such order respecting the appearance of such person as counsel or representative in proceedings before the Secretary as the Secretary finds to be appropriate.

(3) *Failure to appear.* If the petitioner, after being duly notified, fails to appear at the hearing, he shall be deemed to have authorized the dismissal of the proceeding, without further procedure, and with or without prejudice as the judge may determine. In the event that the petitioner appears at the hearing and no representative of the Department appears, the judge shall proceed ex parte to hear the evidence of the petitioner. *Provided,* That failure on the part of such representative of the Department to appear at a hearing shall not be deemed to be waiver of the Department's right to file suggested findings of fact, conclusions and order; to be served with a copy of the judge's initial decision and to appeal to the Secretary with respect thereto.

(c) *Order of proceeding.* Except as may be determined otherwise by the judge, the petitioner shall proceed first at the hearing.

(d) *Evidence*—(1) *In general.* The hearing shall be publicly conducted, and the testimony given at the hearing shall be reported verbatim.

(i) The testimony of witnesses at a hearing shall be upon oath or affirmation and subject to cross-examination.

(ii) Any witness may, in the discretion of the judge, be examined separately and apart from all other witnesses except those who may be parties to the proceeding.

(iii) The judge shall exclude, insofar as practicable, evidence which is immaterial, irrelevant, or unduly repetitious, or which is not of the sort upon which responsible persons are accustomed to rely.

(2) *Objections.* If a party objects to the admission or rejection of any evidence or to the limitation of the scope

of any examination or cross-examination, or any other ruling of the judge, he shall state briefly the grounds of such objection, whereupon an automatic exception will follow which may be pursued in an appeal pursuant to §900.65 by the party adversely affected by the judge's ruling.

(3) *Depositions.* The deposition of any witness shall be admitted, in the manner hereinafter provided in and subject to the provisions of §900.61.

(4) *Affidavits.* Except as is otherwise provided in this subpart, affidavits may be admitted only if the evidence is otherwise admissible and the parties agree (which may be determined by their failure to make timely objections) that affidavits may be used.

(5) *Proof and authentication of official records or documents.* An official record or document, when admissible for any purpose, shall be admissible in evidence without the production of the person who made or prepared the same. Such record or document shall, in the discretion of the judge, be evidenced by an official publication thereof or by a copy attested by the person having legal custody thereof and accompanied by a certificate that such person has the custody.

(6) *Exhibits.* All written statements, charts, tabulations, or similar data offered in evidence at the hearing shall, after identification by the proponent and upon a satisfactory showing of the admissibility of the contents thereof, be numbered as exhibits and received in evidence and made a part of the record. Except where the judge finds that the furnishing of copies is impracticable, a copy of each exhibit, in addition to the original, shall be filed with the judge for the use of each other party to the proceeding. The judge shall advise the parties as to the exact number of copies which will be required to be filed and shall make and have noted on the record the proper distribution of the copies. If the testimony of a witness refers to a statute, or to a report, document, or transcript, the judge, after inquiry relating to the identification of such statute, report, document, or transcript, shall determine whether the same shall be produced at the hearing and physically be made a part of the evidence as an exhibit, or whether it shall be incorporated into the evidence by reference. If relevant and material matter offered in evidence is embraced in a report, document, or transcript containing immaterial or irrelevant matter, such immaterial or irrelevant matter shall be excluded and shall be segregated insofar as practicable, subject to the direction of the judge.

(7) *Official notice.* Official notice will be taken of such matters as are judicially noticed by the courts of the United States and of any other matter of technical, scientific, or commercial fact of established character: *Provided,* That the parties shall be given adequate notice, at the hearing or by reference in the judge's report or the tentative order or otherwise, of matters so noticed, and (except where official notice is taken, for the first time in the proceeding, in the final order) shall be given adequate opportunity to show that such facts are erroneously noticed.

(8) *Offer of proof.* Whenever evidence is excluded from the record, the party offering such evidence may make an offer of proof, which shall be included in the transcript. The offer of proof shall consist of a brief statement describing the evidence to be offered. If the evidence consists of a brief oral statement or of an exhibit, it shall be inserted into the transcript in toto. In such event, it shall be considered a part of the transcript if the Secretary decides that the judge's ruling in excluding the evidence was erroneous. The judge shall not allow the insertion of such evidence in toto if the taking of such evidence will consume a considerable length of time at the hearing. In the latter event, if on appeal the Secretary decides that the judge erred in excluding the evidence, and that such error was substantial, the hearing shall be reopened to permit the taking of such evidence.

(e) *Transcript.* Transcripts of hearings shall be made available to any person at actual cost of duplication.

[25 FR 5907, June 28, 1960, as amended at 38 FR 29798, Oct. 29, 1973; 67 FR 10829, Mar. 11, 2002]

§ 900.61 Depositions.

(a) *Procedure in lieu of deposition.* Before any party may have testimony taken by deposition, said party shall, if practicable, submit to the other party an affidavit which shall set forth the facts to which the witness would testify, if the deposition should be taken. If, after examination of such affidavit, the other party agrees, or (within 10 days after submission of the affidavit) fails to object, that the affidavit may be used in lieu of the deposition, the judge shall admit the affidavit in evidence and shall not order the deposition to be taken.

(b) *Application for taking deposition.* Upon the application of a party to the proceeding, the judge may, at any time after the filing of the moving paper, order, under the facsimile signature of the Secretary, the taking of testimony by deposition. The application shall be in writing and shall be filed with the hearing clerk and shall set forth: (1) The name and address of the proposed deponent; (2) the name and address of the person (referred to hereinafter in this section as the *judge*), qualified under the rules in this part to take depositions, before whom the proposed examination is to be made; (3) the proposed time and place of the examination, which shall be at least 15 days after the date of the mailing of the application; and (4) the reasons why such deposition should be taken.

(c) *Judge's order for taking deposition.* If, after the examination of the application, the judge is of the opinion that the deposition should be taken, he shall order its taking. The order shall be filed with the hearing clerk and shall be served upon the parties and shall state: (1) The time and place of the examination (which shall not be less than 10 days after the filing of the order); (2) the name of the judge before whom the examination is to be made; (3) the name of the deponent. The judge and the time and place need not be the same as those suggested in the application.

(d) *Qualifications of judge.* The deposition shall be taken before the judge, or before a judge authorized by the law of the United States or by the law of the place of the examination to administer oaths, or before a judge authorized by the Secretary to administer oaths.

(e) *Procedure on examination.* (1) The deponent shall be examined under oath or affirmation and shall be subject to cross-examination. The testimony of the deponent shall be recorded by the judge or by some person under his direction and in his presence. In lieu of oral examination, parties may transmit written interrogatories to the judge prior to the examination and the judge shall propound such interrogatories to the deponent.

(2) The applicant must arrange for the examination of the witness either by oral examination or by written interrogatories. If it is found by the judge, upon the protest of a party to the proceeding, that such party has his residence and his place of business more than 100 miles from the place of the examination and that it would constitute an undue hardship upon such party to be represented at the examination, the applicant will be required to conduct the examination by means of interrogatories. When the examination is conducted by means of interrogatories, copies of the interrogatories shall be served upon the other parties to the proceeding at least five days prior to the date set for the examination, and the other parties shall be afforded an opportunity to file with the judge cross-interrogatories at any time prior to the time of the examination.

(f) *Certification by judge.* The judge shall certify on the deposition that the deponent was duly sworn by him and that the deposition is a true record of the deponent's testimony. He shall then securely seal the deposition, together with two copies thereof, in an envelope and mail the same by registered mail to the hearing clerk.

(g) *Use of depositions.* A deposition ordered and taken in accord with the provisions of this section may be used in a proceeding under the act if the judge finds that the evidence is otherwise admissible and (1) that the witness is dead; or (2) that the witness is at a distance greater than 100 miles from the place of hearing, unless it appears that the absence of the witness was procured by the party offering the deposition; or (3) that the witness is unable to attend or testify because of age, sickness, infirmity, or imprisonment; or (4) that the party offering the deposition has endeavored to procure the attendance of the witness by subpena but has been unable to do so; or (5) that such exceptional circumstances exist as to make it desirable, in the interests of justice, to allow the deposition to be used. If a deposition has been taken, and the party upon whose application it was taken refuses to offer it in evidence, the other party may offer the deposition, or any part thereof, in evidence.

§900.62 Subpenas.

(a) *Issuance of subpenas.* The attendance of witnesses and the production of documentary evidence from any place in the United States on behalf of any party to the proceeding may, by subpena, be required at any designated place of hearing. Subpenas may be issued by the Secretary or by the judge, under the facsimile signature of the Secretary, upon a reasonable showing by the applicant of the grounds, necessity, and reasonable scope thereof.

(b) *Application for subpena duces tecum.* Subpenas for the production of documentary evidence, unless issued by the judge upon his own motion, shall be issued only upon a certified written application. Such application shall specify, as exactly as possible, the documents desired and shall show their competency, relevancy, and materiality and the necessity for their production.

(c) *Service of subpenas.* Subpenas may be served (1) by a United States Marshal or his deputy, or (2) by any other person who is not less than 18 years of age, or (3) by registering and mailing a copy of the subpena addressed to the person to be served at his or its last known residence or principal place of business or residence. Proof of service may be made by the return of service on the subpena by the United States Marshal or his deputy; or, if served by an individual other than a United States Marshal or his deputy, by an affidavit of such person stating that he personally served a copy of the subpena upon the person named therein; or, if service was by registered mail, by an affidavit made by the person mailing the subpena that it was mailed as pro-

vided in this paragraph and by the signed return post office receipt: *Provided,* That, if the subpena is issued on behalf of the Department, the return receipt without an affidavit of mailing shall be sufficient proof of service. In making personal service, the person making service shall leave a copy of the subpena with the person subpenaed; the original, bearing or accompanied by the required proof of service, shall be returned to the official who issued the same.

[25 FR 5907, June 28, 1960, as amended 67 FR 10829, Mar. 11, 2002]

§900.63 Fees and mileage.

Witnesses who are subpenaed and who appear in such proceeding, including witnesses whose depositions are taken, shall be paid the same fees and mileage that are paid witnesses in the courts of the United States, and persons taking depositions shall be entitled to the same fees as are paid for like services in the courts of the United States, to be paid by the party at whose request the deposition is taken. Witness fees and mileage shall be paid by the party at whose instance the witnesses appear, and claims therefor, as to witnesses subpenaed on behalf of the Department, shall be proved before the person issuing the subpena, and, as to witnesses subpenaed on behalf of any other party, shall be presented to such party.

§900.64 The Judge's decision.

(a) *Corrections to and certification of transcript.* (1) At such time as the judge may specify, but not later than the time fixed for filing proposed findings of fact, conclusions and order, or briefs, as the case may be, the parties may file with the judge proposed corrections to the transcript.

(2) As soon as practicable after the filing of proposed findings of fact, conclusions and order, or briefs, as the case may be, the judge shall file with the hearing clerk his certificate indicating any corrections to be made in the transcript, and stating that, to the best of his knowledge and belief, the transcript, as corrected, is a true, correct, and complete transcript of the testimony given at the hearing, and that the exhibits are all the exhibits

properly a part of the hearing record. The original of such certificate shall be attached to the original transcript and a copy of such certificate shall be served upon each of the parties by the hearing clerk who shall also enter onto the transcript (without obscuring the text) any correction noted in the certification.

(b) *Proposed findings of fact, conclusions, and orders.* Within 10 days (unless the judge shall have announced at the hearing a shorter or longer period of time) after the transcript has been filed with the hearing clerk, as provided in paragraph (a) of this section, each party may file with the hearing clerk proposed findings of fact, conclusions, and order, based solely upon the evidence of record, and briefs in support thereof.

(c) *Judge's Decision.* The judge, within a reasonable time after the termination of the period allowed for the filing of proposed findings of fact, conclusions,and orders, and briefs in support thereof, shall prepare upon the basis of the record, and shall file with the hearing clerk, his initial decision, a copy of which shall be served by the hearing clerk, upon each of the parties. Such decision shall become final without further proceedings 35 days after the date of service thereof, unless there is an appeal to the Secretary by a party to the proceeding: *Provided, however,* That no decision shall be final for the purpose of judicial review except a final decision issued by the Secretary pursuant to an appeal by a party to the proceeding.

[25 FR 5907, June 28, 1960, as amended at 38 FR 29799, Oct. 29, 1973; 67 FR 10830, Mar. 11, 2002]

§ 900.65 Appeals to Secretary: Transmittal of record.

(a) *Filing of appeal.* Any party who disagrees with a judge's decision or any part thereof, may appeal the decision to the Secretary by filing an appeal petition with the Hearing Clerk within 30 days after service of said decision upon said party. Each issue set forth in the appeal, and the arguments thereon, shall be separately numbered; shall be plainly and concisely stated; and shall contain detailed citations of the record, statutes, regulations and authorities being relied upon in support thereof. The appeal petition shall be served upon the other party to the proceeding by the hearing clerk.

(b) *Argument before Secretary*—(1) *Oral argument.* A party bringing an appeal may request within the prescribed time period for filing such appeal, an opportunity for oral argument before the Secretary. Failure to make such request in writing, within the prescribed time period, shall be deemed a waiver of oral argument. The Secretary, in his discretion, may grant, refuse or limit any request for oral argument on appeal.

(2) *Scope of argument.* Argument to be heard on appeal, whether oral or in a written brief, shall be limited to the issues raised by the appeal, except that if the Secretary determines that additional issues should be argued, the parties shall be given reasonable notice of such determination, so as to permit preparation of adequate arguments on all the issues to be argued.

(c) *Response.* Within 20 days after service of an appeal brought by a party to the proceeding, any other party may file a response in support of or in opposition to such appeal.

(d) *Transmittal of record.* Whenever an appeal is filed by a party to the proceeding, the hearing clerk shall transmit to the Secretary the record of the proceeding. Such record shall include: The pleadings; any motions and requests filed, and the rulings thereon; the transcript of the testimony taken at the hearing, as well as the exhibits filed in connection therewith; any statements filed under the shortened procedure; any documents or papers filed in connection with prehearing conferences; such proposed findings of fact, conclusions, and orders, and briefs in support thereof, as may have been filed in connection with the hearing; the judge's initial decision; and the appeal petition; briefs in support thereof, and responses thereto as may have been filed in the proceeding.

[38 FR 29799, Oct. 29, 1973, as amended at 67 FR 10830, Mar. 11, 2002]

§900.66 Consideration of appeal by the Secretary and issuance of final order.

(a) *Consideration of appeal.* As soon as practicable after the receipt of the record from the hearing clerk, or, in case oral argument was had, as soon as practicable thereafter, the Secretary, upon the basis of and after due consideration of the record, shall rule on the appeal. If the Secretary decides that no change or modification of the judge's decision is warranted, he may adopt the Judge's decision as the final order of the Secretary, preserving any right of the party bringing the appeal to seek judicial review of such decision in the proper forum. At no stage of the proceeding between its institution and the issuance of the order shall the Secretary discuss ex parte the merits of the proceeding with any person who is connected with the proceeding in an advocative or an investigative capacity, or with any representative of such person: *Provided, however,* That the Secretary may discuss the merits of the proceeding with such a person if all parties to the proceeding, or their representatives, have been given an opportunity to be present. If, notwithstanding the foregoing provisions of this section, a memorandum or other communication from any party, or from any person acting on behalf of any party, which relates to the merits of the proceeding, receives the personal attention of the Secretary (or, if an official other than the Secretary is to issue the order, then of such other official) during the pendency of the proceeding, such memorandum or communication shall be regarded as argument made in the proceeding and shall be filed with the hearing clerk, who shall serve a copy thereof upon the opposite party to file a reply thereto.

(b) *Issuance of final order.* A final order issued by the Secretary shall be filed with the hearing clerk, who shall serve it upon the parties: *Provided,* That, if the terms of the order differ substantially from those proposed in the decision of the judge, the Secretary shall, if he deems it advisable to do so, direct that a copy of the order be served upon the parties as a tentative order; and, in such event, opportunity shall be given the parties to file exceptions thereto and written arguments or briefs in support of such exceptions. In such case, if exceptions are filed within a period of time (to be fixed by the Secretary but not to exceed 20 days) following the service of the tentative order, the Secretary shall give consideration, to and shall make such changes in the tentative order as he deems to be appropriate; otherwise, tentative order shall become final, as of the day following the date of expiration of the period fixed for the filing of exceptions.

[38 FR 29799, Oct. 29, 1973]

§900.68 Petitions for reopening hearings; for rehearings or rearguments of proceedings; or for reconsideration of orders.

(a) *Petition requisite*—(1) *Filing; service.* A petition for reopening the hearing to take further evidence, or for rehearing or reargument of the proceeding, or for reconsideration of the order shall be made by petition addressed to the Secretary and filed with the hearing clerk, who immediately shall notify and serve a copy thereof upon the other party to the proceeding. Every such petition shall state specifically the grounds relied upon.

(2) *Petitions to reopen hearings.* A petition to reopen the hearing for the purpose of taking additional evidence may be filed at any time prior to the issuance of the final order. Every such petition shall state briefly the nature and purpose of the evidence to be adduced, shall show that such evidence is not merely cumulative, and shall set forth a good reason why such evidence was not adduced at the hearing.

(3) *Petitions to rehear or reargue proceedings, or to reconsider orders.* A petition to rehear or reargue the proceeding or to reconsider the final order shall be filed within 15 days after the date of the service of such order. Every such petition shall state specifically the matters claimed to have been erroneously decided, and alleged errors must be briefly stated.

(b) *Procedure for disposition of petitions.* Within 10 days following the service of any petition provided for in this section, the other party to the proceeding shall file with the hearing clerk an answer thereto. As soon as

29

practicable thereafter, the Secretary shall announce the decision granting or denying the petition. Unless the Secretary shall determine otherwise, the issuance or operation of the order shall not be stayed pending the decision of the Secretary upon the petition. In the event that any such petition is granted by the Secretary, the applicable rules of practice, as set out elsewhere in this subpart, shall be followed.

[25 FR 5907, June 28, 1960, as amended at 67 FR 10830, Mar. 11, 2002]

§ 900.69 Filing; service; extensions of time; effective date of filing; and computation of time.

(a) *Filing; number of copies.* Except as provided otherwise herein, all documents or papers required or authorized in this subpart to be filed with the hearing clerk shall be filed in quadruplicate: *Provided,* That, if there are more than two parties to the proceeding, a sufficient number of additional copies shall be filed so as to provide for service upon all the parties to the proceeding. Any document or paper, required or authorized in this subpart to be filed with the hearing clerk, shall, during the course of an oral hearing, be filed with the judge.

(b) *Service; proof of service.* Copies of all such papers shall be served upon the parties by the hearing clerk, by the judge, or by some other employee of the Department or by a United States Marshal or his deputy. Service shall be made either (1) by delivering a copy of the document or paper to the individual to be served or to a member of the partnership to be served or to the president, secretary, or other executive officer or any director of the corporation, organization, or association to be served, or to the attorney or agent of record of such individual, partnership, corporation, organization, or association; or (2) by leaving a copy of the document or paper at the principal office or place of business of such individual, partnership, corporation, organization, or association, or of his or its attorney or agent of record; or (3) by registering and mailing a copy of the document or paper, addressed to such individual, partnership, corporation, organization, or association, or to his or its attorney or agent of record, at

his or its last known principal office, place of business, or residence. Proof of service hereunder shall be made by the affidavit of the person who actually made the service. The affidavit contemplated herein shall be filed with the hearing clerk, and the fact of filing thereof shall be noted on the docket of the proceeding.

(c) *Extensions of time.* The time for the filing of any documents or papers required or authorized in this subpart to be filed may be extended upon (1) a written stipulation between the parties, or (2) upon the request of a party, by the judge before the transmittal of the record to the Secretary, or by the Secretary at any other time if, in the judgment of the Secretary or the judge, as the case may be, there is good reason for the extension.

(d) *Effective date of filing.* Any document or paper required or authorized in this subpart to be filed shall be deemed to be filed at the time it is received by the Hearing Clerk.

(e) *Computation of time.* Each day, including Saturdays, Sundays, and legal public holidays, shall be included in computing the time allowed for filing any document or paper: *Provided,* That when the time for filing a document or paper expires on a Saturday, Sunday, or legal public holiday, the time allowed for filing the document or paper shall be extended to include the following business day.

[25 FR 5907, June 28, 1960, as amended at 67 FR 10830, Mar. 11, 2002]

§ 900.70 Applications for interim relief.

(a) *Filing the application.* A person who has filed a petition pursuant to § 900.52 may by separate application filed with the hearing clerk apply to the Secretary for an order postponing the effective date of, or suspending the application of, the marketing order or any provision thereof, or any obligation imposed in connection therewith, pending final determination of the proceeding.

(b) *Contents of the application.* The application shall contain a statement of the facts upon which the relief is requested, including any facts showing irreparable injury. The application must be signed and sworn to by the petitioner and any facts alleged therein which are not within his personal knowledge shall be supported by affidavits of a person or persons having personal knowledge of such facts or by proper documentary evidence thereof.

(c) *Answer to application.* Immediately upon receipt of the application, the hearing clerk shall transmit a copy thereof, together with all supporting papers, to the Administrator, who shall, within 20 days, or such other time fixed by the Secretary, after the filing of the application file an answer thereto with the hearing clerk.

(d) *Contents of answer.* The answer shall contain a statement of the objections, if any, of the Administrator to the application for interim relief, and may be supported by affidavits and documentary evidence.

(e) *Transmittal to Secretary.* Upon receiving the answer of the Administrator or upon the expiration of the time for filing the answer, the hearing clerk shall transmit to the Secretary for his decision all papers filed in connection with the application.

(f) *Hearing and oral argument.* The Secretary may, in his discretion, permit oral argument or the taking of testimony in connection with such application. However, unless written request therefor is filed with the hearing clerk prior to the transmittal of the papers to the Secretary, the parties shall be deemed to have waived oral argument and the taking of testimony.

(g) *Decision by Secretary.* The Secretary may grant or deny the application. Any action taken by the Secretary shall be in the form of an order filed with the hearing clerk and shall contain a brief statement of the reasons for the action taken. The hearing clerk shall cause copies of the order to be served upon the parties.

[25 FR 5907, June 28, 1960, as amended at 67 FR 10830, Mar. 11, 2002]

§900.71 Hearing before Secretary.

The Secretary may act in the place and stead of a judge in any proceeding hereunder. When he so acts the hearing clerk shall transmit the record to the Secretary at the expiration of the period provided for the filing of proposed findings of fact, conclusions and orders, and the Secretary shall thereupon, after due consideration of the record,

issue his final order in the proceeding: *Provided,* That he may issue a tentative order in which event the parties shall be afforded an opportunity to file exceptions before the issuance of the final order.

Subpart E—Supplemental Procedural Requirements for Marketing Orders, Marketing Agreements, and Requirements Covering Fruits, Vegetables, and Nuts

AUTHORITY: 7 U.S.C. 601–674.

SOURCE: 61 FR 20717, May 8, 1996, unless otherwise noted.

§900.80 Words in the singular form.

Words in this subpart in the singular form shall be deemed to import the plural, and vice versa, as the case may demand.

§900.81 Definitions.

As used in this subpart, the terms as defined in the act shall apply with equal force and effect. In addition, unless the context otherwise requires:

(a) The term *Act* means Public Act No. 10, 73 Congress (48 Stat. 31) as amended and as reenacted and amended by the Agricultural Marketing Agreement Act of 1937 (50 Stat. 246), as amended.

(b) The term *Department* means the United States Department of Agriculture.

(c) The term *Secretary* means the Secretary of Agriculture of the United States, or any officer or employee of the Department to whom authority has heretofore been delegated, or to whom authority may hereafter be delegated, to act in his stead.

(d) The term *Administrator* means the Administrator of the Agricultural Marketing Service, with power to redelegate, or any officer or employee of the Department to whom authority has been delegated or may hereafter be delegated to act in his stead.

(e) The term *proceeding* means a proceeding before the Secretary arising under sections 8a, 8b(b), 8c(14), 8e, 10(c) and 10(h).

(f) The term *hearing* means that part of the proceeding which involves the submission of evidence.

(g) The term *marketing agreement* means any marketing agreement or any amendment thereto which may be entered into pursuant to section 8b of the act.

(h) The term *marketing order* means any order or any amendment thereto which may be issued pursuant to section 8c of the act, and after notice and hearing as required by said section.

(i) The term *handler* means any person who, by the terms of a marketing order or marketing agreement, is subject thereto, or to whom a marketing order or marketing agreement is sought to be made applicable.

(j) The term *importer* means any person who, by the terms of section 8e of the act, is subject thereto.

(k) The term *person* means any individual, corporation, partnership, association, or any other business unit.

§ 900.82 Stipulation procedures.

The Administrator, or the Administrator's representative, may, at any time before the issuance of a complaint seeking a civil penalty under the Act, enter into a stipulation with any handler or importer in accordance with the following procedures:

(a) The Administrator, or the Administrator's representative, shall give the handler or importer notice of the alleged violation of the applicable marketing order or marketing agreement, or the requirements issued pursuant to 7 U.S.C. 608b(b) and 7 U.S.C. 608e, and an opportunity for a hearing thereon as provided by the Act;

(b) In agreeing to the proposed stipulation, the handler or importer expressly waives the opportunity for a hearing and agrees to pay a specified civil penalty within a designated time;

(c) The Administrator, or the Administrator's representative, agrees to accept the specified civil penalty in settlement of the particular matter involved if it is paid within the designated time;

(d) In cases where the handler or importer does not pay the specified civil penalty within the designated time, or the handler or importer does not agree to the stipulation, the Administrator

may issue an administrative complaint; and

(e) The civil penalty that the Administrator may have proposed in a stipulation agreement shall have no bearing on the civil penalty amount that the Department may seek in a formal administrative proceeding against the same handler or importer for the same alleged violation.

§ 900.83 Conducting Meetings via Electronic Communication or Otherwise.

Notwithstanding any other provisions of a marketing order in this part, administrative bodies of fruit, vegetable, and specialty crop marketing orders, and their committees/subcommittees may, upon due notice to all members and the public:

(a) Conduct meetings by any means of communication available, electronic or otherwise, that effectively assembles members and the public, and facilitates open communication.

(b) Vote by any means of communication available, electronic or otherwise; *Provided,* That votes cast are verifiable and that quorum and other procedural requirements of each respective marketing order are met.

(c) With the approval of the Secretary, each administrative body may prescribe any additional procedures necessary to carry out the objectives of paragraphs (a) and (b) of this section.

[83 FR 22832, May 17, 2018]

Subpart F—Procedure Governing Meetings To Arbitrate and Mediate Disputes Relating to Sales of Milk or Its Products

AUTHORITY: Sec. 3, 50 Stat. 248; 7 U.S.C. 671.

§ 900.100 Words in the singular form.

Words in this subpart in the singular form shall be deemed to import the plural, and vice versa, as the case may demand.

§ 900.101 Definitions.

As used in this subpart, the terms as defined in the act shall apply with equal force and effect. In addition, unless the context otherwise requires:

(a) The term *act* means section 3 of the Agricultural Marketing Agreement Act of 1937, as amended (50 Stat. 248, as amended; 7 U.S.C. 671);

(b) The term *Department* means the United States Department of Agriculture;

(c) The term *Secretary* means the Secretary of Agriculture of the United States, or any officer or employee of the Department to whom authority has heretofore been delegated, or to whom authority may hereafter be delegated, to act in his stead;

(d) The term *General Counsel* means the General Counsel of the Department;

(e) The term *Administrator* means the Administrator of the Agricultural Marketing Service, with power to redelegate, or any officer or employee of the Department to whom authority has been delegated or may hereafter be delegated to act in his stead;

(f) The term *Service* means the Agricultural Marketing Service;

(g) The term *Division* means the Dairy Division of the Service;

(h) The term *cooperative* means any association, incorporated or otherwise, which is in good faith owned or controlled by producers, or organizations thereof, of milk or its products, and which is bona fide engaged in the collective processing or preparing for market or handling or marketing, in the current of interstate or foreign commerce, of milk or its products;

(i) The term *arbitrator* means any officer or employee of the Service designated by the Administrator, pursuant to the act, to arbitrate a bona fide dispute with reference to the terms and conditions of the sale of milk or its products between a producer cooperative and purchasers, handlers, processors, or distributors of milk or its products;

(j) The term *mediator* means any officer or employee of the Service designated by the Administrator, pursuant to the act, to mediate a bona fide dispute with reference to terms and conditions of the sale of milk or its products between a producer cooperative and purchasers, handlers, processors, or distributors of milk or its products;

(k) The term *hearing clerk* means the hearing clerk, United States Departent of Agriculture, Washington, DC.

[25 FR 5907, June 28, 1960, as amended at 26 FR 7797, Aug. 22, 1961; 28 FR 579, Jan. 23, 1963; 37 FR 8059, Apr. 25, 1972]

§ 900.102   Filing of applications for mediation or arbitration.

All applications for mediation or arbitration, all submissions, and all correspondence regarding mediation or arbitration shall be addressed to the Secretary, attention of the Division.

§ 900.103   Application for mediation.

An application for mediation by a cooperative, shall be in writing and shall include the following information:

(a) Names in full of the parties to the dispute and their addresses;

(b) Description of the cooperative organization and business, including copies of the articles of incorporation or association, by-laws, and membership contract; information regarding the number of shares of outstanding stock and the approximate portion owned by active producers; a statement of the function performed in connection with the collective processing, preparing, handling, or marketing of milk or its products; and data relative to the distribution of membership by States, the distribution by States of plant facilities for collecting, processing, or disposing of milk or its products, and the business operations for the year last past, including the total quantity of milk and its products handled by the applicant and the proportion of that quantity that was sold in States other than the States of production;

(c) Suggested time and place for meeting between parties and mediator.

§ 900.104   Inquiry by the Administrator.

Upon receipt of an application for mediation, the Administrator, through such officers or employees of the Service as he may designate, may make any inquiry which is deemed to be necessary or proper in order to determine whether a bona fide dispute exists.

§ 900.105   Notification.

The Administrator, acting on behalf of the Secretary will notify the applicant as to whether he considers that mediation will effectuate the purpose of the act and as to whether he will mediate.

§ 900.106　Assignment of mediator.

The Director of the Division shall assign a mediator, from the group designated by the Administrator, to act in such capacity.

§ 900.107　Meetings.

All meetings held pursuant to §§ 900.103 to 900.109 shall be held with and under the direction of the mediator.

§ 900.108　Mediator's report.

The mediator, upon the completion of mediation proceedings, shall submit to the Administrator a complete report on such proceedings.

§ 900.109　Mediation agreement.

An agreement arrived at by mediation shall not become effective until approved by the Secretary, and the Secretary will not approve an agreement if there is evidence of fraud, if there is a lack of evidence to support the agreement, or if the agreement provides for any unfair trade practice.

§ 900.110　Application for arbitration.

An application for arbitration by a cooperative shall be in writing and shall contain the following information:

(a) Names in full of the parties to the dispute and their addresses;

(b) The same information required under § 900.103(b);

(c) Concise statement of dispute to be submitted;

(d) Originals or certified copies of all contracts, if any, involved in the dispute, and of correspondence which has passed between the parties and of any other documents or information relied upon;

(e) Dates before which it is desired that the hearing shall be had and the award shall become effective;

(f) Suggested time and place for arbitration hearing.

The applicant shall send a copy of the application to each other party to the dispute.

§ 900.111　Inquiry by the Administrator.

Upon receipt of an application for arbitration, the Administrator, through such officers or employees of the Service as he may designate, may make any inquiry deemed to be necessary or proper in order to determine whether a bona fide dispute exists, to assist the parties in reducing the dispute to well-defined issues, and to select an arbitrator who would be satisfactory to all parties.

§ 900.112　Notification.

The Administrator, acting on behalf of the Secretary, within a reasonable time after the receipt of an application, will notify the applicant as to whether he will grant the application.

§ 900.113　Submission.

(a)(1) Within a reasonable time after the receipt of the Administrator's consent to arbitrate, the parties to the dispute shall file with the Administrator a formal submission, which shall contain the following information:

(i) Names in full of the parties;

(ii) Addresses of the parties to whom all notifications and communications concerning the arbitration shall be sent;

(iii) Description of the organization and businesses of all parties to the dispute, including sufficient information to show that the cooperative is a bona fide one, and that the parties are engaged in activities in the current of interstate or foreign commerce;

(iv) Concise statement of the specific questions submitted and a brief outline of the contentions of each party to the dispute, and a statement as to the period of time during which the award shall be in effect, said period to be not less than thirty days from the effective date of the award;

(v) Name of arbitrator;

(vi) Time and place of arbitration, including street address;

(vii) Stipulation by the parties that they will produce any books, records, and correspondence required by the arbitrator as being necessary to a fair determination of the dispute;

(viii) Agreement by the parties that they will consider the award as final and will comply therewith;

(ix) Stipulation by the parties that arbitration is to take place under rules and regulations issued by the Secretary, and that any such rules and regulations pertaining to mediation and arbitration shall be considered a part of the submission;

(x) Stipulation that a stenographic report of the proceedings must be made.

(2) The submission shall be signed by each party before a notary public, and when the signature is that of an agent of a corporation or cooperative association, the same shall be accompanied by evidence of the authority to sign.

(3) A submission may be withdrawn at any time before the award, and any question held by the arbitrator to be a separable question may be withdrawn before award by agreement of all parties. When any question is so withdrawn, the parties shall file with the arbitrator the agreement on that question reached by the parties, showing all the details thereof, and the arbitrator shall include it in the record of the arbitration.

(b) [Reserved]

§900.114 Designation of arbitrator.

The Administrator, after receiving the submission, will designate one or more persons to act as arbitrator.

§900.115 Hearing.

(a) The arbitrator shall have full discretion to conduct the hearing in such manner as will, in his opinion, enable him to ascertain all the facts in the case.

(b) Parties to the dispute may appear in person or by duly accredited agents and may be represented by counsel.

(c) All relevant and material evidence may be presented. The arbitrator shall not be bound by the legal rules of evidence.

(d) The arbitrator, in the presence of the parties, may require the production of books and records for examination by himself, but not for examination of confidential information by other parties to the dispute unless the party producing the same consents to its examination by the other parties to the dispute.

(e) No evidence offered by one party shall be received except in the presence of all parties unless the parties so agree in a submission specifying the nature of the evidence to be received.

(f) Final determination as to what will be considered confidential information shall be made by the arbitrator.

(g) The arbitrator may request the opinions of economists, marketing specialists, statisticians, lawyers, accountants, and other experts.

(h) When more than two arbitrators are designated to hear a dispute, and they disagree, the award of the majority shall be the final award. If the arbitrators are evenly divided, there shall be no award.

(i) A stenographic record of all the proceedings during an arbitration must be made.

§900.116 Award.

(a) An award shall be made within ten days after the close of the hearing.

(1) The award shall be in writing and shall cover only points of dispute raised in the submission.

(2) The arbitrator, in making the award, may use his own technical knowledge in addition to the evidence submitted by the parties.

(3) The award shall state the period during which it shall be in effect, said period to be not less than thirty days from the effective date thereof; and said period may be extended by agreement among the parties upon notification thereof to the Administrator, unless or until the Administrator withdraws his approval.

(4) The arbitrator shall sign the award in the presence of a notary public, or, when more than one arbitrator is designated the arbitrator shall sign in the presence of each other.

(5) Copies of the award shall be delivered to the parties by the Division.

(b) [Reserved]

§900.117 Approval of award.

The award shall not become effective until approved by the Secretary, and the Secretary will not approve an award if there is evidence of fraud, or evidence of misconduct upon the part of the arbitrator, or lack of evidence to support the award, or if the award provides for any unfair trade practice.

§ 900.118 Costs.

The parties jointly shall pay for the stenographic record. A copy of the record shall be furnished by the parties to the arbitrator and shall be forwarded by him to the Administrator, ultimately to be filed in the office of the hearing clerk. The arbitrator shall not receive compensation for parties to the dispute.

Subpart G—Miscellaneous Requirements

AUTHORITY: Sec. 10, 48 Stat. 37, as amended; 7 U.S.C. 610.

§ 900.200 Definitions.

As used in this subpart, the terms as defined in the Act shall apply with equal force and effect. In addition, unless the context otherwise requires:

(a) The term *Act* means Public Act No. 10, 73d Congress (48 Stat. 31), as amended and as reenacted and amended by the Agricultural Marketing Agreement Act of 1937 (50 Stat. 246, 7 U.S.C. 601), as amended;

(b) The term *Department* means the United States Department of Agriculture;

(c) The term *Secretary* means the Secretary of Agriculture of the United States, or any officer or employee of the Department to whom authority has heretofore been delegated, or to whom authority may hereafter be delegated, to act in his stead;

(d) The term *General Counsel* means the General Counsel of the Department;

(e) The term *Administrator* means the Administrator of the Agricultural Marketing Service, with power to redelegate, or any officer or employee of the Department to whom authority has been delegated or may hereafter be delegated to act in his stead.

(f) The term *mail* means to transmit either electronically or through a postal or other delivery system, information or a package (*e.g.*, letter or envelope) to a recipient.

(g) The term FEDERAL REGISTER means the publication provided for by the Act of July 26, 1935 (49 Stat. 500), and Acts supplementary thereto and amendatory thereof;

(h) The term *marketing agreement* means any marketing agreement or any amendment thereto which may be entered into pursuant to section 8b of the Act;

(i) The term *marketing order* means any order or any amendment thereto which may be issued pursuant to section 8c of the Act;

(j) The term *person* means any individual, corporation, partnership, association, or any other business unit;

(k) The term *official* means the Secretary, any officer, employee, or other person employed or appointed by the Department, and any agency or agent appointed by the Secretary to administer a marketing agreement or a marketing order, and any agent or employee of any such agency or agent;

(l) The term *information* means and includes reports, books, accounts, records, and the facts and information contained therein and required to be furnished to or acquired by any official pursuant to the provisions of any marketing agreement or marketing order.

[25 FR 5907, June 28, 1960, as amended at 26 FR 7796, Aug. 22, 1961; 28 FR 579, Jan. 23, 1963; 37 FR 8059, Apr. 25, 1972; 83 FR 27682, June 14, 2018]

§ 900.201 Investigation and disposition of alleged violations.

Whenever the Administrator has reason to believe that any handler has violated, or is violating, the provisions of any marketing order, he may institute such investigation and, after due notice to such handler, conduct such hearing in order to determine the facts as, in his opinion, are warranted. If, in the opinion of the Administrator and the General Counsel, the facts developed as a result of such investigation or hearing warrant such action, the General Counsel shall refer the matter to the Attorney General for appropriate action.

§ 900.202 Restrictions applicable to Committee personnel.

Members and employees of Federal marketing order boards and committees are immune from prosecution under the United States antitrust laws only insofar as their conduct in administering the respective marketing order is authorized by the Agricultural Marketing Agreement Act of 1937, 7 U.S.C. 601–674, or the provisions of the respective order. Under the antitrust laws, Committee members and employees may not engage in any unauthorized agreement or concerted action that unreasonably restrains United States domestic or foreign commerce. For example, Committee members and employees have no authority to participate, either directly or indirectly, whether on an informal or formal, written or oral basis, in any bilateral or international undertaking or agreement with any competing foreign producer or seller or with any foreign government, agency, or instrumentality acting on behalf of competing foreign producers or sellers to raise, fix, stabilize, or set a floor for commodity prices, or limit the quantity or quality of commodity imported into or exported from the United States. Participation in any such unauthorized agreement or joint undertaking could result in prosecution under the antitrust laws by the United States Department of Justice and/or suit by injured private persons seeking treble damages, and could also result in expulsion of members from the Committee or termination of employment with the Committee.

[80 FR 45396, July 30, 2015]

§900.210 Disclosures of information.

All information in the possession of any official which relates to the business or property of any person, and which was furnished by, or obtained from, such person pursuant to the provisions of any marketing agreement or marketing order, shall be kept confidential and shall not be disclosed, divulged, or made public, unless otherwise expressly provided in said marketing agreement or marketing order, or unless said person authorizes said official, in writing, to disclose such information, except that:

(a) Such information may be disclosed, divulged, or made public if it has been obtained from or furnished by a person who is not the person to whose business or property such information relates or an employee of such latter person, or if such information is otherwise required by law to be furnished to an official;

(b) Such information may be furnished to other officials for use in the regular course of their official duties;

(c) Such information may be combined and published in the form of general statistical studies or data in which the identity of the person furnishing such information or from whom it was obtained shall not be disclosed;

(d) Such information may be disclosed upon lawful demand made by the President or by either House of Congress or any committee thereof, or, if the Secretary determines that such disclosure is not contrary to the public interest, such information may be disclosed in response to a subpena by any court of competent jurisdiction.

(e) Such information may be offered in evidence (whether or not it has been obtained from or furnished by the person against whom it is offered) by or on behalf of the Secretary, the United States, or the official who obtained it or to whom it was furnished, in any administrative hearing held pursuant to section 8c(15)(A) of the Act or in any action, suit, or proceeding, civil or criminal, in which the Secretary or the United States or any such official is a party, and:

(1) Which is instituted (i) for the purpose of enforcing or restraining the violation of any marketing agreement or marketing order, or (ii) for the purpose of collecting any penalty or forfeiture provided for in the Act, or (iii) for the purpose of collecting any monies due under a marketing agreement or marketing order, or

(2) In which the validity of any marketing agreement or marketing order, or any provision of either, is challenged or involved.

(f) Such information may be furnished to the duly constituted authorities of any State, pursuant to a written agreement made under authority of section 10(i) of the Act, to the extent that such information is relevant to transactions within the regulatory jurisdiction of such authorities.

§900.211 Penalties.

Any official who shall have violated the provisions of §900.210 by wilfully divulging, disclosing, or making public any information acquired by or furnished to or in the possession or custody of such official pursuant to the provisions of a marketing agreement or marketing order shall be subject to a penalty of the amount specified at § 3.91(b)(1) (viii) of this title for each offense. (The civil penalty provided in this section is prescribed under the authority contained in sec. 10(c) of the Act (7 U.S.C. 610(c)); this provision is not intended to supersede the provision in section 8d(2) of the Act (7 U.S.C. 608d(2)) for criminal liability and removal from office.)

[25 FR 5907, June 28, 1960, as amended at 75 FR 17560, Apr. 7, 2010]

Subpart H—Procedure for Conduct of Referenda To Determine Producer Approval of Milk Marketing Orders To Be Made Effective Pursuant to Agricultural Marketing Agreement Act of 1937, as Amended

AUTHORITY: Secs. 1–19, 48 Stat. 31, as amended; 7 U.S.C. 601–674.

SOURCE: 30 FR 15412, Dec. 15, 1965, unless otherwise noted.

§ 900.300   General.

Unless otherwise prescribed, the procedure contained in this subpart shall be applicable to each producer referendum conducted for the purpose of ascertaining whether the issuance by the Secretary of a milk marketing order is approved or favored, as required under the applicable provisions of the Agricultural Marketing Agreement Act of 1937, as amended (48 Stat. 31, as amended, 7 U.S.C. 601–674). The procedure in this subpart replaces the procedure for conducting similar referenda (15 FR 5177) issued August 7, 1950.

§ 900.301   Definitions.

As used in this subpart and in all supplementary instructions, forms, and documents, unless the context or subject matter otherwise requires, the following terms shall have the following meanings:

(a) *Act. Act* means Public Act No. 10, 73d Congress (48 Stat. 31), as amended, and as re-enacted and amended by the Agricultural Marketing Agreement Act of 1937 (50 Stat. 246), as amended.

(b) *Department. Department* means the United States Department of Agriculture.

(c) *Secretary. Secretary* means the Secretary of Agriculture of the United States, or any officer or employee of the Department to whom authority has heretofore been delegated, or to whom authority may hereafter be delegated, to act in his stead.

(d) *Administrator. Administrator* means the Administrator of the Agricultural Marketing Service, with power to redelegate, or any officer or employee of the Department to whom authority has been delegated or may hereafter be delegated to act in his stead.

(e) *Person. Person* includes any individual, partnership, corporation, association, and any other business unit.

(f) *Order. Order* means the marketing order (including an amendatory order) with respect to which the Secretary has directed that a referendum be conducted.

(g) *Producer. Producer* means any person who is a dairy farmer and who, during the representative period, met the requirements of the term *producer* as defined in the order had such order been in effect during the representative period.

(h) *Handler. Handler* means any person who, during the representative period, met the requirements of the term *handler* as defined in the order had such order been in effect during the representative period.

(i) *Referendum agent. Referendum agent* means the person designated by the Secretary to conduct the referendum.

(j) *Representative period. Representative period* means the period designated by the Secretary pursuant to section 8c of the Act (7 U.S.C. 608c).

(k) *Cooperative association. Cooperative association* means any association of producers that the administrator has found to be qualified pursuant to section 608c(12) of the Act.

[30 FR 15412, Dec. 15, 1965, as amended at 37 FR 8059, Apr. 25, 1972]

§ 900.302 Associations eligible to vote.

(a) Any association of producers, not previously determined to be a cooperative association may file an application for a determination as to whether it is a cooperative association and thus eligible to vote in a referendum. Such application shall be filed with the Administrator at least 60 days prior to the holding of the referendum: *Provided, however,* That the Administrator may permit the filing of an application in less than 60 days when, in the opinion of the Administrator, such filing would not delay the conduct of the referendum.

(b) Within a time fixed by the referendum agent, but not later than 5 days prior to the final date for balloting, each cooperative association electing to vote shall, upon the request of the referendum agent, furnish to him a certified list showing the name and address of each producer for whom it claims the right to vote and the plant at which such person's milk was received during the representative period.

§ 900.303 Conduct of referendum.

The referendum shall be conducted by mail in the manner prescribed in this subpart. The referendum agent may utilize such personnel or agencies of the Department as are deemed necessary by the Administrator.

§ 900.304 Who may vote.

(a) Each producer shall be entitled to only one vote and to cast one ballot in each referendum; and no person who may claim to be a producer shall be refused a ballot. Each producer casting more than one ballot with conflicting votes shall thereby invalidate all ballots cast by such producer in such referendum. Each ballot cast shall contain a certification by the person casting the ballot that he is a producer.

(b) Except as provided in section 8c(5)(B) of the act, as amended, any cooperative association eligible under § 900.302 may, if it elects to do so, vote and cast one ballot for producers who are members of, stockholders in, or under contract with, such cooperative association. A cooperative association shall submit, with its ballot, a certified copy of the resolution authorizing the casting of the ballot. Each such cooperative association entitled to vote in a referendum casting more than one ballot with conflicting votes shall thereby invalidate all ballots cast by such voter in such referendum.

(c) Voting by proxy or agent, or in any manner, except by the producer or cooperative association will not be permitted; however, a producer which is other than an individual may cast its ballot by a person who is duly authorized and such ballot shall contain a certification by such person that the person on whose behalf the ballot is cast is a producer.

§ 900.305 Duties of referendum agent.

The referendum agent shall also:

(a) For purposes of mailing, prepare a record of producers which will disclose the name of each such person, his address, the name of the handler who received the producer's milk during the representative period, and the name of the cooperative association, if any, which claims the right to vote for the producer. Such record may be compiled from readily available sources, including the following:

(1) Records of the Department;

(2) Producer records supplied by handlers;

(3) Health authority records;

(4) Certifications signed by dairy farmers who claim to be producers;

(5) Any other reliable sources of information which may be available to the referendum agent.

(b) Apply, as a guide, the following criteria in preparing a record of producers:

(1) When the order requires approval by an appropriate health authority before a person meets the definition of producer, only those persons having such approval and who otherwise meet the definition may be regarded as producers. When the definition of producer requires the shipment of milk to a handler or a plant as well as health authority approval, only those persons having such approval and whose milk was received by a handler or at a plant may be regarded as producers.

(2) When the order requries shipment to a handler or to a plant, without regard to health authority approval, a

person may not be regarded as a producer, except as provided in paragraph (b)(6) of this section, unless his name appears on the handler's producer records.

(3) In the case of a producer that is other than an individual, the business unit shall be regarded as the producer.

(4) No person may be included in the record more than once although he may operate more than one farm, hold more than one health authority approval, or appear on more than one handler's producer records.

(5) In the event the health authority records are not available, are inaccurate, or are incomplete, the appearance of the producer's name on a handler's records as an approved producer shall be prima facie evidence of health authority approval.

(6) In the event any handler refuses or fails to make his records available to the referendum agent, a certification signed by the producer shall be regarded by the referendum agent as prima facie evidence that such person is eligible to vote.

(c) Verify the information supplied by each cooperative association which wishes to vote on behalf of producers, as follows:

(1) Examine the records of the cooperative association for the purpose of ascertaining whether each producer claimed by the cooperative association is a member of, stockholder in, or under contract with the cooperative association.

(2) Identify the persons ascertained to be members of, stockholders in, or under contract with a cooperative association which wishes to vote on behalf of its producers with the names of producers which appear on the record compiled pursuant to paragraph (a) of this section.

(3) In determining whether a cooperative association may vote on behalf of a producer the following criteria shall be used:

(i) The cooperative association may vote for each producer who is a member of, stockholder in, or under contract with such cooperative association on the date of the order directing that the referendum be conducted.

(ii) The cooperative association may cast only one ballot for all such producers.

(iii) Whenever more than one cooperative association claims the right to vote for a producer only the cooperative association which furnished evidence satisfactory to the referendum agent that such association was in fact marketing the milk of the producer on the date of the referendum order may vote for such producer.

§ 900.306   Notice of the referendum.

(a) The referendum agent shall at least 5 days prior to the final date for balloting:

(1) Mail to each cooperative association which has elected to cast a ballot on behalf of its producers and to each of all other known producers, a notice of the referendum which will include instructions for completing the ballot, a statement as to the time within which the ballot must be mailed to, and received by, the referendum agent, a copy of the final decision, and a ballot containing a description of the terms and conditions of the order.

(2) Give public notice of the referendum:

(i) By furnishing press releases and other information to available media of public information (including but not limited to press, radio, and television facilities) serving the area, announcing the time within which ballots must be completed and mailed to and received by the referendum agent, eligibility requirements, where additional information may be procured, and other pertinent information; and

(ii) By such other means as said agent may deem advisable.

(b) [Reserved]

§ 900.307   Time for voting.

There shall be no voting except within the time specified by the referendum agent as stated in the notice of the referendum.

§ 900.308   Tabulation of ballots.

(a) *General.* The referendum agent shall verify the information supplied with each ballot. If he ascertains that the person who cast the ballot was eligible to do so, that the ballot is complete and was mailed and received within the prescribed time, the ballot shall be eligible to be counted. If the referendum agent ascertains that the person who cast the ballot was not eligible to do so, or if the producer who cast the ballot was a member of, stockholder in, or under contract with a cooperative association which cast a valid ballot, or if the ballot is not completed or cast in accordance with instructions, or if the ballot was not mailed to or received by the referendum agent within the prescribed time, the ballot shall be marked "disqualified" with a notation on the ballot as to the reason for the disqualification. The total number of ballots cast, including the disqualified ballots, shall be ascertained. The number of eligible ballots cast approving and the number of eligible ballots cast disapproving the issuance of the order shall also be ascertained. The ballots marked "disqualified" shall not be considered as approving or disapproving the issuance of the order, and the persons who cast such ballots shall not be regarded as participating in the referendum.

(b) *Individual-handler pool provisions.* Whenever separate approval of the pooling provisions of the order is required by section 608c(5)(B)(i) of the act, any ballot which approves the issuance of the order and disapproves the pooling provisions, or approves the pooling provisions and disapproves the issuance of the order, shall be disqualified; and the referendum agent shall mark the ballot accordingly.

(c) *Record of results of the referendum.* The referendum agent shall notify the Administrator of the number of eligible ballots cast, the count of the votes, the number of disqualified ballots and the number of producers who were eligible to cast ballots. The referendum agent shall seal the ballots, including those marked "disqualified", the list of eligible voters and tabulation of ballots, and shall transmit to the Administrator a complete detailed report of all action taken in connection with the referendum together with all the ballots cast and all other information furnished to or compiled by the referendum agent.

(d) *Announcement of the results of the referendum.* Announcement of the re-

sults of the referendum will be made only at the direction of the Secretary. The referendum agent, or others who assist in the referendum, shall not disclose the results of the referendum or the total number of ballots cast.

§ 900.309 Confidential information.

The ballots cast, the identity of any person who voted, or the manner in which any person voted and all information furnished to, compiled by, or in the possession of the referendum agent, shall be regarded as confidential.

§ 900.310 Supplementary instructions.

The Administrator is authorized to issue instructions and to prescribe forms and ballots, not inconsistent with the provisions of this subpart, to govern the conduct of referenda by referendum agents.

§ 900.311 Submittals or requests.

Interested persons may secure information or make submittals or requests to the Administrator with respect to the provisions contained in this subpart.

Subpart I—Procedure for Determining the Qualification of Cooperative Milk Marketing Associations

AUTHORITY: Secs. 1–19, 48 Stat. 31, as amended; 7 U.S.C. 601–674.

SOURCE: 32 FR 9821, July 6, 1967, unless otherwise noted.

§ 900.350 General statement.

Cooperative marketing associations apply for qualification by the Secretary under the Federal milk order program for certain privileges and exemptions. These privileges and exemptions are expressed in the Agricultural Marketing Agreement Act of 1937 (50 Stat. 246) as amended, and the milk marketing orders issued pursuant to its provisions.

§ 900.351 Applications for qualification.

Any association of producers may apply for determinations as to whether it is a qualified cooperative association with authority to represent producers in order referendums; has authorization to collect payment from handlers for members' milk; and is rendering specified marketing services to producers. Applicant associations should supply information for these determinations, using as a guide Application Form DA–25. The application form may be obtained from the Dairy Division, Agricultural Marketing Service, United States Department of Agriculture, Washington, DC 20250. Determinations required of the Secretary of Agriculture, or the Administrator of the Agricultural Marketing Service, by delegation are made by the Director of the Dairy Division. Once issued they are valid until amended, suspended or terminated.

§ 900.352　Confidential information.

The documents and other information submitted by an applicant association and otherwise obtained by investigation, examination of books, documents, papers, records, files and facilities, and in reports filed subsequent to initial determinations of qualification, shall be regarded as confidential and shall be governed by § 900.210.

§ 900.353　Qualification standards.

Statutory requirements for qualification of coopertive associations are provided in subsections (5) and (12) of section 608c of the Agricultural Marketing Agreement Act of 1937, as amended (7 U.S.C. 601 *et seq.*). The association must: (a) Be a cooperative marketing association of producers, qualified under the provisions of the Act of Congress of February 18, 1922, as amended, known as the "Capper-Volstead Act," (7 U.S.C. 291, 292); (b) have its entire organization and all of its activities under the control of its members; (c) have full authority in the sale of its members' milk; and (d) be engaged in making collective sales or marketing of milk or milk products for the producers thereof. Qualification for exemption from deductions for marketing service payments under specific marketing orders and payment for milk of members under specific orders shall be determined in accordance with the terms of the respective marketing orders.

§ 900.354　Inspection and investigation.

The Secretary of Agriculture, or his duly authorized representative, shall have the right, at any time after an application is received, to examine all books, documents, papers, records, files and facilities of the association, to verify any of the information submitted and to procure such other information as may be required to determine whether the association is qualified in accordance with its application.

§ 900.355　Annual reporting.

Determinations of qualification for privileges and exemptions are subject to amendment, termination or suspension if the association does not currently meet the qualification standards. An association found to be qualified pursuant to the Act is required to file an annual report after its annual meeting has been held following the close of its fiscal year. Form DA–24 is used for this purpose. The report form is available at the Dairy Division, Agricultural Marketing Service, U.S. Department of Agriculture, Washington, DC 20250. The association is required to file a copy of its report with the Dairy Division at Washington and with the market administrator of each order under which it operates.

§ 900.356　Listing of qualified associations.

A copy of each determination of qualification is furnished to the respective association. Copies are also filed in the Dairy Division, Agricultural Marketing Service, and with the Hearing Clerk, Office of the Secretary, U.S. Department of Agriculture, Washington, DC 20250, where they are available for public inspection. A list of qualified associations engaged in marketing milk under a particular milk marketing order is maintained at the office of the market administrator of the order.

§ 900.357　Denial of application; suspension or revocation of determination of qualification.

Any cooperative association whose application has been wholly or partially denied, or whose determination of qualification has been wholly or partly revoked or suspended, may petition the Secretary for a review of such action. Such petition shall state facts relevant to the matter for which review is sought. After due notice to such cooperative association, the Director of the Dairy Division, or in his absence the Acting Director, shall hold, in the manner hereinafter specified, an informal hearing.

(a) *Notice*. Notice shall be given in writing and shall be mailed to the last known address of the association, or of an officer thereof, at least 3 days before the date set for a hearing. Such notice shall contain: A statement of the time and place of the hearing, said place to be as convenient to the association as can reasonably be arranged, and may contain a statement of the reason for calling the hearing and the nature of the questions upon which evidence is desired or upon which argument may be presented.

(b) *Parties*. Hearings are not to be public and are to be attended only by representatives of the association and of the Government, and such other persons as either the association or the Government desires to have appear for purposes of submitting information or as counsel.

(c) *Conduct of hearing*. The Director or Acting Director of the Dairy Division, or a person designated by him, shall preside at the hearing. The hearing shall be conducted in such manner as will be most conducive to the proper disposition of the matter. Written statements or briefs may be filed by the association within the time specified by the presiding officer.

(d) *Preliminary report*. The presiding officer shall prepare a preliminary report setting forth a recommendation as to what action shall be taken and the basis for such action. A copy of said report shall be served upon the association by mail or in person. The association may file exceptions to said report within 10 days after service thereof.

(e) *Final report*. After due consideration of all the facts and the exceptions, if any, the Director of the Dairy Division shall issue a final report setting forth the action to be taken and the basis for such action.

Subpart J—Procedure for the Conduct of Referenda in Connection With Marketing Orders for Fruits, Vegetables, and Nuts Pursuant to the Agricultural Marketing Agreement Act of 1937, as Amended

AUTHORITY: Secs. 1–19, 48 Stat. 31, as amended; 7 U.S.C. 601–674.

SOURCE: 30 FR 15414, Dec. 15, 1965, unless otherwise noted.

§ 900.400 General.

Referenda for the purpose of ascertaining whether the issuance by the Secretary of Agriculture of a marketing order to regulate the handling of any fruit, vegetable, or nut, or product thereof, or the continuance or termination of such an order, is approved or favored by producers or processors shall, unless supplemented or modified by the Secretary, be conducted in accordance with this subpart.

§ 900.401 Definitions.

(a) *Act* means Public Act No. 10, 73d Congress (48 Stat. 31), as amended, and as reenacted and amended by the Agricultural Marketing Agreement Act of 1937 (50 Stat. 246), as amended (7 U.S.C. 601–674).

(b) *Secretary* means the Secretary of Agriculture of the United States, or any officer or employee of the Department to whom authority has heretofore been delegated, or to whom authority may hereafter be delegated, to act in his stead; and *Department* means the United States Department of Agriculture.

(c) *Administrator* means the Administrator of the Agricultural Marketing Service, with power to redelegate, or any officer or employee of the Department to whom authority has been delegated or may hereafter be delegated to act in his stead.

(d) *Order* means the marketing order (including an amendatory order) with respect to which the Secretary has directed that a referendum be conducted.

(e) *Referendum agent* means the individual or individuals designated by the Secretary to conduct the referendum.

(f) *Representative period* means the period designated by the Secretary pursuant to section 8c of the act (7 U.S.C. 608c).

(g) *Person* means any individual, partnership, corporation, association, or other business unit. For the purpose of this definition, the term *partnership* includes (1) a husband and wife who have title to, or leasehold interest in, land as tenants in common, joint tenants, tenants by the entirety, or, under community property laws, as community property, and (2) so-called *joint ventures,* wherein one or more parties to the agreement, informal or otherwise, contributed capital and others contribute labor, management, equipment, or other services, or any variation of such contributions by two or more parties, so that it results in the growing of the commodity for market and the authority to transfer title to the commodity so produced.

(h) *Producer* means any person defined as a producer in the order who: (1) Owns and farms land, resulting in his ownership of the commodity produced thereon; (2) Rents and farms land, resulting in his ownership of all or a portion of the commodity produced thereon; or (3) Owns land which he does not farm and, as rental for such land, obtains the ownership of a portion of the commodity produced thereon. Ownership of, or leasehold interest in, land and the acquisition, in any manner other than as hereinbefore set forth, of legal title to the commodity grown thereon shall not be deemed to result in such owners or lessees becoming producers.

[30 FR 15414, Dec. 15, 1965, as amended at 37 FR 8059, Apr. 25, 1972]

§ 900.402 Voting.

(a) Each person who is a producer, as defined in this subpart, at the time of the referendum and who also was a producer during the representative period, shall be entitled to only one vote in the referendum, except that: (1) In a landlord-tenant relationship, where in each of the parties is a producer, each such producer shall be entitled to one vote in the referendum; and (2) a cooperative association of producers, bona fide engaged in marketing the commodity or product thereof proposed to be regulated, or in rendering services for or advancing the interest of the producers of such commodity or product, may, if it elects to do so, vote, both by number and total volume, for the producers who are members of, stockholders in, or under contract with such association.

(b) Whenever, as required by the act, processors vote on the issuance of an order, each processor who is engaged in canning or freezing within the production area of the commodity covered by the order shall be entitled to vote in the referendum the quantity of such commodity canned or frozen within the production area for market by him during the representative period determined by the Secretary.

(c) Proxy voting is not authorized but an officer or employee of a corporate producer, processor or cooperative association, or an administrator, executor or trustee of a producing estate may cast a ballot on behalf of such producer, processor, estate, or cooperative association. Any individual so voting in a referendum shall certify that he is an officer or employee of the producer, processor, or cooperative association, or an administrator, executor, or trustee of a producing estate, and that he has the authority to take such action. Upon request of the referendum agent, the individual shall submit adequate evidence of such authority.

(d) Each producer, cooperative association of producers, and processor entitled to vote in a referendum shall be entitled to cast one ballot in the referendum. Each producer, cooperative association of producers, and processor casting more than one ballot with conflicting votes shall thereby invalidate all ballots cast by such producer, cooperative association of producers, or processor in such referendum.

§ 900.403 Instructions.

The referendum agent shall conduct the referendum, in the manner herein provided, under supervision of the Administrator. The Administrator may prescribe additional instructions, not inconsistent with the provisions hereof, to govern the procedure to be followed by the referendum agent. Such agent shall:

(a) Determine the time of commencement and termination of the period of the referendum, and the time prior to which all ballots must be cast.

(b) Determine whether ballots may be cast by mail, at polling places, at meetings of producers or processors, or by any combination of the foregoing.

(c) Provide ballots and related material to be used in the referendum. Ballot material shall provide for recording essential information for ascertaining:

(1) Whether the person voting, or on whose behalf the vote is cast, is an eligible voter, and

(2) The total volume (i) produced for market during the representative period, or (ii) canned or frozen for market during the representative period.

(d) Give reasonable advance notice of the referendum (1) by utilizing without advertising expense available media of public information (including, but not being limited to, press and radio facilities) serving the production area, announcing the dates, places, or methods of voting, eligibility requirements, and other pertinent information, and (2) by such other means as said agent may deem advisable.

(e) Make available to producers and the aforesaid cooperative associations which indicate to the agent their intentions to vote, and to processors when required, instructions on voting, appropriate ballot and certification forms, and, except in the case of a referendum on the termination or continuance of an order, the text of the proposed order and a summary of its terms and conditions: *Provided,* That no person who claims to be qualified to vote shall be refused a ballot.

(f) If ballots are to be cast by mail, cause all the material specified in paragraph (e) of this section to be mailed to each producer (and processor when required) whose name and address is known to the referendum agent.

(g) If ballots are to be cast at polling places or meetings, determine the necessary number of polling or meeting places, designate them, announce the time of each meeting or the hours during which each polling place will be open, provide the material specified in paragraph (e) of this section, and provide for appropriate custody of ballot forms and delivery to the referendum agent of ballots cast.

(h) At the conclusion of the referendum, canvass the ballots, tabulate the results, and, except as otherwise directed, report the outcome to the Administrator and promptly thereafter submit the following:

(1) All ballots received by the agent and appointees, together with a certificate to the effect that the ballots forwarded are all of the ballots cast and received by such persons during the referendum period;

(2) A list of all challenged ballots deemed to be invalid; and

(3) A tabulation of the results of the referendum and a report thereon, including a detailed statement explaining the method used in giving publicity to the referendum and showing other information pertinent to the manner in which the referendum was conducted.

§ 900.404 Subagents.

The referendum agent may appoint any person or persons deemed necessary or desirable to assist said agent in performing his functions hereunder. Each person so appointed may be authorized by said agent to perform, in accordance with the requirements herein set forth, any or all of the following functions (which, in the absence of such appointment, shall be performed by said agent):

(a) Give public notice of the referendum in the manner specified herein;

(b) Preside at a meeting where ballots are to be cast or as poll officer at a polling place;

(c) Distribute ballots and the aforesaid texts to producers (and to processors when required) and receive any ballots which are cast; and

(d) Record the name and address of each person receiving a ballot from, or casting a ballot with, said subagent and inquire into the eligibility of such person to vote in the referendum.

§ 900.405 Ballots.

The referendum agent and his appointees shall accept all ballots cast; but, should they, or any of them, deem that a ballot should be challenged for any reason, said agent or appointee shall endorse above his signature, on said ballot, a statement to the effect that such ballot was challenged, by whom challenged, the reasons therefor, the results of any investigations made with respect thereto, and the disposition thereof. Invalid ballots shall not be counted.

§ 900.406 Referendum report.

Except as otherwise directed, the Administrator shall prepare and submit to the Secretary a report on results of the referendum, the manner in which it was conducted, the extent and kind of public notice given, and other information pertinent to analysis of the referendum and its results.

§ 900.407 Confidential information.

All ballots cast and the contents thereof (whether or not relating to the identity of any person who voted or the manner in which any person voted) and all information furnished to, compiled by, or in possession of, the referendum agent shall be treated as confidential.

Subpart K—Public Information

AUTHORITY: 5 U.S.C. 301, 552.

AVAILABILITY OF PROGRAM INFORMATION, STAFF MANUALS AND INSTRUCTIONS, AND RELATED MATERIAL

§ 900.500 General.

This subpart is issued in accordance with the regulations of the Secretary of Agriculture in part 1, subpart A, of subtitle A of this title (7 CFR 1.1 through 1.16), and appendix A thereto, implementing the Freedom of Information Act (5 U.S.C. 552). The Secretary's regulations, as implemented by the regulations of this subpart, govern the availability of records of AMS to the public.

[40 FR 20267, May 9, 1975]

§ 900.501 Public inspection and copying.

(a) Facilities for public inspection and copying of the indexes and materials required to be made available under § 1.2(a) of this title will be provided by AMS during normal information should be made to the Freedom of

Information Act Officer at the following address:

Freedom of Information Act Officer, Agricultural Marketing Service, United States Department of Agriculture, Washington, DC 20250.

(b) Copies of such material may be obtained in person or by mail. Applicable fees for copies will be charged in accordance with the regulations prescribed by the Director, Office of Operations and Finance, USDA.

[44 FR 39151, July 5, 1979]

§ 900.502 Indexes.

Pursuant to the regulations in § 1.4(b) of this title, AMS will maintain and make available for public inspection and copying current indexes of all material required to be made available in § 1.2(a) of this title. Notice is hereby given that publication of these indexes is unnecessary and impractical, since the material is voluminous and does not change often enough to justify the expense of publication.

[44 FR 39151, July 5, 1979]

§ 900.503 Request for records.

(a) Requests for records under 5 U.S.C. 552(a)(3) shall be made in accordance with § 1.3(a) of this title. Authority to make determinations regarding initial requests in accordance with § 1.4(c) of this title is delegated to the Freedom of Information Act Officer of AMS. Requests should be submitted to the FOIA Officer at the following address:

Freedom of Information Act Officer (FOIA Request). Agricultural Marketing Service, United States Department of Agriculture, Washington, DC 20250.

(b) The request shall identify each record with reasonable specificity as prescribed in § 1.3 of this title.

(c) The FOIA Officer is authorized to receive requests and to exercise the authority to (1) make determinations to grant requests or deny initial requests, (2) extend the administrative deadline, (3) make discretinary release of exempt records, and (4) make determinations regarding charges pursuant to the fee schedule.

[44 FR 39151, July 5, 1979]

§ 900.504 Appeals.

Any person whose request under § 900.503 above is denied shall have the right to appeal such denial in accordance with § 1.3(e) of this title. Appeals shall be addressed to the Administrator, Agricultural Marketing Service, U.S. Department of Agriculture, Washington, DC 20250.

[40 FR 20267, May 9, 1975]

Subpart L—Information Collection

AUTHORITY: 44 U.S.C. Ch. 35.

§ 900.600 General.

This subpart shall contain such requirements as pertain to the information collection provisions under the Paperwork Reduction Act of 1995.

[63 FR 10492, Mar. 4, 1998]

§ 900.601 OMB control numbers assigned pursuant to the Paperwork Reduction Act.

(a) *Purpose.* This section collects and displays the control numbers assigned to information collection requirements by the Office of Management and Budget contained in 7 CFR parts 905 through 998 under the Paperwork Reduction Act of 1995.

(b) *Display.*

7 CFR part where identified and described	Current OMB control No.
905, Florida Oranges, Grapefruit Tangerines, Tangelos	0581–0094
906, Texas Oranges & Grapefruit	0581–0068
911, Florida Limes	0581–0091
915, Florida Avocados	0581–0078
916, California Nectarines	0581–0072
917, California Pears and Peaches	0581–0080
920, California Kiwifruit	0581–0149
922, Washington Apricots	0581–0095
923, Washington Sweet Cherries	0581–0133
924, Washington-Oregon Fresh Prunes	0581–0134
925, S.E. California Desert Grapes	0581–0109
927, Oregon-Washington-California Winter Pears	0581–0089
928, Hawaiian Papayas	0581–0102
929, Cranberries Grown in Designated States	0581–0103
930, Red Tart Cherries	0581–0177
931, Oregon-Washington Bartlett Pears	0581–0092
932, California Olives	0581–0142
945, Idaho-Eastern Oregon Potatoes	0581–0178
946, Washington Potatoes	0581–0178
947, Oregon-California Potatoes	0581–0178
948, Colorado Potatoes	0581–0178
953, Southeastern Potatoes	0581–0178
955, Vidalia Onions	0581–0178
956, Walla Walla Onions	0581–0178
958, Idaho-Oregon Onions	0581–0178

7 CFR part where identified and described	Current OMB control No.
959, South Texas Onions	0581–0178
966, Florida Tomatoes	0581–0178
979, South Texas Melons	0581–0178
981, California Almonds	0581–0071
982, Oregon-Washington Hazelnuts	0581–0178
984, California Walnuts	0581–0178
985, Spearmint Oil	0581–0065
987, California Dates	0581–0178
989, California Raisins	0581–0178
993, California Dried Prunes	0581–0178
997, Domestic Peanuts Not Covered Under the Peanut Marketing Agreement	0581–0163
998, Domestic Peanuts Covered Under the Peanut Marketing Agreement	0581–0067

[63 FR 10492, Mar. 4, 1998]

Subpart M—Assessment of Exemptions

§ 900.700 Exemption from assessments.

(a) This section specifies criteria for identifying persons eligible to obtain an exemption from the portion of the assessment used to fund marketing promotion activities under a marketing order and the procedures for applying for such an exemption under 7 CFR parts 905, 906, 915, 922, 923, 925, 927, 929, 930, 932, 948, 955, 956, 958, 959, 966, 981, 982, 984, 985, 987, 989, 993, and such other parts (included in 7 CFR parts 905 through 998) covering marketing orders for fruits, vegetables, and specialty crops as may be established or amended to include market promotion. For the purposes of this section, the term "assessment period" means fiscal period, fiscal year, crop year, or marketing year as defined under these parts; the term "marketing promotion" means marketing research and development projects or marketing promotion, including paid advertising designed to assist, improve, or promote the marketing, distribution, or consumption of the applicable commodity.

(b) A handler who operates under an approved National Organic Program (7 CFR part 205) (NOP) organic handling system plan and is subject to assessments under a part or parts specified in paragraph (a) of this section may be exempt from the portion of the assessment applicable to marketing promotion, including paid advertising, provided that:

(1) Only agricultural commodities certified as "organic" or "100 percent

47

organic" (as defined in the NOP) are eligible for exemption;

(2) The exemption shall apply to all certified "organic" or "100 percent organic" (as defined in the NOP) products of a handler regardless of whether the agricultural commodity subject to the exemption is handled by a person that also handles conventional or non-organic agricultural products of the same agricultural commodity as that for which the exemption is claimed;

(3) The handler maintains a valid certificate of organic operation as issued under the Organic Foods Production Act of 1990 (7 U.S.C. 6501-6522)(OFPA) and the NOP regulations issued under OFPA (7 CFR part 205);

(4) Any handler so exempted shall continue to be obligated to pay assessments under such part or parts specified that are associated with any agricultural products that do not qualify for an exemption under this section; and

(5) For exempted products, any handler so exempted shall be obligated to pay the portion of the assessment associated with the other authorized activities under such part or parts other than marketing promotion, including paid advertising.

(c) *Assessment exemption application.* (1) To be exempt from paying assessments for these purposes under a part or parts listed in paragraph (a) of this section, the handler shall submit an application to the board or committee established under the applicable part or parts prior to or during the assessment period. This application, Form FV-649, "Certified Organic Handler Application for Exemption from Market Promotion Assessments Paid Under Federal Marketing Orders," shall include:

(i) The date, applicable committee or board, and Federal marketing order number;

(ii) The applicant's full name, company name, address, telephone and fax numbers, and email address;

(iii) Certification that the applicant maintains a valid certificate of organic operation under the OFPA and the NOP;

(iv) Certification that the applicant handles or markets organic products

eligible to be labeled "organic" or "100 percent organic" under the NOP;

(v) Certification that the applicant is otherwise subject to assessments under the Federal marketing order program for which the exemption is requested;

(vi) The number of organic certified producers for whom they handle or market product (including the applicant);

(vii) A requirement that the applicant attach a copy of their certificate of organic operation and all applicable producer certificates of organic operation issued by a USDA-accredited certifying agent under the OFPA and the NOP;

(viii) Certification, as evidenced by signature and date, that all information provided by the applicant is true; and

(ix) Such other information as the committee or board may require, with the approval of the Secretary.

(2) The handler shall file the application with the committee or board, prior to or during the applicable assessment period, and annually thereafter, as long as the handler continues to be eligible for the exemption. If the person complies with the requirements of this section and is eligible for an assessment exemption, the committee or board will approve the exemption request and provide written notification of such to the applicant within 30 days. If the application is disapproved, the committee or board will provide written notification of the reason(s) for such disapproval within the same timeframe.

(3) The exemption will apply at the beginning of the next assessable period following notification of approval of the assessment exemption, in writing, by the committee or board.

(d) *Assessment exemption calculation.* (1) The applicable assessment rate for any handler approved for an exemption shall be computed by dividing the committee's or board's estimated non-marketing promotion expenditures by the committee's or board's estimated total expenditures approved by the Secretary and applying that percentage to the assessment rate applicable to all persons for the assessment period. The modified assessment rate shall then be applied to the quantity of certified "organic" or "100 percent organic" products handled under an approved organic assessment exemption as provided in paragraph (c)(2) of this section. Products handled not subject to an approved organic assessment exemption shall be assessed at the assessment rate applicable to all persons for the assessment period. The committee's or board's estimated non-marketing promotion expenditures shall exclude the direct costs of marketing promotion and the portion of committee's or board's administrative and overhead costs (*e.g.*, salaries, supplies, printing, equipment, rent, contractual expenses, and other applicable costs) to support and administer the marketing promotion activities.

(2) If a committee or board does not plan to conduct any market promotion activities in a fiscal year, the committee or board may submit a certification to that effect to the Secretary, and as long as no assessments for such fiscal year are used for marketing promotion projects, or the administration of projects are funded by a previous fiscal period's assessments, the committee or board may assess all handlers, regardless of their organic status, the full assessment rate applicable to the assessment period.

(3) For each assessment period, the Secretary shall review the portion of the assessment rate applicable to marketing promotion for persons eligible for an exemption and, if appropriate, approve the assessment rate.

(4) When the requirements of this section for exemption no longer apply to a handler, the handler shall inform the committee or board within 30 days and pay the full assessment on all remaining assessable product for all committee or board assessments from the date the handler no longer is eligible to the end of the assessment service.

(5) Within 30 days following the applicable assessment period, the committee or board shall re-compute the applicable assessment rate for handlers exempt under this section based on the actual expenditures incurred during the applicable assessment period. The Secretary shall review, and if appropriate, approve any change in the portion of the assessment rate for market promotion applicable to exempt handlers, and authorize adjustments for any overpayments or collection of underpayments.

[80 FR 82020, Dec. 31, 2015]

PART 905—ORANGES, GRAPEFRUIT, TANGERINES, AND PUMMELOS GROWN IN FLORIDA

Subpart A—Order Regulating Handling

DEFINITIONS

Subpart A—Order Regulating Handling

SOURCE: 22 FR 10734, Dec. 27, 1957, unless otherwise noted. Redesignated at 26 FR 12751, Dec. 30, 1961.

DEFINITIONS

§ 905.1　Secretary.

Secretary means the Secretary of Agriculture of the United States, or any officer or employee of the United States Department of Agriculture to whom authority has heretofore been delegated, or to whom authority may hereafter be delegated, to act in his stead.

[42 FR 59368, Nov. 17, 1977]

§ 905.2　Act.

Act means Public Act No. 10, 73d Congress (May 12, 1933), as amended and as reenacted and amended by the Agricultural Marketing Agreement Act of 1937, as amended. (48 Stat. 31, as amended; 7 U.S.C. 601 *et seq.;* 68 Stat. 906, 1047.)

§ 905.3　Person.

Person means an individual, partnership, corporation, association, business trust, legal representative, or any organized group of individuals.

§ 905.4　Fruit.

Fruit means any or all varieties of the following types of citrus fruits grown in the production area:

(a) Citrus sinensis, Osbeck, commonly called "oranges";

(b) Citrus paradisi, MacFadyen, commonly called "grapefruit";

(c) Citrus reticulata, commonly called "tangerines" or "mandarin";

(d) Citrus maxima Merr (L.); Osbeck, commonly called "pummelo"; and,

(e) "Citrus hybrids" that are hybrids between or among one or more of the four fruits in paragraphs (a) through (d) of this section and the following: Trifoliate orange (Poncirus trifoliata), sour orange (C. aurantium), lemon (C. limon), lime (C. aurantifolia), citron (C. medica), kumquat (Fortunella species), tangelo (C. reticulata x C. paradisi or C. grandis), tangor (C. reticulata x C. sinensis), and varieties of these species. In addition, citrus hybrids include: Tangelo (C. reticulata x C. paradisi or C. grandis), tangor (C. reticulata x C. sinensis), Temple oranges, and varieties thereof

[81 FR 10454, Mar. 1, 2016]

§905.5 Variety.

Variety or *varieties* means any one or more of the following classifications or groupings of fruit:

(a) *Oranges.* (1) Early and Midseason oranges;

(2) Valencia, Lue Gim Gong, and similar late maturing oranges of the Valencia type;

(3) Navel oranges.

(b) *Grapefruit.* (1) Red Grapefruit, to include all shades of color;

(2) White Grapefruit.

(c) *Tangerines and mandarins.* (1) Dancy and similar tangerines;

(2) Robinson tangerines;

(3) Honey tangerines;

(4) Fall-Glo tangerines;

(5) US Early Pride tangerines;

(6) Sunburst tangerines;

(7) W-Murcott tangerines;

(8) Tangors.

(d) *Pummelos.* (1) Hirado Buntan and other pink seeded pummelos;

(2) [Reserved].

(e) *Citrus hybrids*—(1) *Tangelos.* (i) Orlando tangelo;

(ii) Minneola tangelo.

(2) Temple oranges.

(f) *Other varieties of citrus fruits specified in §905.4, including hybrids, as recommended and approved by the Secretary. Provided,* That in order to add any hybrid variety of citrus fruit to be regulated under this provision, such variety must exhibit similar characteristics and be subject to cultural practices common to existing regulated varieties.

[81 FR 10454, Mar. 1, 2016]

§905.6 Producer.

Producer is synonymous with *grower* and means any person who is engaged in the production for market of fruit in the production area and who has a proprietary interest in the fruit so produced.

[42 FR 59368, Nov. 17, 1977]

§905.7 Handler.

Handler is synonymous with *shipper* and means any person (except a common or contract carrier transporting fruit for another person) who, as owner, agent, or otherwise, handles fruit in fresh form, or causes fruit to be handled. Each handler shall be registered with the Committee pursuant to rules recommended by the Committee and approved by the Secretary.

[81 FR 10454, Mar. 1, 2016]

§905.8 Prepare for market.

Prepare for market means to wash, grade, size, or place fruit (whether or not wrapped) into any container whatsoever; but such term shall not include the harvesting of fruit.

§905.9 Handle or ship.

Handle or *ship* means to sell, transport, deliver, pack, prepare for market, grade, or in any other way to place fruit in the current of commerce within the production area or between any point in the production area and any point outside thereof.

[81 FR 10455, Mar. 1, 2016]

§905.10 Carton or standard packed carton.

Carton or standard packed carton means a unit of measure equivalent to four-fifths (⅘) of a United States bushel of fruit, whether in bulk or in any container.

[42 FR 59368, Nov. 17, 1977]

§905.11 Fiscal period.

Fiscal period means the period of time from August 1 of any year until July 31 of the following year, both dates inclusive.

§905.12 Committee.

Committee means the Citrus Administrative Committee established pursuant to §905.19.

[42 FR 59368, Nov. 17, 1977]

§905.13 District.

(a) *Citrus District One* shall include the Counties of Hillsborough, Pinellas, Pasco, Hernando, Citrus, Sumter, and Lake.

(b) *Citrus District Two* shall include the Counties of Osceola, Orange, Seminole, Alachua, Putnam, St. Johns, Flagler, Marion, Levy, Duval, Nassau, Baker, Union, Bradford, Columbia, Clay, Gilchrist, and Suwannee, and County Commissioner, Districts One, Two, and Three of Volusia County, and that part of the Counties of Indian River and Brevard not included in Regulation Area II.

(c) *Citrus District Three* shall include the County of St. Lucie and that part of the Counties of Brevard, Indian River, Martin, and Palm Beach described as lying within Regulation Area II, and County Commissioner's Districts Four and Five of Volusia County.

(d) *Citrus District Four* shall include the Counties of Manatee, Sarasota, Hardee, Highlands, Okeechobee, Glades, De Sota, Charlotte, Lee, Hendry, Collier, Monroe, Dade, Broward, and that part of the Counties of Palm Beach and Martin not included in Regulation Area II.

(e) *Citrus District Five* shall include the County of Polk.

[42 FR 59368, Nov. 17, 1977]

§ 905.14   Redistricting.

(a) The Committee may, with the approval of the Secretary, redefine the districts into which the production area is divided or reapportion or otherwise change the grower membership of districts, or both: *Provided,* That the membership shall consist of at least eight but not more than nine grower members, and any such change shall be based, insofar as practicable, upon the respective averages for the immediately preceding three fiscal periods of:

(1) The number of bearing trees in each district;

(2) The volume of fresh fruit produced in each district;

(3) The total number of acres of citrus in each district; and

(4) Other relevant factors.

(b) Each redistricting or reapportionment shall be announced on or prior to March 1 preceding the effective fiscal period.

[81 FR 10455, Mar. 1, 2016]

§ 905.15   Regulation Area I.

Regulation Area I is defined as the "Interior District", and shall include all that part of the production area not included in Regulation Area II.

[54 FR 37292, Sept. 8, 1989]

§ 905.16   Regulation Area II.

Regulation Area II is defined as the "Indian River District", and shall include that part of the State of Florida particularly described as follows:

Beginning at a point on the shore of the Atlantic Ocean where the line between Flagler and Volusia Counties intersects said shore, thence follow the line between said two counties to the Southwest corner of Section 23, Township 14 South, Range 31 East; thence continue South to the Southwest corner of Section 35, Township 14 South, Range 31 East; thence East to the Northwest corner of Township 15 South, Range 32 East; thence South to the Southwest corner of Township 17 South, Range 32 East; thence East to the Northwest corner of Township 18 South, Range 33 East; thence South to the St. Johns River; thence along the main channel of the St. Johns River and through Lake Harney, Lake Poinsett, Lake Winder, Lake Washington, Sawgrass Lake, and Lake Helen Blazes to the range line between Ranges 35 East and 36 East; thence South to the South line of Brevard County; thence East to the line between Ranges 36 East and 37 East; thence South to the Southwest corner of St. Lucie County; thence East to the line between Ranges 39 East and 40 East; thence South to the South line of Martin County; thence East to the line between Ranges 40 East and 41 East; thence South to the West Palm Beach Canal (also known as the Okeechobee Canal); thence follow said canal eastward to the mouth thereof; thence East to the shore of the Atlantic Ocean; thence Northerly along the shore of the Atlantic Ocean to the point of beginning.

[22 FR 10734, Dec. 27, 1957. Redesignated at 26 FR 12751, Dec. 30, 1961, and further redesignated at 42 FR 59368, Nov. 17, 1977, as amended at 42 FR 59370, Nov. 17, 1977; 54 FR 37292, Sept. 8, 1989]

§ 905.17   Production area.

Production area means that portion of the State of Florida which is bounded by the Suwannee River, the Georgia border, the Atlantic Ocean, and the Gulf of Mexico.

[22 FR 10734, Dec. 27, 1957. Redesignated at 26 FR 12751, Dec. 30, 1961, and further redesignated at 42 FR 59368, Nov. 17, 1977]

§905.18 Improved No. 2 grade and Improved No. 2 Bright grade.

Improved No. 2 grade and *Improved No. 2 Bright grade* means grapefruit meeting all of the respective requirements of the U.S. No. 2 grade and the U.S. No. 2 Bright grade and those requirements of the U.S. No. 1 grade relating to shape (form) and color, as such requirements are set forth in the U.S. Standards for Grades of Florida Grapefruit (§§51.750–51.783 of this title) or as such standards may hereafter be amended.

[31 FR 15060, Dec. 1, 1966. Redesignated at 42 FR 59368, Nov. 17, 1977]

ADMINISTRATIVE BODIES

§905.19 Establishment and membership.

(a) There is hereby established a Citrus Administrative Committee consisting of at least eight but not more than nine grower members, and eight shipper members. Grower members shall be persons who are not shippers or employees of shippers: Provided, that the committee, with the approval of the Secretary, may establish alternative qualifications for such grower members. Shipper members shall be shippers or employees of shippers. The committee may be increased by one non-industry member nominated by the committee and selected by the Secretary. The committee, with approval of the Secretary, shall prescribe qualifications, term of office, and the procedure for nominating the non-industry member.

(b) Each member shall have an alternate who shall have the same qualifications as the member for whom this person is an alternate.

[54 FR 37293, Sept. 8, 1989]

§905.20 Term of office.

The term of office of members and alternate members shall begin on the first day of August of even-numbered years and continue for two years and until their successors are selected and have qualified. The consecutive terms of office of a member shall be limited to two terms. The terms of office of alternate members shall not be so limited. Members, their alternates, and their respective successors shall be nominated and selected by the Secretary as provided in §§905.22 and 905.23.

[81 FR 10455, Mar. 1, 2016]

§905.21 Selection of initial members of the committee.

The initial members of the Citrus Administrative Committee and their respective alternates shall be the members and alternates of the Growers Administrative Committee and the Shippers Advisory Committee serving on the effective date of his amendment. Each member and alternate shall serve until completion of the term for which he was selected and until his successor has been selected and qualified.

[42 FR 59369, Nov. 17, 1977]

§905.22 Nominations.

(a) *Grower members.* (1) The Committee shall give public notice of a meeting of producers in each district to be held not later than June 10th of even-numbered years, for the purpose of making nominations for grower members and alternate grower members. The Committee, with the approval of the Secretary, shall prescribe uniform rules to govern such meetings and the balloting thereat. The chairman of each meeting shall publicly announce at such meeting the names of the persons nominated, and the chairman and secretary of each such meeting shall transmit to the Secretary their certification as to the number of votes so cast, the names of the persons nominated, and such other information as the Secretary may request. All nominations shall be submitted to the Secretary on or before the 20th day of June.

(2) Each nominee shall be a producer in the district from which he or she is nominated. In voting for nominees, each producer shall be entitled to cast one vote for each nominee in each of the districts in which he or she is a producer. At least two of the nominees and their alternates so nominated shall be affiliated with a bona fide cooperative marketing organization.

(b) *Shipper members.* (1) The Committee shall give public notice of a meeting for bona fide cooperative marketing organizations which are handlers, and a meeting for other handlers who are not so affiliated, to be held not later than June 10th of even-numbered years, for the purpose of making nominations for shipper members and their alternates. The Committee, with the approval of the Secretary, shall prescribe uniform rules to govern each such meeting and the balloting thereat. The chairperson of each such meeting shall publicly announce at the meeting the names of the persons nominated and the chairman and secretary of each such meeting shall transmit to the Secretary their certification as to the number of votes cast, the weight by volume of those shipments voted, and such other information as the Secretary may request. All nominations shall be submitted to the Secretary on or before the 20th day of June.

(2) Nomination of at least two members and their alternates shall be made by bona fide cooperative marketing organizations which are handlers. Nominations for not more than six members and their alternates shall be made by handlers who are not so affiliated. In voting for nominees, each handler or his or her authorized representative shall be entitled to cast one vote, which shall be weighted by the volume of fruit by such handler during the then current fiscal period.

(c) Notwithstanding the provisions of paragraphs (a) and (b) of this section, nomination and election of members and alternate members to the Committee may be conducted by mail, electronic mail, or other means according to rules and regulations recommended by the Committee and approved by the Secretary.

[42 FR 59369, Nov. 17, 1977, as amended at 74 FR 46306, Sept. 9, 2009; 81 FR 10455, Mar. 1, 2016]

§ 905.23   Selection.

(a) From the nominations made pursuant to § 905.22(a) or from other qualified persons, the Secretary shall select one member and one alternate member to represent District 2 and two members and two alternate members each to represent Districts 1, 3, 4, and 5 or such other number of members and alternate members from each district as may be prescribed pursuant to § 905.14. At least two such members and their alternates shall be affiliated with bona

fide cooperative marketing organizations.

(b) From the nominations made pursuant to § 905.22(b) or from other qualified persons, the Secretary shall select at least two members and their alternates to represent bona fide cooperative marketing organizations which are handlers, and the remaining members and their alternates to represent handlers who are not so affiliated.

[74 FR 46306, Sept. 9, 2009]

§ 905.27   Failure to nominate.

In the event nominations for a member or alternate member of the committee are not made pursuant to the provisions of §§ 905.22 and 905.25, the Secretary may select such member or alternate member without regard to nominations.

[22 FR 10734, Dec. 27, 1957. Redesignated at 26 FR 12751, Dec. 30, 1961, as amended at 42 FR 59370, Nov. 17, 1977]

§ 905.28   Qualification and acceptance.

Any person nominated to serve as a member or alternate member of the Committee shall, prior to selection by the Secretary, qualify by filing a written qualification and acceptance statement indicating such person's qualifications and willingness to serve in the position for which nominated.

[81 FR 10455, Mar. 1, 2016]

§ 905.29   Inability of members to serve.

(a) An alternate for a member of the committee shall act in the place and stead of such member (1) in his absence, or (2) in the event of his removal, resignation, disqualification, or death, and until a successor for his unexpired term has been selected.

(b) If both a member and his or her respective alternate are unable to attend a committee meeting, such member may designate another alternate to act in his or her place in order to obtain a quorum: *Provided*, That such alternate member represents the same group affiliation as the absent member. If the member is unable to designate such an alternate, the committee members present may designate such alternate.

(c) In the event of the death, removal, resignation, or disqualification of any person selected by the Secretary as a member or an alternate member of the committee, a successor for the unexpired term of such person shall be selected by the Secretary. Such selection may be made without regard to the provisions of this subpart as to nominations.

[22 FR 10734, Dec. 27, 1957. Redesignated at 26 FR 12751, Dec. 30, 1961, as amended at 42 FR 59370, Nov. 17, 1977; 74 FR 46306, Sept. 9, 2009]

§905.30 Powers of the committee.

The committee, in addition to the power to administer the terms and provisions of this subpart, as herein specifically provided, shall have power (a) to make, only to the extent specifically permitted by the provisions contained in this subpart, administrative rules and regulations; (b) to receive, investigate and report to the Secretary complaints of violations of this subpart; and (c) to recommend to the Secretary amendments to this subpart.

[22 FR 10734, Dec. 27, 1957. Redesignated at 26 FR 12751, Dec. 30, 1961, as amended at 42 FR 59370, Nov. 17, 1977]

§905.31 Duties of Citrus Administrative Committee.

It shall be the duty of the Citrus Administrative Committee:

(a) To select a chairman from its membership, and to select such other officers and adopt such rules and regulations for the conduct of its business as it may deem advisable;

(b) To keep minutes, books, and records which will clearly reflect all of its acts and transactions, which minutes, books, and records shall at all times be subject to the examination of the Secretary;

(c) To act as intermediary between the Secretary and the producers and handlers;

(d) To furnish the Secretary with such available information as he may request;

(e) To appoint such employees as it may deem necessary and to determine the salaries and define the duties of such employees;

(f) To cause its books to be audited by one or more certified or registered public accountants at least once for each fiscal period, and at such other times as it deems necessary or as the Secretary may request, and to file with the Secretary copies of all audit reports;

(g) To prepare and publicly issue a monthly statement of financial operations of the committee;

(h) To provide an adequate system for determining the total crop of each variety of fruit, and to make such determinations, including determinations by grade and size, as it may deem necessary, or as may be prescribed by the Secretary, in connection with the administration of this subpart;

(i) To perform such duties in connection with the administration of section 32 of the act to amend the Agricultural Adjustment Act and for other purposes, Public Act No. 320, 74th Congress, as amended, as may from time to time be assigned to it by the Secretary;

[22 FR 10734, Dec. 27, 1957. Redesignated at 26 FR 12751, Dec. 30, 1961, as amended at 30 FR 13934, Nov. 4, 1965; 42 FR 59369, Nov. 17, 1977]

§905.33 Compensation and expenses of committee members.

The members and alternate members of the Committee shall serve without compensation but may be reimbursed for expenses necessarily incurred by them in attending committee meetings and in the performance of their duties under this part.

[42 FR 59369, Nov. 17, 1977]

§905.34 Procedure of committees.

(a) Ten members of the committee shall constitute a quorum.

(b) For any decision or recommendation of the committee to be valid, ten concurring votes, five of which must be grower votes, shall be necessary: *Provided*, That the committee may recommend a regulation restricting the shipment of grapefruit grown in Regulation Area I or Regulation Area II which meets the requirements of the Improved No. 2 grade or the Improved No. 2 Bright grade only upon the affirmative vote of a majority of its members present from the regulation area in which such restriction would apply; and whenever a meeting to consider a recommendation for release of such grade is requested by a majority of the members from the affected area, the committee shall hold a meeting within a reasonable length of time for the purpose of considering such a recommendation. If after such consideration the requesting area majority present continues to favor such release for their area, the request shall be considered a valid recommendation and transmitted to the Secretary. The votes of each member cast for or against any recommendation made pursuant to this subpart shall be duly recorded. Whenever an assembled meeting is held each member must vote in person.

(c) The committee may provide for meeting by telephone, telegraph, or other means of communication, and any vote cast at such a meeting shall be promptly confirmed in writing: *Provided*, That if any assembled meeting is held, all votes shall be cast in person.

(d) The committee shall give the Secretary the same notice of meetings as is given to the members thereof.

[42 FR 59369, Nov. 17, 1977, as amended at 74 FR 46306, Sept. 9, 2009]

§ 905.35 Right of the Secretary.

The members of the committee (including successors and alternates), and any agent or employee appointed or employed by the committee, shall be subject to removal or suspension by the Secretary at any time. Each and every order, regulation, decision, determination, or other act of the committee shall be subject to the continuing right of the Secretary to disapprove of the same at any time and upon his disapproval shall be deemed null and void, except as to acts done in reliance thereon or in compliance therewith.

[22 FR 10734, Dec. 27, 1957. Redesignated at 26 FR 12751, Dec. 30, 1961, as amended at 42 FR 59370, Nov. 17, 1977]

§ 905.36 Funds.

(a) All funds received by the committee pursuant to any provision of this subpart shall be used solely for the purposes herein specified and shall be accounted for in the manner provided in this subpart.

(b) The Secretary may, at any time, require the committee and its members to account for all receipts and disbursements.

(c) Upon the removal or expiration of the term of office of any member of the committee, such member shall account for all receipts and disbursements and deliver all property and funds, together with all books and records, in his possession, to his successor in office, and shall execute such assignments and other instruments as may be necessary or appropriate to vest in such successor full title to all of the property, funds, and claims vested in such member pursuant to this subpart.

[22 FR 10734, Dec. 27, 1957. Redesignated at 26 FR 12751, Dec. 30, 1961, as amended at 42 FR 59370, Nov. 17, 1977]

EXPENSES AND ASSESSMENTS

§ 905.40 Expenses.

The committee is authorized to incur such expenses as the Secretary finds are reasonable and likely to be incurred to carry out the functions of the committee under this subpart during each fiscal period. The funds to cover such expenses shall be acquired by the levying of assessments upon handlers as provided in § 905.41.

[22 FR 10734, Dec. 27, 1957. Redesignated at 26 FR 12751, Dec. 30, 1961, as amended at 42 FR 59370, Nov. 17, 1977]

§ 905.41 Assessments.

(a) Each handler who first handles fruit shall pay to committee, upon demand, such handler's pro rata share of the expenses which the Secretary finds will be incurred by the committee for the maintenance and functioning, during each fiscal period, of the committee established under this subpart. Each such handler's share of such expenses shall be that proportion thereof which the total quantity of fruit shipped by such handler as the first handler thereof during the applicable fiscal period is of the total quantity of fruit so shipped by all handlers during the same fiscal period. The Secretary shall fix the rate of assessment per standard packed carton of fruit to be paid by each such handler. The payment of assessments for the maintenance and functioning of the committee may be required under this part throughout the period it is in effect irrespective of whether particular provisions thereof are suspended or become inoperative.

(b) At any time during or after the fiscal period, the Secretary may increase the rate of assessment so that the sum of money collected pursuant to the provisions of this section shall be adequate to cover the said expenses. Such increase shall be applicable to all fruit shipped during the given fiscal period. In order to provide funds to carry out the functions of the committee established under §905.19, handlers may make advance payment of assessments.

(c) In the case of an extreme emergency, the committee may borrow money on a short term basis to provide funds for the administration of this part. Any such borrowed money shall only be used to meet the committee's current financial obligations, and the committee shall repay all such borrowed money by the end of the next fiscal period from assessment income.

[22 FR 10734, Dec. 27, 1957. Redesignated at 26 FR 12751, Dec. 30, 1961, as amended at 42 FR 59370, Nov. 17, 1977; 54 FR 37293, Sept. 8, 1989]

§905.42 Handler's accounts.

(a) If, at the end of a fiscal period, the assessments collected are in excess of expenses incurred, the Committee, with the approval of the Secretary, may carry over such excess into subsequent fiscal periods as a reserve: *Provided*, That funds already in the reserve do not exceed approximately two fiscal periods' expenses. Such reserve funds may be used (1) to cover any expenses authorized by this part and (2) to cover necessary expenses of liquidation in the event of termination of this part. If any such excess is not retained in a reserve, each handler entitled to a proportionate refund shall be credited with such refund against the operations of the following fiscal period unless he demands payment of the sum due him, in which case such sum shall be paid to him. Upon termination of this part, any funds not required to defray the necessary expenses of liquidation shall be disposed of in such manner as the Secretary may determine to be appropriate: *Provided*, That to the extent practical, such funds shall be returned pro rata to the persons from whom such funds were collected.

(b) The committee may, with the approval of the Secretary, maintain in its own name or in the name of its members a suit against any handler for the collection of such handler's pro rata share of the said expense.

[22 FR 10734, Dec. 27, 1957. Redesignated at 26 FR 12751, Dec. 30, 1961, as amended at 34 FR 12427, July 30, 1969; 42 FR 59371, Nov. 17, 1977; 81 FR 10455, Mar. 1, 2016]

REGULATIONS

§905.50 Marketing policy.

(a) Before making any recommendations pursuant to §905.51 for any variety of fruit, the committee shall, with respect to the regulations permitted by §905.52, submit to the Secretary a detailed report setting forth an advisable marketing policy for such variety for the then current shipping season. Such report shall set forth the proportion of the remainder of the total crop of such variety of fruit (determined by the committee to be available for shipment during the remainder of the shipping season of such variety) deemed advisable by the committee to be shipped during such season.

(b) In determining each such marketing policy and advisable proportion, the committee shall give due consideration to the following factors relating to citrus fruit produced in Florida and in other States:

(1) The available crop of each variety of citrus fruit in Florida, and in other States, including the grades and sizes thereof, which grades and sizes in Florida shall be determined by the committee pursuant to §905.31;

(2) The probable shipments of citrus fruit from other States;

(3) The level and trend in consumer income;

(4) The prospective supplies of competitive commodities; and

(5) Other pertinent factors bearing on the marketing of fruit.

(c) In addition to the foregoing, the committees shall set forth a schedule of proposed regulations for the remainder of the shipping season for each variety of fruit for which recommendations to the Secretary pursuant to

§ 905.51 are contemplated. Such schedules shall recognize the practical operations of harvesting and preparation for market of each variety and the change in grades and sizes thereof as the respective seasons advance. In the event it is deemed advisable to alter such marketing policy or advisable proportion as the shipping season progresses, in view of changed demand and supply conditions with respect to fruit, the said committee shall submit to the Secretary a report thereon.

(d) The committee shall transmit a copy of each marketing policy report or revision thereof to the Secretary and to each producer and handler who files a request therefor. Copies of all such reports shall be maintained in the office of the committee where they shall be available for examination by producers and handlers.

[22 FR 10734, Dec. 27, 1957. Redesignated at 26 FR 12751, Dec. 30, 1961, as amended at 30 FR 13934, Nov. 4, 1965; 42 FR 59371, Nov. 17, 1977]

§ 905.51 Recommendations for regulation.

(a) Whenever the committee deems it advisable to regulate any variety in the manner provided in § 905.52, it shall give due consideration to the following factors relating to the citrus fruit produced in Florida and in other States: (1) Market prices, including prices by grades and sizes of the fruit for which regulation is recommended; (2) maturity, condition, and available supply, including the grade and size thereof in the producing areas; (3) other pertinent market information; and (4) the level and trend in consumer income. The committee shall submit to the Secretary its recommendations and supporting information respecting the factors enumerated in this section.

(b) The committee shall give notice of any meeting to consider the recommendation of regulations pursuant to § 905.52 by mailing a notice of meeting to each handler who has filed his address with committee for this purpose. The committee shall give the same notice of any such recommendation before the time it is recommended that such regulation become effective.

[42 FR 59370, Nov. 17, 1977]

§ 905.52 Issuance of regulations.

(a) Whenever the Secretary shall find from the recommendations and reports of the committee, or from other available information, that to limit the shipment of any variety would tend to effectuate the declared policy of the act, he shall so limit the shipment of such variety during a specified period or periods. Such regulations may:

(1) Limit the shipments of any grade or size, or both, of any variety, in any manner as may be prescribed, and any such limitation may provide that shipments of any variety grown in Regulation Area II shall be limited to grades and sizes different from the grade and size limitations applicable to shipments of the same varieties grown in Regulation Area I: *Provided,* That whenever any such grade or size limitation restricts the shipment of a portion of a specified grade or size of a variety the quantity of such grade or size that may be shipped by a handler during a particular week shall be established as a percentage of the total shipments of such variety by such handler in such prior period established by the committee with the approval of the Secretary, in which he shipped such variety.

(2) Limit the shipment of any variety by establishing and maintaining, only in terms of grades or sizes, or both, minimum standards of quality and maturity;

(3) Limit the shipment of the total quantity of any variety by prohibiting the shipment thereof: Provided, that no such prohibition shall apply to exports or be effective during any fiscal period with respect to any variety other than for one period not exceeding five days during the week in which Thanksgiving Day occurs, and for not more than two periods not exceeding a total of 14 days during the period December 20 to January 20, both dates inclusive.

(4) Establish, prescribe, and fix the size, capacity, weight, dimensions, marking (including labels and stamps), or pack of the container or containers which may be used in the packaging, transportation, sale, shipment, or other handling of fruit.

(5) Provide requirements that may be different for the handling of fruit within the production area, the handling of fruit for export, or for the handling of fruit between the production area and any point outside thereof within the United States.

(6) Any regulations or requirements pertaining to intrastate shipments shall not be implemented unless Florida statutes and regulations regulating such shipments are not in effect.

(b) Prior to the beginning of any such regulations, the Secretary shall notify the committee of the regulation issued by him, and the committee shall notify all handlers by mailing a copy thereof to each handler who has filed his address with said committee for this purpose.

(c) Whenever the Secretary finds from the recommendations and reports of the committee, or from other available information, that a regulation should be modified, suspended, or terminated with respect to any or all shipments of fruit in order to effectuate the declared policy of the act, he shall so modify, suspend, or terminate such regulation. If the Secretary finds that a regulation obstructs or does not tend to effectuate the declared policy of the act, he shall suspend or terminate such regulation. On the same basis, and in like manner, the Secretary may terminate any such modification or suspension.

(d) Whenever any variety is regulated pursuant to paragraph (a)(3) of this section, no such regulation shall be deemed to limit the right of any person to sell, contract to sell, or export such variety but no handler shall otherwise ship any fruit of such variety which was prepared for market during the effective period of such regulation.

[22 FR 10734, Dec. 27, 1957. Redesignated at 26 FR 12751, Dec. 30, 1961, as amended at 34 FR 12427, July 30, 1969; 42 FR 59370, Nov. 17, 1977; 54 FR 37292, Sept. 8, 1989; 81 FR 10455, Mar. 1, 2016]

§905.53 Inspection and certification.

(a) Whenever the handling of a variety of a type of fruit is regulated pursuant to §905.52, each handler who handles any variety of such type of fruit shall, prior to the handling of any lot of such variety, cause such lot to be inspected by the Federal-State Inspection Service and certified by it as meeting all applicable requirements of such regulation: *Provided,* That such inspection and certification shall not be required if the particular lot of fruit previously had been so inspected and certified unless such prior inspection was not performed within such time limitations as may be prescribed pursuant to paragraph (b) of this section. Each handler shall promptly submit, or cause to be submitted, to the committee a copy of each certificate of inspection issued to him covering varieties so handled.

(b) With respect to any variety regulated pursuant to §905.52(a)(4), the committee may prescribe, with the approval of the Secretary, such requirements with respect to time of inspection as it may deem necessary to insure satisfactory condition of the fruit at time of export.

[22 FR 10734, Dec. 27, 1957. Redesignated at 26 FR 12751, Dec. 30, 1961, as amended at 42 FR 59371, Nov. 17, 1977]

§905.54 Marketing, research and development.

The committee may, with the approval of the Secretary, establish, or provide for the establishment of, projects including production research, marketing research and development projects, and marketing promotion including paid advertising, designed to assist, improve, or promote the marketing, distribution, and consumption or efficient production of fruit. The expenses of such projects shall be paid by funds collected pursuant to §905.41. Upon conclusion of each project, but at least annually, the committee shall summarize the program status and accomplishments to its members and the Secretary. A similar report to the committee shall be required of any contracting party on any project carried out under this section. Also, for each project, the contracting party shall be required to maintain records of money received and expenditures, and such shall be available to the committee and the Secretary.

[74 FR 46306, Sept. 9, 2009]

HANDLERS' REPORTS

§ 905.70 Manifest report.

The committee may request information from each handler regarding the variety, grade, and size of each standard packed carton of fruit shipped by him and may require such information to be mailed or delivered to the committee or its duly authorized representative, within 24 hours after such shipment is made, in a manner or by such method as the said committee may prescribe, and upon such forms as may be prepared by it.

[42 FR 59371, Nov. 17, 1977]

§ 905.71 Other information.

Upon request of the committee, made with the approval of the Secretary, every handler shall furnish the committee, in such manner and at such times as it prescribes, such other information as will enable it to perform its duties under this subpart.

[22 FR 10734, Dec. 27, 1957. Redesignated at 26 FR 12751, Dec. 30, 1961, as amended at 42 FR 59371, Nov. 17, 1977]

MISCELLANEOUS PROVISIONS

§ 905.80 Fruit not subject to regulation.

Except as otherwise provided in this section, any person may, without regard to the provisions of §§ 905.52 and 905.53 and the regulations issued thereunder, ship any variety for the following purposes: (a) To a charitable institution for consumption by such institution; (b) to a relief agency for distribution by such agency; (c) to a commercial processor for conversion by such processor into canned or frozen products or into a beverage base; (d) by parcel post; or (e) in such minimum quantities, types of shipments, or for such purposes as the committee with the approval of the Secretary may specify. No assessment shall be levied on fruit so shipped. The committee shall, with the approval of the Secretary, prescribe such rules, regulations, or safeguards as it may deem necessary to prevent varieties handled under the provisions of this section from entering channels of trade for other than the purposes authorized by this section. Such rules, regulations, and safeguards may include the requirements that handlers shall file applications with the committee for authorization to handle a variety pursuant to this section, and that such applications be accompanied by a certification by the intended purchaser or receiver that the variety will not be used for any purpose not authorized by this section.

[22 FR 10734, Dec. 27, 1957. Redesignated at 26 FR 12751, Dec. 30, 1961, as amended at 42 FR 59371, Nov. 17, 1977]

§ 905.81 Compliance.

Except as provided in this part, no person shall ship fruit the shipment of which has been prohibited by the Secretary in accordance with the provisions of this part.

§ 905.82 Effective time.

The provisions of this subpart shall become effective on and after 12:01 a.m., e.s.t., September 1, 1946, and shall continue in force until terminated in one of the ways specified in § 905.83.

§ 905.83 Termination.

(a) The Secretary may at any time terminate the provisions of this part by giving at least one day's notice by means of a press release or in any other manner which he may determine.

(b) The Secretary shall terminate the provisions of this part at the end of any fiscal period whenever he finds that such termination is favored by a majority of producers who, during the preceding fiscal period, have been engaged in the production for market of fruit: *Provided,* That such majority have, during such period, produced for market more than 50 percent of the volume of such fruit produced for market, but such termination shall be effective only if announced on or before July 31 of the then current fiscal period.

(c) The Secretary shall conduct a referendum six years after the effective date of this paragraph and every sixth year thereafter to ascertain whether continuance of this part is favored by producers. The Secretary may terminate the provisions of this part at the end of any fiscal period in which the Secretary has found that continuance of this part is not favored by producers who during a representative period, determined by the Secretary, have been engaged in the production for market of the fruit in the production area. Such termination shall be announced on or before July 31 of the fiscal period.

(d) The provisions of this part shall, in any event, terminate whenever the provisions of the act authorizing it cease to be in effect.

[22 FR 10734, Dec. 27, 1957. Redesignated at 26 FR 12751, Dec. 30, 1961, as amended at 54 FR 37293, Sept. 8, 1989]

§905.84 Proceedings after termination.

(a) Upon the termination of the provisions of this part, the then functioning members · of the committee shall continue as joint trustees, for the purpose of liquidating the affairs of the committee, of all the funds and property then in the possession of or under control of committee, including claims for any funds unpaid or property not delivered at the time of such termination.

(b) The said trustees (1) shall continue in such capacity until discharged by the Secretary, (2) shall, from time to time, account for all receipts and disbursements or deliver all property on hand, together with all books and records of the committee and of the joint trustees, to such person as the Secretary may direct; and (3) shall, upon the request of the Secretary, execute such assignments or other instruments necessary or appropriate to vest in such person full title and right to all of the funds, property, and claims vested in the committee, or the joint trustees pursuant to this part.

(c) Any funds collected pursuant to §905.41, over and above the amounts necessary to meet outstanding obligations and expenses necessarily incurred during the operation of this part and during the liquidation period, shall be returned to handlers as soon as practicable after the termination of this part. The refund to each handler shall be represented by the excess of the amount paid by him over and above his pro rata share of the expenses.

(d) Any person to whom funds, property, or claims have been transferred or delivered by the committee or its members, pursuant to this section, shall be subject to the same obligations imposed upon the members of the committee and upon the said joint trustees.

[22 FR 10734, Dec. 27, 1957. Redesignated at 26 FR 12751, Dec. 30, 1961, as amended at 42 FR 59371, Nov. 17, 1977]

§905.85 Duration of immunities.

The benefits, privileges, and immunities conferred upon any person by virtue of this part shall cease upon its termination, except with respect to acts done under and during the existence of this part.

§905.86 Agents.

The Secretary may, by designation in writing, name any person, including any officer or employee of the Government, or name any bureau or division in the United States Department of Agriculture, to act as his agent or representative in connection with any of the provisions of this part.

§905.87 Derogation.

Nothing contained in this part is, or shall be construed to be in derogation or in modification of the rights of the Secretary or of the United States (a) to exercise any powers granted by the act or otherwise, or (b) in accordance with such powers, to act in the premises whenever such action is deemed advisable.

§905.88 Personal liability.

No member or alternate of the committee nor any employee or agent thereof, shall be held personally responsible, either individually or jointly with others, in any way whatsoever, to any handler or to any other person for errors in judgment, mistakes, or other acts, either of commission or omission, as such member, alternate, or employee, except for acts of dishonesty.

[22 FR 10734, Dec. 27, 1957. Redesignated at 26 FR 12751, Dec. 30, 1961, as amended at 42 FR 59371, Nov. 17, 1977]

§905.89 Separability.

If any provision of this part is declared invalid, or the applicability thereof to any person, circumstance, or thing is held invalid, the validity of the remainder of this part or the applicability thereof to any other person, circumstance, or thing shall not be affected thereby.

Subpart B—Administrative Requirements

SOURCE: 42 FR 59371, Nov. 17, 1977, unless otherwise noted.

§ 905.105 Tangerine and grapefruit classifications.

(a) Pursuant to § 905.5(m), the following classifications of grapefruit are renamed as follows:

(1) Marsh and other seedless grapefruit, excluding pink grapefruit, are renamed as Marsh and other seedless grapefruit, excluding red grapefruit;

(2) Duncan and other seeded grapefruit, excluding pink grapefruit, are renamed as Duncan and other seeded grapefruit, excluding red grapefruit;

(3) Pink seedless grapefruit is renamed as Red seedless grapefruit;

(4) Pink seeded grapefruit is renamed as Red seeded grapefruit.

(b) Pursuant to § 905.5(m), the term *variety* or *varieties* includes Sunburst and Fallglo tangerines.

[56 FR 49132, Sept. 27, 1991, as amended at 63 FR 55500, Oct. 16, 1998]

§ 905.114 Redistricting of citrus districts and reapportionment of grower members.

Pursuant to § 905.14, the citrus districts and membership allotted each district shall be as follows:

(a) Citrus District One shall include the counties of Alachua, Baker, Bradford, Citrus, Clay, Columbia, Duval, Flagler, Gilchrist, Hernando, Hillsborough, Lake, Levy, Marion, Nassau, Orange, Osceola, Pasco, Pinellas, Polk, Putnam, Seminole, St. Johns, Sumter, Suwannee, and Union and County Commissioner's Districts One, Two, and Three of Volusia County, and that part of the counties of Indian River and Brevard not included in Regulation Area II. This district shall have two grower members and alternates.

(b) Citrus District Two shall include the counties of Broward, Charlotte, Collier, Dade, De Soto, Glades, Hardee, Hendry, Highlands, Lee, Manatee, Monroe, Okeechobee, Sarasota, and that part of the counties of Palm Beach and Martin not included in Regulation Area II. This district shall have three grower members and alternates.

(c) Citrus District Three shall include the County of St. Lucie and that part of the counties of Brevard, Indian River, Martin, and Palm Beach described as lying within Regulation Area II, and County Commissioner's Districts Four and Five of Volusia County. This district shall have four grower members and alternates.

[78 FR 13781, Mar. 1, 2013]

§ 905.120 Nomination procedure.

Meetings shall be called by the committee in accordance with the provisions of § 905.22, for the purpose of making nominations for members and alternate members of the Citrus Administrative Committee. The manner of nominating members and alternate members of said committee shall be as follows:

(a) At each such meeting the committee's representative shall announce the requirements as to eligibility for voting for nominees and the procedure for voting, and shall explain the duties of the committee.

(b) A chairman and a secretary of each meeting shall be selected.

(c) At each meeting there shall be presented for nomination and there shall be nominated not less than the number of nominees required under the provisions of § 905.19, all of whom shall have the qualifications as specified in § 905.22.

(d) At the meetings of handlers, any person authorized to represent a handler may cast a ballot for such handler.

(e) At each meeting each eligible person may cast one vote for each of the persons to be nominated to represent the district or group, as the case may be.

(f) Voting may be by written ballot. If written ballots are used, all ballots shall be delivered by the chairman or the secretary of the meeting to the agent of the Secretary. If ballots are not used, the committee's representative shall deliver to the Secretary's agent a listing of each person nominated and a count of the number of votes cast for each nominee for grower member and alternate. Said representative shall also provide the agent the register of eligible voters present at each meeting, a listing of each person nominated, the number of votes cast, and the weight by volume of shipments voted for each nominee for shipper member and alternate.

(g) Up to four grower members may be growers who are also shippers, or growers who are also employees of shippers.

[43 FR 9455, Mar. 8, 1979, as amended at 78 FR 13781, Mar. 1, 2013]

NON-REGULATED FRUIT

§905.140 Gift packages.

Any handler may, without regard to the provisions of §§905.52 and 905.53 and the regulations issued thereunder, ship any varieties for the following purpose and types of shipment:

(a) To any person gift packages containing such varieties: *Provided,* That such packages are individually addressed to such person, and shipped directly to the addressee for use by such person other than for resale; or

(b) to any individual gift package distributor of such varieties to be handled by such distributor: *Provided,* That such person is the original purchaser and the gift packages are individually addressed or marked "not for resale". This exemption does not apply to "commercially handled" shipments for resale.

[58 FR 65539, Dec. 15, 1993]

§905.141 Minimum exemption.

Any shipment of fruit which meets each of the following requirements may be transported from the production area during any one day by any person or by the occupants of one vehicle exempt from the requirements of §§905.52 and 905.53 and regulations issued thereunder:

(a) The shipment does not exceed a total of 15 standard packed cartons (12 bushels) of fruit, either a single fruit or a combination of two or more fruits;

(b) The shipment consists of fruit not for resale; and

(c) Such exempted quantity is not included as a part of a shipment exceeding 15 standard packed cartons (12 bushels) of fruit.

§905.142 Animal feed.

(a) The handling of citrus for animal feed shall be exempt from the provisions of §§905.52 and 905.53 and the regulations issued thereunder under the following conditions:

(1) The handler notifies the committee each fiscal period, prior to such handling of his/her intention to handle such fruit, the quantity he/she anticipates handling and the destination point of each lot of fruit and receives from the committee a special shipping permit for the shipment;

(2) The fruit is used for animal feed and is not offered for resale, disposed of, or in any way handled so as to enter fresh fruit channels;

(3) The quantity does not exceed 1,000 ⅘ bushel cartons per fiscal period or such other quantity as may be specified by the committee;

(4) The fruit is placed in containers of uniform capacity; and

(5) Each shipment is certified by the Federal-State Inspection Service as to the quantity shipped.

(b) [Reserved]

[46 FR 47056, Sept. 24, 1981]

§905.145 Certification of certain shipments.

Whenever a regulation pursuant to §905.52 restricts the shipment of a portion of a specified grade or size of a variety, each handler shipping such variety during the regulation period shall, with respect to each such shipment, certify to the U.S. Department of Agriculture and the committee the quantity of the partially restricted grade or size, or both, contained in such shipment. Such certification shall accompany the manifest of such shipment which the handler furnishes to the Federal-State Inspection Service.

§905.146 Special purpose shipments.

(a) A Special Purpose Shipper is one who handles Florida citrus fruit that is certified by a Florida Department of Agriculture and Consumer Services licensed certifying agent as organically grown under Florida law. In addition, the shipper shall certify that shipments will be limited to outlets handling organically grown fruits. Any such shipments shall be subject to a Certificate of Privilege issued by the committee.

(b) To qualify for a Certificate of Privilege, each such shipper must notify the committee prior to the first shipment of certified organically grown Florida citrus fruit in the fiscal period of the shipper's intent to ship such citrus, submit an application on forms supplied by the committee, and agree to other requirements as set forth in §§ 905.147 and 905.148 inclusive, with respect to such shipments. The shipper shall certify that no claims will be made, written or verbal, concerning any alleged advantages of using, or any alleged superiority of, fruit shipped under a Certificate of Privilege, compared to other Florida produced citrus.

(c) Citrus meeting all other applicable requirements may be handled without regard to grade regulations issued under § 905.52 under the following conditions:

(1) Such fruit meets the requirements of U. S. No. 2 Russet grade and those requirements of U. S. No. 1 grade relating to shape (form), as such requirements are set forth in the revised U. S. Standards for Grades of Florida Oranges and Tangelos (7 CFR 51.1140 through 51.1179), the revised Standards for Florida Tangerines (7 CFR 51.1810 through 51.1837), or the revised U. S. Standards for Grades of Florida Grapefruit (7 CFR 51.750 through 51.784). Such fruit also meets applicable minimum size requirements in effect for domestic shipments of citrus fruits.

(2) All such citrus shall be inspected as required by § 905.53 by the Federal or Federal-state Inspection Service prior to the time such citrus is shipped from the packing facility, and certified as meeting the applicable requirements.

(3) Be reported as required in § 905.148.

[59 FR 26928, May 25, 1994, as amended at 66 FR 229, Jan. 3, 2001]

§ 905.147 Certificate of privilege.

(a) *Application.* Application for Certificate of Privilege by a Special Purpose Shipper shall be made on forms furnished by the committee. Each application may contain, but need not be limited to, the name and address of each handler; a list of certified organic citrus fruit growers, including addresses; a list of receivers; the quantity and variety of citrus to be shipped; a certification to the Secretary of Agriculture and to the committee as to the truthfulness of the information shown thereon; and any other appropriate information or documents deemed necessary by the committee or its duly authorized agents for the purposes stated in § 905.146.

(b) *Approval.* The committee or its duly authorized agents shall give prompt consideration to each application for a Certificate of Privilege. Approval of an application based upon a determination as to whether the information contained therein and other information available to the committee supports approval, shall be evidenced by the issuance of a Certificate of Privilege to the applicant. Each certificate shall expire at the end of the fiscal period.

(c) *Suspension or Denial of Certificate of Privilege.* The committee may investigate the handling of special purpose shipments under Certificates of Privilege to determine whether Special Purpose Shippers are complying with the requirements and regulations applicable to such certificates. Whenever the committee finds that a Special Purpose Shipper or consignee is failing to comply with the requirements and regulations applicable to such certificates, the Certificate of Privilege issued to such Special Purpose Shipper may be suspended or, in the case of an application for the issuance of an initial Certificate of Privilege, may be denied. Such suspension of a certificate shall be for a reasonable period of time as determined by the committee, but in no event shall it extend beyond the end of the current fiscal period. In the case of the denial of an application for the issuance of an initial certificate, such certificate shall be denied until the applicant comes into compliance with the requirements and regulations applicable to such certificates. Prior to suspending or denying an application for a Certificate of Privilege, the committee shall give the shipper or applicant reasonable advance notice in writing of its intention and the facts and reasons therefor, and afford the shipper or applicant an opportunity, either orally or in writing, to present opposing facts and reasons. The shipper or applicant shall be informed of the committee's determination in writing and in a timely manner.

[43 FR 9456, Mar. 8, 1978, as amended at 59 FR 26929, May 25, 1994]

§905.148 Reports of special purpose shipments under certificates of privilege.

(a) Each handler of citrus shipping under Certificates of Privilege shall supply the committee with reports on each shipment as requested by the committee, on forms supplied by the committee, showing the name and address of the shipper or shippers; name and address of the certified organic Florida citrus fruit grower or growers supplying fruit for such shipment; truck or other conveyance identification; the loading point; destination, consignee; the inspection certificate number; and any other information deemed necessary by the committee.

(b) One copy of the report on each shipment shall be forwarded by the shipper to the committee within 10 days after such shipment, and two copies of the report shall accompany each shipment to the receiver. Upon the receipt of each shipment, the receiver shall complete the applicable portion of the form and return one copy to the committee within 10 days and one copy shall be retained by the shipper. Such completion shall contain a certification to the Secretary and the committee that the citrus described shall be distributed in the outlets described. Failure to complete and return such forms will be cause to remove that receiver's name from the committee's list of eligible receivers.

[43 FR 9456, Mar. 8, 1978, as amended at 59 FR 26929, May 25, 1994]

§905.149 Procedure for permitting growers to ship tree run citrus fruit.

(a) *Tree run citrus fruit.* Tree run citrus fruit as referenced in this section is defined in the Florida Department of Citrus (FDOC) regulation 20–35.006, which specifies that "Tree run grade is

that grade of naturally occurring sound and wholesome citrus fruit which has not been separated either as to grade or size after severance from the tree." Wholesomeness is as defined in FDOC regulation 20–62.002. The tree run citrus fruit shipped under this provision also must be from the applying grower's own grove.

(b) *Application.* A grower shall apply to ship tree run fruit using a Grower Tree Run Certificate Application, furnished by the committee. Such application shall contain, but not be limited to: the name, address, and phone number of the grower; legal description of the grove(s) from which citrus will be shipped; variety of citrus produced on the identified grove(s); approximate number of boxes produced on the identified grove(s); and a certification to the U.S. Department of Agriculture and to the committee as to the truthfulness of the information shown thereon; and any other appropriate information or documents deemed necessary by the committee or its duly authorized agents.

(c) *Approval.* The committee or its duly authorized agents shall give prompt consideration to each application for a Grower Tree Run Certificate. Approval of an application will be based upon a determination as to whether the information contained therein and on whether other information available to the committee supports an application's approval. Approval of an application shall be evidenced by the issuance of a Grower Tree Run Certificate to the applicant. Each certificate shall expire at the end of the fiscal period.

(d) *Suspension or denial of a Grower Tree Run Certificate.* The committee may investigate the handling of tree run shipments under a Grower Tree Run Certificate to determine whether growers are complying with the requirements and regulations applicable to such certificates. Whenever the committee finds that a grower is failing to comply with the requirements and regulations applicable to such certificates, the Grower Tree Run Certificate issued to such grower may be suspended or, in the case of an application for the issuance of an initial Grower Tree Run Certificate, may be denied.

Such suspension of a certificate shall be for a reasonable period of time as determined by the committee, but in no event shall it extend beyond the end of the fiscal period. In the case of the denial of an application for the issuance of an initial certificate, such certificate shall be denied until the applicant comes into compliance with the requirements and regulations applicable to such certificates. Prior to suspending or denying an application for a Grower Tree Run Certificate, the committee shall give the grower reasonable advance notice in writing of its intention and the facts and reasons therefor, and afford the grower an opportunity, either orally or in writing, to present opposing facts and reasons. The grower shall be informed of the committee's determination in writing and in a timely manner.

(e) To qualify for a Grower Tree Run Certificate, each such grower must notify the committee prior to the first shipment of tree run Florida citrus fruit of the grower's intent to ship such citrus, submit an application on forms supplied by the committee, and agree to other requirements as set forth in this section with respect to such shipments.

(f) The handling of tree run citrus under a Grower Tree Run Certificate shall be exempt from the provisions of §§ 905.52 and 905.53 and the regulations issued thereunder, under the following conditions:

(1) A grower may only ship up to 150 1⅗ bushel boxes per variety, per shipment.

(2) A grower may only ship up to 3,000 boxes per variety per season.

(3) Each grower shall apply to the Citrus Administrative Committee and receive a Grower Tree Run Certificate prior to shipping their own tree run Florida citrus fruit.

(4) Each grower of citrus shipping under a Grower Tree Run Certificate shall supply the committee with reports on each shipment as requested by the committee, on forms supplied by the committee, providing the following information: The name and address of the grower, along with the grower's Grower Tree Run Certificate number; the legal description of the grove; the variety and amount of citrus shipped;

the date the fruit was shipped; and the truck/trailer license number. A copy of the form will be completed for each shipment. One copy of the report will be forwarded by the grower to the committee office within 10 days after such shipment, and one copy of the report will accompany each shipment and be given to the Road Guard Station.

(5) Each container of tree run fruit shipped under a Grower Tree Run Certificate shall be labeled with or contain the name and address of the grower shipping under the Grower Tree Run Certificate.

[67 FR 62313, Oct. 7, 2002, as amended at 68 FR 52329, Sept. 3, 2003; 69 FR 50269, Aug. 16, 2004]

§ 905.150 Eligibility requirements for public member and alternate member.

(a) The public member shall be neither a producer nor a handler of Florida citrus fruit and shall have no direct financial interest in the production or marketing of citrus fruit (except as a consumer of agricultural products).

(b) The public member should be able to devote sufficient time and express a willingness to attend Committee activities regularly and become familiar with the background and economics of the industry.

(c) The public member must be a resident of the production area.

(d) The public member should be nominated by the Citrus Administrative Committee and should serve a 1-year term which coincides with the term of office of producer and handler members of the Committee.

[43 FR 32397, July 27, 1978]

§ 905.153 Procedure for determining handlers' permitted quantities of red seedless grapefruit when a portion of sizes 48 and 56 of such variety is restricted.

(a) For the purposes of this section, the prior period specified in § 905.52 is hereby established as an average week within the immediately preceding three seasons. Each handler's average week shall be computed by adding the total volume of red seedless grapefruit handled in the immediately preceding three seasons and dividing the total by 99. The average week for handlers with less than three previous seasons of shipments shall be calculated by adding the total volume of shipments for the seasons they did ship red seedless grapefruit, divide by the number of seasons, divide further by 33. If crop conditions limit shipments from any or all of the immediately preceding three season(s), the committee may use a prior season or seasons for the purposes of calculating an average week. New handlers with no record of shipments could ship size 48 and 56 red seedless grapefruit as a percentage of total shipments equal to the percentage applied to other handlers' average week; once such handlers have recorded shipments, their average week shall be calculated as an average of total shipments for the weeks they have shipped red seedless grapefruit during the current season. When used in the regulation of red seedless grapefruit, the term season means the weeks beginning the third Monday in September and ending the first Sunday in the following May. The term *regulation period* means the 22-week period beginning the third Monday in September of the current season.

(b) When a size limitation restricts the shipment of a portion of sizes 48 and 56 red seedless grapefruit during a particular week as provided in §905.52, the committee shall compute the quantity of sizes 48 and 56 red seedless grapefruit that may be shipped by each handler by multiplying the handler's calculated average week shipments of such grapefruit by the percentage established by regulation for red seedless grapefruit for that week. Such set percentage may vary from week to week but shall not be less than 25 percent in any week.

(c) The committee shall notify each handler of the quantity of size 48 and 56 red seedless grapefruit such handler may handle during a particular week.

(d) During any regulation week for which the Secretary has fixed the percentage of sizes 48 and 56 red seedless grapefruit, any person who has received an allotment may handle, in addition to their total allotment available, an amount of size 48 and 56 red seedless grapefruit up to 10 percent greater than their allotment. The quantity of the overshipment shall be deducted from the handler's allotment

for the following week. Overshipments will not be allowed during the last week of regulation. If the handler fails to use his or her entire allotment, the undershipment is not carried over to the following week. Each handler shipping size 48 and/or 56 red seedless grapefruit during the regulation period shall complete and submit to the committee, no later than 2 p.m. of the business day following the shipment, a report of red seedless grapefruit shipments by day for each regulation week.

(e) Any handler may transfer or loan any or all of their shipping allotment (excluding the overshipment allowance) of size 48 and 56 red seedless grapefruit to any other handler. Each handler party to such transfer or loan shall no later than noon on the Wednesday following the regulation week notify the committee so the proper adjustment of records may be made. In each case, the committee shall confirm in writing all such transactions, prior to the following week, to the handlers involved. The committee may act on behalf of handlers wanting to arrange allotment loans or participate in the transfer of allotments.

(f) New handlers with no record of shipments planning to ship red seedless grapefruit covered by any percentage size regulation shall register with the committee prior to the regulation period so their allotments can be properly calculated. Each new handler shall provide on a form furnished by the committee their Florida citrus fruit dealer's license number, their Florida Department of Agriculture and Consumer Services' Fruit and Vegetable Division packinghouse registration number, and the physical location of the packinghouse where the red seedless grapefruit is to be prepared for market. The committee shall notify any new handlers of their allotments prior to the regulation period.

[61 FR 69015, Dec. 31, 1996, as amended at 62 FR 52011, Oct. 6, 1997; 64 FR 51892, Sept. 27, 1999; 67 FR 809, Jan. 8, 2002; 69 FR 50278, Aug. 16, 2004; 70 FR 54242, Sept. 14, 2005]

§905.161 Repacking shipper.

(a) A repacking shipper is a person who repacks and ships citrus fruit grown in the production area in Florida which has been previously inspected and certified as meeting the requirements specified under § 905.52 of the order, and who has obtained a currently valid repacking certificate of privilege issued to him or her by the committee as specified in § 905.162.

(b) Each repacking shipper, to qualify for a repacking certificate of privilege, must notify the committee 10 days prior to his or her first shipment of repacked citrus fruit during a particular fiscal period of his or her intent to ship such citrus fruit, submit an Application for a Repacking Certificate of Privilege form supplied by the committee, and agree to other requirements as set forth in §§ 905.162 and 905.163 inclusive, with respect to such shipments. The repacking shipper shall certify that he or she will only handle previously inspected and certified citrus fruit.

(c) Any repacking shipper who handles citrus fruit shipped under a repacking certificate of privilege must, other order provisions not withstanding, meet the following requirements:

(1) All such citrus fruit must be positive lot identified by the Federal or Federal/State Inspection Service and certified as meeting the applicable requirements for citrus fruit shipped to the domestic market (fruit shipped from the production area to any point outside thereof in the 48 contiguous States and the District of Columbia of the United States), prior to being repacked and shipped by the repacking shipper. Each such citrus fruit shipment shall be accompanied by a Federal-State manifest that certifies the grade and amount of each load of citrus fruit received, which shall be retained by the repacking shipper.

(2) Be reported as required in § 905.163.

(3) The repacking facility used to repack previously inspected and certified citrus fruit by the repacking shipper shall not have operable equipment to wash, brush, wax, or dry citrus fruit.

(4) All citrus fruit handled by a repacking shipper shall be packed in approved Florida Department of Citrus fruit containers.

(5) Each container shipped with such citrus fruit shall be marked with the repacking shipper's repacking certificate of privilege number.

[59 FR 48782, Sept. 23, 1994]

§ 905.162 Repacking certificate of privilege.

(a) *Application.* Application for a repacking certificate of privilege by a repacking shipper shall be made on an Application for a Repacking Certificate of Privilege form supplied by the committee. Each such application shall contain, but need not be limited to, the name, address and Florida citrus fruit dealer license number of the applicant; approximate number of boxes to be handled during the season; the various types of containers to be used to ship the repacked citrus fruit; a certification to the Secretary of Agriculture and to the committee as to the truthfulness of the information shown thereon; and any other appropriate information or documents deemed necessary by the committee or duly authorized agents for the purposes stated in § 905.161.

(b) *Approval.* The committee or its duly authorized agents shall give prompt consideration to each application for a repacking certificate of privilege. Approval of an application based upon a determination as to whether the information contained therein and other information available to the committee supports approval, shall be evidenced by the issuance of a repacking certificate of privilege to the applicant. Each such certificate shall expire at the end of the fiscal period.

(c) *Suspension or denial of certificate of privilege.* The committee may investigate the handling of repacked fresh citrus fruit shipments under certificates of privilege to determine whether repacking shippers are complying with the requirements and regulations applicable to such certificates. Whenever the committee finds that a repacking shipper is failing to comply with the requirements and regulations applicable to such certificates, the certificate of privilege issued to such repacking shipper may be suspended or, in the case of an application for the issuance of an initial certificate of privilege, may be denied. Such suspension of a certificate shall be for a reasonable period of time as determined by the committee, but in no event shall it extend beyond the end of the then current fiscal period. In the case of the denial of an application for the issuance of an initial certificate, such certificate shall be denied until the applicant comes into compliance with the requirements and regulations applicable to such certificates. Prior to suspending or denying an application for a certificate of privilege, the committee shall give the shipper or applicant an opportunity, either orally or in writing, to present opposing facts and reasons. The shipper or applicant shall be informed of the committee's determination in writing and in a timely manner.

[59 FR 48783, Sept. 23, 1994]

§905.163 Reports of shipments under repacking certificate of privilege.

(a) Each repacking shipper who handles citrus fruit under a repacking certificate of privilege shall supply the committee with reports on each shipment as requested by the committee, on a Report of Shipments Under Certificate of Privilege form supplied by the committee, showing the name and address of the repacking shipper; name and address of the handler supplying the inspected and certified citrus fruit for such shipment; number of packages; size and containers; brand; grade; certificate number; and any other information deemed necessary by the committee. Each repacking shipper of citrus fruit shall maintain on file a copy of the Federal-State manifest that certifies the grade and amount of each load of citrus fruit received. These manifests shall be readily available to the committee upon request.

(b) One copy of the Report of Shipments Under Certificate of Privilege form on each shipment shall be forwarded to the committee promptly, one copy of such form shall be retained by the repacking shipper, and one copy of such form shall accompany the shipment. Failure to complete and return such form shall be cause for suspension of the repacking shippers repacking certificate of privilege.

[59 FR 48783, Sept. 23, 1994]

§905.171 Handler supplier report.

Each handler shall furnish a supplier report to the Committee on an annual basis. Such reports shall be made on forms provided by the Committee and shall include the name and business address of each grower whose fruit was shipped or acquired by the handler during the season. Handlers shall submit this report to the Committee not later than June 15 of each season.

[78 FR 32070, May 29, 2013]

Subpart C—Assessment Rate

§905.235 Assessment rate.

On and after August 1, 2018, an assessment rate of $0.015 per 4/5-bushel carton or equivalent is established for Florida citrus covered under the Order.

[84 FR 2049, Feb. 6, 2019]

Subpart D—Grade and Size Requirements

§905.306 Orange, Grapefruit, Tangerine and Tangelo Regulation.

(a) No handler shall ship between the production area and any point outside thereof, in the 48 contiguous States and the District of Columbia of the United States, any variety of fruit listed in column (1) of Table I, except for Ambersweet and Temple, unless such variety meets the applicable minimum grade and size (with tolerances for size as specified in paragraph (c) of this section) specified for such variety in columns (2) and (3) of Table I: *Provided,* That all grapefruit meet the minimum maturity requirements specified in paragraph (e) of this section.

TABLE I

Variety	Regulation period	Minimum grade	Minimum diameter (inches)
(1)	(2)	(3)	(4)
Oranges			
Early and midseason	....	U.S. No. 1	2⁴⁄₁₆
Navel		U.S. No. 1	2⁴⁄₁₆
Valencia and other late type.	September 1–May 14	U.S. No. 1	2⁴⁄₁₆
	May 15–June 14	U.S. No. 1 Golden	2⁴⁄₁₆
	June 15–August 31	U.S. No. 2, External/U.S. No. 1, Internal.	2⁴⁄₁₆
Grapefruit, Seedless	....	U.S. No. 1	3
Tangerines:			
Fallglo	....	U.S. No. 1	2⁵⁄₁₆
Honey	....	Florida No. 1	2⁵⁄₁₆
Sunburst	....	U.S. No. 1	2⁵⁄₁₆
Tangelos	....	U.S. No. 1	2⁸⁄₁₆

(b) No handler shall ship to any destination outside the 48 contiguous States and the District of Columbia of the United States any variety of fruit listed in column (1) of Table II, except for Ambersweet and Temple, unless such variety meets the applicable minimum grade and size (with tolerances for size as specified in paragraph (c) of this section) specified for such variety in columns (2) and (3) of Table II: *Provided,* That all grapefruit meet the minimum maturity requirements specified in paragraph (e) of this section.

TABLE II

Variety	Minimum grade	Minimum diameter (inches)
(1)	(2)	(3)
Oranges	U.S. No. 1	2⁴⁄₁₆
Navels	U.S. No. 1 Golden	2⁴⁄₁₆
Grapefruit, Seedless	U.S. No. 1	3
Tangerines:		
Fallglo	U.S. No. 1	2⁵⁄₁₆
Honey	Florida No. 1	2⁵⁄₁₆
Sunburst	U.S. No. 1	2⁵⁄₁₆
Tangelos	U.S. No. 1	2⁸⁄₁₆

(c) *Size tolerances.* To allow for variations incident to proper sizing in the determination of minimum diameters as prescribed in Tables I and II, not more than 10 percent, by count, of the fruit in any lot of containers may fail to meet the minimum diameter size requirements, and not more than 15 percent, by count, in any individual sample may fail to meet the minimum diameter size requirements specified: *Provided,* That such tolerances for other than Navel and Temple oranges shall be based only on the oranges in the lot measuring 2¹⁴⁄₁₆ inches or smaller in diameter.

(d) Terms used in the marketing order including Improved No. 2 grade for grapefruit, when used herein, mean the same as is given to the terms in the order; Florida No. 1 grade for Honey tangerines means the same as provided in Rule No. 20–35.03 of the Regulations of the Florida Department of Citrus, and terms relating to grade, except Improved No. 2 grade for grapefruit and diameter, shall mean the same as is given to the terms in the revised U. S. Standards for Grades of Florida Oranges and Tangelos (7 CFR 51.1140 through 51.1179), the revised U. S. Standards for Florida Tangerines (7

CFR 51.1810 through 51.1837), or the revised U. S. Standards for Grades of Florida Grapefruit (7 CFR 51.750 through 51.784).

(e)(1) All grapefruit shipped under the order shall meet minimum maturity requirements of 8.0 percent soluble solids (sugars) and 7.5 to 1 solids to acid ratio or shall comply with one of the alternate equivalent soluble solids and solids to acid ratio combinations set forth in Table III: *Provided,* That the minimum ratio shall not drop below 7.2 even if the soluble solids (sugars) reaches a level higher than 9.6.

(2) Notwithstanding the provision of paragraph (e)(1) of this section, for the period December 23, 2004 to July 31, 2005, all grapefruit shipped under the order shall meet minimum maturity requirements of 7.5 percent soluble solids (sugars) and 7.5 to 1 solids to acid ratio or shall comply with one of the alternate equivalent soluble solids and solids to acid ratio combinations set forth in Table III: Provided, That the minimum ratio shall not drop below 7.2 even if the soluble solids (sugars) reaches a level higher than 9.6.

TABLE III

Minimum total solids (sugars), %	Solids to acid minimum ratio
8.0 to (not including) 9.1	7.50 to 1
9.1 to (not including) 9.2	7.45 to 1
9.2 to (not including) 9.3	7.40 to 1
9.3 to (not including) 9.4	7.35 to 1
9.4 to (not including) 9.5	7.30 to 1
9.5 to (not including) 9.6	7.25 to 1
9.6 and greater	7.20 to 1

[46 FR 60171, Dec. 8, 1981]

EDITORIAL NOTE: For FEDERAL REGISTER citations affecting § 905.306, see the List of CFR Sections Affected, which appears in the Finding Aids section of the printed volume and at *www.govinfo.gov.*

§ 905.350 [Reserved]

Subpart E—Interpretations

§ 905.400 Interpretation of certain provisions.

(a) In interpreting the provisions of paragraph (d) of § 905.52, the limitation on shipment of any variety of fruit regulated pursuant to paragraph (a)(3) of that section, which was prepared for market during the effective period of such regulation, shall not be deemed to apply to shipment of such variety which was prepared for market incidentally as part of a lot packed for export and shipped following the period of regulation.

(b) Prior to shipment of any variety of fruit so prepared, the handler shall provide the Citrus Administrative Committee or its designated agent a copy of the shipping manifest applicable to such shipment with a notation thereon that the fruit was packed incidentally as part of a lot packed for export.

[54 FR 46597, Nov. 6, 1989]

PART 906—ORANGES AND GRAPEFRUIT GROWN IN LOWER RIO GRANDE VALLEY IN TEXAS

Subpart A—Order Regulating Handling

DEFINITIONS

906.35   Accounting.

AUTHORITY: 7 U.S.C. 601–674.

Subpart A—Order Regulating Handling

SOURCE: 25 FR 9093, Sept. 22, 1960, unless otherwise noted. Redesignated at 26 FR 12751, Dec. 30, 1961.

DEFINITIONS

§ 906.1   Secretary.

Secretary means the Secretary of Agriculture of the United States, or any employee of the Department to whom authority has heretofore been delegated, or to whom authority may hereafter be delegated, to act in his stead.

§ 906.2   Act.

Act means Public Act No. 10, 73d Congress, as amended and as re-enacted and amended by the Agricultural Marketing Agreement Act of 1937, as amended (sections 1–19, 48 Stat. 31, as amended; 7 U.S.C. 601–674).

§ 906.3   Person.

Person means an individual, partnership, corporation, association, or any other business unit.

§ 906.4   Production area.

Production area means all territory in the counties of Cameron, Hidalgo, and Willacy in the State of Texas.

§ 906.5   Fruit.

Fruit means either or both of the following citrus fruits grown in the production area: (a) Citrus grandis, Osbeck, commonly called grapefruit, and (b) Citrus sinensis, Osbeck, commonly called oranges.

§ 906.6   Handler.

Handler is synonymous with *shipper* and means any person (except a common or contract carrier of fruit owned by another person) who handles fruit or causes fruit to be handled.

(a) *Independent handler.* Independent *handler* means any handler other than a handler that is a cooperative marketing organization.

(b) [Reserved]

§ 906.7   Handle.

Handle or *ship* means to transport or sell fruit, or in any other way to place fruit, in the current of commerce between the production area and any point outside thereof in the United States, Canada, or Mexico.

§906.8 Producer.

Producer means any person engaged in a proprietary capacity in the production of fruit for market.

(a) *Independent producer.* Independent producer means any producer who does not market his fruit through a handler that is a cooperative marketing organization.

(b) [Reserved]

§906.9 Grade and size.

Grade means any one of the established grades of fruit and *size* means any one of the established sizes of fruit as defined and set forth in the applicable U.S. Standards for fruit (§§51.680 through 51.714 and §§51.620 through 51.653) issued by the United States Department of Agriculture, or amendments thereto, or modifications thereof, or variations based thereon recommended by the committee and approved by the Secretary.

§906.10 Pack.

Pack means the specific grade, quality, size, or arrangement of fruit in a particular container or containers.

§906.11 Maturity.

Maturity means various degrees of ripeness for fruit as established by the committee with approval of the Secretary.

§906.12 Container.

Container means any box, bag, crate, hamper, basket, package, bulk carton, or any other type of receptacle used in the packaging, transportation, sale, or other handling of fruit.

§906.13 Variety or varieties.

Variety or varieties means any one or more of the following groupings or classifications of fruit: (a) Navel oranges; (b) Early and Midseason oranges, except Navel oranges; (c) Valencia and similar late type oranges; (d) white seeded grapefruit; (e) white seedless grapefruit; (f) pink and red seeded grapefruit; and (g) pink and red seedless grapefruit.

§906.14 Committee.

Committee means the Texas Valley Citrus Committee, established pursuant to §906.18.

§906.15 Fiscal period.

Fiscal period means the period beginning August 1 and ending July 31 following; or such annual beginning and ending dates as may be approved by the Secretary pursuant to recommendations of the committee.

§906.16 District.

District means any of the geographic divisions of the production area initially established pursuant to §906.20 or as re-established pursuant to §906.21.

COMMITTEE

§906.18 Establishment and membership.

(a) The Texas Valley Citrus Committee, consisting of fifteen (15) members is hereby established. For each member of the committee there shall be an alternate who shall have the same qualifications as the member.

(b) Nine members shall be producers who produce fruit in the district which they represent and are residents of the production area. Two of the producer members shall be producers who market their fruit through cooperative marketing organizations, and seven of the producer members shall be independent producers. Producer members shall not have a proprietary interest in or be employees of a handler organization: *Provided,* That members of a cooperative marketing organization shall not be considered as having a proprietary interest in a handler organization because of such membership.

(c) Six members shall be handlers who are residents of the production area. One handler member shall represent cooperative marketing organizations; five handler members shall represent independent handlers.

§906.19 Term of office.

(a) The term of office of committee members and their respective alternates shall be for three years beginning August 1 and ending July 31: *Provided,* That the term of office of one-third of the initial producer members and alternates and one-third of the initial handler members and alternates shall end July 31, 1961, and the term of office of an identical number of such committee members and alternates shall end July 31, 1962. No member or alternate member shall succeed himself.

(b) Members and alternates shall serve in that capacity during the portion of the term of office for which they are selected and have qualified, and until their respective successors are selected and have qualified. Should a producer member or alternate member change his marketing affiliation during his term of office, he may continue to serve in such capacity during the remainder of such term.

[25 FR 9093, Sept. 22, 1960. Redesignated at 26 FR 12751, Dec. 30, 1961, as amended at 31 FR 10462, Aug. 4, 1966]

§ 906.20 Districts.

For the purpose of determining the basis for selecting producer committee members the following districts of the production area are hereby initially established:

District No. 1: The county of Cameron in the State of Texas;
District No. 2: The county of Hidalgo in the State of Texas; and
District No. 3 The county of Willacy in the State of Texas.

§ 906.21 Redistricting.

The committee may recommend, and pursuant thereto the Secretary may approve, the reapportionment of members among districts, the reapportionment of members between grower and handler members representing cooperative marketing organizations and independent grower and independent handler members, and the re-establishment of districts within the production area. In recommending such changes, the committee shall give consideration to: (a) Shifts in production; (b) the importance of new production in its relation to existing districts; (c) the equitable relationship of committee membership and districts; (d) changes in amount of fruit handled by cooperative marketing organizations in relation to fruit handled by independent handlers; and (e) other relevant factors. No changes in districting or in apportionment of members may become effective in less than 30 days prior to the date on which terms of office begin each year and no recommendations for such redistricting or reapportionment may be made less than six months prior to such date.

§ 906.22 Selection.

(a) From District No. 1 the Secretary shall select initially two producer members and their alternates representing independent producers. From District No. 2 the Secretary shall select initially two producer members and their respective alternates representing producers who market their fruit through cooperative marketing organizations, and four producer members and their respective alternates representing independent producers. From District No. 3 the Secretary shall select initially one producer member and his alternate representing independent producers.

(b) From the production area the Secretary shall select initially six handler members and their respective alternates. One handler member shall represent cooperative marketing organizations and five handler members shall represent independent handlers.

§ 906.23 Nominations.

The Secretary may select the members of the committee and alternates from nominations which may be made in the following manner:

(a) A meeting of producers who are members of cooperative marketing organizations and a meeting of independent producers shall be held for each district having both cooperative and independent producer members and alternates to elect nominees for such positions. For all other districts, meetings of all producers shall be held for such purpose. A meeting of handlers representing cooperative marketing organizations and a meeting of independent handlers shall be held in the production area to elect nominees for handler members and alternates. For nominations to the initial committee, the meetings may be sponsored by the United States Department of Agriculture or by any agency or group requested to do so by such Department.

For nominations for succeeding members and alternates on the committee, the committee shall hold such meetings or cause them to be held prior to June 15 of each year, after the effective date of this subpart.

(b) At each such meeting at least one nomination shall be designated for each position as member and alternate.

(c) Nominations for committee members and alternates following the initial committee shall be supplied to the Secretary not later than July 1 each year.

(d) In districts having both cooperative and independent producer members, only producers who market their fruit through cooperative marketing organizations may participate in designating nominees for members and alternates representing cooperative producers; and only independent producers may participate in designating nominees for members and alternates representing independent producers. In all other districts, all producers may participate in designating the nominees for producer members and alternates. Only handlers representing cooperative marketing organizations may participate in designating nominees for members and alternates representing cooperative handlers; and only independent handlers may participate in designating nominees for members and alternates representing independent handlers. In the event that a person is engaged in producing fruit in more than one district such person shall elect the district within which he may participate, as aforesaid, in designating nominees.

(e) Regardless of the amount of fruit handled by a handler or the number of districts in which a person produces fruit, each person is entitled to cast only one vote on behalf of himself, his agents, subsidiaries, affiliates, and representatives in designating nominees for committee members and alternates. An eligible voter's privilege of casting only one vote shall be construed to permit a voter to cast one vote for each position to be filled. Votes must be cast in person at all nomination meetings.

§ 906.24   Failure to nominate.

If nominations are not made within the time and in the manner specified in § 906.23, the Secretary may, without regard to nominations, select the committee members and alternates, which selection shall be on the basis of the representation provided for in §§ 906.20 through 906.22, inclusive.

§ 906.25   Acceptance.

Any person selected as a committee member or alternate shall qualify by filing a written acceptance with the Secretary within ten days after being notified of such selection.

§ 906.26   Vacancies.

To fill committee vacancies, the Secretary may select such members or alternates from unselected nominees on the current nominee list from the district and group involved, or from nominations made in the manner specified in § 906.23. If the names of nominees to fill any such vacancy are not made available to the Secretary within 30 days after such vacancy occurs, such vacancy may be filled without regard to nominations, which selection shall be made on the basis of representation provided for in §§ 906.20 through 906.22, inclusive.

§ 906.27   Alternate members.

An alternate member of the committee shall act in the place and stead of the member for whom he is an alternate, during such member's absence or when designated to do so by the member for whom he is an alternate. In the event both a member and his alternate are unable to attend a committee meeting, the committee members present may designate another alternate of the same classification (handler or producer, and to the extent practical, independent, or co-op) to serve in such member's place and stead. In the event of the death, removal, resignation, or disqualification of a member, his alternate shall act for him until a successor of such member is selected and has qualified.

[31 FR 10462, Aug. 4, 1966]

§ 906.28 Procedure.

Ten members of the committee shall be necessary to constitute a quorum, six of whom shall be producer members. Ten affirmative votes shall be required to pass any motion or approve any committee action. All votes shall be cast in person.

§ 906.29 Expenses and compensation.

The members of the committee, and alternates, shall serve without compensation; but they may be reimbursed for expenses necessarily incurred by them in the performance of their duties and in the exercise of their powers under this subpart.

§ 906.30 Powers.

The committee shall have the following powers:

(a) To administer the provisions of this part in accordance with its terms;

(b) To make rules and regulations to effectuate the terms and provisions of this part;

(c) To receive, investigate, and report to the Secretary complaints of violation of the provisions of this part; and

(d) To recommend to the Secretary amendments to this part.

§ 906.31 Duties.

It shall be, among other things, the duty of the committee:

(a) At the beginning of each term of office, to meet and organize, to select a chairman and such other officers as may be necessary, to select sub-committees, and to adopt such rules and regulations for the conduct of its business as it may deem advisable;

(b) To act as intermediary between the Secretary and any producer or handler;

(c) To furnish to the Secretary such available information as he may request;

(d) To appoint such employees, agents, and representatives as it may deem necessary and to determine the salaries and define the duties of each such person;

(e) To require adequate fidelity bonds for all persons handling funds;

(f) To investigate from time to time and to assemble data on the growing, harvesting, shipping, and marketing conditions with respect to fruit;

(g) To prepare a marketing policy;

(h) To recommend marketing regulations to the Secretary;

(i) To recommend rules and procedures for, and to make determinations in connection with, issuance of certificates of privilege;

(j) To keep minutes, books, and records which clearly reflect all of the acts and transactions of the committee; and such minutes, books, and records shall be subject to examination at any time by the Secretary or his authorized agent or representative; and minutes of each committee meeting shall be promptly submitted to the Secretary;

(k) At the beginning of each fiscal period, to prepare a budget of its expenses for such fiscal period, together with a report thereon;

(l) To cause the books of the committee to be audited by a competent accountant at least once each fiscal period, and at such other time as the committee may deem necessary or as the Secretary may request (the report of each such audit shall show the receipt and expenditure of funds collected pursuant to this part; a copy of each such report shall be furnished to the Secretary and a copy of each report shall be made available at the principal office of the committee for inspection by producers and handlers); and

(m) To consult, cooperate, and exchange information with other marketing agreement committees and other individuals or agencies in connection with all proper committee activities and objectives under this part.

EXPENSES AND ASSESSMENTS

§ 906.32 Expenses.

The committee is authorized to incur such expenses as the Secretary may find are reasonable and likely to be incurred during each fiscal period for its maintenance and functioning, and for such purposes as the Secretary, pursuant to this subpart, determines to be appropriate. Each handler's share of such expense shall be proportionate to the ratio between the total quantity of fruit handled by him as the first handler thereof during a fiscal period and the total quantity of fruit handled by all handlers as first handlers thereof during such fiscal period.

§906.33 Budget.

At the beginning of each fiscal period and as may be necessary thereafter, the committee shall prepare an estimated budget of income and expenditures necessary for the administration of this part. The committee shall recommend the rate of assessment calculated to provide adequate funds to defray its proposed expenditures. The committee shall present such budget to the Secretary with an accompanying report showing the basis for its estimates and recommendations.

§906.34 Assessments.

(a) The funds to cover the committee's expenses shall be acquired by the levying of assessments upon handlers as provided in this subpart. Each handler who first handles fruit shall, with respect to the fruit so handled by him, pay assessments to the committee upon demand, which assessments shall be in payment of such handler's pro rata share of the committee's expenses.

(b) Assessments shall be levied upon handlers at rates established by the Secretary. Such rates may be established upon the basis of the committee's recommendations and other available information. Such rates may be applied to specified containers used in the production area.

(c) The rate of assessment may be increased at any time by the Secretary if he finds such increase is necessary in order that the money collected shall be adequate to cover the committee's expenses during a given fiscal period. Such increase shall be applicable to all fruit handled during such fiscal period.

(d) The payment of assessments for the maintenance and functining of the committee may be required under this part throughout the period it is in effect irrespective of whether particular provisions of this part are suspended or become inoperative.

§906.35 Accounting.

(a) If, at the end of a fiscal period, the assessments collected are in excess of expenses incurred, such excess shall be accounted for in accordance with one of the following:

(1) If such excess is not retained in a reserve, as provided in paragraph (a)(2) of this section, it shall be refunded proportionately to the persons from whom collected.

(2) The committee, with the approval of the Secretary may carry over such excess into subsequent fiscal periods as a reserve: *Provided,* That funds already in the reserve do not equal approximately 1 fiscal period's expenses. Such reserve funds may be used for any expenses authorized pursuant to §906.32 and for necessary expenses of liquidation in the event of termination of this part. Upon such termination, any funds not required to defray the necessary expenses of liquidation shall be disposed of in such manner as the Secretary may determine to be appropriate. To the extent practical, such funds shall be returned pro rata to the persons from whom such funds were collected.

(b) All funds received by the committee pursuant to the provisions of this part shall be used solely for the purpose specified in this part and shall be accounted for in the manner provided in this part. The Secretary may at any time require the committee and its members to account for all receipts and disbursements.

(c) Upon the removal or expiration of the terms of office of any member of the committee, such member shall account for all receipts and disbursements and deliver all property and funds in his possession to the committee, and shall execute such assignments and other instruments as may be necessary or appropriate to vest in the committee full title to all of the property, funds, and claims vested in such member pursuant to this part.

(d) The committee may make recommendations to the Secretary for one or more of the members thereof, or any other person, to act as a trustee for holding records, funds, or any other committee property during periods of suspension of this subpart, or during any period or periods when regulations are not in effect, and if the Secretary determines such action appropriate, he may direct that such person or persons shall act as trustee or trustees for the committee.

[25 FR 9093, Sept. 22, 1960. Redesignated at 26 FR 12751, Dec. 30, 1961, as amended at 31 FR 10462, Aug. 4, 1966]

RESEARCH AND DEVELOPMENT

§ 906.37 Research and development.

The committee, with the approval of the Secretary, may establish or provide for the establishment of marketing research and development projects, including paid advertising, designed to assist, improve, or promote the marketing, distribution, and consumption of fruit. Any such project for the promotion and advertising of fruit may utilize an identifying mark which shall be made available for use by all handlers in accordance with such terms and conditions as the committee, with the approval of the Secretary, may prescribe. The expenses of such projects shall be paid from funds collected pursuant to § 906.34.

[31 FR 10462, Aug. 4, 1966]

REGULATION

§ 906.38 Marketing policy.

Prior to or at the same time as initial recommendations are made pursuant to § 906.39, the committee shall submit to the Secretary a report setting forth the marketing policy it deems desirable for the industry to follow in shipping fruit from the production area during the ensuing season. Additional reports shall be submitted from time to time if it is deemed advisable by the committee to adopt a new or modified marketing policy because of changes in the demand and supply situation with respect to fruit. The committee shall publicly announce the submission of each marketing policy report and copies thereof shall be available at the committee's office for inspection by any producer or handler. In determining each such marketing policy the committee shall give due consideration to the following:

(a) Market prices of fruit, including prices by grade, size, and quality in different packs, and such prices by foreign competing areas;

(b) Supply of fruit, by grade, size, and quality in the production area, and in other production areas, including foreign production areas;

(c) Trend and level of consumer income;

(d) Marketing conditions affecting fruit prices; and

(e) Other relevant factors.

§ 906.39 Recommendations for regulations.

The committee, upon complying with the requirements of § 906.38, may recommend regulations to the Secretary whenever it finds that such regulations, as are provided for in this subpart, will tend to effectuate the declared policy of the act. The committee shall give notice to handlers of any such recommendation at the same time such recommendation is submitted to the Secretary.

§ 906.40 Issuance of regulations.

The Secretary shall limit the handling of fruit whenever he finds from the recommendation and information submitted by the committee, or from other available information, that such rgulation would tend to effectuate the declared policy of the act. Such regulations may:

(a) Limit the handling of particular grades, sizes, qualities, maturities, or packs of any or all varieties of fruit during a specified period or periods: *Provided*, That specific maturity requirements applicable to the handling of any variety may be prescribed under this section only in the event that appropriate maturity requirements for such variety are not in effect under State authority.

(b) Limit the handling of particular grades, sizes, qualities, or packs of fruit differently for different varieties, for different containers, for different purposes specified in § 906.42, or any combination of the foregoing, during any period.

(c) Limit the handling of fruit by establishing, in terms of grades, sizes, or both, minimum standards of quality and maturity.

(d) Fix the size, weight, capacity, dimensions, or pack of the container or containers which may be used in the packaging, transportation, sale, shipment, or other handling of fruit.

(e) Prohibit the handling (1) of any fruit which does not have marked on each container the grade or the registered grade label of the fruit contained therein; (2) of any grapefruit which does not have marked on each fruit the word *Texas* or other words implying Texas origin, except that the committee may recommend and the Secretary establish a tolerance for grapefruit in any container or lot not so marked; and (3) of any container fruit which is misbranded as to variety.

(f) No regulations may be issued under the provisions of this subpart which allots to individual handlers the quantity of fruit which each handler may ship during any regulation period.

§906.41 Gift fruit shipments.

The handling to any person of gift packages of fruit individually addressed to such person, in quantities aggregating not more than 500 pounds and not for resale, are exempt from the provisions of §§906.34, 906.40, and 906.45, and the regulations issued thereunder, but shall conform to such safeguards as may be established pursuant to §906.43.

§906.42 Shipments for special purposes.

Upon the basis of recommendations and information submitted by the committee, or other available information, the Secretary, whenever he finds that it will tend to effectuate the declared policy of the act, shall modify, suspend, or terminate regulations issued pursuant to §§906.34, 906.40, 906.45, or any combination thereof, in order to facilitate the handling of fruit:

(a) For relief or for charity;

(b) For processing or for manufacture or conversion into specified products; and

(c) In such minimum quantities and for such other purposes as may be specified by the committee with the approval of the Secretary.

§906.43 Notification of regulations.

The Secretary shall notify the committee of any regulations issued or of any modification, suspension, or termination thereof. The committee shall give reasonable notice thereof to handlers.

§906.44 Safeguards.

(a) The committee, with the approval of the Secretary, may prescribe adequate safeguards to prevent the handling of fruit pursuant to §906.41 or §906.42 from entering channels of trade for other than the specific purpose authorized therefor, and rules governing the issuance and the contents of certificates of privilege if such certificates are prescribed as safeguards by the committee. Such safeguards may include requirements that:

(1) Handlers shall file applications with the committee to ship fruit pursuant to §§906.41 and 906.42.

(2) Handlers shall obtain inspection provided by §906.45, or pay the assessment levied pursuant to §906.34, or both, in connection with shipments made under §906.42: *Provided*, That such inspection and assessment requirements shall not apply to fruit handled for canning or freezing.

(3) Handlers shall obtain certificates of privilege from the committee to handle fruit affected or to be affected under the provisions of §§906.41 and 906.42.

(b) The committee may rescind or deny certificates of privilege to any handler if proof is obtained that fruit handled by him for the purposes stated in §§906.41 and 906.42 was handled contrary to the provisions of this part.

(c) The Secretary shall have the right to modify, change, alter, or rescind any safeguards prescribed and any certificates issued by the committee pursuant to the provisions of this section.

(d) The committee shall make reports to the Secretary, as requested, showing the number of applications for such certificates, the quantity of fruit covered by such applications, the number of such applications denied and certificates granted, the quantity of fruit handled under duly issued certificates, and such other information as may be requested.

INSPECTION

§906.45 Inspection and certification.

(a) During any period in which handling of a variety of a type of fruit is regulated pursuant to §§906.34, 906.40, 906.42, or any combination thereof, no handler shall handle any variety of such type of fruit which has not been inspected by an authorized representative of the Federal or Federal-State Inspection Service, unless such handling is relieved from such requirements pursuant to § 906.41 or § 906.42, or both;

(b) Regrading, resorting, or repacking any lot of fruit shall invalidate any prior inspection insofar as the requirements of this section are concerned. No handler shall handle fruit after it has been regraded, resorted, repacked, or in any other way prepared for market, unless each lot of fruit is inspected by an authorized representative of the Federal or Federal-State Inspection Service: *Provided,* That the committee, with the approval of the Secretary, may provide for waiving inspection requirements on any fruit in circumstances where it appears reasonably certain that, after regrading, resorting, or repacking, such fruit meets the applicable quality and other standards then in effect;

(c) Insofar as the requirements of this section are concerned, the length of time for which an inspection certificate is valid may be established by the committee with the approval of the Secretary;

(d) When fruit is inspected in accordance with the requirements of this section a copy of each inspection certificate issued shall be made available to the committee by the inspection service;

(e) The committee may recommend and the Secretary may require that any fruit handled or transported by motor vehicle shall be accompanied by a copy of the inspection certificate issued thereon, which certificate shall be surrendered to such authority as may be designated.

REPORTS

§ 906.51 Reports.

Upon request of the committee, made with the approval of the Secretary, each handler shall furnish to the committee, in such manner and at such time as it may prescribe, such reports and other information as may be necessary for the committee to perform its duties under this part.

(a) Such reports may include, but are not necessarily limited to, the following:

(1) The quantities of fruit received by a handler;

(2) The quantities disposed of by him, segregated as to the respective quantities subject to regulation and not subject to regulation;

(3) The date of each such disposition and the identification of the carrier transporting such fruit;

(4) Identification of the inspection certificates, and the certificates of privilege, if any, pursuant to which the fruit was handled, together with the destination of each lot of fruit handled pursuant to § 906.41.

(b) All such reports shall be held under appropriate protective classification and custody of the committee, or duly appointed employees thereof, so that the information contained therein which may adversely affect the competitive position of any handler in relation to other handlers will not be disclosed. Compilations of general reports from data submitted by handlers is authorized, subject to prohibition of disclosure of individual handlers identities or operations.

(c) Each handler shall maintain for at least two succeeding years such records of the fruit received and disposed of by such handler as may be necessary to verify the reports he submits to the committee pursuant to this section.

MISCELLANEOUS PROVISIONS

§ 906.52 Compliance.

Except as provided in this subpart, no handler shall handle fruit, the handling of which has been prohibited by the Secretary in accordance with provisions of this subpart, or the rules and regulations issued thereunder, and no handler shall handle fruit except in conformity to the provisions of this part.

§ 906.53 Right of the Secretary.

The members of the committee (including successors and alternates), and any agent or employee appointed or employed by the committee, shall be subject to removal or suspension by the Secretary at any time. Each and

80

every order, regulation, decision, determination or other act of the committee shall be subject to the continuing right of the Secretary to disapprove of the same at any time. Upon such disapproval, the disapproved action of the said committee shall be deemed null and void, except as to acts done in reliance thereon or in compliance therewith prior to such disapproval by the Secretary.

§906.54 Effective time.

The provisions of this subpart, or any amendment thereto, shall become effective at such time as the Secretary may declare and shall continue in force until terminated in one of the ways specified in this subpart.

§906.55 Termination.

(a) The Secretary may, at any time, terminate the provisions of this subpart by giving at least one day's notice by means of a press release or in any other manner he may determine.

(b) The Secretary may terminate or suspend the operation of any or all of the provisions of this subpart whenever he finds that such provisions do not tend to effectuate the declared policy.

(c) The Secretary shall terminate the provisions of this subpart at the end of any fiscal period whenever he finds that such termination is favored by a majority of producers who, during a representative period, have been engaged in the production of fruit for market: *Provided*, That such majority has, during such representative period, produced for market more than fifty percent of the volume of such fruit produced for market.

(d) The provisions of this subpart shall, in any event, terminate whenever the provisions of the act authorizing them cease to be in effect.

§906.56 Proceedings after termination.

(a) Upon the termination of the provisions of this subpart the then functioning members of the committee shall, for the purpose of liquidating the affairs of the committee continue as joint trustees of all the funds and property then in the possession of or under control of the committee, including claims for any funds unpaid or property not delivered at the time of such termination. Action by said trusteeship shall require the concurrence of a majority of the said trustees.

(b) The said trustees shall continue in such capacity until discharged by the Secretary; shall, from time to time, account for all receipts and disbursements and deliver all property on hand, together with all books and records of the committee and of the trustees, to such person as the Secretary may direct; and shall, upon the request of the Secretary, execute such assignments or other instruments necessary or appropriate to vest in such person full title and right to all funds, property, and claims vested in the committee or the trustees pursuant to this subpart.

(c) Any person to whom funds, property, or claims have been transferred or delivered by the committee or its members pursuant to this section, shall be subject to the same obligations imposed upon the members of the committee and upon the said trustees.

§906.57 Effect of termination or amendment.

Unless otherwise expressly provided by the Secretary, the termination of this subpart or of any regulation issued pursuant to this subpart, or the issuance of any amendments to either thereof, shall not (a) affect or waive any right, duty, obligation, or liability which shall have arisen or which may thereafter arise in connection with any provision of this subpart or any regulation issued under this subpart, or (b) release or extinguish any violation of this subpart or of any regulations issued under this subpart, or (c) affect or impair any rights or remedies of the Secretary or of any other person with respect to any such violations.

§906.58 Duration of immunities.

The benefits, privileges, and immunities conferred upon any person by virtue of this subpart shall cease upon the termination of this subpart, except with respect to acts done under and during the existence of this subpart.

§906.59 Agents.

The Secretary may, by designation in writing, name any person, including any officer or employee of the United

States, or name any agency in the United States Department of Agriculture, to act as his agent or representative in connection with any of the provisions of this subpart.

§906.60 Derogation.

Nothing contained in this subpart is, or shall be construed to be, in derogation or in modification of the rights of the Secretary or of the United States to exercise any powers granted by the act or otherwise, or in accordance with such powers, to act in the premises whenever such action is deemed advisable.

§906.61 Personal liability.

No member or alternate of the committee or any employee or agent thereof, shall be held personally responsible, either individually or jointly with others, in any way whatsoever, to any handler or to any person for errors in judgment, mistakes, or other acts, either of commission or omission, as such member, alternate, agent, or employee, except for act of dishonesty, willful misconduct, or gross negligence.

§906.62 Separability.

If any provision of this subpart is declared invalid, or the applicability thereof to any person, circumstance, or thing is held invalid, the validity of the remainder of this subpart, or the applicability thereof to any other person, circumstance, or things, shall not be affected thereby.

Subpart B—Administrative Requirements

§906.120 Fruit exempt from regulations.

(a) *Minimum quantity.* Any person or the occupants of any one vehicle may ship fruit from the production area during any one day exempt from the requirements of §§906.34, 906.40, and 906.45, and regulations issued thereunder: *Provided,* That the shipment does not exceed 400 pounds of fruit (either oranges or grapefruit or a combination of both), it consists solely of fruit not for resale, and it is not part of a shipment of fruit exceeding 400 pounds.

(b) *Processing.* The term *processing* as used in §906.42(b) means the manufacture of any orange or grapefruit product which has been converted into sectioned fruit or into fresh juice, or preserved by any commercial process, including canning, freezing, dehydrating, drying, and the addition of chemical substances, or by fermentation. Fruit so processed, if handled in accordance with §906.123, shall be exempt from the provisions of §§906.34 and 906.40.

(c) *Special purpose shipments and safeguards.* (1) Fruit may be handled for relief or charity exempt from the requirements of §§906.34, 906.40, and 906.45 and the regulations issued thereunder: *Provided,* That the fruit shall not be offered for resale, and the handler submits, prior to any such handling, an application to the committee on forms provided by the committee. The application shall contain the name and address of the handler and such other information that the committee may require including, but not limited to, the quantity of fruit involved, license number of the conveyance, and supporting documentation. Approval of the application by the committee shall be evidenced by the issuance of a certificate of privilege to the applicant in accordance with paragraph (d) of this section.

(2) Gift packages of fruit handled pursuant to §906.41 shall be in containers stamped or marked with the handler's name and address.

(3) Fruit may be handled exempt from regulations issued pursuant to §906.40(d), if the following conditions are met:

(i) Each fiscal period the handler submits prior to such handling a written application to the committee on forms provided by the committee. The application shall contain the name and address of the handler, and a description of the container or containers in which such fruit would be handled.

(ii) The fruit grades at least U.S. No. 1.

(iii) The fruit is handled in closed fully telescopic fiberboard cartons with inside dimensions of $16\frac{1}{2} \times 10\frac{3}{4} \times 10\frac{1}{2}$ inches, and the cover and bottom section have a Mullen or Cady test of at least 250 pounds; in six-packs; in 12-packs; in baskets of a capacity of 1 bushel or less; or in any of the containers authorized under §906.340, provided they are stamped or marked *special purpose shipment.*

(iv) Each handler shall file a report with the committee within 1 business day after each shipment handled pursuant to paragraph (c)(3). Such report shall contain the name and address of the handler; date fruit is handled; the number and type of containers and packs in such shipment; the inspection certificate numbers applicable to such shipment; name and address of the purchaser; and the license number of the truck, trailer, or automobile, as the case may be, in which the shipment was loaded.

(4) Oranges and grapefruit grown in the production area may be handled exempt from container and pack regulations issued pursuant to §906.40(d), under the following conditions:

(i) Such oranges and/or grapefruit grown in the production area are mixed with other types of fruit;

(ii) Such oranges and/or grapefruit grown in the production area constitute at least one-third by volume of the contents of any container, and any such container is not larger than a 7/10 bushel carton.

(iii) Such grapefruit grown in the production area grade at least U.S. No. 1, and such oranges grown in the production area grade at least U.S. Combination (with not less than 60 percent, by count, of the oranges in any lot grading at least U.S. No.1).

(d) The committee or its duly authorized agents, shall approve or deny each handler's request to handle fruit under paragraphs (c)(1) and (c)(3) of this section and promptly notify such handler in writing of its decision: *Provided,* That if it approves a handler's request, it shall issue a certificate of privilege as provided in §906.44, but if it denies a request it shall advise the handler why the application was denied. The committee may rescind a certificate of privilege issued to a handler, or deny a certificate of privilege to a handler upon proof satisfactory to the committee that such handler has shipped fruit contrary to the provisions of this part. Such action denying a certificate of privilege shall apply to and not exceed a reasonable period of time as determined by the committee. Any handler who has had a certificate of privilege rescinded or denied may file a written appeal with the committee for reconsideration.

(e) *Terms.* The term *bushel* means a unit of measure equivalent to 2,150.42 cubic inches; the term *six-pack* means any container with a capacity of one-fourth of a bushel, the term *basket* means any container made of interwoven material; the term *closed* means closed in accordance with good commercial practices; and terms relating to grade mean the same as in the U.S. Standards for Grades of Grapefruit (Texas and States other than Florida, California, and Arizona) (7 CFR 51.620 through 51.653), or in the U.S. Standards for Grades of Oranges (Texas and States other than Florida, California, and Arizona) (7 CFR 51.680 through 51.714).

[25 FR 9757, Oct. 12, 1960. Redesignated at 26 FR 12751, Dec. 30, 1961, as amended at 39 FR 44736, Dec. 27, 1974; 40 FR 3286, Jan. 21, 1975; 44 FR 75103, Dec. 19, 1979; 48 FR 50502, Nov. 2, 1983; 49 FR 3173, Jan. 26, 1984; 54 FR 18095, Apr. 27, 1989; 59 FR 50826, Oct. 6, 1994; 59 FR 63693, Dec. 9, 1994; 60 FR 13892, Mar. 15, 1995; 70 FR 51578, Aug. 31, 2005]

§906.121   Reestablishment of districts.

The three districts of the production area specified in §906.20 *Districts* are reestablished as a single district comprising the entire production area.

[34 FR 6651, Apr. 18, 1969]

§906.122   [Reserved]

§906.123   Fruit for processing.

(a) No person shall be granted exemption from regulation to handle oranges and grapefruit for processing unless such fruit is shipped to an approved processor. All such shipments to an approved processor shall be reported to the committee on a form approved by it.

(b) *Approved processor.* Any person who desires to acquire, as an approved processor, fruit for processing, as set forth in §906.120(b), shall, prior thereto, file an application with the committee on a form approved by it, which shall contain, but not be limited to, the following information:

(1) Name and address of applicant;

(2) Location of plant or plants where manufacturing is to take place;

(3) Approximate quantity of fruit used each month;

(4) A statement that the fruit obtained exempt from fresh fruit regulations will not be resold or transferred for resale, directly or indirectly, but will be used only for processing;

(5) A statement agreeing to hold a license issued under the Perishable Agricultural Commodities Act, 1930 (7 U.S.C. 499r), and regulations issued thereunder (7 CFR part 46) when buying Texas oranges and grapefruit for processing;

(6) A statement agreeing to undergo random inspection by the committee;

(7) A statement that the requesting processor has no facilities, equipment, or outlet to repack or sell fruit in fresh form;

(8) A statement agreeing to submit such reports as are required by the committee.

Such application shall be investigated by the committee staff. After such investigation, the staff shall report its findings to the committee at its next meeting or to its delegated subcommittee. Based upon the staff's report and other reliable information, the committee or delegated subcommittee shall approve or disapprove the application and notify the applicant accordingly. If the application is approved, the applicant's name shall be placed upon the list of approved processors.

(c) *Certificate by processors.* Upon request by the committee each approved processor shall submit to the committee on or before the 10th day of each month a report of the oranges and grapefruit used during the preceding calendar month. Each report shall contain a certificate to the United States Department of Agriculture and to the committee as to the truthfulness of the information shown therein.

(d) *Diversion report.* Each handler who ships fruit to processors for processing shall report to the committee on a form approved by it the following information:

(1) Name and address of the processor's place of business where the fruit was shipped;

(2) The net weight of oranges or grapefruit;

(3) Truck license number or rail car initial and number;

(4) Inspection certificate number; and

(5) Such other information as the committee may require.

The handler shall prepare 4 copies of the report and sign them. The original copy shall be submitted to the committee within 7 days. One copy shall be retained by the handler. One copy shall be given to the party transporting the fruit who, upon arrival at the processor's place of business, shall turn it over to the party receiving the fruit with the understanding that the processor will record thereon the actual net weight of the fruit received and forward such copy to the committee office. One copy shall be submitted to the processor along with the invoice.

[39 FR 44736, Dec. 27, 1974, as amended at 54 FR 18095, Apr. 27, 1989]

§ 906.137 Handlers use of identifying marks utilized by the committee in promotional and advertising projects.

(a) Pursuant to § 906.37, the identifying marks "Texasweet", "Sweeter By Nature", "Texas Fancy", and "Texas Choice" shall be available to handlers only under the following terms and conditions:

(1) The identifying marks "Texasweet" and "Sweeter by Nature" may severally or jointly be affixed only to containers of grapefruit or to individual grapefruit comprising a lot which grades at least U.S. No. 1.

(2) The identifying mark "Texas Fancy" may be affixed only to containers of grapefruit or to individual grapefruit comprising a lot which grades at least U.S. No. 1 with no more than 40 percent of the surface of the grapefruit, in the aggregate, affected by discoloration.

(3) The identifying mark "Texas Choice" may be affixed only to containers of grapefruit or to individual grapefruit comprising a lot which grades at least U.S. No. 2, with no more than 60 percent of the surface of the grapefruit, in the aggregate, affected by discoloration.

(4) The identifying marks "Texasweet" and "Sweeter by Nature"

may severally or jointly be affixed only to containers of oranges or to individual oranges comprising a lot which grades at least U.S. Combination, with not less than 60 percent, by count, of the oranges in each container thereof grading at least U.S. No. 1 and the remainder U.S. No. 2.

(5) The identifying mark "Texas Choice" may be affixed only to containers of oranges or to individual oranges comprising a lot which grades at least U.S. No. 2, except that in determining whether the fruit is reasonably well colored the yellow or orange color must predominate over the green color on at least 75 percent of the fruit surface in the aggregate which is not discolored.

(b) When used herein, terms relating to grade shall have the same meaning as is given to the respective term in the U.S. Standards for Grapefruit (Texas and States other than Florida, California, and Arizona) (7 CFR 51.620 through 51.653) and in the U.S. Standards for Oranges (Texas and States other than Florida, California, and Arizona) (7 CFR 51.680 through 51.714).

[33 FR 14069, Sept. 17, 1968, as amended at 53 FR 40398, Oct. 17, 1988; 53 FR 50916, Dec. 19, 1988; 70 FR 51578, Aug. 31, 2005]

§906.151 Reports.

(a) During each fiscal period, each handler shall upon request by the committee file with the committee within the time specified in the request an accurate report showing the total quantity or oranges and the total quantity of grapefruit received by him during such fiscal period or the preceding fiscal period, as requested.

(b) Each handler who sells over 400 pounds of oranges or grapefruit or a combination of both for resale inside the production area shall, for each transaction, report to the committee on a form approved by it the following information:

(1) Name and address of seller;

(2) Name and address of buyer;

(3) Description and quantity of oranges or grapefruit sold;

(4) Destination of fruit;

(5) A statement that the buyer certifies that fruit that is subsequently taken outside the production area for resale will be inspected; and

(6) Such other pertinent information as the committee may require.

(c) The handler shall prepare the report in triplicate. The buyer shall sign the certification statement. The pink copy shall be submitted to the committee within 7 days. The white copy shall be retained by the handler and the canary copy shall be given to the buyer. Such form shall be reviewed by the committee staff and the information compiled for the committee's use.

[34 FR 6651, Apr. 18, 1969, as amended at 61 FR 64255, Dec. 4, 1996; 62 FR 3603, Jan. 24, 1997]

§906.235 Assessment rate.

On and after August 1, 2018, an assessment rate of $0.01 per 7/10-bushel carton or equivalent is established for oranges and grapefruit grown in the Lower Rio Grande Valley in Texas.

[83 FR 55933, Dec. 10, 2018]

Subpart C—Container and Pack Requirements

§906.340 Container, pack, and container marking regulations.

(a) No handler shall handle any variety of oranges or grapefruit grown in the production area unless such fruit is in one of the following containers, and the fruit is packed and the containers are marked as specified in this section:

(1) *Containers.* (i) Closed fiberboard carton with approximate inside dimensions of 13¼ x 10½ x 7¼ inches: *Provided,* That the container has a Mullen or Cady test of at least 200 pounds;

(ii) Closed fully telescopic fiberboard carton with approximate inside dimensions of 16½ x 10¾ x 9½ inches (Standard carton);

(iii) Poly or mesh bags having a capacity of 4, 5, 8, 10, or 18 pounds of fruit;

(iv) Rectangular or octagonal bulk fiberboard crib with approximate dimensions of 46 to 47½ inches in length, 37 to 38 inches in width, and 36 inches in height: *Provided,* That the container has a Mullen or Cady test of at least 1,300 pounds, and that it is used only once for the shipment of citrus fruit: *And Provided further,* That the container may be used to pack any poly or mesh bags authorized in this section, or bulk fruit;

(v) Rectangular or octagonal ⅔ fiberboard crib with approximate dimensions of 46 to 47½ inches in length, 37 to 38 inches in width, and 24 inches in height: *Provided,* That the crib has a Mullen or Cady test of at least 1,300 pounds, and that it is used only once for the shipment of citrus fruit: *And Provided further,* That the container may be used to pack any poly or mesh bags authorized in this section, or bulk fruit;

(vi) Octagonal fiberboard crib with approximate dimensions of 46 to 47½ inches in width, 37 to 38 inches in depth, and 26 to 26½ inches in height: *Provided,* That the crib has a Mullen or Cady test of at least 1,300 pounds, and that it is used only once for the shipment of citrus fruit: *And Provided further,* That the crib may be used to pack any poly or mesh bags authorized in this section, or bulk fruit;

(vii) Fiberboard box holding two layers of fruit, with approximate dimensions of 23 inches in length, 15½ inches in width, and 7 inches in depth;

(viii) Reusable collapsible plastic container with approximate dimensions of 23 inches in length, 15 inches in width, and 7 to 11 inches in depth;

(ix) Reusable collapsible plastic bin with approximate dimensions of 36¾ x 44¾ x 27 inches;

(x) Octagonal bulk triple wall fiberboard crib with approximate dimensions of 37¾ inches in length, 25 inches in width, and 25 inches in height: *Provided,* That the container has a Mullen or Cady test of at least 1,100 pounds: *And Provided further,* That the container may be used to pack any poly or mesh bags authorized in this section, or bulk fruit;

(xi) Bag having the capacity of 15 pounds of fruit, either in a combination ½ poly and ½ mesh bag or mesh bag;

(xii) Reusable collapsible plastic mini bin with approximate dimensions of 39½ inches in length, 24 inches in width, and 30½ inches in height: *Provided,* That the container may be used to pack any poly or mesh bags authorized in this section, or bulk fruit;

(xiii) Bag having the capacity of three pounds of fruit;

(xiv) Standard carton with approximate inside dimensions of 16.375 x 10.6875 x 10.25 inches;

(xv) ⅞ Body master carton with approximate inside dimensions of 19.5385 x 13.125 x 11.625 inches, one piece;

(xvi) Euro ⅝ (5 Down) with approximate inside dimensions of 22.813 x 14.688 x 7.0 up to 7.936 inches;

(xvii) Fiberboard one piece display container with approximate inside dimensions of 23 inches x 15 inches x 9½ up to 10½ inches in depth;

(xviii) Such types and sizes of containers as may be approved by the committee for testing in connection with a research project conducted by or in cooperation with the committee: *Provided,* That the handling of each lot of fruit in such test containers shall be subject to prior approval and under the supervision of the committee.

(2) *Pack regulation—*(i) *Oranges.* (A) Oranges, when packed in any carton, bag, or other container, shall be sized in accordance with the sizes in the following Table I, and meet the requirements of standard pack; and, when in containers not packed according to a definite pattern, shall be sized in accordance with the sizes in Table I and otherwise meet the requirements of standard sizing: Provided, That the packing tolerances in the U.S. Standards for Grades of Oranges (Texas and States other than Florida, California, and Arizona), shall apply to fruit so packed. All fruit packed to size 163 in the following Table I shall be sized in accordance with the sizes in Table I but need not otherwise meet the requirements of standard sizing or standard pack: Provided, That they meet the same tolerances for off-size and pack as defined in the U.S. Standards for Grades of Oranges (Texas and States other than Florida, California, and Arizona):

TABLE I—ORANGES

[⁷⁄₁₀ bushel carton]

Rack size/ number of oranges	Diameter in inches	
	Minimum	Maximum
24	3¹²⁄₁₆	5¹⁄₁₆
32	3⁶⁄₁₆	4⁹⁄₁₆
36	3⁴⁄₁₆	4⁶⁄₁₆
40	3²⁄₁₆	4⁴⁄₁₆
48	2¹⁵⁄₁₆	4
56	2¹³⁄₁₆	3¹³⁄₁₆

TABLE I—ORANGES—Continued

[7/10 bushel carton]

Rack size/ number of oranges	Diameter in inches	
	Minimum	Maximum
64	2¹¹/₁₆	3¹⁰/₁₆
72	2⁹/₁₆	3⁸/₁₆
88	2⁸/₁₆	3⁴/₁₆
113	2⁷/₁₆	3
138	2⁶/₁₆	2¹²/₁₆
163	2³/₁₆	2⁸/₁₆

(B) If 7/10 bushel containers of oranges are marked, the count of fruit in each container shall not be less than the count marked on the container, but may exceed the count marked on the container by not more than 8 percent. When packed in marked containers other than 7/10 bushel, the pack sizes applicable to 7/10 bushel containers shall also apply to such containers.

(ii) *Grapefruit.* (A) Grapefruit, when packed in any carton, bag, or other container, shall be sized in accordance with the sizes in the following Table II, except as otherwise provided in the regulations issued pursuant to this part, and meet the requirements of standard pack; and, when in containers not packed according to a definite pattern, shall be sized in accordance with the sizes in Table II: Provided, That the packing tolerances in the U.S. Standards for Grades of Grapefruit (Texas and States other than Florida, California, and Arizona), shall apply to fruit so packed. All fruit packed to size 64 in the following Table II shall be sized in accordance with the sizes in Table II but need not otherwise meet the requirements of standard pack: Provided, That they meet the same tolerances for off-size and pack as defined in the U.S. Standards for Grades of Grapefruit (Texas and States other than Florida, California, and Arizona).

TABLE II—GRAPEFRUIT

[7/10 Bushel carton]

Pack size/ number of grapefruit	Diameter in inches	
	Minimum	Maximum
18	4¹⁵/₁₆	5⁹/₁₆
23	4⁵/₁₆	5
27	4²/₁₆	4¹²/₁₆
32	3¹⁵/₁₆	4⁸/₁₆
36	3¹³/₁₆	4⁵/₁₆
40	3¹⁰/₁₆	4²/₁₆
48	3⁹/₁₆	3¹⁴/₁₆
56	3⁵/₁₆	3¹⁰/₁₆

TABLE II—GRAPEFRUIT—Continued

[7/10 Bushel carton]

Pack size/ number of grapefruit	Diameter in inches	
	Minimum	Maximum
64	3	3⁸/₁₆

(B) If 7/10 bushel containers of grapefruit are marked, the count of fruit in the container shall not be less than the count marked on the container, but may exceed the count marked on the container by not more than 8 percent. When packed in marked containers other than 7/10 bushel, the pack sizes applicable to 7/10 bushel containers shall also apply to such containers.

(3) *Container grade markings.* Except when the identifying marks "Texas Choice" or "Texas Fancy" are used by handlers pursuant to § 906.137, any container of U.S. No. 2 grade fruit shall be marked to indicate the grade of the fruit in letters and numbers at least three-fourths inch in height: *Provided,* That bags containing five or eight pounds of fruit shall be so marked with letters and numbers at least one-fourth inch in height prominently displayed on the front panel of the bag. The requirements of this paragraph (a)(3) will not be effective until February 16, 1992.

(b) *Nonapplicability.* The provisions of this section shall not apply to gift packages of fruit.

(c) As used herein, terms relating to grade, pack, standard pack, and diameter mean the same as defined in the United States Standards for Grades of Oranges (Texas and States other than Florida, California, and Arizona), (7 CFR 51.680 through 51.714), or in the United States Standards for Grades of Grapefruit (Texas and States other than Florida, California, and Arizona), (7 CFR 51.620 through 51.653); and *closed* means closed in accordance with good commercial practices.

[33 FR 11542, Aug. 14, 1968]

EDITORIAL NOTE: For FEDERAL REGISTER citations affecting § 906.340, see the List of CFR Sections Affected, which appears in the Finding Aids section of the printed volume and at *www.govinfo.gov.*

§ 906.365 Texas Orange and Grapefruit Regulation 34.

(a) No handler shall handle any variety of oranges or grapefruit grown in the production area unless:

(1) Such oranges grade U.S. Fancy, U.S. No. 1, U.S. No. 1 Bright, U.S. No. 1 Bronze, U.S. Combination (with not less than 60 percent, by count, of the oranges in any lot thereof grading at least U.S. No. 1), or U.S. No. 2;

(2) Such oranges are at least pack size 163 with a minimum diameter of 2–3/16 inches;

(3) Such grapefruit grade U.S. Fancy, U.S. No. 1, U.S. No. 1 Bright, or U.S. No. 1 Bronze, or meet the quality requirements of "Texas Fancy" or "Texas Choice" as defined in § 906.137 of this part;

(4) Such grapefruit are at least pack size 64 with a minimum diameter of 3 inches.

(5) An appropriate inspection certificate has been issued for such fruit within 48 hours prior to the time of shipment. No handler may transport by motor vehicle or cause the transportation of any shipment of fruit for which an inspection certificate is required unless each such shipment is accompanied by a copy of the inspection certificate applicable thereto, and a copy of such inspection certificate is surrendered upon request to Texas Department of Agriculture personnel designated by the committee.

(6) The fruit meets all the applicable container and pack requirements effective under this marketing order.

(7) Beginning in 1995, this paragraph (a) is suspended each year from July 1 through August 31 of each year.

(b) Terms relating to grade, pack size, and diameter shall mean the same as in the U.S. Standards for Grades of Oranges (Texas and States other than Florida, California, and Arizona) (7 CFR 51.680 through 51.714) or in the U.S. Standards for Grades of Grapefruit (Texas and States other than Florida, California and Arizona) (7 CFR 51.620 through 51.653).

[47 FR 1266, Jan. 12, 1982, as amended at 51 FR 41070, Nov. 13, 1986; 54 FR 3421, Jan. 24, 1989; 54 FR 41584, Oct. 11, 1989; 56 FR 55983, Oct. 31, 1991; 58 FR 52401, Oct. 8, 1993; 58 FR 54926, Oct. 25, 1993; 59 FR 56383, Nov. 14, 1994; 60 FR 33679, June 29, 1995; 60 FR 54292, Oct. 23, 1995; 61 FR 43141, Aug. 21, 1996; 64 FR 47358, Aug. 31, 1999; 79 FR 11297, 11300, Feb. 28, 2014]

PART 915—AVOCADOS GROWN IN SOUTH FLORIDA

Subpart A—Order Regulating Handling

AUTHORITY: 7 U.S.C. 601–674.

SOURCE: 19 FR 3439, June 11, 1954, unless otherwise noted. Redesignated at 26 FR 12751, Dec. 30, 1961.

Subpart A—Order Regulating Handling

DEFINITIONS

§ 915.1   Secretary.

Secretary means the Secretary of Agriculture of the United States or any officer or employee of the United States Department of Agriculture who is, or may hereafter be, authorized to exercise the powers and perform the duties of the Secretary of Agriculture of the United States.

§ 915.2   Act.

Act means Public Act No. 10, 73d Congress (May 12, 1933), as amended and as reenacted and amended by the Agricultural Marketing Agreement Act of 1937, as amended (48 Stat. 31, as amended, 68 Stat. 906, 1047; 7 U.S.C. 601 *et seq.*).

[19 FR 3439, June 11, 1954, as amended at 20 FR 4177, June 15, 1955. Redesignated at 26 FR 12751, Dec. 30, 1961]

§ 915.3   Person.

Person means an individual, partnership, corporation, association or any other business unit.

§ 915.4   Production area.

Production area means the counties of Brevard, Orange, Lake, Polk, Hillsborough, and Pinellas in the State of Florida, and all of the counties of that State situated south of such counties.

[20 FR 4177, June 15, 1955. Redesignated at 26 FR 12751, Dec. 30, 1961]

§ 915.5   Avocados.

Avocados means all varieties of avocados grown in the production area.

§ 915.6   Fiscal year.

Fiscal year means the twelve-month period ending March 31 of each year.

§ 915.7   Committee.

Committee means the Avocado Administrative Committee established pursuant to § 915.20.

§ 915.8   Grower.

Grower is synonymous with producer and means any person who produces avocados for market and who has a proprietary interest therein: *Provided,* That as used in § 915.22 the term grower shall include only those who have a proprietary interest in the production of 10 or more bearing avocado trees.

[19 FR 3439, June 11, 1954, as amended at 22 FR 3513, May 21, 1957. Redesignated at 26 FR 12751, Dec. 30, 1961]

§ 915.9   Handler.

Handler is synonymous with shipper and means any person (except a common or contract carrier transporting avocados owned by another person) who handles avocados or causes avocados to be handled.

§ 915.10 Handle.

Handle means to sell, consign, deliver, or transport avocados within the production area or between the production area and any point outside thereof: *Provided*, That such term shall not include: (a) The sale or delivery of avocados to a handler, registered as such with the committee in accordance with such rules and regulations as it may prescribe with the approval of the Secretary, who has facilities within the production area for preparing avocados for market; (b) the delivery of avocados to such a handler solely for the purpose of having such avocados prepared for market; or (c) the transportation of avocados by a handler, so registered with the committee, from the grove to his packing facilities within the production area for the purpose of having such avocados prepared for market. In the event a grower sells his avocados to a handler who is not so registered with the committee, such grower shall be the first handler of such avocados.

[19 FR 3439, June 11, 1954, as amended at 20 FR 4177, June 15, 1955. Redesignated at 26 FR 12751, Dec. 30, 1961]

§ 915.11 District.

District means the applicable one of the following described subdivisions of the production area:

(a) *District 1* shall include Miami-Dade County.

(b) *District 2* shall include all of the production area except Miami-Dade County.

[19 FR 3439, June 11, 1954. Redesignated at 26 FR 12751, Dec. 30, 1961, as amended at 73 FR 6837, Feb. 6, 2008]

§ 915.12 Export.

Export means to ship avocados to any destination which is not within the 48 contiguous States of the District of Columbia of the United States or Canada.

[43 FR 39322, Sept. 5, 1978]

ADMINISTRATIVE BODY

§ 915.20 Establishment and membership.

(a) There is hereby established an Avocado Administrative Committee consisting of nine members, each of whom shall have an alternate who shall have the same qualifications as the member for whom he is an alternate. Five of the members and their respective alternates shall be growers who shall not be handlers of avocados produced by others or employees of such handlers. Four of the members and their respective alternates shall be handlers or employees of handlers. The five members of the committee who shall be growers who shall not be handlers of avocados produced by others or employees of such handlers are referred to as "grower" members of the committee; and the four members who shall be handlers or employees of handlers are referred to as "handler" members of the committee. Four of the five grower members shall be producers of avocados in District 1, and one grower member shall be a producer of avocados in District 2. Three of the four handler members shall be handlers, or employees of handlers, of avocados in District 1, and one handler member shall be a handler, or an employee of a handler, of avocados in District 2. No handler or handler organization shall be permitted to have more than one handler member and alternate on the committee from each district: *Provided*, That this requirement may be waived by the Secretary in the event that there are not enough persons available to be nominated and selected to serve on the committee.

(b) The committee may be increased by one public member and an alternate. Persons for the public member positions would be nominated by the committee and selected by the Secretary. The committee, with the approval of the Secretary, shall prescribe qualifications, term of office and the procedure for nominating the public member and alternate.

[19 FR 3439, June 11, 1954. Redesignated at 26 FR 12751, Dec. 30, 1961, as amended at 30 FR 917, Jan. 29, 1965; 43 FR 39322, Sept. 5, 1978; 52 FR 7118, Mar. 9, 1987]

§915.21 Term of office.

The term of office of each member and alternate member of the committee shall begin April 1, and shall terminate March 31 of the following year. Members and alternate members shall serve in such capacities for the portion of the term of office for which they are selected and qualify and until their respective successors are selected and have qualified. The consecutive terms of office of members shall be limited to three terms.

§915.22 Nomination.

(a) *Initial members.* Nominations for each of the five initial grower members and four initial handler members of the committee, together with nominations for the initial alternate members for each position, may be submitted to the Secretary by individual growers and handlers. Such nominations may be made by means of group meetings of the growers and handlers concerned in each district. Such nominations, if made, shall be filed with the Secretary no later than ten calendar days prior to the effective date hereof. In the event nominations for initial members and alternate members of the committee; or the filed pursuant to, and within the time specified in, this section, the Secretary may select such initial members and alternate members without regard to nominations, but selections shall be on the basis of the representation provided for in §915.20.

(b) *Successor members.* (1) The committee shall hold or cause to be held a meeting or meetings of growers and handlers in each district to designate nominees for successor members and alternate members of the committee; or the committee may conduct nominations in Districts 1 and 2 by mail in a manner recommended by the committee and approved by the Secretary. Such nominations shall be submitted to the Secretary by the committee not later than March 1 of each year. The committee shall prescribe procedural rules, not inconsistent with the provisions of this section, for the conduct of nomination.

(2) Only growers may participate in the nomination and election of nominees for grower members and their al-ternates. Each grower shall be entitled to cast only one vote for each nominee to be elected in the district in which he produced avocados. No grower shall participate in the election of nominees in more than one district in any one fiscal year.

(3) Only handlers may participate in the nomination and election of nominees for handler members and their alternates. Each handler shall be entitled to cast only one vote for each nominee to be elected in the district in which such handler handles avocados. Each vote shall be weighted by the volume of avocados shipped by such handler during the immediately preceding twelve-month period, January through December.

[19 FR 3439, June 11, 1954, as amended at 22 FR 3513, May 21, 1957. Redesignated at 26 FR 12751, Dec. 30, 1961, as amended at 40 FR 52605, Nov. 11, 1975; 40 FR 59719, Dec. 30, 1975; 52 FR 7118, Mar. 9, 1987; 73 FR 6837, Feb. 6, 2008]

§915.23 Selection.

From the nominations made pursuant to §915.22, or from other qualified persons, the Secretary shall select the five grower members of the committee, the four handler members of the committee, and an alternate for each such member.

§915.24 Failure to nominate.

If nominations are not made within the time and in the manner prescribed in §915.22, the Secretary may, without regard to nominations, select the members and alternate members of the committee on the basis of the representation provided for in §915.20.

§915.25 Acceptance.

Any person selected by the Secretary as a member or as an alternate member of the committee shall qualify by filing a written acceptance with the Secretary within ten days after being notified of such selection.

§915.26 Vacancies.

To fill any vacancy occasioned by the failure of any person selected as a member or as an alternate member of the committee to qualify, or in the event of the death, removal, resignation, or disqualification of any member or alternate member of the committee, a successor for the unexpired term of such member or alternate member of the committee shall be nominated and selected in the manner specified in §§ 915.22 and 915.23. If the names of nominees to fill any such vacancy are not made available to the Secretary within fifteen days after such vacancy occurs, the Secretary may fill such vacancy without regard to nominations, which selection shall be made on the basis of representation provided for in § 915.20.

§ 915.27 Alternate members.

An alternate member of the committee, during the absence or at the request of the member for whom he is an alternate, shall act in the place and stead of such member. In the event of the death, removal, resignation, or disqualification of a member, his alternate shall act for him until a successor for such member is selected and has qualified. In the event both a member and his alternate are unable to attend a committee meeting, the chairman may designate any alternate who is present and who is not serving for any member to serve in such absent member's place and stead: *Provided,* That only grower alternate members may be so designated to serve for grower members and only handler alternate members may be so designated to serve only for handler members.

[19 FR 3439, June 11, 1954. Redesignated at 26 FR 12751, Dec. 30, 1961, as amended at 35 FR 16627, Oct. 27, 1970]

§ 915.28 Powers.

The committee shall have the following powers:

(a) To administer the provisions of this part in accordance with its terms;

(b) To receive, investigate, and report to the Secretary complaints of violations of the provisions of this part;

(c) To make and adopt rules and regulations to effectuate the terms and provisions of this part; and

(d) To recommend to the Secretary amendments to this part.

§ 915.29 Duties.

The committee shall have, among others, the following duties:

(a) To select a chairman and such other officers as may be necessary, and to define the duties of such officers;

(b) To appoint such employees, agents, and representatives as it may deem necessary, and to determine the compensation and to define the duties of each;

(c) To submit to the Secretary as soon as practicable after the beginning of each fiscal year a budget for such fiscal year, including a report in explanation of the items appearing therein and a recommendation as to the rate of assessment for such fiscal year;

(d) To keep minutes, books, and records which will reflect all of the acts and transactions of the committee and which shall be subject to examination by the Secretary;

(e) To prepare periodic statements of the financial operations of the committee and to make copies of each such statement available to growers and handlers for examination at the office of the committee;

(f) To cause its books to be audited by a certified public accountant at least once each fiscal year, and at such other times as the Secretary may request;

(g) To act as intermediary between the Secretary and any grower or handler;

(h) To investigate growing and maturity conditions of avocados, and to assemble data in connection therewith;

(i) To engage in such research relating to the determination of maturity and grade standards for avocados as may be approved by the Secretary;

(j) To submit to the Secretary such available information as he may request;

(k) To notify, as provided in this part, producers and handlers of all meetings of the committee to consider recommendations for regulation;

(l) To give the Secretary the same notice of meetings of the committee as is given to its members;

(m) To consult with such representatives of growers or groups of growers as may be deemed necessary and to pay the travel expenses incurred by such representatives in attending committee meetings at the request of the committee: *Provided,* That the committee shall not pay the travel expenses of more than three such representatives in connection with any one meeting of the committee; and

(n) To investigate compliance with the provisions of this part.

§915.30 Procedure.

(a) Except as provided in paragraph (c) of this section, six members of the committee, including alternates acting for members, shall constitute a quorum and any decision, recommendation or other action of the committee shall require not less than five concurring votes including one by a handler, or an alternate acting as such: *Provided,* That if the committee is increased by one, the quorum requirement shall be increased to seven and any decision, recommendation or other action of the committee shall require not less than six concurring votes including one by a handler, or an alternate acting as such.

(b) The committee may provide for simultaneous meetings of groups of its members assembled at two or more designated places: *Provided,* That such meetings shall be subject to the establishment of telephone communication between all such groups and the availability of loud speaker receivers for each group so that each member may participate in the discussions and other actions the same as if the committee were assembled in one place.

(c) For any recommendation of the committee for an assessment rate change, a quorum of seven committee members and a two-thirds majority vote of approval of those in attendance is required.

[19 FR 3439, June 11, 1954, as amended at 22 FR 3513, May 21, 1957. Redesignated at 26 FR 12751, Dec. 30, 1961, as amended at 43 FR 39322, Sept. 5, 1978; 73 FR 6837, Feb. 6, 2008]

§915.31 Expenses.

The members of the committee and their respective alternates when performing duties at the direction of the committee, shall be reimbursed for expenses necessarily incurred by them in the performance of their duties under this part.

[43 FR 39323, Sept. 5, 1978]

§915.32 Annual report.

The committee shall, as soon as practicable after the close of each fiscal year, prepare and mail an annual report to the Secretary, and to each handler and grower who requests a copy of the report. This annual report shall contain at least: (a) A complete review, by districts, of the regulatory operations during the fiscal year; (b) an appraisal of the effect of such regulatory operations upon the avocado industry; and (c) any recommendations for changes in the program.

[19 FR 3439, June 11, 1954, as amended at 22 FR 3513, May 21, 1957. Redesignated at 26 FR 12751, Dec. 30, 1961]

EXPENSES AND ASSESSMENTS

§915.40 Expenses.

The committee is authorized to incur such expenses as the Secretary finds are reasonable and likely to be incurred to enable the committee to exercise its powers and perform its duties in accordance with the provisions of this part during each fiscal year. The funds to cover such expenses shall be acquired by the levying of assessments as provided for in §915.41.

[19 FR 3439, June 11, 1954. Redesignated at 26 FR 12751, Dec. 30, 1961, as amended at 35 FR 16627, Oct. 27, 1970]

§915.41 Assessments.

(a) Each person who first handles avocados shall, with respect to the avocados so handled by him, pay to the committee upon demand such person's pro rata share of the expenses which the Secretary finds are reasonable and likely to be incurred by the committee during each fiscal year. Each such person's share of such expenses shall be equal to the ratio between the total quantity of avocados handled by him as the first handler thereof during the applicable fiscal year, and the total quantity of avocados so handled by all persons during the same fiscal year. The payment of assessments for the maintenance and functioning of the committee may be required under this part throughout the period it is in effect irrespective of whether particular provisions thereof are suspended or become inoperative. If a handler does not pay his assessment within the time prescribed by the committee, the unpaid assessment may be subject to an interest charge at rates prescribed by the committee with the approval of the Secretary.

(b) The Secretary shall fix the rate of assessment per 55-pounds of fruit or equivalent in any container or in bulk, to be paid by each such handler. At any time during or after a fiscal year, the Secretary may increase the rate of assessment, in order to secure sufficient funds to cover any later finding by the Secretary relative to the expense which may be incurred. Such increase shall be applied to all fruit handled during the applicable fiscal year. In order to provide funds for the administration of the provisions of this part, the committee may accept the payment of assessments in advance, or borrow money on an emergency short-term basis. The authority of the committee to borrow money is subject to approval of the Secretary and may be used only to meet financial obligations as the obligations occur or to allow the committee to adjust its reserve funds to meet such obligations.

[19 FR 3439, June 11, 1954. Redesignated at 26 FR 12751, Dec. 30, 1961, as amended at 35 FR 16627, Oct. 27, 1970; 40 FR 52605, Nov. 11, 1975; 43 FR 39323, Sept. 5, 1978; 73 FR 6837, Feb. 6, 2008]

§ 915.42　Accounting.

(a) If, at the end of a fiscal year, the assessments collected are in excess of expenses incurred, such excess shall be accounted for as follows:

(1) Except as provided in paragraph (a)(2) of this section, each person entitled to a proportionate refund of the excess assessment shall be credited with such refund against the operation of the following fiscal year unless such person demands repayment thereof, in which event it shall be paid to him: *Provided,* That any sum paid by a person in excess of his pro rata share of the expenses during any fiscal year may be applied by the committee at the end of such fiscal year to any outstanding obligations due the committee from such person.

(2) The Secretary, upon recommendation of the committee, may determine that it is appropriate for the maintenance and functioning of the committee that the funds remaining at the end of a fiscal year which are in excess of the expenses necessary for committee operations during such year may be carried over into following years as a reserve. Such reserve may be established at an amount not to exceed approximately 3 fiscal years' operational expenses. Funds in the reserve may be used to cover the necessary expenses of liquidation, in the event of termination of this part, and to cover the expenses incurred for the maintenance and functioning of the committee during any fiscal year when there is crop failure, or during any period of suspension of any or all of the provisions of this part. Such reserve may also be used by the committee to finance its operations during any fiscal year prior to the time that assessment income is sufficient to cover such expenses and to cover deficits incurred during any fiscal year when income is less than expenses. Upon termination of this part, any funds not required to defray the necessary expenses of liquidation shall be disposed of in such manner as the Secretary may determine to be appropriate: *Provided,* That to the extent practical, such funds shall be returned pro rata to the persons from whom such funds were collected.

(b) All funds received by the committee pursuant to the provisions of this part shall be used solely for the purposes specified in this part, and shall be accounted for in the manner provided in this part. The Secretary may, at any time, require the committee and its members to account for all receipts and disbursements.

[19 FR 3439, June 11, 1954, as amended at 22 FR 3513, May 21, 1957. Redesignated at 26 FR 12751, Dec. 30, 1961, as amended at 35 FR 16628, Oct. 27, 1970; 43 FR 39323, Sept. 5, 1978]

§ 915.43　Contributions.

The committee may accept voluntary contributions. Such contributions shall be free from any encumbrances by the donor and the committee shall retain complete control of their use.

[73 FR 6837, Feb. 6, 2008]

RESEARCH AND DEVELOPMENT

§915.45 Production research, marketing research and development.

The committee may, with the approval of the Secretary, establish or provide for the establishment of production research, marketing research and development projects designed to assist, improve or promote the marketing, distribution, and consumption or efficient production of avocados. Such products may provide for any form of marketing promotion, including paid advertising. The expenses of such projects shall be paid from funds collected pursuant to the applicable provisions of §915.41, or from such other funds as approved by the USDA.

[73 FR 6837, Feb. 6, 2008]

§915.49 Marketing policy.

Each season prior to making any recommendations pursuant to §915.50, the committee shall submit to the Secretary a report setting forth its marketing policy for the ensuing season. Such marketing policy report shall contain information relative to (a) the estimated total production of avocados within the production area; (b) the expected general quality and maturity of avocados in the production area and in competing areas; (c) the expected demand conditions for avocados in different market outlets; (d) the expected shipments of avocados produced in the production area and competing areas; (e) supplies of competing commodities; (f) trend and level of consumer income; (g) other factors having a bearing on the marketing of avocados; and (h) the type of regulations expected to be recommended during the season. In the event it becomes advisable, because of changes in the supply and demand situation for avocados, to modify substantially such marketing policy, the committee shall submit to the Secretary a revised marketing policy report setting forth the information prescribed in this section. The committee shall publicly announce the contents of each marketing policy report and copies thereof shall be maintained in the offices of the committee where they shall be available for examination by growers and handlers.

[40 FR 52605, Nov. 11, 1975]

REGULATIONS

§915.50 Recommendations for regulation.

(a) Whenever the committee deems it advisable to regulate the handling of any variety or varieties of avocados grown in District 1 or District 2 in the manner provided in §915.51, it shall so recommend to the Secretary.

(b) In arriving at its recommendations pursuant to paragraph (a) of this section, the committee shall give consideration to such of the following factors as may be applicable:

(1) The estimated total production and shipments of each variety of avocados, including avocados grown in other areas;

(2) The time of bloom and growing conditions during the development of the crop;

(3) The quality of the avocado crop;

(4) The anticipated demand for avocados; and

(5) Other available information having a bearing on the market for avocados with each recommendation for regulation the committee shall submit to the Secretary the data and information on which such recommendation is predicated, and such other available information as the Secretary may request.

(c) All meetings of the committee held for the purpose of formulating recommendations for regulations shall be open to growers and handlers. The committee shall give notice of such meetings to growers and handlers by mailing such notice to each grower and handler who has filed his address with the committee and requested such notice.

§915.51 Issuance of regulations.

(a) The Secretary shall regulate, in the manner specified in this section, the handling of avocados whenever he finds, from the recommendations and information submitted by the committee or from other available information, that such regulations will tend to effectuate the declared policy of the act. Such regulations may:

(1) Prohibit, prior to such time as shall be specified, the handling of any size or sizes of any variety or varieties of avocados grown in District 1 or District 2.

(2) Prohibit the handling of any variety or varieties of avocados grown in District 1 or District 2 which do not meet such quality and maturity standards as shall be prescribed.

(3) Limit the shipment of the total quantity of avocados by prohibiting the shipment thereof: *Provided,* That no such prohibition shall be effective during any fiscal period, other than for four periods not exceeding six days each immediately prior to, including, or following July 4, Labor Day, Thanksgiving Day, and Christmas Day.

(4) Fix the size, capacity, weight, dimensions, or pack of the container or containers which may be used in the packaging, and the transportation, sale, shipment or other handling of avocados.

(5) Establish and prescribe pack specifications for the grading and packing of any variety or varieties of avocados and require that all avocados handled shall be packed in accordance with such pack specifications, and shall be identified by appropriate labels, seals, stamps, or tags, affixed to the containers by the handler under the supervision of the committee or an inspector of the Federal-State Inspection Service, showing the particular pack specifications of the lot.

(6) Provide that any or all requirements effective pursuant to paragraphs (a)(1), (2), (3), and (4) of this section applicable to the handling of avocados shall be different for the handling of avocados within the production area and for the handling of avocados between the production area and any point outside thereof.

(7) Prescribe requirements, as provided in this paragraph, applicable to exports of any variety of avocados which are different from those applicable to the handling of the same variety to other destinations.

(b) The committee shall be informed immediately of any such regulations issued by the Secretary and the committee shall promptly give notice thereof to growers and handlers.

[19 FR 3439, June 11, 1954, as amended at 20 FR 4177, June 15, 1955; 22 FR 3514, May 21, 1957. Redesignated at 26 FR 12751, Dec. 30, 1961, as amended at 36 FR 14126, July 30, 1971; 40 FR 52606, Nov. 11, 1975; 43 FR 39322, Sept. 5, 1978]

§ 915.52 Modification, suspension, or termination of regulations.

(a) In the event the committee at any time finds that, by reason of changed conditions, any regulations issued pursuant to § 915.51 should be modified, suspended, or terminated, it shall so recommend to the Secretary.

(b) Whenever the Secretary finds, from the recommendations and information submitted by the committee or from other available information, that a regulation should be modified, suspended, or terminated with respect to any or all shipments of avocados in order to effectuate the declared policy of the act, he shall modify, suspend, or terminate such regulation. If the Secretary finds that a regulation obstructs or does not tend to effectuate the declared policy of the act, he shall suspend or terminate such regulation. On the same basis and in like manner the Secretary may terminate any such modification or suspension.

§ 915.53 Exemption certificates.

Whenever a regulation is in effect pursuant to § 915.51(a)(1), the committee shall issue one or more exemption certificates to any person who furnishes proof, satisfactory to the committee, that his avocados of a particular variety are mature prior to the time such variety may be handled under such regulation. Such exemption certificates shall authorize the person to whom the certificates are issued to handle, or have handled, only that portion of his avocados of the particular variety which the committee has determined to be mature. The committee shall adopt, with the approval of the Secretary, procedural rules by which such exemption certificates will be issued and the avocados covered thereunder may be handled. Exemption certificates shall be transferred to the handler of the avocados covered by such certificates at the time the avocados are delivered to such handler.

§915.54 Inspection and certification.

Whenever the handling of any variety of avocados is regulated pursuant to §915.51, each handler who handles avocados shall, prior thereto, cause each lot of avocados handled to be inspected by the Federal-State Inspection Service and certified by it as meeting the applicable requirements of such regulation: *Provided,* That such inspection and certification shall not be required whenever the avocados previously have been so inspected and certified. Promptly thereafter, each such handler shall submit, or cause to be submitted, to the committee a copy of the certificate of inspection with respect to such handling.

§915.55 Avocados not subject to regulations.

Except as otherwise provided in this section, any person may, without regard to the provisions of §§915.41, 915.51, and 915.54, and the regulations issued thereunder, handle avocados (a) for consumption by charitable institutions; (b) for distribution by relief agencies; (c) for commercial processing into products; or (d) in such minimum quantities or types of shipments as the committee, with the approval of the Secretary, may prescribe. The committee shall, with the approval of the Secretary, prescribe such rules, regulations, and safeguards as it may deem necessary to prevent avocados handled under the provisions of this section from entering channels of trade for other than the specific purposes authorized by this section. Such rules, regulations, and safeguards may include the requirements that handlers shall file applications with the committee for authorization to handle avocados pursuant to this section, and that such applications be accompanied by a certification by the intended purchaser or receiver that the avocados will not be used for any purpose not authorized by this section.

[19 FR 3439, June 11, 1954, as amended at 20 FR 4177, June 15, 1955. Redesignated at 26 FR 12751, Dec. 30, 1961]

§915.60 Reports.

(a) Each handler shall furnish to the committee, at such times and for such periods as the committee may designate, certified reports covering, to the extent necessary for the committee to perform its functions, the following:

(1) The quantities of each variety of avocados he received;

(2) A complete record of the quantities disposed of by him, segregated as to varieties and as to the respective quantities subject to regulation and not subject to regulation;

(3) The date of each such disposition and the identification of the carrier transporting such fruit;

(4) Identification of the inspection certificates and the exemption certificates, if any, pursuant to which the fruit was handled, together with the destination of each such exempted disposition, and of all fruit handled pursuant to §915.55; and

(5) The quantity of each variety held by him at the end of the period.

(b) Upon request of the committee, made with the approval of the Secretary, each handler shall furnish to the committee, in such manner and at such times as it may prescribe, such other information as may be necessary to enable the committee to perform its duties under this part.

MISCELLANEOUS PROVISIONS

§915.61 Compliance.

Except as provided in this part, no person shall handle avocados, the shipment of which have been prohibited by the Secretary in accordance with the provisions of this part; and no person shall handle avocados except in conformity with the provisions of this part and the regulations issued under this part.

§915.62 Right of the Secretary.

The members of the committee (including successors and alternates), and any agents, employees, or representatives thereof, shall be subject to removal or suspension by the Secretary at any time. Each and every regulation, decision, determination, or other act of the committee shall be subject to the continuing right of the Secretary to disapprove of the same at any time. Upon such disapproval, the disapproved action of the committee shall be deemed null and void, except as to acts done in reliance thereon or in accordance therewith prior to such disapproval by the Secretary.

§ 915.63 Effective time.

The provisions of this part shall become effective at such time as the Secretary may declare above his signature to this part, and shall continue in force until terminated in one of the ways specified in § 915.64.

§ 915.64 Termination.

(a) The Secretary may at any time terminate the provisions of this part by giving at least one day's notice by means of a press release or in any other manner in which he may determine.

(b) The Secretary shall terminate or suspend the operation of any and all of the provisions of this part whenever he finds that such provisions do not tend to effectuate the declared policy of the act.

(c) The Secretary shall terminate the provisions of this part whenever the Secretary finds by referendum or otherwise that such termination is favored by a majority of the producers: *Provided,* That such majority has, during a representative period determined by the Secretary, produced more than 50 percent of the volume of the avocados produced within the production area: *And Provided further,* That such termination shall be announced by March 15 of the then current fiscal year.

(d) The Secretary shall conduct a referendum as soon as practicable after the end of the fiscal year ending March 31, 1990, and at such time every sixth year thereafter, to ascertain whether continuance of this part is favored by avocado producers. The Secretary may terminate the provisions of this part at the end of any fiscal year in which the Secretary has found that continuance of this part is not favored by producers who, during a representative period determined by the Secretary, have been engaged in the production for market of avocados in the production area: *Provided,* That termination of this part shall be effective only if announced on or before March 15 of the then current fiscal year.

(e) The provisions of this part shall, in any event, terminate whenever the provisions of the Act authorizing them cease to be in effect.

[19 FR 3439, June 11, 1954. Redesignated at 26 FR 12751, Dec. 30, 1961, as amended at 52 FR 7118, Mar. 9, 1987]

§ 915.65 Proceedings after termination.

(a) Upon the termination of the provisions of this part, the committee shall, for the purpose of liquidating the affairs of the committee, continue as trustees of all the funds and property then in its possession, or under its control, including claims for any funds unpaid or property not delivered at the time of such termination.

(b) The said trustees shall (1) continue in such capacity until discharged by the Secretary; (2) from time to time account for all receipts and disbursements and deliver all property on hand, together with all books and records of the committee and of the trustees, to such persons as the Secretary may direct; and (3) upon the request of the Secretary, execute such assignments or other instruments necessary or appropriate to vest in such person, full title and right to all of the funds, property, and claims vested in the committee or the trustees pursuant thereto.

(c) Any person to whom funds, property, or claims have been transferred or delivered, pursuant to this section, shall be subject to the same obligation imposed upon the committee and upon the trustees.

§ 915.66 Effect of termination or amendment.

Unless otherwise expressly provided by the Secretary, the termination of this part or of any regulation issued pursuant to this part, or the issuance of any amendment to either thereof, shall not (a) affect or waive any right, duty, obligation, or liability which shall have arisen or which may thereafter arise in connection with any provision of this part or any regulation issued under this part, or (b) release or extinguish any violation of this part or of any regulation issued under this part, or (c) affect or impair any rights or remedies of the Secretary or of any other person with respect to any such violation.

§915.67 Duration of immunities.

The benefits, privileges, and immunities conferred upon any person by virtue of this part shall cease upon its termination, except with respect to acts done under and during the existence of this part.

§915.68 Agents.

The Secretary may, by designation in writing, name any officer or employee of the United States, or name any agency or division in the United States Department of Agriculture, to act as his agent or representative in connection with any of the provisions of this part.

§915.69 Derogation.

Nothing contained in this part is, or shall be construed to be, in derogation or in modification of the rights of the Secretary or of the United States (a) to exercise any powers granted by the act or otherwise, or (b) in accordance with such powers, to act in the premises whenever such action is deemed advisable.

§915.70 Personal liability.

No member or alternate member of the committee and no employee or agent of the committee shall be held personally responsible, either individually or jointly with others, in any way whatsoever, to any person for errors in judgment, mistakes, or other acts, either of commission or omission, as such member, alternate, employee, or agent, except for acts of dishonesty, willful misconduct, or gross negligence.

§915.71 Separability.

If any provision of this part is declared invalid or the applicability thereof to any person, circumstance, or thing is held invalid, the validity of the remainder of this part or the applicability thereof to any other person, circumstance, or thing shall not be affected thereby.

Subpart B—Administrative Requirements

§915.110 Exemption certificates.

Exemption certificates under §915.53 shall be issued by the Avocado Administrative Committee pursuant to the following rules and regulations:

(a) The grower must make application for exemption on a form supplied by the committee. A separate application must be made for each variety or classification of avocados and shall contain the following:

(1) Name and address of the applicant, and date of application;

(2) District in which the applicant's grove is located;

(3) Regulation from which exemption is requested;

(4) Variety for which exemption is requested;

(5) Location (by county, highway, rural route, distance from nearest town, etc.) of grove from which avocados are to be shipped pursuant to the requested exemption certificate;

(6) Information as to the average size of such avocados and the reasons why applicant believes he is entitled to an exemption certificate; and

(7) Name of the person who will handle any exempted fruit if different than the applicant.

(b) Upon receipt of an application for exemption certificate, the Avocado Administrative Committee shall check all information furnished by the applicant and shall conduct such investigations concerning the maturity of the applicant's avocados as may be necessary to determine whether the application shall be approved or denied.

(c) Approval of the application shall be evidenced by the issuance to the applicant, by the Manager of the Avocado Administrative Committee on its behalf, of one or more exemption certificates which shall authorize the handling of the quantity of the applicant's avocados which the committee has determined is mature.

(d) If the application is denied, the applicant shall be informed of such denial by written notice stating the reasons therefor.

[19 FR 5439, Aug. 26, 1954. Redesignated at 26 FR 12751, Dec. 30, 1961]

§ 915.115 Nomination procedure.

(a) Any grower who resides outside the production area and desires to be represented in a nomination meeting by a duly authorized agent and to have such grower's vote cast by such agent in the nomination and election of nominees for grower members and alternate members to fill positions on the Avocado Administrative Committee, as provided in § 915.22(b)(2), shall submit to the committee, not later than January 20, a written statement containing the following:

(1) Name of grower;

(2) Mailing address;

(3) Location of each avocado grove (either legal or from established landmarks);

(4) Number of avocado trees owned;

(5) Number of 55-pound units of avocados marketed to date during the current season;

(6) Name of the handler of the fruit marketed;

(7) Authorization, including the name and address, of the person who is to represent said grower at the nomination meeting.

(b) Any grower who has not filed the statement as prescribed in paragraph (a) of this section must be present at the nomination meeting to be eligible to have his vote counted in connection with the nomination and election of nominees.

(c) Any grower who, pursuant to the provisions of paragraph (a) of this section, has authorized an agent to cast such grower's vote, may rescind such authorization by appearing at the nomination meeting and exercising his right to vote in person.

[21 FR 78, Jan. 5, 1956. Redesignated at 26 FR 12751, Dec. 30, 1961, as amended at 48 FR 2519, Jan. 20, 1983]

§ 915.120 Handler registration.

(a) Each handler who desires to handle avocados pursuant to the exceptions in § 915.10 shall, prior thereto, register with the committee. Such registration shall be by application for registration filed with the Avocado Administrative Committee on a form, prescribed and furnished by the committee, which shall contain the following information:

(1) Name and address of applicant;

(2) Applicant's principal place(s) of business;

(3) Type of business organization (individual, corporation, partnership, etc.);

(4) If other than an individual, the names and addresses of officers, partners, etc.;

(5) Nature of business (handler, trucker, wholesaler, etc.);

(6) Number of years engaged in avocado business;

(7) Estimated seasonal volume of avocados handled;

(8) Place within production area where the avocados will be prepared for market, and name and address of person responsible for such preparation;

(9) Name and address of three references, one of which shall be a bank;

(10) Certification of accuracy of information furnished; and

(11) An agreement to comply with the provisions of this part.

(b) When the committee receives an application for registration, it shall issue the applicant a certificate of registration, if it determines based upon an investigation that the applicant may be expected to handle avacados in accordance with this part.

(c) If it is determined from the available information that the applicant is not entitled to be registered with the committee, he shall be so informed by written notice stating why the certificate of registration was not issued.

(d) Any certificate of registration issued to a handler pursuant to this section may be canceled by the committee under circumstances which would have justified denial of his application.

(e) The committee shall suspend the certificate of registration issued under this section of any handler who fails to pay assessments or furnish reports as required under this part, and so advise the handler in writing of the suspension and the effective date. The committee shall lift such suspension at such time as the handler pays such assessments and files such reports, and the committee determines that the handler may be expected to handle avacados in the future in accordance with this part.

[19 FR 5439, Aug. 26, 1954. Redesignated at 26 FR 12751, Dec. 30, 1961, as amended at 49 FR 33203, Aug. 22, 1984]

§ 915.140 Avocados not subject to regulation.

(a) *Minimum quantity.* During any one day any handler may handle not to exceed 55 pounds total of avocados exempt from the provisions of §§ 915.41, 915.51, and 915.54, and the regulations issued thereunder: *Provided,* That such exempted quantity shall not be included as part of a shipment exceeding 55 pounds.

(b) *Gift shipments.* Any handler may, exempt from the provisions of §§ 915.41, 915.51, and 915.54, and the regulations issued thereunder, handle avocados in individually addressed gift containers not exceeding 20 pounds net weight for use by the addressee other than for resale.

(c) *Commercial processing into products.* The term *commercial processing into products,* as used in § 915.55(c), means the manufacture of any avocado product which is preserved by any recognized commercial process, including canning, freezing, dehydrating, drying, the addition of chemical substances, or by fermentation.

(d) *Avocados for seed.* Any handler may ship avocados to be used for seed purposes exempt from the provisions of §§ 915.41, 915.51, and 915.54, and the regulations issued thereunder: *Provided,* That such handler shall make application to the committee for an exemption prior to the loading of each shipment and that the receiver of each such shipment shall certify, on a form provided by the committee, that such fruit was used for the intended purpose, and that the residue from the seed separation process will not be allowed to enter fresh channels of trade.

[23 FR 9126, Nov. 26, 1958. Redesignated at 26 FR 12751, Dec. 30, 1961, as amended at 30 FR 10880, Aug. 21, 1965; 36 FR 1191, Jan. 26, 1971; 43 FR 23557, May 31, 1978]

§ 915.141 Handling avocados for commercial processing into products.

(a) No person shall handle any avocados for commercial processing into products unless prior to such handling such person notifies the Avocado Administrative Committee of the proposed handling and provides the committee with the name of the intended processor. If the intended processor's name is not on the Avocado Administrative Committee's current list of approved manufacturers of avocado products, as prescribed in paragraph (b) of this section, or if on the list is suspended, such person shall furnish committee, prior to each such handling, with a statement executed by the intended processor that the avocados will be used for the stated purpose only.

(b) Any person who desires to have his name placed on the Avocado Administrative Committee's list of approved manufacturers of avocado products shall, prior to such listing, submit to the Avocado Administrative Committee an application containing the following information:

(1) Name and address of applicant;

(2) Location of the facilities for commercial processing into products;

(3) Proposed type of avocado product or products to be manufactured from avocados and the proposed commercial process of preservation;

(4) Description of facilities for commercial processing into products;

(5) Quantity of avocados used in commercial processing into products during the previous fiscal year and estimate of the quantity of avocados to be similarly processed during the current fiscal year; (6) expected source of avocados for commercial processing into products; (7) method of transporting avocados and unloading point; (8) Avocado Administrative Committee handler certificate of registration number, if any; (9) a statement that the avocados obtained for commercial processing into products will be used for that purpose only and will not be resold or disposed of in fresh fruit channels; and (10) an agreement to submit such reports as are required by the Avocado Administrative Committee with approval of the Secretary.

(c) Upon receipt of an application for such listing, the Avocado Administrative Committee shall make such investigation as it deems appropriate, and if it appears that the applicant may reasonably be expected to use avocados covered by the application in accordance with, and to comply with, the requirements of paragraph (b) of this section, it shall place the person's name on Avocado Administrative Committee's current list of approved manufacturers of avocado products.

(d) If it is determined by the committee from the available information that the applicant is not entitled to such listing he shall be so informed by written notice stating why his application was denied.

(e) Any such listing pursuant to paragraphs (b) and (c) of this section may be canceled by the committee under circumstances which would have justified denial of this application.

(f) The committee shall suspend the listing of any approved manufacturer who fails to submit reports as prescribed pursuant to the provisions of paragraph (b) of this section. The committee shall advise such manufacturer in writing of the pending suspension and shall specify the time such suspension is to become effective. Upon determination by the committee that the manufacturer has satisfied by such effective time the requirements with respect to the submission of reports the committee shall not make such suspension effective. However, if the suspension is in effect, the committee shall terminate such suspension at such time as it determines that the manufacturer has satisfied the requirements with respect to the submission of reports.

[36 FR 1191, Jan. 26, 1971]

§ 915.142   Reserve fund.

(a) The establishment of a reserve fund at an amount not to exceed approximately 3 fiscal years' operational expenses is appropriate and necessary to the maintenance and functioning of the Avocado Administrative Committee. Such reserve, including funds carried forward from prior fiscal years, shall be used to provide for the maintenance and functioning of the committee in accordance with the provisions of the marketing agreement, as amended, and this part.

(b) Terms used in this section shall have the same meaning as when used in said marketing agreement and order.

[43 FR 39323, Sept. 5, 1978. Redesignated at 45 FR 47653, July 16, 1980]

§ 915.150   Reports.

(a) Each handler shall file with the Avocado Administrative Committee, on a weekly basis, a report of all avocados received by him. Such report shall be on forms prescribed by the committee and shall include: (1) The name and address of the handler; (2) weekly period covered by the report; (3) district in which the avocados were received; and (4) the quantity of each variety of avocados received. Each such report shall be filed with the committee not later than one week after the close of business of the last day of the period covered by the report.

(b) Each handler registered with the Avocado Administrative Committee shall render a report to the committee of the disposition of each lot of noncertified avocados removed from the premises of his handling facilities during each week in which any avocados are handled subject to the provisions of §§ 915.41, 915.51, and 915.54, or exemptions therefrom pursuant to § 915.53. Such report shall be on forms prescribed by the committee and shall include: (1) The quantity; (2) purpose for which removed; (3) date of removal; and (4) the name of the person or firm to which the avocados were delivered or consigned. Each such report shall be signed by the handler or his authorized representative, shall cover the period Sunday through Saturday, and shall be placed in the mail not later than one week after the close of business of the Saturday ending the period covered by the report.

(c) Each handler shall render a report to the Avocado Administrative Committee of each lot of noncertified avocados received from a district other than that in which his handling facilities are located. Such report shall be on forms prescribed by the committee and shall include: (1) The name of the handler; (2) the quantity of avocados received; (3) date received; (4) name and address of the person from whom the avocados were purchased; (5) the district from which the avocados were transferred; and (6) the district to which the avocados were transferred. Each such report shall cover the period Sunday through Saturday and shall be placed in the mail not later than one week after the close of business of the Saturday ending the period covered by the report.

(d) Each handler shall, at the end of the day's operation, report to the committee the number of containers of avocados sold and delivered in the State of Florida in the following containers: (1) ¼ Bushel, (2) ½ Bushel, and (3) ⅘ Bushel. Upon request by the committee, such reports shall be confirmed in writing on a weekly basis on a form prescribed by the committee.

(e) At the time of inspection, each handler shall provide to the Federal-State Inspection Service the quantity and size of containers being packed and inspected for the fresh avocado market. In addition, each handler shall provide the number of avocados packed per container (count per container).

[21 FR 6695, Sept. 6, 1956, as amended at 21 FR 7368, Sept. 27, 1956. Redesignated at 26 FR 12751, Dec. 30, 1961, as amended at 53 FR 1743, Jan. 22, 1988; 70 FR 36470, June 24, 2005]

EFFECTIVE DATE NOTES: 1. At 61 FR 17552, Apr. 22, 1996, in §915.150, paragraph (d) was suspended indefinitely.

2. At 71 FR 76899, Dec. 22, 2006, in §915.150, paragraphs (b), (c), and (d) were suspended indefinitely, effective Dec. 26, 2006.

§915.155 Delinquent assessments.

Each handler shall pay interest of one percent per month on any unpaid assessment balance beginning 30 days after date of billing. Such interest charge is to apply to any unpaid assessments which become due the Avocado Administrative Committee after the effective date of this section.

[40 FR 50024, Oct. 28, 1975]

§915.160 Public member eligibility requirements and nomination procedures.

(a) Public member and alternate member candidates shall not represent an agricultural interest, and shall not have a financial interest in, or be associated with the production, processing, financing, or marketing of avocados.

(b) Public member and alternate member candidates should be able to devote sufficient time to attend committee activities regularly and to familiarize themselves with the background and economics of the avocado industry.

(c) The public member and alternate member shall be a resident of the production area.

(d) The public member and alternate member should be nominated by the Avocado Administrative Committee, and shall serve a one-year term which coincides with the term of the producer and handler members of the committee.

[44 FR 9370, Feb. 13, 1979]

Subpart C—Assessment Rates

§915.235 Assessment rate.

On and after April 1, 2016, an assessment rate of $0.35 per 55-pound container or equivalent is established for avocados grown in South Florida.

[81 FR 38885, June 15, 2016]

Subpart D—Container and Pack Requirements

§915.305 Florida Avocado Container Regulation 5.

(a) No handler shall handle any avocados for the fresh market from the production area to any point outside thereof in containers having a capacity of more than 4 pounds of avocados unless the containers meet the requirements specified in this section: *Provided,* That the containers authorized in this section shall not be used for handling avocados for commercial processing into products pursuant to §915.55(c). All avocados shall be packed in containers of 33, 31, 24, 12, and 8.5 pounds designated net weights and shall conform to all other applicable requirements of this section:

(1) Containers shall not contain less than 33 pounds net weight of avocados, except that for avocados of unnamed varieties, which are avocados that have not been given varietal names, and for Booth 1, Fuchs, and Trapp varieties, such weight shall be not less than 31 pounds. Not more than 10 percent, by count, of the individual containers in any lot may fail to meet the applicable specified weight. No container in any

lot may contain a net weight of avocados exceeding 2 pounds less than the specified net weight; or

(2) Containers shall not contain less than 24 pounds net weight of avocados: *Provided,* That not to exceed 5 percent, by count, of such containers in any lot may fail to meet such weight requirement. All avocados packed at this designated net weight shall be placed in two layers and the net weight of all avocados in any such container shall not be less than 24 pounds: *Provided,* That the requirement as to placing avocados in two layers only shall not apply to such container if each of the avocados therein weighs 14 ounces or less; or

(3) Containers shall not contain less than 12 pounds net weight of avocados: *Provided,* That not to exceed 5 percent, by count, of such containers in any lot may fail to meet such weight requirement. All avocados packed at this designated net weight shall be placed in one layer only and the net weight of all avocados in any such container shall not be less than 12 pounds; or

(4) Containers shall not contain less than 8.5 pounds net weight of avocados: *Provided,* That not to exceed 5 percent, by count, of such containers in any lot may fail to meet such weight requirement. All avocados packed at this designated net weight shall be placed in one layer only and the net weight of all avocados in any such container shall not be less than 8.5 pounds. Such containers shall be for export shipments only.

(5) Such other types and sizes of containers as may be approved by the Avocado Administrative Committee, with the approval of the Secretary, for testing in connection with a research project conducted by or in cooperation with said committee: *Provided,* That the handling of each lot of avocados in such test containers shall be subject to prior approval, and under the supervision of, the Avocado Administrative Committee.

(b) The limitations set forth in paragraph (a) of this section shall not apply to master containers for individual packages of avocados: *Provided,* That the markings or labels, if any, on the individual packages within such master containers do not conflict with the markings or labels on the master container.

(c) No handler shall handle any avocados for the fresh market in 20 bushel plastic field bins to destinations inside the production area.

(d) Avocados handled for the fresh market in containers other than those authorized under §915.305(a) and shipped to destinations within the production area must be packed in 1-bushel containers.

(e) All containers in which the avocados are packed must be new, and clean in appearance, without marks, stains, or other evidence of previous use.

[63 FR 37480, July 13, 1998, as amended at 64 FR 69383, Dec. 13, 1999; 65 FR 15205, Mar. 22, 2000; 70 FR 36470, June 24, 2005; 73 FR 66719, Nov. 12, 2008]

§ 915.306 **Florida avocado grade, pack, and container marking regulation.**

(a) No handler shall handle any variety of avocados grown in the production area unless:

(1) Such avocados grade at least U.S. Combination, except that avocados handled to destinations within the production area grade at least U.S. No. 2.

(2) Such avocados are in containers authorized under §915.305, when handled to points outside the production area.

(3) Such avocados are packed in accordance with standard pack, when handled in containers authorized under §915.305.

(4) Such avocados are in containers marked with a Federal-State Inspection Service lot stamp number, when handled in containers authorized under §915.305: *Provided,* That when inspection occurs after palletization, only all exposed or outside containers of avocados must be plainly marked with the lot stamp number corresponding to the lot inspection conducted by an authorized inspector.

(5) Such avocados are in containers marked with a Federal-State Inspection Service (FSIS) lot stamp number applied to an adhesive tape seal affixed to the container in a manner to prevent the container from being opened and/or the fruit being removed without breaking the seal, when handled in containers other than those authorized under §915.305. The stamp and tape

shall be affixed to the container by the FSIS or by the handler under the supervision of the FSIS. Only stamps and tape which have been approved by the Fresh Products Branch, Fruit and Vegetable Division, Agricultural Marketing Service, U.S. Department of Agriculture, may be used for purposes of stamping and sealing containers to meet these requirements.

(6) Such avocados when handled in containers authorized under §915.305, except for those to export destinations, are marked once with the grade of fruit in letters and numbers at least 1 inch in height on the top or one side of the container, not to include the bottom.

(7) Such avocados when handled in containers other than those authorized under §915.305(a) for shipment to destinations within the production area are marked once with the grade of fruit in letters and numbers at least 3 inches in height on the top or one side of the container, not to include the bottom. Each such container is also to be marked at least once with either the registered handler number assigned to the handler at the time of certification as a registered handler or with the name and address of the handler.

(b) The provisions of paragraphs (a)(2), (a)(3), (a)(4), (a)(5), and (a)(6) of this section shall not apply to individual packages of avocados weighing four pounds or less, net weight, in master containers.

(c) Terms pertaining to grades and standard pack mean the same as those defined in the United States Standards for Florida Avocados (7 CFR 51.3050 through 51.3069).

[50 FR 32553, Aug. 13, 1985, as amended at 56 FR 36080, July 31, 1991; 57 FR 3716, Jan. 31, 1992; 57 FR 48931, Oct. 29, 1992; 60 FR 42770, Aug. 17, 1995; 61 FR 31006, June 19, 1996; 63 FR 37480, July 13, 1998; 73 FR 66719, Nov. 12, 2008; 78 FR 51043, Aug. 20, 2013; 79 FR 67039, Nov. 12, 2014]

§915.332 Florida avocado maturity regulation.

(a) No handler shall handle any variety of avocados, except Hass, Fuerte, Zutano, and Edranol, grown in the production area unless:

(1) Any portion of the skin of the individual avocados has changed to the color normal for that fruit when mature for those varieties which normally change color to any shade of red or purple when mature, except for the Linda variety; or

(2) Such avocados meet the minimum weight or diameter requirements for the Monday nearest each date specified, through the Sunday immediately prior to the nearest Monday of the specified date in the next column, for each variety listed in the following table I: *Provided,* that avocados may not be handled prior to the earliest date specified in column A of such table for the respective variety; *Provided further,* There are no restrictions on size or weight on or after the date specified in column D; *Provided further,* That up to a total of 10 percent, by count to the individual fruit in each lot may weigh less than the minimum specified or be less than the specified diameter, except that no such avocados shall be over 2 ounces lighter than the minimum weight specified for the variety: *Provided further,* That up to double such tolerance shall be permitted for fruit in an individual container in a lot.

TABLE I

Variety	A date	Min. wt.	Min. diam.	B date	Min. wt.	Min. diam.	C date	Min. wt.	Min. diam.	D date
Dr. Dupuis #2	5–30	16	3⁷⁄₁₆	6–13	14	3⁵⁄₁₆	7–04	12	3²⁄₁₆	7–18
Simmons	6–20	16	3⁹⁄₁₆	7–04	14	3⁷⁄₁₆	7–18	12	3¹⁄₁₆	8–01
Pollock	6–20	18	3¹¹⁄₁₆	7–04	16	3⁷⁄₁₆	7–18	14	3⁴⁄₁₆	8–01
Hardee	6–27	16	3²⁄₁₆	7–04	14	2¹⁴⁄₁₆	7–11	12		7–25
Nadir	6–27	14	3³⁄₁₆	7–04	12	3¹⁄₁₆	7–11	10	2¹⁴⁄₁₆	7–18
Ruehle	7–04	18	3¹¹⁄₁₆	7–11	16	3⁹⁄₁₆	8–01	12	3⁵⁄₁₆	8–15
				7–18	14	3⁷⁄₁₆	8–08	10	3³⁄₁₆	
Bernecker	7–18	18	3⁶⁄₁₆	8–01	16	3⁵⁄₁₆	8–15	14	3⁴⁄₁₆	8–29

TABLE I—Continued

Variety	A date	Min. wt.	Min. diam.	B date	Min. wt.	Min. diam.	C date	Min. wt.	Min. diam.	D date
Miguel (P)	7–18	22	$3\frac{13}{16}$	8–01	20	$3\frac{12}{16}$	8–15	18	$3\frac{10}{16}$	8–29
Nesbitt	7–18	22	$3\frac{12}{16}$	8–01	16	$3\frac{5}{16}$	8–08	14	$3\frac{3}{16}$	8–22
Tonnage	8–01	16	$3\frac{9}{16}$	8–15	14	$3\frac{4}{16}$	8–22	12	$3\frac{9}{16}$	8–29
Waldin	8–01	16	$3\frac{9}{16}$	8–15	14	$3\frac{7}{16}$	8–29	12	$3\frac{4}{16}$	9–12
Tower II	8–01	14	$3\frac{6}{16}$	8–15	12	$3\frac{4}{16}$	8–29	10	$3\frac{2}{16}$	9–05
Beta	8–08	18	$3\frac{8}{16}$	8–15	16	$3\frac{5}{16}$	8–29	14	$3\frac{3}{16}$	9–05
Lisa (P)	8–08	12	$3\frac{2}{16}$	8–15	11	3	....			8–22
Black Prince	8–15	28	$4\frac{1}{16}$	8–29	23	$3\frac{14}{16}$	9–12	16	$3\frac{9}{16}$	10–03
Loretta	8–22	30	$4\frac{3}{16}$	9–05	26	$3\frac{15}{16}$	9–19	22	$3\frac{12}{16}$	9–26
Booth 8	8–29	16	$3\frac{9}{16}$	9–12	14	$3\frac{6}{16}$	9–26	12	$3\frac{3}{16}$	10–24
							10–10	10	$3\frac{1}{16}$	
Booth 7	8–29	18	$3\frac{13}{16}$	9–12	16	$3\frac{10}{16}$	9–26	14	$3\frac{8}{16}$	10–10
Booth 5	9–05	14	$3\frac{9}{16}$	9–19	12	$3\frac{6}{16}$	....			10–03
Choquette	9–26	28	$4\frac{4}{16}$	10–10	24	$4\frac{1}{16}$	10–24	20	$3\frac{14}{16}$	11–7
Hall	9–26	26	$3\frac{14}{16}$	10–10	20	$3\frac{9}{16}$	10–24	18	$3\frac{8}{16}$	11–07
Lula	10–03	18	$3\frac{11}{16}$	10–10	14	$3\frac{5}{16}$	10–31	12	$3\frac{3}{16}$	11–14
Monroe	11–07	26	$4\frac{3}{16}$	11–21	24	$4\frac{1}{16}$	12–05	20	$3\frac{14}{16}$	1–02
							12–19	16	$3\frac{9}{16}$	
Arue	5–16	16	....	5–30	14	$3\frac{3}{16}$	6–20	12	....	7–04
Donnie	5–23	16	$3\frac{5}{16}$	6–06	14	$3\frac{4}{16}$	6–20	12	....	7–04
Fuchs	6–06	14	$3\frac{3}{16}$	6–20	12	$3\frac{9}{16}$	....			7–04
K–5	6–13	18	$3\frac{5}{16}$	6–27	14	$3\frac{3}{16}$	....			7–11
West Indian Seedling [1]	6–20	18	....	7–18	16	....	8–22	14	....	9–19
Gorham	7–04	29	$4\frac{5}{16}$	7–18	27	$4\frac{3}{16}$	....			8–15
Biondo	7–11	13	....	....			....			8–15
Petersen	7–11	14	$3\frac{8}{16}$	7–18	12	$3\frac{5}{16}$	7–25	10	$3\frac{2}{16}$	8–08
232	7–18	14	....	8–01	12	....	....			8–15
Pinelli	7–18	18	$3\frac{12}{16}$	8–01	16	$3\frac{10}{16}$	....			8–15
Trapp	7–18	14	$3\frac{10}{16}$	8–01	12	$3\frac{7}{16}$	....			8–15
K–9	8–01	16	$2\frac{14}{16}$	....			....			8–22
Christina	8–01	11	$2\frac{14}{16}$	....			....			8–22
Catalina	8–15	24	....	8–29	22	....	....			9–19
Blair	8–29	16	$3\frac{8}{16}$	9–12	14	$3\frac{5}{16}$	....			10–03
Guatemalan Seedling [2]	9–05	15	....	10–03	13	....	....			12–05
Marcus	9–05	32	$4\frac{12}{16}$	9/19	24	$4\frac{5}{16}$	....			10–31
Brooks 1978	9–05	12	$3\frac{4}{16}$	9–12	10	$3\frac{1}{16}$	9–19	8	$2\frac{14}{16}$	10–10
Rue	9–12	30	$4\frac{3}{16}$	9–19	24	$3\frac{15}{16}$	10–03	18	$3\frac{9}{16}$	10–17
Collinson	9–12	16	$3\frac{10}{16}$	....			....			10–10
Hickson	9–12	12	$3\frac{1}{16}$	9–26	10	$3\frac{9}{16}$	....			10–10
Simpson	9–19	16	$3\frac{8}{16}$	....			....			10–10
Chica	9–19	12	$3\frac{7}{16}$	10–03	10	$3\frac{4}{16}$	....			10–17
Leona	9–26	18	$3\frac{10}{16}$	10–03	16	....	....			10–10
Melendez	9–26	26	$3\frac{14}{16}$	10–10	22	$3\frac{11}{16}$	10–24	18	$3\frac{7}{16}$	11–07
Herman	10–03	16	$3\frac{8}{16}$	10–17	14	$3\frac{5}{16}$	....			10–31
Pinkerton (CP)	10–03	13	$3\frac{3}{16}$	10–17	11	$3\frac{9}{16}$	10–31	9	....	11–14
Taylor	10–10	14	$3\frac{5}{16}$	10–24	12	$3\frac{2}{16}$	....			11–07
Ajax (B–7)	10–10	18	$3\frac{14}{16}$	....			....			10–31
Booth 3	10–10	16	$3\frac{8}{16}$	10–17	14	$3\frac{8}{16}$	....			10–31
Semil 34	10–17	18	$3\frac{10}{16}$	10–31	16	$3\frac{8}{16}$	11–14	14	$3\frac{5}{16}$	11–28
Semil 43	10–24	18	$3\frac{10}{16}$	11–7	16	$3\frac{8}{16}$	11–21	14	$3\frac{5}{16}$	12–05
Booth 1	11–14	16	$3\frac{12}{16}$	11–28	12	$3\frac{8}{16}$	....			12–12
Zio (P)	11–14	12	$3\frac{1}{16}$	11–28	10	$2\frac{14}{16}$	....			12–12
Gossman	11–28	11	$3\frac{1}{16}$	....			....			12–26
Brookslate	12–05	18	$3\frac{13}{16}$	12–12	16	$3\frac{10}{16}$	1–02	12	$3\frac{5}{16}$	1–30
				12–19	14	$3\frac{8}{16}$	1–16	10		
Meya (P)	12–12	13	$3\frac{2}{16}$	12–26	11	$3\frac{9}{16}$	....			1–09
Reed (CP)	12–12	12	$3\frac{4}{16}$	12–26	10	$3\frac{3}{16}$	1–09	9	$3\frac{9}{16}$	1–23

[1] Avocados of the West Indian type varieties and seedlings not listed elsewhere in table I.

[2] Avocados of the Guatemalan type varieties and seedlings, hybrid varieties and seedlings, and unidentified seedlings not listed elsewhere in table I.

(3) Avocados which fail to meet the maturity requirements specified in this section must be maintained under the supervision of the Federal or Federal-State Inspection Service using the Positive Lot Identification program, and when presented for reinspection, must meet the maturity requirements which correspond to the date of the original inspection.

(b) The term *diameter* means the greatest dimension measured at a right angle to a straight line from the stem to the blossom end of the fruit.

[59 FR 30869, June 16, 1994, as amended at 64 FR 53185, Oct. 1, 1999; 71 FR 11294, Mar. 7, 2006; 73 FR 26945, May 12, 2008; 79 FR 55353, Sept. 16, 2014]

PART 917—FRESH PEARS AND PEACHES GROWN IN CALIFORNIA

Subpart A—Order Regulating Handling

AUTHORITY: 7 U.S.C. 601–674.

Subpart A—Order Regulating Handling

SOURCE: 41 FR 17528, Apr. 27, 1976, unless otherwise noted.

DEFINITIONS

§ 917.1 Secretary.

Secretary means the Secretary of Agriculture of the United States, or any officer or employee of the Department to whom authority has heretofore been delegated, or to whom authority may hereafter be delegated, to act in his stead.

EFFECTIVE DATE NOTE: At 76 FR 66605, Oct. 27, 2011, § 917.1 was suspended indefinitely, effective Oct. 28, 2011.

§ 917.2 Act.

Act means Public Act No. 10, 73d Congress (May 12, 1933), as amended, and as reenacted and amended by the Agricultural Marketing Agreement Act of 1937, as amended (48 Stat. 31, as amended; 7 U.S.C. 601–674).

EFFECTIVE DATE NOTE: At 76 FR 66605, Oct. 27, 2011, § 917.2 was suspended indefinitely, effective Oct. 28, 2011.

§ 917.3 Person.

Person means an individual, partnership, corporation, association, or any other business unit.

EFFECTIVE DATE NOTE: At 76 FR 66605, Oct. 27, 2011, § 917.3 was suspended indefinitely, effective Oct. 28, 2011.

§ 917.4 Fruit.

Fruit means the edible product of the following kinds of trees:

(a) All varieties of pears except Beurre Hardy, Beurre D'Anjou, Bosc, Winter Nelis, Doyenne du Comice, Beurre Easter, and Beurre Clairgeau.

(b) [Reserved]

[71 FR 41351, July 21, 2006, as amended at 76 FR 66605, Oct. 27, 2011]

EFFECTIVE DATE NOTE: At 76 FR 66605, Oct. 27, 2011, § 917.4 was suspended indefinitely, effective Oct. 28, 2011.

§ 917.5 Grower.

Grower is synonymous with producer and means any person who produces fruit for market in fresh form, and who has a proprietary interest therein.

[71 FR 41351, July 21, 2006, as amended at 76 FR 66605, Oct. 27, 2011]

EFFECTIVE DATE NOTE: At 76 FR 66605, Oct. 27, 2011, § 917.5 was suspended indefinitely, effective Oct. 28, 2011.

§ 917.6 Handle.

Handle and ship are synonymous and mean to sell, consign, deliver or transport fruit or to cause fruit to be sold, consigned, delivered or transported between the production area and any point outside thereof, or within the production area: *Provided,* That the term handle shall not include the sale of fruit on the tree, the transportation within the production area of fruit from the orchard where grown to a packing facility located within such area for preparation for market, or the delivery of such fruit to such packing facility for such preparation.

[71 FR 41351, July 21, 2006, as amended at 76 FR 66605, Oct. 27, 2011]

EFFECTIVE DATE NOTE: At 76 FR 66605, Oct. 27, 2011, § 917.6 was suspended indefinitely, effective Oct. 28, 2011.

§ 917.7 Handler.

Handler is synonymous with shipper and means any person (except a common or contract carrier transporting fruit owned by another person) who handles fruit.

EFFECTIVE DATE NOTE: At 76 FR 66605, Oct. 27, 2011, § 917.7 was suspended indefinitely, effective Oct. 28, 2011.

§ 917.9 Fiscal period.

Fiscal period is synonymous with fiscal year and means the 12-month period ending on the last day of February of each year, or such other period that may be approved by the Secretary pursuant to recommendations by the committee.

EFFECTIVE DATE NOTE: At 76 FR 66605, Oct. 27, 2011, § 917.9 was suspended indefinitely, effective Oct. 28, 2011.

§ 917.11 Production area.

Production area means the State of California.

EFFECTIVE DATE NOTE: At 76 FR 66605, Oct. 27, 2011, §917.11 was suspended indefinitely, effective Oct. 28, 2011.

§917.12 Container.

Container means a box, bag, crate, lug, basket, carton, package, or any other type of receptacle used in the packaging or handling of fruit.

EFFECTIVE DATE NOTE: At 76 FR 66605, Oct. 27, 2011, §917.12 was suspended indefinitely, effective Oct. 28, 2011.

§917.13 Pack.

Pack means the specific arrangement, size, weight, count, or grade of a quantity of fruit in a particular type and size of container or any combination thereof.

EFFECTIVE DATE NOTE: At 76 FR 66605, Oct. 27, 2011, §917.13 was suspended indefinitely, effective Oct. 28, 2011.

§917.14 District.

District means any of the following subdivisions of the State of California:

(a) *North Sacramento Valley District* includes and consists of Glenn County, Shasta County, Tehama County, Modoc County, Siskiyou County, Lassen County, Plumas County, and Colusa County.

(b) *Central Sacramento Valley District* includes and consists of Sutter County, Butte County, Yuba County, and Sierra County.

(c) *Sacramento River District* includes and consists of Sacramento County, that portion of Yolo County east of a straight line from the northwest corner of Sacramento County to the Northeast corner of Solano County, and that portion of Solano County east of a straight line from the northeast corner of Solano County to the town of Rio Vista.

(d) *El Dorado District* includes and consists of El Dorado County.

(e) *Placer-Colfax District* includes and consists of Nevada and Placer Counties.

(f) *Solano District* includes and consists of that portion of Yolo County not included in the Sacramento River District, and that portion of Solano County not included in the Sacramento River District.

(g) *Contra Costa District* includes and consists of Contra Costa County.

(h) *Santa Clara District* includes and consists of Alameda County, Monterey County, Santa Clara County, San Mateo County, Santa Cruz County, and San Benito County.

(i) *Lake District* includes and consists of Lake County.

(j) *Mendocino District* includes and consists of Mendocino County, Humboldt County, Trinity County, and Del Norte County.

(k) *South Coast District* includes and consists of San Luis Obispo County, Santa Barbara County, and Ventura County.

(l) *Stockton District* includes and consists of San Joaquin County, Amador County, Calaveras County, and Alpine County.

(m) *Stanislaus District* includes and consists of Merced County, Stanislaus County, Tuolumne County, and Mariposa County.

(n) *Fresno District* includes and consists of Madera County, Fresno County, and Mono County.

(o) *Tulare District* includes and consists of Tulare County and Kings County.

(p) *Kern District* includes and consists of that portion of Kern County west of the Tehachapi Mountains.

(q) *Tehachapi District* includes and consists of that portion of Kern County not included in Kern District, and Inyo County.

(r) *Southern California District* includes and consists of San Bernardino County, Orange County, San Diego County, Imperial County, Riverside County, and Los Angeles County.

(s) *North Bay District* includes and consists of Sonoma County, Napa County, and Marin County.

[41 FR 17528, Apr. 27, 1976, as amended at 71 FR 41351, July 21, 2006]

EFFECTIVE DATE NOTE: At 76 FR 66605, Oct. 27, 2011, §917.14 was suspended indefinitely, effective Oct. 28, 2011.

§917.15 Representation area.

Representation area means any one of the districts or groups of districts which are designated for nominating members and alternate members to the commodity committees under § 917.21 or as changed pursuant to § 917.35(g).

[56 FR 46369, Sept. 12, 1991, as amended at 76 FR 66605, Oct. 27, 2011]

EFFECTIVE DATE NOTE: At 76 FR 66605, Oct. 27, 2011, § 917.15 was suspended indefinitely, effective Oct. 28, 2011.

ADMINISTRATIVE BODIES

§ 917.16 Designation of Control Committee.

A Control Committee is hereby established consisting of 12 shipper members and 13 commodity committee members, and the members shall be selected in accordance with the provisions of § 917.17 through § 917.19. The members shall be selected annually for a term ending on the last day of February, and said members shall serve until their respective successors are selected and have qualified.

EFFECTIVE DATE NOTE: At 76 FR 66605, Oct. 27, 2011, § 917.16 was suspended indefinitely, effective Oct. 28, 2011.

§ 917.17 Nomination of shipper members of the Control Committee.

Nominations for the 12 members of the Control Committee to represent shippers shall be made in the following manner:

(a) By February 1 of each year the Control Committee shall announce a time and place for a meeting of all shippers of fruit and shall conduct the election of nominees at such meeting. At said election meeting the shippers present shall select a nominee for each of the shipper member positions on the Control Committee. Each shipper shall cast only one vote.

(b) No shipper shall be entitled to participate in the nomination of members of the Control Committee, or be eligible for membership on such committee, if such shipper has failed to pay the assessments, due to be paid by him pursuant to the provisions of § 917.37.

EFFECTIVE DATE NOTE: At 76 FR 66605, Oct. 27, 2011, § 917.17 was suspended indefinitely, effective Oct. 28, 2011.

§ 917.18 Nomination of commodity committee members of the Control Committee.

Nominations for the 13 members of the Control Committee to represent the commodity committees shall be made in the following manner:

(a) A nomination for one member shall be made by each commodity committee selected pursuant to § 917.25. Nominations for the remaining members shall be made by the respective commodity committees as provided in this section. The number of remaining members which each respective commodity committee shall be entitled to nominate shall be based upon the proportion that the previous three fiscal periods' shipments of the respective fruit is of the total shipments of all fruit to which this part is applicable during such periods. In the event provisions of this part are terminated or suspended as to any fruit, nominations of members to the Control Committee shall be composed of representatives of any remaining fruit.The apportionment shall be determined as aforesaid.In the event provisions of this part are terminated or suspended as to any fruit, the members of the commodity committee of the remaining fruit shall have all the powers, duties, and functions given to the Control Committee under this part and sections of this part pertaining to the designation of the Control Committee shall be terminated or suspended.

(b) A person nominated by any commodity committee for membership on the Control Committee shall be an individual person who is a member or alternate member of the commodity committee that nominates him/her. Each member of each commodity committee shall have only one vote in the selection of nominees for membership on the Control Committee.

EFFECTIVE DATE NOTE: At 76 FR 66605, Oct. 27, 2011, § 917.18 was suspended indefinitely, effective Oct. 28, 2011.

§ 917.19 Selection of members of the Control Committee.

From the nominations made pursuant to § 917.17, or from other persons, the Secretary shall select the shipper members of the Control Committee. From the nominations made pursuant to § 917.18, or from other persons, the Secretary shall select the commodity committee members of the Control Committee. Any person selected as member of the Control Committee

110

shall qualify by filing with the Secretary a written acceptance of the appointment.

EFFECTIVE DATE NOTE: At 76 FR 66605, Oct. 27, 2011, §917.19 was suspended indefinitely, effective Oct. 28, 2011.

§917.20 Designation of members of commodity committees.

There is hereby established a Pear Commodity Committee consisting of 13 members. Each commodity committee may be increased by one public member nominated by the respective commodity committee and selected by the Secretary. The members of each said committee shall be selected biennially for a term ending on the last day of February of odd numbered years, and such members shall serve until their respective successors are selected and have qualified. The members of each commodity committee shall be selected in accordance with the provisions of §917.25.

[76 FR 66605, Oct. 27, 2011]

EFFECTIVE DATE NOTE: At 76 FR 66605, Oct. 27, 2011, §917.20 was suspended indefinitely, effective Oct. 28, 2011.

§917.21 Nomination of Pear Commodity Committee members.

Nominations for membership on the Pear Commodity Committee shall be made by the growers of pears in the respective representation areas as follows:

(a) North Sacramento Valley District and the Central Sacramento Valley District one nominee.

(b) Sacramento River District, Stockton District, Stanislaus District, Contra Costa District, Santa Clara District, and Solano District four nominees.

(c) Placer-Colfax District one nominee.

(d) Lake District four nominees.

(e) Mendocino District and the North Bay District one nominee.

(f) El Dorado District one nominee.

(g) All of the production area not included in paragraphs (a) through (f) of this section one nominee.

EFFECTIVE DATE NOTE: At 59 FR 10055, Mar. 3, 1994, §917.21 was suspended, effective Apr. 4, 1994.

§917.24 Procedure for nominating members of various commodity committees.

(a) The Control Committee shall hold or cause to be held not later than February 15 for pears of each odd numbered year a meeting or meetings of the growers of the fruits in each representation area set forth in §917.21. These meetings shall be supervised by the Control Committee, which shall prescribe such procedures as shall be reasonable and fair to all persons concerned.

(b) With respect to each commodity committee, only growers of the particular fruit who are present at such nomination meetings or represented at such meetings by duly authorized employees may participate in the nomination and election of nominees for commodity committee members and alternates. Each such grower, including employees of such grower, shall be entitled to cast but one vote for each position to be filled for the representation area in which he produces such fruit.

(c) A particular grower, including employees of such growers, shall be eligible for membership as principle or alternate to fill only one position on a commodity committee. A grower nominated for membership on the Pear Commodity Committee must have produced at least 51 percent of the pears shipped by him during the previous fiscal period, or he must represent an organization which produced at least 51 percent of the pears shipped by it during such period.

[76 FR 66605, Oct. 27, 2011]

EFFECTIVE DATE NOTE: At 76 FR 66605, Oct. 27, 2011, §917.24 was suspended indefinitely, effective Oct. 28, 2011.

§917.25 Selection of members of various commodity committees.

(a) The Secretary shall select the members of each commodity committee from nominations made by growers, as provided in §§917.21 through 917.24, or from among other eligible persons. Any person selected as a member of a commodity committee shall qualify by filing with the Secretary a written acceptance of the appointment.

(b) [Reserved]

[41 FR 17528, Apr. 27, 1976, as amended at 76 FR 66605, Oct. 27, 2011]

EFFECTIVE DATE NOTE: At 76 FR 66605, Oct. 27, 2011, § 917.25 was suspended indefinitely, effective Oct. 28, 2011.

§ 917.26 Failure to nominate.

If nominations are not made within the time and in the manner prescribed in §§ 917.21 through 917.24, the Secretary may, without regard to nominations, select the member and alternate members of commodity committees on the basis of representation provided in § 917.21. In the event nominations are not made for membership on the Control Committee, pursuant to the provisions of §§ 917.17 and 917.18, by May 1 of each year, the Secretary may select such members without waiting for nominees to be designated.

[41 FR 17528, Apr. 27, 1976, as amended at 56 FR 46369, Sept. 12, 1991; 76 FR 66605, Oct. 27, 2011]

EFFECTIVE DATE NOTE: At 76 FR 66605, Oct. 27, 2011, § 917.26 was suspended indefinitely, effective Oct. 28, 2011.

§ 917.27 Alternates.

There shall be an alternate for each member of the Control Committee, and an alternate for each member of each commodity committee. Each such alternate shall possess the same qualifications, shall be nominated and selected in the same manner and shall hold office for the same term, as the member for whom he is alternate. An alternate shall, in the event of such member's absence at a meeting of the committee, act in the place and stead of such member; and, in the event of such member's removal, resignation, disqualification, or death, the alternate for such member shall, until a successor for the unexpired term of said member has been selected, act in the place and stead of said member. In the event both a member and his alternate are unable to attend a meeting the member or the committee members present may designate any other alternative to serve in such member's place and stead provided such action is necessary to secure a quorum.

EFFECTIVE DATE NOTE: At 76 FR 66605, Oct. 27, 2011, § 917.27 was suspended indefinitely, effective Oct. 28, 2011.

§ 917.28 Procedure for filling vacancies on committees.

To fill any vacancy on the Control Committee or on any of the commodity committees occasioned by the failure of any person selected as a member or as an alternate member to qualify, or in the event of the death, removal, resignation, or disqualification of any member or alternate member, a successor for the unexpired term of such member or alternate shall be nominated and selected in the manner specified in §§ 917.17 through 917.19 and §§ 917.21 through 917.25. If the names of nominees to fill any such vacancy are not made available to the Secretary within a reasonable time after such vacancy occurs, the Secretary may fill such vacancy without regard to nominations on the basis of representation provided for in §§ 917.16 and 917.21.

[41 FR 17528, Apr. 27, 1976, as amended at 56 FR 46369, Sept. 12, 1991; 76 FR 66605, Oct. 27, 2011]

EFFECTIVE DATE NOTE: At 76 FR 66605, Oct. 27, 2011, § 917.28 was suspended indefinitely, effective Oct. 28, 2011.

§ 917.29 Organization of committees.

(a) A majority of all of the members of the Control Committee shall constitute a quorum, and any action of the Control Committee shall require the concurrence of the majority of all members present at the meeting.

(b) A quorum of the Pear Commodity Committee shall consist of nine members.

(c) The Control Committee and each commodity committee shall give to the Secretary the same notice of each meeting that is given to the members of the respective committee.

(d) The Control Committee or any commodity committee may, upon due notice to all of the members of the respective committee, vote by letter, telegraph or telephone: *Provided,* That any member voting by telephone shall promptly thereafter confirm in writing his/her vote so cast.

[41 FR 17528, Apr. 27, 1976, as amended at 56 FR 46369, Sept. 12, 1991; 71 FR 41352, July 21, 2006; 76 FR 66605, Oct. 27, 2011]

EFFECTIVE DATE NOTE: At 76 FR 66605, Oct. 27, 2011, §917.29 was suspended indefinitely, effective Oct. 28, 2011.

§917.30 Removal and disapproval.

The members of the Control Committee, including their respective successors and alternates, and the members of each commodity committee, including their respective successors and alternates, and any agent or employee appointed or employed by the Control Committee and the members of any other committee established pursuant to the provisions of this subpart shall be subject to removal or suspension at any time by the Secretary. Each regulation, decision, determination, or other act of the Control Committee, or any commodity committee, or any other committee established pursuant to the provisions of this subpart, shall be subject to the continuing right of the Secretary to disapprove of the same at any time; and, upon such disapproval, each such regulation, decision, determination, or other act, shall be deemed null and void except as to acts done in reliance thereon or in compliance therewith prior to such disapproval by the Secretary.

EFFECTIVE DATE NOTE: At 76 FR 66605, Oct. 27, 2011, §917.30 was suspended indefinitely, effective Oct. 28, 2011.

§917.31 Compensation and expenses.

All committee members shall serve without compensation, but said members, and their respective alternates, shall be reimbursed for expenses necessarily incurred in the performance of their duties. At its discretion any committee may request the attendance of one or more alternates at any or all meetings, notwithstanding the expected or actual presence of the respective members, and may pay expenses as aforesaid.

EFFECTIVE DATE NOTE: At 76 FR 66605, Oct. 27, 2011, §917.31 was suspended indefinitely, effective Oct. 28, 2011.

§917.32 Funds and other property.

(a) All funds received by the Control Committee, pursuant to the provisions of this part, shall be used solely for the purpose specified in this part; and the Secretary may require the Control Committee and its members to account for all receipts and disbursements.

(b) Upon the resignation, removal, or expiration of the term of any member or employee of the Control Committee, or of any member of any commodity committee, all books, records, funds, and other property in his possession belonging to the Control Committee or any commodity committee shall be delivered to the Control Committee or to his successor in office; and such assignments and other instruments shall be executed as may be necessary to vest in the Control Committee full title to all the books, records, funds, and other property in the possession or under the control of such member or employee, pursuant to the provisions of this part.

(c) The Control Committee may, with the approval of the Secretary, maintain in its own name, or in the name of its members, a suit against any shipper for the collection of such shipper's pro rata share of expenses, pursuant to the provisions of this part.

EFFECTIVE DATE NOTE: At 76 FR 66605, Oct. 27, 2011, §917.32 was suspended indefinitely, effective Oct. 28, 2011.

§917.33 Powers of Control Committee.

The Control Committee shall have the following powers:

(a) To administer, as specifically provided in this part, the terms and provisions of this part.

(b) To make administrative rules and regulations in accordance with and to effectuate the terms and provisions of this part.

(c) To receive, investigate, and report to the Secretary complaints of violations of the provisions of this part.

(d) To recommend to the Secretary amendments to this part.

EFFECTIVE DATE NOTE: At 76 FR 66605, Oct. 27, 2011, §917.33 was suspended indefinitely, effective Oct. 28, 2011.

§917.34 Duties of Control Committee.

The Control Committee shall have the following duties:

(a) To act as intermediary between the Secretary and any grower or shippers.

(b) To keep minute books and records which will clearly reflect all of the acts and transactions of said Control Committee; and such minute books and records shall be subject at any time to examination by the Secretary or by such person as may be designated by the Secretary.

(c) To investigate, from time to time, and assemble data on the growing, shipping, and marketing conditions respecting fruit, as defined in § 917.4; to engage in such research and service activities in connection with the handling of such fruit as may be approved, from time to time, by the Secretary; and to furnish to the Secretary such available information as may be requested.

(d) To appoint such employees, agents, and representatives as it may deem necessary, and to determine the compensation and define the duties of each.

(e) To develop and provide the commodity committees data on shared expenses to facilitate equitable apportionment of such expenses in the development of budgets.

(f) To confer with representatives of shippers and growers of fruit produced in other states and areas with respect to the formulation or operation of marketing agreements providing for the regulation of shipments among the several states and areas in the United States in which such fruit is grown.

(g) With the approval of the Secretary establish procedures for the selection and appointment of a public member and alternate to each of the commodity committees.

(h) To establish and define the duties of additional committees or subcommittees to assist in the performance of any of the duties and functions of the Control Committee.

(i) To defend all legal proceedings against any committee members (individually or as members) or any officers or employees of such committees arising out of any act or omission made in good faith pursuant to the provisions of this part.

(j) To cause the books of the Control Committee to be audited by a competent accountant at least once each fiscal period and at such other time or times as the Control Committee may deem necessary or as the Secretary may request. Such audit shall indicate whether the funds have been received and expended in accordance with the provisions of this part.

(k) To appoint nomination committees if it deems proper for any or each nomination meeting held pursuant to § 917.21. Such nomination committees would canvas prospective members and alternate members to the commodity committees to determine their eligibility and willingness to serve and present a slate of nominees to the meeting or meetings. The presentation of nominees by the nominating committee at these meetings shall not exclude the right of any grower to nominate any eligible person at such meeting.

[41 FR 17528, Apr. 27, 1976, as amended at 56 FR 46369, Sept. 12, 1991; 76 FR 66605, Oct. 27, 2011]

EFFECTIVE DATE NOTE: At 76 FR 66605, Oct. 27, 2011, § 917.34 was suspended indefinitely, effective Oct. 28, 2011.

§ 917.35 Powers and duties of each commodity committee.

Each commodity committee shall have the following powers and duties:

(a) With regard to the respective fruit for which it was established, to establish production research and marketing research and development projects as authorized under § 917.39, to recommend to the Secretary regulation of shipments pursuant to the provisions of this part, and to possess such other powers and exercise such other duties as will properly effectuate the purpose of this part: *Provided, however,* That the Pear Commodity Committee shall approve actions under § 917.39 and make said recommendations pursuant to §§ 917.40 through 917.43 only upon the affirmative vote of not less than nine members of said committee.

(b) To make such rules and regulations with respect to fruit for which it was established as may be necessary to effectuate the terms and provisions of this part.

(c) To forward to the Control Committee and to the Secretary a record of the minutes of each meeting of the commodity committee.

(d) To establish such other committees to aid the commodity committee in the performance of its duties under this part as may be deemed advisable.

(e) Each season prior to any recommendation to the Secretary for a regulation of shipments pursuant to §§917.40 through 917.43 to determine the marketing policy to be followed for the respective commodity during the ensuing fiscal period and to submit such policy to the Secretary, said policy report to contain, among other provisions, information relative to the estimated total production and shipments of the fruit by districts, information as to the expected general quality and size of fruit, possible or expected demand conditions of different market outlets, supplies of competitive commodities, such analysis of the foregoing factors and conditions as the committee deems appropriate, and the type of regulations of shipments expected to be recommended for the respective fruit.

(f) To submit as soon as practicable after the beginning of each fiscal year to the Secretary, for his approval, a budget of its expenses for such fiscal period, including its proportional share of the expenses of the Control Committee and an explanation of the items therein, and a recommendation as to the rate of assessment for the respective fruit for which the commodity committee was established.

(g) With the approval of the Secretary, to redefine the Districts into which the State of California has been divided under §917.14 or change the representation of any representation area affecting the respective commodity committee: *Provided, however,* That if any such changes are made, representation on any such committee from the various representation areas shall be based, so far as practicable, upon the proportionate quantity of the respective fruit shipped from the respective representation area during the preceding three fiscal periods: *Provided further,* That the commodity committees shall follow the principle, so far as practicable, of assigning a member position on the commodity committees to any representation area from which five percent of regulated shipments have originated during such periods.

[41 FR 17528, Apr. 27, 1976, as amended at 56 FR 46369, Sept. 12, 1991; 71 FR 41352, July 21, 2006; 76 FR 66605, Oct. 27, 2011]

EFFECTIVE DATE NOTE: At 76 FR 66605, Oct. 27, 2011, §917.35 was suspended indefinitely, effective Oct. 28, 2011.

EXPENSES AND ASSESSMENTS

§917.36 Expenses.

Each commodity committee is authorized to incur such expenses as the Secretary finds are reasonable and are likely to be incurred by the said commodity committee during each fiscal period for the maintenance and functioning of such committee, including its proportionate share of the expenses of the Control Committee; and for such research and service activities relating to handling of the fruit for which the commodity committee was established as the Secretary may determine to be appropriate. The funds to cover such expenses shall be acquired by the levying of assessments as provided in §917.37.

EFFECTIVE DATE NOTE: At 76 FR 66605, Oct. 27, 2011, §917.36 was suspended indefinitely, effective Oct. 28, 2011.

§917.37 Assessments.

(a) As his/her pro rata share of the expenses which the Secretary finds are reasonable and are likely to be incurred by the commodity committees during a fiscal period, each handler shall pay to the Control Committee, upon demand, assessments on all fruit handled by him/her. The payment of assessments for the maintenance and functioning of the committees may be required under this part throughout the period it is in effect irrespective of whether particular provisions thereof are suspended or become inoperative.

(b) The Secretary shall fix the respective rate of assessment, which handlers shall pay with respect to each fruit during each fiscal period in an amount designed to secure sufficient funds to cover the respective expenses, which may be incurred during such period. At any time during or after the fiscal period, the Secretary may increase the rates of assessment in order to secure funds to cover any later findings by the Secretary relative to such expenses, and such increase shall apply to all fruit shipped during the fiscal period.

(c) In order to provide funds to carry out the functions of the commodity committee prior to commencement of shipments in any season, shippers may make advance payments of assessments, which advance payments shall be credited to such shippers and the assessments of such shippers shall be adjusted so that such assessments are based upon the quantity of fruit shipped by such shippers during such season. Any shipper who ships fruit for the account of a grower may deduct, from the account of sale covering such shipment or shipments, the amount of assessments levied on said fruit shipped for the account of such grower.

[71 FR 41352, July 21, 2006, as amended at 76 FR 66605, Oct. 27, 2011]

EFFECTIVE DATE NOTE: At 76 FR 66605, Oct. 27, 2011, § 917.37 was suspended indefinitely, effective Oct. 28, 2011.

§ 917.38   Accounting.

If, at the end of a fiscal period the assessments collected are in excess of expenses incurred, each commodity committee, with the approval of the Secretary, may carry over such excess into subsequent fiscal periods as a reserve: *Provided,* That funds already in the reserve do not exceed approximately one fiscal period's expenses. Such reserve funds may be used (1) to cover any expenses authorized by this part and (2) to cover necessary expenses of liquidation in the event of termination of this part. If any such excess is not retained in a reserve, each handler entitled to a proportionate refund shall be credited with such refund against the operations of the following fiscal period or be paid such refund. Upon termination of this part, any funds not required to defray the necessary expenses of liquidation shall be disposed of in such manner as the Secretary may determine to be appropriate: *Provided,* That, to the extent practical, such funds shall be returned pro rata to the persons from whom such funds were collected.

EFFECTIVE DATE NOTE: At 76 FR 66605, Oct. 27, 2011, § 917.38 was suspended indefinitely, effective Oct. 28, 2011.

RESEARCH

§ 917.39   Production research, market research and development.

The committees, with the approval of the Secretary, may establish or provide for the establishment of production research, marketing research, and development projects designed to assist, improve, or promote the marketing, distribution, and consumption or efficient production of fruit. Such projects may provide for any form of marketing promotion including paid advertising. The expenses of such projects shall be paid from funds collected pursuant to § 917.37.

EFFECTIVE DATE NOTE: At 76 FR 66605, Oct. 27, 2011, § 917.39 was suspended indefinitely, effective Oct. 28, 2011.

REGULATIONS

§ 917.40   Recommendations for regulations.

(a) Whenever a commodity committee deems it advisable to regulate the handling of any variety or varieties of fruit in the manner provided in § 917.41, it shall so recommend to the Secretary.

(b) In arriving at its recommendations for regulation pursuant to paragraph (a) of this section, the commodity committee shall give consideration to current information with respect to the factors affecting the supply and demand for such fruit during the period or periods when it is proposed that such regulation should be made effective. With each such recommendation for regulation, the commodity committee shall submit to the Secretary the data and information on which such recommendation is predicated and such other available information as the Secretary may request.

EFFECTIVE DATE NOTE: At 76 FR 66605, Oct. 27, 2011, § 917.40 was suspended indefinitely, effective Oct. 28, 2011.

§ 917.41   Issuance of regulations.

(a) The Secretary shall regulate, in the manner specified in this section, the handling of any variety or varieties of fruit whenever he finds, from the recommendations and information submitted by the commodity committee, or from other available information, that such regulations will tend to effectuate the declared policy of the act. Such regulations may:

(1) Limit, during any period or periods, the total quantity of any grade, size, quality, maturity, or pack, or any combination thereof, of any variety or varieties of fruit;

(2) Limit the shipment of any variety or varieties of fruit by establishing, in terms of grades, sizes, or both, minimum standards of quality and maturity during any period when season average prices are expected to exceed the parity level;

(3) Fix the size, capacity, weight, dimensions, markings, or pack of the container, or containers, which may be used in the packaging or handling of any fruit.

(b) The commodity committee shall be informed immediately of any such regulation issued by the Secretary, and the commodity committee shall promptly give notice thereof to handlers.

EFFECTIVE DATE NOTE: At 76 FR 66605, Oct. 27, 2011, § 917.41 was suspended indefinitely, effective Oct. 28, 2011.

§ 917.42 Modification, suspension, or termination of regulations.

(a) In the event the commodity committee at any time finds that, by reason of changed conditions, any regulations issued pursuant to § 917.41 should be modified, suspended, or terminated, it shall so recommend to the Secretary.

(b) Whenever the Secretary finds, from the recommendations and information submitted by the commodity committee or from other available information, that a regulation should be modified, suspended, or terminated with respect to any or all shipments of fruit in order to effectuate the declared policy of the act, he shall modify, suspend, or terminate such regulation. If the Secretary finds that a regulation obstructs or does not tend to effectuate the declared policy of the act, he shall suspend or terminate such regulation. On the same basis and in like manner the Secretary may terminate any such modification or suspension.

EFFECTIVE DATE NOTE: At 76 FR 66605, Oct. 27, 2011, § 917.42 was suspended indefinitely, effective Oct. 28, 2011.

§ 917.43 Special purpose shipments.

(a) Except as otherwise provided in this section, any person may, without regard to the provisions of §§ 917.37, 917.41, and 917.42, and the regulations issued thereunder, handle fruit (1) for consumption by charitable institutions; (2) for distribution by relief agencies; or (3) for commercial processing into products.

(b) Upon the basis of recommendations and information submitted by the commodity committee, or from other available information, the Secretary may relieve from any or all requirements, under or established pursuant to § 917.41, § 917.42, § 917.45, or § 917.37, the handling of fruit; (1) To designated market areas outside the continental United States; (2) for such specified purposes (including shipments to facilitate the conduct of marketing research and development projects established pursuant to § 917.39); or (3) in such minimum quantities or types of shipments, as may be prescribed.

(c) The commodity committee shall, with the approval of the Secretary, prescribe such rules, regulations, and safeguards as it may deem necessary to prevent fruit handled under the provisions of this section from entering the channels of trade for other than the specified purposes authorized by this section. Such rules, regulations, and safeguards may include the requirements that handlers shall file applications and receive approval from the commodity committee for authorization to handle fruit pursuant to this section, and that such applications be accompanied by a certification by the intended purchaser or receiver that the fruit will not be used for any purpose not authorized by this section.

EFFECTIVE DATE NOTE: At 76 FR 66605, Oct. 27, 2011, § 917.43 was suspended indefinitely, effective Oct. 28, 2011.

§ 917.45 Inspection and certification.

(a) Whenever the handling of any variety of a particular fruit is regulated pursuant to § 917.41 or § 917.42, each handler who handles such fruit shall, prior thereto, cause such fruit to be inspected by the Federal or Federal-State Inspection Service: *Provided,* That inspection and certification shall not be required if such fruit has previously been so inspected and certified. Promptly after inspection and certification, each such handler shall submit, or cause to be submitted, to the commodity committee a copy of the certificate of inspection issued with respect to such fruit. The commodity committees may, with the approval of the Secretary, prescribe rules and regulations waiving the inspection requirements of this section where it is determined that inspection is not available: *Provided*, That all shipments made under such waiver shall comply with all regulations in effect.

(b) The Control Committee may enter into an agreement with the Federal and Federal-State Inspection Services with respect to the costs of the inspection required by paragraph (a) of this section, for any or all fruits, and may collect from handlers their respective pro rata shares of such costs.

EFFECTIVE DATE NOTE: At 76 FR 66605, Oct. 27, 2011, §917.45 was suspended indefinitely, effective Oct. 28, 2011.

REPORTS

§917.50   Reports.

(a) Each handler shall furnish to the Manager of the Control Committee, at such times and for such periods as the Control Committee or the commodity committees may designate, certified reports covering, to the extent necessary for the committees to perform their functions, each shipment of fruits as follows:

(1) The name of the shipper and the shipping point;

(2) The car or truck license number (or name of the trucker), and identification of the carrier;

(3) The date and time of departure;

(4) The number and type of containers in the shipment;

(5) The quantities shipped, showing separately the variety, grade, and size of the fruit;

(6) The destination;

(7) Identification of the inspection certificate or waiver pursuant to which the fruit was handled;

(8) The price per package at which sold, including specific and detailed information relative to all discounts, allowances, rebates, or other adjustments thereof.

(b) Upon request of any committee, made with the approval of the Secretary, each handler shall furnish to the Manager of the Control Committee, in such manner and at such times as it may prescribe, such other information as may be necessary to enable the committee to perform its duties under this part.

(c) Each handler shall maintain for at least two succeeding fiscal years, such records of the fruits received and disposed of by him as may be necessary to verify the reports he submits to the committee pursuant to this section.

(d) All reports and records submitted by handlers pursuant to the provisions of this section shall be received by, and at all times be in custody of, one or more designated employees of the Control Committee. No such employee shall disclose to any person, other than the Secretary upon request therefor, data or information obtained or extracted from such reports and records which might affect the trade position, financial condition, or business operation of the particular handler from whom received: *Provided*, That such data and information may be combined, and made available to any person, in the form of general reports in which the identities of the individual handlers furnishing the information are not disclosed and may be revealed to any extent necessary to effect compliance with the provisions of this part and the regulations issued thereunder.

EFFECTIVE DATE NOTE: At 76 FR 66605, Oct. 27, 2011, §917.50 was suspended indefinitely, effective Oct. 28, 2011.

MISCELLANEOUS PROVISIONS

§917.60   Effective time.

The provisions of this part and of any amendment thereto, shall become effective at such time as the Secretary may declare above his signature and shall continue in force until terminated in one of the ways specified in §917.61.

EFFECTIVE DATE NOTE: At 76 FR 66605, Oct. 27, 2011, §917.60 was suspended indefinitely, effective Oct. 28, 2011.

§917.61 Termination.

(a) The Secretary may at any time terminate the provisions of this part by giving at least one day's notice by means of a press release or in any other manner in which he may determine.

(b) The Secretary shall terminate or suspend the operation of any and all of the provisions of this part whenever he finds that such provisions do not tend to effectuate the declared policy of the act.

(c) The Secretary shall terminate the provisions of this part or the applicability of the provisions of this part as to a particular fruit whenever he finds by referendum or otherwise that such termination is favored by a majority of the growers of the fruit: *Provided,* That such majority has during the current fiscal period produced more than 50 percent of the volume of the fruit which was produced within the production area for shipment in fresh form. Such termination shall become effective on the first day of March subsequent to the announcement therof by the Secretary.

(d) The Control Committee shall consider all petitions from growers submitted to it for termination of this part provided such petitions are received by the Control Committee prior to October 1 of the then current fiscal period. Upon recommendation of the Control Committee, received not later than December 1 of the then current fiscal period, the Secretary shall conduct a referendum among the growers of the particular kind of fruit prior to February 15 of such fiscal period to ascertain whether continuance of this part is favored by producers.

(e) The Secretary shall conduct a referendum within the period beginning December 1, 1974, and ending February 15, 1975, to ascertain whether continuance of this part as to any fruit included in this part is favored by the growers. The Secretary shall conduct such a referendum within the same period of every fourth fiscal period thereafter.

(f) The provisions of this part shall, in any event, terminate whenever the provisions of the act authorizing them cease to be in effect.

EFFECTIVE DATE NOTE: At 76 FR 66605, Oct. 27, 2011, §917.61 was suspended indefinitely, effective Oct. 28, 2011.

§917.62 Proceedings after termination.

(a) Upon the termination of the provisions of this part pertaining to any fruit or fruits, the Control Committee then functioning shall for the purpose of liquidating the affairs of the Control Committee with respect to such fruit continue as trustee of all the funds and property then in its possession, or under its control, including claims for any funds unpaid or property not delivered at the time of such termination.

(b) The said trustees shall (1) continue in such capacity until discharged by the Secretary; (2) from time to time account for all receipts and disbursements and deliver all property on hand, together with all books and records of the committee and of the trustees, to such persons as the Secretary may direct; and (3) upon the request of the Secretary, execute such assignments or other instruments necessary or appropriate to vest in such person, full title and right to all funds, property, and claims vested in the Control Committee or the trustees pursuant thereto.

(c) Any person to whom funds, property, or claims have been transferred or delivered, pursuant to this section, shall be subject to the same obligation imposed upon the Control Committee and upon the trustees.

EFFECTIVE DATE NOTE: At 76 FR 66605, Oct. 27, 2011, §917.62 was suspended indefinitely, effective Oct. 28, 2011.

§917.63 Effect of termination or amendment.

Unless otherwise expressly provided by the Secretary, the termination of this subpart or of any regulation issued pursuant to this subpart, or the issuance of any amendment to either thereof, shall not (a) affect or waive any right, duty, obligation, or liability which shall have arisen or which may thereafter arise in connection with any provision of this subpart or any regulation issued under this subpart, or (b) release or extinguish any violation of this subpart or of any regulation issued under this subpart, or (c) affect or impair any rights or remedies of the Secretary or of any other person with respect to any such violation.

EFFECTIVE DATE NOTE: At 76 FR 66605, Oct. 27, 2011, § 917.63 was suspended indefinitely, effective Oct. 28, 2011.

§ 917.64 Compliance.

Each shipper shall comply with all regulations. No shipper shall ship fruit in violation of the provisions of this part or in violation of any regulation issued by the Secretary pursuant to the provisions of this part.

EFFECTIVE DATE NOTE: At 76 FR 66605, Oct. 27, 2011, § 917.64 was suspended indefinitely, effective Oct. 28, 2011.

§ 917.65 Duration of immunities.

The benefits, privileges, and immunities conferred by virtue of the provisions of this subpart shall cease upon its termination except with respect to acts done under and during the time the provisions of this part are in force and effect.

EFFECTIVE DATE NOTE: At 76 FR 66605, Oct. 27, 2011, § 917.65 was suspended indefinitely, effective Oct. 28, 2011.

§ 917.66 Agents.

The Secretary may by a designation in writing name any person, including any officer or employee of the Government or any agency or Division in the United States Department of Agriculture, to act as his agent or representative in connection with any of the provisions of this part.

EFFECTIVE DATE NOTE: At 76 FR 66605, Oct. 27, 2011, § 917.66 was suspended indefinitely, effective Oct. 28, 2011.

§ 917.67 Derogation.

Nothing contained in this part is or shall be construed to be in derogation or in modification of the rights of the Secretary or of the United States to exercise any powers granted by the act or otherwise, and in accordance with such powers to act in the premises whenever such action is deemed advisable.

EFFECTIVE DATE NOTE: At 76 FR 66605, Oct. 27, 2011, § 917.67 was suspended indefinitely, effective Oct. 28, 2011.

§ 917.68 Liability of committee members.

No members of the Control Committee, any commodity committee, or other committee, or any subcommittee, or any employee of the Control Committee shall be held personally responsible, either individually or jointly with others, in any way whatsoever, to any shipper or any other person for errors in judgment, mistakes, or other acts, either of commission or omission, as such member or employee, except for acts of dishonesty.

EFFECTIVE DATE NOTE: At 76 FR 66605, Oct. 27, 2011, § 917.68 was suspended indefinitely, effective Oct. 28, 2011.

§ 917.69 Separability.

If any provision of this part is declared invalid or the applicability thereof to any person, circumstance, thing, or any particular kind of fruit is held invalid, the validity of the remainder of this part or the applicability thereof to any other person, circumstance, thing, or kind of fruit shall not be affected thereby.

EFFECTIVE DATE NOTE: At 76 FR 66605, Oct. 27, 2011, § 917.69 was suspended indefinitely, effective Oct. 28, 2011.

Subpart B—Administrative Requirements

SOURCE: 16 FR 12776, Dec. 20, 1951, unless otherwise noted. Redesignated at 26 FR 12751, Dec. 30, 1961.

DEFINITIONS

§ 917.100 Order.

Order means Marketing Order No. 917, as amended (this part 917), regulating the handling of fresh pears grown in the State of California.

[31 FR 7476, May 5, 1966, as amended at 56 FR 46369, Sept. 12, 1991; 76 FR 66605, Oct. 27, 2011]

EFFECTIVE DATE NOTE: At 76 FR 66605, Oct. 27, 2011, § 917.100 was suspended indefinitely, effective Oct. 28, 2011.

§ 917.101 Marketing agreement.

Marketing agreement means Marketing Agreement No. 85 as amended.

EFFECTIVE DATE NOTE: At 76 FR 66605, Oct. 27, 2011, §917.101 was suspended indefinitely, effective Oct. 28, 2011.

§917.103 Terms.

All other terms used in this subpart shall have the same meaning as when used in the marketing agreement and order.

[18 FR 712, Feb. 4, 1953. Redesignated at 26 FR 12751, Dec. 30, 1961]

EFFECTIVE DATE NOTE: At 76 FR 66605, Oct. 27, 2011, §917.103 was suspended indefinitely, effective Oct. 28, 2011.

GENERAL

§917.110 Communications.

Unless otherwise prescribed in this subpart, or in the marketing agreement and order, or required by the Control Committee, or a particular commodity committee, all reports, applications, submittals, requests, and communications in connection with the marketing agreement and order shall be addressed as follows:

California Tree Fruit Agreement, P.O. Box 968, Reedley, CA, 93654–0968.

[63 FR 16041, Apr. 1, 1998, as amended at 71 FR 78041, Dec. 28, 2006]

EFFECTIVE DATE NOTE: At 76 FR 66605, Oct. 27, 2011, §917.110 was suspended indefinitely, effective Oct. 28, 2011.

ADMINISTRATIVE BODIES

§917.115 Nomination of shipper members for the Control Committee.

(a) All shippers who, prior to February 1 of the then current year, have not advised the manager of the Control Committee in writing of their participation in the formation of an elective body shall be notified promptly by the manager after that date, by mail, of the time and place for a meeting of such shippers to elect nominees for shipper membership on the Control Committee.

(b) The chairman of the then existing Control Committee shall schedule a meeting of shippers in the month of February of the then current year, for the purpose of making nominations to the shipper membership of the Control Committee; and such chairman is authorized to appoint a member of the Control Committee to act as chairman of the meeting and to conduct the election.

EFFECTIVE DATE NOTE: At 76 FR 66605, Oct. 27, 2011, §917.115 was suspended indefinitely, effective Oct. 28, 2011.

§917.119 Procedure for nominating members for various Commodity Committees; meetings.

(a) The manager of the then existing Control Committee shall arrange for, and publicize, meetings of growers to nominate members for the different commodity committees, and each such meeting shall be attended by one or more employees of the Control Committee. Members of the Agricultural Extension Service of the University of California may be authorized by the manager to assist in calling such meetings and advise growers, on their respective mailing lists, of such meetings.

(b) Growers assembled at any such meetings may select a chairman and secretary, but in the event none of the aforesaid employees of the Control Committee is selected as secretary of the meeting, one such employee shall, nevertheless, record all nominations made.

(c) The nominations at any meeting shall be conducted according to Robert's rules of order. However, voting may be by secret ballot or by acclamation in accordance with the desire of the majority of the growers attending the meeting.

(d) No individual, whether representing a corporation or otherwise, may cast more than one vote for each nominee to be selected at the meeting where such individual is eligible to participate in the selection of nominees for members and alternate members of the Commodity Committees.

[16 FR 12776, Dec. 20, 1951, as amended at 24 FR 470, Jan. 21, 1959. Redesignated at 26 FR 12751, Dec. 30, 1961, as amended at 71 FR 78041, Dec. 28, 2006; 76 FR 66605, Oct. 27, 2011]

EFFECTIVE DATE NOTE: At 76 FR 66605, Oct. 27, 2011, §917.119 was suspended indefinitely, effective Oct. 28, 2011.

§917.121 Changes in nomination of Pear Commodity Committee members.

Nominations for membership on the Pear Commodity Committee shall be made by the growers of pears in the respective representation areas as follows:

(a) North Sacramento Valley District, Central Sacramento Valley District, Placer-Colfax District, El Dorado District, and all of the production area not included in paragraphs (b) through (d) of this section, one nominee.

(b) Sacramento River District, Stockton District, Stanislaus District, Contra Costa District, Santa Clara District and Solano District, three nominees.

(c) Lake District, six nominees.

(d) Mendocino District and North Bay District, three nominees.

[52 FR 12513, Apr. 17, 1987]

EFFECTIVE DATE NOTE: At 59 FR 10056, Mar. 3, 1994, § 917.121 was suspended, effective Apr. 4, 1994.

§ 917.122 Qualification requirements and nomination procedure for public members of Commodity Committees.

(a) Public members shall not have a financial interest in or be associated with production, processing, financing, or marketing (except as consumers) of the commodities regulated under this part.

(b) Public members should be able to devote sufficient time and express a willingness to attend committee activities regularly, and to familiarize themselves with the background and economics of the industry.

(c) Public members must be residents of California.

(d) Public members should be nominated by each Commodity Committee and should serve a two-year term which coincides with the term of office of grower members of Commodity Committees.

[42 FR 3625, Jan. 19, 1977, as amended at 43 FR 58355, Dec. 14, 1978]

EFFECTIVE DATE NOTE: At 76 FR 66605, Oct. 27, 2011, § 917.122 was suspended indefinitely, effective Oct. 28, 2011.

REGULATION BY GRADES, SIZES, AND MINIMUM STANDARDS OF QUALITY AND MATURITY

§ 917.143 Exemptions.

(a) *Waivers.* A handler may handle fruit without inspection and certification, as prescribed under § 917.45, if all the following conditions are met:

(1) The handler requests the Federal-State Inspection Service to provide inspection during its regular working hours at least two hours in advance of the time when inspection is needed. The request need not be in writing but it shall be confirmed immediately in writing on a waiver form supplied by the inspection service;

(2) The Federal-State Inspection Service advises the handler that it is not practicable to provide inspection at the time and place designated by the handler. Such advice may be verbal but it shall be confirmed in writing by the Federal-State Inspection Service by execution of the waiver form on which the handler submitted his written request. A confirmed copy thereof shall be forwarded by the inspection service to the office of the Control Committee;

(3) The Federal-State Inspection Service furnishes the handler with the number of the waiver which shall cover the fruit on which inspection is requested;

(4) When so instructed, the handler plainly and conspicuously marks one end of each container with the letter W and the waiver number supplied by the Federal-State Inspection Service. The letter W and the number so marked shall be not less than one-half inch in height.

(b) *Minimum quantities.* Notwithstanding any other provisions of this section, pears may be handled without regard to the provisions of §§ 917.37, 917.41, 917.42, 917.45 and 917.50 under the following conditions:

(1) Such pears meet the grade requirements set forth in Articles 35, 38, and 34, respectively of the Food and Agriculture Code of California.

(2) Such pears are for home use and not for resale.

(3) The shipment does not exceed 200 pounds of pears to any one vehicle during any one day.

(4) Such pears are handled by the person who produces them; and the handling takes place (i) on the premises where grown, (ii) at a packinghouse or retail stand nearby which is operated by said handler, or (iii) at a certified farmers market in compliance with section 1392 of the regulations of the

California Department of Food and Agriculture: *Provided,* That the exemption for certified farmers markets shall not apply to fruit sorted out by a handler unless such fruit is packed in containers clearly and legibly marked to show that the fruit contained therein is only to be sold at a certified farmers market, and the handler complies with regulations established under §§917.37, 917.41(a)(1), 917.45, and 917.50, except that such fruit may be handled to such markets if the fruit fails to meet the applicable grade only on account of being soft and overripe.

[31 FR 7476, May 24, 1966, as amended at 41 FR 22071, June 1, 1976; 41 FR 28509, July 12, 1976; 42 FR 22875, May 5, 1977; 47 FR 30452, July 14, 1982; 49 FR 36361, Sept. 17, 1984; 53 FR 18818, May 25, 1988; 56 FR 46369, Sept. 12, 1991; 76 FR 66606, Oct. 27, 2011]

EFFECTIVE DATE NOTE: At 76 FR 66606, Oct. 27, 2011, §917.143 was suspended indefinitely, effective Oct. 28, 2011.

§917.149   Special purpose shipments.

Any person may file a request with the Pear Commodity Committee to transport pears to a packing facility located in the State of Oregon without inspection and certification prior to such transporting. The committee may approve such a request subject to the following terms and conditions:

(a) Approval shall be requested by the person prior to transporting the pears out of the area of production.

(b) Such person shall file with the committee, in such manner as required, reports showing, among other things, the date and quantity of pears comprising each shipment of pears transported to Oregon and the disposition thereof.

(c) All such pears shall be of the person's own production and the packing facility to which they are transported must be owned and operated by that person.

(d) All such pears shall be inspected and certified, as required by §917.45, by the Federal or Federal-State Inspection Service prior to the time such pears are shipped from the packing facility. Any pears shipped to any such facility which, upon inspection, do not meet the requirements of the then effective grade, size, or quality regulations, may be shipped, or handled, within the State, for consumption by any charitable institution or for distribution by any relief agency or for conversion into products. Prior to any such shipment or handling, there shall first have been submitted to the committee proof satisfactory to the committee that the pears will not be handled contrary to the requirements of the marketing agreement and order. Such proof shall include a written certificate, executed by both the handler and the intended receiver, stating that the pears will not be used for any purpose not authorized by this section.

[41 FR 31180, July 27, 1976]

EFFECTIVE DATE NOTE: At 59 FR 10056, Mar. 3, 1994, §917.149 was suspended, effective Apr. 4, 1994.

REGULATION OF DAILY SHIPMENTS

REPORTS

§917.176   Pears.

(a) *Report of daily packout.* When requested by the Pear Commodity Committee, each shipper who ships pears shall furnish to the manager of the Control Committee or when designated to the Federal-State Inspection Service a report of the number of packages by container type, by variety and by district of origin, which the shipper packed during the preceding day.

(b) *Recapitulation of shipments.* When requested by the Pear Commodity Committee, each shipper of pears shall furnish to the manager of the Control Committee a recapitulation of his shipments. The recapitulation shall show:

(1) The name of the shipper,

(2) The shipping point,

(3) The district of origin,

(4) The variety, and

(5) The number of packages, by size, for each container type.

(c) *Report of pears held in storage.* Each shipper who has pears under refrigeration in a storage warehouse shall upon request, file with the manager of the Control Committee within the time specified in the request an accurate report containing the following information:

(1) The name and address of the shipper; and

(2) The total quantity, as of the date specified in the request, of pears in

storage outside of the State of California and in storage in the State of California.

[39 FR 27117, July 25, 1974]

EFFECTIVE DATE NOTE: At 59 FR 10056, Mar. 3, 1994, § 917.176 was suspended, effective Apr. 4, 1994.

Subpart C—Grade and Size Requirements

§ 917.461   Pear Regulation 12.

(a) No handler shall ship:

(1) Bartlett or Max-Red (Max-Red Bartlett, Red Bartlett) varieties of pears which do not grade at least U.S. Combination with not less than 80 percent, by count, of the pears grading at least U.S. No. 1: *Provided,* That for the 1992 crop year, no handler shall ship organic pears of these varieties unless they grade at least U.S. Combination with not less than 50 percent, by count, grading at least U.S. No. 1 and the remainder grading at least U.S. No. 2, except that russeting shall not be scored as a defect for such organic pears. Handlers who intend to ship organic pears in accordance with this paragraph shall provide, upon request of the committee, with the approval of the Secretary, information to indicate that the pears were grown in accordance with the provisions of paragraph (b)(5) of this section.

(2) Any box or container, including consumer packages in master containers and consumer packages not in master containers, of Bartlett or Max-Red (Max-Red Bartlett, Red Bartlett) varieties of pears unless such pears are of a size not smaller than the size known commercially as size 165;

(3) Any box or container, other than consumer packages in master containers and consumer packages not in master containers, of Bartlett or Max-Red (Max-Red Bartlett, Red Bartlett) varieties of pears unless such box or container is stamped or otherwise marked, in plain sight and in plain letters, on one outside end with the name of the variety;

(4) Bartlett or Max-Red (Max-Red Bartlett, Red Bartlett) varieties of pears, when packed in closed containers, other than consumer packages in master containers and consumer packages not in master containers, unless such box or container conforms to the requirement of standard pack, except that such pears may be fairly tightly packed;

(5) Bartlett or Max-Red (Max-Red Bartlett, Red Bartlett) varieties of pears, when packed in other than a closed container, unless such pears do not vary more than ⅜ inch in their transverse diameter for counts 120 or less, and ¼ inch for counts 135 to 165, inclusive: *Provided,* That 10 percent of the containers in any lot may fail to meet the requirements of this subparagraph: *Provided further,* That such varieties of pears shipped in bulk bin containers containing 300 pounds or more of pears shall be exempt from the requirements in this subparagraph.

(6) Any volume-filled box or container of Bartlett or Max-Red (Max-Red Bartlett, Red Bartlett) varieties of pears (not packed in rows and not wrap packed), other than consumer packages in master containers and consumer packages not in master containers, unless (i) such boxes or containers are well filled with pears fairly uniform in size; (ii) such pears are packed fairly tight; (iii) there is an approved top pad in each box or container that will cover the fruit with no more than ¼ inch between the pad and any side or end of the box or container; and (iv) the top of the box or container shall be securely fastened to the bottom: Provided, That 10 percent of the boxes or containers in any lot may fail to meet the requirements of this paragraph.

(7) Each master container, when filled with pears packed in consumer packages, shall bear on one outside end in plain sight and plain letters the varietal name and size description of the contents; the number of consumer packages packed in the master container; the net weight of each consumer package; and the name and address, including zip code, of the handler.

(8) Each individual consumer package shall bear the name and address, including the zip code, of the handler and the net weight of the contents. When a consumer package is not shipped in a master container, it must also bear the varietal name, number

and size description of pears contained in the package.

(b) *Definitions.* (1) *Size known commercially as size 165* means a size of pear that will pack a standard pear box, packed in accordance with the specifications of standard pack, with 165 pears and that one-half of the count size designated, representative of the size of the pears in the box or container, shall weigh at least 22 pounds.

(2) *Standard pear box* means the container so designated in § 1380.19 of the regulations of the California Department of Food and Agriculture.

(3) *U.S. No. 1, U.S. No. 2, U.S. Combination,* and *Standard Pack* mean the same as defined in the United States Standards for Summer and Fall Pears (7 CFR 51.1260 to 51.1280).

(4) *Approved top pad* shall mean a pad of wood-type excelsior construction, fairly uniform in thickness, weighing at least 160 pounds per 1,000 square feet (e.g., an 11 inch by 17 inch pad will weigh at least 21 pounds per 100 pads) or an equivalent made of material other than wood excelsior approved by the committee.

(5) *Organic pears* means pears which are produced, harvested, distributed, stored, processed and packaged without application of synthetically compounded fertilizers, pesticides, or growth regulators. In addition, no synthetically compounded fertilizers, pesticides, or growth regulators shall be applied by the grower to the field or area in which the pears are grown for 12 months prior to the appearance of flower buds and throughout the entire growing and harvest season for pears.

(6) *Consumer package* means a package holding 15 pounds or less net weight of pears.

[46 FR 48116, Oct. 1, 1981, as amended at 47 FR 34116, Aug. 6, 1982; 54 FR 32796, Aug. 10, 1989; 55 FR 25958, June 26, 1990; 56 FR 32063, July 15, 1991; 57 FR 31093, July 14, 1992]

EFFECTIVE DATE NOTE: At 59 FR 10056, Mar. 3, 1994, § 917.461 was suspended, effective Apr. 4, 1994.

PART 920—KIWIFRUIT GROWN IN CALIFORNIA

DEFINITIONS

Sec.

920.302   Grade, size, pack, and container regulations.
920.303   Container marking regulations.

AUTHORITY: 7 U.S.C. 601–674.

SOURCE: 49 FR 39658, Oct. 10, 1984, unless otherwise noted.

DEFINITIONS

§ 920.1   Secretary.

Secretary means the Secretary of Agriculture of the United States, or any officer or employee of the Department of whom authority has heretofore been delegated, or to whom authority may hereafter be delegated.

§ 920.2   Act.

Act means Public Act No. 10, 73d Congress (May 12, 1933), as amended and as reenacted and amended by the Agricultural Marketing Agreement Act of 1937, as amended (48 Stat. 31, as amended; 7 U.S.C. 601 *et seq.*).

§ 920.3   Person.

Person means an individual, partnership, corporation, association or any other business unit.

§ 920.4   Production area.

Production area means the State of California.

§ 920.5   Kiwifruit.

Kiwifruit means all varieties of kiwifruit, or kiwi grown in the production area.

[58 FR 65102, Dec. 13, 1993]

§ 920.6   Varieties.

Varieties means and includes all classifications or subdivisions of kiwifruit.

§ 920.7   Fiscal period.

Fiscal period is synonymous with fiscal year and means a 12-month period beginning on August 1 of one year and ending on the last day of July of the following year or such other period as the committee, with the approval of the Secretary, may prescribe.

§ 920.8   Committee.

Committee means the Kiwifruit Administrative Committee established pursuant to § 920.20.

§ 920.9   Grower.

Grower is synonymous with producer and means any person who produces kiwifruit for the fresh market and who has a proprietary interest therein.

§ 920.10   Handler.

Handler is synonymous with shipper and means any person (except a common or contract carrier transporting kiwifruit owned by another person) who handles kiwifruit.

§ 920.11   Handle.

Handle and ship are synonymous and mean to sell, consign, deliver, or transport kiwifruit, or to cause kiwifruit to be sold, consigned, delivered, or transported, between the production area and any point outside thereof, or within the production area: *Provided,* That the term handle shall not include the sale of kiwifruit on the vine, the transportation within the production area of kiwifruit from the vineyard where grown to a packing facility located within such area for preparation for market, or the delivery of such kiwifruit to such packing facility for such preparation.

§ 920.12   District.

District means the applicable one of the following described subdivisions of the production area or such other subdivision as may be prescribed pursuant to § 920.31:

(a) *District 1* shall include Butte, Sutter, and Yuba Counties.

(b) *District 2* shall include Tulare County.

(c) *District 3* shall include all counties within the production area not included in Districts 1 and 2.

[75 FR 37291, June 29, 2010]

§ 920.13   Pack.

Pack means the specific arrangement, size, weight, count, or grade of a quantity of kiwifruit in a particular type and size of container, or any combination thereof.

§ 920.14   Container.

Container means a box, bag, crate, lug, basket, carton, package, or any other type of receptacle used in the packaging or handling of kiwifruit.

ADMINISTRATIVE BODY

§920.20 Establishment and membership.

There is hereby established a Kiwifruit Administrative Committee consisting of 12 members, each of whom shall have an alternate who shall have the same qualifications as the member for whom he or she is an alternate. The 12-member committee shall be made up of the following: One public member (and alternate), and eleven members (and alternates). With the exception of the public member and alternate, all members and their respective alternates shall be growers or employees of growers. In accordance with §920.31(1), district representation on the committee shall be based upon the previous five-year average production in the district and shall be established so as to provide an equitable relationship between membership and districts. The committee may, with the approval of the Secretary, provide such other allocation of membership as may be necessary to assure equitable representation.

[75 FR 37291, June 29, 2010]

§920.21 Term of office.

The term of office of each member and alternate member of the committee shall be for two years from the date of their selection and until their successors are selected. The terms of office shall begin on August 1 and end on the last day of July, or such other dates as the committee may recommend and the Secretary approve. Members may serve up to three consecutive 2-year terms not to exceed 6 consecutive years as members. Alternate members may serve up to three consecutive 2-year terms not to exceed 6 consecutive years as alternate members.

[76 FR 4202, Jan. 25, 2011]

§920.22 Nomination.

(a) Except as provided in paragraph (b) of this section, the committee shall hold, or cause to be held, not later than June 1 of each year in which nominations are made, or such other date as may be specified by the Secretary, a meeting or meetings of growers in each district for the purpose of designating nominees to serve as grower members and alternates on the committee. Any such meetings shall be supervised by the committee, which shall prescribe such procedures as shall be reasonable and fair to all persons concerned.

(b) Nominations in any or all districts may be conducted by mail in a manner recommended by the committee and approved by the Secretary.

(c) Only growers may participate in the nomination of grower members and their alternates. Each grower shall be entitled to cast only one vote for each position to be filled in the district in which such grower produces kiwifruit. No grower shall participate in the election of nominees in more than one district in any one fiscal year.

(d) A particular grower shall be eligible for membership as member or alternate member to fill only one position on the committee.

(e) The public member and alternate shall be nominated by the grower members of the committee.

[57 FR 1219, Jan. 12, 1992, as amended at 75 FR 37291, June 29, 2010]

§920.23 Selection.

From the nominations made pursuant to §920.22, or from other qualified persons, the Secretary shall select the 12 members of the committee and an alternate for each such member, with the exception of the public member and alternate member, who shall be selected by the Secretary in his discretion.

§920.24 Failure to nominate.

If nominations are not made within the time and in the manner prescribed in §920.22, the Secretary may, without regard to nominations, select the members and alternate members of the committee on the basis of the representation provided for in §920.20.

§920.25 Acceptance.

Each person to be selected by the Secretary as a member or as an alternate member of the committee shall, prior to such selection, qualify by advising the Secretary that he/she agrees to serve in the position for which nominated for selection.

§ 920.26 Vacancies.

To fill any vacancy occasioned by the failure of any person selected as a member or as an alternate member of the committee to qualify, or in the event of the death, removal, resignation, or disqualification of any member or alternate member of the committee, a successor for the unexpired term of such member or alternate member of the committee shall be nominated and selected, or, in the case of the public member and alternate, selected by the Secretary in his discretion, in the manner specified in §§ 920.22 and 920.23. If the names of nominees to fill any such vacancy are not made available to the Secretary within a reasonable time after such vacancy occurs, the Secretary may fill such vacancy without regard to nominations, which selection shall be made on the basis of representation provided for in § 920.20.

§ 920.27 Alternate members.

An alternate member of the committee, during the absence of the member for whom that individual is an alternate, shall act in the place and stead of such member and perform such other duties as assigned. In the event both a member and his or her alternate are unable to attend a committee meeting, the committee may designate any other alternate member from the same district to serve in such member's place and stead. In the event of the death, removal, resignation, or disqualification of a member, the alternate of such member shall act for him or her until a successor for such member is selected and has qualified.

[79 FR 30445, May 28, 2014]

§ 920.30 Powers.

The committee shall have the following powers:

(a) To administer the provisions of this part in accordance with its terms;

(b) To receive, investigate, and report to the Secretary complaints of violations of the provisions of this part;

(c) To make and adopt rules and regulations to effectuate the terms and provisions of this part; and

(d) To recommend to the Secretary amendments to this part.

§ 920.31 Duties.

The committee shall have, among others, the following duties:

(a) To select a chairperson and such other officers as may be necessary, and to define the duties of such officers;

(b) To appoint such employees, agents and representatives as it may deem necessary, and to determine compensation and to define the duties of each;

(c) To submit to the Secretary as soon as practicable after the beginning of each fiscal period a budget for such fiscal period, including a report in explanation of the items appearing therein and a recommendation as to the rate of assessment for such period;

(d) To keep minutes, books and records which will reflect all of the acts and transactions of the committee and which shall be subject to examination by the Secretary;

(e) To prepare periodic statements of the financial operations of the committee and to make copies of each such statement available to growers and handlers for examination at the office of the committee;

(f) To cause its books to be audited by a public accountant at least once each fiscal year and at such times as the Secretary may request;

(g) To act as intermediary between the Secretary and any grower or handler;

(h) To investigate and assemble data on the growing, handling and marketing conditions with respect to kiwifruit;

(i) To submit to the Secretary the same notice of meetings of the committee as is given to its members;

(j) To submit to the Secretary such available information as may be requested;

(k) To investigate compliance with the provisions of this part;

(l) With the approval of the Secretary, to redefine the districts into which the production area is divided and to reapportion the representation of any district on the committee: *Provided*, That any such changes shall reflect, insofar as practicable, shifts in kiwifruit production within the districts and the production area.

§920.32 Procedure.

(a) Eight members of the committee, or alternates acting for members, shall constitute a quorum and any action of the committee shall require the concurring vote of the majority of those present: *Provided,* That actions of the committee with respect to expenses and assessments, production and postharvest research, market research and development, or recommendations for regulations pursuant to §§920.50 through 920.55, of this part shall require at least eight concurring votes.

(b) Committee meetings may be assembled or held by telephone, video conference, or other means of communication. The committee may vote by telephone, facsimile, or other means of communication. Votes by members or alternates present at assembled meetings shall be cast in person. Votes by members or alternates participating by telephone or other means of communication shall be by roll call; *Provided,* That a video conference shall be considered an assembled meeting, and votes by those participating through video conference shall be considered as cast in person.

[49 FR 39658, Oct. 10, 1984, as amended at 75 FR 37291, June 29, 2010; 79 FR 30445, May 28, 2014]

§920.33 Expenses and compensation.

(a) Except for the public member and alternate, the members of the committee, and alternates when acting as members, shall serve without compensation but shall be reimbursed for expenses necessarily incurred by them in the performance of their duties under this part: *Provided,* That the committee at its discretion may request the attendance of one or more alternates, including the public alternate, at any or all meetings notwithstanding the expected or actual presence of the respective members and may pay expenses as aforesaid.

(b) The public member and alternate shall be reimbursed for expenses necessarily incurred by them in the performance of their duties under this part, and shall receive per diem compensation established by the committee.

§920.34 Annual report.

The committee shall, as soon as is practicable after the close of each marketing season, prepare and mail an annual report to the Secretary and make a copy available to each grower and handler who requests a copy of the report.

EXPENSES AND ASSESSMENTS

§920.40 Expenses.

The committee is authorized to incur such expenses as the Secretary finds are reasonable and likely to be incurred by the committee for its maintenance and functioning and to enable it to exercise its powers and perform its duties in accordance with the provisions of this part. The funds to cover such expenses shall be acquired in the manner prescribed in §920.41.

§920.41 Assessments.

(a) As his or her pro rata share of the expenses which the Secretary finds are reasonable and likely to be incurred by the committee during a fiscal period, each person who first handles kiwifruit during such period shall pay to the committee, upon demand, assessments on all kiwifruit so handled. The payment of assessments for the maintenance and functioning of the committee may be required under this part throughout the period it is in effect, irrespective of whether particular provisions thereof are suspended or become inoperative. If a handler does not pay any assessment within the time prescribed by the committee, the assessment may be subject to an interest or late payment charge, or both, as may be established by the Secretary upon recommendation of the committee.

(b) The Secretary shall fix the rate of assessment to be paid by each such person during a fiscal period in an amount designed to secure sufficient funds to cover the expenses which may be incurred during such period and to accumulate and maintain a reserve fund equal to approximately one fiscal period's expenses. At any time during or after the fiscal period, the Secretary may increase the rate of assessment in order to secure sufficient funds to cover any later finding by the Secretary relative to the expenses which may be incurred: *Provided,* That any assessment, excluding any amount collected pursuant to § 920.55(c), must be limited to a maximum assessment rate of three and one-half cents per flat, or the equivalent thereof. The Secretary may increase this maximum rate in each succeeding year after the initial year of order operation by the Consumer Price Index (cost of living) for California as published by the Bureau of Labor Statistics. Such increase shall be applied to all kiwifruit handled during the applicable fiscal period. In order to provide funds for the administration of the provisions of this part during the first part of a fiscal period before sufficient operating income is available from assessments on the current year's shipments, the committee may accept the payment of assessments in advance, and may also borrow money for such purposes.

[49 FR 39658, Oct. 10, 1984, as amended at 57 FR 1220, Jan. 12, 1992]

§ 920.42 Accounting.

(a) If, at the end of a fiscal period, the assessments collected are in excess of expenses incurred, such excess shall be accounted for in accordance with one of the following:

(1) If such excess is not retained in a reserve, as provided in paragraph (a)(2) of this section, it shall be refunded proportionately to the persons from whom it was collected: *Provided,* That any sum paid by a person in excess of his or her pro rata share of the expenses during any fiscal period may be applied by the committee at the end of such fiscal period to any outstanding obligations due the committee from such person.

(2) The committee, with the approval of the Secretary, may carry over such excess into subsequent fiscal periods as a reserve: *Provided,* That funds already in the reserve do not equal approximately one fiscal period's expenses. Such reserve funds may be used: (i) To defray expenses, during any fiscal period, prior to the time assessment income is insufficient to cover such expenses; (ii) to cover deficits incurred during any fiscal year when assessment income is less than expenses; (iii) to defray expenses incurred during any period when any or all provisions of this part are suspended or are inoper-

ative; and, (iv) to cover necessary expenses of liquidation in the event of termination of this part. Upon such termination, any funds not required to defray the necessary expenses of liquidation shall be disposed of in such manner as the Secretary may determine to be appropriate: *Provided,* That to the extent practical, such funds shall be returned pro rata to the persons from whom such funds were collected.

(b) All funds received by the committee pursuant to the provisions of this part shall be used solely for the purpose specified in this part and shall be accounted for in the manner provided in this part. The Secretary may at any time require the committee and its members to account for all receipts and disbursements.

(c) Upon the removal or expiration of the term of office of any member of the committee, such member shall account for all receipts and disbursements and deliver all property and funds in his or her possession to the committee, and shall execute such assignments and other instruments as may be necessary or appropriate to vest in the committee full title to all of the property, funds, and claims vested in such member pursuant to this part.

§ 920.45 Contributions.

The committee may accept voluntary contributions, but these shall only be used to pay expenses incurred pursuant to § 920.47 and § 920.48. Furthermore, such contributions shall be free from any encumbrances by the donor, and the committee shall retain complete control of their use.

[79 FR 30445, May 28, 2014]

§ 920.47 Production and postharvest research.

The committee, with the approval of the Secretary, may establish or provide for the establishment of projects involving research designed to assist or improve the efficient production and postharvest handling of kiwifruit.

[79 FR 30445, May 28, 2014]

§ 920.48 Market research and development.

The committee, with the approval of the Secretary, may establish or provide for the establishment of marketing research and development projects designed to assist, improve, or promote the marketing, distribution, and consumption of kiwifruit.

[79 FR 30445, May 28, 2014]

REGULATIONS

§ 920.50 Marketing policy.

(a) Each season prior to making any recommendations pursuant to § 920.51, the committee shall submit to the Secretary a report setting forth its marketing policy for the ensuing marketing season. Such marketing policy report shall contain information relative to:

(1) The estimated total production of kiwifruit within the production area;

(2) The expected general quality and size of kiwifruit in the production area and in other areas;

(3) The expected demand conditions for kiwifruit in different market outlets;

(4) The expected shipments of kiwifruit produced in the production area and in areas outside the production area;

(5) Supplies of competing commodities;

(6) Trend and level of consumer income;

(7) Other factors having a bearing on the marketing of kiwifruit; and

(8) The type of regulations expected to be recommended during the marketing season.

(b) [Reserved]

§ 920.51 Recommendations for regulation.

(a) Whenever the committee deems it advisable to regulate the handling of any variety or varieties of kiwifruit in the manner provided in § 920.52, it shall so recommend to the Secretary.

(b) In arriving at its recommendations for regulation pursuant to paragraph (a) of this section, the committee shall give consideration to current information with respect to the factors affecting the supply and demand for kiwifruit during the period or periods when it is proposed that such regulations should be made effective. With each such recommendation for regulation, the committee shall submit to the Secretary the data and information on which such recommendation is predicated and such other available information as the Secretary may request.

§ 920.52 Issuance of regulations.

(a) The Secretary shall regulate, in the manner specified in this section, the handling of kiwifruit whenever the Secretary finds, from the recommendations and information submitted by the committee, or from other available information, that such regulations will tend to effectuate the declared policy of the act. Such regulations may:

(1) Limit, during any period or periods, the shipment of any particular grade, size, quality, maturity, or pack, or any combination thereof, of any variety or varieties of kiwifruit grown in the production area;

(2) Limit the shipment of kiwifruit by establishing, in terms of grades, sizes, or both, minimum standards of quality and maturity during any period when season average prices are expected to exceed the parity level;

(3) Fix the size, capacity, weight, dimensions, markings, or pack of the container, or containers, which may be used in the packaging or handling of kiwifruit.

(b) The committee shall be informed immediately of any such regulation issued by the Secretary and the committee shall promptly give notice thereof to handlers.

§ 920.53 Modification, suspension, or termination of regulations.

(a) In the event the committee at any time finds that, by reason of changed conditions, any regulations issued pursuant to § 920.52 should be modified, suspended, or terminated, it shall so recommend to the Secretary.

(b) Whenever the Secretary finds from the recommendations and information submitted by the committee or from other available information, that a regulation should be modified, suspended, or terminated with respect to any or all shipments of kiwifruit in order to effectuate the declared policy of the act, the Secretary shall modify, suspend, or terminate such regulation. If the Secretary finds that a regulation obstructs or does not tend to effectuate the declared policy of the act, the Secretary shall suspend or terminate such regulation. On the same basis and in like manner the Secretary may terminate any such modification or suspension.

§ 920.54 Special purpose shipments.

(a) Except as otherwise provided in this section, any person may, without regard to the provisions of §§ 920.41, 920.52, 920.53 and 920.55 and the regulations issued thereunder, handle kiwifruit:

(1) For consumption by charitable institutions;

(2) For distribution by relief agencies; or

(3) For commercial processing into products.

(b) Upon the basis of recommendations and information submitted by the committee, or from other available information, the Secretary may relieve from any or all requirements, under or established pursuant to § 920.41, § 920.52, § 920.53 or § 920.55, the handling of kiwifruit:

(1) To designated market areas;

(2) For such specified purposes (including shipments to facilitate the conduct of marketing research and development projects); or,

(3) In such minimum quantities or types of shipments, as may be prescribed.

(c) The committee shall, with the approval of the Secretary, prescribe such rules, regulations, and safeguards as it may deem necessary to prevent kiwifruit handled under the provisions of this section from entering the channels of trade for other than the specific purposes authorized by this section. Such rules, regulations, and safeguards may include the requirements that handlers shall file applications and receive approval from the committee for authorization to handle kiwifruit pursuant to this section, and that such applications be accompanied by a certification by the intended purchaser or receiver that the kiwifruit will not be used for any purpose not authorized by this section.

§ 920.55 Inspection and certification.

(a) Whenever the handling of any variety of kiwifruit is regulated pursuant to § 920.52, or § 920.53, each handler who handles kiwifruit shall, prior thereto, cause such kiwifruit to be inspected by the Federal or Federal-State Inspection Service and certified as meeting the applicable requirements of such regulation: *Provided,* That inspection and certification shall not be required for kiwifruit which previously have been so inspected and certified if such prior inspection was performed within such period as may be established pursuant to paragraph (b) of this section. Promptly after inspection and certification, each such handler shall submit, or cause to be submitted, to the committee a copy of the certificate of inspection issued with respect to such kiwifruit. The committee may, with the approval of the Secretary, prescribe rules and regulations waiving the inspection requirements of this section where it is determined that inspection is not available: *Provided,* That all shipments made under such waiver shall comply with all regulations in effect.

(b) The committee may, with the approval of the Secretary, establish a period prior to shipment during which the inspection required by this section must be performed.

(c) The committee may enter into an agreement with the Federal and Federal-State Inspection Services with respect to the costs of the inspection required by paragraph (a) of this section, and may collect from handlers their respective pro rata shares of such costs.

REPORTS

§ 920.60 Reports.

(a) Each handler shall furnish to the committee, at such times and for such periods as the committee may designate, certified reports covering, to the extent necessary for the committee to perform its functions, each shipment of kiwifruit as follows:

(1) The name of the shipper and the shipping point;

(2) The car or truck license number (or name of the trucker), and identification of the carrier;

(3) The date and time of departure;

(4) The number and type of containers in the shipment;

(5) The quantities shipped, showing separately the variety, size and grade of the fruit;

(6) The destination;

(7) Identification of the inspection certificate or waiver pursuant to which the fruit was handled.

(b) Upon request of the committee, made with the approval of the Secretary, each handler shall furnish to the committee, in such manner and at such times as it may prescribe, such other information as may be necessary to enable the committee to perform its duties under this part.

(c) Each handler shall maintain for at least two succeeding fiscal years, such records of the kiwifruit received and disposed of by such handler as may be necessary to verify the reports submitted to the committee pursuant to this section.

(d) All reports and records submitted by handlers pursuant to the provisions of this section shall be received by, and at all times be in custody of, one or more designated employees of the committee. No such employee shall disclose to any person, other than the Secretary upon request therefor, data or information obtained or extracted from such reports and records which might affect the trade position, financial condition, or business operation of the particular handler from whom received: *Provided,* That such data and information may be combined, and made available to any person, in the form of general reports in which the identities of the individual handler furnishing the information, is not disclosed but may be revealed to any extent necessary to effect compliance with the provisions of this part and the regulations issued thereunder.

MISCELLANEOUS PROVISIONS

§ 920.61   Compliance.

(a) Except as provided in this part, no person shall handle kiwifruit, the shipment of which has been prohibited by the Secretary in accordance with the provisions of this part; and no person shall handle kiwifruit except in conformity with the provisions of this part and the regulations issued under this part.

(b) For the purpose of checking and verifying reports filed by handlers, the committee, through its duly authorized representatives shall have access to any handler's premises during regular business hours, and shall be permitted at any such times to inspect such premises and any kiwifruit held by such handler, and any and all records of the handler with respect to his or her acquisition, sales, uses and shipments of kiwifruit. Each handler shall furnish all labor and equipment necessary to make such inspections.

§ 920.62   Right of the Secretary.

The members of the committee (including successors and alternates), and any agents, employees, or representatives thereof, shall be subject to removal or suspension by the Secretary at any time. Each and every regulation, decision, determination, or other act of the committee shall be subject to the continuing right of the Secretary to disapprove of the same at any time. Upon such disapproval, the disapproved action of the committee shall be deemed null and void, except as to acts done in reliance thereon or in accordance therewith prior to such disapproval by the Secretary.

§ 920.63   Termination.

(a) The Secretary may at any time terminate the provisions of this part by giving at least one day's notice by means of a press release or in any other manner in which the Secretary may determine.

(b) The Secretary shall terminate or suspend the operation of any and all of the provisions of this part whenever the Secretary finds that such provisions do not tend to effectuate the declared policy of the act.

(c) The Secretary shall terminate the provisions of this part whenever the Secretary finds by referendum or otherwise that such termination is favored by a majority of the growers: *Provided,* That such majority has, during the current marketing season, produced more than 50 percent of the volume of the kiwifruit which were produced within the production area for shipment in fresh form. Such termination shall become effective on the first day of August subsequent to the announcement thereof by the Secretary.

(d) The committee shall consider all petitions from growers submitted to it for termination of this part provided such petitions are received by the committee prior to February 1 of the then current fiscal period. Upon recommendation of the committee received not later than April 1 of the then current fiscal period, the Secretary shall conduct a referendum among the growers prior to July 15 of such fiscal period to ascertain whether continuance of this part is favored by producers.

(e) The Secretary shall conduct a referendum within the period beginning May 15, 1990, and ending July 15, 1990, to ascertain whether continuance of this part is favored by the growers as set forth in paragraph (c) of this section. The Secretary shall conduct such a referendum within the same period of every sixth fiscal period thereafter.

(f) The provisions of this part shall, in any event, terminate whenever the provisions of the act authorizing them cease to be in effect.

§ 920.64 Proceeding after termination.

(a) Upon the termination of the provisions of this part, the committee shall, for the purpose of liquidating the affairs of the committee, continue as trustee of all the funds and property then in its possession, or under its control, including claims for any funds unpaid or property not delivered at the time of such termination.

(b) The said trustees shall:

(1) Continue in such capacity until discharged by the Secretary;

(2) From time to time account for all receipts and disbursements and deliver all property on hand, together with all books and records of the committee and of the trustees, to such persons as the Secretary may direct; and

(3) Upon the request of the Secretary, execute such assignments or other instruments necessary or appropriate to vest in such person, full title and right to all of the funds, property, and claims vested in the committee of the trustees pursuant thereto.

(c) Any person to whom funds, property, or claims have been transferred or delivered, pursuant to this section, shall be subject to the same obligation imposed upon the committee and upon the trustees.

§ 920.65 Effect of termination or amendment.

Unless otherwise expressly provided by the Secretary, the termination of this part or of any regulation issued pursuant to this part, or the issuance of any amendment to either thereof, shall not (a) affect or waive any right, duty, obligation, or liability which shall have arisen or which may thereafter arise in connection with any provision of this part or any regulation issued under this part, or (b) release or extinguish any violation of this part or of any regulation issued under this part, or (c) affect or impair any rights or remedies of the Secretary or of any other person with respect to any such violation.

§ 920.66 Duration of immunities.

The benefits, priviliges, and immunities conferred upon any person by virtue of this part shall cease upon its termination, except with respect to acts done under and during the existence of this part.

§ 920.67 Agents.

The Secretary may, by designation in writing, name any officer or employee of the United States, or name any agency or division in the United States Department of Agriculture, to act as the Secretary's agent or representative in connection with any of the provisions of this part.

§ 920.68 Derogation.

Nothing contained in this part is, or shall be construed to be, in derogation or in modification of the rights of the Secretary or of the United States (a) to exercise any powers granted by the act or otherwise, or (b) in accordance with such powers, to act in the premises whenever such action is deemed advisable.

§ 920.69 Personal liability.

No member or alternate member of the committee and no employee or agent of the committee shall be held personally responsible, either individually or jointly with others, in any way whatsoever, to any person for errors in judgment, mistakes, or other acts, either of commission or omission, as such member, alternate, employee or agent, except for acts of dishonesty, willful misconduct, or gross negligence.

§920.70 Separability.

If any provision of this part is declared invalid or the applicability thereof to any person, circumstance, or thing is held invalid, the validity of the remainder of this part or the applicability thereof to any other person, circumstance, or thing shall not be affected thereby.

§920.110 Exemptions.

(a) *Waivers.* A handler may handle kiwifruit without inspection and certification, as prescribed under §920.55, if all shipments made under such waivers comply with all regulations in effect, and all the following conditions are met:

(1) The handler requests the Federal-State Inspection Service to provide inspection during its regular working hours at least 4 hours in advance of the time when inspection is needed. The request need not be in writing but it shall be confirmed immediately in writing by the inspection service.

(2) The Federal-State Inspection Service advises the handler that it is not practicable to provide inspection at the time and place designated by the handler. This advice may be verbal but it shall be confirmed in writing by the Federal-State Inspection Service. A confirmed copy thereof shall be forwarded by the inspection service to the office of the Kiwifruit Administrative Committee.

(3) The Federal-State Inspection Service furnishes the handler with the waiver number which shall cover the kiwifruit on which inspection is requested.

(4) When instructed to do so, the handler plainly and conspicuously marks the end of each container with the letter "W" and the waiver number assigned by the Federal-State Inspection Service. The letter "W" and the number shall not be less than one-half inch in height.

(b) *Minimum quantities.* Notwithstanding any other provision of this section, kiwifruit may be handled without regard to the provision of §§920.41, 920.52, 920.55 and 920.60 under the following conditions:

(1) Such kiwifruit are for home use and not for resale.

(2) The total weight of such kiwifruit sold to all persons collectively in any one vehicle during any one day does not exceed 200 pounds.

(3) Such kiwifruit are handled by the person who produced them and, the handling takes place: (i) On the premises where grown, (ii) at a packing house, or retail stand (roadside stand, flea market or any other outlet approved by the committee) which is operated by said handler, or (iii) at a Certified Farmers Market.

[50 FR 4856, Feb. 4, 1985, as amended at 53 FR 34035, Sept. 2, 1988]

§920.112 Late payments.

Pursuant to §920.41(a), interest will be charged at a 1.5 percent monthly simple interest rate. Assessments for kiwifruit shall be deemed late if not received within 30 days of invoice, or such other later time period as specified by the committee. A 10 percent late charge will be assessed when payment becomes 30 days late. Interest and late payment charges shall be applied only to the overdue assessment.

[62 FR 45295, Aug. 27, 1997]

§920.122 Nomination procedures.

(a) The manner of nominating grower members and alternate members to the committee shall be as follows:

(1) The committee's mailing of an approved nomination form to all kiwifruit growers of record shall constitute notice of nominations. All eligible kiwifruit growers may nominate themselves or any other eligible kiwifruit grower to vacant committee positions for the nominee's district. Completed nomination forms shall be returned to the office of the committee by a date specified by the committee and approved by an agent of the Secretary. Nomination forms shall provide for names of nominees, as well as the

nominating grower's name, address, telephone number, and signature. Incomplete nominations forms will not be considered valid.

(2) For each district involved in the current year's nominations, committee staff, with the Secretary's oversight, shall establish a slate of candidates with the names of all qualified nominees received. Persons submitting invalid nomination forms shall be notified of such by the committee. Within a reasonable time period, a ballot containing a slate of candidates shall be mailed to all growers of record within the respective district represented by such candidates. The committee shall provide a reasonable period of time to growers to cast votes on the candidates and return the completed ballots for tallying.

(3) To be eligible to vote, growers must be producing kiwifruit during the crop year nominations are held and within the district represented by the candidates on the ballot. A grower may only vote for candidates from one district and may only cast one ballot. Growers may also cast votes for eligible candidates who do not appear on the ballots by writing in the name of such candidates on the ballot. Each ballot shall provide for a voter eligibility certification which must include the voter's name, address, telephone number, and signature, as well as the name(s) of all handlers which handled the current season's crop. At the discretion of the Secretary, the ballots may include other background information about each candidate.

(4) In order to be valid, ballots must be executed in accordance with the instructions set forth on the ballot, and are to be returned to the Secretary's agent who will tally the ballots with such assistance from the committee as may be requested by the agent. Such ballots shall be postmarked by a date specified by the committee and approved by an agent of the Secretary.

(5) The names of the persons receiving the highest total number of votes for a particular position shall be submitted to the Secretary as the nominees for such positions. In the event of a tie vote, a ballot containing only the names of the candidates receiving the tied vote shall be mailed to all growers in the affected district.

(b) In the event of a vacancy as specified in § 920.26, the committee shall utilize the same procedure as prescribed in § 920.122(a)(1) through (a)(5) to fill such vacancy.

[57 FR 62160, Dec. 30, 1992]

§ 920.160 Reports.

(a) When requested by the Kiwifruit Administrative Committee, each shipper who ships kiwifruit, except as provided in paragraph (e) of this section, shall furnish a report of shipment and inventory data to the committee no later than the fifth day of the month following such shipment, or such other later time established by the committee: *Provided,* That each shipper who ships less than 10,000 trays, or the equivalent thereof, per fiscal year and has qualified with the committee shall furnish such report of shipment and inventory data to the committee twice per fiscal year. The first report shall be due no later than January 5 and the final report no later than the fifth day of the following month after such shipment is completed for the season, or such other later times established by the committee. Such report shall show:

(1) The reporting period;

(2) The name and other identification of the shipper;

(3) The number of containers by type and weight by shipment destination category;

(b) *Kiwifruit Inventory Shipping System (KISS) form.* Each handler, except such handlers that ship less than 10,000 trays, or the equivalent thereof, per season and have qualified with the committee, shall file with the committee the initial Kiwifruit Inventory Shipment System (KISS) form, which consists of three sections "KISS/Add Inventory," "KISS/Deduct Inventory," and "KISS/Shipment," on or before November 5th, or such other later time as the committee may establish. Subsequent KISS forms, including all three sections, shall be filed with the committee by the fifth day and again by the twentieth day of each calendar month, or such other later time as the committee may establish, and will contain the following information:

(1) The beginning inventory of the handler by size and container type;

(2) The quantity of fruit the handler lost in repack and repacked into other container types;

(3) The total domestic and export shipments of the handler by size and container type; and

(4) Any other adjustments which increase or decrease posted handler inventory.

(c) *Handler report of returned fruit.* After fruit is returned to a grower, each handler shall file with the committee, no later than five days from the date the fruit is returned, or such other time as the committee may establish, a Return Receipt of Kiwifruit to Grower Form.

(d) *KISS Price/Shipment report.* Each handler who ships 100,000 or more trays, or the equivalent thereof, per season, shall file the KISS Price/Shipment report with the committee. Handlers are not required to report organic kiwifruit shipments on this report. The handler shall file the report weekly following the first week he or she makes shipments and shall continue filing reports until he or she submits a final report for the season. Each such report shall be filed with the committee no later than 5:00 p.m. (the close of business) on the Tuesday immediately following the shipping week. For the purpose of this subsection, the shipping week is defined as Sunday through Saturday. The report shall show:

(1) The company name, contact person, and phone number of the handler;

(2) Weekly period covered by the report;

(3) Total fresh market shipments and gross f.o.b. sales of kiwifruit by pack style and size; and

(4) Total fresh market shipments and gross f.o.b. sales to export markets by pack style and size.

(e) Handlers who file the KISS Price/Shipment report specified in paragraph (d) of this section are exempt from filing the shipping report specified in paragraph (a) of this section and the KISS/Shipment report specified in paragraph (b) of this section.

(f) Each handler shall file annually with the Committee an End-of-Season F.O.B. Sales Report, due within 30 days after such handler has completed current season shipments, reporting gross f.o.b. sales value and number of containers by pack style and size for fresh market shipments for the season. The report shall also show the company name, contact person, and phone number of the handler.

(g) Each handler shall file annually with the Committee a Final Packout Report, due within 30 days after such handler has completed current season shipments, reporting total containers shipped, by pack style for fresh market shipments, for each grower entity during the season. The report shall also include the grower entity and farm name, mailing address, the county in which the farm is located, and total acreage for each reported grower entity. Also, the report shall show the company name, contact person, and phone number of the handler.

[50 FR 4856, Feb. 4, 1985, as amended at 52 FR 37130, Oct. 5, 1987; 59 FR 53565, Oct. 25, 1994; 61 FR 51576, Oct. 3, 1996; 67 FR 54332, Aug. 22, 2002; 74 FR 46309, Sept. 9, 2009; 76 FR 80211, Dec. 23, 2011]

§ 920.213 Assessment rate.

On and after August 1, 2018, an assessment rate of $0.025 per 9-kilo volume-fill container or equivalent of kiwifruit is established for kiwifruit grown in California.

[83 FR 66079, Dec. 26, 2018]

§ 920.302 Grade, size, pack, and container regulations.

(a) No handler shall ship any kiwifruit unless such kiwifruit meet the following requirements:

(1) *Grade requirements.* Fresh shipments of kiwifruit shall be at least KAC No. 1 quality.

(2) *Size Requirements.* Such kiwifruit shall be at least a minimum Size 45. Size 45 is defined as a maximum of 55 pieces of fruit in an 8-pound sample.

(3) *Maturity requirements.* Such kiwifruit shall have a minimum of 6.2 percent soluble solids at the time of inspection.

(4) *Pack Requirements.* (i) Kiwifruit packed in containers with cell compartments, cardboard fillers, or molded trays shall be of proper size for the cells, fillers, or molds in which they are packed. Such fruit shall be fairly uniform in size.

(ii) (A) Kiwifruit packed in any container shall be subject to the size designation, maximum number of fruit per 8-pound sample, and the size variation tolerance specified as follows:

SIZE DESIGNATION AND SIZE VARIATION CHART

Column 1 size designation	Column 2 maximum number of fruit per 8-pound sample	Column 3 size variation tolerance (diameter)
18 or larger	25	½-inch (12.7 mm).
20	27	½-inch (12.7 mm).
23	30	½-inch (12.7 mm).
25	32	½-inch (12.7 mm).
27/28	35	½-inch (12.7 mm).
30	39	½-inch (12.7 mm).
33	43	⅜-inch (9.5 mm).
36	46	⅜-inch (9.5 mm).
39	49	⅜-inch (9.5 mm).
42	53	⅜-inch (9.5 mm).
45 or smaller	55	¼-inch (6.4 mm).

(B) The average weight of all sample units in a lot must weigh at least 8 pounds, but no sample unit may be more than 4 ounces less than 8 pounds.

(C) Not more than 10 percent, by count, of the containers in any lot and not more than 5 percent, by count, of kiwifruit in any container, (except that for Sizes 42 and 45 kiwifruit, the tolerance, by count, in any one container, may not be more than 25 percent) may fail to meet the size variation requirements of this paragraph.

(iii) All volume fill containers of kiwifruit designated by weight shall hold 19.8-pounds (9-kilograms) net weight of kiwifruit unless such containers hold less than 15 pounds or more than 35 pounds net weight of kiwifruit.

(b) *Definitions.* The term *KAC No. 1 quality* means kiwifruit that meets the requirements of the U.S. No. 1 grade as defined in the United States Standards for Grades of Kiwifruit (7 CFR 51.2335 through 51.2340) except that the kiwifruit shall be "not badly misshapen," and an additional tolerance of 16 percent is provided for kiwifruit that is "badly misshapen," and except that all varieties of kiwifruit are exempt from the "tightly packed" standard as defined in § 51.2338(a) of the U.S. Standards for Grades of Kiwifruit. The terms *fairly uniform in size and diameter* mean the same as defined in the U.S. Standards for Grades of Kiwifruit.

(c) *Exemptions.* Any person may handle kiwifruit without regard to the provisions of this section provided that such kiwifruit is handled for (1) consumption by charitable institutions; (2) distribution by relief agencies; or (3) commercial processing into products. For the purposes of this section, *commercial processing into products* means that the kiwifruit is physically altered in form or chemical composition through freezing, canning, dehydrating, pulping, juicing, or heating of the product. The act of slicing, dicing, or peeling shall not be considered commercial processing into products.

[50 FR 36568, Sept. 9, 1985]

EDITORIAL NOTE: For FEDERAL REGISTER citations affecting § 920.302, see the List of CFR Sections Affected, which appears in the Finding Aids section of the printed volume and at *www.govinfo.gov.*

§ 920.303 **Container marking regulations.**

No handler shall ship any kiwifruit except in accordance with the following terms and conditions:

(a) Each package or container of kiwifruit shall bear on at least one outside principal display panel in plain sight and in plain letters, the word *kiwifruit*, the name of the variety (if other than the Hayward variety), if known or, when the variety is not known, the words *unknown variety.*

(b) Each package or container of kiwifruit shall bear on one outside principal display panel in plain sight and in plain letters the name and address (including the city, state, and zip code) of the shipper.

(c) Each package or container of kiwifruit shall bear on one outside principal display panel in plain sight and in plain letters the following information regarding the quantity of kiwifruit packed within the container:

(1) The quantity shall be indicated in terms of count and size for kiwifruit packed in cell compartments, cardboard fillers, or molded trays, and the contents shall conform to the count.

(2) The quantity shall be indicated in terms of the size designation and either the net weight for volume-fill containers packed by weight or the count for volume-fill containers packed by count.

(3) For bulk containers or individual consumer packages not within a master container, the quantity shall be indicated in terms of the size designation and net weight, or in terms of the size designation and count.

(4) Master containers, which hold more than one individual package, must be properly marked with the quantity of the contents. The size designation must also be indicated.

(5) The quantity shall be indicated in terms of either net weight or count (or both) for individual consumer packages within a master container. If count is used, it must be accompanied by the size designation.

(6) Designations of size, count, and net weight on each container shall be accompanied by the words *size, count,* or *net weight* as applicable.

(d) Except as provided in paragraph (f) of this section, containers of kiwifruit must be positive lot identified prior to shipment in accordance with the following requirements. All exposed or outside containers of kiwifruit, but not less than 75 percent of the total containers on the pallet, shall be positive lot identified with a plain mark corresponding to the lot inspection conducted by an authorized inspector, except for individual consumer packages within a master container and containers that are being directly loaded into a vehicle for export shipment under the supervision of the Federal or Federal-State Inspection

Service. Individual consumer packages of kiwifruit placed directly on a pallet shall have all outside or exposed packages on a pallet positive lot identified with a plain mark corresponding to the lot inspection conducted by an authorized inspector or have one inspection label placed on each side of the pallet. Reusable plastic containers of kiwifruit, placed on a pallet, shall be positive lot identified in accordance with Federal or Federal-State Inspection Service procedures and shall have required information on the cards of the individual containers, as provided in this section of the regulations.

(e) As used in this section, the term *principal display panel* means that part of the package or container most likely to be displayed, presented, shown or examined under normal or customary conditions of display and purchase.

(f) Kiwifruit that has been inspected and certified, and is subsequently placed into new containers, does not have to be positive lot identified, as prescribed in paragraph (d) of this section: *Provided,* That:

(1) Such kiwifruit is of the same grade and size as originally inspected; and

(2) The handler requests a verification number from the Federal or Federal-State Inspection Service prior to shipment; plainly marks one end of each container with such number and the letter "R," both of which shall be at least one-half inch in height; and submits a Kiwifruit Verification Form to the Federal or Federal-State Inspection Service within 3 business days of such request. The handler shall provide the following information on the Kiwifruit Verification Form.

(i) From the original inspection:

(A) The positive lot identification numbers;

(B) The identity of the handler;

(C) The inspection certificate numbers;

(D) The grade and size of the kiwifruit;

(E) The number and type of containers; and

(F) The handler's brand; and

(ii) On the kiwifruit placed into new containers:

(A) The number and type of containers; and

(B) The applicable brand.

[58 FR 43246, Aug. 16, 1993, as amended at 61 FR 13395, Mar. 27, 1996; 64 FR 41019, July 29, 1999; 69 FR 54199, Sept. 8, 2004; 71 FR 58249, Oct. 3, 2006]

PART 922—APRICOTS GROWN IN DESIGNATED COUNTIES IN WASHINGTON

Subpart A—Order Regulating Handling

Subpart B—Container Exemption; Waivers of Inspection and Certification

Subpart C—Assessment Rate

Subpart D—Container Requirements

Subpart E—Grade and Size Requirements

AUTHORITY: 7 U.S.C. 601–674.

Subpart A—Order Regulating Handling

SOURCE: 22 FR 3514, May 21, 1957, unless otherwise noted. Redesignated at 26 FR 12751, Dec. 30, 1961.

DEFINITIONS

§ 922.1   Secretary.

Secretary means the Secretary of Agriculture of the United States, or any officer or employee of the Department to whom authority has heretofore been delegated, or to whom authority may hereafter be delegated, to act in his stead.

§ 922.2   Act.

Act means Public Act No. 10, 73d Congress (May 12, 1933), as amended and as reenacted and amended by the Agricultural Marketing Agreement Act of 1937, as amended (48 Stat. 31, as amended; 7 U.S.C. 601 *et seq.;* 68 Stat. 906, 1047).

§922.3 Person.

Person means an individual, partnership, corporation, association, or any other business unit.

§922.4 Production area.

Production area means the counties of Okanogan, Chelan, Kittitas, Yakima, and Klickitat in the State of Washington and all of the counties in Washington lying east thereof.

[27 FR 5188, June 2, 1962]

§922.5 Apricots.

Apricots means all varieties of apricots, grown in the production area, classified botanically as Prunus armeniaca.

§922.6 Varieties.

Varieties means and includes all classifications or subdivisions of Prunus armeniaca.

§922.7 Fiscal period.

Fiscal period is synonymous with fiscal year and means the 12-month period ending on March 31 of each year or such other period that may be approved by the Secretary pursuant to recommendations by the committee.

§922.8 Committee.

Committee means the Washington Apricot Marketing Committee established pursuant to §922.20.

§922.9 Grade.

Grade means any one of the officially established grades of apricots as defined and set forth in:

(a) United States Standards for Apricots (§§51.2925 to 51.2932 of this title) or amendments thereto, or modifications thereof, or variations based thereon;

(b) Standards for apricots issued by the State of Washington or amendments thereto, or modifications thereof, or variations based thereon.

§922.10 Size.

Size means the greatest diameter, measured through the center of the apricot, at right angles to a line running from the stem to the blossom end,

or such other specification as may be established by the committee with the approval of the Secretary.

§922.11 Grower.

Grower is synonymous with producer and means any person who produces apricots for market and who has a proprietary interest therein: *Provided,* That a grower who is also a handler must have produced not less than 51 percent of the apricots handled by him during the previous season in order to qualify as a grower under §§922.20, 922.22, and 922.23.

§922.12 Handler.

Handler is synonymous with shipper and means any person (except a common or contract carrier transporting apricots owned by another person) who handles apricots.

§922.13 Handle.

Handle or *ship* means to sell, consign, deliver, or transport apricots within the production area or between the production area and any point outside thereof: *Provided,* That the term "handle" shall not include the transportation within the production area of apricots from the orchard where grown to a packing facility located within such area for preparation for market.

[27 FR 5188, June 2, 1962]

§922.14 District.

District means the applicable one of the following described subdivisions of the production area, or such other subdivisions as may be prescribed pursuant to §922.31(m):

(a) *District 1* shall include all counties within the production area not included in District 2.

(b) *District 2* shall include the Counties of Yakima, Benton, and Klickitat.

[22 FR 3514, May 21, 1957. Redesignated at 26 FR 12751, Dec. 30, 1961, as amended at 27 FR 5188, June 2, 1962]

§922.15 Export.

Export means to ship apricots beyond the continental boundaries of the United States.

§ 922.16 Pack.

Pack means the specific arrangement, size, weight, count, or grade of a quantity of apricots in a particular type and size of container, or any combination thereof.

§ 922.17 Container.

Container means a box, bag, crate, lug, basket, carton, package, or any other type of receptacle used in the packaging or handling of apricots.

ADMINISTRATIVE BODY

§ 922.20 Establishment and membership.

There is hereby established a Washington Apricot Marketing Committee consisting of twelve members, each of whom shall have an alternate who shall have the same qualifications as the member for whom he is an alternate. Eight of the members and their respective alternates shall be growers or officers or employees of corporate growers. Four of the members and their respective alternates shall be handlers, or officers or employees of corporate handlers. The eight members of the committee who are growers or employees or officers of corporate growers are referred to in this part as "grower members" of the committee; and the four members of the committee who shall be handlers, or officers or employees of corporate handlers, are referred to in this part as "handler members" of the committee. Four of the grower members and their respective alternates shall be producers of apricots in District 1, and four of the grower members and their respective alternates shall be producers of apricots in District 2. Two of the handler members and their respective alternates shall be handlers of apricots in District 1, and two of the handler members with their respective alternates shall be handlers of apricots in District 2.

§ 922.21 Term of office.

The term of office of each member and alternate member of the committee shall be for 2 years beginning April 1 and ending March 31: *Provided,* That the terms of office of one-half the initial members and alternates shall end March 31, 1958. Members and alternate members shall serve in such capacities for the portion of the term of office for which they are selected and have qualified and until their respective successors are selected and have qualified. The terms of office of successor members and alternates shall be so determined that one-half of the total committee membership ends each March 31.

§ 922.22 Nomination.

(a) *Initial members.* Nominations for each of the eight initial grower members and four initial handler members of the committee, together with nominations for the initial alternate members for each position may be submitted to the Secretary by individual growers and handlers. Such nominations may be made by means of group meetings of the growers and handlers concerned in each district. Such nominations, if made, shall be filed with the Secretary no later than the effective date of this part. In the event nominations for initial members and alternate members of the committee are not filed pursuant to, and within the time specified, in this section, the Secretary may select such initial members and alternate members without regard to nominations, but selections shall be on the basis of the representation provided for in § 922.20.

(b) *Successor members.* (1) The committee shall hold or cause to be held, not later than March 1 of each year, a meeting or meetings of growers and handlers in each district for the purpose of designating nominees for successor members and alternate members of the committee. At each such meeting a chairman and a secretary shall be selected by the growers and handlers eligible to participate therein. The chairman shall announce at the meeting the number of votes cast for each person nominated for member or alternate member and shall submit promptly to the committee a complete report concerning such meeting. The committee shall, in turn, promptly submit a copy of each such report to the Secretary.

(2) Only growers, including duly authorized officers or employees of corporate growers, who are present at

such nomination meetings may participate in the nomination and election of nominees for grower members and their alternates. Each grower shall be entitled to cast only one vote for each nominee to be elected in the district in which he produces apricots. No grower shall participate in the election of nominees in more than one district in any one fiscal year. If qualified, a person may vote either as a grower or as a handler but not as both.

(3) Only handlers, including duly authorized officers or employees of corporate handlers, who are present at such nomination meetings, may participate in the nomination and election of nominees for handler members and their alternates. Each handler shall be entitled to cast only one vote for each nominee to be elected in the district in which he handles apricots. No handler shall participate in the election of nominees in more than one district in any one fiscal year. If qualified, a person may vote either as a grower or as a handler but not as both.

§922.23 Selection.

From the nominations made pursuant to §922.22, or from other qualified persons, the Secretary shall select the eight grower members of the committee, the four handler members of the committee, and an alternate for each such member.

§922.24 Failure to nominate.

If nominations are not made within the time and in the manner prescribed in §922.22, the Secretary may, without regard to nominations, select the members and alternate members of the committee on the basis of the representation provided for in §922.20.

§922.25 Acceptance.

Any person selected by the Secretary as a member or as an alternate member of the committee shall qualify by filing a written acceptance with the Secretary promptly after being notified of such selection.

§922.26 Vacancies.

To fill any vacancy occasioned by the failure of any person selected as a member or as an alternate member of the committee to qualify, or in the event of the death, removal, resignation, or disqualification of any member or alternate member of the committee, a successor for the unexpired term of such member or alternate member of the committee shall be nominated and selected in the manner specified in §§922.22 and 922.23. If the names of nominees to fill any such vacancy are not made available to the Secretary within a reasonable time after such vacancy occurs, the Secretary may fill such vacancy without regard to nominations, which selection shall be made on the basis of representation provided for in §922.20.

§922.27 Alternate members.

An alternate member of the committee, during the absence or at the request of the member for whom he is an alternate, shall act in the place and stead of such member and perform such other duties as assigned. In the event of the death, removal, resignation, or disqualification of a member, his alternate shall act for him until a successor for such member is selected and has qualified. In the event both a member of the committee and his alternate are unable to attend a committee meeting, the member or the committee may designate any other alternate member from the same district and group (handler or grower) to serve in such member's place and stead.

§922.30 Powers.

The committee shall have the following powers:

(a) To administer the provisions of this part in accordance with its terms;

(b) To receive, investigate, and report to the Secretary complaints of violations of the provisions of this part;

(c) To make and adopt rules and regulations to effectuate the terms and provisions of this part; and

(d) To recommend to the Secretary amendments to this part.

§922.31 Duties.

The committee shall have, among others, the following duties:

(a) To select a chairman and such other officers as may be necessary, and to define the duties of such officers;

(b) To appoint such employees, agents, and representatives as it may deem necessary, and to determine compensation and to define the duties of each;

(c) To submit to the Secretary as soon as practicable after the beginning of each fiscal period a budget for such fiscal period, including a report in explanation of the items appearing therein and a recommendation as to the rate of assessment for such period;

(d) To keep minutes, books, and records which will reflect all of the acts and transactions of the committee and which shall be subject to examination by the Secretary;

(e) To prepare periodic statements of the financial operations of the committee and to make copies of each such statement available to growers and handlers for examination at the office of the committee;

(f) To cause its books to be audited by a competent accountant at least once each fiscal year and at such times as the Secretary may request;

(g) To act as intermediary between the Secretary and any grower or handler;

(h) To investigate and assemble data on the growing, handling, and marketing conditions with respect to apricots;

(i) To submit to the Secretary such available information as he may request;

(j) To notify producers and handlers of all meetings of the committee to consider recommendations for regulations;

(k) To give the Secretary the same notice of meetings of the committee as is given to its members;

(l) To investigate compliance with the provisions of this part;

(m) With the approval of the Secretary, to redefine the districts into which the production area is divided, and to reapportion the representation of any district on the committee: *Provided,* That any such changes shall reflect, insofar as practicable, shifts in apricot production within the districts and the production area.

§ 922.32　Procedure.

(a) Eight members of the committee, including alternates acting for members, shall constitute a quorum; and any action of the committee shall require the concurring vote of at least 7 members: *Provided,* That when two-thirds of the membership present is greater than 7, such requirement shall be two-thirds of such membership.

(b) The committee may provide for simultaneous meetings of groups of its members assembled at two or more designated places: *Provided,* That such meetings shall be subject to the establishment of communication between all such groups and the availability of loud speaker receivers for each group so that each member may participate in the discussions and other actions the same as if the committee were assembled in one place. Any such meeting shall be considered as an assembled meeting.

(c) The committee may vote by telegraph, telephone, or other means of communication, and any votes so cast shall be confirmed promptly in writing: *Provided,* That if an assembled meeting is held, all votes shall be cast in person.

§ 922.33　Expenses and compensation.

The members of the committee, and alternates when acting as members, shall be reimbursed for expenses necessarily incurred by them in the performance of their duties under this part and may also receive compensation, as determined by the committee, which shall not exceed $10 per day or portion thereof spent in performing such duties: *Provided,* That at its discretion the committee may request the attendance of one or more alternates at any or all meetings, notwithstanding the expected or actual presence of the respective members, and may pay expenses and compensation, as aforesaid.

§ 922.34　Annual report.

The committee shall, prior to the last day of each fiscal period, prepare and mail an annual report to the Secretary and make a copy available to each handler and grower who requests a copy of the report. This annual report shall contain at least:

(a) A complete review of the regulatory operations during the fiscal period;

(b) An appraisal of the effect of such regulatory operations upon the apricot industry; and

(c) Any recommendations for changes in the program.

EXPENSES AND ASSESSMENTS

§922.40 Expenses.

The committee is authorized to incur such expenses as the Secretary finds are reasonable and likely to be incurred by the committee to enable it to exercise its powers and perform its duties in accordance with the provisions of this part during each fiscal period. The funds to cover such expenses shall be acquired by the levying of assessments as prescribed in §922.41.

§922.41 Assessments.

(a) Each person who first handles apricots shall, with respect to the apricots so handled by him, pay to the committee upon demand such person's pro rata share of the expenses which the Secretary finds will be incurred by the committee during each fiscal period. Each such person's share of such expenses shall be equal to the ratio between the total quantity of apricots handled by him as the first handler thereof during the applicable fiscal period and the total quantity of apricots so handled by all persons during the same fiscal period. The payment of assessments for the maintenance and functioning of the committee may be required under this part throughout the period it is in effect irrespective of whether particular provisions thereof are suspended or become inoperative.

(b) The Secretary shall fix the rate of assessment to be paid by each such person. At any time during or after the fiscal period, the Secretary may increase the rate of assessment in order to secure sufficient funds to cover any later finding by the Secretary relative to the expenses which may be incurred. Such increase shall be applied to all apricots handled during the applicable fiscal period. In order to provide funds for the administration of the provisions of this part during the first part of a fiscal period before sufficient operating income is available from assessments on the current year's shipments, the committee may accept the payment of assessments in advance, and may also borrow money for such purpose.

§922.42 Accounting.

(a) If, at the end of a fiscal period, the assessments collected are in excess of expenses incurred, such excess shall be accounted for as follows:

(1) Except as provided in paragraph (a)(2) of this section each person entitled to a proportionate refund of any excess assessment shall be credited with such refund against the operation of the following fiscal period unless such person demands repayment thereof, in which event it shall be paid to him: *Provided,* That any sum paid by a person in excess of his pro rata share of the expenses during any fiscal period may be applied by the committee at the end of such fiscal period to any outstanding obligations due the committee from such person.

(2) The Secretary, upon recommendation of the committee, may determine that it is appropriate for the maintenance and functioning of the committee that the funds remaining at the end of a fiscal period which are in excess of the expenses necessary for committee operations during such period, may be carried over into following periods as a reserve. Such reserve may be established at an amount not to exceed approximately one fiscal period's operational expenses; and such reserve may be used to cover the necessary expenses of liquidation, in the event of termination of this part, and to cover the expenses incurred for the maintenance and functioning of the committee during any fiscal period when there is a crop failure, or during any period of suspension of any or all of the provisions of this part. Such reserve may also be used by the committee to finance its operations, during any fiscal period, prior to the time that assessment income is sufficient to cover such expenses; but any of the reserve funds so used shall be returned to the reserve as soon as assessment income is available for this purpose. Upon termination of this part, any funds not required to defray the necessary expenses of liquidation shall be disposed of in such manner as the Secretary may determine to be appropriate: *Provided,* That

to the extent practical, such funds shall be returned pro rata to the persons from whom such funds were collected.

(b) All funds received by the committee pursuant to the provisions of this part shall be used solely for the purposes specified in this part and shall be accounted for in the manner provided in this part. The Secretary may at any time require the committee and its members to account for all receipts and disbursements.

(c) Upon the removal or expiration of the term of office of any member of the committee, such member shall account for all receipts and disbursements and deliver all property and funds in his possession to his successor in office, and shall execute such assignments and other instruments as may be necessary or appropriate to vest in such successor full title to all of the property, funds, and claims vested in such member pursuant to this part.

RESEARCH

§ 922.45 Marketing research and development.

The committee, with the approval of the Secretary, may establish or provide for the establishment of marketing research and development projects designed to assist, improve, or promote the marketing, distribution, and consumption of apricots. The expense of such projects shall be paid from funds collected pursuant to § 922.41.

REGULATIONS

§ 922.50 Marketing policy.

(a) Each season prior to making any recommendations pursuant to § 922.51, the committee shall submit to the Secretary a report setting forth its marketing policy for the ensuing season. Such marketing policy report shall contain information relative to:

(1) The estimated total production of apricots within the production area;

(2) The expected general quality and size of apricots in the production area and in other areas;

(3) The expected demand conditions for apricots in different market outlets;

(4) The expected shipments of apricots produced in the production area

and in areas outside the production area;

(5) Supplies of competing commodities;

(6) Trend and level of consumer income;

(7) Other factors having a bearing on the marketing of apricots; and

(8) The type of regulations expected to be recommended during the season.

(b) In the event it becomes advisable, because of changes in the supply and demand situation for apricots, to modify substantially such marketing policy, the committee shall submit to the Secretary a revised marketing policy report setting forth the information prescribed in this section. The committee shall publicly announce the contents of each marketing policy report, including each revised marketing policy report, and copies thereof shall be maintained in the office of the committee where they shall be available for examination by growers and handlers.

§ 922.51 Recommendations for regulation.

(a) Whenever the committee deems it advisable to regulate the handling of any variety or varieties of apricots in the manner provided in § 922.52, it shall so recommend to the Secretary.

(b) In arriving at its recommendations for regulation pursuant to paragraph (a) of this section, the committee shall give consideration to current information with respect to the factors affecting the supply and demand for apricots during the period or periods when it is proposed that such regulation should be made effective. With each such recommendation for regulation, the committee shall submit to the Secretary the data and information on which such recommendation is predicated and such other available information as the Secretary may request.

§ 922.52 Issuance of regulations.

(a) The Secretary shall regulate, in the manner specified in this section, the handling of apricots whenever he finds from the recommendations and information submitted by the committee, or from other available information, that such regulations will tend to effectuate the declared policy of the act. Such regulations may:

(1) Limit, during any period or periods, the shipment of any particular grade, size, quality, maturity, or pack, or any combination thereof, of any variety or varieties of apricots grown in any district or districts of the production area;

(2) Limit the shipment of apricots by establishing, in terms of grades, sizes, or both, minimum standards of quality and maturity during any period when season average prices are expected to exceed the parity level;

(3) Fix the size, capacity, weight, dimensions, markings, or pack of the container, or containers, which may be used in the packaging or handling of apricots.

(b) The committee shall be informed immediately of any such regulation issued by the Secretary, and the committee shall promptly give notice thereof to growers and handlers.

[22 FR 3514, May 21, 1957. Redesignated at 26 FR 12751, Dec. 30, 1961, as amended at 27 FR 5188, June 2, 1962]

§922.53 Modification, suspension, or termination of regulations.

(a) In the event the committee at any time finds that, by reason of changed conditions, any regulations issued pursuant to §922.52 should be modified, suspended, or terminated, it shall so recommend to the Secretary.

(b) Whenever the Secretary finds, from the recommendations and information submitted by the committee or from other available information, that a regulation should be modified, suspended, or terminated with respect to any or all shipments of apricots in order to effectuate the declared policy of the act, he shall modify, suspend or terminate such regulation. On the same basis and in like manner the Secretary may terminate any such modification or suspension. If the Secretary finds that a regulation obstructs or does not tend to effectuate the declared policy of the act, he shall suspend or terminate such regulation. On the same basis and in like manner the Secretary may terminate any such suspension.

§922.54 Special purpose shipments.

(a) Except as otherwise provided in this section, any person may, without regard to the provisions of §§922.41, 922.52, 922.53, and 922.55, and the regulations issued thereunder, handle apricots (1) for consumption by charitable institutions; (2) for distribution by relief agencies; or (3) for commercial processing into products.

(b) Upon the basis of recommendations and information submitted by the committee, or from other available information, the Secretary may relieve from any or all requirements, under or established pursuant to §922.41, §922.52, §922.53, or §922.55, the handling of apricots in such minimum quantities, or types of shipments, or for such specified purposes (including shipments to facilitate the conduct of marketing research and development projects established pursuant to §922.45), as the committee, with approval of the Secretary, may prescribe,

(c) The committee shall, with the approval of the Secretary, prescribe such rules, regulations, and safeguards as it may deem necessary to prevent apricots handled under the provisions of this section from entering the channels of trade for other than the specific purposes authorized by this section. Such rules, regulations, and safeguards may include the requirements that handlers shall file applications and receive approval from the committee for authorization to handle apricots pursuant to this section, and that such applications be accompanied by a certification by the intended purchaser or receiver that the apricots will not be used for any purpose not authorized by this section.

§922.55 Inspection and certification.

Whenever the handling of any variety of apricots is regulated pursuant to §922.52 or §922.53, each handler who handles apricots shall, prior thereto, cause such apricots to be inspected by the Federal-State Inspection Service and certified by it as meeting the applicable requirements of such regulation: *Provided,* That inspection and certification shall be required for apricots which previously have been so inspected and certified only if such apricots have been regraded, resorted, repackaged, or in any other way further prepared for market. Promptly after inspection and certification, each such handler shall submit, or cause to be submitted, to the committee a copy of the certificate of inspection issued with respect to such apricots. The committee may, with the approval of the Secretary, prescribe rules and regulations modifying the inspection requirements of this section as to time and place such inspection shall be performed whenever it is determined it would not be practical to perform the required inspection at a particular location: *Provided,* That all such shipments shall comply with all regulations in effect.

[22 FR 3514, May 21, 1957. Redesignated at 26 FR 12751, Dec. 30, 1961, as amended at 27 FR 5188, June 2, 1962]

REPORTS

§ 922.60   Reports.

(a) Upon request of the committee, made with the approval of the Secretary, each handler shall furnish to the committee, in such manner and at such time as it may prescribe, such reports and other information as may be necessary for the committee to perform its duties under this part. Such reports may include, but are not necessarily limited to, the following:

(1) The quantities of each variety of apricots received by a handler,

(2) The quantities disposed of by him segregated as to the respective quantities subject to regulation and not subject to regulation;

(3) The date of each such disposition and the identification of the carrier transporting such apricots, and

(4) The destination of each such shipment.

(b) All such reports shall be held under appropriate protective classification and custody by the committee, or duly appointed employees thereof, so that the information contained therein which may adversely affect the competitive position of any handler in relation to other handlers will not be disclosed. Compilations of general reports from data submitted by handlers is authorized, subject to the prohibition of disclosure of individual handler's identities or operations.

(c) Each handler shall maintain for at least two succeeding years such records of the apricots received, and of apricots disposed of, by such handler as may be necessary to verify reports pursuant to this section.

MISCELLANEOUS PROVISIONS

§ 922.61   Compliance.

Except as provided in this part, no person shall handle apricots, the shipment of which has been prohibited by the Secretary in accordance with the provisions of this part; and no person shall handle apricots except in conformity with the provisions of this part.

§ 922.62   Right of the Secretary.

The members of the committee (including successors and alternates), and any agents, employees, or representatives thereof, shall be subject to removal or suspension by the Secretary at any time. Each and every regulation, decision, determination, or other act of the committee shall be subject to the continuing right of the Secretary to disapprove of the same at any time. Upon such disapproval, the disapproved action of the committee shall be deemed null and void, except as to acts done in reliance thereon or in accordance therewith prior to such disapproval by the Secretary.

§ 922.63   Effective time.

The provisions of this part, and of any amendment thereto, shall become effective at such time as the Secretary may declare above his signature to this part, and shall continue in force until terminated in one of the ways specified in § 922.64.

§ 922.64   Termination.

(a) The Secretary may at any time terminate the provisions of this part by giving at least one day's notice by means of a press release or in any other manner in which he may determine.

(b) The Secretary shall terminate or suspend the operation of any and all of the provisions of this part whenever he finds that such provisions do not tend to effectuate the declared policy of the act.

(c) The Secretary shall terminate the provisions of this part at the end of any fiscal period whenever he finds that continuance is not favored by the majority of producers who, during a representative period determined by the Secretary, were engaged in the production area in the production of apricots for market: *Provided,* That such majority has produced for market during such period more than 50 percent of the volume of apricots produced for market in the production area; but such termination shall be effective only if announced on or before March 31 of the then current fiscal period.

(d) The provisions of this part shall, in any event, terminate whenever the provisions of the act authorizing them cease to be in effect.

§922.65 Proceedings after termination.

(a) Upon the termination of the provisions of this part, the committee shall, for the purpose of liquidating the affairs of the committee, continue as trustees of all the funds and property then in its possession, or under its control, including claims for any funds unpaid or property not delivered at the time of such termination.

(b) The said trustees shall (1) continue in such capacity until discharged by the Secretary; (2) from time to time account for all receipts and disbursements and deliver all property on hand, together with all books and records of the committee and of the trustees, to such person as the Secretary may direct; and (3) upon the request of the Secretary, execute such assignments or other instruments necessary or appropriate to vest in such person, full title and right to all of the funds, property, and claims vested in the committee or the trustees pursuant hereto.

(c) Any person to whom funds, property, or claims have been transferred or delivered pursuant to this section shall be subject to the same obligation imposed upon the committee and upon the trustees.

§922.66 Effect of termination or amendment.

Unless otherwise expressly provided by the Secretary, the termination of this subpart or of any regulation issued pursuant to this subpart, or the issuance of any amendment to either thereof, shall not (a) affect or waive any right, duty, obligation, or liability which shall have arisen or which may thereafter arise in connection with any provision of this subpart or any regulation issued under this subpart, or (b) release or extinguish any violation of this subpart or of any regulation issued under this subpart, or (c) affect or impair any rights or remedies of the Secretary or of any other person with respect to any such violation.

§922.67 Duration of immunities.

The benefits, privileges, and immunities conferred upon any person by virtue of this subpart shall cease upon the termination of this subpart, except with respect to acts done under and during the existence of this subpart.

§922.68 Agents.

The Secretary may, by designation in writing, name any officer or employee of the United States, or name any agency or division in the United States Department of Agriculture, to act as his agent or representative in connection with any of the provisions of this part.

§922.69 Derogation.

Nothing contained in the provisions of this part is, or shall be construed to be, in derogation or in modification of the rights of the Secretary or of the United States (a) to exercise any powers granted by the act or otherwise, or (b) in accordance with such powers, to act in the premises whenever such action is deemed advisable.

§922.70 Personal liability.

No member or alternate member of the committee and no employee or agent of the committee shall be held personally responsible, either individually or jointly with others, in any way whatsoever, to any person for errors in judgment, mistakes, or other act, either of commission or omission, as such member, alternate, employee, or agent, except for acts of dishonesty, willful misconduct, or gross negligence.

§ 922.71   Separability.

If any provision of this part is declared invalid, or the applicability thereof to any person, circumstance, or thing is held invalid, the validity of the remainder of this part or the applicability thereof to any other person, circumstance, or thing shall not be affected thereby.

Subpart B—Container Exemption; Waivers of Inspection and Certification

§ 922.110   Container exemption.

Whenever container limitations are effective pursuant to § 922.52, a handler may make test shipments of apricots in experimental containers, approved by the committee, subject to the following:

(a) Test shipments shall be made only in connection with a container research project, or projects, being conducted by or in cooperation with the Washington Apricot Marketing Committee.

(b) The handler shall first make application to, and receive a permit from, the Washington Apricot Marketing Committee on a form of the committee to handle each experimental container proposed to be used by the handler for test shipments. Such application shall contain the following information:

(1) Name and address of the applicant and date of application;

(2) Description of the container, including size, weight, inside dimensions, and type of pack;

(3) Quantity of such containers proposed to be shipped.

(c) Approval of the application shall be evidenced by the issuance to the applicant by the committee of a permit which shall authorize the handling of apricots in such quantity of experimental containers as the committee may approve.

(d) With respect to each test shipment of apricots handled in experimental containers, the handler shall, prior to such handling, advise the committee as to (1) the number and type of the container or containers in the test shipment, (2) identification of the carrier, (3) name and address of the re-

ceiver, and (4) expected time of arrival at destination.

(e) Terms used in this section shall have the same meaning as when used in said marketing agreement and order (§§ 922.1 to 922.71).

[23 FR 4781, June 28, 1958. Redesignated at 26 FR 12751, Dec. 30, 1961]

§ 922.111   Waiver of inspection and certification.

(a) *Application.* Any handler (including a grower-handler packing and handling apricots of his own production), whose packing facilities are located in an area where a Washington State Horticultural Division Inspection Office or Federal-State Inspector is not readily available to perform the required inspection may, prior to shipment, apply to the Committee for a permit authorizing a waiver of inspection. Applications shall be made on forms furnished by the Committee and shall contain such information as the Committee may require including: Name and address of applicant, location of packing facility, distance of packing facility from the nearest inspection office, period (by approximate beginning and ending dates) during which applicant expects to ship apricots, estimated quantity of apricots applicant expects to ship to fresh market during such period, manner in which the majority of applicant's fruit will be marketed (i.e., transported by applicant to market, sold at orchard to truckers, etc.), areas or markets to which applicant expects to ship the majority of his apricots. The application shall also contain an agreement by applicant (1) not to ship or handle any apricots unless such apricots meet the grade, size, maturity, container, and all other requirements of the amended marketing agreement and order in effect at time of handling, (2) to report periodically to the Committee on reporting forms furnished by the Committee the following information on each shipment: quantity, variety, grade, minimum size, container, date of shipment, destination, name and address of buyer or receiver, and such other information as the Committee may specify, (3) to pay applicable assessments on each shipment, (4) to have or cause to have each shipment of apricots inspected when such shipment is transported to a market or through a location en route to market where an inspector is available, and (5) to comply with such other safeguards as the Committee may prescribe.

(b) *Issuance of permit.* Whenever the Committee finds and determines from the information contained in the application or from other proof satisfactory to the Committee that the applicant is entitled to a waiver from the inspection requirements of the amended marketing agreement and order at time of shipment, the Committee shall issue a permit authorizing the applicant to ship apricots in accordance with these administrative regulations and the terms and conditions of such permit.

[29 FR 9526, July 14, 1964]

EFFECTIVE DATE NOTE: At 78 FR 62966, Oct. 23, 2013, §922.111 was suspended, effective Oct. 24, 2013.

§922.142 Reserve fund.

(a) The establishment of a reserve fund of an amount not greater than approximately one fiscal year's operational expenses is appropriate and necessary to the maintenance and functioning of the Washington Apricot Marketing Committee. The committee is hereby authorized to carry forward in the aforesaid reserve $5,765.09 which are excess assessment funds from the fiscal period ended March 31, 1960, and $787.61 which are excess assessment funds from the fiscal period ended March 31, 1961. Such reserve shall be used in accordance with the provisions of §922.42 of the said marketing agreement and order (§§922.1 to 922.71).

(b) Terms used in this section shall have the same meaning as given to the respective term in said marketing agreement and order.

[26 FR 8664, Sept. 16, 1961. Redesignated at 26 FR 12751, Dec. 30, 1961, and further redesignated at 44 FR 73010, Dec. 17, 1979]

Subpart C—Assessment Rate

§922.235 Assessment rate.

On and after April 1, 2017, an assessment rate of $1.00 per ton is established for Washington apricots handled in the production area.

[82 FR 43299, Sept. 15, 2017]

Subpart D—Container Requirements

§922.306 Apricot Regulation 6.

(a) No handler shall handle any apricots unless such apricots are:

(1) In open containers or telescope fiberboard cartons and the net weight of the apricots is not less than 28 pounds; or

(2) In closed containers containing not less than 14 pounds, net weight, of apricots: *Provided,* That when the apricots are packed in such containers they are row-faced or tray-packed; or

(3) In closed containers that are marked "12 pounds net weight" and contain not less than 12 pounds, net weight, of apricots which are of random size and are not row-faced; or

(4) In closed containers containing not less than 24 pounds, net weight, of apricots when packed loose in such containers; or

(5) If exported to Canada, in any of the containers specified in this paragraph (a) or in containers having inside dimensions of 16⅛ × 11½ inches with 4¾-inch end pieces and 3¾-inch side pieces.

(b) Notwithstanding any other provisions of this section, any individual shipment of apricots which, in the aggregate, does not exceed 500 pounds, net weight, may be handled without regard to the requirements specified in this section or in §922.41 or §922.55.

(c) All apricots handled are also subject to all applicable grade, size, quality, maturity and pack regulations which are in effect pursuant to this part.

(d) The terms *handler, handle* and *apricots* shall have the same meaning as when used in the amended marketing agreement and order.

[59 FR 30673, June 15, 1994, as amended at 63 FR 32718, June 16, 1998]

EFFECTIVE DATE NOTE: At 72 FR 16265, Apr. 4, 2007, §922.306 was suspended indefinitely, effective Apr. 1, 2007.

Subpart E—Grade and Size Requirements

§922.321 Apricot Regulation 21.

(a) On and after August 1, 1981, no handler shall handle any container of apricots unless such apricots meet the following applicable requirements, or are handled in accordance with paragraph (a)(3) of this section:

(1) *Minimum grade and maturity requirements.* Such apricots that grade not less than Washington No. 1 and are at least reasonably uniform in color: *Provided,* That such apricots of the Moorpark variety in open containers shall be generally well matured.

(2) *Minimum size requirements.* Such apricots measure not less than 1⅝ inches in diameter except that apricots of the Blenheim, Blenril, and Tilton varieties may measure not less than 1¼ inches: *Provided,* That not more than 10 percent, by count, of such apricots may fail to meet the applicable minimum diameter requirements.

(3) Notwithstanding any other provision of this section, any individual shipment of apricots which meets each of the following requirements may be handled without regard to the provisions of this paragraph, of §922.41 (Assessments), and of §922.55 (Inspection and Certification):

(i) The shipment consists of apricots sold for home use and not for resale;

(ii) The shipment does not, in the aggregate, exceed 500 pounds, net weight, of apricots; and

(iii) Each container is stamped or marked with the words "not for resale" in letters at least one-half inch in height.

(b) The terms *diameter* and *Washington No. 1* shall have the same meaning as when used in the State of Washington Department of Agriculture Standards for Apricots, effective May 31, 1966; *reasonably uniform in color* means that the apricots in the individual container do not show sufficient variation in color to materially affect the general appearance of the apricots; and *generally well matured* means that with respect to not less than 90 percent, by count, of the apricots in any lot of containers, and not less than 85 percent, by count, of such apricots in any container in such lot, at least 40 percent of the surface area of the fruit is at least as yellow as Shade 3 on U.S. Department of Agriculture Standard

Ground Color Chart of Apples and Pears in Western States.

[46 FR 38668, July 29, 1981, as amended at 54 FR 26186, June 22, 1989; 60 FR 32430, June 22, 1995; 61 FR 30497, June 17, 1996; 75 FR 72935, Nov. 29, 2010]

EFFECTIVE DATE NOTE: At 78 FR 62966, Oct. 23, 2013, §922.321 was suspended, effective Oct. 24, 2013.

PART 923—SWEET CHERRIES GROWN IN DESIGNATED COUNTIES IN WASHINGTON

Subpart A—Order Regulating Handling

AUTHORITY: 7 U.S.C. 601–674.

SOURCE: 22 FR 3835, June 1, 1957, unless otherwise noted. Redesignated at 26 FR 12751, Dec. 30, 1961.

Subpart A—Order Regulating Handling ≤

DEFINITIONS

§923.1 Secretary.

Secretary means the Secretary of Agriculture of the United States, or any officer or employee of the Department to whom authority has heretofore been delegated, or to whom authority may hereafter be delegated, to act in his stead.

§923.2 Act.

Act means Public Act No. 10, 73d Congress (May 12, 1933), as amended and as reenacted and amended by the Agricultural Marketing Agreement Act of 1937, as amended (48 Stat. 31, as amended; 7 U.S.C. 601 *et seq.*; 68 Stat. 906, 1047).

§923.3 Person.

Person means an individual, partnership, corporation, association, or any other business unit.

§923.4 Production area.

Production area means the counties of Okanogan, Chelan, Kittitas, Yakima, Klickitat in the State of Washington and all of the counties in Washington lying east thereof.

[66 FR 58356, Nov. 21, 2001]

§923.5 Cherries.

Cherries means all varieties of sweet cherries grown in the production area, classified botanically as Prunus avium.

§923.6 Varieties.

Varieties means and includes all classifications or subdivisions of Prunus avium.

§923.7 Fiscal period.

Fiscal period is synonymous with fiscal year and means the 12-month period ending on March 31 of each year or such other period that may be approved by the Secretary pursuant to recommendations by the committee.

§923.8 Committee.

Committee means the Washington Cherry Marketing Committee established pursuant to §923.20.

§923.9 Grade.

Grade means any one of the officially established grades of cherries as defined and set forth in:

(a) United States Standards for Sweet Cherries (§§51.2646 to 51.2660 of this title) or amendments thereto, or modifications thereof, or variations based thereon;

(b) Standards for sweet cherries issued by the State of Washington or amendments thereto, or modifications thereof, or variations based thereon.

§923.10 Size.

Size means the greatest diameter, measured through the center of the cherry, at right angles to a line running from the stem to the blossom end, or such other specification as may be established by the committee with the approval of the Secretary.

§923.11 Grower.

Grower is synonymous with producer and means any person who produces cherries for market and who has a proprietary interest therein.

§ 923.12 Handler.

Handler is synonymous with shipper and means any person (except a common or contract carrier transporting cherries owned by another person) who handles cherries.

§ 923.13 Handle.

Handle and *ship* are synonymous and mean to sell, consign, deliver, or transport cherries or cause the sale, consignment, delivery, or transportation of cherries or in any other way to place cherries, or cause cherries to be placed, in the current of the commerce from any point within the production area to any point outside thereof: *Provided,* That the term *handle* shall not include the transportation within the production area of cherries from the orchard where grown to a packing facility located within such area for preparation for market, or the delivery of such cherries to such packing facility for such preparation.

§ 923.14 District.

District means the applicable one of the following described subdivisions of the production area, or such other subdivisions as may be prescribed pursuant to § 923.31(m):

(a) *District 1* shall include the Counties of Chelan, Okanogan, Douglas, Grant, Lincoln, Spokane, Pend Oreille, Stevens, and Ferry.

(b) *District 2* shall include the counties of Kittitas, Yakima, Klickitat, Benton, Adams, Franklin, Walla Walla, Whitman, Columbia, Garfield and Asotin.

[22 FR 3835, June 1, 1957. Redesignated at 26 FR 12751, Dec. 30, 1961, as amended at 66 FR 58356, Nov. 21, 2001]

§ 923.15 Export.

Export means to ship cherries beyond the continental boundaries of the United States.

§ 923.16 Pack.

Pack means the specific arrangement, size, weight, count, or grade of a quantity of cherries in a particular type and size of container, or any combination thereof.

§ 923.17 Container.

Container means a box, bag, crate, lug, basket, carton, package, or any other type of receptacle used in the packaging or handling of cherries.

ADMINISTRATIVE BODY

§ 923.20 Establishment and membership.

There is hereby established a Washington Cherry Marketing Committee consisting of sixteen members, each of whom shall have an alternate who shall have the same qualifications as the member for whom he is an alternate. Ten of the members and their respective alternates shall be growers or officers or employees of corporate growers. Six of the members and their respective alternates shall be handlers, or officers or employees of handlers. The ten members of the committee who are growers or employees or officers of corporate growers are referred to in this part as "grower members" of the committee; and the six members of the committee who shall be handlers, or officers or employees of handlers, are referred to in this part as "handler members" of the committee. Five of the grower members and their respective alternates shall be producers of cherries in District 1, and five of the grower members and their respective alternates shall be producers of cherries in District 2. Three of the handler members and their respective alternates shall be handlers of cherries in District 1, and three of the handler members and their respective alternates shall be handlers of cherries in District 2.

[22 FR 3835, June 1, 1957. Redesignated at 26 FR 12751, Dec. 30, 1961, as amended at 66 FR 58356, Nov. 21, 2001]

§ 923.21 Term of office.

The term of office of each member and alternate member of the committee shall be for two years beginning April 1 and ending March 31. Members and alternate members shall serve in such capacities for the portion of the term of office for which they are selected and have qualified and until their respective successors are selected and have qualified. Committee members shall not serve more than three consecutive terms. Members who have served for three consecutive terms must leave the committee for at least one year before becoming eligible to serve again.

[70 FR 44252, Aug. 2, 2005]

§923.22 Nomination.

(a) *Initial members.* Nominations for each of the ten initial grower members and five initial handler members of the committee, together with nominations for the initial alternate members for each position, may be submitted to the Secretary by individual growers and handlers. Such nominations may be made by means of group meetings of the growers and handlers concerned in each district. Such nominations, if made, shall be filed with the Secretary no later than the effective date of this part. In the event nominations for initial members and alternate members of the committee are not filed pursuant to, and within the time specified, in this section, the Secretary may select such initial members and alternate members without regard to nominations, but selections shall be on the basis of the representation provided for in §923.20.

(b) *Successor members.* (1) The committee shall hold or cause to be held, not later than March 1 of each year, a meeting or meetings of growers and handlers in each district for the purpose of designating nominees for successor members and alternate members of the committee. At each such meeting a chairman and a secretary shall be selected by the growers and handlers eligible to participate therein. The chairman shall announce at the meeting the number of votes cast for each person nominated for member or alternate member and shall submit promptly to the committee a complete report concerning such meeting. The committee shall, in turn, promptly submit a copy of each such report to the Secretary.

(2) Only growers, including duly authorized officers or employees of corporate growers, who are present at such nomination meetings may participate in the nomination and election of nominees for grower members and their alternates. Each grower shall be entitled to cast only one vote for each nominee to be elected in the district in which he produces cherries. No grower shall participate in the election of nominees in more than one district in any one fiscal year. If a person is both a grower and a handler of cherries, such person may vote either as a grower or as a handler but not as both.

(3) Only handlers, including duly authorized officers or employees of handlers, who are present at such nomination meetings, may participate in the nomination and election of nominees for handler members and their alternates. Each handler shall be entitled to cast only one vote for each nominee to be elected in the district in which he handles cherries. No handler shall participate in the election of nominees in more than one district in any one fiscal year. If a person is both a grower and a handler of cherries, such person may vote either as a grower or as a handler but not as both.

§923.23 Selection.

From the nominations made pursuant to §923.22, or from other qualified persons, the Secretary shall select the ten grower members of the committee, the five handler members of the committee, and an alternate for each member.

§923.24 Failure to nominate.

If nominations are not made within the time and in the manner prescribed in §923.22, the Secretary may, without regard to nominations, select the members and alternate members of the committee on the basis of the representation provided for in §923.20.

§923.25 Acceptance.

Any person prior to selection as a member or an alternate member of the committee shall qualify by filing with USDA a written acceptance of willingness to serve on the committee.

[66 FR 58356, Nov. 21, 2001]

§923.26 Vacancies.

To fill any vacancy occasioned by the failure of any person selected as a member or as an alternate member of the committee to qualify, or in the event of the death, removal, resignation, or disqualification of any member or alternate member of the committee, a successor for the unexpired term of such member or alternate member of the committee shall be nominated and selected in the manner specified in §§ 923.22 and 923.23. If the names of nominees to fill any such vacancy are not made available to the Secretary within a reasonable time after such vacancy occurs, the Secretary may fill such vacancy without regard to nominations, which selection shall be made on the basis of representation provided for in § 923.20.

§ 923.27 Alternate members.

An alternate member of the committee, during the absence or at the request of the member for whom he is an alternate, shall act in the place and stead of such member and perform such other duties as assigned. In the event of the death, removal, resignation, or disqualification of a member, his alternate shall act for him until a successor for such member is selected and has qualified. In the event both a member of the committee and his alternate are unable to attend a committee meeting, the member of the committee may designate any other alternate member from the same district and group (handler or grower) to serve in such member's place and stead.

§ 923.30 Powers.

The committee shall have the following powers:

(a) To administer the provisions of this part in accordance with its terms;

(b) To receive, investigate, and report to the Secretary complaints of violations of the provisions of this part;

(c) To make and adopt rules and regulations to effectuate the terms and provisions of this part; and

(d) To recommend to the Secretary amendments to this part.

§ 923.31 Duties.

The committee shall have, among others, the following duties:

(a) To select a chairman and such other officers as may be necessary, and to define the duties of such officers;

(b) To appoint such employees, agents, and representatives as it may deem necessary, and to determine the compensation and to define the duties of each;

(c) To submit to the Secretary as soon as practicable after the beginning of each fiscal period a budget for such fiscal period, including a report in explanation of the items appearing therein and a recommendation as to the rate of assessment for such period;

(d) To keep minutes, books, and records which will reflect all of the acts and transactions of the committee and which shall be subject to examination by the Secretary;

(e) To prepare periodic statements of the financial operations of the committee and to make copies of each such statement available to growers and handlers for examination at the office of the committee;

(f) To cause its books to be audited by a competent accountant at least once each fiscal year and at such time as the Secretary may request;

(g) To act as intermediary between the Secretary and any grower or handler;

(h) To investigate and assemble data on the growing, handling, and marketing conditions with respect to cherries;

(i) To submit to the Secretary such available information as he may request;

(j) To notify producers and handlers of all meetings of the committee to consider recommendations for regulations;

(k) To give the Secretary the same notice of meetings of the committee as is given to its members;

(l) To investigate compliance with the provisions of this part;

(m) With the approval of the Secretary, to redefine the districts into which the production area is divided, and to reapportion the representation of any district on the committee: *Provided,* That any such changes shall reflect, insofar as practicable, shifts in cherry production within the districts and the production area.

§ 923.32 Procedure.

(a) Twelve members of the committee, including alternates acting for members, shall constitute a quorum; and any action of the committee shall require the concurring vote of at least nine members.

(b) The committee may provide for simultaneous meetings of groups of its members assembled at two or more designated places: *Provided,* That such meetings shall be subject to the establishment of communication between all such groups and the availability of loud speaker receivers for each group so that each member may participate in the discussions and other actions the same as if the committee were assembled in one place. Any such meeting shall be considered as an assembled meeting.

(c) The committee may vote by telegraph, telephone, or other means of communication, and any votes so cast shall be confirmed promptly in writing: *Provided,* That if an assembled meeting is held, all votes shall be cast in person.

§923.33   Expenses and compensation.

The members of the committee, and alternates when acting as members, shall be reimbursed for expenses necessarily incurred by them in the performance of their duties under this part and may also receive compensation, as determined by the committee, which shall not exceed $10 per day or portion thereof spent in performing such duties: *Provided,* That at its discretion the committee may request the attendance of one or more alternates at any or all meetings, notwithstanding the expected or actual presence of the respective members, and may pay expenses and compensation, as aforesaid.

§923.34   Annual report.

The committee shall, prior to the last day of each fiscal period, prepare and mail an annual report to the Secretary and make a copy available to each handler and grower who requests a copy of the report. This annual report shall contain at least: (a) A complete review of the regulatory operations during the fiscal period; (b) an appraisal of the effect of such regulatory operations upon the cherry industry; and (c) any recommendations for changes in the program.

EXPENSES AND ASSESSMENTS

§923.40   Expenses.

The committee is authorized to incur such expenses as the Secretary finds are reasonable and likely to be incurred by the committee to enable it to exercise its powers and perform its duties in accordance with the provisions of this part during each fiscal period. The funds to cover such expenses shall be acquired by the levying of assessments as prescribed in §923.41.

§923.41   Assessments.

(a) Each person who first handles cherries shall, with respect to the cherries so handled by him, pay to the committee upon demand such person's pro rata share of the expenses which the Secretary finds will be incurred by the committee during each fiscal period. Each such person's share of such expenses shall be equal to the ratio between the total quantity of cherries handled by him as the first handler thereof during the applicable fiscal period and the total quantity of cherries so handled by all persons during the same fiscal period. The payment of assessments for the maintenance and functioning of the committee may be required under this part throughout the period it is in effect irrespective of whether particular provisions thereof are suspended or become inoperative.

(b) The Secretary shall fix the rate of assessment to be paid by each such person. At any time during or after the fiscal period, the Secretary may increase the rate of assessment in order to secure sufficient funds to cover any later finding by the Secretary relative to the expenses which may be incurred. Such increase shall be applied to all cherries handled during the applicable fiscal period. In order to provide funds for the administration of the provisions of this part during the first part of a fiscal period before sufficient operating income is available from assessments on the current year's shipments, the committee may accept the payment of assessments in advance, and may also borrow money for such purpose.

(c) If a handler does not pay any assessment within the time prescribed by the committee, the assessment may be subject to an interest or late payment charge, or both, as may be established by USDA as recommended by the committee.

[22 FR 3835, June 1, 1957. Redesignated at 26 FR 12751, Dec. 30, 1961, as amended at 66 FR 58356, Nov. 21, 2001.]

§ 923.42   Accounting.

(a) If, at the end of a fiscal period, the assessments collected are in excess of expenses incurred, such excess shall be accounted for as follows:

(1) Except as provided in paragraph (a)(2) of this section, each person entitled to a proportionate refund of any excess assessment shall be credited with such refund against the operation of the following fiscal period unless such person demands repayment thereof, in which event it shall be paid to him: *Provided,* That any sum paid by a person in excess of his pro rata share of the expenses during any fiscal period may be applied by the committee at the end of such fiscal period to any outstanding obligations due the committee from such person.

(2) The Secretary, upon recommendation of the committee, may determine that it is appropriate for the maintenance and functioning of the committee that the funds remaining at the end of a fiscal period which are in excess of the expenses necessary for committee operations during such period may be carried over into following periods as a reserve. Such reserve may be established at an amount not to exceed approximately one fiscal period's operational expenses; and such reserve may be used to cover the necessary expenses of liquidation, in the event of termination of this part, and to cover the expenses incurred for the maintenance and functioning of the committee during any fiscal period when there is a crop failure, or during any period of suspension of any or all of the provisions of this part. Such reserve may also be used by the committee to finance its operations, during any fiscal period, prior to the time that assessment income is sufficient to cover such expenses; but any of the reserve funds so used shall be returned to the reserve as soon as assessment income is available for this purpose. Upon termination of this part, any funds not required to

defray the necessary expenses of liquidation shall be disposed of in such manner as the Secretary may determine to be appropriate: *Provided,* That to the extent practical, such funds shall be returned pro rata to the persons from whom such funds were collected.

(b) All funds received by the committee pursuant to the provisions of this part shall be used solely for the purposes specified in this part and shall be accounted for in the manner provided in this part. The Secretary may at any time require the committee and its members to account for all receipts and disbursements.

(c) Upon the removal or expiration of the term of office of any member of the committee, such member shall account for all receipts and disbursements and deliver all property and funds in his possession to his successor in office, and shall execute such assignments and other instruments as may be necessary or appropriate to vest in such successor full title to all of the property, funds, and claims vested in such member pursuant to this part.

§ 923.43   Contributions.

The committee may accept voluntary contributions but these shall only be used to pay expenses incurred pursuant to § 923.45. Furthermore, such contributions shall be free from any encumbrances by the donor and the committee shall retain complete control of their use.

[70 FR 44252, Aug. 2, 2005]

RESEARCH

§ 923.45   Marketing research and development.

The committee, with the approval of the Secretary, may establish or provide for the establishment of marketing research and development projects designed to assist, improve, or promote the marketing, distribution, and consumption of cherries. The expense of such projects shall be paid from funds collected pursuant to § 923.41.

§923.50 Marketing policy.

(a) Each season prior to making any recommendations pursuant to §923.51, the committee shall submit to the Secretary a report setting forth its marketing policy for the ensuing season. Such marketing policy report shall contain information relative to:

(1) The estimated total production of cherries within the production area;

(2) The expected general quality and size of cherries in the production area and in other areas;

(3) The expected demand conditions for cherries in different market outlets;

(4) The expected shipments of cherries produced in the production area and in areas outside the production area;

(5) Supplies of competing commodities;

(6) Trend and level of consumer income;

(7) Other factors having a bearing on the marketing of cherries; and

(8) The type of regulations expected to be recommended during the season.

(b) In the event it becomes advisable, because of changes in the supply and demand situation for cherries, to modify substantially such marketing policy, the committee shall submit to the Secretary a revised marketing policy report setting forth the information prescribed in this section. The committee shall publicly announce the contents of each marketing policy report, including each revised marketing policy report, and copies thereof shall be maintained in the office of the committee where they shall be available for examination by growers and handlers.

§923.51 Recommendations for regulation.

(a) Whenever the committee deems it advisable to regulate the handling of any variety or varieties of cherries in the manner provided in §923.52, it shall so recommend to the Secretary.

(b) In arriving at its recommendations for regulation pursuant to paragraph (a) of this section, the committee shall give consideration to current information with respect to the factors affecting the supply and demand for cherries during the period or periods when it is proposed that such regulation should be made effective. With each such recommendation for regulation, the committee shall submit to the Secretary the data and information on which such recommendation is predicated and such other available information as the Secretary may request.

§923.52 Issuance of regulations.

(a) The Secretary shall regulate, in the manner specified in this section, the handling of cherries whenever he finds, from the recommendations and information submitted by the committee, or from other available information, that such regulations will tend to effectuate the declared policy of the act. Such regulations may:

(1) Limit, during any period or periods, the shipment of any particular grade, size, quality, maturity, or pack, or any combination thereof, of any variety or varieties of cherries grown in any district or districts of the production area;

(2) Limit the shipment of cherries by establishing, in terms of grades, sizes, or both, minimum standards of quality and maturity during any period when season average prices are expected to exceed the parity level;

(3) Fix the size, capacity, weight, dimensions, markings, or pack of the container, or containers, which may be used in the packaging or handling of cherries.

(b) The committee shall be informed immediately of any such regulation issued by the Secretary, and the committee shall promptly give notice thereof to growers and handlers.

[22 FR 3835, June 1, 1957. Redesignated at 26 FR 12751, Dec. 30, 1961, as amended at 66 FR 58356, Nov. 21, 2001]

§923.53 Modification, suspension, or termination of regulations.

(a) In the event the committee at any time finds that, by reason of changed conditions, any regulations issued pursuant to §923.52 should be modified, suspended, or terminated, it shall so recommend to the Secretary.

(b) Whenever the Secretary finds, from the recommendations and information submitted by the committee or from other available information, that a regulation should be modified, suspended, or terminated with respect to any or all shipments of cherries in order to effectuate the declared policy of the act, he shall modify, suspend, or terminate such regulation. On the same basis and in like manner the Secretary may terminate any such modification or suspension. If the Secretary finds that a regulation obstructs or does not tend to effectuate the declared policy of the act, he shall suspend or terminate such regulation. On the same basis and in like manner the Secretary may terminate any such suspension.

§ 923.54 Special purpose shipments.

(a) Except as otherwise provided in this section, any person may, without regard to the provisions of §§ 923.41, 923.52, 923.53, and 923.55, and the regulations issued thereunder, handle cherries (1) for consumption by charitable institutions; (2) for distribution by relief agencies; or (3) for commercial processing into products.

(b) Upon the basis of recommendations and information submitted by the committee, or from other available information, the Secretary may relieve from any or all requirements, under or established pursuant to § 923.41, § 923.52, § 923.53, or § 923.55, the handling of cherries in such minimum quantities, or types of shipments, or for such specified purposes, as the committee, with approval of the Secretary, may prescribe. Specified purposes under this section may include shipments of cherries for grading or packing to specified locations outside the production area and shipments to facilitate the conduct of marketing research and development projects established pursuant to § 923.45.

(c) The committee shall, with the approval of the Secretary, prescribe such rules, regulations, and safeguards as it may deem necessary to prevent cherries handled under the provisions of this section from entering the channels of trade for other than the specific purposes authorized by this section. Such rules, regulations, and safeguards may

include the requirements that handlers shall file applications and receive approval from the committee for authorization to handle cherries pursuant to this section, and that such applications be accompanied by a certification by the intended purchaser or receiver that the cherries will not be used for any purpose not authorized by this section. The committee may rescind or deny to any packing facility the special purpose shipment certificate if proof satisfactory to the committee is obtained that cherries shipped for the purpose stated in this section were handled contrary to the provisions of this section.

[22 FR 3835, June 1, 1957. Redesignated at 26 FR 12751, Dec. 30, 1961, as amended at 66 FR 58356, Nov. 21, 2001]

§ 923.55 Inspection and certification.

Whenever the handling of any variety of cherries is regulated pursuant to § 923.52 or § 923.53, each handler who handles cherries shall, prior thereto, cause such cherries to be inspected by the Federal-State Inspection Service, and certified by it as meeting the applicable requirements of such regulation: *Provided*, That inspection and certification shall be required for cherries which previously have been so inspected and certified only if such cherries have been regraded, resorted, repackaged, or in any other way further prepared for market. Promptly after inspection and certification, each such handler shall submit, or cause to be submitted, to the committee a copy of the certificate of inspection issued with respect to such cherries.

REPORTS

§ 923.60 Reports.

(a) Upon request of the committee, made with the approval of the Secretary, each handler shall furnish to the committee, in such manner and at such time as it may prescribe, such reports and other information as may be necessary for the committee to perform its duties under this part. Such reports may include, but are not necessarily limited to, the following:

(1) The quantities of each variety of cherries received by a handler;

(2) The quantities disposed of by him, segregated as to the respective quantities subject to regulation and not subject to regulation;

(3) The date of each such disposition and the identification of the carrier transporting such cherries, and

(4) The destination of each shipment of such cherries.

(b) All such reports shall be held under appropriate protective classification and custody by the committee, or duly appointed employees thereof, so that the information contained therein which may adversely affect the competitive position of any handler in relation to other handlers will not be disclosed. Compilations of general reports from data submitted by handlers are authorized, subject to the prohibition of disclosure of individual handler's identities or operations.

(c) Each handler shall maintain for at least two succeeding years such records of the cherries received, and of cherries disposed of, by such handler as may be necessary to verify reports pursuant to this section.

MISCELLANEOUS PROVISIONS

§923.61 Compliance.

Except as provided in this part, no person shall handle cherries, the shipment of which has been prohibited by the Secretary in accordance with the provisions of this part: And no person shall handle cherries except in conformity with the provisions of this part.

§923.62 Right of the Secretary.

The members of the committee (including successors and alternates), and any agents, employees, or representatives thereof, shall be subject to removal or suspension by the Secretary at any time. Each and every regulation, decision, determination, or other act of the committee shall be subject to the continuing right of the Secretary to disapprove of the same at any time. Upon such disapproval, the disapproved action of the committee shall be deemed null and void, except as to acts done in reliance thereon or in accordance therewith prior to such disapproval by the Secretary.

§923.63 Effective time.

The provisions of this part, and of any amendment thereto, shall become effective at such time as the Secretary may declare above his signature to this part, and shall continue in force until terminated in one of the ways specified in §923.64.

§923.64 Termination.

(a) The Secretary may at any time terminate the provisions of this part by giving at least one day's notice by means of a press release or in any other manner in which he may determine.

(b) The Secretary shall terminate or suspend the operation of any and all of the provisions of this part whenever he finds that such provisions do not tend to effectuate the declared policy of the act.

(c) The Secretary shall terminate the provisions of this part whenever it is found that such termination is favored by a majority of growers who, during a representative period, have been engaged in the production of cherries: *Provided*, that such majority has, during such representative period, produced for market more than 50 percent of the volume of such cherries produced for market.

(d) The Secretary shall conduct a referendum six years after the effective date of this section and every sixth year thereafter, to ascertain whether continuance of this subpart is favored by growers. The Secretary may terminate the provisions of this subpart at the end of any fiscal period in which the Secretary has found that continuance of this subpart is not favored by growers who, during a representative period determined by the Secretary, have been engaged in the production of cherries in the production area.

(e) The provisions of this part shall, in any event, terminate whenever the provisions of the act authorizing them cease to be in effect.

[22 FR 3835, June 1, 1957, as amended at 70 FR 44252, Aug. 2, 2005]

§923.65 Proceedings after termination.

(a) Upon the termination of the provisions of this part, the committee shall, for the purpose of liquidating the affairs of the committee, continue as trustees of all the funds and property then in its possession, or under its control, including claims for any funds unpaid or property not delivered at the time of such termination.

(b) The said trustees shall (1) continue in such capacity until discharged by the Secretary; (2) from time to time account for all receipts and disbursements and deliver all property on hand, together with all books and records of the committee and of the trustees, to such person as the Secretary may direct; and (3) upon the request of the Secretary, execute such assignments or other instruments necessary or appropriate to vest in such person, full title and right to all of the funds, property, and claims vested in the committee or the trustees pursuant hereto.

(c) Any person to whom funds, property, or claims have been transferred or delivered pursuant to this section shall be subject to the same obligation imposed upon the committee and upon the trustees.

§ 923.66 Effect of termination or amendment.

Unless otherwise expressly provided by the Secretary, the termination of this subpart or of any regulation issued pursuant to this subpart, or the issuance of any amendment to either thereof, shall not (a) affect or waive any right, duty, obligation, or liability which shall have arisen or which may thereafter arise in connection with any provision of this subpart or any regulation issued under this subpart, or (b) release or extinguish any violation of this subpart or of any regulation issued under this subpart, or (c) affect or impair any rights or remedies of the Secretary or of any other person with respect to any such violation.

§ 923.67 Duration of immunities.

The benefits, privileges, and immunities conferred upon any person by virtue of this subpart shall cease upon the termination of this subpart, except with respect to acts done under and during the existence of this subpart.

§ 923.68 Agents.

The Secretary may, by designation in writing, name any officer or employee of the United States, or name any agency or division in the United States Department of Agriculture, to act as his agent or representative in connection with any of the provisions of this part.

§ 923.69 Derogation.

Nothing contained in the provisions of this part is, or shall be construed to be, in derogation or in modification of the rights of the Secretary or of the United States (a) to exercise any powers granted by the act or otherwise, or (b) in accordance with such powers, to act in the premises whenever such action is deemed advisable.

§ 923.70 Personal liability.

No member or alternate member of the committee and no employee or agent of the committee shall be held personally responsible, either individually or jointly with others, in any way whatsoever, to any person for errors in judgment, mistakes, or other acts, either of commission, or omission, as such member, alternate, employee, or agent, except for acts of dishonesty, willful misconduct, or gross negligence.

§ 923.71 Separability.

If any provision of this part is declared invalid, or the applicability thereof to any person, circumstance, or thing is held invalid, the validity of the remainder of this part or the applicability thereof to any other person, circumstance, or thing shall not be affected thereby.

§ 923.142 Reserve fund.

(a) The establishment of a reserve fund of an amount which shall not exceed approximately 1 fiscal year's operational expenses is appropriate and necessary to the maintenance and functioning of the Washington Cherry Marketing Committee. The committee is authorized to expend any funds in such reserve for expenses authorized pursuant to § 923.42.

(b) Terms used in this section shall have the same meaning as given to the respective term in said marketing agreement and order.

[33 FR 9147, June 21, 1968. Redesignated at 44 FR 73011, Dec. 17, 1979]

§923.236 Assessment rate.

On and after April 1, 2019, an assessment rate of $0.20 per ton is established for the Washington Cherry Marketing Committee.

[84 FR 65265, Nov. 27, 2019]

GRADE, SIZE, CONTAINER AND PACK REGULATION

§923.322 Washington cherry handling regulation.

(a) *Grade.* No handler shall handle, except as otherwise provided in this section, any lot of cherries, except cherries of the Rainier, Royal Anne, and similar varieties, commonly referred to as "light sweet cherries" unless such cherries grade at least Washington No. 1 grade except that the following tolerances, by count, of the cherries in the lot shall apply in lieu of the tolerances for defects provided in the Washington State Standards for Grades of Sweet Cherries: *Provided,* That a total of 10 percent for defects including in this amount not more than 5 percent, by count, of the cherries in the lot, for serious damage, and including in this latter amount not more than one percent, by count, of the cherries in the lot, for cherries affected by decay: *Provided further,* That the contents of individual packages in the lot are not limited as to the percentage of defects but the total of the defects of the entire lot shall be within the tolerances specified.

(b) *Size.* No handler shall handle, except as otherwise provided in this section, any lot of cherries unless such cherries meet the following minimum size requirements:

(1) For the Rainier variety and similar varieties commonly referred to as "lightly colored sweet cherries," at least 90 percent, by count, of the cherries in any lot shall measure not less than $^{61}/_{64}$-inch in diameter and not more than 5 percent, by count, may be less than $^{57}/_{64}$-inch in diameter.

(2) For all other varieties, at least 90 percent, by count, of the cherries in any lot shall measure not less than $^{54}/_{64}$ inch in diameter and not more than 5 percent, by count, may be less than $^{52}/_{64}$ inch in diameter.

(c) *Maturity.* No handler shall handle, except as otherwise provided in this section, any lot of Rainier cherries or other varieties of "lightly colored sweet cherries" unless such cherries meet a minimum of 17 percent soluble solids as determined from a composite sample by refractometer prior to packing, at time of packing, or at time of shipment: *Provided,* That individual lots shall not be combined with other lots to meet soluble solids requirements.

(d) *Pack.* (1) When containers of cherries are marked with a row count/row size designation the row count/row size marked shall be one of those shown in Column 1 of the following table and at least 90 percent, by count, of the cherries in any lot shall be not smaller than the corresponding diameter shown in Column 2 of such table: *Provided,* That the content of individual containers in the lot are not limited as to the percentage of undersize; but the total of undersize of the entire lot shall be within the tolerance specified.

TABLE

Column 1, row count/row size	Column 2 diameter (inches)
8	$^{84}/_{64}$
8½	$^{79}/_{64}$
9	$^{75}/_{64}$
9½	$^{71}/_{64}$
10	$^{67}/_{64}$
10½	$^{64}/_{64}$
11	$^{61}/_{64}$
11½	$^{57}/_{64}$
12	$^{54}/_{64}$

(2) When containers of cherries are marked with a minimum diameter, at least 95 percent, by count, of the cherries in any lot and at least 90 percent, by count, of the cherries in any container, shall be not smaller than such minimum diameter.

(e) *Light sweet cherries marked as premium.* No handler shall handle, except as otherwise provided in this section, any package or container of Rainier cherries or other varieties of lightly colored sweet cherries marked as premium except in accordance with the following:

(1) *Quality.* 90 percent, by count, of such cherries in any lot must exhibit a pink-to-red surface blush and, for any given sample, not more than 20 percent

of the cherries shall be absent a pink-to-red surface blush.

(2) *Pack.* At least 90 percent, by count, of the cherries in any lot shall measure not less than $^{64}/_{64}$ inch ($10\frac{1}{2}$ row) in diameter and not more than 5 percent, by count, may be less than $^{61}/_{64}$ inch (11-row) in diameter.

(f) *Grading or packing cherries outside the production area.* (1) Persons desiring to ship or receive cherries for grading or packing outside the production area shall apply to the committee on a "Shippers/Receivers Application for Special Purpose Shipment Certificate" form, and receive approval from the Committee. The application shall contain the following: (i) Name, address, telephone number, and signature of applicant;

(ii) Certification by the applicant that cherries graded and packed outside the production area shall be inspected by the Federal-State Inspection Service and shall meet the grade, size, maturity, and pack requirements of this section prior to shipment; and

(iii) Such other information as the committee may require.

(2) Each approved applicant shall furnish to the committee, at the close of business every Friday, a report containing the following information on a "Special Purpose Shipment Report" form:

(i) Name, address, telephone number, and signature of applicant;

(ii) Names of growers and handlers of such cherries;

(iii) The total quantity of each variety of cherries; and

(iv) Such other information as the committee may require.

(3) The committee may rescind or deny to any applicant its approval of the "Shippers/Receivers Application for Special Purpose Shipment Certificate" if proof satisfactory to the committee is obtained that any cherries shipped or received by such applicant for grading or packing were handled contrary to the provisions of this section.

(g) *Exceptions.* Any individual shipment of cherries which meets each of the following requirements may be handled without regard to the provisions of paragraphs (a), (b), (c), (d), and (e) of this section, and of §§ 923.41 and 923.55.

(1) The shipment consists of cherries sold for home use and not for resale;

(2) The shipment does not, in the aggregate, exceed 100 pounds, net weight, of cherries; and

(3) Each container is stamped or marked with the words *not for resale* in letters at least one-half inch in height.

(h) *Definitions.* When used herein, *Washington No. 1* and *diameter* shall have the same meaning as when used in the Washington State Standards for Grades of Sweet Cherries (Order 1550 effective April 29, 1978, WAC 16–414–050); *face packed* means that cherries in the top layer in any container are so placed that the stem ends are pointing downward toward the bottom of the container; *row count/row size* means the number of cherries of a uniform size necessary to pack row-faced across a $10\frac{1}{2}$ inch inside width container or comparable number of cherries when packed loose in a container.

[47 FR 31538, July 21, 1982]

EDITORIAL NOTE: For FEDERAL REGISTER citations affecting § 923.322, see the List of CFR Sections Affected, which appears in the Finding Aids section of the printed volume and at *www.govinfo.gov.*

PART 925—GRAPES GROWN IN A DESIGNATED AREA OF SOUTHEASTERN CALIFORNIA

Subpart A—Order Regulating Handling

AUTHORITY: 7 U.S.C. 601–674.

SOURCE: 45 FR 40566, June 16, 1980, unless otherwise noted.

Subpart A—Order Regulating Handling

DEFINITIONS

§925.1 Secretary.

Secretary means the Secretary of Agriculture of the United States, or any officer or employee of the Department to whom authority has heretofore been delegated, or to whom authority may hereafter be delegated.

§925.2 Act.

Act means Public Act No. 10, 73d Congress (May 12, 1933), as amended and as reenacted and amended by the Agricultural Marketing Agreement Act of 1937, as amended (48 Stat. 31, as amended; 7 U.S.C. 601–674).

§925.3 Person.

Person means an individual, partnership, corporation, association, or any other business unit.

§925.4 Grapes.

Grapes means any variety of vinifera species table grapes grown in the production area.

§925.5 Production area.

Production area means Imperial County, California, and that part of Riverside County and San Diego County, California, situated east of a line drawn due north and south through the Post Office in White Water, California.

§925.6 Varieties.

Varieties means and includes all classifications or subdivisions of Vitis vinifera table grapes.

§925.7 Producer.

Producer is synonymous with *grower* and means any person who produces grapes for the fresh market and who has a proprietary interest therein.

§925.8 Handler.

Handler is synonymous with *shipper* and means any person (except a common or contract carrier of grapes owned by another person) who handles grapes or causes grapes to be handled.

§ 925.10 Handle.

Handle is synonymous with *ship* and means to pack, sell, deliver (including delivery to a storage facility), transport, or in any way to place grapes in the current of commerce within the production area or between the production area and any point outside thereof: *Provided,* That such term shall not include the sale of grapes on the vine and except when regulations are effective pursuant to § 925.52(a)(5) shall not include the transportation or delivery of grapes to a packinghouse within the production area for preparation for market.

§ 925.11 Pack.

Pack means the specific arrangement, weight, grade or size, including the uniformity thereof, of the grapes within a container: *Provided,* That when used in or with respect to § 925.52(a)(5) such term shall mean to place grapes into containers for shipment to market as fresh grapes.

§ 925.12 Fiscal period.

Fiscal period is synonymous with *fiscal year* and means the 12 month period beginning on December 1 of one year and ending the last day of November of the following year or such other period as the committee, with the approval of the Secretary, may prescribe.

§ 925.13 Container.

Container means any lug, box, bag, crate, carton, or any other receptacle used in packing grapes for shipment as fresh grapes, and includes the dimensions, capacity, weight, marking, and any pads, liners, lids, and any or all appurtenances thereto or parts thereof. The term applies, in the case of grapes packed in consumer packages, to the master receptacle and to any and all packages therein.

§ 925.14 Committee.

Committee means the California Desert Grape Administrative Committee established under § 925.20.

ADMINISTRATIVE BODY

§ 925.20 Establishment and membership.

(a) There is hereby established a California Desert Grape Administrative Committee consisting of 12 members, each of whom shall have an alternate who shall have the same qualifications as the member. Five of the members and their alternates shall be producers or officers or employees of producers (producer members). Five of the members and their alternates shall be handlers or officers or employees of handlers (handler members). One member and alternate shall be either a producer or handler or officer or employee thereof. One member and alternate shall represent the public.

(b) Not more than two members and not more than two alternate members shall be affiliated with the same handler entity.

(c) The committee may, with the approval of the Secretary, provide such other allocation of producer or handler membership, or both, as may be necessary to assure equitable representation.

§ 925.21 Term of office.

The term of office of the members and alternates shall be four fiscal periods. Each member and alternate shall serve in such capacities for the portion of the term of office for which they are selected and have qualified and until their respective successors are selected and have qualified.

[45 FR 40566, June 16, 1980, as amended at 81 FR 44761, July 11, 2016]

§ 925.22 Nomination.

(a) *Initial members.* Nominations for each of the initial members, together with nominations for the initial alternate members for each position, may be submitted to the Secretary by the Committee responsible for promulgation of this part. Such nominations may be made by means of a meeting of the growers and a meeting of the handlers. Such nominations, if made, shall be filed with the Secretary no later than the effective date of this part. In the event nominations for initial members and alternate members of the committee are not filed pursuant to, and within the time specified in, this section, the Secretary may select such initial members and alternate members without regard to nominations, but selections shall be on the basis of the representation provided in §925.20.

(b) *Successor members.* The Secretary shall cause to be held, not later than November 15, of each year, meetings of producers and handlers for the purpose of making nominations for members and alternate members of the committee.

(c) Only producers, including duly authorized officers or employees of producers, who are present at such nomination meetings, may participate in the nomination and election of nominees for producer members and their alternates. Each producer entity shall be entitled to cast only one vote. If a person is both a producer and a handler of grapes, such person may participate in both producer and handler nominations.

(d) Only handlers, including duly authorized officers or employees of handlers, who are present at such nomination meetings, may participate in the nomination and election of nominees for handler members and their alternates. Each handler entity shall be entitled to cast only one vote.

(e) One member and alternate member shall be nominated by a vote of both producers and handlers and may be of either group.

(f) The public member and alternate member shall be nominated by the committee. The committee shall prescribe, with the approval of the Secretary, procedures for the nomination of the public member and qualification requirements for such member.

§925.23 Selection.

The Secretary shall select members and alternate members of the committee from persons nominated pursuant to §925.22 or from other qualified persons.

§925.24 Failure to nominate.

If nominations are not made within the time and in the manner specified in §925.22 the Secretary may select the members and alternate members of the committee without regard to nomina-

tions on the basis of the representation provided for in §925.20.

§925.25 Qualification and acceptance.

Any person selected as a member or alternate member of the Committee shall, prior to such selection, qualify by filing a qualifications questionnaire advising the Secretary that he or she agrees to serve in the position for which nominated.

[81 FR 44761, July 11, 2016]

§925.26 Vacancies.

To fill any vacancy occasioned by the failure of any person selected as a member or as an alternate member of the committee to qualify, or in the event of the death, removal, resignation, or disqualification of any member or alternate member of the committee, a successor for the unexpired term of such member or alternate member of the committee shall be nominated and selected in the manner specified in §§925.22 and 925.23. If the names of the nominees to fill any such vacancy are not made available to the Secretary within a reasonable time after such vacancy occurs, the Secretary may fill such vacancy without regard to nominations, which selection shall be made on the basis of the representation provided for in §925.20.

§925.27 Alternate members.

An alternate member shall act in the place of the member during such member's absence or at such member's request, and may be assigned other program duties by the chairman or the committee. In the event of the death, removal, resignation, or disqualification of a member the alternate shall act for the member until a successor for such member is selected and has qualified. In the event that both a member and that member's alternate are unable to attend a committee meeting, the member or committee members present may designate any other alternate to serve in such member's place at the meeting if such action is necessary to secure a quorum: *Provided,* That not more than two members or alternates acting for members who are affiliated with the same handler entity shall serve as members at the same meeting.

§ 925.28 Powers.

The committee shall have the following powers:

(a) To administer the provisions of this part in accordance with its terms;

(b) To receive, investigate, and report to the Secretary complaints of violations of the provisions of this part;

(c) To make and adopt rules and regulations to effectuate the terms and provisions of this part; and

(d) To recommend to the Secretary amendments to this part.

§ 925.29 Duties.

The committees shall have, among others, the following duties:

(a) To select a chairman and such other officers as may be necessary, and to define the duties of such officers;

(b) To appoint such employees, agents, and representatives as it may deem necessary, and to determine compensation and to define the duties of each;

(c) To submit to the Secretary as soon as practicable after the beginning of each fiscal period a budget for such period, including a report in explanation of the items appearing therein and a recommendation as to the rate of assessment for such period;

(d) To keep minutes, books, and records, which will reflect all of the acts and transactions of the committee and which shall be subject to examination by the Secretary;

(e) To prepare periodic statements of the financial operations of the committee and to make copies of each such statement available to growers and handlers for examination at the office of the committee;

(f) To cause its books to be audited by a competent public accountant at least once each fiscal period and at such times as the Secretary may request;

(g) To act as intermediary between the Secretary and any grower or handler;

(h) To investigate and assemble data on the growing, handling, and marketing conditions with respect to grapes;

(i) To submit to the Secretary the same notice of meetings of the committee as is given to its members;

(j) To submit to the Secretary such available information as may be requested; and

(k) To investigate compliance with the provisions of this part.

§ 925.30 Procedure.

(a) Eight members of the committee shall constitute a quorum and any action of the committee shall require at least eight concurring votes;

(b) The committee may vote by telephone, telegraph, or other means of communications; and any votes so cast shall be confirmed promptly in writing: *Provided,* That if an assembled meeting is held, all votes shall be cast in person.

§ 925.31 Compensation and expenses.

The members of the committee, and alternates when acting as members, shall serve without compensation but may be reimbursed for expenses necessarily incurred by them in the performance of their duties under this part: *Provided,* That the committee at its discretion may request the attendance of one or more alternates at any or all meetings notwithstanding the expected or actual presence of the respective members and may pay expenses as aforesaid.

§ 925.32 Annual report.

The committee should, as soon as practicable, after the close of each fiscal period, prepare and mail an annual report to the Secretary and make a copy available to each grower and handler who requests a copy of the report.

EXPENSES AND ASSESSMENTS

§ 925.40 Expenses.

The committee is authorized to incur such expenses as the Secretary finds are reasonable and likely to be incurred by the committee for its maintenance and functioning and to enable it to exercise its powers and perform its duties in accordance with the provisions of this part. The funds to cover such expenses shall be acquired in the manner prescribed in § 925.41.

§925.41 Assessments.

(a) Each person who first handles grapes shall pay to the committee, upon demand, such handler's pro rata share of the expenses which the Secretary finds are reasonable and likely to be incurred by the committee during a fiscal period. The payment of assessments for the maintenance and functioning of the committee may be required under this part throughout the period it is in effect irrespective of whether particular provisions thereof are suspended or become inoperative.

(b) The Secretary shall fix the rate of assessment to be paid by each such person during a fiscal period in an amount designed to secure sufficient funds to cover the expenses which may be incurred during such period and to accumulate and maintain a reserve fund equal to approximately one fiscal period's expenses. At any time during or after a fiscal period, the Secretary may increase the rate of assessment in order to secure sufficient funds to cover any later findings by the Secretary relative to the expenses which may be incurred. Such increase shall be applied to all grapes handled during the applicable fiscal period. In order to provide funds for the administration of the provisions of this part during the first part of a fiscal period before sufficient operating income is available from assessments in the current period's shipments, the committee may accept the payment of assessments in advance, and may also borrow money for such purpose.

(c) Any assessment not paid by a handler within a period of time prescribed by the committee may be subject to an interest or late payment charge, or both. The period of time, rate of interest, and late payment charge shall be recommended by the committee and approved by the Secretary. Subsequent to such approval, all assessments not paid within the prescribed time shall be subject to the interest or late payment charge, or both.

§925.42 Accounting.

(a) If, at the end of a fiscal period, the assessments collected are in excess of expenses incurred, such excess shall be accounted for in accordance with one of the following:

(1) If such excess is not retained in a reserve, as provided in paragraph (d)(2) of this section, it shall be refunded proportionately to the persons from whom it was collected: *Provided,* That any sum paid by a person in excess of that person's pro rata share of the expenses during any fiscal period may be applied by the committee at the end of such fiscal period to any outstanding obligations due the committee from such person.

(2) The committee, with the approval of the Secretary, may carry over such excess into subsequent fiscal periods as a reserve: *Provided,* That funds in the reserve shall not exceed approximately one fiscal period's expenses. Such reserve funds may be used: (i) To defray expenses, during any fiscal period, prior to the time the assessment income is sufficient to cover such expenses; (ii) to cover deficits incurred during any fiscal period when assessment income is less than expenses; (iii) to defray expenses incurred during any period when any or all provisions of this part are suspended or are inoperative; or (iv) to cover necessary expenses of liquidation in the event of termination of this part. Upon such termination, any funds not required to defray the necessary expenses of liquidation shall be disposed of in such manner as the Secretary may determine to be appropriate: *Provided,* That to the extent practicable such funds shall be returned pro rata to the persons from whom such funds were collected.

(b) All funds received by the committee under this part shall be used solely for the purpose specified in this part and shall be accounted for in the manner provided in this part. The Secretary may at any time require the committee and its members to account for all receipts and disbursements.

(c) Upon the removal or expiration of the term of office of any member of the committee, such member shall account for all receipts and disbursements and deliver all property and funds in such member's possession to the committee, and shall execute such assignments and other instruments as may be necessary

169

or appropriate to vest in the committee full title to all of the property, funds, and claims vested in such member pursuant to this part.

RESEARCH AND MARKET DEVELOPMENT

§ 925.45 Production research and market research and development.

The committee, with the approval of the Secretary, may establish or provide for the establishment of production research, marketing research and development projects designed to assist, improve or promote the marketing, distribution and consumption or the efficient production of grapes. The expense of such projects shall be paid from funds collected pursuant to this part.

REGULATIONS

§ 925.50 Marketing policy.

Each season prior to making any recommendation pursuant to § 925.51 the committee shall submit to the Secretary a report setting forth its marketing policy for the ensuing marketing season. Such marketing policy report shall contain information relative to:

(a) The estimated total shipments of grapes produced within the production area;

(b) The expected general quality of grapes in the production area;

(c) The expected demand conditions for grapes;

(d) The probable prices for grapes;

(e) Supplies of competing commodities, including foreign produced grapes;

(f) Trend and level of consumer income;

(g) Other factors having a bearing on the marketing of grapes; and

(h) The type of regulations expected to be recommended during the marketing season.

§ 925.51 Recommendation for regulation.

Upon complying with the requirements of § 925.50 the committee may recommend regulations to the Secretary whenever the committee deems that such regulations as are provided in § 925.52 will tend to effectuate the declared policy of the act.

§ 925.52 Issuance of regulations.

(a) The Secretary shall regulate, in the manner specified in this section, the handling of grapes upon finding from the recommendations and information submitted by the committee, or from other available information, that such regulation would tend to effectuate the declared policy of the act. Such regulation may:

(1) Limit the handling of any grade, size, quality, maturity, or pack, or any combination thereof, of any or all varieties of grapes during any period or periods;

(2) Limit the handling of any grade, size, quality, maturity, or pack of grapes differently for different varieties, or any combination of the foregoing during any period or periods;

(3) Limit the handling of grapes by establishing in terms of grades, sizes, or both, minimum standards of quality and maturity during any period when season average prices are expected to exceed the parity level;

(4) Fix the size, capacity, weight, dimensions, markings, materials, or pack of the container which may be used in handling of grapes;

(5) Establish holidays by prohibiting the packing of all varieties of grapes during a specified period or periods.

(b) No handler shall handle grapes that were packed during any period when such packing was prohibited by any regulation issued under paragraph (a)(5) of this section unless such grapes are handled under § 925.54.

§ 925.53 Modification, suspension, or termination of regulations.

(a) In the event the committee at any time finds that, by reason of changed conditions, any regulations issued pursuant to § 925.52 should be modified, suspended, or terminated, it shall so recommend to the Secretary.

(b) Whenever the Secretary finds from the recommendations and information submitted by the committee or from other available information that a regulation should be modified, suspended, or terminated with respect to any or all shipments of grapes in order to effectuate the declared policy of the act, the Secretary shall modify, suspend, or terminate such regulation. If the Secretary finds that a regulation obstructs or does not tend to effectuate the declared policy of the act, the Secretary shall suspend or terminate such regulation. On the same basis and in like manner the Secretary may terminate any such modification or suspension.

§925.54 Special purpose shipments.

(a) Regulations in effect pursuant to §925.41, §925.52, or §925.55 may be modified, suspended, or terminated to facilitate handling of grapes for purposes which may be recommended by the committee and approved by the Secretary.

(b) The committee shall, with the approval of the Secretary, prescribe such rules, regulations, and safeguards as it may deem necessary to prevent grapes handled under the provisions of this section from entering the channels of trade for other than the specific purposes authorized by this section.

INSPECTION AND CERTIFICATION

§925.55 Inspection and certification.

(a) Whenever the handling of any variety of grapes is regulated pursuant to §925.52, each handler who handles grapes shall, prior thereto, cause such grapes to be inspected by the Federal or Federal-State Inspection Service and certified as meeting the applicable requirements of such regulation: *Provided,* That inspection and certification shall not be required for grapes which previously have been so inspected and certified if such prior inspection was performed within such period as may be established pursuant to paragraph (b) of this section. Promptly after the inspection and certification each such handler shall submit, or cause to be submitted, to the committee a copy of the certificate of inspection issued with respect to such grapes.

(b) The committee may, with the approval of the Secretary, establish a period prior to shipment during which the inspection required by this section must be performed.

(c) The committee may enter into an agreement with the Federal and Federal-State Inspection Services with respect to the costs of the inspection required by paragraph (a) of this section,

and may collect from handlers their respective pro rata share of such costs.

REPORTS

§925.60 Reports.

(a) Each handler shall furnish to the committee, at such times and for such periods as the committee may designate, certified reports covering, to the extent necessary for the committee to perform its functions, each shipment of grapes as follows:

(1) The name of the shipper and the shipping point;

(2) The car or truck license number (or name of the trucker), and identification of the carrier;

(3) The date and time of departure;

(4) The variety;

(5) The number and type of containers in the shipment;

(6) The destination; and

(7) Identification of the inspection certificate pursuant to which the grapes were handled.

(b) Upon request of the committee, made with the approval of the Secretary, each handler shall furnish to the committee, in such manner and at such times as it may prescribe, such other information as may be necessary to enable the committee to perform its duties under this part.

(c) Each handler shall maintain for at least two succeeding fiscal periods after the end of the fiscal period in which the transactions occurred, such records of the grapes received and disposed of by such handler as may be necessary to verify the reports such handler submits to the committee pursuant to this section.

(d) All reports and records submitted by handlers pursuant to the provisions of this section shall be received by, and at all times be in custody of one or more designated employees of the committee. No such employee shall disclose to any person, other than the Secretary upon request therefor, data or information obtained or extracted from such reports and records which might affect the trade position, financial condition, or business operation of the particular handler from whom received: *Provided,* That such data and information may be combined, and made available to any person, in the form of

171

general reports in which the identities of the individual handlers furnishing the information are not disclosed and may be revealed to any extent necessary to effect compliance with the provisions of this part and the regulations issued thereunder.

MISCELLANEOUS PROVISIONS

§ 925.61 Compliance.

Except as provided in this part, no handler shall handle grapes except in conformity with the provisions of this part and the regulations issued thereunder.

§ 925.62 Right of the Secretary.

The members of the committee (including successors and alternates) and any agents, employees, or representatives thereof, shall be subject to removal or suspension by the Secretary at any time. Each and every regulation, decision, determination, or other act of the committee shall be subject to the continuing right of the Secretary to disapprove of the same at any time. Upon such disapproval, the disapproved action of the committee shall be deemed null and void, except as to acts done in reliance thereon or in accordance therewith prior to such disapproval by the Secretary.

§ 925.63 Termination.

(a) The Secretary shall terminate or suspend the operation of any and all of the provisions of this part whenever the Secretary finds that such provisions do not tend to effectuate the declared policy of the act.

(b) The Secretary shall terminate the provisions of this part whenever it is found by referendum or otherwise that such termination is favored by a majority of the growers: *Provided,* That such majority has during the current marketing season produced more than 50 percent of the volume of grapes which were produced within the production area for shipment in fresh form. Such termination shall become effective on the first day of December subsequent to the announcement thereof by the Secretary.

(c) The provisions of this part shall, in any event, terminate whenever the provisions of the act authorizing them cease to be in effect.

§ 925.64 Proceedings after termination.

(a) Upon the termination of the provisions of this part, the committee shall, for the purpose of liquidating the affairs of the committee, continue as trustees of all the funds and property then in its possession, or under its control, including claims for any funds unpaid or property not delivered at the time of such termination. Any action by said trustees shall require the concurrence of a majority of the trustees.

(b) The said trustees shall:

(1) Continue in such capacity until discharged by the Secretary;

(2) From time to time account for all receipts and disbursements and deliver all property on hand, together with all books and records of the committee and of the trustees, to such persons as the Secretary may direct;

(3) Upon the request of the Secretary, execute such assignments or other instruments necessary or appropriate to vest in such person, full title and right to all of the funds, property, and claims vested in the committee or the trustees pursuant thereto.

(c) Any person to whom funds, property, or claims have been transferred or delivered, pursuant to this section, shall be subject to the same obligation imposed upon the committee and upon the trustees.

§ 925.65 Effect of termination or amendment.

Unless otherwise expressly provided by the Secretary, the termination of this part or any regulation issued pursuant to this part, or the issuance of any amendment to either thereof, shall not:

(a) Affect or waive any right, duty, obligation, or liability which shall have arisen or which may thereafter arise in connection with any provision of this part or any regulation issued under this part; or

(b) Release or extinguish any violation of this part or any regulation issued under this part; or

(c) Affect or impair any rights or remedies of the Secretary or any other person with respect to any such violation.

§925.66 Duration of immunities.

The benefits, privileges, and immunities conferred upon any person by virtue of this part shall cease upon its termination, except with respect to acts done under and during the existence of this part.

§925.67 Derogation.

Nothing contained in this part is, or shall be construed to be, in derogation or in modification of the rights of the Secretary or of the United States: (a) To exercise any powers granted by the act or otherwise; or (b) in accordance with such powers, to act in the premises whenever such action is deemed advisable.

§925.68 Personal liability.

No member or alternate member of the committee and no employee or agent of the committee shall be held personally responsible, either individually or jointly with others, in any way whatsoever, to any person for errors in judgment, mistakes, or other acts, either of commission or omission, as such member, alternate, employee, or agent, except for acts of dishonesty, willful misconduct, or gross negligence.

§925.69 Separability.

If any provision of this part is declared invalid or the applicability thereof to any person, circumstance, or thing is held invalid, the validity of the remainder of this part or the applicability thereof to any other person, circumstance, or thing shall not be affected thereby.

(Secs. 1–19, 48 Stat. 31, as amended (7 U.S.C. 601–674))

Subpart B—Administrative Requirements

§925.112 Fiscal period.

Beginning January 1, 1988, *fiscal period* will mean January 1 through December 31 of each year.

[52 FR 27538, July 22, 1987]

§925.141 Late payments.

(a) The committee shall impose a late payment charge of 5 percent on the unpaid balance on any handler whose assessment has not been received in the committee's office, or the envelope containing the payment legibly postmarked by the U.S. Postal Service, within 45 days of the invoice date shown on the handler's assessment statement.

(b) In addition to that specified in paragraph (a) of this section, the committee shall impose an interest charge on any handler whose assessment payment has not been received in the committee's office, or the envelope containing the payment legibly postmarked by the U.S. Postal Service, within 45 days of the invoice date. The rate of 1½ percent per month shall be applied to the unpaid balance and late payment charge for the number of days all or any part of the assessment specified in the handler's assessment statement is delinquent beyond the 45 day period.

(c) The committee, upon receipt of a late payment, shall promptly notify the handler (by registered mail) of any late payment charge and/or interest charge due as provided in paragraphs (a) and (b) of this section. If such charges are not paid, or the envelope containing payment is not legibly postmarked by the U.S. Postal Service, within 45 days of the date of such notification, late payment and interest charges as provided in paragraphs (a) and (b) of this section will accrue on the unpaid amount.

[57 FR 24352, June 9, 1992]

§925.160 Reports.

(a) When requested by the California Desert Grape Administrative Committee, each shipper who ships grapes, shall furnish an end-of-season grape shipment report (CDGAC-3) to the Committee no later than 10 days after the last day of shipment for the season or such later time the Committee deems appropriate. Such reports shall show the reporting period, the name and other identification of the shipper and grower, the invoice number, shipping date, varietal name, shipment destination (city and state), and the number of lugs shipped (pounds).

(b) When requested by the California Desert Grape Administrative Committee (CDGAC), each shipper who

173

ships grapes shall furnish to the committee at such time as the committee shall require, an annual grape acreage survey (CDGAC Form 7), which shall include, but is not limited to, the following: The applicable year in which the report is requested; the names of the shipper (handler) who will handle the grapes and the grower who produces them; the location of each vineyard; the variety or varieties grown in each vineyard; and the bearing, nonbearing, and total acres of each vineyard.

(c) Handlers that donate grapes to charitable organizations pursuant to § 925.304(c) shall submit a completed Food Donation Form (CDGAC Form No. 8) to the Committee within 2 days of receipt by the charitable organization. Such form shall include the following: The name of the producer; the name of the handler; loading location and date; inspection location and date; Variety(s) Federal State Inspection Service (FSIS) Certificate number(s); lug weight (pounds); number of lugs; label; signature of person responsible for loading at handling facility; recipient charity name; how many lugs received; signature of responsible charity recipient and date received. Any such grapes shall not be used for resale.

[69 FR 21692, Apr. 22, 2004, as amended at 72 FR 29840, May 30, 2007; 81 FR 24458, Apr. 26, 2016]

Subpart C—Assessment Rates

§ 925.215   Assessment rate.

On and after January 1, 2018, an assessment rate of $0.020 per 18-pound lug is established for grapes grown in a designated area of southeastern California.

[83 FR 21167, May 9, 2018]

§ 925.304   California Desert Grape Regulation 6.

During the period April 10 through July 10 each year, no person shall pack or repack any variety of grapes except Emperor, Almeria, Calmeria, and Ribier varieties, on any Saturday, Sunday, Memorial Day, or the observed Independence Day holiday, unless approved in accordance with paragraph (e) of this section, nor handle any variety of grapes except Emperor, Calmeria, Almeria, and Ribier varieties, unless such grapes meet the requirements specified in this section.

(a) *Grade, size, and maturity.* Except as provided in paragraphs (a)(3) and (4) of this section, such grapes shall meet the minimum grade and size requirements established in paragraphs (a)(1) or (2) of this section.

(1) U.S. No. 1 Table, as set forth in the United States Standards for Grades of Table Grapes (European or Vinifera Type 7 CFR 51.880 through 51.914), with the exception of the tolerance percentage for bunch size when packed in individual consumer clamshell packages weighing 5 pounds or less: Provided that not more than 20 percent of the weight of such containers may consist of single clusters weighing less than one-quarter pound, but with at least five berries each; or

(2) U.S. No. 1 Institutional, with the exception of the tolerance percentage for bunch size. Such tolerance shall be 33 percent instead of 4 percent as is required to meet U.S. No. 1 Institutional grade. Grapes meeting these quality requirements may be marked "DGAC No. 1 Institutional" but shall not be marked "Institutional Pack."

(3) Grapes of the Perlette variety shall meet the minimum berry size requirement of ten-sixteenths of an inch;

(4) Grapes of the Flame Seedless variety shall meet the minimum berry size requirement of ten-sixteenths of an inch and shall be considered mature if the juice meets or exceeds 16.5 percent soluble solids, or contains not less than 15 percent soluble solids and the soluble solids are equal to or in excess of 20 parts to every part acid contained in juice in accordance with applicable sampling and testing procedures specified in sections 1436.3, 1436.5, 1436.6, 1436.7, 1436.12, and 1436.17 of Article 25 of Title 3: California Code of Regulations (CCR).

(b) *Container and pack.* (1) Such grapes shall be packed in one of the following containers, which are new and clean, and otherwise meet the requirements of sections 1380.14, and 1380.19(n), 1436.37, and 1436.38 of Title 3: California Code of Regulations, except that reusable plastic containers may be reused if such containers are clean:

CONTAINER DESCRIPTIONS IN INCHES

Container	Depth	Width	Length
28 Sawdust Pack	7¾ (inside)	14¹⁵/₁₆ (inside)	18⅝ (inside)
38J Polystyrene Lug	6¾ (inside)	12½ (inside)	15⅜ (inside)
38K Standard Grape	4½–8½ (inside)	13½–14½ (outside)	16⅝–17½ (outside)
38L Grape Lug	7⅝ (inside)	13¹¹/₁₆ (outside)	16 (outside)
38M Grape Lug	4¼–5¾ (inside)	15⅜–16 (outside)	23½–24 (outside)
38Q Polystyrene Lug	6¼–8¼ (inside)	11¼ (inside)	18⅛ (inside)
38R Grape Lug	4–7 (inside)	15¾–16 (outside)	19¹¹/₁₆–20 (outside)
38S Grape Lug	5–9 (inside)	11¹¹/₁₆–12 (outside)	19¹¹/₁₆—20 (outside)
38T Grape Lug	5½–7½ (inside)	13⅛–13¹⁵/₁₆ (outside)	15⁵/₁₆–16 (outside)
38U Grape Lug	6³/₁₆–7 (inside)	13¹¹/₁₆ (outside)	20½ (outside)
38 V Grape Lug	5¾ (inside)	14 (outside)	16 (outside)
CP Grape Lug	3¹⁵/₁₆–4¾ (inside)	15¾–15⁹/₁₆ (outside)	23½–23¾ (outside)
CP1 Grape Lug	4¾–5 (inside)	19½–20 (outside)	23¾–24 (outside)

(ii) Containers with a net weight of 5 kilograms (approximately 11 pounds) shall be for export only.

(iii) Such other types and sizes of containers as may be approved by the Committee for experimental or research purposes.

(2) The minimum net weight of grapes in any such containers, except for containers containing grapes packed in sawdust, cork, excelsior or similar packing material, or packed in bags or wrapped in plastic or paper, and containers authorized in paragraph (b)(1)(iii) of this section, shall be 20 pounds based on the average net weight of grapes in a representative sample of containers. Grapes in any such containers packed in bags, or wrapped in plastic or paper prior to being placed in these containers shall meet a minimum net weight of 18 pounds based on the average net weight of grapes in a representative sample of containers: *Provided,* That grapes packed in master containers containing individual consumer packages are exempt from container marking requirements and minimum net weight requirements. Containers of grapes other than master containers containing individual consumer packages shall be marked with the minimum net weight of 20 or 18 pounds.

(3) Such containers of grapes shall be plainly marked with the minimum net weight of grapes contained therein (with numbers and letters at least one-fourth inch in height), the name of the variety of the grapes and the name of the shipper, as provided in §§ 1436.30 and 1359 of Title 3: California Code of Regulations.

(4) Such containers of grapes shall be plainly marked with the lot stamp number corresponding to the lot inspection conducted by an authorized inspector, except that such requirement shall not apply to containers in the center tier of a lot palletized in a 3 box by 3 box pallet configuration: *Provided,* That pallets of reusable plastic containers shall have the lot stamp number stamped on two USDA-approved pallet tags, each affixed to opposite sides of the pallet of containers, in addition to other required information on the cards of the individual containers.

(c) *Donation to charitable organizations.* Handlers of grapes failing to meet the requirements of § 925.55 and paragraph (a) of this section may donate such grapes to charitable organizations. Any such grapes shall not be used for resale. Handlers donating such grapes to a charitable organization shall submit a completed Food Donation Form, CDGAC Form No. 8, as required in § 925.160(c), within 2 days of receipt by the intended charity.

(d) *Organically grown grapes.* Organically grown grapes (defined to mean grapes which have been grown for market as natural grapes by performing all the normal cultural practices, but not using any inorganic fertilizers or agricultural chemicals including insecticides, herbicides, and growth regulators, except sulfur) need not meet the minimum individual berry size requirements of this section if the following conditions and safeguards are met: (1)

175

The handler of such grapes has registered and certified with the committee on a date specified by the committee the location of the vineyard, the acreage and variety of grapes, and such other information as may be needed by the committee to carry out these provisions; (2) each container of organically grown grapes bears the words "organically grown" on one outside end of the container in plain letters in addition to requirements specified under paragraph (b)(3) of this section.

(e) *By-product grapes.* The handling of grapes for processing (raisins, crushing and other by-products) is exempt from requirements specified in paragraphs (a), (b), and (c) of this section if the committee determines that the person handling such grapes has secured the appropriate permit or order from the County Agricultural Commissioner, and the by-product plant or packing plant to which the grapes are shipped has adequate facilities for commercial processing, grading, packing or manufacturing of by-products for resale.

(f) *Suspension of packing holidays.* Upon recommendation of the committee and approval of the Secretary, the prohibition against packing or repacking grapes on any Saturday, Sunday or on Memorial Day or Independence Day holidays of each year, may be modified or suspended to permit the handling of grapes provided such handling complies with procedures and safeguards specified by the committee as follows:

(1) All requests for suspension of a packing holiday shall be in writing, shall state the reasons the suspension is being requested, and shall be submitted to the Committee manager by noon on Wednesday or at least 3 days prior to the requested suspension date;

(2) Upon receipt of a written request, the Committee manager shall promptly give reasonable notice to producers and handlers and to the Secretary that an assembled Committee meeting will be held to discuss the request(s). The representative of the Secretary shall attend the meeting via speakerphone or in person, and all votes of the Committee members shall be cast in person;

(3) The Committee members shall consider marketing conditions (i.e.,

supplies of competing commodities to include quantities in inventory, the expected demand conditions for grapes in different markets, and any pertinent documents which provide data on market conditions), weather conditions, labor shortages, the size of the crop remaining to be marketed, and other pertinent factors in reaching a decision to suspend packing holidays;

(4) Once a vote is taken, any documents utilized during the meeting will be forwarded immediately to the Secretary's representative and a summary of the Committee's action and reasons for recommending approval or disapproval will be prepared and also forwarded by the committee; and

(5) The Secretary's representative shall notify the Committee manager of approval or disapproval of the request prior to commencement of the suspended packing holiday and the Committee manager shall notify handlers and producers accordingly.

(g) Certain maturity, container, and pack requirements cited in this regulation are specified in the Title 3: California Code of Regulations and are incorporated by reference. Copies of such requirements are available from Ronald L. Cioffi, Chief, Marketing Order Administration Branch, F&V, AMS, USDA, Washington, DC 20090–6456, telephone (202) 720–2491. They are also available for inspection at the National Archives and Records Administration (NARA). For information on the availability of this material at NARA, call 202–741–6030, or go to: *http:// www.archives.gov/federal_register/ code_of_federal_regulations/ ibr_locations.html.* This incorporation by reference was approved by the Director of the Federal Register. These materials are incorporated as they existed on the date of the approval and a notice of any change in these materials will be published in the FEDERAL REGISTER.

(h) The Federal or Federal-State Inspection Service, F&V, AMS, USDA, is the governmental inspection service for certifying the grade, size, quality, and maturity of table grapes grown in the production area. The inspection and certification services will be available upon application in accordance with the rules and regulations governing inspections and certification of fresh fruits, vegetables, and other products (7 CFR part 51); except that all persons who request such inspection and certification must provide adequate facilities in which the inspections may be conducted and also provide the necessary equipment and incidental supplies that are considered as standard requirements for providing fresh inspection under Federal or Federal-State inspection procedures.

[51 FR 12501, Apr. 11, 1986]

EDITORIAL NOTE: For FEDERAL REGISTER citations affecting § 925.304, see the List of CFR Sections Affected, which appears in the Finding Aids section of the printed volume and at *www.govinfo.gov*.

PART 926—DATA COLLECTION, REPORTING AND RECORDKEEPING REQUIREMENTS APPLICABLE TO CRANBERRIES NOT SUBJECT TO THE CRANBERRY MARKETING ORDER

AUTHORITY: 7 U.S.C. 601–674.

SOURCE: 70 FR 1999, Jan. 12, 2005, unless otherwise noted.

§ 926.1   Secretary.

Secretary means the Secretary of Agriculture of the United States or any officer or employee of the United States Department of Agriculture who is, or who may hereafter be authorized to act in her/his stead.

EFFECTIVE DATE NOTE: At 71 FR 78046, Dec. 28, 2006, § 926.1 was suspended indefinitely, effective Dec. 29, 2006.

§ 926.2   Act.

Act means Public Act No. 10, 73d Congress [May 12, 1933], as amended, and as reenacted and amended by the Agricultural Marketing Agreement Act of 1937, as amended (Secs. 1–19, 48 Stat. 31, as amended; 7 U.S.C. 601 *et seq.*).

EFFECTIVE DATE NOTE: At 71 FR 78046, Dec. 28, 2006, § 926.2 was suspended indefinitely, effective Dec. 29, 2006.

§ 926.3   Person.

Person means an individual, partnership, corporation, association, or any other business unit.

EFFECTIVE DATE NOTE: At 71 FR 78046, Dec. 28, 2006, § 926.3 was suspended indefinitely, effective Dec. 29, 2006.

§ 926.4   Cranberries.

Cranberries means all varieties of the fruit Vaccinium Macrocarpon and Vaccinium oxycoccus, known as cranberries.

EFFECTIVE DATE NOTE: At 71 FR 78046, Dec. 28, 2006, § 926.4 was suspended indefinitely, effective Dec. 29, 2006.

§ 926.5   Fiscal period.

Fiscal period is synonymous with fiscal year and crop year and means the 12-month period beginning September 1 and ending August 31 of the following year.

EFFECTIVE DATE NOTE: At 71 FR 78046, Dec. 28, 2006, § 926.5 was suspended indefinitely, effective Dec. 29, 2006.

§ 926.6   Committee.

Committee means the Cranberry Marketing Committee, which is hereby authorized by USDA to collect information on sales, acquisitions, and inventories of cranberries and cranberry products under this part. The Committee is established pursuant to the Federal cranberry marketing order regulating the handling of cranberries grown in the States of Massachusetts, Rhode Island, Connecticut, New Jersey,

Wisconsin, Michigan, Minnesota, Oregon, Washington, and Long Island in the State of New York (7 CFR part 929).

EFFECTIVE DATE NOTE: At 71 FR 78046, Dec. 28, 2006, § 926.6 was suspended indefinitely, effective Dec. 29, 2006.

§ 926.7 Producer.

Producer is synonymous with grower and means any person who produces cranberries for market and has a proprietary interest therein.

EFFECTIVE DATE NOTE: At 71 FR 78046, Dec. 28, 2006, § 926.7 was suspended indefinitely, effective Dec. 29, 2006.

§ 926.8 Handler.

Handler means any person who handles cranberries and is not subject to the reporting requirements of Part 929.

EFFECTIVE DATE NOTE: At 71 FR 78046, Dec. 28, 2006, § 926.8 was suspended indefinitely, effective Dec. 29, 2006.

§ 926.9 Handle.

Handle means to can, freeze, dehydrate, acquire, sell, consign, deliver, or transport (except as a common or contract carrier of cranberries owned by another person) fresh or processed cranberries produced within or outside the United States or in any other way to place fresh or processed cranberries into the current of commerce within or outside the United States. This term includes all initial and subsequent handling of cranberries or processed cranberries up to, but not including, the retail level.

EFFECTIVE DATE NOTE: At 71 FR 78046, Dec. 28, 2006, § 926.9 was suspended indefinitely, effective Dec. 29, 2006.

§ 926.10 Acquire.

Acquire means to obtain cranberries by any means whatsoever for the purpose of handling cranberries.

EFFECTIVE DATE NOTE: At 71 FR 78046, Dec. 28, 2006, § 926.10 was suspended indefinitely, effective Dec. 29, 2006.

§ 926.11 Processed cranberries or cranberry products.

Processed cranberries or cranberry products means cranberries which have been converted from fresh cranberries into canned, frozen, or dehydrated cranberries or other cranberry products by any commercial process.

EFFECTIVE DATE NOTE: At 71 FR 78046, Dec. 28, 2006, § 926.11 was suspended indefinitely, effective Dec. 29, 2006.

§ 926.12 Producer-handler.

Producer-handler means any person who is a producer of cranberries for market and handles such cranberries.

EFFECTIVE DATE NOTE: At 71 FR 78046, Dec. 28, 2006, § 926.12 was suspended indefinitely, effective Dec. 29, 2006.

§ 926.13 Processor.

Processor means any person who receives or acquires fresh or frozen cranberries or cranberries in the form of concentrate from handlers, producer-handlers, importers, brokers or other processors and uses such cranberries or concentrate, with or without other ingredients, in the production of a product for market.

EFFECTIVE DATE NOTE: At 71 FR 78046, Dec. 28, 2006, § 926.13 was suspended indefinitely, effective Dec. 29, 2006.

§ 926.14 Broker.

Broker means any person who acts as an agent of the buyer or seller and negotiates the sale or purchase of cranberries or cranberry products.

EFFECTIVE DATE NOTE: At 71 FR 78046, Dec. 28, 2006, § 926.14 was suspended indefinitely, effective Dec. 29, 2006.

§ 926.15 Importer.

Importer means any person who causes cranberries or cranberry products produced outside the United States to be brought into the United States with the intent of entering the cranberries or cranberry products into the current of commerce.

EFFECTIVE DATE NOTE: At 71 FR 78046, Dec. 28, 2006, § 926.15 was suspended indefinitely, effective Dec. 29, 2006.

§ 926.16 Reports.

(a) Each handler, producer-handler, processor, broker, and importer engaged in handling or importing cranberries or cranberry products who is not subject to the reporting requirements of the Federal cranberry marketing order, (7 CFR Part 926) shall, in accordance with Sec. 926.17, file promptly with the Committee reports of sales, acquisitions, and inventory information on fresh cranberries and cranberry products using forms supplied by the Committee.

(b) Upon the request of the Committee, with the approval of the Secretary, each handler, producer-handler, processor, broker, and importer engaged in handling or importing cranberries or cranberry products who is not subject to the Federal cranberry marketing order (7 CFR Part 926) shall furnish to the Committee such other information with respect to fresh cranberries and cranberry products acquired and disposed of by such entity as may be necessary to meet the objectives of the Act.

EFFECTIVE DATE NOTE: At 71 FR 78046, Dec. 28, 2006, §926.16 was suspended indefinitely, effective Dec. 29, 2006.

§926.17 Reporting requirements.

Handlers, producer-handlers, importers, processors, and brokers not subject to the Federal cranberry marketing order (7 CFR part 926) shall be required to submit four times annually, for each fiscal period reports regarding sales, acquisitions, movement for further processing, and dispositions of fresh cranberries and cranberry products using forms supplied by the Committee. An Importer Cranberry Inventory Report Form shall be required to be completed by importers and brokers. This report shall indicate the name, address, variety acquired, the amount sold to and received by brokers, processors, and handlers, and the beginning and ending inventories of cranberries held by the importer for each applicable fiscal period. A Handler/Processor Cranberry Inventory Report Form shall be completed by handlers, producer-handlers, and processors and shall indicate the name, address, variety acquired, domestic/foreign sales, acquisitions, and beginning and ending inventories.

EFFECTIVE DATE NOTE: At 71 FR 78046, Dec. 28, 2006, §926.17 was suspended indefinitely, effective Dec. 29, 2006.

§926.18 Records.

Each handler, producer-handler, processor, broker, and importer shall maintain such records of all fresh cranberries and cranberry products acquired, imported, handled, withheld from handling, and otherwise disposed of during the fiscal period to substantiate the required reports. All such records shall be maintained for not less than three years after the termination of the fiscal year in which the transactions occurred or for such lesser period as the Committee may direct.

EFFECTIVE DATE NOTE: At 71 FR 78046, Dec. 28, 2006, §926.18 was suspended indefinitely, effective Dec. 29, 2006.

§926.19 Confidential information.

All reports and records furnished or submitted pursuant to this part which include data or information constituting a trade secret or disclosing the trade position or financial condition, or business operations from whom received, shall be in the custody and control of the authorized agents of the Committee, who shall disclose such information to no person other than the Secretary.

EFFECTIVE DATE NOTE: At 71 FR 78046, Dec. 28, 2006, §926.19 was suspended indefinitely, effective Dec. 29, 2006.

§926.20 Verification of reports and records.

For the purpose of assuring compliance and checking and verifying records and reports required to be filed by handlers, producer-handlers, processors, brokers, and importers, USDA or the Committee, through its duly authorized agents, shall have access to any premises where applicable records are maintained, where cranberries and cranberry products are received, acquired, stored, handled, and otherwise disposed of and, at any time during reasonable business hours, shall be permitted to inspect such handler, producer-handler, processor, broker, and importer premises, and any and all records of such handlers, producer-handlers, processors, brokers, and importers. The Committee's authorized agents shall be the manager of the Committee and other staff under the supervision of the Committee manager.

EFFECTIVE DATE NOTE: At 71 FR 78046, Dec. 28, 2006, §926.20 was suspended indefinitely, effective Dec. 29, 2006.

§ 926.21 Suspension or termination.

The provisions of this part shall be suspended or terminated whenever there is no longer a Federal cranberry marketing order in effect.

EFFECTIVE DATE NOTE: At 71 FR 78046, Dec. 28, 2006, § 926.21 was suspended indefinitely, effective Dec. 29, 2006.

PART 927—PEARS GROWN IN OREGON AND WASHINGTON

Subpart A—Order Regulating Handling

Subpart B—Administrative Provisions

AUTHORITY: 7 U.S.C. 601–674.

SOURCE: 70 FR 29392, May 20, 2005, unless otherwise noted.

Subpart A—Order Regulating Handling

DEFINITIONS

§ 927.1 Secretary.

Secretary means the Secretary of Agriculture of the United States, or any officer or employee of the Department of Agriculture who has been delegated, or to whom authority may hereafter be delegated, the authority to act for the Secretary.

§ 927.2 Act.

Act means Public Act No. 10, 73d Congress (May 12, 1933), as amended and as reenacted and amended by the Agricultural Marketing Agreement Act of 1937, as amended (48 Stat. 31, as amended; 7 U.S.C. 601 *et seq.*).

§ 927.3 Person.

Person means an individual partnership, corporation, association, legal representative, or any other business unit.

§ 927.4 Pears.

(a) *Pears* means and includes any and all varieties or subvarieties of pears with the genus *Pyrus* that are produced in the production area and are classified as:

(1) Summer/fall pears including Bartlett and Starkrimson pears;

(2) Winter pears including Beurre D'Anjou, Beurre Bosc, Doyenne du Comice, Concorde, Forelle, Winter Nelis, Packham, Seckel, and Taylor's Gold pears; and

(3) Other pears including any or all other varieties or subvarieties of pears not classified as summer/fall or winter pears.

(b) The Fresh Pear Committee and/or the Processed Pear Committee, with the approval of the Secretary, may recognize new or delete obsolete varieties or subvarieties for each category.

§ 927.5 Size.

Size means the number of pears which can be packed in a 44-pound net weight standard box or container equivalent, or as "size" means the greatest transverse diameter of the pear taken at right angles to a line running from the stem to the blossom end, or such other specifications more specifically defined in a regulation issued under this part.

§ 927.6 Grower.

Grower is synonymous with producer and means any person engaged in the production of pears, either as owner or as tenant.

§ 927.7 Handler.

Handler is synonymous with shipper and means any person (except a common or contract carrier transporting pears owned by another person) who, as owner, agent, broker, or otherwise, ships or handles pears, or causes pears to be shipped or handled by rail, truck, boat, or any other means whatsoever.

§ 927.8 Ship or handle.

Ship or handle means to sell, deliver, consign, transport or ship pears within the production area or between the production area and any point outside thereof, including receiving pears for processing: *Provided*, That the term "handle" shall not include the transportation of pear shipments within the production area from the orchard where grown to a packing facility located within the production area for preparation for market or delivery for processing.

§ 927.9 Fiscal period.

Fiscal period means the period beginning July 1 of any year and ending June 30 of the following year or such may be approved by the Secretary pursuant to a joint recommendation by the Fresh Pear Committee and the Processed Pear Committee.

§ 927.10 Production area.

Production area means and includes the States of Oregon and Washington.

§ 927.11 District.

District means the applicable one of the following-described subdivisions of the production area covered by the provisions of this subpart:

(a) For the purpose of committee representation, administration and application of provisions of this subpart as applicable to pears for the fresh market, districts shall be defined as follows:

(1) *Medford District* shall include all the counties in the State of Oregon except for Hood River and Wasco counties.

(2) *Mid-Columbia District* shall include Hood River and Wasco counties in the State of Oregon, and the counties of Skamania and Klickitat in the State of Washington.

(3) *Wenatchee District* shall include the counties of King, Chelan, Okanogan, Douglas, Grant, Lincoln, and Spokane in the State of Washington, and all other counties in Washington lying north thereof.

(4) *Yakima District* shall include all of the State of Washington, not included in the Wenatchee District or in the Mid-Columbia District.

(b) For the purpose of committee representation, administration and application of provisions of this subpart as applicable to pears for processing, districts shall be defined as follows:

(1) The State of Washington.

(2) The State of Oregon.

(c) The Secretary, upon recommendation of the Fresh Pear Committee or the Processed Pear Committee, may reestablish districts within the production area.

§ 927.12   Export market.

Export market means any destination which is not within the 50 states, or the District of Columbia, of the United States.

§ 927.13   Subvariety.

Subvariety means and includes any mutation, sport, or other derivation of any of the varieties covered in § 927.4 which is recognized by the Fresh Pear Committee or the Processed Pear Committee and approved by the Secretary. Recognition of a subvariety shall include classification within a varietal group for the purposes of votes conducted under § 927.52.

§ 927.14   Processor.

Processor means any person who as owner, agent, broker, or otherwise, commercially processes pears in the production area.

§ 927.15   Process.

Process means to can, concentrate, freeze, dehydrate, press or puree pears, or in any other way convert pears commercially into a processed product.

ADMINISTRATIVE BODIES

§ 927.20   Establishment and membership.

There are hereby established two committees to administer the terms and provisions of this subpart as specifically provided in §§ 927.20 through 927.35:

(a) A Fresh Pear Committee, consisting of 13 individual persons as its members is established to administer order provisions relating to the handling of pears for the fresh market. Six members of the Fresh Pear Committee shall be growers, six members shall be handlers, and one member shall represent the public. For each member there shall be two alternates, designated as the "first alternate" and the "second alternate," respectively. Each district shall be represented by one grower member and one handler member, except that the Mid-Columbia District and the Wenatchee District shall be represented by two grower members and two handler members.

(b) A Processed Pear Committee consisting of 10 members is established to administer order provisions relating to the handling of pears for processing. Three members of the Processed Pear Committee shall be growers, three members shall be handlers, three members shall be processors, and one member shall represent the public. For each member there shall be two alternates, designated as the "first alternate" and the "second alternate," respectively. District 1, the State of Washington, shall be represented by two grower members, two handler members and two processor members. District 2, the State of Oregon, shall be represented by one grower member, one handler member and one processor member.

(c) The Secretary, upon recommendation of the Fresh Pear Committee or the Processed Pear Committee may reapportion members among districts, may change the number of members and alternates, and may change the composition by changing the ratio of members, including their alternates. In recommending any such changes, the following shall be considered:

(1) Shifts in pear acreage within districts and within the production area during recent years;

(2) The importance of new pear production in its relation to existing districts;

(3) The equitable relationship between membership and districts;

(4) Economies to result for growers in promoting efficient administration due to redistricting or reapportionment of members within districts; and

(5) Other relevant factors.

§ 927.21 Nomination and selection of members and their respective alternates.

Grower members and their respective alternates for each district shall be selected by the Secretary from nominees elected by the growers in such district. Handler members and their respective alternates for each district shall be selected by the Secretary from nominees elected by the handlers in such district. Processor members and their respective alternates shall be selected by the Secretary from nominees elected by the processors. Public members for each committee shall be nominated by the Fresh Pear Committee and the Processed Pear Committee, each independently, and selected by the Secretary. The Fresh Pear Committee and the Processed Pear Committee may, each independently, prescribe such additional qualifications, administrative rules and procedures for selection for each candidate as it deems necessary and as the Secretary approves.

§ 927.22 Meetings for election of nominees.

(a) Nominations for members of the Fresh Pear Committee and their alternates shall be made at meetings of growers and handlers held in each of the districts designated in § 927.11 at such times and places designated by the Fresh Pear Committee.

(b) Nominations for grower and handler members of the Processed Pear Committee and their alternates shall be made at meetings of growers and handlers held in each of the districts designated in § 927.11 at such times and places designated by the Processed Pear Committee. Nominations for processor members of the Processed Pear Committee and their alternates shall be made at a meeting of processors at such time and place designated by the Processed Pear Committee.

§ 927.23 Voting.

Only growers in attendance at meetings for election of nominees shall participate in the nomination of grower members and their alternates, and only handlers in attendance at meetings for election of nominees shall participate in the nomination of handler members and their alternates, and only processors in attendance for election of nominees shall participate in the nomination of processor members and their alternates. A grower may participate only in the election held in the district in which he or she produces pears, and a handler may participate only in the election held in the district in which he or she handles pears. Each person may vote as a grower, handler or processor, but not a combination thereof. Each grower, handler and processor shall be entitled to cast one vote, on behalf of himself, his agents, partners, affiliates, subsidiaries, and representatives, for each nominee to be elected.

§ 927.24 Eligibility for membership.

Each grower member and each of his or her alternates shall be a grower, or an officer or employee of a corporate or LLC grower, who grows pears in the district in which and for which he or she is nominated and selected. Each handler member and each of his or her alternates shall be a handler, or an officer or employee of a handler, handling pears in the district in and for which he or she is nominated and selected. Each processor member and each of their alternates shall be a processor, or an officer or employee of a processor, who processes pears in the production area.

§ 927.25 Failure to nominate.

In the event nominations are not made pursuant to §§ 927.21 and 927.22 on or before June 1 of any year, the Secretary may select members and alternates for members without regard to nominations.

§ 927.26 Qualifications.

Any person prior to or within 15 days after selection as a member or as an alternate for a member of the Fresh Pear Committee or the Processed Pear Committee shall qualify by filing with the Secretary a written acceptance of the person's willingness to serve.

§ 927.27 Term of office.

The term of office of each member and alternate member of the Fresh Pear Committee and the Processed Pear Committee shall be for two years beginning July 1 and ending June 30: *Provided,* That the terms of office of one-half the initial members and alternates shall end June 30, 2006; and that beginning with the 2005–2006 fiscal period, no member shall serve more than three consecutive two-year terms unless specifically exempted by the Secretary. Members and alternate members shall serve in such capacities for the portion of the term of office for which they are selected and have qualified and until their respective successors are selected and have qualified. The terms of office of successor members and alternates shall be so determined that one-half of the total committee membership ends each June 30.

§ 927.28 Alternates for members.

The first alternate for a member shall act in the place and stead of the member for whom he or she is an alternate during such member's absence. In the event of the death, removal, resignation, or disqualification of a member, his or her first alternate shall act as a member until a successor for the member is selected and has qualified. The second alternate for a member shall serve in the place and stead of the member for whom he or she is an alternate whenever both the member and his or her first alternate are unable to serve. In the event that a member of the Fresh Pear Committee or the Processed Pear Committee and both that member's alternates are unable to attend a meeting, the member may designate any other alternate member from the same group (handler, processor, or grower) to serve in that member's place and stead.

§ 927.29 Vacancies.

To fill any vacancy occasioned by the failure of any person selected as a member or as an alternate for a member of the Fresh Pear Committee or the Processed Pear Committee to qualify, or in the event of death, removal, resignation, or disqualification of any qualified member or qualified alternate for a member, a successor for his or her unexpired term shall be nominated and selected in the manner set forth in §§ 927.20 to 927.35. If nominations to fill any such vacancy are not made within 20 days after such vacancy occurs, the Secretary may fill such vacancy without regard to nominations.

§ 927.30 Compensation and expenses.

The members and alternates for members shall serve without compensation, but may be reimbursed for expenses necessarily incurred by them in the performance of their respective duties.

§ 927.31 Powers.

The Fresh Pear Committee and the Processed Pear Committee shall have the following powers to exercise each independently:

(a) To administer, as specifically provided in §§ 927.20 to 927.35, the terms and provisions of this subpart:

(b) To make administrative rules and regulations in accordance with, and to effectuate, the terms and provisions of this subpart; and

(c) To receive, investigate, and report to the Secretary complaints of violations of the provisions of this subpart.

§ 927.32 Duties.

The duties of the Fresh Pear Committee and the Processed Pear Committee, each independently, shall be as follows:

(a) To act as intermediary between the Secretary and any grower, handler or processor;

(b) To keep minutes, books, and records which will reflect clearly all of the acts and transactions. The minutes, books, and records shall be subject at any time to examination by the Secretary or by such person as may be designated by the Secretary;

(c) To investigate, from time to time, and to assemble data on the growing, harvesting, shipping, and marketing conditions relative to pears, and to furnish to the Secretary such available information as may be requested;

(d) To perform such duties as may be assigned to it from time to time by the Secretary in connection with the administration of section 32 of the Act to amend the Agricultural Adjustment Act, and for other purposes, Public Act No. 320, 74th Congress, approved August 24, 1935 (49 Stat. 774), as amended;

(e) To cause the books to be audited by one or more competent accountants at the end of each fiscal year and at such other times as the Fresh Pear Committee or the Processed Pear Committee may deem necessary or as the Secretary may request, and to file with the Secretary copies of any and all audit reports made;

(f) To appoint such employees agents, and representatives as it may deem necessary, and to determine the compensation and define the duties of each;

(g) To give the Secretary, or the designated agent of the Secretary, the same notice of meetings as is given to the members of the Fresh Pear Committee or the Processed Pear Committee;

(h) To select a chairman of the Fresh Pear Committee or the Processed Pear Committee and, from time to time, such other officers as it may deem advisable and to define the duties of each; and

(i) To submit to the Secretary as soon as practicable after the beginning of each fiscal period, a budget for such fiscal year, including a report in explanation of the items appearing therein and a recommendation as to the rate of assessment for such period.

§927.33 **Procedure.**

(a) *Quorum and voting.* A quorum at a meeting of the Fresh Pear Committee or the Processed Pear Committee shall consist of 75 percent of the number of committee members, or alternates then serving in the place of any members, respectively. Except as otherwise provided in §927.52, all decisions of the Fresh Pear Committee or the Processed Pear Committee at any meeting shall require the concurring vote of at least 75 percent of those members present, including alternates then serving in the place of any members.

(b) *Mail voting.* The Fresh Pear Committee or the Processed Pear Committee may provide for members voting by mail, telecopier or other electronic means, telephone, or telegraph, upon due notice to all members. Promptly after voting by telephone or telegraph, each member thus voting shall confirm in writing, the vote so cast.

§927.34 **Right of the Secretary.**

The members and alternates for members and any agent or employee appointed or employed by the Fresh Pear Committee or the Processed Pear Committee shall be subject to removal or suspension by the Secretary at any time. Each and every regulation, decision, determination, or other act shall be subject to the continuing right of the Secretary to disapprove of the same at any time, and, upon such disapproval, shall be deemed null and void, except as to acts done in reliance thereon or in compliance therewith prior to such disapproval by the Secretary.

§927.35 **Funds and other property.**

(a) All funds received pursuant to any of the provisions of this subpart shall be used solely for the purposes specified in this subpart, and the Secretary may require the Fresh Pear Committee or the Processed Pear Committee and its members to account for all receipts and disbursements.

(b) Upon the death, resignation, removal, disqualification, or expiration of the term of office of any member or employee, all books, records, funds, and other property in his or her possession belonging to the Fresh Pear Committee or the Processed Pear Committee shall be delivered to his or her successor in office or to the Fresh Pear Committee or Processed Pear Committee, and such assignments and other instruments shall be executed as may be necessary to vest in such successor or in the Fresh Pear Committee or Processed Pear Committee full title to all the books, records, funds, and other property in the possession or under the control of such member or employee pursuant to this subpart.

EXPENSES AND ASSESSMENTS

§ 927.40 Expenses.

The Fresh Pear Committee and the Processed Pear Committee are authorized, each independently, to incur such expenses as the Secretary finds may be necessary to carry out their functions under this subpart. The funds to cover such expenses shall be acquired by the levying of assessments as provided in § 927.41.

§ 927.41 Assessments.

(a) Assessments will be levied only upon handlers who first handle pears. Each handler shall pay assessments on all pears handled by such handler as the pro rata share of the expenses which the Secretary finds are reasonable and likely to be incurred by the Fresh Pear Committee or the Processed Pear Committee during a fiscal period. The payment of assessments for the maintenance and functioning of the Fresh Pear Committee or the Processed Pear Committee may be required under this part throughout the period such assessments are payable irrespective of whether particular provisions thereof are suspended or become inoperative.

(b)(1) Based upon a recommendation of the Fresh Pear Committee or other available data, the Secretary shall fix three base rates of assessment for pears that handlers shall pay on pears handled for the fresh market during each fiscal period. Such base rates shall include one rate of assessment for any or all varieties or subvarieties of pears classified as summer/fall; one rate of assessment for any or all varieties or subvarieties of pears, classified as winter; and one rate of assessment for any or all varieties or subvarieties of pears classified as other. Upon recommendation of the Fresh Pear Committee or other available data, the Secretary may also fix supplemental rates of assessment on individual varieties or subvarieties categorized within the assessment classifications in this paragraph (b)(1) to secure sufficient funds to provide for projects authorized under § 927.47. At any time during the

fiscal period when it is determined on the basis of a Fresh Pear Committee recommendation or other information that different rates are necessary for fresh pears or for any varieties or subvarieties, the Secretary may modify those rates of assessment and such new rate shall apply to any or all varieties or subvarieties that are shipped during the fiscal period for fresh market.

(2) Based upon a recommendation of the Processed Pear Committee or other available data, the Secretary shall fix three base rates of assessment for pears that handlers shall pay on pears handled for processing during each fiscal period. Such base rates shall include one rate of assessment for any or all varieties or subvarieties of pears classified as summer/fall; one rate of assessment for any or all varieties or subvarieties of pears, classified as winter; and one rate of assessment for any or all varieties or subvarieties of pears classified as other. Upon recommendation of the Processed Pear Committee or other available data, the Secretary may also fix supplemental rates of assessment on individual varieties or subvarieties categorized within the assessment classifications defined in paragraph (b)(1) of this section to secure sufficient funds to provide for projects authorized under § 927.47. At any time during the fiscal period when it is determined on the basis of a Processed Pear Committee recommendation or other information that different rates are necessary for pears for processing or for any varieties or subvarieties, the Secretary may modify those rates of assessment and such new rate shall apply to any or all varieties or subvarieties of pears that are shipped during the fiscal period for processing.

(c) Based on the recommendation of the Fresh Pear Committee, the Processed Pear Committee or other available data, the Secretary may establish additional base rates of assessments, or change or modify the base rate classifications defined in paragraphs (a) and (b) of this section.

(d) The Fresh Pear Committee or the Processed Pear Committee may impose a late payment charge on any handler who fails to pay any assessment within the time prescribed. In the event the handler thereafter fails to pay the amount outstanding, including the late payment charge, within the prescribed time, the Fresh Pear Committee or the Processed Pear Committee may impose an additional charge in the form of interest on such outstanding amount. The Fresh Pear Committee or the Processed Pear Committee, with the approval of the Secretary, shall prescribe the amount of such late payment charge and rate of interest.

(e) In order to provide funds to carry out the functions of the Fresh Pear Committee or the Processed Pear Committee prior to commencement of shipments in any season, handlers may make advance payments of assessments, which advance payments shall be credited to such handlers and the assessments of such handlers shall be adjusted so that such assessments are based upon the quantity of each variety or subvariety of pears handled by such handlers during such season. Further, payment discounts may be authorized by the Fresh Pear Committee or the Processed Pear Committee upon the approval of the Secretary to handlers making such advance assessment payments.

§927.42 Accounting.

(a) If, at the end of a fiscal period, the assessments collected are in excess of expenses incurred, the Fresh Pear Committee or the Processed Pear Committee may carryover such excess into subsequent fiscal periods as a reserve: *Provided,* That funds already in the reserve do not exceed approximately one fiscal period's expenses. Such reserve may be used to cover any expense authorized under this part and to cover necessary expenses of liquidation in the event of termination of this part. Any such excess not retained in a reserve or applied to any outstanding obligation of the person from whom it was collected shall be refunded proportionately to the persons from whom it was collected. Upon termination of this part, any funds not required to defray the necessary expenses of liquidation shall be disposed of in such manner as the Secretary may determine to be appropriate: *Provided,* That to the extent practical, such funds shall be returned pro rata to the persons from whom such funds were collected.

(b) All funds received pursuant to the provisions of this part shall be used solely for the purpose specified in this part and shall be accounted for in the manner provided in this part. The Secretary may at any time require the Fresh Pear Committee or the Processed Pear Committee and its members to account for all receipts and disbursements.

§927.43 Use of funds.

From the funds acquired pursuant to §927.41 the Fresh Pear Committee and the Processed Pear Committee, each independently, shall pay the salaries of its employees, if any, and pay the expenses necessarily incurred in the performance of the duties of the Fresh Pear Committee or the Processed Pear Committee.

§927.44 [Reserved]

§927.45 Contributions.

The Fresh Pear Committee or the Processed Pear Committee may accept voluntary contributions, but these shall only be used to pay expenses incurred pursuant to §927.47. Furthermore, such contributions shall be free from any encumbrances by the donor, and the Fresh Pear Committee or the Processed Pear Committee shall retain complete control of their use.

RESEARCH AND DEVELOPMENT

§927.47 Research and development.

The Fresh Pear Committee or the Processed Pear Committee, with the approval of the Secretary, may establish or provide for the establishment of production and post-harvest research, or marketing research and development projects designed to assist, improve, or promote the marketing, distribution, and consumption of pears. Such projects may provide for any form of marketing promotion, including paid advertising. The expense of such projects shall be paid from funds collected pursuant to §§927.41 and 927.45. Expenditures for a particular variety or subvariety of pears shall approximate the amount of assessments and voluntary contributions collected for that variety or subvariety of pears.

REGULATION OF SHIPMENTS

§ 927.50 Marketing policy.

(a) It shall be the duty of the Fresh Pear Committee to investigate, from time to time, supply and demand conditions relative to pears and each grade, size, and quality of each variety or subvariety thereof. Such investigations shall be with respect to the following:

(1) Estimated production of each variety or subvariety of pears and of each grade, size, and quality thereof;

(2) Prospective supplies and prices of pears and other fruits, both in fresh and processed form, which are competitive to the marketing of pears;

(3) Prospective exports of pears and imports of pears from other producing areas;

(4) Probable harvesting period for each variety or subvariety of pears;

(5) The trend and level of consumer income;

(6) General economic conditions; and

(7) Other relevant factors.

(b) On or before August 1 of each year, the Fresh Pear Committee shall recommend regulations to the Secretary if it finds, on the basis of the investigations specified in this section, that such regulation as is provided in § 927.51 will tend to effectuate the declared policy of the act.

(c) In the event the Fresh Pear Committee at any time finds that by reason of changed conditions any regulation issued pursuant to § 927.51 should be modified, suspended, or terminated, it shall so recommend to the Secretary.

§ 927.51 Issuance of regulations; and modification, suspension, or termination thereof.

(a) Whenever the Secretary finds, from the recommendations and information submitted by the Fresh Pear Committee, or from other available information, that regulation, in the manner specified in this section, of the shipment of fresh pears would tend to effectuate the declared policy of the act, he or she shall so limit the shipment of such pears during a specified period or periods. Such regulation may:

(1) Limit the total quantity of any grade, size, quality, or combinations thereof, of any variety or subvariety of pears grown in any district and may prescribe different requirements applicable to shipments to different export markets;

(2) Limit, during any period or periods, the shipment of any particular grade, size, quality, or any combination thereof, of any variety or subvariety, of pears grown in any district or districts of the production area; and

(3) Provide a method, through rules and regulation issued pursuant to this part, for fixing markings on the container or containers, which may be used in the packaging or handling of pears, including appropriate logo or other container markings to identify the contents thereof.

(b) Whenever the Secretary finds, from the recommendations and information submitted by the Fresh Pear Committee, or from other available information, that a regulation should be modified, suspended, or terminated with respect to any or all shipments of fresh pears grown in any district in order to effectuate the declared policy of the act, he or she shall so modify, suspend, or terminate such regulation. If the Secretary finds, from the recommendations and information submitted by the Fresh Pear Committee, or from other available information, that a regulation obstructs or does not tend to effectuate the declared policy of the act, he or she shall suspend or terminate such regulation. On the same basis and in like manner, the Secretary may terminate any such modification or suspension.

§ 927.52 Prerequisites to recommendations.

(a) Decisions of the Fresh Pear Committee or the Processed Pear Committee with respect to any recommendations to the Secretary pursuant to the establishment or modification of a supplemental rate of assessment for an individual variety or subvariety of pears shall be made by affirmative vote of not less than 75 percent of the applicable total number of votes, computed in the manner described in paragraph (b) of this section, of all members. Decisions of the Fresh

Pear Committee pursuant to the provisions of §927.50 shall be made by an affirmative vote of not less than 80 percent of the applicable total number of votes, computed in the manner prescribed in paragraph (b) of this section, of all members.

(b) With respect to a particular variety or subvariety of pears, the applicable total number of votes shall be the aggregate of the votes allotted to the members in accordance with the following: Each member shall have one vote as an individual and, in addition, shall have a vote equal to the percentage of the vote of the district represented by such member; and such district vote shall be computed as soon as practical after the beginning of each fiscal period on either:

(1) The basis of one vote for each 25,000 boxes (except 2,500 boxes for varieties or subvarieties with less than 200,000 standard boxes or container equivalents) of the average quantity of such variety or subvariety produced in the particular district and shipped therefrom during the immediately preceding three fiscal periods; or

(2) Such other basis as the Fresh Pear Committee or the Processed Pear Committee may recommend and the Secretary may approve. The votes so allotted to a member may be cast by such member on each recommendation relative to the variety or subvariety of pears on which such votes were computed.

§927.53 **Notification.**

(a) The Fresh Pear Committee shall give prompt notice to growers and handlers of each recommendation to the Secretary pursuant to the provisions of §927.50.

(b) The Secretary shall immediately notify the Fresh Pear Committee of the issuance of each regulation and of each modification, suspension, or termination of a regulation and the Fresh Pear Committee shall give prompt notice thereof to growers and handlers.

§927.54 **[Reserved]**

INSPECTION

§927.60 **Inspection and certification.**

(a) Handlers shall ship only fresh pears inspected by the Federal-State Inspection Service or under a program developed by the Federal-State Inspection Service: except, that such inspection and certification of shipments of pears may be performed by such other inspection service as the Fresh Pear Committee, with the approval of the Secretary, may designate. Promptly after shipment of any pears, the handler shall submit, or cause to be submitted, to the Fresh Pear Committee a copy of the inspection certificate issued on such shipment.

(b) Any handler may ship pears, on any one conveyance and in such quantity as the committee, with the approval of the Secretary, may prescribe, exempt from the inspection and certification requirements of paragraph (a) of this section.

(c) The Fresh Pear Committee may, with the approval of the Secretary, prescribe rules and regulations modifying or eliminating the requirement for mandatory inspection and certification of shipments: Provided, That an adequate method of ensuring compliance with quality and size requirements is developed.

EXCEPTIONS

§927.65 **Exemption from regulation.**

(a) Nothing contained in this subpart shall limit or authorize the limitation of shipment of pears for consumption by charitable institutions or distribution by relief agencies, nor shall any assessment be computed on pears so shipped. The Fresh Pear Committee or the Processed Pear Committee may prescribe regulations to prevent pears shipped for either of such purposes from entering commercial channels of trade contrary to the provisions of this subpart.

(b) The Fresh Pear Committee or the Processed Pear Committee may prescribe rules and regulations, to become effective upon the approval of the Secretary, whereby quantities of pears or

189

types of pear shipments may be exempted from any or all provisions of this subpart.

MISCELLANEOUS PROVISIONS

§ 927.70 Reports.

(a) Upon the request of the Fresh Pear Committee or the Processed Pear Committee, and subject to the approval of the Secretary, each handler shall furnish to the aforesaid committee, respectively, in such manner and at such times as it prescribes, such information as will enable it to perform its duties under this subpart.

(b) All such reports shall be held under appropriate protective classification and custody by the Fresh Pear Committee or the Processed Pear Committee, or duly appointed employees thereof, so that the information contained therein which may adversely affect the competitive position of any handler in relation to other handlers will not be disclosed. Compilations of general reports from data submitted by handlers are authorized subject to the prohibition of disclosure of individual handler's identities or operations.

(c) Each handler shall maintain for at least two succeeding years such records of the pears received and of pears disposed of, by such handler as may be necessary to verify reports pursuant to this section.

§ 927.71 Compliance.

Except as provided in § 927.65, no handler shall ship any pears contrary to the applicable restrictions and limitations specified in, or effective pursuant to, the provisions of this subpart.

§ 927.72 Duration of immunities.

The benefits, privileges, and immunities conferred by virtue of this subpart shall cease upon termination hereof, except with respect to acts done under and during the existence of this subpart.

§ 927.73 Separability.

If any provision of this subpart is declared invalid, or the applicability thereof to any person, circumstance, or thing is held invalid, the validity of the remaining provisions and the applicability thereof to any other person, circumstance, or thing shall not be affected thereby.

§ 927.74 Derogation.

Nothing contained in this subpart is or shall be construed to be in derogation of, or in modification of, the rights of the Secretary or of the United States to exercise any powers granted by the act or otherwise, or, in accordance with such powers, to act in the premises whenever such action is deemed advisable.

§ 927.75 Liability.

No member or alternate for a member of the Fresh Pear Committee or the Processed Pear Committee, nor any employee or agent thereof, shall be held personally responsible, either individually or jointly with others, in any way whatsoever, to any party under this subpart or to any other person for errors in judgment, mistakes, or other acts, either of commission or omission, as such member, alternate for a member, agent or employee, except for acts of dishonesty, willful misconduct, or gross negligence.

§ 927.76 Agents.

The Secretary may name, by designation in writing, any person, including any officer or employee of the Government or any bureau or division in the Department of Agriculture to act as his or her agent or representative in connection with any of the provisions of this subpart.

§ 927.77 Effective time.

The provisions of this subpart and of any amendment thereto shall become effective at such time as the Secretary may declare, and shall continue in force until terminated in one of the ways specified in § 927.78.

§ 927.78 Termination.

(a) The Secretary may at any time terminate this subpart.

(b) The Secretary shall terminate or suspend the operation of any or all of the provisions of this subpart whenever he or she finds that such operation obstructs or does not tend to effectuate the declared policy of the act.

(c) The Secretary shall terminate the provisions of this subpart applicable to fresh pears for market or pears for processing at the end of any fiscal period whenever the Secretary finds, by referendum or otherwise, that such termination is favored by a majority of growers of fresh pears for market or pears for processing, respectively: *Provided*, That such majority has during such period produced more than 50 percent of the volume of fresh pears for market or pears for processing, respectively, in the production area. Such termination shall be effective only if announced on or before the last day of the then current fiscal period.

(d) The Secretary shall conduct a referendum within every six-year period beginning on May 21, 2005, to ascertain whether continuance of the provisions of this subpart applicable to fresh pears for market or pears for processing are favored by producers of pears for the fresh market and pears for processing, respectively. The Secretary may terminate the provisions of this subpart at the end of any fiscal period in which the Secretary has found that continuance of this subpart is not favored by producers who, during a representative period determined by the Secretary, have been engaged in the production of fresh pears for market or pears for processing in the production area: *Provided*, That termination of the order shall be effective only if announced on or before the last day of the then current fiscal period.

(e) The provisions of this part shall, in any event, terminate whenever the provisions of the act authorizing them cease to be in effect.

§927.79 Proceedings after termination.

(a) Upon the termination of this subpart, the members of the Fresh Pear Committee or the Processed Pear Committee then functioning shall continue as joint trustees for the purpose of liquidating all funds and property then in the possession or under the control of the Fresh Pear Committee or the Processed Pear Committee, including claims for any funds unpaid or property not delivered at the time of such termination.

(b) The joint trustees shall continue in such capacity until discharged by the Secretary; from time to time account for all receipts and disbursements; deliver all funds and property on hand, together with all books and records of the Fresh Pear Committee or the Processed Pear Committee and of the joint trustees, to such person as the Secretary shall direct; and, upon the request of the Secretary, execute such assignments or other instruments necessary and appropriate to vest in such person full title and right to all of the funds, property, or claims vested in the Fresh Pear Committee or the Processed Pear Committee or in said joint trustees.

(c) Any funds collected pursuant to this subpart and held by such joint trustees or such person over and above the amounts necessary to meet outstanding obligations and the expenses necessarily incurred by the joint trustees or such other person in the performance of their duties under this subpart, as soon as practicable after the termination hereof, shall be returned to the handlers pro rata in proportion to their contributions thereto.

(d) Any person to whom funds, property, or claims have been transferred or delivered by the Fresh Pear Committee or the Processed Pear Committee or its members, upon direction of the Secretary, as provided in this section, shall be subject to the same obligations and duties with respect to said funds, property, or claims as are imposed upon the members or upon said joint trustees.

§927.80 Amendments.

Amendments to this subpart may be proposed from time to time by the Fresh Pear Committee or the Processed Pear Committee or by the Secretary.

Subpart B—Administrative Provisions

SOURCE: 70 FR 59625, Oct. 13, 2005, unless otherwise noted.

DEFINITIONS

§ 927.100 Terms.

Each term used in this subpart shall have the same meaning as when used in the marketing order.

[70 FR 59625, Oct. 13, 2005, as amended at 71 FR 7676, Feb. 14, 2006]

§ 927.101 [Reserved]

§ 927.102 Order.

Order means Marketing Order No. 927, as amended (§§ 927.1 to 927.81), regulating the handling of pears grown in the States of Oregon and Washington.

[71 FR 7676, Feb. 14, 2006]

§ 927.103 Organically produced pears.

Organically produced pears means pears that have been certified by an organic certification organization currently registered with the Oregon or Washington State Departments of Agriculture, or such certifying organization accredited under the National Organic Program.

COMMUNICATIONS

§ 927.105 Communications.

Unless otherwise prescribed in this subpart or in the order, or required by the Fresh Pear Committee or the Processed Pear Committee, all reports, applications, submittals, requests, inspection certificates, and communications in connection with the order shall be forwarded to: Fresh Pear Committee, 4382 SE International Way, Suite A, Milwaukie OR 97222–4635 and or the Processed Pear Committee, 105 South 18th Street, Suite 205, Yakima WA 98901.

[71 FR 7676, Feb. 14, 2006]

EXEMPTIONS AND SAFEGUARDS

§ 927.120 Pears for charitable or by-product purposes.

Pears which do not meet the requirements of the then effective grade, size, or quality regulations shall not be shipped or handled for consumption by any charitable institution or for distribution by any relief agency or for conversion into any by-product, unless there first shall have been delivered to the manager of the Fresh Pear Committee a certificate executed by the intended receiver and user of said pears showing, to the manager's satisfaction, that said pears actually will be used for one or more of the aforesaid purposes.

[70 FR 59625, Oct. 13, 2005, as amended at 71 FR 7676, Feb. 14, 2006]

§ 927.121 Pears for gift purposes.

There are exempted from the provisions of the order any and all pears which, in individual gift packages, are shipped directly to, or which are shipped for distribution without resale to, an individual person as the consumer thereof, and any and all pears which, in individual gift packages are shipped directly to, or are shipped for distribution without resale to, a purchaser who will use these pears solely for gift purposes and not for sale.

[70 FR 59625, Oct. 13, 2005, as amended at 71 FR 7676, Feb. 14, 2006]

§ 927.122 Consumer direct pear sales.

Notwithstanding any other provision of this section, fresh pears may be handled without regard to the provisions of §§ 927.41, 927.51, 927.60, and 927.70 under the following conditions:

(a) Such pears are sold in person and sold directly to consumers on the premises where grown, at packing facilities, at roadside stands, or at farmers' markets.

(b) Such pears are for home use only and are not for resale.

(c) The total quantity of such pears sold to each consumer during any single transaction does not exceed 220 pounds.

[76 FR 4204, Jan. 25, 2011]

§ 927.123 Interest and late payment charges.

Payments received more than 45 days after the date on which they are due shall be considered delinquent and subject to a late payment charge of $25.00 or 2 percent of the total due, whichever is greater. Payments received more than 60 days after the date on which they are due shall be subject to a 1½ percent interest charge per month, until final payment is made and interest shall be applied to the total unpaid balance, including the late payment charge and any accumulated interest. Any amount paid shall be credited when the payment is received in the Fresh Pear Committee or Processed Pear Committee office.

[70 FR 59625, Oct. 13, 2005, as amended at 71 FR 7676, Feb. 14, 2006]

REPORTS

§927.125 Fresh pear reports.

(a) Each handler shall furnish to the Fresh Pear Committee, as of every other Friday or at such other times established by the Fresh Pear Committee, a "Handler's Statement of Fresh Pear Shipments" containing the following information:

(1) The quantity of each variety or subvariety of fresh pears shipped by that handler during the preceding two weeks;

(2) The assessment payment due and enclosed;

(3) The date of each shipment;

(4) The ultimate destination by city and state or city and country;

(5) The name and address of such handler; and

(6) Other information as may be requested by the Fresh Pear Committee.

(b) Each handler shall furnish to the Fresh Pear Committee, each Friday during the shipping season or at such other times established by the Fresh Pear Committee, a "Handler's Packout Report" containing the following information:

(1) The projected total quantity of the packout of each variety or subvariety;

(2) The quantity to date of the packout of each variety or subvariety;

(3) The quantity of each variety or subvariety loose in storage;

(4) The quantity of the packout in controlled atmosphere (C.A.) storage and the quantity in C.A. storage which is sold;

(5) The quantity of each variety or subvariety shipped;

(6) The name and address of such handler; and

(7) Other information as may be requested by the Fresh Pear Committee.

(c) Each handler shall furnish to the Fresh Pear Committee, upon request, the "Pear Size and Grade Storage Report" containing the quantity of specific grades and sizes of fresh pears in regular and C.A. storage by variety or subvariety, and such other information as may be requested from the Fresh Pear Committee for the time period specified.

(d) Each handler who has shipped less than 2,500 44-pound net weight standard boxes or container equivalents of fresh pears during any reporting period of the shipping season may, in lieu of reporting as provided in (a) and (b) of this section, report as follows:

(1) At completion of harvest, on the next reporting date, furnish to the Fresh Pear Committee a "Handlers Packout Report';

(2) After unreported shipments total 2,500 44-pound net weight standard boxes or container equivalents of fresh pears, furnish to the Fresh Pear Committee a "Handler's Statement of Fresh Pear Shipments" and a "Handler's Packout Report" on the next reporting date;

(3) After completion of all shipments from regular storage (i.e. non-C.A. storage), furnish to the Fresh Pear Committee a "Handler's Statement of Fresh Pear Shipments" and a "Handler's Packout Report" on the next reporting date;

(4) At mid-season for C.A. storage, at a date established by the Fresh Pear Committee, furnish to the Fresh Pear Committee a "Handler's Statement of Fresh Pear Shipments", and a "Handler's Packout Report'; and

(5) At the completion of all seasonal pear shipments, furnish to the Fresh Pear Committee a "Handler's Statement of Fresh Pear Shipments" and a "Handler's Packout Report", on the next reporting date. Each of these reports shall be marked "final report" and include an explanation of the actual shipments versus the original estimate, if different.

(e) Each handler shall specify on each bill of lading covering each shipment, the variety or subvariety and quantity of all pears included in that shipment.

[71 FR 7677, Feb. 14, 2006]

§927.126 Processed pear reports.

(a) Each handler shall furnish to the Processed Pear Committee annually on a date established by the Processed Pear Committee the "Processed Pear

Assessment Report" containing the following information:

(1) The name of the processor(s) or firm(s) to whom pears were sold;

(2) The quantity of each variety or subvariety of pears shipped by that handler;

(3) The crop year covered in the report;

(4) The assessment payment due and enclosed;

(5) The name and address of such handler; and

(6) Other information as may be requested by the Processed Pear Committee.

(b) Each handler shall specify on each bill of lading covering each shipment, the variety or subvariety and quantity of all pears included in that shipment.

[71 FR 7677, Feb. 14, 2006]

§ 927.142 [Reserved]

ADMINISTRATIVE BODIES

§ 927.150 Reapportionment of the Processed Pear Committee.

Pursuant to § 927.20(c), on or after July 1, 2019, the 10-member Processed Pear Committee is reapportioned and shall consist of three grower members, three handler members, three processor members, and one member representing the public. For each member there shall be an alternate. District 1, the State of Washington, shall be represented by two grower members and two handler members. District 2, the State of Oregon, shall be represented by one grower member and one handler member. Processor members may be from District 1, District 2, or from both districts.

[84 FR 9222, Mar. 14, 2019]

ASSESSMENT RATE

§ 927.236 Fresh pear assessment rate.

On and after July 1, 2018, the following base rates of assessment for fresh pears are established for the Fresh Pear Committee:

(a) $0.463 per 44-pound net weight standard box or container equivalent for any or all varieties or subvarieties of fresh pears classified as "summer/fall";

(b) $0.463 per 44-pound net weight standard box or container equivalent for any or all varieties or subvarieties of fresh pears classified as "winter"; and

(c) $0.000 per 44-pound net weight standard box or container equivalent for any or all varieties or subvarieties of fresh pears classified as "other".

[71 FR 7677, Feb. 14, 2006, as amended at 76 FR 54078, Aug. 31, 2011; 78 FR 24035, Apr. 24, 2013; 83 FR 56257, Nov. 13, 2018]

§ 927.237 Processed pear assessment rate.

On and after July 1, 2018, the following base rates of assessment for pears for processing are established for the Processed Pear Committee:

(a) $7.15 per ton for any or all varieties or subvarieties of pears for canning classified as "summer/fall" excluding pears for other methods of processing;

(b) $0.00 per ton for any or all varieties or subvarieties of pears for processing classified as "winter"; and

(c) $0.00 per ton for any or all varieties or subvarieties of pears for processing classified as "other".

[71 FR 7677, Feb. 14, 2006, as amended at 76 FR 53813, Aug. 30, 2011; 77 FR 72199, Dec. 5, 2012; 83 FR 591, Jan. 5, 2018; 83 FR 62451, Dec. 4, 2018]

§ 927.316 Handling regulation.

During the period August 15 through November 1, no person shall handle any fresh Beurre D'Anjou variety of pears for shipments to North America (Continental United States, Mexico, or Canada), unless such pears meet the following requirements:

(a) Fresh Beurre D'Anjou variety of pears shall have a certification by the Federal-State Inspection Service, issued prior to shipment, showing that the core/pulp temperature of such pears has been lowered to 35 degrees Fahrenheit or less and any such pears have an average pressure test of 14 pounds or less. The handler shall submit, or cause to be submitted, a copy of the certificate issued on the shipment to the Fresh Pear Committee.

(b) Each handler may ship on any one conveyance 8,800 pounds or less of fresh

Beurre D'Anjou variety of pears without regard to the quality and inspection requirements in paragraph (a) of this section.

[71 FR 7677, Feb. 14, 2006]

PART 929—CRANBERRIES GROWN IN STATES OF MASSACHUSETTS, RHODE ISLAND, CONNECTICUT, NEW JERSEY, WISCONSIN, MICHIGAN, MINNESOTA, OREGON, WASHINGTON, AND LONG ISLAND IN THE STATE OF NEW YORK

Subpart A—Order Regulating Handling

DEFINITIONS

AUTHORITY: 7 U.S.C. 601–674.

SOURCE: 27 FR 8101, Aug. 15, 1962, unless otherwise noted.

Subpart A—Order Regulating Handling

DEFINITIONS

§ 929.1   Secretary.

Secretary means the Secretary of Agriculture of the United States, or any officer or employee of the United States Department of Agriculture to whom authority has heretofore been delegated, or to whom authority may hereafter be delegated, to act in his stead.

§ 929.2   Act.

Act means Public Act No. 10, 73d Congress (May 12, 1933), as amended, and as reenacted and amended by the Agricultural Marketing Agreement Act of 1937, as amended (secs. 1–19, 48 Stat. 31, as amended; 7 U.S.C. 601–674).

§ 929.3   Person.

Person means an individual, partnership, corporation, association, or any other business unit.

§ 929.4   Production area.

Production area means the States of Massachusetts, Rhode Island, Connecticut, New Jersey, Wisconsin, Michigan, Minnesota, Oregon, Washington, and Long Island in the State of New York.

§ 929.5   Cranberries.

Cranberries means all varieties of the fruit Vaccinium Macrocarpon, known as cranberries, grown in the production area.

§ 929.6   Fiscal period.

Fiscal period is synonymous with *fiscal year* and *crop year* and means the 12-month period beginning September 1 of 1 year and ending August 31 of the following year.

[33 FR 11640, Aug. 16, 1968]

§ 929.7   Committee.

Committee means the Cranberry Marketing Committee established pursuant to § 929.20.

§ 929.8   Grower.

Grower is synonymous with producer and means any person who produces cranberries for market and who has a proprietary interest therein.

§ 929.9   Handler.

Handler means any person who handles cranberries.

§ 929.10   Handle.

(a) *Handle* means:

(1) To can, freeze, or dehydrate cranberries within the production area or;

(2) To sell, consign, deliver, or transport (except as a common or contract carrier of cranberries owned by another person) fresh cranberries or in any other way to place fresh cranberries in the current of commerce within the production area or between the production area and any point outside thereof.

(b) The term handle shall not include:

(1) The sale of non harvested cranberries;

(2) The delivery of cranberries by the grower thereof to a handler having packing or processing facilities located within the production area;

(3) The transportation of cranberries from the bog where grown to a packing or processing facility located within the production area; or

(4) The cold storage or freezing of excess or restricted cranberries for the purpose of temporary storage during periods when an annual allotment percentage and/or a handler withholding program is in effect prior to their disposal, pursuant to §§ 929.54 or 929.59.

[57 FR 38748, Aug. 27, 1992, as amended at 70 FR 7640, Feb. 15, 2005]

§929.11 To can, freeze, or dehydrate.

To can, freeze, or dehydrate means to convert cranberries into canned, frozen, or dehydrated cranberries or other cranberry products by any commercial process.

§929.12 Acquire.

Acquire means to obtain cranberries by any means whatsoever for the purpose of handling such cranberries.

§929.13 Sales history.

Sales history means the number of barrels of cranberries established for a grower by the committee pursuant to §929.48.

[57 FR 38748, Aug. 27, 1992]

§929.14 Marketable quantity.

Marketable quantity means for a crop year the number of pounds of cranberries necessary to meet the total market demand and to provide for an adequate carryover.

[33 FR 11640, Aug. 16, 1968]

§929.15 Annual allotment.

A grower's annual allotment for a particular crop year is the number of barrels of cranberries determined by multiplying such grower's sales history by the allotment percentage established pursuant to §929.49 for such crop year.

[57 FR 38748, Aug. 27, 1992]

§929.17 Barrel.

Barrel means a quantity of cranberries equivalent to 100 pounds of cranberries.

[57 FR 38748, Aug. 27, 1992]

ADMINISTRATIVE BODY

§929.20 Establishment and membership.

(a) There is hereby established a Cranberry Marketing Committee consisting of 13 grower members, and 9 grower alternate members. Except as hereafter provided, members and alternate members shall be growers or employees, agents, or duly authorized representatives of growers.

(b) The committee shall include one public member and one public alternate member nominated by the committee and selected by the Secretary. The public member and public alternate member shall not be a cranberry grower, processor, handler, or have a financial interest in the production, sales, marketing or distribution of cranberries or cranberry products. The committee, with the approval of the Secretary, shall prescribe qualifications and procedures for nominating the public member and public alternate member.

(c) Members shall represent each of the following subdivisions of the production areas in the number specified in Table 1. Members shall reside in the designated district of the production area from which they are nominated and selected. Provided, that there shall also be one member-at-large who may be nominated from any of the marketing order districts.

District 1: The States of Massachusetts, Rhode Island, and Connecticut;
District 2: The State of New Jersey and Long Island in the State of New York.
District 3: The States of Wisconsin, Michigan, and Minnesota.
District 4: The States of Oregon and Washington.

TABLE 1

Districts	Major cooperative	Major cooperative	Other than major	Other than major
			Members	Alternates
1	2	1	2	1
2	1	1	1	1
3	2	1	2	1
4	1	1	1	1
Any	1 member-at-large			

(d) *Disclosure of unregulated production.* All grower nominees and alternate grower nominees of the committee shall disclose any financial interest in the production of cranberries that are not subject to regulation by this part.

(e) The committee may establish, with the approval of the Secretary, rules and regulations for the implementation and operation of this section.

[69 FR 18806, Apr. 9, 2004]

§ 929.21 Term of office.

(a) The term of office for each member and alternate member of the committee shall be for two years, beginning on August 1 of each even-numbered year and ending on the second succeeding July 31. *Provided,* That following adoption of this amendment, the term of office for the initial members and alternates shall also include any time served prior to August 1 of the first even numbered year served. Members and alternate members shall serve the term of office for which they are selected and have been qualified or until their respective successors are selected and have been qualified.

(b) Beginning on August 1 of the even-numbered year following the adoption of this amendment, committee members shall be limited to three consecutive terms. This limitation on tenure shall not include service on the committee prior to the adoption of this amendment or service on the committee by the initial members prior to August 1 of the first even-numbered year served and shall not apply to alternate members.

(c) Members who have served three consecutive terms must leave the committee for at least one full term before becoming eligible to serve again unless specifically exempted by the Secretary. The consecutive terms of office for alternate members shall not be so limited.

[69 FR 18806, Apr. 9, 2004]

§ 929.22 Nomination.

(a) *Initial members.* As soon as practicable after adoption of this amendment, the committee shall hold nominations in accordance with this section. The names and addresses of all nominees shall be submitted to the Secretary for selection as soon as the nomination process is complete. Nominees selected for the initial Committee, following adoption of this amendment, shall serve a minimum of one two-year term beginning on August 1 of the first even-numbered year served.

(b) *Successor members.* Beginning on June 1 of the even-numbered year following the adoption of this amendment, the committee shall hold nominations in accordance with this section.

(c) Whenever any cooperative marketing organization handles more than fifty percent of the total volume of cranberries produced during the fiscal period in which nominations for membership on the committee are made, such cooperative or growers affiliated therewith shall nominate:

(1) Six qualified persons for members and four qualified persons for alternate members of the committee. These members and alternate members shall be referred to as the major cooperative members and alternate members. Nominee(s) for major cooperative member and major cooperative alternate member shall represent growers from each of the marketing order districts designated in § 929.20.

(2) A seventh major cooperative member shall be referred to as the major cooperative member-at-large. The major cooperative member-at-large may be nominated from any of the marketing order districts.

(3) Six qualified persons for members and four qualified persons for alternate members of the committee shall be nominated by those growers who market their cranberries through entities other than the major cooperative marketing organization. Nominees for member and alternate member representing entities other than the major cooperative marketing organization shall represent growers from each of the marketing order districts as designated in § 929.20(c).

(d) Whenever any major cooperative marketing organization handles 50 percent or less of the total volume of cranberries produced during the fiscal period in which nominations for membership on the committee are made, the major cooperative or growers affiliated therewith, shall nominate:

(1) Six qualified persons for major cooperative members and four qualified persons for major cooperative alternate members of the committee. Nominees for member and alternate member shall represent growers from each of the marketing order districts as designated in §929.20(c).

(2) Six qualified persons for members and four qualified persons for alternate members of the committee shall be nominated by those growers who market their cranberries through entities other than the major cooperative marketing organization. Nominees for member and alternate member shall represent growers from each of the marketing order districts as designated in §929.20(c).

(3) A seventh member nominee shall be referred to as the member-at-large representing entities other than the major cooperative marketing organization. The member-at-large may be nominated from any of the marketing order districts.

(e) Nominations of qualified member nominees representing entities other than the major cooperative marketing organization shall be made through a call for nominations sent to all eligible growers residing within each of the marketing order districts. The call for such nominations shall be by such means as are recommended by the committee and approved by the Secretary.

(1) The names of all eligible nominees from each district received by the committee, by such date and in such form as recommended by the committee and approved by the Secretary, will appear on the nomination ballot for that district.

(2) Election of the member nominees and alternate member nominees shall be conducted by mail ballot.

(3) Eligible growers shall participate in the election of nominees from the district in which they reside.

(4) When voting for member nominees, each eligible grower shall be entitled to cast one vote on behalf of him/herself.

(5) The nominee receiving the highest number of votes cast in districts two and four shall be the member nominee representing entities other than the major cooperative marketing organization from that district. The nominee receiving the second highest number of votes cast in districts two and four shall be the alternate member representing entities other than the major cooperative marketing organization from that district.

(6) The nominees receiving the highest and second highest number of votes cast in districts one and three shall be the member nominees representing entities other than the major cooperative marketing organization from that district. The nominee receiving the third highest number of votes cast in districts one and three shall be the alternate member representing entities other than the major cooperative marketing organization from that district.

(f) Nominations for the member-at-large representing entities other than the major cooperative marketing organization shall be made through a call for nominations sent to all eligible growers residing within the marketing order districts. The call for such nominations shall be by such means as recommended by the committee and approved by the Secretary.

(1) Election of the member-at-large shall be held by mail ballot sent to all eligible growers in the marketing order districts by such date and in such form as recommended by the committee and approved by the Secretary.

(2) Eligible growers casting ballots may vote for a member-at-large nominee from marketing order districts other than where they produce cranberries.

(3) When voting for the member-at-large nominee, each eligible grower shall be entitled to cast one vote on behalf of him/herself.

(4) The nominee receiving the highest number of votes cast shall be designated the member-at-large nominee representing entities other than the major cooperative marketing organization. The nominee receiving the second

highest number of votes cast shall be declared the alternate member-at-large nominee representing entities other than the major cooperative marketing organization.

(g) The committee may request that growers provide their federal tax identification number(s) in order to determine voting eligibility.

(h) The names and addresses of all successor member nominees shall be submitted to the Secretary for selection no later than July 1 of each even-numbered year.

(i) The committee, with the approval of the Secretary, may issue rules and regulations to carry out the provisions or to change the procedures of this section.

[69 FR 18807, Apr. 9, 2004]

§ 929.23 Selection.

(a) From nominations made pursuant to § 929.22(b), the Secretary shall select members and alternate members to the committee on the basis of the representation provided for in § 929.20 and in paragraph (b) or (c) of this section.

(b) Whenever any cooperative marketing organization handles more than 50 percent of the total volume of cranberries produced during the fiscal year in which nominations for membership on the committee are made, the Secretary shall select:

(1) Six major cooperative members and four major cooperative alternate members from nominations made pursuant to § 929.22(c)(1).

(2) One major cooperative member-at-large from nominations made pursuant to § 929.22(c)(2), and

(3) Six members and four alternate members from growers who market their cranberries through other than the major cooperative marketing organization made pursuant to § 929.22(c)(3).

(c) Whenever any major cooperative marketing organization handles 50 percent or less of the total volume of cranberries produced during the fiscal year in which nominations for membership on the committee are made, the Secretary shall select:

(1) Six major cooperative members and four major cooperative alternate members from nominations made pursuant to § 929.22(d)(1).

(2) Six members and four alternate members from nominations made pursuant to § 929.22(d)(2).

(3) One member-at-large representing entities other than the major cooperative marketing organization from nominations made pursuant to § 929.22(d)(3).

[69 FR 18808, Apr. 9, 2004]

§ 929.24 Failure to nominate.

If nominations are not made within the time and in the manner prescribed in § 929.22, the Secretary may, without regard to nominations, select the members and alternate members of the committee on the basis of representation provided for in §§ 929.20 and 929.23.

§ 929.25 Acceptance.

Any person selected by the Secretary as a member or as an alternate member of the committee shall qualify by filing a written acceptance with the Secretary promptly after being notified of such selection.

§ 929.26 Vacancies.

To fill any vacancy occasioned by the failure of any person selected as a member or as an alternate member of the committee to qualify, or in the event of the death, removal, resignation, or disqualification of any member or alternate member of the committee, a successor for the unexpired term of such member or alternate member of the committee shall be nominated and selected in the manner specified in §§ 929.22 and 929.23. If the names of nominees to fill any such vacancy are not made available to the Secretary within a reasonable time after such vacancy occurs, the Secretary may fill such vacancy without regard to nominations, which selection shall be made on the basis of representation provided for in §§ 929.20 and 929.23.

§ 929.27 Alternate members.

An alternate member of the committee shall act in the place and stead of a member during the absence of such member and may perform such other duties as assigned. In the event of the death, removal, resignation, or disqualification of a member, an alternate shall act for him/her until a successor for such member is selected and has qualified. In the event both a member and alternate member from the same marketing order district are unable to attend a committee meeting, the committee may designate any other alternate member to serve in such member's place and stead at that meeting provided that:

(a) An alternate member representing the major cooperative shall not serve in place of a member representing other than the major cooperative or the public member.

(b) An alternate member representing other than the major cooperative shall not serve in place of a major cooperative member or the public member.

(c) A public alternate member shall not serve in place of any industry member.

[69 FR 18808, Apr. 9, 2004]

§929.28 Redistricting and Reapportionment.

(a) The committee, with the approval of the Secretary, may reestablish districts within the production area and reapportion membership among the districts. In recommending such changes, the committee shall give consideration to:

(1) The relative volume of cranberries produced within each district.

(2) The relative number of cranberry producers within each district.

(3) Cranberry acreage within each district.

(4) Other relevant factors.

(b) The committee may establish, with the approval of the Secretary, rules and regulations for the implementation and operation of this section.

[70 FR 7641, Feb. 15, 2005]

§929.30 Powers.

The committee shall have the following powers:

(a) To administer the provisions of this part in accordance with its terms;

(b) To receive, investigate, and report to the Secretary complaints of violations of the provisions of this part;

(c) To make and adopt rules and regulations to effectuate the terms and provisions of this part; and

(d) To recommend to the Secretary amendments to this part.

§929.31 Duties.

The committee shall have, among others, the following duties:

(a) To select a chairman and such other officers as may be necessary, and to define the duties of such officers;

(b) To appoint such employees, agents, and representatives as it may deem necessary and to determine the compensation and to define the duties of each;

(c) To submit to the Secretary as soon as practicable after the beginning of each fiscal period a budget for such fiscal period, including a report in explanation of the items appearing therein and a recommendation as to the rate of assessment for such period;

(d) To keep minutes, books, and records which will reflect all of the acts and transactions of the committee and which shall be subject to examination by the Secretary;

(e) To prepare periodic statements of the financial operations of the committee and to make copies of each such statement available to growers and handlers for examination at the office of the committee;

(f) To cause its books to be audited by a competent public accountant at least once each fiscal year and at such times as the Secretary may request;

(g) To act as intermediary between the Secretary and any grower or handler;

(h) To investigate and assemble data on the growing, handling, and marketing conditions with respect to cranberries;

(i) To submit to the Secretary the same notice of meetings of the committee as is given to its members;

(j) To submit to the Secretary such available information as he may request; and

(k) To investigate compliance with the provisions of this part.

§929.32 Procedure.

(a) Ten members of the committee, or alternates acting for members, shall constitute a quorum. All actions of the committee shall require at least ten concurring votes: Provided, if the public member or the public alternate member acting in the place and stead of the public member, is present at a meeting, then eleven members shall constitute a quorum. Any action of the committee on which the public member votes shall require eleven concurring votes. If the public member abstains from voting on any particular matter, ten concurring votes shall be required for an action of the committee.

(b) The committee may vote by mail, telephone, fax, telegraph, or other electronic means; Provided that any votes cast by telephone shall be confirmed promptly in writing. Voting by proxy, mail, telephone, fax, telegraph, or other electronic means shall not be permitted at any assembled meeting of the committee.

(c) All assembled meetings of the committee shall be open to growers and handlers. The committee shall publish notice of all meetings in such manner as it deems appropriate.

[69 FR 18808, Apr. 9, 2004]

§ 929.33   Expenses and compensation.

The members of the committee, and alternates when acting as members, shall serve without compensation but shall be reimbursed for necessary expenses, as approved by the committee, incurred by them in the performance of their duties under this part. The committee at its discretion may request the attendance of one or more alternates at any or all meetings, notwithstanding the expected or actual presence of the respective members, and may pay expenses, as aforesaid.

EXPENSES AND ASSESSMENTS

§ 929.40   Expenses.

The committee is authorized to incur such expenses as the Secretary finds are reasonable and likely to be incurred by the committee for its maintenance and functioning and to enable it to exercise its powers and perform its duties in accordance with the provisions hereof. The funds to cover such expenses shall be paid to the committee by handlers in the manner prescribed in § 929.41.

§ 929.41   Assessments.

(a) As a handler's pro rate share of the expenses which the Secretary finds are reasonable and likely to be incurred by the committee during a fiscal period, a handler shall pay to the committee assessments on all cranberries acquired as the first handler thereof during such period, except as provided in § 929.55: *Provided,* That no handler shall pay assessments on excess cranberries as provided in § 929.57. The payment of assessments for the maintenance and functioning of the committee may be required under this part throughout the period it is in effect, irrespective of whether particular provisions thereof are suspended or become inoperative.

(b) The Secretary shall fix the rate of assessment to be paid by each handler during a fiscal period in an amount designated to secure funds sufficient to cover the expenses which may be incurred during such period and to accumulate and maintain a reserve fund equal to approximately one fiscal period's expenses. At any time during or after the fiscal period, the Secretary may increase the assessment rate in order to secure funds sufficient to cover any later finding by the Secretary relative to the expenses which may be incurred. Such increase shall be applied to all cranberries acquired during the applicable fiscal period. In order to provide funds for the administration of the provisions of this part during the first part of a fiscal year, before sufficient operating income is available from assessments, the committee may accept the payment of assessments in advance and may also borrow money for such purposes.

(c) If a handler does not pay such assessment within the period of time prescribed by the committee, the assessment may be increased by either a late payment charge, or an interest charge, or both, at rates prescribed by the committee, with the approval of the Secretary.

[57 FR 38748, Aug. 27, 1992]

§ 929.42   Accounting.

(a) If, at the end of a fiscal period, the assessments collected are in excess of expenses incurred, the committee, with the approval of the Secretary, may carryover such excess into subsequent fiscal periods as a reserve: *Provided,* That funds already in the reserve do not exceed approximately one fiscal period's expenses. Such reserve funds may be used (1) to cover any expenses authorized by this part and (2) to cover necessary expenses of liquidation in the event of termination of this part. If any such excess is not retained in a reserve, it shall be refunded proportionately to the handlers from whom the excess was collected. Upon termination of this part, any funds not required to defray the necessary expenses of liquidation shall be disposed of in such manner as the Secretary may determine to be appropriate; *Provided,* That to the extent practical, such funds shall be returned pro rata to the persons from whom such funds were collected.

(b) All funds received by the committee pursuant to the provisions of this part shall be used solely for the purpose specified in this part and shall be accounted for in the manner provided in this part. The Secretary may at any time require the committee and its members to account for all receipts and disbursements.

§929.43 Contributions.

The Committee may accept voluntary contributions to pay expenses incurred pursuant to §929.45, Research and development. Such contributions may only be accepted if they are sourced from domestic contributors and are free from any encumbrances or restrictions on their use by the donor. The Cranberry Marketing Committee shall retain complete control of their use.

[84 FR 9939, Mar. 19, 2019]

RESEARCH

§929.45 Research and development.

(a) The committee, with the approval of the Secretary, may establish or provide for the establishment of production research, marketing research, and market development projects, including paid advertising, designed to assist, improve, or promote the marketing, distribution, consumption, or efficient production of cranberries. The expense of such projects shall be paid from funds collected pursuant to §929.41, or from such other funds as approved by the Secretary.

(b) The committee may, with the approval of the Secretary, establish rules and regulations as necessary for the implementation and operation of this section.

[57 FR 38748, Aug. 27, 1992, as amended at 70 FR 7641, Feb. 15, 2005]

REGULATIONS

§929.46 Marketing policy.

Each season prior to making any recommendation pursuant to §929.51, the committee shall submit to the Secretary a report setting forth its marketing policy for the crop year. Such marketing policy shall contain the following information for the current crop year:

(a) The estimated total production of cranberries;

(b) The expected general quality of such cranberry production;

(c) The estimated carryover, as of September 1, of frozen cranberries and other cranberry products;

(d) The expected demand conditions for cranberries in different market outlets;

(e) The recommended desirable total marketable quantity of cranberries including a recommended adequate carryover into the following crop year of frozen cranberries and other cranberry products;

(f) Other factors having a bearing on the marketing of cranberries.

[70 FR 7641, Feb. 15, 2005]

§929.48 Sales history.

(a) A sales history for each grower shall be computed by the committee in the following manner:

(1) For growers with acreage with 6 or more years of sales history, the sales history shall be computed using an average of the highest four of the most recent six years of sales.

(2) For growers with 5 years of sales history from acreage planted or replanted 2 years prior to the first harvest on that acreage, the sales history is computed by averaging the highest 4 of the 5 years.

(3) For growers with 5 years of sales history from acreage planted or replanted 1 year prior to the first harvest on that acreage, the sales history is computed by averaging the highest 4 of the 5 years and in a year prior to a year of a producer allotment volume regulation shall be adjusted as provided in paragraph (a)(6) of this section.

(4) For a grower with 4 years or less of sales history, the sales history shall be computed by dividing the total sales from that acreage by 4 and in a year prior to a year of a producer allotment volume regulation shall be adjusted as provided in paragraph (a)(6) of this section.

(5) For growers with acreage having no sales history, or for the first harvest of replanted acres, the sales history will be the average first year yields (depending on whether first harvested 1 or 2 years after planting or replanting) as established by the committee and multiplied by the number of acres.

(6) In a year prior to a year of a producer allotment volume regulation, in addition to the sales history computed in accordance with paragraphs (a)(3) and (a)(4) of this section, additional sales history shall be assigned to growers using the formula $x = (a - b)c$. The letter "x" constitutes the additional number of barrels to be added to the grower's sales history. The value "a" is the expected yield for the forthcoming year harvested acreage as established by the committee. The value "b" is the total sales from that acreage as established by the committee divided by four. The value "c" is the number of acres planted or replanted in the specified year. For acreage with five years of sales history: a = the expected yield for the forthcoming sixth year harvested acreage (as established by the committee); b = an average of the most recent 4 years of expected yields (as established by the committee); and c = the number of acres with 5 years of sales history.

(b) A new sales history shall be calculated for each grower after each crop year, using the formulas established in paragraph (a) of this section, or such other formula(s) as determined by the committee, with the approval of the Secretary.

(c) The committee, with the approval of the Secretary, may adopt regulations to change the number and identity of years to be used in computing sales histories, including the number of years to be used in computing the average. The committee may establish, with the approval of the Secretary, rules and regulations necessary for the implementation and operation of this section.

(d) Sales histories, starting with the crop year following adoption of this part, shall be calculated separately for fresh and processed cranberries. The amount of fresh fruit sales history may be calculated based on either the delivered weight of the barrels paid for by the handler (excluding trash and unusable fruit) or on the weight of the fruit paid for by the handler after cleaning and sorting for the retail market. Handlers using the former calculation shall allocate delivered fresh fruit subsequently used for processing to growers' processing sales. Fresh fruit sales history, in whole or in part, may be added to process fruit sales history with the approval of the committee in the event that the grower's fruit does not qualify as fresh fruit at delivery.

(e) The committee may recommend rules and regulations, with the approval of the Secretary, to adjust a grower's sales history to compensate for catastrophic events that impact the grower's crop.

[70 FR 7641, Feb. 15, 2005]

929.49 Marketable quantity, allotment percentage, and annual allotment.

(a) *Marketable quantity and allotment percentage.* If the Secretary finds, from the recommendation of the committee or from other available information, that limiting the quantity of cranberries purchased from or handled on behalf of growers during a crop year would tend to effectuate the declared policy of the Act, the Secretary shall determine and establish a marketable quantity for that crop year.

(b) The marketable quantity shall be apportioned among growers by applying the allotment percentage to each grower's sales history, established pursuant to §929.48. Such allotment percentage shall be established by the

Secretary and shall equal the marketable quantity divided by the total of all growers' sales histories including the estimated total sales history for new growers. Except as provided in paragraph (g) of this section, no handler shall purchase or handle on behalf of any grower cranberries not within such grower's annual allotment.

(c) In any crop year in which the production of cranberries is estimated by the committee to be equal to or less than its recommended marketable quantity, the committee may recommend that the Secretary increase or suspend the allotment percentage applicable to that year. In the event it is found that market demand is greater than the marketable quantity previously set, the committee may recommend that the Secretary increase such quantity.

(d) *Issuance of annual allotments.* The committee shall require all growers to qualify for such allotment by filing with the committee a form wherein growers include the following information:

(1) The amount of acreage which will be harvested;

(2) A copy of any lease agreement covering cranberry acreage;

(3) The name of the handler(s) to whom their annual allotment will be delivered;

(4) Such other information as may be necessary for the implementation and operation of this section.

(e) On or before such date as determined by the committee, with the approval of the Secretary, the committee shall issue to each grower an annual allotment determined by applying the allotment percentage established pursuant to paragraph (b) of this section to the grower's sales history.

(f) On or before such date as determined by the committee, with the approval of the Secretary, in which an allotment percentage is established by the Secretary, the committee shall notify each handler of the annual allotment that can be handled for each grower whose total crop will be delivered to that handler. In cases where a grower delivers a crop to more than one handler, the grower must specify how the annual allotment will be apportioned among the handlers. If a grower does not specify how their annual allotment is to be apportioned among the handlers, the Committee will apportion such annual allotment equally among those handlers they are delivering their crop to.

(g) Growers who do not produce cranberries equal to their computed annual allotment shall transfer their unused allotment to such growers' handlers unless it is transferred to another grower in accordance with §929.50(b) or if it is not assigned in accordance with paragraph (i) of this section. The handler shall equitably allocate the unused annual allotment to growers with excess cranberries who deliver to such handler. Unused annual allotment remaining after all such transfers have occurred shall be reported and transferred to the committee by such date as established by the committee with the approval of the Secretary.

(h) Handlers who receive cranberries more than the sum of their growers' annual allotments have "excess cranberries," pursuant to §929.59, and shall so notify the committee. Handlers who have remaining unused allotment pursuant to paragraph (g) of this section are "deficient" and shall so notify the committee. The committee shall allocate unused allotment to all handlers having excess cranberries, proportional to each handler's total allotment.

(i) Growers who decide not to grow a crop, during any crop year in which a volume regulation is in effect, may choose not to assign their allotment to a handler.

(j) The committee may establish, with the approval of the Secretary, rules and regulations necessary for the implementation and operation of this section.

[70 FR 7641, Feb. 15, 2005]

§929.50 Transfers of Sales Histories and Annual Allotments.

(a) *Leases and sales of cranberry acreage—(1) Total or partial lease of cranberry acreage.* When total or partial lease of cranberry acreage occurs, sales history attributable to the acreage being leased shall remain with the lessor.

(2) *Total sale of cranberry acreage.* When there is a sale of a grower's total

cranberry producing acreage, the committee shall transfer all owned acreage and all associated sales history to such acreage to the buyer. The seller and buyer shall file a sales transfer form providing the committee with such information as may be requested so that the buyer will have immediate access to the sales history computation process.

(3) *Partial sale of cranberry acreage.* When less than the total cranberry producing acreage is sold, sales history associated with that portion of the acreage being sold shall be transferred with the acreage. The seller shall provide the committee with a sales transfer form containing, but not limited to the distribution of acreage and the percentage of sales history, as defined in § 929.48(a)(1), attributable to the acreage being sold.

(4) No sale of cranberry acreage shall be recognized unless the committee is notified in writing.

(b) *Allotment transfers.* During a year of volume regulation, a grower may transfer all or part of his/her allotment to another grower. If a lease is in effect the lessee shall receive allotment from lessor attributable to the acreage leased. *Provided,* That the transferred allotment shall remain assigned to the same handler and that the transfer shall take place prior to a date to be recommended by the Committee and approved by the Secretary. Transfers of allotment between growers having different handlers may occur with the consent of both handlers.

(c) The committee may establish, with the approval of the Secretary, rules and regulations, as needed, for the implementation and operation of this section.

[70 FR 7642, Feb. 15, 2005]

§ 929.51 Recommendations for regulation.

(a) Except as otherwise provided in paragraph (b) of this section, if the committee deems it advisable to regulate the handling of cranberries in the manner provided in § 929.52, it shall so recommend to the Secretary by the following appropriate dates:

(1) An allotment percentage regulation must be recommended by no later than March 1;

(2) A handler withholding program must be recommended by not later than August 31. Such recommendation shall include the free and restricted percentages for the crop year;

(3) If both programs are recommended in the same year, the Committee shall submit with its recommendation an economic analysis to the USDA prior to March 1 of the year in which the programs are recommended.

(b) An exception to the requirement in paragraph (a)(1) of this section may be made in a crop year in which, due to unforeseen circumstances, a producer allotment regulation is deemed necessary subsequent to the March 1 deadline.

(c) In arriving at its recommendations for regulation pursuant to paragraph (a) of this section, the committee shall give consideration to current information with respect to the factors affecting the supply of and demand for cranberries during the period when it is proposed that such regulation should be imposed. With each such recommendation for regulation, the committee shall submit to the Secretary the data and information on which such recommendation is based and any other information the Secretary may request.

[70 FR 7642, Feb. 15, 2005]

§ 929.52 Issuance of regulations.

(a) The Secretary shall regulate, in the manner specified in this section, the handling of cranberries whenever the Secretary finds, from the recommendations and information submitted by the committee, or from other available information, that such regulation will tend to effectuate the declared policy of the Act. Such regulation shall limit the total quantity of cranberries which may be handled during any fiscal period by fixing the free and restricted percentages, applied to cranberries acquired by handlers in accordance with § 929.54, and/or by establishing an allotment percentage in accordance with § 929.49.

(b) The committee shall be informed immediately of any such regulation issued by the Secretary, and the committee shall promptly give notice thereof to handlers.

[70 FR 7643, Feb. 15, 2005]

§929.53 Modification, suspension, or termination of regulations.

(a) In the event the committee at any time finds that, by reason of changed conditions, any regulations issued pursuant to §929.52 should be modified, suspended, or terminated, it shall so recommend to the Secretary.

(b) Whenever the Secretary finds, from the recommendations and information submitted by the committee or from other available information, that a regulation should be modified, suspended, or terminated in order to effectuate the declared policy of the act, he shall modify, suspend, or terminate such regulation: *Provided*, That no such modification shall increase the restricted percentage previously established for the then current fiscal year. If the Secretary finds that a regulation obstructs or does not tend to effectuate the declared policy of the act, he shall suspend or terminate such regulation.

§929.54 Withholding.

(a) Whenever the Secretary has fixed the free and restricted percentages for any fiscal period, as provided for in §929.52(a), each handler shall withhold from handling a portion of the cranberries acquired during such period. The withheld portion shall be equal to the restricted percentage multiplied by the volume of marketable cranberries acquired. Such withholding requirements shall not apply to any lot of cranberries for which such withholding requirement previously has been met by another handler in accordance with §929.55.

(b) The committee, with the approval of the Secretary, shall prescribe the manner in which, and date or dates during the fiscal period by which, handlers shall have complied with the withholding requirements specified in paragraph (a) of this section.

(c) Withheld cranberries may meet such standards of grade, size, quality, or condition as the committee, with the approval of the Secretary, may prescribe. The Federal or Federal-State Inspection Service may inspect all such cranberries. A certificate of such inspection shall be issued which shall include the name and address of the handler, the number and type of containers in the lot, the location where the lot is stored, identification marks (including lot stamp, if used), and the quantity of cranberries in such lot that meet the prescribed standards. Promptly after inspection and certification, each such handler shall submit to the committee a copy of the certificate of inspection issued with respect to such cranberries.

(d) Any handler who withholds from handling a quantity of cranberries in excess of that required pursuant to paragraph (a) of this section shall have such excess quantity credited toward the next fiscal year's withholding obligation, if any—provided that such credit shall be applicable only if the restricted percentage established pursuant to §929.52 was modified pursuant to §929.53; to the extent such excess was disposed of prior to such modification; and after such handler furnishes the committee with such information as it prescribes regarding such withholding and disposition.

(e) The Committee, with the approval of the Secretary, may establish rules and regulations necessary and incidental to the administration of this section.

[70 FR 7643, Feb. 15, 2005]

§929.55 Interhandler transfer.

(a) Transfer of cranberries from one handler to another may be made without prior notice to the committee, except during a period when a volume regulation has been established. If such transfer is made between handlers who have packing or processing facilities located within the production area, the assessment and withholding obligations provided under this part shall be assumed by the handler who agrees to meet such obligation. If such transfer is to a handler whose packing or processing facilities are outside of the production area, such assessment and withholding obligation shall be met by the handler residing within the production area.

(b) All handlers shall report all such transfers to the committee on a form provided by the committee four times a year or at other such times as may be recommended by the committee and approved by the Secretary.

(c) The committee may establish, with the approval of the Secretary, rules and regulations necessary for the implementation and operation of this section.

[38 FR 29801, Oct. 29, 1973, as amended at 57 FR 38750, Aug. 27, 1992]

§ 929.56 Special provisions relating to withheld (restricted) cranberries.

(a) A handler shall make a written request to the committee for the release of all or part of the cranberries that the handler is withholding from handling pursuant to § 929.54(a). Each request shall state the quantity of cranberries for which release is requested and shall provide such additional information as the committee may require. Handlers may replace the quantity of withheld cranberries requested for release as provided under either paragraph (b) or (c) of this section.

(b) The handler may contract with another handler for an amount of free cranberries to be converted to restricted cranberries that is equal to the volume of cranberries that the handler wishes to have converted from his own restricted cranberries to free cranberries.

(1) The handlers involved in such an agreement shall provide the committee with such information as may be requested prior to the release of any restricted cranberries.

(2) The committee shall establish guidelines to ensure that all necessary documentation is provided to the committee, including but not limited to, the amount of cranberries being converted and the identities of the handlers assuming the responsibility for withholding and disposing of the free cranberries being converted to restricted cranberries.

(3) Cranberries converted to replace released cranberries may be required to be inspected and meet such standards as may be prescribed for withheld cranberries prior to disposal.

(4) Transactions and agreements negotiated between handlers shall include all costs associated with such transactions including the purchase of the free cranberries to be converted to restricted cranberries and all costs associated with inspection (if applicable) and disposal of such restricted cranberries. No costs shall be incurred by the committee other than for the normal activities associated with the implementation and operation of a volume regulation program.

(5) Free cranberries belonging to one handler and converted to restricted cranberries on the behalf of another handler shall be reported to the committee in such manner as prescribed by the committee.

(c) Except as otherwise directed by the Secretary, as near as practicable to the beginning of the marketing season of each fiscal period with respect to which the marketing policy proposes regulation pursuant to § 929.52(a), the committee shall determine the amount per barrel each handler shall deposit with the committee for it to release to him, in accordance with this section, all or part of the cranberries he is withholding; and the committee shall give notice of such amount of deposit to handlers. Such notice shall state the period during which such amount of deposit shall be in effect. Whenever committee determines that, by reason of changed conditions or other factors, a different amount should therefore be deposited for the release of withheld cranberries, it shall give notice to handlers of the new amount and the effective period thereof. Each determination as to the amount of deposit shall be on the basis of the committee's evaluation of the following factors:

(1) The prices at which growers are selling cranberries to handlers,

(2) The prices at which handlers are selling fresh market cranberries to dealers,

(3) The prices at which cranberries are being sold for processing in products,

(4) The prices at which handlers are selling cranberry concentrate,

(5) The prices the committee has paid to purchase cranberries to replace released cranberries in accordance with this section, and

(6) The costs incurred by growers in producing cranberries.

(7) Each request for release of withheld cranberries shall include, in addition to all other information as may be prescribed by the committee, the quantity of cranberries the release is requested and shall be accompanied by a deposit (a cashier's or certified check made payable to the Cranberry Marketing Committee) in an amount equal to the twenty percent of the amount determined by multiplying the number of barrels stated in the request by the then effective amount per barrel as determined in paragraph (c).

(8) Subsequent deposits equal to, but not less than, the ten percent of the remaining outstanding balance shall be payable to the committee on a monthly basis commencing on January 1, and concluding by no later than August 31 of the fiscal period.

(9) If the committee determines such a release request is properly filled out, is accompanied by the required deposit, and contains a certification that the handler is withholding such cranberries, it shall release to such handler the quantity of cranberries specified in his request.

(d) Funds deposited for the release of withheld cranberries, pursuant to paragraph (c) of this section, shall be used by the committee to purchase from handlers unrestricted (free percentage) cranberries in an aggregate amount as nearly equal to, but not in excess of, the total quantity of the released cranberries as it is possible to purchase to replace the released cranberries.

(e) All handlers shall be given an equal opportunity to participate in such purchase of unrestricted (free percentage) cranberries. If a larger quantity is offered than can be purchased, the purchases shall be made at the lowest price possible. If two or more handlers offer unrestricted (free percentage) cranberries at the same price, purchases from such handlers shall be in proportion to the quantity of their respective offerings insofar as such division is practicable. The committee shall dispose of cranberries purchased as restricted cranberries in accordance with § 929.57. Any funds received by the committee for cranberries so disposed of, which are in excess of the costs incurred by the committee in making

such disposition, will accrue to the committee's general fund.

(f) In the event any portion of the funds deposited with the committee pursuant to paragraph (c) of this section cannot, for reasons beyond the committee's control, be expended to purchase unrestricted (free percentage) cranberries to replace those withheld cranberries requested to be released, such unexpended funds shall, after deducting expenses incurred by the committee, be refunded to the handler who deposited the funds. The handler shall equitably distribute such refund among the growers delivering to such handler.

(g) Inspection for restricted (withheld) cranberries released to a handler is not required.

(h) The committee may establish, with the approval of the Secretary, rules and regulations for the implementation of this section. Such rules and regulations may include, but are not limited to, revisions in the payment schedule specified in paragraphs (c)(7) and (c)(8) of this section.

[70 FR 7643, Feb. 15, 2005]

§ 929.57 Outlets for restricted cranberries.

(a) Except as provided in this section and in § 929.56, cranberries withheld from handling may be disposed of only through diversion to such outlets as the committee, with the approval of the Secretary, finds are noncompetitive to outlets for unrestricted (free percentage) cranberries.

(b) The storage and disposition of all cranberries withheld from handling shall be subject to the supervision and accounting control of the committee.

§ 929.58 Exemptions.

(a) Upon the basis of the recommendation and information submitted by the committee, or from other available information, the Secretary may relieve from any or all requirements pursuant to this part the handling of cranberries in such minimum quantities as the committee, with the approval of the Secretary, may prescribe.

(b) Upon the basis of the recommendation and information submitted by the committee, or from

209

other available information, the Secretary may relieve from any or all requirements pursuant to this part the handling of such forms or types of cranberries as the committee, with the approval of the Secretary, may prescribe. Forms of cranberries could include cranberries intended for fresh sales or organically grown cranberries.

(c) The committee, with the approval of the Secretary, shall prescribe such rules, regulations, and safeguards as it may deem necessary to ensure that cranberries handled under the provisions of this section are handled only as authorized.

[70 FR 7644, Feb. 15, 2005]

§ 929.59 Excess cranberries.

(a) Whenever the Secretary establishes an allotment percentage pursuant to § 929.52, handlers shall be notified by the committee of such allotment percentage and shall withhold from handling such cranberries in excess of the total of their growers' annual allotments obtained during such period. Such withheld cranberries shall be defined as "excess cranberries" after all unused allotment has been allocated.

(1) Excess cranberries received by a handler shall be made available for inspection by the committee or its representatives from the time they are received until final disposition is completed. Such excess cranberries shall be identified in such manner as the committee may specify in its rules and regulations with the approval of the Secretary.

(2) All matters dealing with handler-held excess cranberries shall be in accordance with such rules and regulations established by the committee, with the approval of the Secretary.

(b) Prior to January 1, or such other date as recommended by the committee and approved by the Secretary, handlers holding excess cranberries shall submit to the committee a written plan outlining procedures for the systematic disposal of such cranberries in the outlets prescribed in § 929.61.

(c) Prior to March 1, or such other date as recommended by the committee and approved by the Secretary,

all excess cranberries shall be disposed of pursuant to § 929.61.

[57 FR 38750, Aug. 27, 1992]

REPORTS AND RECORDS

§ 929.60 Handling for special purposes.

Regulations in effect pursuant to § 929.10, § 929.41, § 929.47, § 929.48, § 929.49, § 929.51, § 929.52, or § 929.53 or any combination thereof, may be modified, suspended, or terminated to facilitate handling of excess cranberries for the following purposes:

(a) Charitable institutions;

(b) Research and development projects described pursuant to § 929.61;

(c) Any nonhuman food use;

(d) Foreign markets, except Canada; and

(e) Other purposes which may be recommended by the committee and approved by the Secretary.

[57 FR 38750, Aug. 27, 1992]

§ 929.61 Outlets for excess cranberries.

(a) *Noncommercial outlets.* Excess cranberries may be disposed of in noncommercial outlets that the committee finds, with the approval of the Secretary, meet the requirements outlined in paragraph (c) of this section. Noncommercial outlets include, but are not limited to:

(1) Charitable institutions; and

(2) Research and development projects.

(b) *Noncompetitive outlets.* Excess cranberries may be sold in outlets that the committee finds, with the approval of the Secretary, are noncompetitive with established markets for regulated cranberries and meet the requirements outlined in paragraph (c) of this section. Noncompetitive outlets include but are not limited to:

(1) Any nonhuman food use; and

(2) Other outlets established by the committee with the approval of the Secretary.

(c) *Requirements.* The handler disposing of or selling excess cranberries into noncompetitive or noncommercial outlets shall meet the following requirements, as applicable:

(1) *Charitable institutions.* A statement from the charitable institution shall be submitted to the committee

showing the quantity of cranberries received and certifying that the institution will consume the cranberries;

(2) *Research and development projects.* A report shall be given to the committee describing the project, quantity of cranberries contributed, and date of disposition;

(3) *Nonhuman food use.* Notification shall be given to the committee at least 48 hours prior to such disposition;

(4) *Other outlets established by the committee with the approval of the Secretary.* A report shall be given to the committee describing the project, quantity of cranberries contributed, and date of disposition.

(d) The storage and disposition of all excess cranberries withheld from handling shall be subject to the supervision and accounting control of the committee.

(e) The committee, with the approval of the Secretary, may establish rules and regulations for the implementation and operation of this section.

[70 FR 7644, Feb. 15, 2005]

§929.62 Reports.

(a) *Grower report.* Each grower shall file a report with the committee by January 15 of each crop year, or such other date as determined by the committee, with the approval of the Secretary, indicating the following:

(1) Total acreage harvested and whether owned or leased.

(2) Total commercial cranberry sales in barrels from such acreage.

(3) Amount of acreage either in production, but not harvested or taken out of production and the reason(s) why.

(4) Amount of new or replanted acreage coming into production.

(5) Name of the handler(s) to whom commercial cranberry sales were made.

(6) Such other information as may be needed for implementation and operation of this section.

(b) *Inventory.* Each handler engaged in the handling of cranberries or cranberry products shall, upon request of the committee, file promptly with the committee a certified report, showing such information as the committee shall specify with respect to any cranberries and cranberry products which were held by them on such date as the committee may designate.

(c) *Receipts.* Each handler shall, upon request of the committee, file promptly with the committee a certified report as to each quantity of cranberries acquired during such period as may be specified, and the place of production.

(d) *Handling reports.* Each handler shall, upon request of the committee, file promptly with the committee a certified report as to the quantity of cranberries handled during any designated period or periods.

(e) *Withheld and excess cranberries.* Each handler shall, upon request of the committee, file promptly with the committee a certified report showing, for such period as the committee may specify, the total quantity of cranberries withheld from handling or held in excess, in accordance with §§929.49 and 929.54, the portion of such withheld or excess cranberries on hand, and the quantity and manner of disposition of any such withheld or excess cranberries disposed of.

(f) *Other reports.* Upon the request of the committee, with the approval of the Secretary, each handler shall furnish to the committee such other information with respect to the cranberries and cranberry products acquired and disposed of by such person as may be necessary to enable the committee to exercise its powers and perform its duties under this part.

(g) The committee may establish, with the approval of the Secretary, rules and regulations for the implementation and operation of this section.

[70 FR 7644, Feb. 15, 2005]

§929.63 Records.

Each handler shall maintain such records of all cranberries acquired, withheld from handling, handled, and otherwise disposed of as will substantiate the required reports and as may be prescribed by the committee. All such records shall be maintained for not less than three years after the termination of the crop year in which the transactions occurred or for such lesser period as the committee may direct.

[27 FR 8101, Aug. 15, 1962. Redesignated at 57 FR 38750, Aug. 27, 1992]

§ 929.64   Verification of reports and records.

The committee, through its duly authorized agents, during reasonable business hours, shall have access to any handler's premises where applicable records are maintained for the purpose of assuring compliance and checking and verifying records and reports filed by such handler.

[70 FR 7645, Feb. 15, 2005]

§ 929.65   Confidential information.

All reports and records furnished or submitted by handlers to the committee and its authorized agents which include data or information constituting a trade secret or disclosing the trade position, financial condition, or business operations of the particular handler from whom received, shall be received by and at all times kept in the custody and under the control of one or more employees of the committee, who shall disclose such information to no person other than the Secretary.

[27 FR 8101, Aug. 15, 1962. Redesignated at 57 FR 38750, Aug. 27, 1992]

MISCELLANEOUS PROVISIONS

§ 929.66   Compliance.

Except as provided in this part, no person shall handle cranberries, the handling of which has been prohibited by the Secretary in accordance with the provisions of this part; and no person shall acquire or handle cranberries except in conformity with the provisions of this part and the regulations issued hereunder.

[27 FR 8101, Aug. 15, 1962. Redesignated at 57 FR 38750, Aug. 27, 1992]

§ 929.67   Right of the Secretary.

The members of the committee (including successors and alternates), and any agents, employees, or representatives thereof, shall be subject to removal or suspension by the Secretary at any time. Each and every regulation, decision, determination, or other act of the committee shall be subject to the continuing right of the Secretary to disapprove of the same at any time. Upon such disapproval, the disapproved action of the committee shall be deemed null and void, except as to acts done in reliance thereon or in accordance therewith prior to such disapproval by the Secretary.

[27 FR 8101, Aug. 15, 1962. Redesignated at 57 FR 38750, Aug. 27, 1992]

§ 929.68   Effective time.

The provisions of this part, and of any amendment thereto, shall become effective at such time as the Secretary may declare above his signature and shall continue in force until terminated in one of the ways specified in § 929.68.

[27 FR 8101, Aug. 15, 1962. Redesignated at 57 FR 38750, Aug. 27, 1992]

§ 929.69   Termination.

(a) The Secretary may at any time terminate the provisions of this part by giving at least one day's notice by means of a press release or in any other manner in which he may determine.

(b) The Secretary shall terminate or suspend the operation of any or all of the provisions of this part whenever he finds that such provisions do not tend to effectuate the declared policy of the act.

(c) The Secretary shall terminate the provisions of this part whenever he finds by referendum or otherwise that such termination is favored by a majority of the growers: *Provided,* That such majority has, during the current fiscal year, produced more than 50 percent of the volume of the cranberries which were produced within the production area. Such termination shall become effective on the last day of July subsequent to the announcement thereof by the Secretary.

(d) The Secretary shall conduct a referendum during the month of May 1975 to ascertain whether continuance of this part is favored by the growers as set forth in paragraph (c) of this section. The Secretary shall conduct such a referendum during the month of May of every fourth year thereafter.

(e) The provisions of this part shall, in any event, terminate whenever the provisions of the act authorizing them cease to be effective.

[27 FR 8101, Aug. 15, 1962, as amended at 33 FR 11642, Aug. 16, 1968. Redesignated at 57 FR 38750, Aug. 27, 1992]

§929.70 Proceedings after termination.

(a) Upon the termination of the provisions of this part, the committee shall, for the purpose of liquidating the affairs of the committee, continue as trustees of all the funds and property then in its possession, or under its control, including claims for any funds unpaid or property not delivered at the time of such termination.

(b) The said trustees shall (1) continue in such capacity until discharged by the Secretary; (2) from time to time account for all receipts and disbursements and deliver all property on hand, together with all books and records of the committee and of the trustees, to such persons as the Secretary may direct; and (3) upon the request of the Secretary, execute such assignments or other instruments necessary or appropriate to vest in such person full title and right to all of the funds, property, and claims vested in the committee or the trustees pursuant thereto.

(c) Any person to whom funds, property, or claims have been transferred or delivered, pursuant to this section, shall be subject to the same obligation imposed upon the committee and upon the trustees.

[27 FR 8101, Aug. 15, 1962. Redesignated at 57 FR 38750, Aug. 27, 1992]

§929.71 Effect of termination or amendment.

Unless otherwise expressly provided by the Secretary, the termination of this part or of any regulation issued pursuant to this part, or the issuance of any amendment to either thereof, shall not (a) affect or waive any right, duty, obligation, or liability which shall have arisen or which may thereafter arise in connection with any provision of this part or any regulation issued hereunder, or (b) release or extinguish any violation of this part or any regulation issued hereunder, or (c) affect or impair any rights or remedies of the Secretary or of any other person with respect to any such violation.

[27 FR 8101, Aug. 15, 1962. Redesignated at 57 FR 38750, Aug. 27, 1992]

§929.72 Duration of immunities.

The benefits, privileges, and immunities conferred upon any person by virtue of this part shall cease upon its termination, except with respect to acts done under and during the existence of this part.

[27 FR 8101, Aug. 15, 1962. Redesignated at 57 FR 38750, Aug. 27, 1992]

§929.73 Agents.

The Secretary may, by designation in writing, name any officer or employee of the United States, or name any agency or division in the United States Department of Agriculture, to act as his agent or representative in connection with any of the provisions of this part.

[27 FR 8101, Aug. 15, 1962. Redesignated at 57 FR 38750, Aug. 27, 1992]

§929.74 Derogation.

Nothing contained in this part is, or shall be construed to be, in derogation or in modification of the rights of the Secretary or of the United States (a) to exercise any powers granted by the act or otherwise, or (b) in accordance with such powers, to act in the premises whenever such action is deemed advisable.

[27 FR 8101, Aug. 15, 1962. Redesignated at 57 FR 38750, Aug. 27, 1992]

§929.75 Personal liability.

No member or alternate member of the committee and no employee or agent of the committee shall be held personally responsible, either individually or jointly with others, in any way whatsoever, to any person for errors in judgment, mistakes, or other acts, either of commission or ommission, as such member, alternate, employee, or agent, except for acts of dishonesty, willful misconduct, or gross negligence.

[27 FR 8101, Aug. 15, 1962. Redesignated at 57 FR 38750, Aug. 27, 1992]

§929.76 Separability.

If any provision of this part is declared invalid or the applicability thereof to any person, circumstance, or thing is held invalid, the validity of the remainder of this part or the applicability thereof to any other person, circumstance, or thing shall not be affected thereby.

[27 FR 8101, Aug. 15, 1962. Redesignated at 57 FR 38750, Aug. 27, 1992]

Subpart B—Administrative Requirements

§ 929.101   Minimum exemption.

The requirements of § 929.41 *Assessments* and § 929.54 *Withholding* shall not apply to any handler in a fiscal year during which the handler handles notmore than a total of 300 barrels of cranberries.

[53 FR 12374, Apr. 14, 1988]

§ 929.102   Procedure to determine quantity of screened cranberries in unscreened lots.

The determination pursuant to § 929.54 of the quantity of screened cranberries contained in an unscreened lot shall be made in accordance with the following procedure and on the basis of a sample of representative boxes comprising no less than 2 percent of the cranberries in the lot:

(a) The cranberries in the sample are cleaned to remove chaff, and the boxes of cleaned berries are weighed. The weight of the boxes themselves is then deducted to determine the weight of the cleaned berries. The weight of the cleaned berries is divided by the number of boxes in the sample to obtain the net weight of cleaned cranberries per box. The net weight is multiplied by the number of boxes in the lot to obtain the net weight of the berries in the unscreened lot.

(b) The cleaned berries are run through a separator, having a $\frac{9}{32}$ inch screen, and with the bounce boards in the lowest position.

(c) The berries from the lower three bounce boards are rerun through the separator.

(d) The berries from the upper four bounce boards are thoroughly mixed and a random cupful (approximately 1 pint) is used to determine, from a count of the sound and unsound berries, the percentage of sound berries in the lot.

(e) Such percentage is adjusted by increasing it by 5 percentage points but not to exceed a total of 100 percent. (This increase makes the sample comparable to lots of screened cranberries, as such lots generally contain an average of 5 percent unsound berries.)

(f) The net weight, as determined in accordance with paragraph (a) of this section, of the berries in the unscreened lot, is multiplied by the adjusted percentage to obtain the quantity of screened cranberries in the unscreened lot.

[28 FR 11611, Oct. 31, 1963]

§ 929.103   Inspection procedure.

(a) Inspection of withheld cranberries shall be limited to any plant, storage facility, or other location, within the production area where facilities suitable to the inspection service are available for sampling, weighing, and inspection of cranberries.

(b) The handler offering any lot of cranberries for inspection shall furnish the necessary labor and pay the costs of moving, weighing, and otherwise making available the sample the inspector chooses for inspection. The size of the sample shall be determined by the inspector. In the case of inspection of a lot of unscreened cranberries, the sample shall be screened by the handler under the supervision of the inspector and the inspection certificate shall show the quantity of cranberries in such lot which meets the requirements established for withheld cranberries pursuant to § 929.54(c).

[28 FR 11611, Oct. 31, 1963]

§ 929.104   Outlets for excess cranberries.

(a) In accordance with § 929.61, excess cranberries may be diverted only to the following noncommercial or noncompetitive outlets:

(1) Foreign countries, except Canada.

(2) Charitable institutions.

(3) Any nonhuman food use.

(4) Research and development projects approved by the committee dealing with the development of foreign and domestic markets, including, but not limited to dehydration, radiation, freeze drying, or freezing of cranberries.

(b) [Reserved][

[65 FR 42614, July 11, 2000, as amended at 66 FR 34351, June 27, 2001; 83 FR 46075, Sept. 12, 2018]

§929.105 Reporting.

(a) Each report required to be filed with the committee pursuant to §§929.6 and 929.48 shall be mailed to the committee office or delivered to that office. If the report is mailed, it shall be deemed filed when postmarked.

(b) Certified reports shall be filed with the committee, on a form provided by the committee, by each handler not later than January 20, May 20, and July 20 of each fiscal period and by September 20 of the succeeding fiscal period showing:

(1) The total quantity of cranberries the handler acquired and the total quantity of cranberries and *Vaccinium oxycoccus* cranberries the handler handled from the beginning of the reporting period indicated through December 31, April 30, June 30, and August 31, respectively, and

(2) The respective quantities of cranberries and *Vaccinium oxycoccus* cranberries and cranberry products and *Vaccinium oxycoccus* cranberry products held by the handler on January 1, May 1, June 30, and August 31 of each fiscal period.

(c) Beginning with crop year 2018–19, the due date for the grower report required under §929.62(a) is changed to March 1.

[53 FR 12374, Apr. 14, 1988, as amended at 61 FR 30498, June 17, 1996; 62 FR 916, Jan. 7, 1997; 75 FR 20516, Apr. 20, 2010; 77 FR 52597, Aug. 30, 2012; 83 FR 46075, Sept. 12, 2018]

§929.106 Fiscal period.

The fiscal period specified in §929.6 of this part which began September 1, 1968, and ends on August 31, 1969, is changed to include the period of August 1, through August 31, 1968. Thereafter, the fiscal period will begin on September 1 and end on August 31 of the following year.

[33 FR 16492, Nov. 13, 1968]

§929.107 Conversion.

During a year of volume regulation, cranberry concentrate and other processed products made from excess or restricted cranberries harvested in that year may be diverted according to the provisions of this part. Any handler disposing of concentrate or other processed products must report the whole-berry equivalent to the Committee so that all excess or restricted cranberries are accounted for and reported per rules and regulations in effect. Table 1-Conversion Table provides a conversion rate for concentrate to barrels of whole berries based on Brix average by production region. Should requests be made to use other processed products for diversion, conversion rates for those products would be provided by the Committee based on information provided by the requesting handler.

TABLE 1 TO §929.107—CONVERSION TABLE

Region	Brix average	Concentrate yield for one barrel of cranberries
Oregon	9.8	1.91 gallons 50 Brix concentrate.
Washington	9.3	1.81 gallons 50 Brix concentrate.
New Jersey	8.8	1.72 gallons 50 Brix concentrate.
Wisconsin	8.7	1.70 gallons 50 Brix concentrate.
Massachusetts	8.4	1.64 gallons 50 Brix concentrate.
All others	8.7	1.70 gallons 50 Brix concentrate.

[83 FR 14357, Apr. 4, 2018]

§929.108 Outlets for restricted cranberries.

In accordance with §929.57, restricted cranberries may be diverted only to the following noncommercial or noncompetitive outlets:

(a) Foreign countries, except Canada, provided that restricted cranberries diverted under this provision may not be converted into canned, frozen, or dehydrated cranberries or other cranberry products by any commercial process, prior to diversion;

(b) Charitable institutions;

(c) Any nonhuman food use, or;

(d) Research and development projects approved by the Committee dealing with the development of foreign and domestic markets, including, but not limited to dehydration radiation, freeze drying, or freezing of cranberries.

[83 FR 14357, Apr. 4, 2018]

§ 929.110 Transfers or sales of cranberry acreage.

(a) Sales or transfers of cranberry acreage shall be reported by the transferor and transferee to the committee, in writing, on forms provided by the committee. Completed forms shall be sent to the committee office not later than 30 days after the transaction has occurred.

(b) Upon transfer of all or a portion of a growers' acreage, the committee shall be provided with certain information on the forms it will provide to the parties. The transferor and transferee must provide the following information:

(1) Crop records for the acreage involved;

(2) Annual production and sales for each crop year on the acreage involved, either in total, or for each individual parcel; and

(3) Such other information as the committee deems necessary.

(c) Cranberry acreage sold or transferred shall be recognized in connection with the issuance of sales history as follows:

(1) If a grower sells all of the acreage comprising the entity, all prior sales history shall accrue to the purchaser;

(2) If a grower sells only a portion of the acreage comprising the entity from which prior sales have been made, the purchaser and the seller must agree as to the amount of sales history attributed to each portion and shall provide, on a form provided by the committee, sufficient information so that sales are shown separately by crop year. However, the sales history attributed to each portion shall not exceed the total sales history, as determined by the committee, for such acreage at the time of transfer.

(d) During a year of regulation, all transfers of growers' sales histories for partial or total leases of acreage shall be received in the Committee office by close of business on July 31.

[59 FR 36023, July 15, 1994, as amended at 66 FR 34351, June 27, 2001]

§ 929.125 Committee review procedures.

Growers may request, and the Committee may grant, a review of determinations made by the Committee pursuant to section 929.48, in accordance with the following procedures:

(a) If a grower is dissatisfied with a determination made by the Committee which affects such grower, the grower may submit to the Committee within 30 days after receipt of the Committee's determination of sales history, a request for a review by an appeals subcommittee composed of two independent and two cooperative representatives, as well as a public member. Such appeals subcommittee shall be appointed by the Chairman of the Committee. Such grower may forward with the request any pertinent material for consideration of such grower's appeal.

(b) The subcommittee shall review the information submitted by the grower and render a decision within 30 days of receipt of such appeal. The subcommittee shall notify the grower of its decision, accompanied by the reasons for its conclusions and findings.

(c) The grower may further appeal to the Secretary, within 15 days after notification of the subcommittee's findings, if such grower is not satisfied with the appeals subcommittee's decision. The Committee shall forward a file with all pertinent information related to the grower's appeal. The Secretary shall inform the grower and all interested parties of the Secretary's decision. All decisions by the Secretary are final.

[66 FR 34351, June 27, 2001]

§ 929.142 Reserve.

(a) It is necessary and appropriate to establish and maintain a reserve in an amount not to exceed approximately one fiscal period's operational expenses to be used in accordance with the provisions of §929.42 of the marketing agreement and this part, and

(b) Assessments collected for each of the fiscal periods ended July 31, 1963; July 31, 1965; July 31, 1966; and July 31, 1967, were in excess of expenses for such periods. The committee is hereby authorized to place excess funds in said reserve.

[28 FR 11052, Oct. 16, 1963, as amended at 32 FR 13253, Sept. 20, 1967. Redesignated at 44 FR 73011, Dec. 17, 1979]

§929.149 Determination of sales history.

A sales history for each grower shall be computed by the Committee in the following manner when a producer allotment volume regulation is in effect.

(a) For each grower with acreage with 6 or more years of sales history, a new sales history shall be computed using an average of the highest 4 of the most recent 6 years of sales. If the grower has acreage with 5 years of sales history and such acreage was planted more than 6 years ago, a new sales history shall be computed by averaging the highest 4 of the 5 years.

(b) For growers whose acreage has 5 years of sales history and was planted 6 years ago or later, the sales history shall be computed by averaging the highest 4 of the 5 years and shall be adjusted as provided in paragraph (d). For growers whose acreage has 4 years of sales history, the sales history shall be computed by averaging all 4 years and shall be adjusted as provided in paragraph (d). For growers whose acreage has 1 to 3 years of sales history, the sales history shall be computed by dividing the total years sales by 4 and shall be adjusted as provided in paragraph (d).

(c) For growers with acreage with no sales history or for the first harvest of re-planted acres, the sales history will be 75 barrels per acre for acres planted or re-planted 1 year ago and first harvested in the current crop year and 156 barrels per acre for acres planted or re-planted 2 years ago and first harvested in the current crop year.

(d) In addition to the sales history computed in accordance with paragraphs (a) and (b) of this section, additional sales history shall be assigned to growers with acreage planted in the last 6 years. The additional sales histories depending on the date the acreage is planted are shown in Table 1.

TABLE 1—ADDITIONAL SALES HISTORY ASSIGNED TO ACREAGE

Date planted	Additional current crop year sales history per acre
6 years ago	49
5 years ago	117
4 years ago	157
3 years ago	183
2 years ago	156
1 year ago	75

[66 FR 34351, June 27, 2001, as amended at 78 FR 51046, Aug. 20, 2013; 80 FR 37533, July 1, 2015]

§929.150 Transfer or assignment of sales history.

(a) If indebtedness is incurred with regard to the acreage to which the cranberries are attributed, and on which a sales history is established, the sales history holder may transfer or assign the sales history solely as security for the loan. During the existence of such indebtedness no further transfer or assignment of sales history by the sales history holder shall be recognized by the committee unless the lender agrees thereto: Provided, That a copy of such loan agreement or assignment shall be filed with the committee before any right expressed therein, with regard to the sales history, shall be recognized by the committee under this paragraph (a).

(b) This regulation shall not in any way be construed to affect the right of the Secretary of Agriculture to amend, modify or terminate this regulation, or the marketing order under which it is issued as provided by law.

[34 FR 705, Jan. 17, 1969, as amended at 59 FR 36023, July 15, 1994]

§929.152 Delinquent assessments.

There shall be a late payment charge of five percent and an interest charge of 1½ percent per month applied to any assessment not received at the committee's office before the end of the month in which such assessment was first invoiced to the handler: Provided, That if an assessment is first invoiced later than the 15th of the month, no late payment or interest charge shall be levied if such assessment is received at the committee office by the end of the following month in which the assessment was first invoiced to the handler.

[60 FR 2, Jan. 3, 1995]

§ 929.157 Handler diversion.

(a) *Methods of diversion.* Handlers may divert cranberries by disposing of cranberries or cranberry products. Diversion by disposal may take place prior to placing the cranberries into the processing line or after processing. Handlers may also divert cranberries or cranberry products to approved, noncompetitive outlets for withheld fruit. Whole berries or processed products diverted must come from the current crop year. Any information collected of a confidential and/or proprietary nature would be held in confidence pursuant to § 929.65.

(1) *Diversion through disposal.* This type of diversion is to be carried out under the supervision of the Committee, and the cost of such supervision is to be paid by the handler. Handlers shall notify the Committee of their intent to dispose of cranberries or cranberry products using Form CMC–DISP as specified in § 929.162(c). Following notification, a Committee inspector will meet with the handler to verify the documentation provided and, when possible, witness the destruction. The Committee inspector may request receipts, visual proof, or any other information needed to support the disposal as reported. Once the verification process has been completed, the Committee inspector will sign the certification section of Form CMC–DISP and return it to the Committee.

(2) *Diversion through noncompetitive outlets.* To divert cranberries or cranberry products to a noncompetitive outlet, handlers must apply to the Committee using Form CMC–OUT as specified in § 929.162(d) prior to each disposal activity of this type. The Committee will review the information and approve or disapprove the diversion request. Once the cranberries or cranberry products are delivered to the approved noncompetitive outlets, the Committee must receive satisfactory

documentation of the transaction using Form CMC–CONF as specified in § 929.162(e).

(b) *Committee notification and handler plan.* Any handler intending to divert cranberries or cranberry products pursuant to § 929.54 must notify the Committee of such intent and provide a plan by June 1 that shows how the handler intends to meet the restricted percentage obligation. The handler shall submit this plan using Form CMC–JUNE as specified in the reporting requirements under § 929.162(a). The handler will have until August 31 to fulfill the plan, by which time the handler shall submit a final report detailing how the restricted percentage obligation was met using Form CMC–AUG as specified in § 929.162(b).

(c) *Request for review.* (1) If a handler is dissatisfied with a determination made by the Committee which affects such handler, the handler may submit to the Committee within 30 days after receipt of the Committee's determination, a request for a review by an appeals subcommittee composed of two independent growers and two cooperative representatives, as well as a public member. The appeals subcommittee shall be appointed by the Committee chairperson. The handler may forward with the request any pertinent materials for consideration of the appeal.

(2) The subcommittee shall review the information submitted by the handler and render a decision within 30 days of receipt of such appeal. The subcommittee shall notify the handler of its decision, accompanied by the reasons for its conclusions and findings.

(3) The handler may further appeal to the Secretary, within 15 days after notification of the subcommittee's findings, if such handler is not satisfied with the appeals subcommittee's decision. The Committee shall forward a file to the Secretary with all pertinent information related to the handler's appeal. The Secretary shall inform the handler and all interested parties of the Secretary's decision. All decisions by the Secretary are final.

[83 FR 32197, July 12, 2018]

§ 929.158 Exemptions.

If fresh and organically-grown cranberries are exempted from the volume regulation as recommended by the Committee and approved by the Secretary, the following provisions to these exemptions shall apply:

(a) Sales of packed-out cranberries intended for sales to consumers in fresh form shall be exempt from volume regulation provisions. Fresh cranberries are also sold dry in bulk boxes generally weighing less than 30 pounds. Fresh cranberries intended for retail markets are not sold wet. If any such fresh cranberries are diverted into processing outlets, the exemption no longer applies. Growers who intend to handle fresh fruit shall notify the committee of their intent to sell over 300 barrels of fresh fruit.

(b) Sales of organically-grown cranberries are exempt from volume regulation provisions. In order to receive an exemption for organic cranberry sales, such cranberries must be certified as such by a third party organic certifying organization acceptable to the committee.

(c) Handlers shall qualify for the exemptions in paragraphs (a) and (b) of this section by filing the amount of packed-out fresh or organic cranberry sales on the grower acquisition form.

[66 FR 34351, June 27, 2001]

§929.159 Excess cranberries.

(a) Beginning with crop year 2018–19, handlers holding excess cranberries shall submit to the Committee a written plan outlining procedures for the systematic disposal of such cranberries as specified in §929.59(b) by March 1.

(b) Beginning with crop year 2018–19, all excess cranberries shall be diverted as specified in §929.59(c) prior to August 31.

[83 FR 46075, Sept. 12, 2018]

§929.160 Public member eligibility requirements and nomination procedures.

(a) Public member and alternate member candidates shall not represent an agricultural interest and shall not have a financial interest in, or be associated with the production, processing, financing, or marketing of cranberries.

(b) Public member and alternate member candidates should be able to devote sufficient time to attend committee activities regularly and to familiarize themselves with the background and economies of the cranberry industry.

(c) Names of candidates together with evidence of qualification for public membership on the Cranberry Marketing Committee shall be submitted to the committee at its business office.

(d) Questionnaires shall be sent by the committee to those persons submitted as candidates to determine their eligibility and interest in becoming a public member.

(e) The names of persons nominated by the committee for the public member and alternate positions shall be submitted to the Secretary with such information as deemed pertinent by the committee or as requested by the Secretary.

(f) Public members shall serve a two-year term which coincides with the term of office of industry members of the committee.

[44 FR 16884, Mar. 20, 1979, as amended at 53 FR 12374, Apr. 14, 1988]

§929.161 Nomination and balloting procedures for candidates other than the major cooperative marketing organization.

(a) During the nomination process, each eligible candidate shall indicate if he/she is seeking a position on the Committee as a member or alternate member.

(b) Ballots provided by the Committee shall include the names of those candidates seeking member positions on the Committee and those seeking alternate member positions.

(c) All ballots shall be received by a date designated by the Committee office staff. Votes for member positions and alternate member positions shall be tabulated separately. In districts entitled to one member, the successful candidate shall be the person receiving the highest number of votes as a member or alternate member. In districts entitled to two members, the successful candidates shall be those receiving the highest and second highest number of votes as members or alternate members. Those names shall then be forwarded to the Secretary for selection.

[75 FR 18395, Apr. 12, 2010]

§ 929.162　Handler diversion reports.

(a) *Handler withholding report.* Handlers shall submit to the Committee, by June 1, a handler withholding report. The report shall be submitted using Form CMC–JUN and contain the following information:

(1) The name and address of the handler;

(2) The amount of cranberries acquired;

(3) The amount of cranberries withheld by disposal;

(4) The amount of cranberries diverted to noncompetitive outlets;

(5) The form of cranberry products withheld; and

(6) The total withholding obligation.

(b) *Handler Withholding Final Report.* Handlers shall submit to the Committee, by August 31, a final handler withholding report. The final report shall be submitted using Form CMC–AUG and contain the following information:

(1) The name and address of the handler;

(2) The seasonal total of cranberries acquired;

(3) The seasonal total of cranberries withheld by disposal;

(4) The seasonal total of cranberries diverted to noncompetitive outlets;

(5) The form of cranberry products withheld during the season; and

(6) The total withholding obligation.

(c) *Handler disposal certification.* Handlers shall submit to the Committee Form CMC–DISP for each lot of cranberries or cranberry products to be diverted through disposal. The form shall contain the following information:

(1) Name and address of the handler;

(2) Marketable cranberries in whole fruit or processed cranberries converted to whole fruit equivalent disposed of in this lot;

(3) Form of cranberries;

(4) Volume if in processed form;

(5) Lot details;

(6) Disposal site and method; and

(7) Inspector certification of the completion of the disposal.

(d) *Handler application for outlets for withheld fruit.* Handlers shall submit to the Committee Form CMC–OUT for approval for each lot of cranberries or cranberry products to be diverted to noncompetitive outlets in accordance with § 929.57. The form shall contain the following information:

(1) Name and address of the handler;

(2) Project type;

(3) Product form;

(4) Quantity of cranberries in whole fruit or processed cranberries converted to whole fruit equivalent diverted;

(5) A description of the project and how the cranberries will be used.

(e) *Third-party confirmation of receipt of withheld fruit.* Handlers shall submit to the Committee Form CMC–CONF for each diversion to a noncompetitive outlet to verify the receipt of the cranberries or cranberry product by the approved outlet. The form shall contain the following information:

(1) Name and address of the handler;

(2) Project type;

(3) Product form;

(4) Quantity of cranberries in whole fruit or processed cranberries converted to whole fruit equivalent utilized; and

(5) Confirmation or documentation of receipt from the receiving outlet.

(f) *Handler withholding appeal.* Handlers may appeal a determination made by the Committee relating to a handler withholding regulation using the appeals process outlined in § 929.157(c) and Form CMC–APPL, which shall contain the following information:

(1) Name and address of the handler;

(2) Reason for appeal; and

(3) Information in support of appeal.

[83 FR 32197, July 12, 2018]

Subpart C—Assessment Rate

§ 929.236　Assessment rate.

On and after September 1, 2006, an assessment rate of $.28 per barrel is established for cranberries.

[72 FR 14654, Mar. 29, 2007]

§ 929.250　Marketable quantity and allotment percentage for the 2000–2001 crop year.

The marketable quantity for the 2000–2001 crop year is set at 5.468 million barrels and the allotment percentage is designated at 85 percent. The marketable quantity may be adjusted to retain the 85 percent allotment percentage if the total industry sales history increases due to established growers receiving additional sales history on acreage with four years sales or less.

[65 FR 42615, July 11, 2000]

§ 929.251 Marketable quantity and allotment percentage for the 2001–2002 crop year.

The marketable quantity for the 2001–2002 crop year is set at 4.6 million barrels and the allotment percentage is designated at 65 percent. Fresh and organically grown fruit shall be exempt from the volume regulation provisions of this section.

[66 FR 34352, June 27, 2001]

§ 929.252 Free and restricted percentages for the 2017–18 crop year.

(a) The percentages for cranberries handled by handlers during the crop year beginning on September 1, 2017, which shall be free and restricted, respectively are designated as follows: Free percentage, 85 percent and restricted percentage, 15 percent.

(b) Handlers have the option to process restricted cranberries into dehydrated cranberries or other processed products. Handlers also have the option to divert concentrate or other processed products as provided in § 929.107 to account for up to 50 percent of their restriction.

(c) Organically grown fruit shall be exempt from the volume regulation requirements of this section. Small handlers who process less than 125,000 barrels during the 2017–18 fiscal year are exempt from the restriction. Any handlers who do not have carryover inventory at the end of the 2017–18 fiscal year are also exempt.

[83 FR 14357, Apr. 4, 2018]

§ 929.253 Marketable quantity and allotment percentage for the 2018–19 crop year.

(a) The marketable quantity for the 2018–19 crop year is set at 7.275 million barrels and the allotment percentage is designated at 75 percent.

(b) Organically grown fruit shall be exempt from the volume regulation requirements of this section. Small han-

dlers who processed less than 125,000 barrels during the 2017–18 fiscal year are exempt from the volume regulation requirements of this section. Any handler who did not have carryover inventory at the end of the 2017–18 fiscal year is also exempt from the volume regulation requirements of this section.

(c) Handlers have the option to process up to 50 percent of the excess cranberries received over their growers' allotments into dehydrated cranberries or other processed products. Handlers utilizing this option shall divert an amount of 2018–19 processed products equivalent to the volume of excess cranberries processed as provided for in § 929.107. The remaining volume of excess cranberries must be diverted as whole fruit.

[83 FR 46075, Sept. 12, 2018]

PART 930—TART CHERRIES GROWN IN THE STATES OF MICHIGAN, NEW YORK, PENNSYLVANIA, OREGON, UTAH, WASHINGTON, AND WISCONSIN

Subpart A—Order Regulating Handling

DEFINITIONS

AUTHORITY: 7 U.S.C. 601–674.

SOURCE: 61 FR 49942, Sept. 24, 1996, unless otherwise noted.

Subpart A—Order Regulating Handling

DEFINITIONS

§ 930.1   Act.

Act means Public Act No. 10, 73d Congress (May 12, 1933), as amended, and as reenacted and amended by the Agricultural Marketing Agreement Act of 1937, as amended (48 Stat. 31, as amended, 68 Stat. 906, 1047; 7 U.S.C. 601 *et seq.*).

§ 930.2   Board.

Board means the Cherry Industry Administrative Board established pursuant to § 930.20.

§ 930.3   Cherries.

Cherries means all tart/sour cherry varieties grown in the production area classified botanically as *Prunus cerasas,* or hybrids of *Prunus cerasas* by *Prunus avium,* or *Prunus cerasas* by *Prunus fruticosa.*

§ 930.4   Crop year.

Crop year means the 12-month period beginning on July 1 of any year and ending on June 30 of the following year, or such other period as the Board, with the approval of the Secretary, may establish.

§ 930.5   Department or USDA.

Department or *USDA* means the United States Department of Agriculture.

§930.6 District.

District means one of the subdivisions of the production area described in §930.20(c), or such other subdivisions as may be established pursuant to §930.21, or any subdivision added pursuant to §930.52.

§930.7 Fiscal period.

Fiscal period is synonymous with fiscal year and means the 12-month period beginning on July 1 of any year and ending on June 30 of the following year, or such other period as the Board, with the approval of the Secretary, may establish: *Provided*, that the initial fiscal period shall begin on the effective date of this part.

§930.8 Free market tonnage percentage cherries.

Free market tonnage percentage cherries means that proportion of cherries handled in a crop year which are free to be marketed in normal commercial outlets in that crop year under any volume regulation established pursuant to §930.50 or §930.51 and, in the absence of a restricted percentage being established for a crop year pursuant to §930.50 or §930.51, means all cherries received by handlers in that crop year.

§930.9 Grower.

Grower is synonymous with *producer* and means any person who produces cherries to be marketed in canned, frozen, or other processed form and who has a proprietary interest therein: Provided that, the term *grower* shall not include a person who produces cherries to be marketed exclusively for the fresh market in an unpitted condition.

§930.10 Handle.

Handle means the process to brine, can, concentrate, freeze, dehydrate, pit, press or puree cherries, or in any other way convert cherries commercially into a processed product, or divert cherries pursuant to §930.59, or to otherwise place cherries into the current of commerce within the production area or from the area to points outside thereof: *Provided*, That the term handle shall not include:

(a) The brining, canning, concentrating, freezing, dehydration, pitting, pressing or the converting, in any other way, of cherries into a processed product for home use and not for resale.

(b) The transportation within the production area of cherries from the orchard where grown to a processing facility located within such area for preparation for market.

(c) The delivery of such cherries to such processing facility for such preparation.

(d) The sale or transportation of cherries by a grower to a handler of record within the production area.

(e) The sale of cherries in the fresh market in an unpitted condition.

[61 FR 49942, Sept. 24, 1996, as amended at 77 FR 33306, June 6, 2012]

§930.11 Handler.

Handler means any person who first handles cherries or causes cherries to be handled for his or her own account.

§930.12 Person.

Person means an individual, partnership, corporation, association, or any other business unit.

§930.13 Primary inventory reserve.

Primary inventory reserve means that portion of handled cherries that are placed into handlers' inventories in accordance with any restricted percentage established pursuant to §930.50 or §930.51.

§930.14 Production area.

Production area means the States of Michigan, New York, Pennsylvania, Oregon, Utah, Washington and Wisconsin.

§930.15 Restricted percentage cherries.

Restricted percentage cherries means that proportion of cherries handled in a crop year which must be either placed into handlers' inventories in accordance with §930.55 or §930.57 or otherwise diverted in accordance with §930.59 and thereby withheld from marketing in normal commercial outlets under any volume regulation established pursuant to §930.50 or §930.51.

§ 930.16 Sales constituency.

Sales constituency means a common marketing organization or brokerage firm or individual representing a group of handlers and growers. An organization which receives consignments of cherries and does not direct where the consigned cherries are sold is not a sales constituency.

[66 FR 35896, July 10, 2001]

§ 930.17 Secondary inventory reserve.

Secondary inventory reserve means any portion of handled cherries voluntarily placed into inventory by a handler under § 930.57.

§ 930.18 Secretary.

Secretary means the Secretary of Agriculture of the United States, or any officer or employee of the U.S. Department of Agriculture to whom authority has heretofore been delegated, or to whom authority may hereafter be delegated, to act in the Secretary's stead.

ADMINISTRATIVE BODY

§ 930.20 Establishment and membership.

(a) There is hereby established a Cherry Industry Administrative Board, the membership of which shall be calculated in accordance with paragraph (b) of this section. The number of Board members may vary, depending upon the production levels of the districts. All but one of these members shall be qualified growers and handlers selected pursuant to this part, each of whom shall have an alternate having the same qualifications as the member for whom the person is an alternate. One member of the Board shall be a public member who, along with his or her alternate, shall be elected by the Board from the general public.

(b) District representation on the Board shall be based upon the previous three-year average production in the district and shall be established as follows:

(1) Up to and including 10 million pounds shall have 1 member;

(2) Greater than 10 and up to and including 40 million pounds shall have 2 members;

(3) Greater than 40 and up to and including 80 million pounds shall have 3 members; and

(4) Greater than 80 million pounds shall have 4 members; and

(5) Allocation of the seats in each district shall be as follows but subject to the provisions of paragraphs (d), (e) and (f) of this section:

District type	Grower members	or	Handler members
Up to and including 10 million pounds	1		1
More than 10 and up to 40 million pounds	1		1
More than 40 and up to 80 million pounds	1		2
More than 80 million pounds	2		2

(c) Upon the adoption of this part, the production area shall be divided into the following described subdivisions for purposes of this section:

District 1—Northern Michigan: that portion of the State of Michigan which is north of a line drawn along the northern boundary of Mason County and extended east to Lake Huron.

District 2—Central Michigan: that portion of the State of Michigan which is south of District 1 and north of a line drawn along the northern boundary of Allegan County and extended east to Lake St. Clair.

District 3—Southern Michigan: That portion of the State of Michigan not included in Districts 1 and 2.

District 4—The State of New York.

District 5—The State of Oregon.

District 6—The State of Pennsylvania.

District 7—The State of Utah.

District 8—The State of Washington.

District 9—The State of Wisconsin.

(d) The ratio of grower to handler representation in districts with three members shall alternate each time the term of a Board member from the representative group having two seats expires. During the initial period of the order, the ratio shall be as designated in paragraph (b) of this section.

(e) Board members from districts with one seat may be either grower or handler members and will be nominated and elected as outlined in § 930.23.

(f) If the 3-year average production of a district changes so that a different number of seats should be allocated to the district, then the Board will be re-established by the Secretary, and such seats will be filled according to the applicable provisions of this part. Each district's 3-year average production shall be recalculated annually as soon as possible after each season's final production figures are known.

(g) In order to achieve a fair and balanced representation on the Board, and to prevent any one sales constituency from gaining control of the Board, not more than one Board member may be from, or affiliated with, a single sales constituency in those districts having more than one seat on the Board; *Provided,* That this prohibition shall not apply in a district where such a conflict cannot be avoided. There is no prohibition on the number of Board members from differing districts that may be elected from a single sales constituency which may have operations in more than one district. However, as provided in §930.23, a handler or grower may only nominate Board members and vote in one district.

(h) Subject to the approval of the Secretary, the Board shall at its first meeting and annually thereafter elect from among any of its members a chairperson and a vice-chairperson and may elect other appropriate officers.

(i) The Board, with the approval of the Secretary, may establish rules and regulation's necessary and incidental to the administration of this section.

[61 FR 49942, Sept. 24, 1996, as amended at 67 FR 51713, Aug. 8, 2002; 75 FR 33677, June 15, 2010]

§930.21 Reestablishment.

Districts, subdivisions of districts, and the distribution of representation among growers and handlers within a respective district or subdivision thereof, or among the subdivision of districts, may be reestablished by the Secretary, subject to the provisions of §930.23, based upon recommendations by the Board. In recommending any such changes, the Board shall consider:

(a) The relative importance of producing areas;

(b) Relative production;

(c) The geographic locations of producing areas as they would affect the efficiency of administration of this part;

(d) Shifts in cherry production within the districts and the production area;

(e) Changes in the proportion and role of growers and handlers within the districts; and

(f) Other relevant factors.

§930.22 Term of office.

The term of office of each member and alternate member of the Board shall be for three fiscal years: Provided that, of the nine initial members and alternates from the combination of Districts 1, 2 and 3, one-third of such initial members and alternates shall serve only one fiscal year, one-third of such members and alternates shall serve only two fiscal years, one-third of such members and alternates shall serve three fiscal years; and one-half of the initial members and alternates from Districts 4 and 7 shall serve only one fiscal year, and one-half of such initial members and alternates shall serve two fiscal years (determination of which of the initial members and their alternates shall serve for 1 fiscal year, 2 fiscal years, or 3 fiscal years, in both instances, shall be by lot). Members and alternate members shall serve in such capacity for the portion of the term of office for which they are selected and have qualified until their respective successors are selected, have qualified and are appointed. The consecutive terms of office of grower, handler and public members and alternate members shall be limited to two 3-year terms, excluding any initial term lasting less than 3 years. The term of office of a member and alternate member for the same seat shall be the same. If this part becomes effective on a date such that the initial fiscal period is less than six months in duration, then the tolling of time for purposes of this subsection shall not begin until the beginning of the first 12-month fiscal period.

§930.23 Nomination and election.

(a) *Forms and ballots.* Nomination and election of initial and successor members and alternate members of the Board shall be conducted through petition forms and election ballots distributed to all eligible growers and handlers via the U.S. Postal Service or other means, as determined by the Secretary. Similar petition forms and election ballots shall be used for both members and alternate members and

225

any requirements for election of a member shall apply to the election of an alternate.

(b) *Nomination.* (1) In order for the name of a grower nominee to appear on an election ballot, the nominee's name must be submitted with a petition form, to be supplied by the Secretary or the Board, which, except in District 8, contains at least five signatures of growers, other than the nominee, from the nominee's district who are eligible to vote in the referendum. Grower petition forms in District 8 must be signed by only two growers, other than the nominee, from the nominee's district.

(2) In order for the name of a handler nominee to appear on an election ballot, the nominee's name must be submitted with a petition form, to be supplied by the Secretary or the Board, which contains the signature of one or more handler(s), other than the nominee, from the nominee's district who is or are eligible to vote in the election and that handle(s) a combined total of no less than five percent (5%) of the average production, as that term is used §930.20, handled in the district. *Provided,* that this requirement shall not apply if its application would result in a sales constituency conflict as provided in §930.20(g). The requirement that the petition form be signed by a handler other than the nominee shall not apply in any district where fewer than two handlers are eligible to vote.

(3) Only growers, including duly authorized officers or employees of growers, who are eligible to serve as grower members of the Board shall participate in the nomination of grower members and alternate grower members of the Board. No grower shall participate in the submission of nominees in more than one district during any fiscal period. If a grower produces cherries in more than one district, that grower may select in which district he or she wishes to participate in the nominations and election process and shall notify the Secretary or the Board of such selection. A grower may not participate in the nomination process in one district and the election process in a second district in the same election cycle.

(4) Only handlers, including duly authorized officers or employees of handlers, who are eligible to serve as handler members of the Board shall participate in the nomination of handler members and alternate handler members of the Board. No handler shall participate in the selection of nominees in more than one district during any fiscal period. If a handler handles cherries in more than one district, that handler may select in which district he or she wishes to participate in the nominations and election process and shall notify the Secretary or the Board of such selection. A handler may not participate in the nominations process in one district and the elections process in a second district in the same election cycle. If a person is a grower and a grower-handler only because some or all of his or her cherries were custom packed, but he or she does not own or lease and operate a processing facility, such person may vote only as a grower.

(5) In districts entitled to only one Board member, both growers and handlers may be nominated for the district's Board seat. Grower and handler nominations must follow the petition procedures outlined in paragraphs (b)(1) and (b)(2) of this section.

(6) All eligible growers and handlers in all districts may submit the names of the nominees for the public member and alternate public member of the Board.

(7) After the appointment of the initial Board, the Secretary or the Board shall announce at least 180 days in advance when a Board member's term is expiring and shall solicit nominations for that position in the manner described in this section. Nominations for such position should be submitted to the Secretary or the Board not less than 120 days prior to the expiration of such term.

(c) *Election.* (1) After receiving nominations, the Secretary or the Board shall distribute ballots via the U.S. Postal Service or other means, as determined by the Secretary, to all eligible growers and handlers containing the names of the nominees by district for the respective seats on the Board, excluding the public voting member seat. The ballots will clearly indicate that growers and handlers may only rank or otherwise vote for nominees in their own district.

(2) Except as provided in paragraph (c)(4) of this section, only growers, including duly authorized officers or employees of growers, who are eligible to serve as grower members of the Board shall participate in the election of grower members and alternate grower members of the Board. No grower shall participate in the election of Board members in more than one district during any fiscal period. If a grower produces cherries in more than one district, the grower must vote in the same district in which he or she chose to participate in the nominations process under paragraph (b)(3) of this section. However, if the grower did not participate in the nominations process, he or she may select in which district he or she wishes to vote and shall notify the Secretary or the Board of such selection.

(3)(i) Except as provided in paragraph (c)(4) of this section, only handlers, including duly authorized officers or employees of handlers, who are eligible to serve as handler members of the Board shall participate in the election of handler members and alternate handler members of the Board. No handler shall participate in the election of Board members in more than one district during any fiscal period. If a handler does handle cherries in more than one district, he or she must vote in the same district in which the handler elected to participate in the nominations process under paragraph (b)(4) of this section. However, if a handler did not participate in the nominations process, that handler may select in which district he or she chooses to vote and shall notify the Secretary or the Board of such selection. If a person is a grower and a grower-handler only because some or all of his or her cherries were custom packed, but he or she does not own or lease and operate a processing facility, such person may vote only as a grower.

(ii) To be seated as a handler representative in any district, the successful candidate must receive the support of handler(s) that handled a combined total of no less than five percent (5%), of the average production, as that term is used in §930.20, handled in the district; *Provided,* that this paragraph shall not apply if its application would

result in a sales constituency conflict as provided in §930.20(g).

(4) In districts entitled to only one Board member, growers and handlers may vote for either the grower or handler nominee(s) for the single seat allocated to those districts.

(d) The members of the Board appointed by the Secretary pursuant to §930.24 shall, at the first meeting and whenever necessary thereafter, by at least a two-thirds vote of the entire Board, select individuals to serve as the public member and alternate public member of the Board from the list of nominees received from growers and handlers pursuant to paragraph (b) of this section or from other persons nominated by the Board. The persons selected shall be subject to appointment by the Secretary under §930.24.

(e) The Board, with the approval of the Secretary, may establish rules and regulations necessary and incidental to the administration of this section.

[61 FR 49942, Sept. 24, 1996, as amended at 75 FR 33677, June 15, 2010]

§930.24 Appointment.

The selection of nominees made pursuant to elections conducted under §930.23(c) shall be submitted to the Secretary in a format which indicates the nominees by district, with the nominee receiving the highest number of votes at the top and the number of votes received being clearly indicated. The Secretary shall appoint from those nominees or from other qualified individuals, the grower and handler members of the Board and an alternate for each such member on the basis of the representation provided for in §930.20 or as provided for in any reapportionment or reestablishment undertaken pursuant to §930.21. The public member and alternate public member are nominated by the Board pursuant to §930.23(d) and shall also be subject to appointment by the Secretary. The Secretary shall appoint from nominees by the Board or from other qualified individuals the public member and the alternate public member.

§930.25 Failure to nominate.

If nominations are not made within the time and in the manner prescribed in §930.23, the Secretary may, without regard to nominations, select the members and alternate members of the Board on the basis of the representation provided for in § 930.20 or as provided for in any reapportionment or reestablishment undertaken pursuant to § 930.21.

§ 930.26 Acceptance.

Each person to be appointed by the Secretary as a member or as an alternate member of the Board shall, prior to such appointment, qualify by advising the Secretary that he/she agrees to serve in the position for which nominated for selection.

§ 930.27 Vacancies.

To fill any vacancy occasioned by the failure of any person appointed as a member or as an alternate member of the Board to qualify, or in the event of the death, removal, resignation, or disqualification of any member or alternate member of the Board, a successor for the unexpired term of such member or alternate member of the Board shall be appointed by the Secretary from the most recent list of nominations for the Board made by growers and handlers, from nominations made by the Board, or from other qualified individuals. Any nominations made by the Board to fill a vacancy must be received by the Secretary within 90 days of the effective date of the vacancy. Board members wishing to resign from the Board must do so in writing to the Secretary.

§ 930.28 Alternate members.

An alternate member of the Board, during the absence of the member for whom that member serves as an alternate, shall act in the place and stead of such member and perform such other duties as assigned. However, if a member is in attendance at a meeting of the Board, an alternate member may not act in the place and stead of such member. In the event a member and his or her alternate are absent from a meeting of the Board, such member may designate, in writing and prior to the meeting, another alternate to act in his or her place: *Provided*, that such alternate represents the same group (grower or handler) as the member. In the event of the death, removal, resignation or disqualification of a member, the alternate shall act for the member until a successor is appointed and has qualified.

[67 FR 51714, Aug. 8, 2002]

§ 930.29 Eligibility for membership on Cherry Industry Administrative Board.

(a) Each grower member and each grower alternate member of the Board shall be a grower, or an officer or employee of a grower, in the district for which nominated or appointed.

(b) Each handler member and each handler alternate member of the Board shall be a handler, or an officer or employee of a handler, who owns, leases, and operates a cherry processing facility in the district for which nominated or appointed.

(c) The public member and alternate public member of the Board shall be prohibited from having any financial interest in the cherry industry and shall possess such additional qualifications as may be established by regulation.

§ 930.30 Powers.

The Board shall have the following powers:

(a) To administer this part in accordance with its terms and provisions;

(b) To make rules and regulations to effectuate the terms and provisions of this part;

(c) To receive, investigate, and report to the Secretary complaints of violations of this part; and

(d) To recommend to the Secretary amendments to this part.

§ 930.31 Duties.

The Board shall have, among others, the following duties:

(a) To select such officers, including a chairperson and vice-chairperson, as may be necessary, and to define the duties of such officers and the duties of the chairperson and the vice-chairperson;

(b) To employ or contract with such persons or agents as the Board deems necessary and to determine the duties and compensation of such persons or agents;

(c) To select such committees and subcommittees as may be necessary;

(d) To adopt bylaws and to adopt such rules for the conduct of its business as it may deem advisable;

(e) To submit to the Secretary a budget for each fiscal period, prior to the beginning of such period, including a report explaining the items appearing therein and a recommendation as to the rates of assessments for such period;

(f) To keep minutes, books, and records which will reflect all of the acts and transactions of the Board and which shall be subject to examination by the Secretary;

(g) To prepare periodic statements of the financial operations of the Board and to make copies of each statement available to growers and handlers for examination at the office of the Board;

(h) To cause its financial statements to be audited by a certified public accountant at least once each fiscal year and at such times as the Secretary may request. Such audit shall include an examination of the receipt of assessments and the disbursement of all funds. The Board shall provide the Secretary with a copy of all audits and shall make copies of such audits, after the removal of any confidential individual grower or handler information that may be contained in them, available to growers and handlers for examination at the offices of the Board;

(i) To act as intermediary between the Secretary and any grower or handler with respect to the operations of this part;

(j) To investigate and assemble data on the growing, handling, and marketing conditions with respect to cherries;

(k) To apprise the Secretary of all Board meetings in a timely manner;

(l) To submit to the Secretary such available information as the Secretary may request;

(m) To investigate compliance with the provisions of this part;

(n) To develop and submit an annual marketing policy for approval by the Secretary containing the optimum supply of cherries for the crop year established pursuant to §930.50 and recommending such action(s) necessary to achieve such optimum supply;

(o) To implement volume regulations established under §930.50 and issued by the Secretary under §930.51, including the release of any inventory reserves;

(p) To provide thorough communication to growers and handlers regarding the activities of the Board and to respond to industry inquiries about Board activities;

(q) To oversee the collection of assessments levied under this part;

(r) To enter into contracts or agreements with such persons and organizations as the Board may approve for the development and conduct of activities, including research and promotion activities, authorized under this part or for the provision of services required by this part and for the payment of the cost thereof with funds collected through assessments pursuant to §930.41 and income from such assessments. Contracts or agreements for any plan or project shall provide that:

(1) The contractors shall develop and submit to the Board a plan or project together with a budget(s) which shall show the estimated cost to be incurred for such plan or project;

(2) Any contract or agreement for a plan or project and any plan or project adopted by the Board shall only become effective upon approval by the Secretary; and

(3) Every such contracting party shall keep accurate records of all of its transactions and make periodic reports to the Board of activities conducted and an accounting for funds received and expended, and such other reports as the Secretary or the Board may require. The Secretary or employees of the Board may audit periodically the records of the contracting party;

(s) Pending disbursement consistent with its budget, to invest, with the approval of the Secretary, and in accordance with applicable Departmental policies, funds collected through assessments authorized under §930.41 and income from such assessments;

(t) To establish standards or grade requirements for cherries for frozen and canned cherry products, subject to the approval of the Secretary;

(u) To borrow such funds, subject to the approval of the Secretary and not to exceed the expected expenses of one fiscal year, as are necessary for administering its responsibilities and obligations under this part; and

(v) To establish, with the approval of the Secretary, such rules and procedures relative to administration of this subpart as may be consistent with the provisions contained in this subpart and as may be necessary to accomplish the purposes of the Act and the efficient administration of this subpart.

§ 930.32   Procedure.

(a) Two-thirds of the members of the Board, including alternates acting for absent members, shall constitute a quorum. For any action of the Board to pass, at least two-thirds of the entire Board must vote in support of such action.

(b) The Board may provide through its own rules and regulations, subject to approval by the Secretary, for simultaneous meetings of groups of its members assembled at different locations and for votes to be conducted by telephone or other means of communication. Votes so cast shall be promptly confirmed in writing.

(c) All meetings of the Board are open to the public, although the Board may hold portions of meetings in executive session for the consideration of certain business. The Board will establish, with the approval of the Secretary, a means of advanced notification of growers and handlers of Board meetings.

[61 FR 49942, Sept. 24, 1996, as amended at 67 FR 51714, Aug. 8, 2002]

§ 930.33   Expenses and compensation.

Except for the public member and alternate public member who shall receive such compensation as the Board may establish and the Secretary may approve, the members of the Board, and alternates when acting as members, shall serve without compensation but shall be reimbursed for necessary and reasonable expenses, as approved by the Board, incurred by them in the performance of their duties under this part. The Board at its discretion may request the attendance of one or more alternates at any or all meetings, notwithstanding the expected or actual presence of the respective member(s), and may pay the expenses of such alternates.

EXPENSES AND ASSESSMENTS

§ 930.40   Expenses.

The Board is authorized to incur such expenses as the Secretary finds are reasonable and likely to be incurred for its maintenance and functioning and to enable it to exercise its powers and perform its duties in accordance with the provisions of this part. The funds to cover such expenses shall be acquired by the levying of assessments as provided in § 930.41.

§ 930.41   Assessments.

(a) An assessment may be levied upon handlers annually under this part to cover the administrative costs of the Board, costs of inspection, and any research, development and promotion activities initiated by the Board under § 930.48.

(b) Each part of an assessment intended to cover the costs of each activity in paragraph (a) of this section, must be identified and approved by the Board and the Secretary, and any notification or other statement regarding assessments provided to handlers must contain such information.

(c) As a pro rata share of the administrative, inspection, research, development, and promotion expenses which the Secretary finds reasonable and likely to be incurred by the Board during a fiscal period, each handler shall pay to the Board assessments on all cherries handled, as the handler thereof, during such period: *Provided,* a handler shall be exempt from any assessment only on the tonnage of handled cherries that either are diverted by destruction at the handler's facilities according to § 930.59 or are cherries represented by grower diversion certificates issued pursuant to § 930.58(b) and acquired by handlers as described in § 930.59.

(d) The Secretary, after consideration of the recommendation of the Board, shall fix the rate of assessment to be paid by each handler during the fiscal period in an amount designed to secure sufficient funds to cover the expenses which may be approved and incurred during such period or subsequent period as provided in paragraph (c) of this section. At any time during or after the fiscal period, the Secretary may increase the rate of assessment in order to secure sufficient funds to cover any later finding by the Secretary relative to the expenses which may be incurred. Such increase shall be applied to all cherries handled during the applicable fiscal period. In order to provide funds for the administration of the provisions of this part during the first part of a fiscal period before sufficient operating income is available from assessments, the Board may accept the payment of assessments in advance, and may borrow money for such purposes.

(e) Assessments not paid within a time prescribed by the Board may be made subject to interest or late payment charges, or both. The period of time, rate of interest, and late payment charge will be as recommended by the Board and approved by the Secretary: *Provided,* That when interest or late payment charges are in effect, they shall be applied to all assessments not paid within the prescribed period of time.

(f) Assessments shall be calculated on the basis of pounds of cherries handled. The established assessment rate may be uniform, or may vary dependent on the product the cherries are used to manufacture. In recommending annual assessment rates, the Board shall consider:

(1) The differences in the number of pounds of cherries utilized for various cherry products; and

(2) The relative market values of such cherry products.

(g) The Board, with the approval of the Secretary, may establish rules and regulations necessary and incidental to the administration of this section.

[61 FR 49942, Sept. 24, 1996, as amended at 67 FR 51714, Aug. 8, 2002]

§ 930.42 Accounting.

(a) If, at the end of a fiscal period, the assessments collected are in excess of expenses incurred, the Board, with the approval of the Secretary, may carry over all or any portion of such excess into subsequent fiscal periods as a reserve. Such reserve funds may be used to cover any expenses authorized by this part, and to cover necessary expenses of liquidation in the event of termination of this part. If any such excess is not retained in a reserve, it shall be refunded proportionately to the handlers from whom the excess was collected. Without an additional reserve level approved by the Secretary, the amount held in reserve may not exceed approximately one year's operational expenses. Upon termination of this part, any funds not required to defray the necessary expenses of liquidation shall be disposed of in such a manner as the Secretary may determine to be appropriate: *Provided,* That to the extent practicable, such funds shall be returned pro rata to the persons from whom such funds were collected.

(b) All funds received by the Board pursuant to the provisions of this part shall be used solely for the purpose specified in this part and shall be accounted for in the manner provided in this part. The Secretary may at any time require the Board and its members to account for all receipts and disbursements.

QUALITY CONTROL

§ 930.44 Quality control.

(a) *Quality standards.* The Board may establish, with the approval of the Secretary, such minimum quality and inspection requirements applicable to cherries as will contribute to orderly marketing or be in the public interest. If such requirements are adopted, no handler shall process cherries into manufactured products or sell manufactured products in the current of commerce unless such cherries and/or such cherries used in the manufacture of products meet the applicable requirements as evidenced by certification acceptable to the Board. The Board, with the approval of the Secretary, may establish rules and regulations necessary and incidental to the administration of this section.

(b) *Inspection and certification.* Whenever the handling of any cherries requires inspection pursuant to this part, each handler who handles cherries shall cause such cherries to be inspected by the appropriate division of USDA, and certified by it as meeting the applicable requirements of such regulation: *Provided,* That inspection and certification shall be required for cherries which previously have been so inspected and certified only if such cherries have been regraded, resorted, repackaged, or in any other way further prepared for market. Promptly after inspection and certification, each such handler shall submit, or cause to be submitted, to the Board a copy of the certificate of inspection issued with respect to such cherries.

RESEARCH, MARKET DEVELOPMENT AND PROMOTION

§ 930.48 Research, market development and promotion.

The Board, with the approval of the Secretary, may establish or provide for the establishment of production and processing research, market research and development, and/or promotional activities, including paid advertising, designed to assist, improve or promote the efficient production and processing, marketing, distribution, and consumption of cherries subject to this part. The expense of such projects shall be paid from funds collected pursuant to this part and the income from such funds.

REGULATIONS

§ 930.50 Marketing policy.

(a) *Optimum supply.* On or about July 1 of each crop year, the Board shall hold a meeting to review sales data, inventory data, current crop forecasts and market conditions in order to establish an optimum supply level for the crop year. The optimum supply volume shall be calculated as 100 percent of the average sales of the prior three years reduced by average sales that represent dispositions of exempt cherries and restricted percentage cherries qualifying for diversion credit for the same three years, unless the Board determines that it is necessary to recommend otherwise with respect to sales of exempt and restricted percentage cherries, to which shall be added a desirable carry-out inventory not to exceed 20 million pounds or such other amount as the Board, with the approval of the Secretary, may establish. This optimum supply volume shall be announced by the Board in accordance with paragraph (h) of this section.

(b) *Preliminary percentages.* On or about July 1 of each crop year, the Board shall establish a preliminary free market tonnage percentage which shall be calculated as follows: from the optimum supply computed in paragraph (a) of this section, the Board shall deduct the carry-in inventory to determine the tonnage requirements (adjusted to a raw fruit equivalent) for the current crop year which will be subtracted from the current year USDA crop forecast or by an average of such other crop estimates the Board votes to use. If the resulting number is positive, this would represent the estimated overproduction which would be the restricted tonnage. This restricted tonnage would then be divided by the sum of the crop forecast(s) for the regulated districts to obtain a preliminary restricted percentage, rounded to the nearest whole number, for the regulated districts. If subtracting the current crop year requirement, computed in the first sentence from the current crop forecast, results in a negative number, the Board shall establish a preliminary free market tonnage percentage of 100 percent with a preliminary restricted percentage of zero. The Board shall announce these preliminary percentages in accordance with paragraph (h) of this section.

(c) *Interim percentages.* Between July 1 and September 15 of each crop year, the Board may modify the preliminary free market tonnage and restricted percentages to adjust to the actual pack occurring in the industry. The Board shall announce any interim percentages in accordance with paragraph (h) of this section.

(d) *Final percentages.* No later than September 15 of each crop year, the Board shall review the most current information available including, but not limited to, processed production and grower diversions of cherries during the current crop year. The Board shall make such adjustments as are necessary between free and restricted tonnage to achieve the optimum supply and recommend such final free market tonnage and restricted percentages to the Secretary and announce them in accordance with paragraph (h) of this section. The difference between any final free market tonnage percentage designated by the Secretary and 100 percent shall be the final restricted percentage. With its recommendation, the Board shall report on its consideration of the factors in paragraph (e) of this section.

(e) *Factors.* When computing preliminary and interim percentages, or determining final percentages for recommendation to the Secretary, the Board shall give consideration to the following factors:

(1) The estimated total production of cherries;

(2) The estimated size of the crop to be handled;

(3) The expected general quality of such cherry production;

(4) The expected carryover as of July 1 of canned and frozen cherries and other cherry products;

(5) The expected demand conditions for cherries in different market segments;

(6) Supplies of competing commodities;

(7) An analysis of economic factors having a bearing on the marketing of cherries;

(8) The estimated tonnage held by handlers in primary or secondary inventory reserves;

(9) Any estimated release of primary or secondary inventory reserve cherries during the crop year; and

(10) The quantity of grower-diverted cherries during the crop year.

(f) *Modification.* In the event the Board subsequently deems it advisable to modify its marketing policy, because of national emergency, crop failure, or other major change in economic conditions, it shall hold a meeting for that purpose, and file a report thereof with the Secretary within 5 days (exclusive of Saturdays, Sundays, and holidays) after the holding of such meeting, which report shall show the Board's recommended modification and the basis therefor.

(g) *Additional tonnage to sell as free tonnage.* In addition, the Board, in years when restricted percentages are established, shall make available tonnage equivalent to an additional 10 percent, if available, of the average sales of the prior 3 years, as defined in paragraph (a) of this section, for market expansion.

(h) *Publicity.* The Board shall promptly give reasonable publicity to growers and handlers of each meeting to consider a marketing policy or any modification thereof, and each such meeting shall be open to them and to the public. Similar publicity shall be given to growers and handlers of each marketing policy report or modification thereof, filed with the Secretary and of the Secretary's action thereon. Copies of all marketing policy reports shall be maintained in the office of the Board, where they shall be made available for examination. The Board shall notify handlers, and give reasonable publicity to growers, of its computation of the optimum supply, preliminary percentages, and interim percentages and shall notify handlers of the Secretary's action on final percentages by registered or certified mail.

(i) *Restricted percentages.* Restricted percentage requirements established under paragraphs (b), (c), or (d) of this section may be fulfilled by handlers by either establishing an inventory reserve in accordance with § 930.55 or § 930.57 or by diversion of product in accordance with § 930.59. In years where required, the Board shall establish a maximum percentage of the restricted quantity which may be established as a primary inventory reserve such that the total primary inventory reserve does not exceed 50-million pounds; *Provided,* That such 50-million-pound quantity may be changed upon recommendation of the Board and approval of the Secretary. Any such change shall be recommended by the Board on or before September 30 of any crop year to become effective for the following crop year, and the quantity may be changed no more than one time per crop year. Handlers will be permitted to divert (at plant or with grower diversion certificates) as much of the restricted percentage requirement as they deem appropriate, but may not establish a primary inventory reserve in excess of the percentage established by the Board for restricted cherries. In the event handlers wish to establish inventory reserve in excess of this amount, they may do so, in which case it will be classified as a secondary inventory reserve and will be regulated accordingly.

(j) *Inventory Reserve Release.* In years when inventory reserve cherries are available and when the expected availability of cherries from the current crop plus expected carryin inventory does not fulfill the optimum supply, the Board shall release not later than November 1st of the current crop year such volume from the inventory reserve as will satisfy the optimum supply.

(k) The Board, with the approval of the Secretary, may establish rules and regulations necessary and incidental to the administration of this section.

[61 FR 49942, Sept. 24, 1996, as amended at 66 FR 35896, July 10, 2001; 67 FR 51714, Aug. 8, 2002; 75 FR 33677, June 15, 2010; 77 FR 33306, June 6, 2012]

§ 930.51 Issuance of volume regulations.

(a) Whenever the Secretary finds, from the recommendation and supporting information supplied by the Board, that to designate final free market tonnage and restricted percentages for any cherries acquired by handlers during the crop year will tend to effectuate the declared policy of the Act, the Secretary shall designate such percentages. Such regulation designating such percentage shall fix the free market tonnage and restricted percentages, totaling 100 percent, which shall be applied in accordance with this section, §§ 930.55, 930.57 and 930.59 to cherries grown in regulated districts, as determined under § 930.52, and handled during such fiscal period.

(b) The Board shall be informed immediately of any such regulation issued by the Secretary, and the Board shall promptly give notice thereof to handlers.

(c) That portion of a handler's cherries that are restricted percentage cherries is the product of the restricted percentage imposed under paragraph (a) of this section multiplied by the tonnage of cherries, originating in a regulated district, handled, including those diverted according to § 930.59, by that handler in that fiscal year.

(d) The Board, with the approval of the Secretary, shall develop rules and regulations which shall provide guidelines for handlers in complying with any restricted tonnage requirements, including, but not limited to, a grace period of at least 30 days to segregate

and appropriately document any tonnage they wish to place in the inventory reserve and to assemble any applicable diversion certificates.

[61 FR 49942, Sept. 24, 1996, as amended at 67 FR 51714, Aug. 8, 2002]

§ 930.52 Establishment of districts subject to volume regulations.

(a) The districts in which handlers shall be subject to any volume regulations implemented in accordance with this part shall be those districts in which the average annual production of cherries over the prior 3 years has exceeded 6 million pounds. Handlers shall become subject to volume regulation implemented in accordance with this part in the crop year that follows any 3-year period in which the 6-million pound average production requirement is exceeded in that district.

(b) Handlers in districts which are not subject to volume regulation would only be so regulated to the extent that they handled cherries which were grown in a district subject to regulation as specified in paragraph (a) of this section. In such a case, the handler must place in inventory reserve pursuant to § 930.55 or § 930.57 or divert pursuant to § 930.59 the required restricted percentage of the crop originating in the regulated district.

(c) Handlers in districts not meeting the production requirement described in paragraph (a) of this section in a given year would not be subject to volume regulation in the next crop year.

(d) Any district producing a crop which is less than 50 percent of the average annual processed production in that district in the previous five years would be exempt from any volume regulation if, in that year, a restricted percentage is established.

(e) The Board, with the approval of the Secretary, may establish rules and regulations necessary and incidental to the administration of this section.

[61 FR 49942, Sept. 24, 1996, as amended at 67 FR 51714, Aug. 8, 2002]

§ 930.53 Modification, suspension, or termination of regulations.

(a) In the event the Board at any time finds that, by reason of changed conditions, any regulations issued pursuant to § 930.44 or § 930.51 should be modified, suspended, or terminated, it shall so recommend to the Secretary.

(b) Whenever the Secretary finds, from the recommendations and information submitted by the Board or from other available information, that a regulation issued pursuant to § 930.44 or § 930.51 should be modified, suspended or terminated with respect to any or all shipments of cherries in order to effectuate the declared policy of the Act, the Secretary shall modify, suspend, or terminate such regulation.

§ 930.54 Prohibition on the use or disposition of inventory reserve cherries.

Cherries that are placed in inventory reserve pursuant to the requirements of § 930.50, § 930.51, § 930.55, or § 930.57 shall not be used or disposed of by any handler or any other person except as provided in § 930.50 or in paragraphs (a), (b), or (c) of this section.

(a) If the Board determines that the total available supplies for use in commercial outlets are less than the amount needed to meet the demand in such outlets, the Board may recommend to the Secretary that a portion or all of the primary and/or secondary inventory reserve cherries be released for such use.

(b) The Board may recommend to the Secretary that a portion or all of the primary and/or secondary inventory reserve cherries be released for sale in certain designated markets. Such designated markets may be defined in terms of the use or form of the cherries.

(c) Cherries in the primary and/or secondary inventory reserve may be used at any time for uses exempt from regulation under § 930.62.

(d) Should the volume of cherries held in the primary inventory reserves and, subsequently, the secondary inventory reserves reach a minimum amount, which level will be established by the Secretary upon recommendation from the Board, the products held in the respective reserves shall be released from the reserves and made available to the handlers as free tonnage.

[67 FR 51714, Aug. 8, 2002, as amended at 75 FR 33678, June 15, 2010]

EDITORIAL NOTE: At 66 FR 232, Jan. 3, 2001, in § 930.54 paragraph (a), the word "normal" was suspended indefinitely.

§ 930.55 Primary inventory reserves.

(a) Whenever the Secretary has fixed the free market tonnage and restricted percentages for any fiscal period, as provided for in § 930.51(a), each handler in a regulated district shall place in his or her primary inventory reserve for such period, at such time, and in such manner, as the Board may prescribe, or otherwise divert, according to § 930.59, a portion of the cherries acquired during such period.

(b) The form of the cherries, frozen, canned in any form, dried, or concentrated juice, placed in the primary inventory reserve is at the option of the handler. The product(s) placed by the handler in the primary inventory reserve must have been produced in either the current or the preceding two crop years. Except as may be limited by § 930.50(i) or as may be permitted pursuant to §§ 930.59 and 930.62, such inventory reserve portion shall be equal to the sum of the products obtained by multiplying the weight or volume of the cherries in each lot of cherries acquired during the fiscal period by the then effective restricted percentage fixed by the Secretary; *Provided,* That in converting cherries in each lot to the form chosen by the handler, the inventory reserve obligations shall be adjusted in accordance with uniform rules adopted by the Board in terms of raw fruit equivalent.

(c) Inventory reserve cherries shall meet such standards of grade, quality, or condition as the Board, with the approval of the Secretary, may establish. All such cherries shall be inspected by USDA. A certificate of such inspection shall be issued which shall show, among other things, the name and address of the handler, the number and type of containers in the lot, the grade of the product, the location where the lot is stored, identification marks (can codes or lot stamp), and a certification that the cherries meet the prescribed standards. Promptly after inspection and certification, each such handler shall submit, or cause to be submitted, to the Board, at the place designated by the Board, a copy of the certificate of inspection issued with respect to such cherries.

(d) Handlers shall be compensated for inspection costs incurred on cherries placed in the primary inventory reserve. All reporting of cherries placed in, rotated in and out, or released from an inventory reserve shall be in accordance with rules and procedures established by the Board, with the approval of the Secretary. The Board could, with the approval of the Secretary, also limit the number of inspections of reserve cherries being rotated into inventory reserves for which the Board would be financially liable.

(e) Except as provided in § 930.54, handlers may not sell inventory reserve cherries prior to their official release by the Board. Handlers may rotate cherries in their inventory reserves with prior notification to the Board. All cherries rotated into the inventory reserve must meet the applicable inspection requirements.

[61 FR 49942, Sept. 24, 1996, as amended at 75 FR 33678, June 15, 2010]

§ 930.56 Off-premise inventory reserve.

Any handler may, upon notification to the Board, arrange to hold inventory reserve, of his or her own production or which was purchased, on the premises of another handler or in an approved commercial storage facility in the same manner as though the inventory reserve were on the handler's own premises.

§ 930.57 Secondary inventory reserve.

(a) In the event the inventory reserve established under § 930.55 of this part is at its maximum volume, and the Board has announced, in accordance with § 930.50, that volume regulation will be necessary to maintain an orderly supply of quality cherries for the market, handlers in a regulated district may elect to place in a secondary inventory reserve all or a portion of the cherries the volume regulation would otherwise require them to divert in accordance with § 930.59.

(b) Should any handler in a regulated district exercise his or her right to establish a secondary inventory reserve under paragraph (a) of this section, all costs of maintaining that reserve, as well as inspection costs, will be the responsibility of the individual handler.

(c) The secondary inventory reserve shall be established in accordance with §§ 930.55 (b) and (c) and such other rules and regulations which the Board, with the approval of the Secretary, may establish.

(d) The Board shall retain control over the release of any cherries from the secondary inventory reserve. No cherries may be released from the secondary reserve until all cherries in any primary inventory reserve established under § 930.55 have been released. Any release of the secondary inventory reserve shall be in accordance with the annual marketing policy and with § 930.54.

§ 930.58 Grower diversion privilege.

(a) *In general.* Any grower may voluntarily elect to divert, in accordance with the provisions of this section, all or a portion of the cherries which otherwise, upon delivery to a handler, would become restricted percentage cherries. Upon such diversion and compliance with the provisions of this section, the Board shall issue to the diverting grower a grower diversion certificate which such grower may deliver to a handler. Any grower diversions completed in accordance with this section, but which are undertaken in districts subsequently exempted by the Board from volume regulation under § 930.52(d), shall qualify for diversion credit.

(b) *Eligible diversion.* Grower diversion certificates shall be issued to growers only if the cherries are diverted in accordance with the following terms and conditions or such other terms and conditions that the Board, with the approval of the Secretary, may establish. Diversion may take such of the following forms which the Board, with the approval of the Secretary, may designate: uses exempt under § 930.62; nonhuman food uses; or other uses, including diversion by leaving such cherries unharvested.

(c) *Application/mapping.* The Board, with the approval of the Secretary, shall develop rules and regulations providing for the diversion of cherries by growers. Such regulations may include, among other things:

(1) The form and content of applications and agreements relating to the diversion, including provisions for supervision and compensation; and

(2) Provisions for mapping areas in which cherries will be left unharvested.

(d) *Diversion certificate.* If the Board approves the application it shall so notify the applicant and conduct such supervision of the applicant's diversion of cherries as may be necessary to assure that the cherries have been diverted. After the diversion has been accomplished, the Board shall issue to the diverting grower a diversion certificate stating the weight of cherries diverted. Where diversion is carried out by leaving the cherries unharvested, the Board shall estimate the weight of cherries diverted on the basis of such uniform rule prescribed in rules and regulations as the Board, with the approval of the Secretary, may recommend to implement this section.

[61 FR 49942, Sept. 24, 1996, as amended at 67 FR 51715, Aug. 8, 2002; 77 FR 33306, June 6, 2012]

§930.59 Handler diversion privilege.

(a) *In general.* Handlers handling cherries harvested in a regulated district may fulfill any restricted percentage requirement in full or in part by acquiring diversion certificates or by voluntarily diverting cherries or cherry products in a program approved by the Board, rather than placing cherries in an inventory reserve. Upon voluntary diversion and compliance with the provisions of this section, the Board shall issue to the diverting handler a handler diversion certificate which shall satisfy any restricted percentage or diversion requirement to the extent of the Board or Department inspected weight of the cherries diverted.

(b) *Eligible diversion.* Handler diversion certificates shall be issued to handlers only if the cherries are diverted in accordance with the following terms and conditions or such other terms and conditions that the Board, with the approval of the Secretary, may establish. Such diversion may take place in any form which the Board, with the approval of the Secretary, may designate. Tart cherry juice and juice concentrate may receive diversion credit but only if diverted in forms approved under the terms of this section. Such forms may include, but are not limited to:

(1) Contribution to a Board-approved food bank or other approved charitable organization;

(2) Use for new product and new market development;

(3) Export to designated destinations; or

(4) Other uses or disposition, including destruction of the cherries at the handler's facilities.

(c) *Notification.* The handler electing to divert cherries through means authorized under this section shall first notify the Board of such election. Such notification shall describe in detail the manner in which the handler proposes to divert cherries including, if the diversion is to be by means of destruction of the cherries, a detailed description of the means of destruction and ultimate disposition of the cherries. It shall also contain an agreement that the proposed diversion is to be carried out under the supervision of the Board and that the cost of such supervision is to be paid by the handler. Uniform fees for such supervision may be established by the Board, pursuant to rules and regulations approved by the Secretary.

(d) *Diversion certificate.* The Board shall conduct such supervision of the handler's diversion of cherries under paragraph (c) of this section as may be necessary to assure that the cherries are diverted as authorized. After the diversion has been completed, the Board shall issue to the diverting handler a handler diversion certificate indicating the weight of cherries which may be used to offset any restricted percentage requirement.

(e) *Transfer of certificates.* Within such restrictions as may be prescribed in rules and regulations, including but not limited to procedures for transfer of diversion credit and limitations on the type of certification eligible for transfer, a handler who acquires diversion certificates representing diverted cherries during any crop year may transfer such certificates to another handler or handlers. The Board must be notified in writing whenever such transfers take place during a crop year.

(f) The Board, with the approval of the Secretary, may establish rules and regulations necessary and incidental to the administration of this section.

[67 FR 51715, Aug. 8, 2002]

§ 930.60   Equity holders.

(a) *Inventory reserve ownership.* The inventory reserve shall be the sole responsibility of the handlers who place products into the inventory reserve. A handler's equity in the primary inventory reserve may be transferred to another person upon notification to the Board.

(b) *Agreements with growers.* Individual handlers are encouraged to have written agreements with growers who deliver their cherries to the handler as to how any restricted percentage cherries delivered to the handler will be handled and what share, if any, the grower will have in the eventual sale of any inventory reserve cherries.

(c) *Rulemaking authority.* The Board, with the approval of the Secretary, may adopt rules and regulations necessary and incidental to the administration of this section.

§ 930.61   Handler compensation.

Each handler handling cherries from a regulated district that is subject to volume regulations shall be compensated by the Board for inspection relating to the primary inventory reserve as the Board may deem to be appropriate. The Board, with the approval of the Secretary, may establish such rules and regulations as are necessary and incidental to the administration of this section.

§ 930.62   Exempt uses.

(a) The Board, with the approval of the Secretary, may exempt from the provisions of § 930.41, § 930.44, § 940.51, § 930.53, or § 930.55 through § 930.57 cherries for designated uses. Such uses may include, but are not limited to:

(1) New product and new market development;

(2) Export to designated destinations;

(3) Experimental purposes; or

(4) For any other use designated by the Board, including cherries processed into products for markets for which less than 5 percent of the preceding 5-year average production of cherries were utilized.

(b) The Board, with the approval of the Secretary, shall prescribe such rules, regulations, and safeguards as it may deem necessary to ensure that cherries handled under the provisions of this section are handled only as authorized.

(c) Diversion certificates shall not be issued for cherries which are used for exempt purposes; *Provided,* that growers engaging in such activities under the authority of § 930.58 shall be issued diversion certificates for such activities.

[67 FR 51715, Aug. 8, 2002]

§ 930.63   Deferment of restricted obligation.

(a) *Bonding.* The Board, with the approval of the Secretary, may require handlers to secure bonds on deferred inventory reserve tonnage. Handlers may, in order to comply with the requirements of §§ 930.50 and 930.51 and regulations issued thereunder, secure bonds on restricted percentage cherries to temporarily defer the date that inventory reserve cherries must be held to any date requested by the handler. This date shall be not later than 60 days prior to the end of that crop year. Such deferment shall be conditioned upon the voluntary execution and delivery by the handler to the Board of a written undertaking within thirty (30) days after the Secretary announces final restricted percentage under § 930.51. Such written undertaking shall be secured by a bond or bonds with a surety or sureties acceptable to the Board that on or prior to the acceptable deferred date the handler will have fully satisfied the restricted percentage amount required by § 930.51.

(b) *Rulemaking authority.* The Board, with the approval of the Secretary, may adopt rules and regulations necessary and incidental to the administration of this section.

REPORTS AND RECORDS

§ 930.70   Reports.

(a) *Weekly production, monthly sales, and inventory data.* Each handler shall, upon request of the Board, file promptly with the Board, reports showing weekly production data; monthly sales and inventory data; and such other information, including the volume of any cherries placed in or released from a primary or secondary inventory reserve or diverted, as the Board shall specify with respect to any cherries handled by the handler. Such information may be provided to the Board members in summary or aggregated form only without any reference to the individual sources of the information.

(b) *Other reports.* Upon the request of the Board, with the approval of the Secretary, each handler shall furnish to the Board such other information with respect to the cherries acquired, handled, stored and disposed of by such handler as may be necessary to enable the Board to exercise its powers and perform its duties under this part.

(c) *Protection of proprietary information.* Under no circumstances shall any information or reports be made available to the Board members, or to any person designated by the Board or by the Secretary, which will reveal the proprietary information of an individual handler.

§930.71 Records.

Each handler shall maintain such records of all cherries acquired, handled, stored or sold, or otherwise disposed of as will substantiate the required reports and as may be prescribed by the Board. All such records shall be maintained for not less than two years after the termination of the fiscal year in which the transactions occurred or for such lesser period as the Board may direct with the approval of the Secretary.

§930.72 Verification of reports and records.

For the purpose of assuring compliance and checking and verifying the reports filed by handlers, the Secretary and the Board, through its duly authorized agents, shall have access to any premises where applicable records are maintained, where cherries are received, stored, or handled, and, at any time during reasonable business hours, shall be permitted to inspect such handlers premises and any and all records of such handlers with respect to matters within the purview of this part.

§930.73 Confidential information.

All reports and records furnished or submitted by handlers to the Board and its authorized agents which include data or information constituting a trade secret or disclosing trade position, financial condition, or business operations of the particular handler from whom received, shall be received by and at all times kept in the custody and under the control of one or more employees of the Board or its agent, who shall disclose such information to no person other than the Secretary.

MISCELLANEOUS PROVISIONS

§930.80 Compliance.

Except as provided in this part, no person may handle cherries, the handling of which has been prohibited by the Secretary under this part, and no person shall handle cherries except in conformity with the provisions of this part and the regulations issued hereunder. No person may handle any cherries for which a diversion certificate has been issued other than as provided in §§930.58(b) and 930.59(b).

§930.81 Right of the Secretary.

Members of the Board (including successors and alternates), and any agents, employees, or representatives thereof, shall be subject to removal or suspension by the Secretary at any time. Each regulation, decision, determination, or other act of the Board shall be subject to the Secretary's disapproval at any time. Upon such disapproval, the disapproved action of the Board shall be deemed null and void, except as to acts done in reliance thereon or in accordance therewith prior to such disapproval by the Secretary.

§930.82 Effective time.

The provisions of this part, and of any amendment thereto, shall become effective at such time as the Secretary may declare, and shall continue in force until terminated, or suspended.

§930.83 Termination.

(a) The Secretary may, at any time, terminate any or all of the provisions of this part by giving at least 1 day's notice by means of a press notice or in any other manner in which the Secretary may determine.

(b) The Secretary shall terminate or suspend the operation of any or all of the provisions of this part whenever the Secretary finds that such provisions do not tend to effectuate the declared policy of the Act.

(c) The Secretary shall terminate the provisions of this part whenever the Secretary finds by referendum or otherwise that such termination is favored by a majority of the growers and processors: *Provided,* That such majority has, during the current fiscal year, produced or canned and frozen more than 50 percent of the volume of the cherries which were produced or processed within the production area. Such termination shall become effective on the last day of June subsequent to the announcement thereof by the Secretary.

(d) The Secretary shall conduct a referendum within the month of March of every sixth year after the effective date of this part to ascertain whether continuation of this part is favored by the growers and processors. The Secretary may terminate the provisions of this part at the end of any fiscal period in which the Secretary has found that continuance is not favored by a majority of growers and processors who, during a representative period determined by the Secretary, have been engaged in the production or processing of tart cherries in the production area. Such termination shall be announced on or before the end of the fiscal period.

(e) The provisions of this part shall, in any event, terminate whenever the provisions of the Act authorizing them cease to be in effect.

§ 930.84 Proceedings after termination.

(a) Upon the termination of the provisions of this part, the then functioning members of the Board shall, for the purpose of liquidating the affairs of the Board, continue as trustees of all the funds and property then in its possession, or under its control, including claims for any funds unpaid or property not delivered at the time of such termination.

(b) The said trustees shall:

(1) continue in such capacity until discharged by the Secretary;

(2) from time to time account for all receipts and disbursements and deliver all property on hand, together with all books and records of the Board and of the trustees, to such person as the Secretary may direct; and

(3) upon the request of the Secretary, execute such assignments or other instruments necessary or appropriate to vest in such person full title and right to all of the funds, property, and claims vested in the Board or in the trustees pursuant to this part.

(c) Any person to whom funds, property, and claims have been transferred or delivered, pursuant to this section, shall be subject to the same obligations imposed upon the Board and upon the trustees.

§ 930.85 Effect of termination or amendment.

Unless otherwise expressly provided by the Secretary, the termination of this part or of any regulation issued pursuant to this part, or the issuance of any amendment to either thereof, shall not:

(a) Affect or waive any right, duty, obligation, or liability which shall have risen or which may thereafter arise in connection with any provision of this part or any regulation issued thereunder;

(b) Release or extinguish any violation of this part or any regulation issued thereunder;

(c) Affect or impair any rights or remedies of the Secretary or any other person with respect to any such violation.

§ 930.86 Duration of immunities.

The benefits, privileges, and immunities conferred upon any person by virtue of this part shall cease upon its termination, except with respect to acts done under and during the existence of this part.

§ 930.87 Agents.

The Secretary may, by designation in writing, name any officer or employee of the United States, or name any agency or division in the U.S. Department of Agriculture, to act as the Secretary's agent or representative in connection with any provisions of this part.

§930.88 Derogation.

Nothing contained in this part is, or shall be construed to be, in derogation or in modification of the rights of the Secretary or of the United States to exercise any powers granted by the Act or otherwise, or, in accordance with such powers, to act in the premises whenever such action is deemed advisable.

§930.89 Personal liability.

No member or alternate member of the Board and no employee or agent of the Board shall be held personally responsible, either individually or jointly with others, in any way whatsoever, to any person for errors in judgment, mistakes, or other acts, either of commission or omission, as such member, alternate member, employee, or agent, except for acts of dishonesty, willful misconduct, or gross negligence.

§930.90 Separability.

If any provision of this part is declared invalid or the applicability thereof to any person, circumstance, or thing is held invalid, the validity of the remainder of this part or the applicability thereof to any other person, circumstance, or thing shall not be affected thereby.

§930.91 Amendments.

Amendments to this subpart may be proposed, from time to time, by the Board or by the Secretary.

§930.100 Grower diversion certificates.

(a) In accordance with paragraph (b) of this section, the Board may, for the 1997 crop year, issue diversion certificates to growers, in districts subject to volume regulation (Northwest Michigan, Central Michigan, New York, and Utah) who have voluntarily elected to divert in the orchard all or a portion of their 1997 tart cherry production which otherwise, upon delivery to handlers, would become restricted percentage cherries. Growers may offer the diversion certificate to handlers in lieu of delivering cherries.

(b) *Terms and conditions.* To be eligible to receive diversion credit, growers voluntarily choosing to divert cherries must meet the following terms and conditions:

(1) In order to receive a certificate, a grower must demonstrate, to the satisfaction of the Board, that rows or trees which were selected for diversion were not harvested. Trees six years old or younger do not qualify for diversion.

(2) The grower must furnish the Board with a total harvested production amount so the Board can calculate the amount of grower diversion tonnage to be placed on the diversion certificate. The Board will confirm the grower's production amount with information provided by handlers (to which the grower delivers cherries) on Board Form Number Two.

(3) The grower must agree to allow a Board compliance officer to visit the grower's orchard to confirm that diversion has actually taken place.

(c) *Calculation of diversion amounts.* The weight of cherries diverted and left unharvested shall be calculated by the Board after growers furnish the Board with the necessary information concerning their production. After verification of the volume of cherries diverted, the Board shall calculate the amounts of grower diversion tonnage to be placed on the diversion certificates and issue such certificates to growers. Such amounts shall be determined as follows:

(1) For whole block diversion, the weight of a harvested sample of 5 percent of each diverted block, provided by the grower, will be used to calculate the total volume of diverted cherries to be credited on the diversion certificate. For example, a grower farms 1,000 acres and elects to whole block divert a 200 acre block. If 5 percent of the harvested trees in the block diverted yield 80,000 pounds of cherries, the grower would receive a diversion certificate for 1,600,000 pounds (80,000 pounds divided by 5 percent (.05) yields 1,600,000 pounds). The rest of the block would remain unharvested.

(2) For random row diversion, such estimated volume would be calculated by applying the percentage of the grower's production diverted to the actual average volume per acre of cherries produced and harvested. For example, a grower farms 1,000 acres and

elects to divert 20 percent of the harvestable acreage (200 acres). The grower harvests the remaining 800 acres and obtains 6,400,000 pounds of cherries, which represents a yield per acre of 8,000 pounds. Such grower would receive a diversion certificate for 1,600,000 pounds of cherries (8,000 lbs multiplied by the 20 percent of the total acreage diverted; in this instance, 200 acres).

[62 FR 44883, Aug. 25, 1997, as amended at 63 FR 20023, Apr. 22, 1998]

§ 930.107 Fiscal period.

Pursuant to § 930.7, fiscal period shall mean the period beginning October 1 and ending September 30 of each year.

[73 FR 75929, Dec. 15, 2008]

§ 930.120 Board membership.

When the production level from a district falls below the thresholds stated in § 930.20(b)(5), members of the specific district will make a recommendation to the Board as to who should be removed from the Board and the Board shall submit a recommendation to the Secretary for approval. If the recommendation is not made by the Board within a reasonable time, the Secretary may select the member and alternate to be removed.

[71 FR 16985, Apr. 5, 2006]

Subpart B—Administrative Requirements

§ 930.133 Compensation rate.

A compensation rate of $250 per meeting shall be paid to the public member and to the alternate public member when attending Board meetings. Such compensation is a per meeting rate. For example, if a Board meeting is convened and lasts one or two days or only four hours, the public member and/or alternate public member attending the meeting would receive $250 each.

[63 FR 33528, June 19, 1998]

§ 930.141 Delinquent assessments.

(a) Pursuant to § 930.41, the Board shall impose an interest charge on any handler whose assessment payment has not been received by October 1 of each crop year. The interest rate shall be a rate of one percent per month and shall be applied to the unpaid assessment balance not paid by the October 1 due date. In addition to the interest charge, the Board shall impose a late payment charge on any handler whose assessment payment has not been received within 90 days from the due date of October 1. The late payment charge shall be 10 percent of the unpaid balance.

(b) [Reserved]

[62 FR 55150, Oct. 23, 1997, as amended at 63 FR 14024, Mar. 24, 1998]

§ 930.150 Primary inventory reserve.

Beginning July 1, 2012, the primary inventory reserve may not exceed 100 million pounds.

[77 FR 40253, July 9, 2012]

§ 930.151 Desirable carry-out inventory.

For the 2016 crop year, the desirable carry-out inventory, for the purposes of determining an optimum supply volume, will be 57 million pounds.

[82 FR 28755, June 26, 2017]

§ 930.154 Release of inventory reserve cherries.

(a) As provided in § 930.54, the Board may recommend a release of a portion or all of the primary and/or secondary reserve cherries. The total available reserves will be determined at the beginning of the crop year. The primary reserve as defined in §§ 930.55 and 930.150 must be depleted before the secondary reserve can be released. If a release is recommended, the recommended volume shall be apportioned to handlers on the basis of each handler's proportion of the total volume handled in the preceding three crop years.

(b) If a handler has less volume in reserve than is apportioned, the excess volume shall be reapportioned to those who still have volume in reserve until the total release is complete.

[81 FR 63679, Sept. 16, 2016]

§ 930.158 Grower diversion and grower diversion certificates.

(a) *Grower diversion certificates.* The Board may issue diversion certificates to growers in districts subject to volume regulation who have voluntarily elected to divert in the orchard all or a portion of their tart cherry production which otherwise, upon delivery to handlers, would become restricted percentage cherries. Growers may offer the diversion certificate to handlers in lieu of delivering cherries. Handlers may redeem diversion certificates with the Board through June 30 of each crop year. After June 30 of the crop year that crop year's grower diversion certificates are no longer valid. Cherries that have reached a harvestable, marketable condition will be eligible for diversion. Diversion will not be granted to growers whose fruit was destroyed before it set and/or matured on the tree, or whose fruit is unmarketable. If marketable fruit were to be damaged or destroyed by acts of nature such as storms or hail diversion credit could be granted. To be considered marketable for the purposes of this section, sampled fruit may not exceed a 5 percent tolerance for insects or a 7 percent tolerance for rot.

(b) *Application and mapping for diversion.* Any grower desiring to divert cherries using methods other than in-orchard tank shall submit a map of the orchard or orchards to be diverted, along with a completed Grower Diversion Application, to the Board by April 15 of each crop year. The application includes a statement which must be signed by the grower which states that the grower agrees to comply with the regulations established for a tart cherry diversion program. Each map shall contain the grower's name and number assigned by the Board, the grower's address, block name or number when appropriate, location of orchard or orchards and other information which may be necessary to accomplish the desired diversion. On or before July 1, the grower should inform the Board of such grower's intention to divert in-orchard and what type of diversion will be used. The four types of diversion are random row diversion, whole block diversion, partial block diversion and in-orchard tank diversion. A grower who informs the Board about the type of diversion he or she wishes to use by July 1 can elect to use any diversion method or combination of diversion methods.

Only random row or in-orchard tank diversion methods may be used if the Board is not so informed by July 1. Trees that are four years or younger do not qualify for diversion. Annual resubmissions of either the map or application will no longer be required. Growers will only submit a new application and map if they are participating in the grower diversion program for the first time. Growers will need only to submit a new orchard map if he/she adds a new block of trees to the orchard or changes the orchard layout differently from the map previously submitted to the Board.

(1) *Random row diversion.* Using the orchard map furnished by the grower, the Board will randomly select rows of trees within the orchard to be diverted. The amount of cherries to be diverted will be based on the preliminary restricted percentage amount established pursuant to §930.50. A grower may elect a different percentage amount; however, the grower needs to inform the Board as soon as possible after the preliminary percentages are announced of this other amount, but in no event shall this be less than seven days in advance of harvest. The designated rows indicated by the map must not be harvested. After completing harvest of the remaining rows in the orchard, the grower must notify the Board and/or the Board's compliance officer. A compliance officer will then be allowed to observe the grower's orchard to assure that the selected rows have not been harvested. The grower must inform the Board of the total production of the orchard to calculate the tonnage that was diverted.

(2) *Whole block diversion.* Based on maps supplied by the grower, a sampling procedure will be used to determine the amount of cherries in the orchard to be diverted. A block is defined as rows that run in the same direction, are similar in age, and have definable boundaries. The Board will require a number of tree sites to be sampled depending on the size of the block. A tree site is a planted tree or an area where a tree was planted and may have been uprooted or died. If a block has 5 rows or less, or 200 or less tree sites, 3 rows would be randomly chosen to be sampled, if a block has 6 to 15 rows, or 201–

400 tree sites, 4 rows would be randomly chosen to be sampled, and if a block has 16 or more rows and greater than 400 tree sites, 5 rows would be randomly chosen to be sampled. The Board's compliance officer will apply the sampling procedure (based on the number of rows or the number of tree sites) which results in the fewest number of tree sites required to be sampled. From each of the rows to be sampled, ten contiguous tree sites will be sampled. Only trees more than five years old will be harvested for the sample. For example, if it is determined that five rows are to be sampled, 10 contiguous tree sites in each of the five rows will be subject to harvest. Trees within the 10 sites which are more than five years old will be harvested. The harvested tonnage will be converted to a volume that represents the entire block of cherries. If, for example, a total of 4,600 pounds is harvested from the sample tree sites and this total is divided by 50 tree sites a yield of 92 pounds per tree site is obtained. To find the total yield for the block, the 92 pounds per tree site yield is multiplied by the 880 tree sites that were mapped in the block and that equals 80,960 pounds for that block. The compliance officer would be allowed access to the block to oversee the sampling process and to confirm that the block has been diverted.

(3) *Partial block diversion.* Partial block diversion will also be accomplished using maps supplied by the grower. Sampling will be done as in whole block diversion except that only partial blocks would be selected and sampled. Growers may divert up to five partial blocks, or 50 percent of a grower's total number of blocks per year. Such block(s) must be mapped and will be sampled as described under whole block diversion. Rows used in partial block diversion must be contiguous.

(4) *In-orchard tank diversion.* Growers wishing to in-orchard tank divert must pick the cherries to be diverted and place them in harvesting tanks. A compliance officer would then probe the tanks for volume measurement and observe the destruction of the cherries on the grower's premises. Growers wishing to take advantage of this option must have at least 10 tanks ready for diversion. The compliance officer has up to five days to come to the grower's premises to observe the diversion after being contacted.

(c) *Compliance.* Growers who voluntarily participate in the grower diversion program must sign and file with the Board a Grower Diversion Application. By signing the application, a grower agrees to the terms and conditions of the grower diversion program as contained in these regulations. To be eligible to receive diversion credit, growers voluntarily choosing to divert cherries must meet the following terms and conditions:

(1) In order to receive a certificate, a grower must demonstrate, to the satisfaction of the Board, that rows or trees which were selected for diversion were not harvested. Trees four years old or younger do not qualify for diversion.

(2) The grower must furnish the Board with a total harvested production amount so the Board can calculate the amount of grower diversion tonnage to be placed on the diversion certificate. The Board will confirm the grower's production amount with information provided by handlers (to which the grower delivers cherries) on Board form Number Two.

(3)(i) The grower must agree to allow a Board compliance officer to visit the grower's orchard to confirm that diversion has actually taken place. If the terms and conditions for whole block, partial block or in-orchard tank diversion are not completed,, the Board shall not issue the grower a diversion certificate.

(ii) If a grower who chooses random row diversion harvests rows that were designated not to be harvested, the grower should inform the Board immediately of the error. The grower will then be required to divert twice the amount (rows or trees) incorrectly harvested to correct the mistake. The grower will still receive a diversion certificate equal to the original requested amount. However, in instances where a grower is at the end of harvesting the orchard and fails to divert a complete block or specified rows, the Board shall multiply by two the difference between the original diversion amount and the actual diverted amount. The Board shall subtract that amount from the diversion application amount. Thus, the grower would receive a grower diversion certificate equal to a portion of the originally requested amount. If the grower does not inform the Board of such errors, the grower will not receive a diversion certificate.

[63 FR 33528, June 19, 1998, as amended at 64 FR 30232, June 7, 1999;; 71 FR 66098, Nov. 13, 2006; 76 FR 65360, Oct. 21, 2011; 78 FR 46496, Aug. 1, 2013; 84 FR 53008, Oct. 4, 2019]

EFFECTIVE DATE NOTE: At 76 FR 65360, Oct. 21, 2011, in §930.158, paragraphs (b)(1) and (c)(3)(ii) were suspended indefinitely.

§930.159 Handler diversion.

(a) *Methods of diversion.* Handlers may divert cherries by redeeming grower diversion certificates, by destroying cherries at handlers' facilities (at-plant), by diverting cherry products accidentally or voluntarily destroyed, by donating cherries or cherry products to charitable organizations or by using cherries or cherry products for exempt purposes under §930.162, including export to countries other than Canada, and Mexico. Once diversion has taken place, handlers will receive diversion certificates stating the weight of cherries diverted. Diversion credit may be used to fulfill any restricted percentage requirement in full or in part. Any information of a confidential and/or proprietary nature included in this application would be held in confidence pursuant to §930.73 of the order.

(b) *Board notification and handler plan.* Any handler intending to divert cherries or cherry products pursuant to §930.59 of the order (except through exempt uses under §930.62 of the order) must notify the Board of such intent and provide a plan by October 1 which shows how the handler intends to meet the restricted percentage obligation, except that, for the 1997–98 season only, the deadline is February 5, 1998. The Board may extend this date in individual cases pursuant to a written request showing good cause why the plan cannot be provided by the due date. A handler will have one year to fulfill such plan. The details of the plan shall include, but not be limited to, the name and address of the handler, the total product processed at-plant, product diverted at-plant, in-orchard diversion certificates to be redeemed, anticipated donations to charitable outlets, disposition to exempt outlets or uses and detailed plans for how and where such disposition will be made, and inventory reserve amount. It shall also contain an agreement that the proposed diversion is to be carried out under the supervision of the Board and that the cost of such supervision is to be paid by the handler. Supervision of diversion by means other than destruction of the cherries at a handler's facility will be subject to supervision as found necessary by the Board. USDA inspectors or Board employees will supervise diversion of cherry products at the current hourly rate under USDA's inspection fee schedule (7 CFR 52.42). Any cherries not diverted in accordance with the handler's plan will be placed into the secondary inventory reserve or the primary inventory reserve if a secondary inventory reserve has not been established.

(c) *At-plant diversion.* Diversion by disposal at-plant may take place prior to placing the cherries into the processing line, or after processing, but before a finished product is manufactured. Such diversion will take place under the supervision of USDA Inspection Service or Board employee inspectors. USDA inspectors or Board employees or Board agents will supervise diversion of cherry products at-plant at the current hourly rate under USDA's inspection fee schedule (7 CFR 52.42).

(d) *Diversion of finished products.* Handlers may be granted diversion credit for finished tart cherry products that are accidentally destroyed or voluntarily destroyed by the handler. To receive diversion credit under this option the cherry products must be owned by the handler at the time of accidental or voluntary destruction, be a marketable product at the time of processing, be included in the handler's end of the year handler plan, and have been assigned a Raw Product Equivalent (RPE) by the handler to determine the volume of cherries. In addition, the accidental or voluntary destruction and disposition of the product must be verified by either a USDA inspector or Board agent or employee who witnesses the disposition of the accidentally or voluntarily destroyed product. Products will be considered as accidentally destroyed if they sustain damage which renders them unacceptable in normal market channels. Products which are voluntarily destroyed must have deteriorated in condition to such an extent that they are not acceptable for use in normal market channels.

(e) *Contributions to approved charitable organizations.* When diverting by donating cherries or cherry products to charitable organizations, handlers should follow the requirements specified herein. For contributions to qualify for diversion credit, the contributed product should be marked clearly "NOT FOR RESALE". The receiving organization must be approved by the Board as a qualified recipient of contributions of tart cherry products. Such organizations must be tax-exempt, must not sell the donated products and must be noncompetitive with other tart cherry industry sales outlets. Once products are donated to an organization, the Board must receive satisfactory documentation of the transaction. Handlers should provide the Board with information on how the product was used and the volume of product used.

(f) *Grower diversion certificates.* To satisfy restricted percentage obligations by redeeming grower diversion certificates handlers must present to the Board grower diversion certificates obtained from growers who have diverted cherries by non-harvest, and who have been issued diversion certificates by the Board in accordance with the applicable rules and regulations governing the issuance of grower diversion certificates. For this crop year July 1, 1997, through June 30, 1998, grower diversion certificates will be valid until February 5, 1998.

[63 FR 404, Jan. 6, 1998, as amended at 63 FR 20019, Apr. 22, 1998; 64 FR 9268, Feb. 25, 1999; 64 FR 33009, June 21, 1999; 65 FR 35267, June 2, 2000; 69 FR 41385, July 9, 2004; 71 FR 16985, Apr. 5, 2006; 78 FR 46496, Aug. 1, 2013]

§ 930.162   Exemptions.

(a) *General.* Tart cherries which are used for the purpose of new product development, for new market development and market expansion, for the development of export markets, for experimental purposes, for export to countries other than Canada, and Mexico, or which are donated to charitable organizations may be granted an exemption by the Board and will be exempt from §§ 930.41, 930.44, 930.51, 930.53, and §§ 930.55 through 930.57, subject to the following terms and conditions. Any information received of a confidential and/or proprietary nature included in this application will be protected from disclosure pursuant to § 930.73 of the order.

(b) *Definitions.* The terms in paragraph (a) of this section shall have the following meaning:

(1) *New product development.* This term includes the development of new tart cherry products or of foods or other products in which tart cherries or tart cherry products are incorporated which are not presently being produced on a commercial basis. New product development can also include the production or processing of a tart cherry product using a technique not presently being utilized commercially in the tart cherry industry; an end product of the processing of raw tart cherries done by the industry at pack time either for resale or for re-manufacturing which has not been manufactured previously by the industry; or a processed, value-added item that includes tart cherry products as an ingredient which has never been marketed to consumers either by a handler within the industry or by a food manufacturer. In addition, the maximum duration of any credit activity is five years from the date of the first shipment.

(2) *New market development and market expansion.* This includes the development of markets for tart cherry products which are not commercially established markets and which are not competitive with commercial outlets presently utilized by the tart cherry industry (including the development of new export markets): *Provided,* That these markets are a geographic area into which tart cherries or products derived from them have not been previously sold. The term "market expansion", includes activities that incrementally expand the sale of either tart cherries or the products in which tart cherries are an ingredient, such as, but not limited to: Expansions of the geographic areas into which tart cherries or tart

cherry products are marketed; product line extensions; significant improvements to or revisions of existing products; packaging innovations; segmentation of markets along geographic, demographic, or other definable characteristics; and product repositionings. In addition, shipments of tart cherries or tart cherry products in new market development and market expansion outlets are eligible for handler diversion credit for a period of five years from the handler's date of the first shipment into such outlets.

(3) *Development of export markets.* The sale of cherries or cherry products, including the development of sales for new or different tart cherry products or the expansion of sales for existing tart cherry products, to countries other than Canada, and Mexico.

(4) *Experimental purposes.* The use of cherries or cherry products in preliminary and/or developmental activities intended to result in new products, new applications and/or new markets for tart cherry products. Any exemption for experimental work shall be limited in scope, duration and volume based on information supplied by the applicant at the time a request for exemption is made. In no case shall an individual exemption for experimental purposes last longer than five years or exceed 100,000 pounds raw product equivalent of tart cherries.

(c) *Obtaining approval for exempt uses.* In order to receive exemptions for cherries or cherry products utilized for exempt purposes, handlers must apply to the Board for a new exemption or for renewal of an existing exemption by November 1 for the next succeeding year, except for the 1997 year only, handlers may apply through February 5, 1998. A handler shall have one crop year to dispose of cherries or cherry products to exempt outlets approved by the Board, unless granted a renewal. Handlers applying to the Board for a new exemption or for renewal of an existing exemption are subject to the following conditions:

(1) When applying to the Board for an exemption for new product development, handlers must detail the nature of their new product, how it differs from current, existing products and the anticipated short and long term sales volume for the exemption. It will be the Board staff's responsibility to analyze and investigate any request and upon completion of that analysis authorize or deny the exemption.

(2) When applying to the Board for an exemption for new market development, handlers must detail the nature of their new market, how it differs from current, existing markets and the anticipated short and long term sales volume for the exemption. It will be the Board staff's responsibility to analyze and investigate any request and upon completion of that analysis authorize or deny the exemption.

(3) When applying to the Board for an exemption for the use of domestic tart cherry products in markets not currently served by the domestic industry, handlers may provide a verifiable statement from the buyer of its intent to use domestic tart cherry products to the Board staff for review in lieu of review by the subcommittee as detailed in paragraph (d) of this section. A verifiable statement is defined as a written statement from the buyer that it will use domestic tart cherries in products or markets not currently supplied by domestic sources, which will be reviewed and documented by Board staff.

(4) When applying to the Board for an exemption for the development of export markets for tart cherries or cherry products (including juice and juice concentrate) in countries other than Canada and Mexico, including the expansion of sales in existing export markets, handlers must detail the nature of their product, specify whether such product differs from current products being sold in export markets, and estimate the anticipated short and long term sales volumes for the requested exemption.

(5) When applying to the Board for an exemption for experimental purposes, handlers must indicate the preliminary and/or developmental experimental activity. Such experimental purposes should be intended to result in new products, new applications and/or new markets for existing tart cherry products. Any exemption for experimental work shall be limited in scope, duration and volume which the proposing

party shall specify at the time a request for exemption is made. In no case shall an exemption for experimental purposes last longer than five years or exceed 100,000 pounds raw product equivalent per handler of tart cherries during the duration of the experiment.

(6) To be eligible for new product, new market development and market expansion diversion exemptions, a handler must demonstrate involvement in the activity for which the exemptions are sought. The requesting handler must either be or have been involved in development of the product, the market, or market expansion activities for which the exemptions are sought or have had financial involvement in the activities. This involvement must be demonstrated and established to the satisfaction of the Board by the handler requesting the exemptions.

(d) *Review of applications.* A Board appointed subcommittee shall review applications for exemption or renewal of exemption and either approve or deny the exemption. The subcommittee shall consist of three members and one alternate, each having no handler affiliation but knowledge of the tart cherry industry, one of whom shall be the public member or the alternate public member if available to serve. Any denial of an application for exemption or renewal of an existing exemption shall be served on the applicant by certified mail and shall state the reasons for the denial. Within 10 days after the receipt of a denial, the applicant may file an appeal, in writing, with the Deputy Administrator, Specialty Crops Program, supported by any arguments and evidence the applicant may wish to offer as to why the application for exemption or renewal of exemption should have been approved. The Deputy Administrator, upon consideration of such appeal, will take such action as deemed appropriate with respect to the application for exemption or renewal of exemption.

(e) *Progress report.* Each handler that is granted an exemption must submit to the Board an annual progress report, due May 1 of each crop year. The progress report shall include the results of the exemption activity (comparison of intended activity with actual activity) for the year in its entirety, the volume of exempted fruit, an analysis of the success of the exemption program, and such other information as the Board may request.

(f) *Diversion credit; failure to meet terms and conditions of exemption.* Handler diversion certificates for exempt uses shall be issued to handlers provided that terms and conditions applicable to exempt uses are satisfied. Diversion certificates will not be issued to handlers for any volume of tart cherry products for which such terms and conditions are not satisfied and such cherries would be subject to all of the terms and conditions of §§ 930.41, 930.44, 930.51, 930.53, and §§ 930.55 through 930.57.

(g) *Failure to meet terms and conditions for exemption.* Upon termination of an exemption, any volume of tart cherry products that were granted an exemption but were not utilized for the authorized exempt purpose would be subject to all of the terms and conditions of §§ 930.41, 930.44, 930.51, 930.53, and §§ 930.55 through 930.57.

(h) *Extensions and transfers.* (1) If no shipments are made within the first year of any approved exemption project from the date of approval, new applications for a similar project (same market or product) are eligible for approval; *provided that,* handlers with an approved exemption project have the opportunity to apply to the subcommittee for a six-month extension of this time period.

(2) For projects granted extensions, if no shipment is made prior to the end of the extension period, new applications for the same market or project are eligible for approval.

[63 FR 405, Jan. 6, 1998, as amended at 65 FR 35267, June 2, 2000; 66 FR 39413, July 31, 2001; 69 FR 34553, June 22, 2004; 71 FR 16986, Apr. 5, 2006; 77 FR 40253, July 9, 2012; 80 FR 68427, Nov. 5, 2015; 83 FR 31447, July 6, 2018]

§ 930.163 Deferment of restricted obligation.

A handler may obtain a surety bond on restricted percentage cherries to be posted to temporarily defer the date that inventory reserves must be held. The surety bond must be posted at two times the market value of the quantity of cherries for which the holding obligation is being deferred. The Board can temporarily defer the date inventory reserve cherries must be held to any date requested by the handler. However, this date shall not be later than 60 days prior to the end of the crop year. The deferment shall be conditioned on the execution and delivery by the handler to the Board of a written undertaking within 30 days after the Secretary announces the final restricted percentage under § 930.51. The written undertaking (required to be secured by a bond or bonds with a surety or sureties acceptable to the Board) must guarantee that on or prior to the deferment date requested by the handler the handler will have fully satisfied the restricted percentage obligation. On or prior to the deferment date requested by the handler, the handler will have to fully satisfy the restricted percentage obligation. In the event, a handler has posted the surety bond, reached the deferment date deadline and does not have cherries in the inventory reserve to cover his/her inventory reserve obligation, the bond will be forfeited to the Board. The Board will then buy cherries to fulfill that handler's obligation.

[66 FR 35891, July 10, 2001]

Subpart C—Assessment Rate

§ 930.200 Assessment rate.

On and after October 1, 2016, the assessment rate imposed on handlers shall be $0.0075 per pound of tart cherries grown in the production area and utilized in the production of tart cherry products. Included in this rate is $0.0065 per pound of tart cherries to cover the cost of the research and promotion program and $0.001 per pound of tart cherries to cover administrative expenses.

[82 FR 41829, Sept. 5, 2017]

§ 930.256 Free and restricted percentages for the 2018–19 crop year.

The percentages for tart cherries handled by handlers during the crop year beginning on July 1, 2018, which shall be free and restricted, respectively, are designated as follows: Free percentage, 73 percent and restricted percentage, 27 percent.

[84 FR 53008, Oct. 4, 2019]

PART 932—OLIVES GROWN IN CALIFORNIA

Subpart A—Order Regulating Handling

AUTHORITY: 7 U.S.C. 601–674.

SOURCE: 30 FR 12629, Oct. 2, 1965, unless otherwise noted.

Subpart A—Order Regulating Handling

DEFINITIONS

§ 932.1 Secretary.

Secretary means the Secretary of Agriculture of the United States, or any officer or employee of the U.S. Department of Agriculture who is or who may hereafter be authorized to exercise the powers or to perform the duties of the Secretary of Agriculture.

§ 932.2 Act.

Act means Public Act No. 10, 73d Congress (May 12, 1933) as amended and as reenacted and amended by the Agricultural Marketing Agreement Act of 1937, as amended (48 Stat. 31, as amended; 7 U.S.C. 601–674).

§ 932.3 Person.

Person includes an individual, partnership, corporation, association, or any other business unit.

§ 932.4 Area.

Area means the State of California.

§ 932.5 Olives.

Olives means the fruit of any variety of the species olea europaea, whether or not processed, grown within the area.

§ 932.6 Variety group 1.

Variety group 1 means the following varieties and any mutations, sports, or other derivations of such varieties: Aghizi Shami, Amellau, Ascolano, Ascolano dura, Azapa, Balady, Barouni, Carydolia, Cucco, Gigante di Cerignola, Gordale, Grosane, Jahlut, Polymorpha, Prunara, Ropades, Sevillano, Saint Agostino, Tafahi, and Touffahi.

§ 932.7 Variety group 2.

Variety group 2 means the following varieties and any mutations, sports, or other derivations of such varieties: Manzanillo, Mission, Nevadillo, Obliza, Redding Picholine.

§ 932.8 Natural condition olives.

Natural condition olives means olives in their fresh harvested state, whether or not placed in a water or other preserving medium.

[33 FR 11266, Aug. 8, 1968]

§ 932.9 Packaged olives.

Packaged olives means (a) processed olives in hermetically sealed containers and heat sterilized under pressure, otherwise known as *canned ripe olives* and including the three distinct types, *ripe, green ripe,* and *tree-ripened*; or (b) olives, packed in brine, and which have been fermented and cured, otherwise known as *green olives.*

§932.10 Lot.

Lot means the total net weight of natural condition olives of any one variety delivered to a handler at any one time.

§932.11 Grade.

Grade means the classification of olives as to quality according to the grading specifications established pursuant to the provisions of this part.

§932.12 Size.

Size means the number of whole olives contained in a pound and may be referred to in terms of size ranges.

§932.13 Size-grade.

Size-grade means to classify olives, or to cause olives to be classified, by sample or otherwise, into separate size designations.

§932.14 Process.

Process means to change olives in any way from their natural condition by any commercial process.

§932.15 Handler.

Handler means any person who handles olives.

§932.16 Handle.

Handle means to: (a) Size-grade olives, (b) process olives, or (c) use processed olives in the production of packaged olives, within the production area, or (d) ship packaged olives from the area to any point outside thereof or within the area: *Provided,* This term shall not include natural condition olives acquired and (1) used for olive oil, salt cured oil coated olives (also variously referred to as "Greek Olives," "Greek Style Olives," or "Oil Cured Olives"), or Silician Style Olives, or (2) shipped to fresh market outlets.

[36 FR 20356, Oct. 21, 1971]

§932.17 Producer.

Producer means any person engaged in a proprietary capacity in the production of olives for market as packaged olives.

§932.18 Committee.

Committee means the California Olive Committee established pursuant to §932.25.

[47 FR 32906, July 30, 1982]

§932.19 Crop year and fiscal year.

(a) *Crop year* means the 12-month period beginning on August 1 of each year and ending on July 31 of the following year or such other period that may be recommended by the committee and approved by the Secretary.

(b) *Fiscal year* means the 12-month period beginning on January 1 and ending on December 31 of each year or such other period that may be recommended by the committee and approved by the Secretary.

[47 FR 32906, July 30, 1982]

§932.20 Part and subpart.

Part means the Order Regulating the Handling of Olives Grown in California and all rules and regulations, and supplementary orders issued thereunder. The aforesaid Order Regulating the Handling of Olives Grown in California shall be a *subpart* of such part.

§932.21 District.

District means any of the following geographical areas of the State of California:

(a) *District 1* shall include the counties of Glenn, Tehama, and Shasta.

(b) *District 2* shall include the counties of Mono, Mariposa, Merced, San Benito, Monterey, Madera, Fresno, Tulare, and all counties to the south thereof.

(c) *District 3* shall include all counties not included in Districts 1 and 2.

§932.22 Sublot.

Sublot means a quantity of olives resulting from the separation by the handler of a lot into two or more parts.

[36 FR 20356, Oct. 21, 1971]

§932.23 Undersize olives and limited use size olives.

Undersize olives means olives of a size which, pursuant to §932.51(a)(3), shall

251

be disposed of in noncanning use; and *limited use size olives* means processed olives of any size which, pursuant to § 932.52(a)(3), is authorized for limited use.

[36 FR 20356, Oct. 21, 1971, as amended at 47 FR 32906, July 30, 1982]

§ 932.23a Limited use.

Limited use means the use of processed olives in the production of packaged olives of the halved, segmented (wedged), sliced, or chopped styles, as defined in the U.S. Standards for Grades of Canned Ripe Olives (7 CFR part 52) or subsequent amendments thereto, including modifications of the requirements for such styles pursuant to this part, and such additional styles (and the requirements applicable thereto) as may be specified pursuant to § 932.52(a)(7).

[47 FR 32906, July 30, 1982]

§ 932.24 Noncanning use.

Noncanning use means the use of olives other than in the production of canned ripe olives, and is the authorized outlet for undersize olives and the limited use size olives which, pursuant to § 932.52(b), are not permitted for limited use in any crop year in which limited use is restricted to less than the available quantity of limited use size olives.

[36 FR 20356, Oct. 21, 1971]

OLIVE ADMINISTRATIVE COMMITTEE

§ 932.25 Establishment and membership.

A California Olive Committee consisting of 16 members, is hereby established to administer the terms and provisions of this part. Each member shall have an alternate who meets the same qualifications as the member. Eight of the members and their alternates shall be producers or officers or employees of producers, and eight of the members and their alternates shall be handlers or directors, officers, or employees of handlers. The eight members of the committee who are producers or officers or employees of producers are referred to in this subpart as "producer members" of the committee; and the eight members of the committee who are handlers or directors, officers, or employees of handlers are referred to in this subpart as "handler members" of the committee. The committee may be increased by one public member who shall not be a producer or handler of olives nor an officer or employee or director of any producer or handler of olives. District representation of the producer members shall be two from District 1, four from District 2, and two from District 3. Allocation of the handler members shall be four members to represent cooperative marketing organizations, herein referred to as "cooperative handlers", and four members to represent handlers who are not cooperative marketing organizations, herein referred to as "independent handlers": *Provided,* That whenever during the crop year in which nominations are made and in the preceding crop year, the cooperative handlers or the independent handlers handled as first handler 65 percent or more of the total quantity of olives so handled by all handlers, allocation shall be five members to represent the group which so handled 65 percent or more of such olives and three members to represent the group which handled 35 percent or less. The public member and alternate public member shall be selected from any place within the area. The committee may, with the approval of the Secretary, provide such other allocation of producer or handler membership, or both, as may be necessary to assure equitable representation.

[47 FR 32907, July 30, 1982]

§ 932.26 Term of office.

The term of office of members and alternate members of the committee shall be 2 years beginning on June 1 and ending on May 31 of odd numbered years: *Provided,* That the term of office of initial members and alternate members shall begin on the effective date of this subpart. Each such member and alternate member shall serve during that portion of the term of office for which he is selected and has qualified and shall continue to serve until his successor is selected and has qualified.

§ 932.27 Selection.

Selection of members of the committee, and their respective alternates, shall be made in the appropriate numbers specified in §932.25 by the Secretary from nominees nominated pursuant to this part or, in the discretion of the Secretary, from other persons eligible for nominations for such positions.

§932.28 Eligibility.

Each producer member of the committee shall, at the time of selection and during the member's term of office, be a producer in the district for which selected, and except for producers who are members of cooperative handlers shall not be engaged in the handling of olives either in a proprietary capacity, or as a director, officer, or employee. Each handler member of the committee shall, at the time of selection and during the member's term of office, be a handler in the group that the member represents or a director, officer, or employee of such handler. The public member and alternate public member of the committee shall not at the time of selection and during the term of office be engaged in or have a financial interest in the commercial production, marketing, buying, grading, or processing of olives, nor shall such member or alternate be an officer, director, member, or employee of any firm engaged in such activities.

[47 FR 32907, July 30, 1982]

§932.29 Nominations.

(a) *Producer members.* (1) Nominations for producer members of the committee, and their respective alternates, may be conducted according to the following procedures, or other procedures recommended by the committee and approved by the Secretary:

(i) Meetings shall be held in each producer district for the purpose of selecting candidates for the member and alternate member nominations;

(ii) Those candidates selected at the producer meetings shall be nominated by mail balloting of producers in that district;

(iii) The committee shall adopt, with approval of the Secretary, appropriate procedures to be observed for conducting producer nominations by mail: *Provided,* That the names of nominees shall be submitted to the Secretary

prior to April 16 of the year in which nominations are made.

(2) Only producers, including duly authorized officers or employees of producers, shall participate in the nomination of producer members and alternate members. Each producer shall be entitled to cast only one vote for each nominee to be selected in the district in which the producer produces olives. No producer shall participate in the selection of nominees in more than one district. If a producer produces olives in more than one district, such producer shall select the district in which such producer will so participate and notify the committee of such choice.

(b) *Handler members.* (1) At a meeting or meetings called by the committee, the cooperative handlers shall nominate a qualified person for each member position and a qualified person for each alternate member position allocated to cooperative handlers as provided in §932.25.

(2) At a meeting or meetings called by the committee, the independent handlers shall nominate a qualified person for each member position and a qualified person for each alternate member position allocated to independent handlers as provided in §932.25.

(3) Each handler shall be entitled to cast only one vote for each nominee for cooperative handler member or alternate member or independent handler member or alternate member, as the case may be, which vote shall be weighed by the tonnages of olives handled by such handler during the crop year in which nominations are made and in the previous crop year.

(c) *Public member.* Nominations for the public member and alternate public member of the committee shall be submitted to the Secretary prior to April 16 of the year in which nominations are made. The committee shall prescribe procedures for the selection and voting for each candidate.

[33 FR 11266, Aug. 8, 1968, as amended at 47 FR 32907, July 30, 1982]

§932.30 Alternates.

An alternate for a member of the committee shall act in the place and stead of such member (a) during such member's absence, and (b) in the event of such member's removal, resignation, disqualification or death, until a successor for such member's unexpired term has been selected and has qualified. Except as otherwise specifically provided in this subpart, the provisions of this part applicable to members also apply to alternate members. The committee or the chariman of the committee may request one or more alternates to attend any or all meetings notwithstanding the expected or actual attendance of the respective member or members.

[47 FR 32907, July 30, 1982]

§ 932.31 Failure to nominate.

If nominations for any position on the committee are not received by the Secretary by May 1 of the year in which nominations are to be made, the Secretary may select an eligible individual without regard to nomination.

§ 932.32 Acceptance.

Any person selected by the Secretary as a member or as an alternate member of the committee shall qualify by filing a written acceptance with the Secretary promptly after being notified of such selection.

§ 932.33 Vacancies.

To fill any vacancy occasioned by the failure of any person selected as a member, or as an alternate member of the committee to qualify, or in the event of the removal, resignation, disqualification, or death of any member or alternate member, a successor for such person's unexpired term shall be nominated and selected in the manner set forth in § 932.29 insofar as such provisions are applicable. If nomination to fill any such vacancy is not made within 60 calendar days after such vacancy occurs, the Secretary may fill such vacancy without regard to nominations, but on the basis of the applicable representations and qualifications set forth in §§ 932.25, 932.27, and 932.28.

§ 932.34 Powers.

The committee shall have the following powers:

(a) To administer this subpart in accordance with its terms and provisions;

(b) To make rules and regulations to effectuate the terms and provisions of this subpart;

(c) To receive, investigate, and report to the Secretary complaints of violations of the provisions of this subpart; and

(d) To recommend to the Secretary amendments to this subpart.

§ 932.35 Duties.

The committee shall have, among others, the following duties:

(a) To act as intermediary between the Secretary and any producer or handler;

(b) To keep minutes, books, and other records, which shall clearly reflect all of its acts and transactions, and such minutes, books, and other records shall be subject to examination by the Secretary at any time;

(c) To make, subject to approval by the Secretary, scientific and other studies, and assemble data on the producing, handling, shipping, and marketing conditions relative to olives, which are necessary in connection with the performance of its official duties;

(d) To submit to the Secretary such available information with respect to olives as he may request or as the committee may deem desirable and pertinent;

(e) To select, from among its members, a chairman and other officers, and to adopt such rules and regulations for the conduct of its business as it may deem advisable;

(f) To appoint or employ such other persons as it may deem necessary, and to determine the salaries and define the duties of each such person;

(g) To submit to the Secretary, prior to the beginning of each fiscal year and not later than December 15, a budget of the anticipated expenses of the committee and the proposed assessment rate for such fiscal year, together with a report thereon.

(h) To cause the books of the committee to be audited by one or more certified public accountants at least once each fiscal year, and at such other times as the committee may deem necessary or as the Secretary may request. The report of each such audit shall show, among other things, the receipts and expenditures of funds, and at least two copies of each such audit report shall be submitted to the Secretary.

(i) To prepare monthly statements of its financial operations and make such statements, together with the minutes of its meetings, available at the office of the committee for inspection by any producer or handler, and to submit copies of such statements and minutes to the Secretary;

(j) To give reasonable advance notice of each meeting by mail addressed to each member, and such notice shall be given as widespread publicity as practicable. The same notice of meetings given to members shall be given to the Secretary;

(k) With the approval of the Secretary, to redefine the districts into which the area has been divided in §932.21 and to reapportion the membership in accordance therewith: *Provided,* That any such changes reflect insofar as practicable shifts in olive acreage within the districts and area, the numbers of growers in the districts, the tonnage produced, and are equitable as to producers; and

(l) To investigate compliance with the provisions of this part.

[30 FR 12629, Oct. 2, 1965, as amended at 33 FR 11266, Aug. 8, 1968; 47 FR 32907, July 30, 1982]

§932.36 Procedure.

Decisions of the committee shall be by majority vote of the members present and voting, and a quorum must be present: *Provided,* That decisions requiring a recommendation to the Secretary on matters pertaining to grade and size regulations shall require at least 10 affirmative votes, at least 5 of which must be from producer members and at least 5 of which must be from handler members and, if the committee is increased by the addition of a public member, at least 11 affirmative votes shall be required, at least 5 of which must be from producer members and at least 5 of which must be from handler members. A quorum shall consist of at least 10 members of whom at least 5 shall be producer members and at least 5 shall be handler members and, if the committee is increased by the addition of a public member, a quorum shall consist of at least 11 members of which at least 5 shall be producer members

and at least 5 shall be handler members. Except in case of an emergency, a minimum of 5 days advance notice shall be given with respect to any meeting of the committee. In case of an emergency, to be determined within the discretion of the chairman of the committee, as much advance notice of a meeting as is practicable in the circumstances shall be given. The committee may vote by mail or telegram upon due notice to all members, but any proposition to be so voted upon first shall be explained accurately, fully, and identically by mail or telegram to all members. When voted on by such method, at least 14 affirmative votes, of which seven shall be producer member votes and seven shall be handler member votes, shall be required for adoption and, if the committee is increased by the addition of a public member, votes by mail or telegram shall require at least 15 affirmative votes, of which at least 7 shall be producer member votes and at least 7 shall be handler member votes. The committee may recommend for the Secretary's approval changes in the number of affirmative votes required for adoption of any proposition voted upon by means of a mail or telegram ballot: *Provided,* That the number of affirmative votes required for adoption shall not be less than ten, and in any case an equal number of producer member and handler member votes shall be required for adoption and, if the committee is increased by the addition of a public member, the number of affirmative votes required for adoption shall be increased by one.

[47 FR 32908, July 30, 1982]

§932.37 Compensation and expenses.

The members of the committee and alternates when acting as members or at the request of the committee or its chairman shall serve without compensation, but shall be reimbursed for necessary expenses, as approved by the committee, incurred by them in the performance of their duties under this part.

[47 FR 32908, July 30, 1982]

EXPENSES AND ASSESSMENTS

§ 932.38 Expenses.

The committee is authorized to incur such expenses as the Secretary finds are reasonable and likely to be incurred by the committee for its maintenance and functioning and to enable it to exercise its powers and perform its duties in accordance with the provisions of this part. The funds to cover such expenses shall be acquired in the manner prescribed in § 932.39.

§ 932.39 Assessments.

(a) As each handler's pro rata share of the expenses which the Secretary finds are reasonable and likely to be incurred by the committee during a fiscal year, each handler who first handles olives during the current crop year shall pay to the committee, upon demand, assessments less any amounts which may be credited pursuant to § 932.45, on all olives to be used in the production of packaged olives, including olives to be used in canned ripe olives of the "tree-ripened" type or green olives when such are regulated as packaged olives pursuant to § 932.52. The payment of assessments for maintenance and functioning of the committee may be required under this part throughout the period it is in effect irrespective of whether particular provisions thereof are suspended or become inoperative.

(b) The Secretary shall fix the rate of assessment to be paid by each such handler during a fiscal year in an amount designed to secure sufficient funds to cover the expenses which may be incurred during such period. At any time during or after the fiscal year, the Secretary may increase the rate of assessment in order to secure sufficient funds to cover any later finding by the Secretary relative to the expenses which may be incurred. Such increase shall be applied to all olives handled during the applicable crop year. In order to provide funds for the administration of the provisions of this part during the first part of a fiscal year before sufficient operation income is available from assessments, the committee may accept the payment of assessments in advance, and may also borrow money for such purpose.

(c) Any assessment not paid by a handler within a period of time prescribed by the committee may be subject to an interest or late payment charge, or both. The period of time, rate of interest and late payment charge shall be as recommended by the committee and approved by the Secretary. Subsequent to such approval, all assessments not paid within the presecribed period of time shall be subject to an interest or late payment charge or both.

[47 FR 32908, July 30, 1982, as amended at 47 FR 51093, Nov. 12, 1982]

§ 932.40 Accounting.

(a) If, at the end of a fiscal year, the assessments collected are in excess of expenses incurred, such excess shall be accounted for in accordance with one of the following:

(1) If such excess is not retained in a reserve as provided in paragraph (a)(2) of this section, the committee shall refund or credit to handler accounts the aforesaid excess. Each handler's share of such excess funds shall be the amount of assessments such handler has paid in excess of such handler's pro rata share of the actual net expenses of the committee for such fiscal year. Excess funds may be used temporarily by the committee to defray expenses of the subsequent fiscal year: *Provided,* That each handler's share of such excess shall be made available to the handler by the committee within five months after the end of the fiscal year.

(2) The committee, with the approval of the Secretary, may carry over such excess into subsequent fiscal years as a reserve: *Provided,* That funds already in the reserve do not exceed approximately one fiscal year's expenses. Such reserve funds may be used for any expenses authorized pursuant to § 932.38 and for necessary expenses of liquidation in the event of termination of this part. Upon such termination, any funds not required to defray the necessary expenses of liquidation shall be disposed of in such manner as the Secretary may determine to be appropriate: *Provided,* That to the extent practicable, such funds shall be returned pro rata to the persons from whom such funds were collected.

(b) All funds received by the committee pursuant to the provisions of this part shall be used solely for the purpose specified in this part and shall be accounted for in the manner provided in this part. The Secretary may at any time require the committee and its members to account for all receipts and disbursements.

(c) Upon the removal or expiration of the term of office of any member of the committee, such member shall account for all receipts and disbursements and deliver all property and funds in his possession to the committee, and shall execute such assignments and other instruments as may be necessary or appropriate to vest in the committee full title to all of the property, funds, and claims vested in such member pursuant to this part.

[30 FR 12629, Oct. 2, 1965, as amended at 47 FR 32908, July 30, 1982]

RESEARCH AND DEVELOPMENT

§932.45 Production research and marketing research and development projects.

(a) The following activities of the committee are authorized under this section.

(1) The committee may, with the approval of the Secretary, establish or provide for the establishment of production research, and marketing research and development projects designed to assist, improve or promote the marketing, distribution, and consumption or efficient production of California olives. Such projects may provide for any marketing research and development projects designed to assist, improve, or promote the marketing, distribution, and consumption or efficient production of California olives. Such projects may provide for any form of marketing promotion including paid advertising. The expenses of such research and projects shall be paid from funds collected pursuant to §932.39 or from voluntary contributions. Voluntary contributions may be accepted by the committee only to pay the expenses of such projects: *Provided,* That the committee shall retain complete control over the use of such contributions which shall be free from any encumbrances.

(2) The committee, with the approval of the Secretary, may provide for crediting a portion of a handler's direct expenditures for paid brand advertising for olives. Such expenditures may include, but are not limited to, money spent for advertising space in magazines, newspapers, outdoor media and transit or time charges for radio and television. No handler shall receive credit in excess of such handler's pro rata share of the total monies allotted by the committee for brand advertising credit. Each advertisement must be published, broadcast or displayed during the fiscal year for which credit is requested. Before any creditable brand advertising may be undertaken pursuant to this paragraph (a)(2) of this section, the Secretary, upon recommendation by the committee, shall prescribe appropriate rules and regulations as are necessary to effectively regulate such activity.

(b) In recommending marketing research and development projects pursuant to this section, the committee shall give consideration to the following factors:

(1) The expected supply of olives in relation to market requirements;

(2) The supply situation among competing areas and commodities; and

(3) The need for marketing research with respect to any marketing development activity and the need for a coordinated effort with USDA's Plentiful Food Program.

(c) In recommending production research projects pursuant to this section, the committee shall give consideration to the extent and need for assistance to, and improvement of California olive production.

(d) If the committee should conclude that a program of production research, marketing research, or development should be undertaken or continued pursuant to this section in any fiscal year, it shall submit the following for the approval of the Secretary:

(1) Its recommendations as to funds to be obtained pursuant to §932.39 or voluntary contributions;

(2) Its recommendations as to any production research or marketing research project; and

(3) Its recommendation as to promotion activity and paid advertising.

257

(e) The committee shall, as soon as practicable, prepare and mail reports on current production research and marketing research and development projects to the Secretary and make a copy of such reports available at the committee office for examination by producers, handlers, or other interested parties.

[36 FR 20356, Oct. 21, 1971, as amended at 47 FR 32908, July 30, 1982; 47 FR 51093, Nov. 12, 1982]

REGULATIONS

§ 932.50   Report of marketing policy.

At least 14 days prior to the start of each crop year (except that this period may be shortened by the committee not more than 5 days if warranted), the committee shall hold a meeting for the purpose of formulating a marketing policy for the coming crop year: *Provided*, That with respect to the 1982–83 crop year the committee shall hold a meeting for such purpose as soon as practicable. The committee shall prepare and submit to the Secretary promptly after each such meeting, a report setting forth its recommended marketing policy for the ensuing crop year. In the event it becomes advisable to modify such policy, because of changed supply, demand, or other conditions, the committee shall formulate a new policy and shall submit a report thereon to the Secretary. In developing the marketing policy, the committee shall give consideration to the handler carryover, production, probable quality and composition of olive sizes in the crop, trade demand, probable imports, whether producer prices are likely to exceed parity, the probable assessable tonnage and such other factors as may have a bearing on the marketing of olives or the administration of this part. Notice of the committee's marketing policy, and of any modifications thereof, shall be given promptly by reasonable publicity to producers and handlers.

[30 FR 12629, Oct. 2, 1965, as amended at 47 FR 32908, July 30, 1982]

§ 932.51   Incoming regulations.

(a) *Minimum standards for natural condition olives.* (1) Except as otherwise provided in this section, no handler shall process any lot of natural condition olives for use in the production of packaged olives which has not first been:

(i) Weighed on scales sealed by the State of California Department of Weights and Measures, an official certified weight certificate issued thereon, and a copy of such certificate furnished to the Federal or Federal-State Inspection Service and the committee; and

(ii) Size-graded, either by sample or by lot, under the supervision of any such inspection service and classified into separate size designations and a certification issued with respect thereto by such inspection service. Such size designations shall be in accordance with those set forth in the U.S. Standards for Grades of Canned Ripe Olives (7 CFR part 52) or subsequent amendments thereto, or such sizes as may be recommended by the committee and established by the Secretary: *Provided*, That, for the purpose of this part, the size designations in said standards shall be deemed to include the following additional size designations.

Designation(s)	Approximate count (per pound)	Average count range (per pound)
Subpetite	....	181 and up.
Petite	160	141–180, inclusive.
Extra Large Sevillano "L"	82	76–88, inclusive.
Extra Large Sevillano "C"	70	65–75, inclusive.

Provided further, That the additional size designations may be renamed and/or modified as recommended by the committee and approved by the Secretary. Such certification shall show, in addition to the quantities by weight of the olives in the lot that are classified as being in each size or size designation the quantity of olives classified as culls by the handler: *Provided*, That when the Secretary, upon the recommendation of the committee, issues a definition of and classification for "culls", the aforesaid quantity of culls shall be determined on the basis of such definition and in accordance with such classification.

(2) Each handler may satisfy the incoming and outgoing size requirements for any lot of olives under the conditions set forth in subdivisions (i), (ii),

and (iii) of this paragraph: *Provided*, That any such lot shall be kept intact under surveillance by the inspection services:

(i) When the Secretary authorizes use of limited size olives for limited use styles during any crop year, any lot of limited use size olives may be used in the production of packaged olives for limited use styles without an outgoing inspection if such olives are within the following average count range for that variety group, and meet such further size requirements as recommended by the committee with the approval of the Secretary:

Variety	Average count range (per pound)
Group 1, except Ascolano, Barouni, and St. Agostino.	76–88, inclusive.
Group 1, Ascolano, Barouni and St. Agostino.	89–140, inclusive.
Group 2, except Obliza	141–180, inclusive.
Group 2, Obliza	128–140, inclusive.

Provided, That the varietal groupings and/or average count ranges may be changed, and additional size certification procedures and requirements may be established as recommended by the committee and approved by the Secretary;

(ii) When limited use size olives are not authorized for limited use styles during any crop year, any lot of the minimum canning size olives may be used in the production of packaged olives for limited use styles without an outgoing inspection for size if such olives are within the following average count range for that variety group, and meet such further size requirements as recommended by the committee with approval of the Secretary:

Variety	Average count range (per pound)
Group 1, except Ascolano, Barouni, and St. Agostino.	65–75, inclusive.
Group 1, Ascolano, Barouni and St. Agostino.	65–88, inclusive.
Group 2, except Obliza	128–140, inclusive.
Group 2, Obliza	106–121, inclusive.

Provided, That for whole and whole pitted styles of olives an additional size grading is required after processing, prior to canning, and those olives that fail to meet the requirements in §932.52 may be used in limited use styles. *Provided further*, That the varietal

groupings, average count ranges, and/or other size requirements may be changed or modified as recommended by the committee and approved by the Secretary;

(iii) The committee may recommend, subject to approval by the Secretary, size certification procedures for olives used in the production of canned whole or pitted styles of olives: *Provided*, That if size certification for canned whole or pitted styles is implemented, marketing order sizes shall be adopted and size requirements in the U.S. Grade Standards shall not apply. Size certification of such styles shall be applicable to any or all sizes of olives recommended by the committee and approved by the Secretary pursuant to §932.52(a)(2). Size certification procedures recommended to the Secretary may include but are not limited to the establishment of average count ranges, acceptable count ranges, and approximate counts (midpoints) for each variety or variety group.

(3) Each handler shall, under the supervision of any such inspection service, dispose of into noncanning use an aggregate quantity of olives, comparable in size and characteristics and equal to the quantities shown on the certification for each lot to be:

(i) Variety Group 1 olives, except the Ascolano, Barouni, and St. Agostino varieties, of a size which individually weigh less than $1/90$ pound;

(ii) Variety Group 1 olives of the Ascolano, Barouni, and St. Agostino varieties of a size which individually weigh less than $1/140$ pound;

(iii) Variety Group 2 olives, except the Obliza variety, of a size which individually weigh less than $1/180$ pound;

(iv) Variety Group 2 olives of the Obliza variety of a size which individually weigh less than $1/140$ pound;

(v) Such other sizes for the foregoing variety groups as are not authorized for limited use pursuant to §932.52; and

(vi) Olives classified as culls.

(4) Notwithstanding the provisions of paragraph (a)(3) of this section, a handler may (i) meet any deficit in such handler's undersize obligation in one variety by disposing of, under supervision of the inspection service, as other than canned ripe olives, an equal quantity of undersize olives, of any

other variety, or by so disposing of an equal quantity of olives of that or any other variety of sizes larger than undersize of a quality better than culls, and (ii) meet any deficit in such handler's cull obligation in one variety by so disposing of an equal quantity of cull olives of any other variety, or by so disposing of an equal quantity of olives of any variety of sizes larger than undersize of a quality better than culls.

(5) Each handler shall hold at all times a quantity of olives equal to the quantities required in paragraph (a)(3) of this section, less any quantity previously disposed of as specified in such subparagraph.

(b) Whenever a handler receives a lot of natural condition olives, or makes a separation resulting in a sublot, solely for use in the production of green olives or canned ripe olives of the "tree-ripened" type, he may handle such lot or sublot without regard to the provisions of this section and § 932.52 only if (1) he notifies the committee upon receiving such a lot or making such a separation; (2) the identity of all such lots and sublots of olives is maintained by keeping them separate and apart from other olives he receives; (3) the packaged olives produced from such lots and sublots after processing are canned ripe olives of the "tree-ripened" type or green olives; and (4) there are no outgoing regulations pursuant to § 932.52 then applicable to packaged olives that are canned ripe olives of the "tree-ripened" type or green olives.

[30 FR 12629, Oct. 2, 1965, as amended at 33 FR 11267, Aug. 8, 1968; 36 FR 20356, Oct. 21, 1971; 47 FR 32909, July 30, 1982]

Effective Date Notes: 1. At 56 FR 49669, Oct. 1, 1991, in § 932.51, paragraphs (a)(3) (i), (ii), (iii), (iv) and the words "for the foregoing variety groups" in paragraph (a)(3)(v) were suspended indefinitely.

2. At 81 FR 46569, July 18, 2016, in § 932.51 paragraphs (a)(1)(ii) through (a)(5) were suspended indefinitely.

§ 932.52　Outgoing regulations.

(a) *Minimum standards for packaged olives.* No handler shall use processed olives in the production of packaged olives or ship such packaged olives unless they have first been inspected as required pursuant to § 932.53 and meet

each of the following applicable requirements:

(1) Canned ripe olives, other than those of the "tree-ripened" type, shall grade at least U.S. Grade C as such grade is defined in the U.S. Standards for Grades of Canned Ripe Olives (7 CFR part 52) or subsequent amendments thereto, or as modified by the committee, with approval of the Secretary, for purposes of this part.

(2) Except as provided in § 932.51(a) (1) and (2), canned whole ripe olives, other than those of the "tree-ripened" type, shall conform to the single size designations set forth in the U.S. Standards for Grades of Canned Ripe Olives (7 CFR part 52) or subsequent amendments thereto, or as modified by the committee, with the approval of the Secretary, and shall be of a size not smaller than the following applicable size requirements, tolerances and percentages: *Provided,* That the Secretary, on the basis of a recommendation of the committee or other available information, may change such sizes, tolerances or percentages:

(i) With respect to variety group 1 olives, except the Ascolano, Barouni, and St. Agostino varieties, the individual fruits shall each weigh no less than $\frac{1}{75}$ pound, except that (A) for olives of the extra large size designation, not more than 25 percent, by count, of such olives may weigh less than $\frac{1}{75}$ pound each including not more than 10 percent, by count, of such olives that weigh less than $\frac{1}{82}$ pound each; and (B) for olives of any designation except the extra large size, not more than 5 percent, by count, of such olives may weigh less than $\frac{1}{75}$ pound each;

(ii) With respect to variety group 1 olives of the Ascolano, Barouni and St. Agostino varieties, the individual fruits shall each weigh not less than $\frac{1}{88}$ pound except that (A) for olives of the extra large size designation, not more than 25 percent, by count, of such olives may weigh less than $\frac{1}{88}$ pound each including not more than 10 percent, by count, of such olives that weigh less than $\frac{1}{88}$ pound each, and (B) for olives of any size designation, except the extra large size, not more than 5 percent, by count, of such olives may weigh less than $\frac{1}{88}$ pound each;

(iii) With respect to variety group 2 olives, except the Obliza variety, the individual fruits shall each weigh not less than $\frac{1}{140}$ pound except that (A) for olives of the small size designation, not more than 35 percent, by count, of such olives may weigh less than $\frac{1}{140}$ pound each including not more than 7 percent, by count, of such olives that weigh less than $\frac{1}{160}$ pound each; and (B) for olives of any size designation, except the small size, not more than 5 percent, by count, of such olives may weigh less than $\frac{1}{140}$ pound each; and

(iv) With respect to Variety Group 2 olives of the Obliza variety, the individual fruits shall each weigh not less than $\frac{1}{121}$ pound except that (a) for olives of the medium size designation, not more than 35 percent, by count, of such olives may weigh less than $\frac{1}{121}$ pound each including not more than 7 percent, by count, of such olives that weigh less than $\frac{1}{135}$ pound each; and (b) for olives of any size designation, except the medium size, not more than 5 percent, by count, of such olives may weigh less than $\frac{1}{121}$ pound each.

(3) Subject to the provisions set forth in paragraph (a)(4) of this section and §932.51(a) (1) and (2), processed olives to be used in the production of canned pitted ripe olives, other than those of the "tree-ripened" type, shall meet the same requirements as prescribed pursuant to paragraph (a)(2) of this section: *Provided,* That olives smaller than those so prescribed, as recommended annually by the committee and approved by the Secretary, may be authorized for limited use but any such limited use size olives so used shall be not smaller than the following applicable minimum size: *Provided further,* That each such minimum size may also include a size tolerance (specified as a percent) as recommended by the committee and approved by the Secretary.

(i) Variety Group 1 olives, except the Ascolano, Barouni, and St. Agostino varieties, of a size which individually weigh $\frac{1}{90}$ pound;

(ii) Variety Group 1 olives of the Ascolano, Barouni, or St. Agostino varieties, of a size which individually weigh $\frac{1}{140}$ pound;

(iii) Variety Group 2 olives, except the Obliza variety, of a size which individually weigh $\frac{1}{180}$ pound;

(iv) Variety Group 2 olives of the Obliza variety, of a size which individually weigh $\frac{1}{140}$ pound.

(4) The Secretary may, upon recommendation of the committee, restrict the total quantity of limited use size olives for limited use during any crop year. Such restricted quantity shall be apportioned among the handlers by applying a percentage, established annually by the Secretary upon recommendation by the committee, to each handler's total receipts of limited use size olives during such crop year.

(5) Canned ripe olives of the "tree-ripened" type and green olives shall meet such grade, size, and pack requirements as may be established by the Secretary based upon the recommendation of the committee or other available information.

(6) The size designations used in this section mean the size designations described in (a)(1)(ii) of §932.51.

(7) For the purposes of this part the committee may, with the approval of the Secretary, specify the styles of olives, including the requirements with respect thereto, for limited use.

(b) *Disposition requirements for limited use size olives.* (1) The requirements of this paragraph are in addition to and not in substitution of the requirements of §932.51(a)(5).

(2) Each handler shall, under the supervision of the Processed Products Branch, USDA, or the Federal or Federal-State Inspection Service, dispose of limited use size olives into limited use or into noncanning use: *Provided,* That whenever a handler's use of limited use size olives is restricted pursuant to §932.52(a)(4), such handler shall dispose of into noncanning use that quantity of such limited use size olives which is in excess of the quantity permitted for limited use.

(3) Notwithstanding the provisions of paragraph (b)(2) of this section, a handler may meet any deficit in his obligation to dispose of limited use size olives into noncanning use pursuant to this paragraph by disposing of, under supervision of the inspection service, an equivalent quantity of olives of a size larger than the limited use size and of a quality better than culls.

(4) Each handler shall hold at all times a quantity of olives eligible to

meet the disposition requirements of this paragraph less any quantity previously disposed of as specified in paragraphs (b) (2) and (3) of this section.

[36 FR 20357, Oct. 21, 1971, as amended at 47 FR 32910, July 30, 1982]

EFFECTIVE DATE NOTE: At 56 FR 49669, Oct. 1, 1991, in § 932.52, in paragraph (a)(3) introductory text and paragraphs (a)(3)(i) through (a)(3)(iv) the words "but any such limited use size olives so used shall be not smaller than the following applicable minimum size: *Provided further,* That each such minimum size may also include a size tolerance (specified as a percent) as recommended by the committee and approved by the Secretary" were suspended indefinitely.

§ 932.53　Inspection and certification.

(a) Each handler shall have the olives such handler handles inspected and certified as for conformance with all applicable requirements pursuant to §§ 932.51 and 932.52 with respect to such handling. Inspection and certification for conformance with the requirements of § 932.51 shall be by the Federal or Federal-State Inspection Service, including certification as to size, and inspection for conformance with the requirements of § 932.52 shall be by the Processed Products Branch, USDA, except that the disposition of olives, other than as canned ripe olives, in accordance with the requirements of § 932.51(a)(3) may be under the supervision of any of such inspection services. A copy of each certification by the said inspection services, pursuant to the provisions of this section, shall be furnished to the committee.

(b) The committee may enter into an agreement with either or both of said inspection services with respect to the costs of the inspection required by this section and may collect from handlers their respective pro rata share of such costs.

[30 FR 12629, Oct. 2, 1965, as amended at 47 FR 32910, July 30, 1982]

§ 932.54　Transfers.

Transfers within the area of olives from one handler to another for further handling within the area are permitted. Whenever such a transfer of olives is made, the transferring handler shall comply with all applicable regulations up to the time of such transfer, and the receiving handler shall comply with all applicable regulations subsequent to such transfer: *Provided,* That the disposition obligations referable to transferred natural condition olives pursuant to § 932.51(a)(3) may be transferred along with the olives, in which event the receiving handler shall comply with the disposition obligations. Transfers of olives from within the area to any point outside the area shall be subject to such requirements with respect to inspection, holding, disposition, and reporting as may be established by the Secretary on the basis of recommendations by the committee or other available information.

[33 FR 11267, Aug. 8, 1968, as amended at 36 FR 20357, Oct. 21, 1971; 47 FR 32910, July 30, 1982]

§ 932.55　Exemption.

(a) The provisions of this subpart shall not be applicable to processed olives on hand on the effective date of this subpart but only if the identity of such olives is maintained and such olives are not commingled with olives processed after such effective date in the production of packaged olives. However, olives on hand on such effective date that are commingled with olives processed after such date and are used in the production of packaged olives shall be subject to all relevant provisions applicable to the handling of packaged olives.

(b) Upon the basis of the recommendation submitted by the committee or from other available information, the Secretary may relieve from any or all requirements under this part the handling of olives in such minimum quantities, in such types of shipments, or for such specified purposes (including shipments to facilitate the conduct of marketing research and development projects established pursuant to § 932.45) as the committee with the approval of the Secretary may prescribe.

(c) The committee, with the approval of the Secretary, shall prescribe rules, regulations, and safeguards as it may deem necessary to ensure that olives exempted under the provisions of this section are handled only as authorized.

[30 FR 12629, Oct. 2, 1965, as amended at 33 FR 11267, Aug. 8, 1968]

REPORTS AND RECORDS

§932.60 Reports of acquisitions, sales, uses, shipments and creditable brand advertising.

(a) Each handler shall file such reports of his acquisitions, sales, uses, and shipments of olives, as may be requested by the committee.

(b) Upon the request of the committee, each handler shall furnish such other reports and information as are needed to enable the committee to perform its functions under this part.

(c) Each handler shall file such reports of creditable brand advertising as recommended by the committee and approved by the Secretary.

[30 FR 12629, Oct. 2, 1965, as amended at 47 FR 51094, Nov. 12, 1982]

§932.61 Records.

Each handler shall maintain such records of olives acquired, held, and disposed of by such handler as may be prescribed by the committee and needed by it to perform its functions under this subpart. Such records shall be retained for at least two years beyond the crop year in which the transaction occurred. The committee, with the approval of the Secretary, may prescribe rules and regulations to include under this section handler records that detail advertising and promotion activities which the committee may need to perform its functions under §932.45(a).

[47 FR 51094, Nov. 12, 1982]

§932.62 Verification of reports.

For the purpose of checking and verifying reports filed by handlers, the committee, through its duly authorized representatives, shall have access to any handler's premises during regular business hours, and shall be permitted at any such time to: (a) Inspect such premises and any olives held by such handler, and any and all records of the handler with respect to such handler's acquisition, sales, uses and shipments of olives; and (b) inspect any and all records of such handler with respect to advertising and promotion activities subject to §932.45(a) and maintained by the handler pursuant to §932.61. Each handler shall furnish all labor and equipment necessary to make such inspections.

[47 FR 51094, Nov. 12, 1982]

§932.63 Confidential information.

All reports and information submitted by handlers pursuant to the provisions of this part shall be received by, and at all times be in the custody of one or more designated employees of the committee. No such employees shall disclose to any person, other than the Secretary upon request therefor, data, or information obtained or extracted from such reports and records which might affect the trade position, financial condition, or business operation of the particular handler from whom received: *Provided,* That such data and information may be combined, and made available in the form of general reports in which the identities of the individual handlers furnishing the information is not disclosed.

MISCELLANEOUS PROVISIONS

§932.65 Compliance.

Except as provided in this part, no person shall handle olives, the handling of which has been prohibited by the Secretary in accordance with the provisions of this part, and no person shall handle olives except in conformity with the provisions of this part and the regulations issued hereunder.

§932.66 Right of the Secretary.

The members of the committee (including successors and alternates) and any agents or employees appointed or employed by the committee, shall be subject to removal or suspension at any time by the Secretary. Each and every order, regulation, determination, decision, or other act of the committee shall be subject to the continuing right of the Secretary to disapprove of the same at any time. Upon such disapproval, such disapproved action shall be deemed null and void except as to acts done in reliance thereon or in compliance therewith prior to such disapproval by the Secretary.

§ 932.67 Effective time.

The provisions of this subpart, as well as any amendments to this subpart, shall become effective at such time as the Secretary may declare, above his signature, and shall continue in force until terminated in one of the ways specified in § 932.68.

§ 932.68 Termination.

(a) The Secretary may, at any time, terminate the provisions of this subpart by giving at least one day's notice by means of a press release or in any other manner which he may determine.

(b) The Secretary shall terminate or suspend the operation of any or all of the provisions of this subpart whenever he finds such provisions do not tend to effectuate the declared policy of the act.

(c) The Secretary shall terminate the provisions of this subpart at the end of any crop year whenever the Secretary finds that such termination is favored by a majority of producers who, during a representative period determined by the Secretary, have been engaged in the area in the production of olives for market as packaged olives: *Provided,* That such majority have during such representative period produced for market more than 50 percent of the volume of such olives produced for market, but such termination shall be effective only if announced on or before July 15 of the then current crop year.

[30 FR 12629, Oct. 2, 1965, as amended at 47 FR 32910, July 30, 1982]

§ 932.69 Proceedings after termination.

(a) Upon the termination of the provisions of this subpart, the members of the committee then functioning shall continue as joint trustees, for the purpose of liquidating the affairs of the committee, of all funds and property then in the possession or under the control of the committee including claims for any funds unpaid or property not delivered at the time of such termination. Action by such trustee shall require the concurrence of a majority of the trustees.

(b) Said trustees shall continue in such capacity until discharged by the Secretary; shall, from time to time, account for all receipts and disburse- ments, and deliver all property on hand, together with all books and records of the committee and the joint trustees, to such person as the Secretary may direct; and shall, upon the request of the Secretary, execute such assignments or other instruments necessary or appropriate to vest in such person full title and right to all of the funds, property, and claims vested in the committee or the joint trustees.

(c) Any person to whom funds, property, or claims have been transferred or delivered by the committee or the joint trustees, pursuant to this section, shall be subject to the same obligations imposed upon the members of the said committee and upon said joint trustees.

§ 932.70 Effect of termination or amendment.

Unless otherwise expressly provided by the Secretary, the termination of this subpart or any regulation issued pursuant thereto, or the issuance of any amendment to either thereof, shall not (a) affect or waive any right, duty, obligation, or liability which shall have arisen, or which may thereafter arise, in connection with any provision of this subpart, or any regulation issued thereunder; (b) release or extinguish any violation of this subpart or of any regulation issued thereunder; or (c) affect or impair any rights or remedies of the Secretary or any other person with respect to any such violation.

§ 932.71 Duration of immunities.

The benefits, privileges, and immunities conferred upon any person by virtue of this subpart shall cease upon the termination of this subpart, except with respect to acts done under and during the existence of this subpart.

§ 932.72 Agents.

The Secretary may, by a designation in writing, name any person, including any officer or employee of the U.S. Government or name any service or division in the U.S. Department of Agriculture, to act as his agent or representative in connection with any of the provisions of this subpart.

§932.73 Derogation.

Nothing contained in this subpart is or shall be construed to be, in derogation or in modification of the rights of the Secretary or of the United States to exercise any powers granted by the act or otherwise, or, in accordance with such powers, to act in the premises whenever such action is deemed advisable.

§932.74 Personal liability.

No member or alternate member of the committee or any employee or agent thereof shall be held personally responsible, either individually or jointly with others, in any way whatsoever, to any person, for errors in judgment, mistakes, or other acts either of commission or omission, as such member, alternate member, employee, or agent, except for acts of dis- honesty.

§932.75 Separability.

If any provision of this subpart is declared invalid or the applicability thereof to any person, circumstance, or thing is held invalid, the validity of the remainder of this subpart or the applicability thereof to any other person, circumstance, or thing shall not be affected thereby.

Subpart B—Administrative Requirements

§932.108 Noncanning olives.

Noncanning olives means those olives which, pursuant to the requirements of §932.51(a)(2), are to be disposed of as other than canned ripe olives.

[31 FR 12634, Sept. 27, 1966]

§932.109 Canned ripe olives of the tree-ripened type.

(a) *Canned ripe olives of the tree-ripened type* means packaged olives, not oxidized in processing, that are prepared from a lot or sublot of natural condition olives of advanced maturity which:

(1) Range in color from pinkish red, with some greenish cast, to black; and

(2) Have not more than 10 percent, by count, of *off-color* olives (*off-color* means those olives whose greenish cast covers more than 50 percent of the surface of the individual olives).

(b) [Reserved]

[40 FR 38146, Aug. 27, 1975]

§932.121 Producer districts.

Pursuant to the authority in §932.35(k), commencing with the term of office beginning June 1, 2005, district means any of the following geographical areas of the State of California:

(a) District 1 shall include the counties of Alpine, Tuolumne, Stanislaus, Santa Clara, Santa Cruz, and all counties north thereof.

(b) District 2 shall include the counties of Mono, Mariposa, Merced, San Benito, Monterey and all counties south thereof.

[70 FR 6326, Feb. 7, 2005]

§932.125 Producer representation on the committee.

Pursuant to the authority in §§932.25 and 932.35(k), commencing with the term of office beginning June 1, 2005, representation shall be apportioned as follows:

(a) District 1 shall be represented by three producer members and alternates.

(b) District 2 shall be represented by five producer members and alternates.

[70 FR 6326, Feb. 7, 2005]

§932.129 Nomination procedures for producer members.

Members and alternate members on the Committee who represent producers shall be nominated in accordance with the procedures specified in either paragraph (a) or paragraph (b) of this section as the Committee may determine.

(a) *Mail ballot voting.* (1) The Committee shall schedule a meeting, prior to March 1 of each odd-numbered year, in each producing district for the purpose of selecting candidates for member and alternate member nominations. A notice of such meetings will be mailed to each producer of record in each district. The nomination process is as follows:

(i) Any person who produces olives in a particular district may offer the

265

name of any producer from that district as a candidate for either a member or alternate member position in said district.

(ii) A producer, who produces olives in more than one district, can be selected as a candidate for a member or alternate member position in only one district.

(iii) The Committee will notify by mail producers who are selected as candidates but are not in attendance at such meetings. Such producers have the right to decline such listing on the ballot within 7 days of mailing such notice.

(iv) In the event a producer cannot attend a meeting but wishes to be included on the ballot, that producer may notify the Committee office in writing no later than 7 days after the date of the nomination meeting for the producer's district and request that the producer's name be included on the ballot.

(v) In the event that no candidates or an insufficient number of candidates are selected at such meetings for the producer members and alternates in the respective districts, the Committee will give written notice to producers in said district that additional names may be submitted for the specified position(s).

(2) Following such meetings, and no later than March 15 of each odd-numbered year, the Committee shall prepare and mail a ballot to each producer that delivered olives during that crop year in each district.

(i) A producer who produces olives in more than one district must choose the district in which the producer will vote and notify the Committee of that choice. If the Committee is not notified and more than one ballot is received from such a producer, the first ballot received will be counted. Candidates may only vote in the district in which they are seeking nomination.

(ii) Each ballot will list separately the names of candidates for the member positions and the names of candidates for the alternate member positions for said district.

(iii) A ballot will be mailed to producers of record to give them an opportunity to vote. Committee records will be used to determine the list of producers eligible to cast ballots. However, any producer who is not identified in such records may receive a ballot if the Committee determines that such producer is eligible to participate in nominations in that district.

(iv) A producer may cast a vote for as many candidates as there are member or alternate positions in said district.

(v) The candidate on each list, as prescribed in paragraph (a)(2)(ii) of this section, who receives the most votes will be the nominee for the first position, and until all positions for that district are filled, the candidates receiving the second, third and fourth highest number of votes will be the nominees for the second, third and fourth position respectively.

(vi) In the event of a tie which would result in elimination of a tied candidate, a second ballot with the names of those tied candidates will be mailed to producers in said district for another vote.

(b) *Nomination meetings.* In lieu of the mail ballot nomination procedure specified in paragraph (a) of this section, the Committee may schedule nomination meetings. In such an event, the following procedure will apply:

(1) Prior to March 15 of each odd-numbered year, the Committee shall schedule a nomination meeting to be held in each district for the purpose of obtaining nominees for producer members and alternate members for such district.

(2) Nominations for members and balloting thereon shall precede nominations and balloting for alternate members.

(3) The candidate for each position who receives the highest number of votes shall be the nominee for the position: *Provided,* That such candidate receives a majority of the ballots cast. If no candidate receives such a majority, the two candidates who received the highest number of votes shall participate in a run-off balloting to determine which is the nominee.

(c) For the purposes of this section, a producer is a person engaged in a proprietary capacity as a single business unit in the production of olives for market as packaged olives and includes

266

an individual (owner-operated), partnership, corporation, association, institution, or other legal business unit.

(d) *Determination of producer eligibility.* (1) Only producers (including duly authorized officers or employees of producers) who produced olives within the district shall participate in the nomination and election of producer members and alternates.

(2) Each producer (as defined in paragraph (c) of this section) shall be entitled to cast only one vote for each position.

(3) A producer having olive acreage in more than one district may participate in nominations and elections in only one district. The district in which the producer wishes to participate shall be the producer's choice.

(4) Any member of a producer's family (husband, wife, son or daughter) may vote on behalf of an owner-operated, landlord-tenant, family enterprise, or other farming unit.

(5) Any authorized officer or employee of a corporation which is a producer may vote.

(6) Any authorized member of a partnership which is a producer may vote.

(7) Power of attorney (proxies) for voting purposes are not accepted.

[48 FR 24312, June 1, 1983, as amended at 54 FR 46222, Nov. 2, 1989]

§932.130 Public member and alternate public member eligibility requirements and nomination procedures.

(a) *Eligibility requirements.* (1) The public member and alternate public member shall not be a producer, handler, or family member (husband, wife, son or daughter) of a producer or handler of olives and shall have no direct financial interest in, nor be engaged in, the commercial production, marketing, buying, grading or processing of olives; nor shall they be either an officer, director, or employee, or family member of an officer, director, or employee of any firm engaged in such activities.

(2) The public member and alternate public member should be able to devote sufficient time and must express a willingness to attend subcommittee and committee activities regularly and to familiarize themselves with the background and economics of the olive industry.

(3) The public member and alternate public member must be residents of California.

(b) *Nomination procedures.* (1) Prior to April 16 of the year in which nominations are made, the Committee will recommend to the Secretary a public member and alternate public member for the Committee for a two-year term of office beginning June 1 and ending May 31 of odd numbered years.

(2) The Committee will solicit, interview and recommend to the Secretary its nominees for public member and alternate public member.

(3) A majority vote is required in Committee actions concerning the nomination of the public member and alternate public member.

[48 FR 24313, June 1, 1983]

§932.136 Use of communication technology.

The Committee may conduct meetings by any means of audio and/or audiovisual communication technology available that effectively assembles members and alternates, and facilitates open communication; *Provided,* That, quorum and voting requirements specified in §932.36 for physically assembled meetings shall apply. The Committee may also vote electronically; *Provided,* That, such voting shall be subject to the same requirements specified for mail voting in §932.36.

[84 FR 4308, Feb. 15, 2019]

§932.139 Late payment and interest charges.

(a) The committee shall impose a late payment charge on any handler whose assessment has not been received in the committee's office, or the envelope containing the payment legibly postmarked by the U.S. Postal Service, within 30 days of the invoice date shown on the handler's assessment statement. The late payment charge shall be five percent of the unpaid balance.

(b) In addition to that specified in paragraph (a) of this section, the committee shall impose an interest charge

267

on any handler whose assessment payment has not been received in the committee's office, or the envelope containing the payment legibly postmarked by the U.S. Postal Service, within 30 days of the invoice date. The interest charge shall be the current commerical prime rate of the committee's bank plus two percent which shall be applied to the unpaid balance and late payment charge for the number of days all or any part of the assessment specified in the handler's assessment statement is delinquent beyond the 30 day payment period.

(c) The committee, upon receipt of a late payment equal to or greater than the assessment specified on the handler's assessment statement, shall promptly notify the handler (by registered mail) of any late payment charge and/or interest due as provided in paragraphs (a) and (b) of this section. If such charges are not paid, or the envelope containing payment is not legibly postmarked by the U.S. Postal Service, within 30 days of the date on such notification, late payment and interest charges as provided in paragraphs (a) and (b) of this section will accrue on the unpaid amount.

[49 FR 29210, July 19, 1984]

§ 932.149 Modified minimum quality requirements for specified styles of canned olives of the ripe type.

(a) Except as otherwise provided in this section, the minimum quality requirements prescribed in § 932.52(a)(1) are modified as follows, for specified styles of canned olives of the ripe type:

(1) Canned whole and pitted olives of the ripe type shall meet the minimum quality requirements as prescribed in table 1 of this section;

(2) Canned sliced, segmented (wedged), and halved olives of the ripe type shall meet the minimum quality requirements as prescribed in table 2 of this section;

(3) Canned chopped olives of the ripe type shall meet the minimum quality requirements as prescribed in table 3 of this section; and shall be practically free from identifiable units of pit caps, end slices, and slices ("practically free from identifiable units" means that not more than 10 percent, by weight, of the unit of chopped style olives may be

identifiable pit caps, end slices, or slices); and,

(4) Canned broken pitted olives of the ripe type shall meet the minimum quality requirements as prescribed in table 4 of this section;

(5) A lot of canned ripe olives is considered to meet the requirements of this section if all or most of the sample units meet the requirements specified in tables 1 through 4 of this section: *Provided,* That the number of sample units which do not meet the requirements specified in tables 1 through 4 of this section does not exceed the acceptance number prescribed for in the sample size provided in table I of 7 CFR 52.38: *Provided further,* That there is no off flavor in any sample unit.

TABLE 1—WHOLE AND PITTED STYLE

[Defects by count per 50 olives]

FLAVOR	Reasonably good; no "off" flavor
FLAVOR (Green Ripe Type)	Free from objectionable flavors of any kind
SALOMETER	Acceptable Range in degrees: 3.0 to 14.0
COLOR	Reasonably uniform with not less than 60% having a color equal or darker than the USDA Composite Color Standard for Ripe Type
CHARACTER	Not more than 5 soft units or 2 excessively soft units
UNIFORMITY OF SIZE	60%, by visual inspection, of the most uniform in size. The diameter of the largest does not exceed the smallest by more than 4mm
DEFECTS:	
Pitter Damage (Pitted Style Only).	15
Major Blemishes	5
Major Wrinkles	5
Pits and Pit Fragments (Pitted Style Only).	Not more than 1.3% average by count
Major Stems	Not more than 3
HEVM	Not more than 1 unit per sample
Mutilated	Not more than 3
Mechanical Damage.	Not more than 5
Split Pits or Misshapen.	Not more than 5

TABLE 2—SLICED, SEGMENTED (WEDGED), AND HALVED STYLES

[Defects by count per 255 grams]

FLAVOR	Reasonably good; no "off" flavor
SALOMETER	Acceptable Range in degrees: 3.0 to 14.0
COLOR	Reasonably uniform with no units lighter than the USDA Composite Color Standard for Ripe Type
CHARACTER	Not more than 13 grams excessively soft
DEFECTS:	

TABLE 2—SLICED, SEGMENTED (WEDGED), AND HALVED STYLES—Continued

[Defects by count per 255 grams]

Pits and Pit Fragments.	Average of not more than 1 by count per 300 grams
Major Stems	Not more than 3
HEVM	Not more than 2 units per sample
Broken Pieces and End Caps.	Not more than 125 grams by weight

TABLE 3—CHOPPED STYLE

[Defects by count per 255 grams]

FLAVOR	Reasonably good; no "off" flavor
SALOMETER	Acceptable Range in degrees: 3.0 to 14.0
COLOR	Reasonably uniform with no units lighter than the USDA Composite Color Standard for Ripe Type
DEFECTS:	
Pits and Pit Fragments.	Average of not more than 1 by count per 300 grams
Major Stems	Not more than 3
HEVM	Not more than 2 units per sample

TABLE 4—BROKEN PITTED STYLE

[Defects by count per 255 grams]

FLAVOR	Reasonably good; no "off" flavor
SALOMETER	Acceptable Range in degrees: 3.0 to 14.0
COLOR	Reasonably uniform with no units lighter than the USDA Composite Color Standard for Ripe Type
CHARACTER	Not more than 13 grams excessively soft
DEFECTS:	
Pits and Pit Fragments.	Average of not more than 1 by count per 300 grams
Major Stems	Not more than 3
HEVM	Not more than 2 units per sample

(b) Terms used in this section shall have the same meaning as are given to the respective terms in the current U.S. Standards for Grades of Canned Ripe Olives (7 CFR part 52): *Provided,* That the definition of "broken pitted olives" is as follows: "Broken pitted olives" consist of large pieces that may have been broken in pitting but have not been sliced or cut.

[62 FR 1242, Jan. 9, 1997]

§ 932.150 **Modified minimum quality requirements for canned green ripe olives.**

The minimum quality requirements prescribed in § 932.52 (a)(1) of this part are hereby modified with respect to canned green ripe olives so that no requirements shall be applicable with respect to color and blemishes of such olives.

[62 FR 1244, Jan. 9, 1997]

§ 932.151 **Incoming regulations.**

(a) *Inspection stations.* Natural condition olives shall be weighed only at inspection stations which shall be a plant of a handler or other place having facilities for weighing such olives: *Provided,* That such location and facilities are satisfactory to the committee: *Provided further,* That upon prior application to, and approval by, the committee, a handler may receive olives at an inspection station other than the one where the lot was weighed.

(b) *Lot identification.* (1) Immediately upon receipt of each lot of natural condition olives, the handler shall complete Form COC 3A or 3C, weight and grade report or such other lot identification form as may be approved by the committee, which shall contain at least the following:

(i) Lot number;

(ii) Date;

(iii) Variety; and

(iv) Number and type containers.

(2) The handler shall maintain identity of such lot of olives with its corresponding lot weight and grade report.

(c) *Weighing.* Each lot of natural condition olives shall be separately weighed to determine the net weight of olives.

(d) *Handler incoming responsibility*—(1) *General.* The handler is responsible for the proper performance of all actions connected with the identification of lots of olives, the weighing of boxes or bins, the taking of samples, and the furnishing of necessary personnel for the carrying out of such actions.

(2) *Certification.* (i) For each lot of olives that are weighed, the handler shall complete Form COC–3A or 3C, weight and grade report, which shall contain at least the following:

(A) Name of handler;

(B) Name of producer;

(C) County of production;

(D) Applicable lot number;

(E) Weight certificate number;

(F) Net weight;

(G) Number and type of containers;

(H) Date received;

(I) Time received; and

(J) Weight of sample.

269

(ii) The completed Form COC–3A or 3C shall be furnished to the committee, which shall certify thereon that the lot was weighed as required by § 932.51 if in accordance with the facts.

(e) *Disposition of noncanning olives*—(1)(i) *Notification and inspection of noncanning olives.* Prior to disposition of noncanning olives the handler shall complete Form COC–5, report of limited and undersize and cull olives inspection and disposition, which shall contain the following:

(A) Type and number of containers;

(B) Type of olives (undersize or culls);

(C) Net weight;

(D) Variety;

(E) Outlet (green olives, olive oil, etc.); and

(F) Consignee.

(ii) Before disposition of such olives, the completed Form COC–5 shall be furnished to the committee.

(2) *Control and surveillance.* Noncanning olives that have been reported on Form COC–5 shall, unless such olives are disposed of immediately after receipt, be identified by fixing to each bin or pallet of boxes a COC control card which may be obtained from the committee. Such olives shall be kept separate and apart from other olives in the handler's possession and shall be disposed of only in the outlet shown on Form COC–5.

(3) *Time period for disposition.* All required disposition of noncanning olives shall be completed not later than September 30 of the crop year following the one in which the obligation is incurred or such later date that a handler may specify in a notice filed with the committee at least 15 days prior to September 15 of such subsequent crop year: *Provided,* That such notice shows that such handler has a sufficient quantity of olives held in storage to meet his obligation and such later date is not later than the date when he will have completed his disposition of olives of the crop year of obligation.

(4) *Olives not subject to incoming regulation requirements.* Except as otherwise prescribed in § 932.51(b), any lot of olives to be used solely in the production of green olives or canned ripe olives of the "tree ripened" type shall not be subject to incoming regulation: *Pro-*

vided, That the applicable requirements of § 932.51(b) are met and the handler notifies the committee, in writing, that such lot is to be so used. Notice may be given by writing on the weight certificate "Lot to be used solely for use in the production of green olives or tree ripened olives" and a copy of such weight certificate given to the committee.

(f) *Partially exempted lots.* (1) Pursuant to § 932.55, any handler may process any lot of natural condition olives for use in the production of packaged olives which has not first been weighed as an individual lot as required by § 932.51(a)(1)(i) but was combined with any other lot or lots of natural condition olives, only if:

(i) All the olives in the combined lot are delivered to the handler in the same day;

(ii) The total net weight of the olives delivered to the handler by any person in such day does not exceed 500 pounds;

(iii) Each such person had authorized combination of his lot with other lots; and

(iv) The combined lot of the natural condition olives is weighed as required by § 932.51(a)(1)(i) prior to processing the olives.

(2) Whenever the natural condition olives in partially exempt individual lots are combined with other such olives as provided in paragraph (f)(1) of this section, the provision of the section applicable on individual lots shall apply instead to a combined lot.

(3) Each such handler shall file with the committee a weekly report showing for each day of the week the respective quantity in combined lots together with each person's authorization for combining lots. The report shall be filed upon a form supplied by the committee.

(g) *Additional Marketing Order Size Designations.* Pursuant to the authority in § 932.51(a)(1)(ii), the following additional size designations are established:

Designation(s)	Approximate count (per pound)	Average count range (per pound)
Subpetite	....	181 and up.
Petite	166	141–180, inclusive.

Designation(s)	Approximate count (per pound)	Average count range (per pound)
Extra Large Sevillano "L"	86	76–90, inclusive.
Extra Large Sevillano "C"	70	65–75, inclusive.

[31 FR 12635, Sept. 27, 1966, as amended at 33 FR 15631, Oct. 23, 1968; 34 FR 15389, Oct. 2, 1969; 49 FR 34440, Aug. 31, 1984; 49 FR 44448, Nov. 7, 1984; 52 FR 38224, Oct. 15, 1987; 52 FR 49346, Dec. 31, 1987; 81 FR 46569, July 18, 2016]

§ 932.152   Outgoing regulations.

(a) *Inspection stations.* Processed olives shall be sampled and inspected only at an inspection station which shall be any olive processing plant having facilities for in-line or lot inspection which are satisfactory to the Inspection Service and the Committee; or an olive processing plant which has an approved Quality Assurance Program in effect.

(b) *Inspection—General.* Inspection of packaged olives for conformance with § 932.52 shall be by a Quality Assurance Program approved by the Processed Products Branch (PPB), USDA; or by in-line or lot inspection. A PPB approved Quality Assurance Program shall be pursuant to a Quality Assurance contract as referred to in § 52.2.

(c) *Certification.* (1) Each handler shall furnish daily to the Inspection Service a copy of a pack report for the preceding work day which shall contain at least the following:

(i) The total number of cases of packaged olives;

(ii) Number of cans per case;

(iii) Can size;

(iv) Can code;

(v) Variety;

(vi) Fruit size; and

(vii) Style.

(2) The Inspection Service shall issue for each day's pack a signed certificate covering the quantities of such packaged olives which meet all applicable minimum quality and size requirements. Each such certificate shall contain at least the following:

(i) Date;

(ii) Place of inspection;

(iii) Name and address of handler;

(iv) Can code;

(v) Variety;

(vi) Fruit size;

(vii) Can size;

(viii) Style;

(ix) Total number of cases;

(x) Number of cans per case;

(xi) And statement that packaged olives meet the effective minimum quality requirements for canned ripe olives as warranted by the facts.

(d) *Olives which fail to meet minimum quality and size requirements.* (1) Whenever any portion of a handler's daily pack of packaged olives fails to meet all applicable minimum quality and size requirements, the Inspection Service shall issue a signed report covering such olives. Each such report shall contain at least the following:

(i) Date;

(ii) Place of inspection;

(iii) Name and address of handler;

(iv) Can code;

(v) Variety;

(vi) Fruit size;

(vii) Can size;

(viii) Style;

(ix) Total number of cases;

(x) Number of cans per case; and

(xi) Reason why the applicable requirements were not met.

(2) All such packaged olives shall be kept separate and apart from other packaged olives and shall be so identified by control cards or other means satisfactory to the Inspection Service and the committee that their identity is readily apparent. Such packaged olives may be reprocessed under supervision of the Inspection Service. Any such packaged olives that are not so reprocessed may be disposed of only in accordance with § 932.155.

(e) *Examination of certain olives received for use in the production of canned ripe olives of the tree-ripened type.* Pursuant to § 932.51(b), whenever a handler receives a lot of natural condition olives or makes a separation resulting in a sublot, solely for use in the production of canned ripe olives of the tree-ripened type he shall, at the time of receiving such lot or making such separation, notify the committee or the Inspection Service of the lot so received or the sublot so created which shall then be subject to examination by the committee, or by the Inspection Service if so designated by the committee, to assure that the olives in such lot or sublot comply with the specifications

271

set forth in § 932.109. Each such handler shall identify all such lots and sublots of natural condition olives and keep them separate and apart from other olives received. Such identification and separation shall be maintained throughout the processing and production of such olives as canned ripe olives of the tree-ripened type.

(f) *Size designations.* (1) In lieu of the size designations specified in § 932.52(a)(2), except as provided in § 932.51(a) (1) and (2), canned whole ripe olives, other than those of the "tree-ripened" type, shall conform to the marketing order size designations listed in table 1 contained herein, and shall be of a size not smaller than the applicable size requirements, tolerances, and percentages listed in paragraph (h) of this section.

TABLE I—CANNED WHOLE RIPE OLIVE SIZES AVERAGE COUNT RANGES
[Per Pound]

Size designation	Variety group 1		Variety group 2	
	Except Ascolano, Barouni, St. Agostino	Ascolano, Barouni, St. Agostino	Obliza	Except Obliza
Small	N.A.	N.A.	N.A.	128–140
Medium	N.A.	N.A.	106–127	106–127
Large	N.A.	91–105	91–105	91–105
Extra Large	65–75	65–90	65–90	65–90
Jumbo	47–60	47–60	47–60	47–60
Colossal	33–46	33–46	33–46	33–46
Sup. Colossal	(1)	(1)	(1)	(1)

1 32 or fewer.
N.A.—Not Applicable.

(2) The size of the canned whole olives shall conform with the applicable count per pound range indicated in table I of paragraph (f)(1) of this section. When the count per pound of whole olives falls between two count ranges, the size designation shall be that of the smaller size. The average count for canned whole ripe olives is determined from all containers in the sample and is calculated on the basis of the drained weight of the olives.

(3) Pitted olives must meet the size requirements for canned whole olives specified in paragraphs (f)(1) and (f)(2) of this section prior to pitting, or must meet the size designations specified in § 52.3754 of the U.S. Standards for Grades of Canned Ripe Olives subsequent to pitting, subject to the following minimum size requirements:

(i) Variety group 1 olives, except Ascolano, Barouni, and St. Agostino varieties, shall be at least "Extra Large;"

(ii) Variety group 1 olives of the Ascolano, Barouni, and St. Agostino varieties shall be at least "Large;"

(iii) Variety group 2 olives, except the Obliza variety, shall be at least "Small;"

(iv) Variety group 2 olives of the Obliza variety shall be at least "Medium."

(g) *Size Certification.* (1) When limited-use size olives for limited-use styles are authorized during a crop year and a handler elects to have olives sized pursuant to § 932.51(a)(2)(i), any lot of limited-use size olives may be used in the production of packaged olives for limited-use styles if such olives are within the average count range in table II contained herein for that variety group, and meet such further midpoint or acceptable count requirements for the average count range in each size as approved by the committee.

TABLE II—LIMITED USE SIZE OLIVES

Variety	Average count range (per pound)
Group 1, except Ascolano, Barouni, and St. Agostino.	76–90, inclusive.
Group 1, Ascolano, Barouni, and St. Agostino.	106–140, inclusive.
Group 2, except Obliza	141–180, inclusive.
Group 2, Obliza	128–140, inclusive.

(2) When limited-use size olives are not authorized for limited-use styles during a crop year and a handler elects

to have olives sized pursuant to §932.51(a)(2)(ii), any lot of canning-sized olives may be used in the production of packaged olives for whole, pitted, or limited-use styles if such olives are within the average count range in table III contained herein for that variety group, and meet such further midpoint or acceptable count requirements for the average count range in each size as approved by the committee.

TABLE III—CANNED WHOLE RIPE OLIVE SIZES AVERAGE COUNT RANGES

[Per Pound]

Size designation	Variety group 1		Variety group 2	
	Except Ascolano, Barouni, St. Agostino	Ascolano, Barouni, St. Agostino	Obliza	Except Obliza
Small	N.A.	N.A.	N.A.	128–140
Medium	N.A.	N.A.	106–127	106–127
Large	N.A.	91–105	91–105	91–105
Ex. Large	65–75	65–90	65–90	65–90
Jumbo	47–60	47–60	47–60	47–60
Colossal	33–46	33–46	33–46	33–46
Sup. Colossal	(¹)	(¹)	(¹)	(¹)

¹ 32 or fewer.
N.A.—Not Applicable.

(h) Canned whole ripe olives, other than those of the "tree-ripened" type, shall be of a size not smaller than the following applicable size requirements, tolerances and percentages:

(1) With respect to variety group 1 olives, except Ascolano, Barouni, and St. Agostino varieties, the individual fruits shall each weigh no less than 1/75 pound, except that

(i) For olives of the extra large size designation, not more than 25 percent, by count, of such olives may weigh less than 1/75 pound each including not more than 10 percent, by count, of such olives that weigh less than 1/82 pound each; and

(ii) For olives of any designation except the extra large size, not more than 5 percent, by count, of such olives may weigh less than 1/75 pound each;

(2) With respect to variety group 1 olives of the Ascolano, Barouni, and St. Agostino varieties, the individual fruits shall each weigh not less than 1/105 pound, except that

(i) For olives of the large size designation, not more than 25 percent, by count, of such olives may weigh less than 1/105 pound each including not more than 10 percent, by count, of such olives that weigh less than 1/116 pound each; and

(ii) For olives of any designation except the large size, not more than 5

percent, by count, of such olives may weigh less than 1/105 pound each;

(3) With respect to variety group 2 olives, except the Obliza variety, the individual fruits shall each weigh not less than 1/140 pound, except that

(i) For olives of the small size designation, not more than 35 percent by count, of such olives may weigh less than 1/140 pound each including not more than 7 percent, by count, of such olives that weigh less than 1/160 pound each; and

(ii) For olives of any designation except the small size, not more than 5 percent, by count, of such olives may weigh less than 1/140 pound each;

(4) With respect to variety group 2 olives of the Obliza variety, the individual fruit shall each weigh not less than 1/127 pound, except that

(i) For olives of the medium size designation, not more than 35 percent, by count, of such olives may weigh less than 1/127 pound each including not more than 7 percent, by count, of such olives that weigh less than 1/135 pound each; and

(ii) For olives of any designation except the medium size, not more than 5 percent, by count, of such olives may weigh less than 1/127 pound each.

[31 FR 12635, Sept. 27, 1966, as amended at 33 FR 15632, Oct. 23, 1968; 36 FR 24795, Dec. 23, 1971; 48 FR 54212, Dec. 1, 1983; 52 FR 38224, Oct. 15, 1987; 52 FR 49346, Dec. 31, 1987; 57 FR 36353, Aug. 13, 1992; 59 FR 38106, July 27, 1994; 59 FR 55341, Nov. 7, 1994; 62 FR 1244, Jan. 9, 1997]

§ 932.153 Establishment of minimum quality and size requirements for processed olives for limited uses.

(a) *Minimum quality requirements.* On or after August 1, 1996, any handler may use processed olives of the respective variety group in the production of limited use styles of canned ripe olives if such olives were processed after July 31, 1996, and meet the minimum quality requirements specified in § 932.52(a)(1) as modified by § 932.149.

(b) *Sizes.* On and after August 1, 1996, any handler may use processed olives in the production of limited-use styles of canned ripe olives if such olives were harvested after August 1, 1996, and meet the following requirements:

(1) The processed olives shall be identified and kept separate and apart from any olives harvested before August 1, 1996.

(2) Variety Group 1 olives, except the Ascolano, Barouni, or St. Agostino varieties, shall be of a size which individually weigh at least 1/105 pound: *Provided,* That no more than 35 percent of the olives in any lot or sublot may be smaller than 1/105 pound.

(3) Variety Group 1 olives of the Ascolano, Barouni, or St. Agostino varieties shall be of a size which individually weigh at least 1/180 pound: *Provided,* That no more than 35 percent of the olives in any lot or sublot may be smaller than 1/180 pound.

(4) Variety Group 2 olives, except the Obliza variety, shall be of a size which individually weigh at least 1/205 pound: *Provided,* That not to exceed 35 percent of the olives in any lot or sublot may be smaller than 1/205 pound.

(5) Variety Group 2 olives of the Obliza variety shall be of a size which individually weigh at least 1/180 pound: *Provided,* That not to exceed 35 percent

of the olives in any lot or sublot may be smaller than 1/180 pound.

[61 FR 40510, Aug. 5, 1996, as amended at 62 FR 1244, Jan. 9, 1997]

§ 932.154 Handler transfer.

(a) Except as hereinafter provided in paragraph (b) of this section, Form COC–6 "Report of Interhandler Transfer" shall be completed by the transfering handler for all lots of processed, but not packaged, olives transferred to another handler within the area and for all lots and sublots of natural condition olives transferred to another handler within the area or shipped to destinations outside the area except fresh market outlets. For natural condition and processed, but not packaged, olives transferred between handlers within the area, two completed copies of said form, signed by the transferring handler, shall accompany the lot or sublot to the receiving handler who shall certify on both copies as to receipt of the olives and forward one copy to the committee within 10 days following receipt of the olives. For natural condition olives transferred by a handler to a destination outside the area, except fresh market outlets, two copies of said form shall be completed by the transferring handler with the words *Outside the Area* included in the upper right corner of the form and one copy shall be returned to the committee within 10 days following transfer of the olives. The completed form shall contain at least the following information: (1) Name and address of both the transferor and transferee; (2) date of transfer; (3) condition (natural, processed but not packaged); (4) weight, number and size of each type of container; (5) variety; and (6) other identification (undersize olives, culls, style, etc.).

(b) Undersize or cull olives that are transferred from one handler to another and for which the transferring handler desires credit toward satisfaction of his obligation under § 932.51(a)(2) need only be accompanied by two copies of Form COC–5, report of limited and undersize and cull olives inspection and disposition: *Provided,* That such transfers are carried out under the supervision of the Inspection Service.

(c) No handler may ship any lot or sublot of natural condition olives to a destination outside the area, except fresh market outlets, unless such olives have first been size-graded and meet the disposition and holding requirements applicable under paragraphs (a) (2) and (4) of § 932.51. The size of such transferred olives shall be verified, prior to transfer, by certification issued to the transferring handler by the appropriate inspection service (Federal or Federal-State Inspection Service or the Processed Products Branch, USDA).

[31 FR 12636, Sept. 27, 1966, as amended at 36 FR 24795, Dec. 23, 1971; 49 FR 34440, Aug. 31, 1984; 49 FR 44448, Nov. 7, 1984]

§ 932.155 Special purpose shipments.

(a) The disposition of packaged olives covered by § 932.152(d) which are not reprocessed, and new packaged olive products covered under paragraph (b) of this section which have not been disposed of by the end of the test market period, shall be handled in conformity with the applicable provisions of this paragraph.

(1) Under the supervision of the Inspection Service, such packaged olives may be disposed of for use in the production of olive oil or dumped.

(2) Such packaged olives may be disposed of to a charitable organization for use by such organization, provided the following conditions are met:

(i) Any handler who wishes to so dispose of olives shall first file a written application with, and obtain written approval thereof, from the committee. Each such application shall contain at least:

(A) The name and address of the handler and the charitable organization;

(B) The physical location of the charitable organization's facilities;

(C) The quantity, in cases, the variety, size, can size, and can code of the packaged olives; and

(D) A certification from the charitable organization that such olives will be used by the organization and will not be sold.

(ii) Prior to approval, the committee shall perform such verification of the accuracy of the information on the application as it deems necessary. The committee may deny any application if it finds that the required information is incomplete or incorrect, or has reason to believe that the intended receiver is not a charitable organization, or that the handler or the organization has disposed of packaged olives contrary to a previously approved application. The committee shall notify the applicant and the organization in writing of its approval, or denial, of the application. Any such approval shall continue in effect so long as the packaged olives covered thereby are disposed of consistent with this section. The committee shall notify the handler and the organization of each such termination of approval. The handler shall furnish the committee, upon demand, such evidence of disposition of the packaged olives covered by an approved application as may be satisfactory to the committee.

(b) In accordance with the provisions of § 932.55(b), packaged olives to be used in marketing development projects may be handled without regard to § 932.149 provided the following conditions are met. Such olives must be identified to the satisfaction of the Inspection Service and kept separate from other packaged olives. The handler shall submit to the committee for its approval "COC Form 155" at least 10 working days prior to the shipment of such packaged olives to test markets, and report progress or changes to the committee, as requested. The applicant handler shall provide the following information on COC Form 155:

(1) The quantity of olives to be utilized (limited to not more than five percent of the handler's crop year acquisitions);

(2) Specific market outlet;

(3) Flavorings or other ingredients added to the olives;

(4) Style of olives used;

(5) Type of olives used, either black or green ripe;

(6) Container sizes;

(7) Varieties used, whether Ascolano, Barouni, Manzanillo, Mission, Sevillano, etc.;

(8) Sizes of olives utilized;

(9) Approximate dates when the new product will be packaged;

(10) Name and address of requesting handler;

(11) Place of inspection;

(12) Certification that all assessment and reporting requirements in effect under the marketing order will be met prior to shipment;

(13) Certification that all such fruit will be kept separate from other packaged olives and will be so identified by control cards or other means acceptable to the Inspection Service;

(14) Purpose and nature of the request, whether for test marketing, evaluation, market research, etc.; and

(15) An estimate of the amount of time required to complete the test. The committee shall promptly approve or deny the application, and may add limitations to any such approval. Upon approval, the applicant handler shall notify the Inspection Service. Packaged olives so identified and remaining unused at the end of the approved test-market period shall be disposed of according to paragraph (a) of this section.

(c) In accordance with the provisions of § 932.55(b), any handler may use processed olives in the production of packaged olives for repackaging, and ship packaged olives for repackaging, if the packaged olives meet the minimum quality requirements, except for the requirement that the packaged olives possess a reasonably good flavor: *Provided,* That the failure to possess a reasonably good flavor is due only to excessive sodium chloride.

[33 FR 15632, Oct. 23, 1968, as amended at 39 FR 38221, Oct. 30, 1974; 62 FR 1244, Jan. 9, 1997; 65 FR 4575, Jan. 31, 2000]

§ 932.159 Reallocation of handler membership.

Pursuant to § 932.25, handler representation on the Committee is reallocated to provide that the two handlers who handled the largest and second largest total volume of olives during the crop year in which nominations are made and in the preceding crop year shall each be represented by four members and four alternate members.

[65 FR 62994, Oct. 20, 2000]

§ 932.161 Reports.

(a) *Reports of olives received.* Each handler shall submit to the committee, on a form provided by the committee, for each week (Sunday through Satur-

day, or such other 7-day period for which the handler has submitted a request and received approval from the committee) and not later than the fourth day after the close of such week, a report showing by size designation and culls the respective quantities of each variety of olives received. In addition thereto, he shall also report the seasonal totals to date of the report.

(b) *Sales reports.* (1) Each handler shall submit to the committee, on COC Form 21 as provided by the committee, for each month and not later than the 15th day following the end of that month, a report showing the handler's total sales of packaged olives to commercial outlets in the United States, to governmental agencies, and to foreign countries. Such sales shall be reported in the following categories:

(i) Whole and whole pitted styles of canned ripe olives in consumer size containers;

(ii) Whole and whole pitted styles of canned ripe olives in institutional size containers;

(iii) Chopped style of canned ripe olives in all types of containers; and

(iv) Halved, segmented (wedged), and sliced styles of canned ripe olives in all types of containers.

The quantity in each category shall be reported in terms of the equivalent number of cases of 24 No. 300 (300 × 407) size cans.

(2) Each handler shall submit to the committee, on a form provided by the committee, for each month and not more than 15 days after the end of such month, a report showing the total quantity of packaged olives of the ripe and green ripe types sold during the month. Such reports shall include the following information, as applicable:

(i) With respect to the whole, pitted, and broken pitted styles of packaged olives of the ripe or green ripe type, each style shall be reported separately on COC Form 29a in terms of the quantity of each size of olives as designated on the form. Such quantity, or quantities, shall be reported in terms of the total amount packaged in each of the container sizes listed on said form except that the committee may require such reporting in terms of the equivalent number of cases of 24 No. 300 (300

× 407) size cans. Each handler shall report separately the total monthly sales of packaged olives of the green ripe type.

(ii) Limited use styles of packaged olives of the ripe or green ripe type shall be reported in terms of the quantity of each style packaged in each of the container sizes listed on COC Form 29b except that the committee may require such reporting in terms of the equivalent number of cases of 24 No. 300 (300 × 407) size cans.

(c) *Report of handler's utilization of limited size olives.* Each handler shall submit to the committee, on a form provided by the committee, upon completion of the handler's canning season, but not later than August 1st of each crop year, a report showing the quantities of limited canning size olives used in (1) halved; (2) segmented (wedged); (3) sliced; (4) chopped; (5) acidified; (6) Spanish olives; (7) Sicilian style olives; (8) Greek style olives; (9) olive oil; (10) olives dumped; and (11) any other use (specify such use).

(d) *Packaged olive inventory reports.* Each handler shall submit an inventory report to the committee, on a form provided by the committee, not later than the 15th day of each month showing the total quantity of packaged olives of the ripe and green ripe types held in storage at all locations on the last day of the preceding month. Such reports shall contain the following information, as applicable:

(1) With respect to the whole, pitted, and broken pitted styles of packaged ripe or green ripe type olives, each style shall be reported separately on COC Form 27a in terms of the packaged quantity of each size designated on the form. Such quantity, or quantities, shall be reported in terms of the total amount packaged in each of the container sizes listed on said form except that the committee may require such reporting in terms of the equivalent number of cases of 24 No. 300 (300 × 407) size cans. Each handler shall report separately the total quantity of any packaged olives of the green ripe type held in storage at all locations.

(2) Halved, sliced, segmented (wedged), and chopped styles of packaged olives of the ripe or green ripe type shall be reported in terms of the

quantity of each style packaged in each of the container sizes listed on COC Form 27b except that the committee may require such reporting in terms of the equivalent number of cases of 24 No. 300 (300 × 407) size cans.

(e) *Processed olive bulk inventory reports.* Each handler shall submit an inventory report to the committee, on a form provided by the committee, not later than the 15th day of each month showing the total quantity of processed olives of the ripe and green ripe types held in bulk storage at all locations on the last day of the preceding month. Such reports shall contain the following information, as applicable:

(1) The total tonnage of processed olives of the ripe and green ripe types, held in storage by the handler, which are of any size that may be used in the production of packaged olives of the whole or the pitted styles shall be reported on COC Form 27c in terms of the total quantity of each size designated on the form.

(2) The total tonnage of processed olives of the ripe and green ripe types, held in storage by the handler, which are of sizes that may be used in the production of packaged olives of the halved, sliced, segmented (wedged), or chopped style shall be reported on COC Form 27b.

(f) *Packout reports.* Each handler shall submit to the committee, on a form provided by the committee, for each month and not more than 15 days after the end of such month, a report showing the total production of packaged olives of the ripe and green ripe types. Such reports shall include the following information, as applicable:

(1) With respect to the whole, pitted, and broken pitted styles of packaged olives of the ripe or the green ripe types, each style shall be reported separately on COC Form 28a in terms of the total quantity of each size of olives as designated on the form. Such quantity, or quantities, shall be reported in terms of the total amount packaged in each of the container sizes listed on said form except that the committee may require such reporting in terms of the equivalent number of cases of 24 No. 300 (300 × 407) size cans. Each handler shall report separately the total

monthly production of packaged olives of the green ripe type.

(2) Halved, sliced, segmented (wedged), and chopped styles of packaged olives of the ripe or green ripe type shall be reported in terms of the quantity of each style packaged in each of the container sizes listed on COC Form 28b except that the committee may require such reporting in terms of the equivalent number of cases of 24 No. 300 (300 × 407) size cans.

[33 FR 15632, Oct. 23, 1968, as amended at 36 FR 24795, Dec. 23, 1971; 47 FR 13118, Mar. 29, 1982; 49 FR 34440, 34441, Aug. 31, 1984; 49 FR 44448, Nov. 7, 1984]

§ 932.230   Assessment rate.

On and after January 1, 2019, an assessment rate of $44.00 per ton is established for California olives.

[84 FR 33829, July 16, 2019]

PART 944—FRUITS; IMPORT REGULATIONS

AUTHORITY: 7 U.S.C. 601–674.

§ 944.28   Avocado Import Grade Regulation.

(a) Pursuant to section 8e of the Act and Part 944—Fruits; Import Regulations, the importation into the United States of any avocados is prohibited unless such avocados grade at least U.S. No. 2, as such grade is defined in the United States Standards for Grades of Florida Avocados (7 CFR 51.3050 through 51.3069).

(b) The Federal or Federal-State Inspection Service, Fruit and Vegetable Division, Agricultural Marketing Service, United States Department of Agriculture, is designated as the government inspection service for certifying the grade, size, quality, and maturity of avocados that are imported into the United States. Inspection by the Federal or Federal-State Inspection Service with evidence thereof in the form of an official inspection certificate, issued by the respective service, applicable to the particular shipment of avocados, is required on all imports. The inspection and certification services will be available upon application in accordance with the rules and regulations governing inspection and certification of fresh fruits, vegetables, and other products (7 CFR part 51) and in accordance with the regulations designating inspection services and procedure for obtaining inspection and certification (7 CFR 944.400).

(c) The term *importation* means release from custody of the United States Customs Service. The term *commercial processing into products* means the manufacture of avocado product which is preserved by any recognized commercial process, including canning, freezing, dehydrating, drying, the addition of chemical substances, or by fermentation.

(d) Any person may import up to 55 pounds of avocados exempt from the requirements specified in this section.

(e) Any lot or portion thereof which fails to meet the import requirements, and is not being imported for purposes of consumption by charitable institutions, distribution by relief agencies, seed, or commercial processing into products; prior to or after reconditioning may be exported or disposed of under the supervision of the Federal or Federal-State Inspection Service with the costs of certifying the disposal of such lot borne by the importer.

(f) The grade, size, and quality requirements of this section shall not be applicable to avocados imported for consumption by charitable institutions, distribution by relief agencies, seed, or commercial processing into products, but shall be subject to the safeguard provisions contained in §944.350.

[50 FR 21032, May 22, 1985, as amended at 58 FR 69185, Dec. 30, 1993; 79 FR 67039, Nov. 12, 2014]

§944.31 Avocado import maturity regulation.

(a) Pursuant to section 8e [7 U.S.C. 608e–1] of the Agricultural Marketing Agreement Act of 1937, as amended [7 U.S.C 601–674], and Part 944—Fruits; Import Regulations, the importation into the United States of any avocados, except the Hass, Fuerte, Zutano, and Edranol varieties, is prohibited unless:

(1) Any portion of the skin of the individual avocados has changed to the color normal for that fruit when mature for those varieties which normally change color to any shade of red or purple when mature, except for the Linda variety; or

(2) Such avocados meet the minimum weight or diameter requirements for the Monday nearest each date specified, through the Sunday immediately prior to the nearest Monday of the specified date in the next column, for each variety listed in the following table I: Provided, that avocados may not be handled prior to the earliest date specified in column A of such table for the respective variety; Provided further, There are no restrictions on size or weight on or after the date specified in column D; Provided further, That up to a total of 10 percent, by count to the individual fruit in each lot may weigh less than the minimum specified or be less than the specified diameter, except that no such avocados shall be over 2 ounces lighter than the minimum weight specified for the variety: Provided further, That up to double such tolerance shall be permitted for fruit in an individual container in a lot.

TABLE 1

Variety	A date	Min. wt.	Min. diam.	B date	Min. wt.	Min. diam.	C date	Min. wt.	Min. diam.	D date
Dr. Dupuis										
#2	5–30	16	$3^7/16$	6–13	14	$3^5/16$	7–04	12	$3^2/16$	7–18
Simmons ...	6–20	16	$3^9/16$	7–04	14	$3^7/16$	7–18	12	$3^1/16$	8–01
Pollock	6–20	18	$3^{11}/16$	7–04	16	$3^7/16$	7–18	14	$3^4/16$	8–01
Hardee	6–27	16	$3^2/16$	7–04	14	$2^{14}/16$	7–11	12	....	7–25
Nadir	6–27	14	$3^3/16$	7–04	12	$3^1/16$	7–11	10	$2^{14}/16$	7–18
Ruehle	7–04	18	$3^{11}/16$	7–11	16	$3^9/16$	8–01	12	$3^5/16$	8–15
				7–18	14	$3^7/16$	8–08	10	$3^3/16$	
Bernecker ..	7–18	18	$3^6/16$	8–01	16	$3^5/16$	8–15	14	$3^4/16$	8–29
Miguel (P)	7–18	22	$3^{13}/16$	8–01	20	$3^{12}/16$	8–15	18	$3^{10}/16$	8–29
Nesbitt	7–18	22	$3^{12}/16$	8–01	16	$3^5/16$	8–08	14	$3^3/16$	8–22
Tonnage	8–01	16	$3^6/16$	8–15	14	$3^4/16$	8–22	12	$3^2/16$	8–29
Waldin	8–01	16	$3^9/16$	8–15	14	$3^7/16$	8–29	12	$3^4/16$	9–12
Tower II	8–01	14	$3^6/16$	8–15	12	$3^4/16$	8–29	10	$3^2/16$	9–05
Beta	8–08	18	$3^8/16$	8–15	16	$3^5/16$	8–29	14	$3^3/16$	9–05
Lisa (P)	8–08	12	$3^2/16$	8–15	11	$3^0/16$	....	....	....	8–22
Black										
Prince	8–15	28	$4^1/16$	8–29	23	$3^{14}/16$	9–12	16	$3^9/16$	10–03
Loretta	8–22	30	$4^3/16$	9–05	26	$3^{15}/16$	9–19	22	$3^{12}/16$	9–26
Booth 8	8–29	16	$3^8/16$	9–12	14	$3^6/16$	9–26	12	$3^3/16$	10–24
							10–10	10	$3^1/16$	
Booth 7	8–29	18	$3^{13}/16$	9–12	16	$3^{10}/16$	9–26	14	$3^8/16$	10–10
Booth 5	9–05	14	$3^9/16$	9–19	12	$3^6/16$	....	....	....	10–03
Choquette	9–26	28	$4^4/16$	10–10	24	$4^1/16$	10–24	20	$3^{14}/16$	11–7
Hall	9–26	26	$3^{14}/16$	10–10	20	$3^9/16$	10–24	18	$3^8/16$	11–07
Lula	10–03	18	$3^{11}/16$	10–10	14	$3^6/16$	10–31	12	$3^3/16$	11–14
Monroe	11–07	26	$4^3/16$	11–21	24	$4^1/16$	12–05	20	$3^{14}/16$	1–02
							12–19	16	$3^9/16$	
Arue	5–16	16	....	5–30	14	$3^3/16$	6–20	12	....	7–04
Donnie	5–23	16	$3^5/16$	6–06	14	$3^4/16$	6–20	12	....	7–04
Fuchs	6–06	14	$3^3/16$	6–20	12	$3^9/16$	....	....	....	7–04
K–5	6–13	18	$3^5/16$	6–27	14	$3^3/16$	....	....	....	7–11
West Indian										
Seedling [1]	6–20	18	....	7–18	16	....	8–22	14	....	9–19
Gorham	7–04	29	$4^5/16$	7–18	27	$4^3/16$	....	....	....	8–15
Biondo	7–11	13	....	....	....	....	....	....	....	8–15
Petersen	7–11	14	$3^8/16$	7–18	12	$3^5/16$	7–25	10	$3^2/16$	8–08
232	7–18	14	....	8–01	12	....	....	....	....	8–15

TABLE 1—Continued

Variety	A date	Min. wt.	Min. diam.	B date	Min. wt.	Min. diam.	C date	Min. wt.	Min. diam.	D date
Pinelli	7–18	18	3$^{12}/_{16}$	8–01	16	3$^{10}/_{16}$	....	....	....	8–15
Trapp	7–18	14	3$^{10}/_{16}$	8–01	12	3$^7/_{16}$	....	....	....	8–15
K–9	8–01	16	....	....	....	....	....	....	....	8–22
Christina	8–01	11	2$^{14}/_{16}$	....	....	....	....	....	....	8–22
Catalina	8–15	24	....	8–29	22	....	....	....	....	9–19
Blair	8–29	16	3$^8/_{16}$	9–12	14	3$^5/_{16}$	....	....	....	10–03
Guatemalan Seedling[2]	9–05	15	....	10–03	13	....	....	....	....	12–05
Marcus	9–05	32	4$^{12}/_{16}$	9–19	24	4$^5/_{16}$	....	....	....	10–31
Brooks 1978	9–05	12	3$^4/_{16}$	9–12	10	3$^1/_{16}$	9–19	8	2$^{14}/_{16}$	10–10
Rue	9–12	30	4$^3/_{16}$	9–19	24	3$^{15}/_{16}$	10–03	18	3$^9/_{16}$	10–17
Collinson ...	9–12	16	3$^{10}/_{16}$	....	....	....	....	....	....	10–10
Hickson	9–12	12	3$^1/_{16}$	9–26	10	3$^0/_{16}$	....	....	....	10–10
Simpson	9–19	16	3$^9/_{16}$	....	....	....	....	....	....	10–10
Chica	9–19	12	3$^7/_{16}$	10–03	10	3$^4/_{16}$	....	....	....	10–17
Leona	9–26	18	3$^{10}/_{16}$	10–03	16	....	....	....	....	10–10
Melendez ..	9–26	26	3$^{14}/_{16}$	10–10	22	3$^{11}/_{16}$	10–24	18	3$^7/_{16}$	11–07
Herman	10–03	16	3$^9/_{16}$	10–17	14	3$^8/_{16}$	....	....	....	10–31
Pinkerton (CP)	10–03	13	3$^3/_{16}$	10–17	11	3$^0/_{16}$	10–31	9	....	11–14
Taylor	10–10	14	3$^5/_{16}$	10–24	12	3$^2/_{16}$	....	....	....	11–07
Ajax (B–7) ...	10–10	18	3$^{14}/_{16}$	....	....	....	....	....	....	10–31
Booth 3	10–10	16	3$^8/_{16}$	10–17	14	3$^6/_{16}$	....	....	....	10–31
Semil 34	10–17	18	3$^{10}/_{16}$	10–31	16	3$^8/_{16}$	11–14	14	3$^5/_{16}$	11–28
Semil 43	10–24	18	3$^{10}/_{16}$	11–7	16	3$^8/_{16}$	11–21	14	3$^5/_{16}$	12–05
Booth 1	11–14	16	3$^{12}/_{16}$	11–28	12	3$^8/_{16}$	....	....	....	12–12
Zio (P)	11–14	12	3$^1/_{16}$	11–28	10	2$^{14}/_{16}$	....	....	....	12–12
Gossman ...	11–28	11	3$^1/_{16}$	....	....	....	....	....	....	12–26
Brookslate	12–05	18	3$^{13}/_{16}$	12–12	16	3$^{10}/_{16}$	1–02	12	3$^5/_{16}$	1–30
				12–19	14	3$^8/_{16}$	1–16	10		
Meya (P) ...	12–12	13	3$^2/_{16}$	12–26	11	3$^0/_{16}$	....	....	....	1–09
Reed (CP) ...	12–12	12	3$^4/_{16}$	12–26	10	3$^3/_{16}$	1–09	9	3$^9/_{16}$	1–23

[1] Avocados of the West Indian type varieties and seedlings not listed elsewhere in table 1.
[2] Avocados of the Guatemalan type varieties and seedlings, hybrid varieties and seedlings, and unidentified seedlings not listed elsewhere in table I.

(3) Avocados which fail to meet the maturity requirements specified in this section must be maintained under the supervision of the Federal or Federal-State Inspection Service using the Positive Lot Identification program, and when presented for reinspection, must meet the maturity requirements which correspond to the date of the original inspection.

(b) The term *diameter* means the greatest dimension measured at a right angle to a straight line from the stem to the blossom end of the fruit.

(c) The term *importation* means release from custody of the United States Customs Service. The term *commercial processing into products* means the manufacture of avocado product which is preserved by any recognized commercial process, including canning, freezing, dehydrating, drying, the addition of chemical substances, or by fermentation.

(d) Any person may import up to 55 pounds of avocados exempt from the requirements specified in this section.

(e) The Federal or Federal-State Inspection Service, Fruit and Vegetable Division, Agricultural Marketing Service, United States Department of Agriculture, is designated as the governmental inspection service for certifying the grade, size, quality, and maturity of avocados imported into the United States. Inspection by the Federal or Federal-State Inspection Service with evidence thereof in the form of an official inspection certificate, issued by the respective service, applicable to the particular shipment of avocados, is required on all such imports. The inspection and certification services will be available upon application in accordance with the Regulations Governing Inspection, Certification and Standards of Fresh Fruits, Vegetables, and Other Products (7 CFR part 51), and in accordance with the regulation designating inspection services and procedure for obtaining inspection and certification (7 CFR 944.400).

(f) Any lot or portion thereof which fails to meet the import requirements, and is not being imported for purposes of consumption by charitable institutions, distribution by relief agencies, seed, or commercial processing into products; prior to or after reconditioning may be exported or disposed of under the supervision of the Federal or Federal-State Inspection Service with the costs of certifying the disposal of such lot borne by the importer.

(g) The maturity requirements of this section shall not be applicable to avocados imported for consumption by charitable institutions, distribution by relief agencies, seed, or commercial processing into products, but such avocados shall be subject to the safeguard provisions contained in §944.350.

[59 FR 30871, June 16, 1994, as amended at 61 FR 13058, Mar. 26, 1996; 64 FR 53186, Oct. 1, 1999; 73 FR 26945, May 12, 2008; 79 FR 55354, Sept. 16, 2014]

§944.106 **Grapefruit import regulation.**

(a) Pursuant to Section 8e [7 U.S.C. Section 608e–1] of the Agricultural Marketing Agreement Act of 1937, as amended [7 U.S.C. 601–674], and Part 944—Fruits; Import Regulations, the importation into the United States of any grapefruit is prohibited unless such grapefruit meet the following minimum grade and size requirements for each specified grapefruit classification:

Grapefruit classification	Minimum grade	Minimum diameter (inches)
(1)	(2)	(3)
Grapefruit, seedless	U.S. No. 1	3

(b) The term *grapefruit* is defined as citrus paradisi, MacFadyen.

(c) Terms and tolerances pertaining to grade and size requirements, which are defined in the United States Standards for Grades of Florida Grapefruit (7 CFR 51.750–51.784), and in Marketing Order No. 905 (7 CFR 905.18 and 905.306(a) through (d)), shall be applicable herein.

(d) The Federal or Federal-State Inspection Service, Specialty Crops Program, Agricultural Marketing Service, United States Department of Agriculture, is designated as the governmental inspection service for certifying the grade, size, quality, and maturity of grapefruit imported into the United States. Inspection by the Federal or Federal-State Inspection Service with evidence thereof in the form of an official inspection certificate, issued by the respective service, applicable to the particular shipment of grapefruit, is required on all such imports. The inspection and certification services will be available upon application in accordance with the Regulations Governing Inspection, Certification and Standards of Fresh Fruits, Vegetables, and Other Products (7 CFR part 51), and in accordance with the regulation designating inspection services and procedure for obtaining inspection and certification (7 CFR 944.400).

(e) Any lot or portion thereof which fails to meet the import requirements, and is not being imported for purposes of consumption by charitable institutions, distribution by relief agencies, animal feed, or commercial processing into canned or frozen products or into a beverage base; prior to or after reconditioning may be exported or disposed of under the supervision of the Federal or Federal-State Inspection Service with the costs of certifying the disposal of said lot borne by the importer.

(f) Any person may import up to 15 standard packed cartons (12 bushels) of grapefruit exempt from the requirements specified in this section.

(g) Any grapefruit which fail to meet the import requirements prior to or after reconditioning may be exported or disposed of under the supervision of the Federal or Federal-State Inspection Service with the costs of certifying the disposal of such grapefruit borne by the importer.

(h) The Secretary has determined that grapefruit imported into the United States are in most direct competition with grapefruit grown in Florida regulated under Marketing Order No. 905 (7 CFR part 905).

(i) The grade, size, quality, and maturity requirements of this section shall not be applicable to grapefruit imported for consumption by charitable institutions, distribution by relief agencies, animal feed, or commercial processing into canned or frozen products or into a beverage base, but shall be subject to the safeguard provisions contained in § 944.350.

[58 FR 39430, July 23, 1993, as amended at 58 FR 59934, Nov. 12, 1993; 58 FR 69185, Dec. 30, 1993; 59 FR 56380, Nov. 14, 1994; 60 FR 58499, Nov. 28, 1995; 61 FR 64253, Dec. 4, 1996; 63 FR 62923, Nov. 10, 1998; 64 FR 58762, Nov. 1, 1999; 66 FR 229, Jan. 3, 2001; 68 FR 53024, Sept. 10, 2003; 74 FR 15644, Apr. 7, 2009; 82 FR 55308, Nov. 21, 2017; 84 FR 16201, Apr. 18, 2019]

§ 944.312 Orange import regulation.

(a) Pursuant to section 8e (7 U.S.C. 608e–1) of the Agricultural Marketing Agreement Act of 1937, as amended (7 U.S.C 601–674), and part 944—Fruits; Import Regulations, the importation into the United States of any oranges is prohibited unless such oranges grade at least U.S. No. 2, and they are at least 2³⁄₁₆ inches in diameter. Effective July 1 through August 31 of each year this parageaph is suspended.

(b) The term *oranges* is defined as Citrus sinensis, Osbeck.

(c) The term *importation* means release from custody of the United States Customs Service. The term *processing* means the manufacture of any orange product which has been converted into sectioned fruit or into fresh juice, or preserved by any commercial process, including canning, freezing, dehydrating, drying, and the addition of chemical substances, or by fermentation.

(d) Terms and tolerances pertaining to grade and size requirements, which are defined in the United States Standards for Grades of Oranges (Texas and States other than Florida, California, and Arizona) (7 CFR 51.680–51.714), shall be applicable herein.

(e) Any person may import up to 400 pounds a day of oranges exempt from the requirements specified in this section.

(f) The Federal or Federal-State Inspection Service, Fruit and Vegetable Division, Agricultural Marketing Service, United States Department of Agriculture, is designated as the governmental inspection service for certifying the grade, size, quality, and maturity of oranges imported into the United States. Inspection by the Federal or Federal-State Inspection Service with evidence thereof in the form of an official inspection certificate, issued by the respective service, applicable to the particular shipment of oranges, is required on all such imports. The inspection and certification services will be available upon application in accordance with the Regulations Governing Inspection, Certification and Standards of Fresh Fruits, Vegetables, and Other Products (7 CFR part 51), and in accordance with the regulation designating inspection services and procedure for obtaining inspection and certification (7 CFR 944.400).

(g) Any oranges which fail to meet the import requirements, and are not being imported for purposes of consumption by charitable institutions, distribution by relief agencies, or processing into products; prior to or after reconditioning may be exported or disposed of under the supervision of the Federal or Federal-State Inspection Service with the costs of certifying the disposal of such oranges borne by the importer.

(h) The grade, size, quality, and maturity requirements of this section shall not be applicable to oranges imported for consumption by charitable institutions, distribution by relief agencies, or processing into products, but shall be subject to the safeguard provisions contained in § 944.350, *Provided that:* oranges, imported as exempt under this regulation, cannot be shipped to processors who have facilities, equipment, or outlets to repack or sell fruit in fresh form.

(i) The Secretary has determined that oranges imported into the United States are in most direct competition with oranges grown in Texas regulated under Marketing Order No. 906.

[59 FR 25792, May 18, 1994, as amended at 60 FR 33679, June 29, 1995; 61 FR 13059, Mar. 26, 1996; 79 FR 11300, Feb. 28, 2014]

§ 944.350 **Safeguard procedures for avocados, grapefruit, kiwifruit, olives, oranges, prune variety plums (fresh prunes), and table grapes, exempt from grade, size, quality, and maturity requirements.**

(a) Each person who imports or receives any of the commodities listed in paragraphs (a)(1) through (5) of this section shall file (electronically or paper) an "Importer's Exempt Commodity Form" (FV–6) with the Marketing Order and Agreement Division, Fruit and Vegetable Program, AMS, USDA. A "person who imports" may include a customs broker, acting as an importer's representative (hereinafter referred to as "importer"). A copy of the completed form (electronic or paper) shall be provided to the U.S. Customs and Border Protection. If a paper form is used, a copy of the form shall accompany the lot to the exempt outlet specified on the form. Any lot of any commodity offered for inspection and, all or a portion thereof, subsequently imported as exempt under this provision shall also be reported on an FV–6 form. Such form (electronic or paper) shall be provided to the Marketing Order and Agreement Division in accordance with paragraph (d) of this section. The applicable commodities are:

(1) Avocados, grapefruit, kiwifruit, olives, oranges, prune variety plums (fresh prunes) and table grapes for consumption by charitable institutions or distribution by relief agencies;

(2) Avocados, grapefruit, kiwifruit, oranges, prune variety plums (fresh prunes), and table grapes for processing;

(3) Olives for processing into oil;

(4) Grapefruit for animal feed; or

(5) Avocados for seed.

(b) *Certification of exempt use.* (1) Each importer of an exempt commodity as specified in paragraph (a) of this section shall certify on the FV–6 form (electronic or paper) as to the intended exempt outlet (*e.g.*, processing, charity, livestock feed). If certification is made

using a paper FV–6 form, the importer shall provide a handwritten signature on the form.

(2) Each receiver of an exempt commodity as specified in paragraph (a) of this section shall also receive a copy of the associated FV–6 form (electronic or paper) filed by the importer. Within two days of receipt of the exempt lot, the receiver shall certify on the form (electronic or paper) that such lot has been received and will be utilized in the exempt outlet as certified by the importer. If certification is made using a paper FV–6 form, the receiver shall provide a handwritten signature on the form.

(c) It is the responsibility of the importer to notify the Marketing Order and Agreement Division of any lot of exempt commodity rejected by a receiver, shipped to an alternative exempt receiver, exported, or otherwise destroyed. In such cases, a second FV–6 form must be filed by the importer, providing sufficient information to determine ultimate disposition of the exempt lot, and such disposition shall be so certified by the final receiver.

(d) All FV–6 forms and other correspondence regarding entry of exempt commodities must be submitted electronically, by mail, or by fax to the Marketing Order and Agreement Division, Fruit and Vegetable Program, AMS, USDA, 1400 Independence Avenue SW, STOP 0237, Washington, DC 20250–0237; telephone (202) 720–2491; email *ComplianceInfo@ams.usda.gov;* or fax (202) 720–5698.

[80 FR 15677, Mar. 24, 2015, as amended at 81 FR 24458, Apr. 26, 2016]

§ 944.400 **Designated inspection services and procedure for obtaining inspection and certification of imported avocados, grapefruit, kiwifruit, oranges, prune variety plums (fresh prunes), and table grapes regulated under section 8e of the Agricultural Marketing Agreement Act of 1937, as amended.**

(a) The Federal or Federal-State Inspection Service, Fruit and Vegetable Division, Agricultural Marketing Service, United States Department of Agriculture is hereby designated as the governmental inspection service for the purpose of certifying the grade, size, quality, and maturity of avocados, grapefruit, nectarines, oranges, prune variety plums (fresh prunes), and table grapes that are imported into the United States. Agriculture and Agri-Food Canada is also designated as a governmental inspection service for the purpose of certifying grade, size, quality and maturity of prune variety plums (fresh prunes) only. Inspection by the Federal or Federal-State Inspection Service or the Agriculture and Agri-Food Canada, with appropriate evidence thereof in the form of an official inspection certificate, issued by the respective services, applicable to the particular shipment of the specified fruit, is required on all imports. Inspection and certification by the Federal or Federal-State Inspection Service will be available upon application in accordance with the Regulations Governing Inspection, Certification and Standards for Fresh Fruits, Vegetables, and Other Products (7 CFR part 51) but, since inspectors are not located in the immediate vicinity of some of the small ports of entry, such as those in southern California, importers of avocados, grapefruit, nectarines, oranges, prune variety plums (fresh prunes), and table grapes should make arrangements for inspection through the applicable one of the following offices, at least the specified number of the days prior to the time when the fruit will be imported:

PORTS, OFFICES AND ADVANCE NOTICE

Eastern Region

In Alabama, Officer In Charge, Post Office Box 244, Mobile, AL 36601, PH: 205–690–6154, or

In Jacksonville, Florida, Officer In Charge, Unit 8, 3335 N. Edgewood Ave., Jacksonville, FL 32205, PH: 904–354–5983, or

In Miami, Florida, Officer In Charge, 1350 N.W. 12th Ave., RM. 530, Miami FL 33136, PH: 305–324–6116, or

In Maryland, Officer In Charge, Maryland Wholesale Produce Market—Building B Unit 13, Jessup, MD 20794, PH: 301–799–5899, or

In Massachusetts, Officer In Charge, Boston Terminal Market, Room 1, 34 Market Street, Everett, MA 02149, PH: 617–389–2480, or

In Buffalo, New York, Officer In Charge, 176 Niagara Frontier Food Terminal—Rm. 7, Buffalo, NY 14206, PH: 716–824–1585, or

In New Jersey, Officer In Charge, Federal Building, RM. 839, 970 Broad Street, Newark, NJ 07102, PH: 201–645–2208, or

In New York, New York, Officer In Charge, Room 28–A, Hunts Point Market, Bronx, NY 10474, PH: 212–991–7669, or

In Pennsylvania, Officer In Charge, 293 Produce Building, 3301 S. Galloway Street, Philadelphia, PA 19148, PH: 215–336–0845, or

In Virginia, Officer In Charge, 3661 Virginia Beach Blvd., Norfolk, VA 23502, PH: 804–441–6218, or

In Puerto Rico, Officer In Charge, Post Office Box 9112, Santurce, PR 00908, PH: 809–783–2230.

All other Eastern Port of entry: Regional Director Skyline Office Building, 5205 Leesburg Pike—Suite 806, Falls Church, VA 22041, PH: 703–756–6781.

Central Region

In Louisiana, Officer In Charge, 5027 U.S. Postal Service Building, 701 Loyola Avenue, New Orleans, LA 70113, PH: 504–589–6741, or

In Michigan, Officer In Charge, 90 Detroit Union Produce, 7201 West Fort Street, Detroit, MI 48209, PH: 313–226–6059, or

In Minnesota, Officer In Charge, Agriculture Building Room 226, 90 West Plato Boulevard, St. Paul, MN 55107, PH: 612–296–8557, or

In El Paso, Officer In Charge, 6070 Gateway East, Suite 410, El Paso, TX 79905, PH: 915–543–7723, or

In Houston, Texas, Officer In Charge, 3100 Produce Row, Room 14, Houston, TX 77023, PH: 713–923–2557.

All other Texas Ports: Officer In Charge, Post Office Box 107, San Juan, TX 78589, PH: 512–787–4091.

All other Central Ports of Entry: Regional Director, Room 1012, 610 South Canal Street, Chicago, IL 60607, PH: 312–353–6225.

Western Region

In Arizona, Officer In Charge, Post Office Box 1485, Nogales, AZ 85621, PH: 602–281–0783, or

In Los Angeles, California, Officer In Charge, Wholesale Terminal Bldg., Room 271, 784 South Central Avenue, Los Angeles, CA 90021, PH: 213–688–2489, or

In San Francisco, California, Officer In Charge, P.O. Box 4266, Burlingame, CA 94010, PH: 415–876–1093 & 1094, or

In Hawaii, Officer In Charge, P.O. Box 22159, Pawaa Substation, Honolulu, HI 96822, PH: 808–548–7147, or

In Oregon, Officer In Charge, Cascade Plaza, Suite 125, 2828 S.W. Corbett, Portland, OR 97201, PH: 503–229–6161, or

In Washington, Officer In Charge, 5507 Sixth Avenue South, Seattle, WA 98108, PH: 206–764–3500, or

In New Mexico, Officer In Charge, New Mexico Market and Development Branch, New Mexico Department of Agriculture, P.O. Box 5600, Las Cruces, NM 88003, PH: 505–646–4929.

All other Western Ports of Entry: Regional Director, P.O. Box 214287, Sacramento, CA 95821, PH: 916–484–4952 & 3549.

Headquarters: Washington, DC: Chief, Fresh Products Branch, Fruit and Vegetable Division, AMS, Room 2052–S. Bldg., U.S. Department of Agriculture, Washington, DC 20250, PH: 202–447–5870.

Notification:

Port Offices—at least one (1) day;
Regional Director—at least two (2) days;
Headquarters—at least three (3) days.

(b) Inspection certificates shall cover only the quantity of fruit that is being imported at a particular port of entry by a particular importer.

(c) The inspection performed, and certificates issued, by the Federal or Federal-State Inspection Service shall be in accordance with the rules and regulations of the Department governing the inspection and certification of fresh fruits, vegetables, and other products (7 CFR part 51). The cost of any inspection and certification shall be borne by the applicant therefor.

(d) Each inspection certificate issued with respect to any of the specified fruits to be imported into the United States shall set forth among other things:

(1) The name and place of inspection;

(2) The name of the shipper, or applicant;

(3) The Customs entry number pertaining to the lot or shipment covered by the certificate;

(4) The commodity inspected;

(5) The quantity of the commodity covered by the certificate;

(6) The principal identifying marks on the container;

(7) The railroad car initials and number, the truck and the trailer license number, the name of the vessel, the name of the air carrier, or other identification of the shipment; and

(8) The following statement if the facts warrant: Meets U.S. import requirements under section 8e of the Agricultural Marketing Agreement Act of 1937, as amended.

[48 FR 44459, Sept. 29, 1983, as amended at 56 FR 10504, Mar. 13, 1991; 61 FR 40958, Aug. 7, 1996; 68 FR 10347, Mar. 5, 2003; 74 FR 2808, Jan. 16, 2009]

§ 944.401 Olive Regulation 1.

(a) *Definitions.* (1) *Canned ripe olives* means olives in hermetically sealed containers and heat sterilized under pressure, of the two distinct types "ripe" and "green-ripe" as defined in the current U.S. Standards for Grades of Canned Ripe Olives. The term does not include Spanish-style green olives.

(2) *Spanish-style green olives* means olives packed in brine and which have been fermented and cured, otherwise known as "green olives."

(3) *Variety group 1* means the following varieties and any mutations, sports, or other derivations of such varieties: Aghizi Shami, Amellau Ascolano, Ascolano dura, Azapa, Balady, Barouni, Carydolia, Cucco, Gigante di Cerignola, Gordale, Grosane, Jahlut, Polymorpha, Prunara, Ropades, Sevillano, St. Agostino, Tafahi, and Touffahi.

(4) *Variety group 2* means the following varieties and any mutations, sports, or other derivations of such varieties: Manzanillo, Mission, Nevadillo, Obliza, and Redding Picholine.

(5) *USDA Inspector* means an inspector of the Processed Products Branch, Fruit and Vegetable Division, Agricultural Marketing Service, U.S. Department of Agriculture, or any other duly authorized employee of the Department.

(6) *Importation* means release from custody of the U.S. Bureau of Customs.

(7) *Limited use* means the use of processed olives in the production of packaged olives of the halved, segmented (wedged), sliced, or chopped styles, as defined in said standards.

(8) Terms used in this section shall have the same meaning as are given to the respective terms in the current U.S. Standards for Grades of Canned Ripe Olives (7 CFR part 52) including the terms *size, character, defects* and *ripe type: Provided,* That the definition of *broken pitted olives* is as follows: "Broken pitted olives" consist of large

285

pieces that may have been broken in pitting but have not been sliced or cut.

(b) The importation into the United States of any canned ripe olives is prohibited unless such olives are inspected and meet the following applicable requirements: *Provided,* That olives imported in bulk form and used in the production of any canned ripe olives are subject to such applicable requirements and the additional requirements in paragraph (b)(12) of this section.

(1) *Minimum quality requirements.* Canned ripe olives shall meet the following quality requirements, except that no requirements shall be applicable with respect to color and blemishes for canned green ripe olives:

(i) Canned whole and pitted olives of the ripe type shall meet the minimum quality requirements prescribed in table 1 of this section;

(ii) Canned sliced, segmented (wedged), and halved olives of the ripe type shall meet the minimum quality requirements prescribed in table 2 of this section;

(iii) Canned chopped olives of the ripe type shall meet the minimum quality requirements prescribed in table 3 of this section and shall be practically free from identifiable units of pit caps, end slices, and slices ("practically free from identifiable units" means that not more than 10 percent, by weight, of the unit of chopped style olives may be identifiable pit caps, end slices, or slices); and

(iv) Canned broken pitted olives of the ripe type shall meet the minimum quality requirements prescribed in table 4 of this section, *Provided,* That broken pitted olives consist of large pieces that may have been broken in pitting but have not been sliced or cut.

(v) A lot of canned ripe olives is considered to meet the requirements of this section if all or most of the sample units meet the requirements specified in tables 1 through 4 of this section: *Provided,* That the number of sample units which do not meet the requirements specified in tables 1 through 4 of this section does not exceed the acceptance number prescribed for in the sample size provided in table I of 7 CFR 52.38: *Provided further,* That there is no off flavor in any sample unit.

TABLE 1—WHOLE AND PITTED STYLE
[Defects by count per 50 olives]

FLAVOR	Reasonably good; no "off" flavor
FLAVOR (Green Ripe Type).	Free from objectionable flavors of any kind
SALOMETER	Acceptable range in degrees: 3.0 to 14.0
COLOR	Reasonably uniform with not less than 60% having a color equal or darker than the USDA Composite Color Standard for Ripe Type
CHARACTER	Not more than 5 soft units or 2 excessively soft units
UNIFORMITY OF SIZE.	60%, by visual inspection, of the most uniform in size. The diameter of the largest does not exceed the smallest by more than 4mm
DEFECTS:.	
Pitter Damage (Pitted Style Only).	15
Major Blemishes	5
Major Wrinkles	5
Pits and Pit Fragments (Pitted Style Only).	Not more than 1.3% average by count
Major Stems	Not more than 3
HEVM	Not more than 1 unit per sample
Mutilated	Not more than 3
Mechanical Damage.	Not more than 5
Split Pits or Misshapen.	Not more than 5

TABLE 2—SLICED, SEGMENTED (WEDGED), AND HALVED STYLES
[Defects by count per 255]

FLAVOR	Reasonably good; no "off" flavor
SALOMETER	Acceptable range in degrees: 3.0 to 14.0
COLOR	Reasonably uniform with no units lighter than the USDA Composite Color Standard for Ripe Type
CHARACTER	Not more than 13 grams excessively soft
DEFECTS:	
Pits and Pit Fragments.	Average of not more than 1 by count per 300 grams
Major Stems	Not more than 3
HEVM	Not more than 2 units per sample
Broken Pieces and End Caps.	Not more than 125 grams by weight

TABLE 3—CHOPPED STYLE
[Defects by count per 255 grams]

FLAVOR	Reasonably good; no "off" flavor
SALOMETER	Acceptable range in degrees: 3.0 to 14.0
COLOR	Reasonably uniform with no units lighter than the USDA Composite Color Standard for Ripe Type
DEFECTS:	
Pits and Pit Fragments.	Average of not more than 1 by count per 300 grams
Major Stems	Not more than 3
HEVM	Not more than 2 units per sample

TABLE 4—BROKEN PITTED STYLE

[Defects by count per 255 grams]

FLAVOR	Reasonably good; no "off" flavor
SALOMETER	Acceptable range in degrees: 3.0 to 14.0
COLOR	Reasonably uniform with no units lighter than the USDA Composite Color Standard for Ripe Type
CHARACTER	Not more than 13 grams excessively soft
DEFECTS:	
Pits and Pit Fragments.	Average of not more than 1 by count per 300 grams
Major Stems	Not more than 3
HEVM	Not more than 2 units per sample

(2) Canned whole ripe olives of Variety Group 1, except the Ascolano, Barouni, and St. Agostino varieties, shall be of such a size that not more than 25 percent, by count, of the olives may weigh less than 1/75 pound (6.0 grams) each, except that not more than 10 percent, by count, of the olives may weigh less than 1/82 pound (5.5 grams) each;

(3) Canned whole ripe Variety Group 1 olives, of the Ascolano, Barouni, and St. Agostino varieties, shall be of such size that not more than 25 percent, by count, of the olives may weigh less than 1/105 pound (4.3 grams) each except that not more than 10 percent, by count, of the olives may weigh less than 1/116 pound (3.9 grams) each;

(4) Canned whole ripe olives of Variety Group 2, except the Obliza variety, shall be of such a size that not more than 35 percent, by count, of the olives may weigh less than 1/140 pound (3.2 grams) each except that not more than 7 percent, by count, of the olives may weigh less than 1/160 pound (2.8 grams) each;

(5) Canned whole ripe Variety Group 2 olives, of the Obliza variety, shall be of such a size that not more than 35 percent, by count, of the olives may weigh less than 1/127 pound (3.5 grams) each except that not more than 7 percent, by count, of the olives may weigh less than 1/135 pound (3.3 grams) each;

(6) Canned whole ripe olives not identifiable as to variety or variety group shall be of such a size that not more than 35 percent, by count, of the olives may weigh less than 1/140 pound (3.2 grams) each except that not more than 7 percent, by count, of the olives may weigh less than 1/160 pound (2.8 grams) each;

(7) Canned pitted ripe olives of Variety Group 1, except the Ascolano, Barouni, and St. Agostino varieties, shall be at least "Extra Large" as defined in §52.3754 of the U.S. Standards for Grades of Canned Ripe Olives.

(8) Canned pitted ripe Variety Group 1 olives of the Ascolano, Barouni, and St. Agostino varieties shall be at least "Large" as defined in §52.3754 of the U.S. Standards for Grades of Canned Ripe Olives.

(9) Canned pitted ripe olives of Variety Group 2, except the Obliza variety, shall be at least "Small" as defined in §52.3754 of the U.S. Standards for Grades of Canned Ripe Olives.

(10) Canned pitted ripe Variety Group 2 olives of the Obliza variety shall be at least "Medium" as defined in §52.3754 of the U.S. Standards for Grades of Canned Ripe Olives.

(11) Canned pitted ripe olives not identifiable as to variety or variety group shall be at least "Small" as defined in §52.3754 of the U.S. Standards for Grades of Canned Ripe Olives.

(12) Imported bulk olives when used in the production of canned ripe olives must be inspected and certified as prescribed in this section. Imported bulk olives which do not meet the applicable minimum size requirements specified in paragraphs (b)(2) through (b)(11) of this section may be imported after August 1, 1996, for limited-use, but any such olives so used shall not be smaller than the following applicable minimum size:

(i) Whole ripe olives of Variety Group 1, except Ascolano, Barouni, or St. Agostino varieties, of a size that not more than 35 percent of the olives, by count, may be smaller than 1/105 pound (4.3 grams) each.

(ii) Whole ripe olives of Variety Group 1 of the Ascolano, Barouni, or St. Agostino varieties, of a size that not more than 35 percent of the olives, by count, may be smaller than 1/180 pound (2.5 grams) each.

(iii) Whole ripe olives of Variety Group 2, except the Obliza variety, of a size that not more than 35 percent of the olives, by count, may be smaller than 1/205 pound (2.2 grams) each.

(iv) Whole ripe olives of Variety Group 2 of the Obliza variety of a size that not more than 35 percent of the olives, by count, may be smaller than ⅟₁₈₀ pound (2.5 grams) each.

(v) Whole ripe olives not identifiable as to variety or variety group of a size that not more than 35 percent of olives, by count, may be smaller than ⅟₂₀₅ pound (2.2 grams) each.

(c) The Processed Products Branch, Fruit and Vegetable Division, Agricultural Marketing Service, U.S. Department of Agriculture, is hereby designated as the governmental inspection service for the purpose of certifying the grade and size of processed olives from imported bulk lots for use in canned ripe olives and the grade and size of imported canned ripe olives. Inspection by said inspection service with appropriate evidence thereof in the form of an official inspection certificate, issued by the service and applicable to the particular lot of olives, is required. With respect to imported bulk olives, inspection and certification shall be completed prior to use as packaged ripe olives. With respect to canned ripe olives, inspection and certification shall be completed prior to importation. Any lot of olives which fails to meet the import requirements and is not being imported for purposes of contribution to a charitable organization or processing into oil may be exported or disposed of under the supervision of the Processed Products Branch, Fruit and Vegetable Division, AMS, USDA, with the cost of certifying the disposal borne by the importer. Such inspection and certification services will be available, upon application, in accordance with the applicable regulations governing the inspection and certification of Processed Fruits and Vegetables, Processed Products Thereof, and Certain Other Processed Food Products (part 52 of this title). Application for inspection of canned ripe olives shall be made not less than 10 days prior to the time when the olives will be imported. Since inspectors are not located in the immediate vicinity of some of the small ports of entry, importers of canned ripe olives shall make arrangements for inspection through the following office at least 10 days prior to the time when the olives will be imported: Processed Products Branch, USDA, AMS, F&V Division,

P.O. Box 96456, Room 0726–S, Washington, DC 20090–6456, telephone (202) 720–5021, fax (202) 690–1527. Application for inspection of processed bulk olives shall be made not less than 3 days prior to use in the production of canned ripe olives. Such application shall be made through one of the following offices: Regional Director, Eastern Regional Office, 800 Roosevelt Road, Building A, suite 380 Glen Ellyn, IL 60137, telephone (708) 790–6937/8/9, fax (708) 469–5162; or Regional Director, Western Regional Office, 2202 Monterey Street, suite 102–C, Fresno, CA 93721, telephone (209) 487–5891, fax (209) 487–5900.

(d) Inspection certificates shall cover only (1) the quantity of canned ripe olives that is being imported at a particular port of entry by a particular importer or (2) the quantity of canned ripe olives processed from a lot or sublot of imported bulk olives.

(e) Inspection shall be performed by USDA inspectors in accordance with said regulations governing the inspection and certification of processed fruits and vegetables and related products (part 52 of this title). The cost of each such inspection and related certification shall be borne by the applicant therefore. Applicants shall provide USDA inspectors with the entry number and such other identifying information for each lot as the inspector may request.

(f) Notwithstanding any other provisions of this regulation, any importation of canned ripe olives or olives imported in bulk for use in the production of canned ripe olives which, in the aggregate, does not exceed 100 pounds drained weight may be imported without regard to the requirements of this section.

(g) It is hereby determined, on the basis of the information currently available, that the minimum quality requirements and size requirements set forth in this part are comparable to those applicable to California canned ripe olives.

(h) No provisions of this section shall supersede the restrictions or prohibitions on canned ripe olives under the provisions of the Federal Food, Drug, and Cosmetic Act, or any other applicable laws or regulations or the need to comply with applicable food and sanitary regulations of city, county, State, or Federal agencies.

(i) Each inspection certificate issued with respect to canned ripe olives to be imported into the United States and canned ripe olives processed from a lot or sublot of imported bulk olives shall set forth among other things:

(1) The date and place of inspection;

(2) The name of the shipper or applicant;

(3) The Customs entry number pertaining to the lot or shipment covered by the certificate;

(4) The commodity inspected;

(5) The quantity of the commodity covered by the certificate;

(6) The principal identifying marks on the container;

(7) The railroad car initials and number, the truck and the trailer license number, the name of the vessel, or other identification of the shipment;

(8) The Consumption Entry Number for Canned Ripe Olives; and

(9) The following statement if the facts warrant: Meets the U.S. import requirements under section 8e of the Agricultural Marketing Agreement Act of 1937, as amended.

(j) The minimum quality, size, and maturity requirements of this section shall not be applicable to olives imported for charitable organizations or processing for oil, but shall be subject to the safeguard provisions contained in §944.350.

[47 FR 51349, Nov. 15, 1982, as amended at 49 FR 34441, Aug. 31, 1984; 49 FR 44448, Nov. 7, 1984; 52 FR 38225, Oct. 15, 1987; 56 FR 49671, Oct. 1, 1991; 57 FR 36355, Aug. 13, 1992; 58 FR 69186, Dec. 30, 1993; 59 FR 38106, July 27, 1994; 59 FR 46910, Sept. 13, 1994; 60 FR 42774, Aug. 17, 1995; 61 FR 13059, Mar. 26, 1996; 61 FR 40510, Aug. 5, 1996; 62 FR 1244, Jan. 9, 1997; 74 FR 2808, Jan. 16, 2009; 81 FR 87412, Dec. 5, 2016]

§944.503 Table Grape Import Regulation 4.

(a)(1) Pursuant to section 8e of the Act and Part 944—Fruits, Import Regulations, and except as provided in paragraphs (a)(1)(iii) and (iv) of this section, the importation into the United States of any variety of Vinifera species table grapes, except Emperor, Calmeria, Almeria, and Ribier varieties, is prohibited unless such grapes meet the minimum grade and size requirements established in paragraphs (a)(1)(i) or (ii) of this section.

(i) U.S. No. 1 Table, as set forth in the United States Standards for Grades of Table Grapes (European or Vinifera Type 7 CFR 51.880 through 51.914), with the exception of the tolerance percentage for bunch size when packed in individual consumer clamshell packages weighing 5 pounds or less: not more than 20 percent of the weight of such containers may consist of single clusters weighing less than one-quarter pound, but with at least five berries each; or

(ii) U.S. No. 1 Institutional, with the exception of the tolerance percentage for bunch size. Such tolerance shall be 33 percent instead of 4 percent as is required to meet U.S. No. 1 Institutional grade. Grapes meeting these quality requirements may be marked "DGAC No. 1 Institutional" but shall not be marked "Institutional Pack."

(iii) Grapes of the Perlette variety shall meet the minimum berry size requirement of ten-sixteenths of an inch, and

(iv) Grapes of the Flame Seedless variety shall meet the minimum berry size requirement of ten-sixteenths of an inch (1.5875 centimeters) and shall be considered mature if the juice meets or exceeds 16.5 percent soluble solids, or the juice contains not less than 15 percent soluble solids and the soluble solids are equal to or in excess of 20 parts to every part acid contained in the juice, in accordance with applicable sampling and testing procedures specified in sections 1436.3, 1436.5, 1436.6, 1436.7, 1436.12, and 1436.17 of Article 25 of Title 3: California Code of Regulations (CCR).

(2) Such minimum maturity standards are incorporated by reference, copies of which are available from Ronald L. Cioffi, Chief, Marketing Order Administration Branch, F&V, AMS, USDA, Washington, DC 20090-6456, telephone (202) 720-2491. They are also available for inspection at the National Archives and Records Administration (NARA). For information on the availability of this material at NARA, call 202–741–6030, or go to: *http://www.archives.gov/federal_register/code_of_federal_regulations/ibr_locations.html.* This incorporation

by reference was approved by the Director of the Federal Register. These materials are incorporated as they exist on the date of approval and a notice of any change in these materials will be published in the FEDERAL REGISTER.

(3) All regulated varieties of grapes offered for importation shall be subject to the grape import requirements contained in this section effective April 10 through July 10.

(b) The Federal or Federal-State Inspection Service, F&V, AMS, USDA, is designated as the governmental inspection service for certifying the grade, size, quality, and maturity of table grapes that are imported into the United States. Inspection by the Federal or Federal-State Inspection Service with evidence thereof in the form of an official inspection certificate, issued by the respective service, applicable to the particular shipment of table grapes, is required on all imports. The inspection and certification services will be available upon application in accordance with the rules and regulations governing inspection and certification of fresh fruits, vegetables, and other products (7 CFR part 51) and in accordance with the Procedure for Requesting Inspection and designating the Agencies to Perform Requested Inspection and Certification (7 CFR 944.400).

(c) The term *importation* means release from custody of the United States Customs Service.

(d) Any lot or portion thereof which fails to meet the import requirements, and is not being imported for purposes of processing or donation to charitable organizations, prior to or after reconditioning may be exported or disposed of under the supervision of the Federal or Federal-State Inspection Service with the costs of certifying the disposal of said lot borne by the importer.

(e) The grade, size, quality, and maturity requirements of this section shall not be applicable to grapes imported for processing or donation to charitable organizations, but shall be subject to the safeguard provisions contained in § 944.350.

[51 FR 12502, Apr. 11, 1986, as amended at 51 FR 13209, Apr. 18, 1986; 52 FR 31979, Aug. 25, 1987; 53 FR 22128, June 14, 1988; 58 FR 21537, Apr. 22, 1993; 58 FR 69186, Dec. 30, 1993; 59 FR 67619, 67620, Dec. 30, 1994; 60 FR 33681, June 29, 1995; 63 FR 28480, May 26, 1998; 69 FR 18801, Apr. 9, 2004; 74 FR 3419, Jan. 21, 2009; 74 FR 11277, Mar. 17, 2009; 75 FR 17034, Apr. 5, 2010; 80 FR 68424, Nov. 5, 2015; 81 FR 24459, Apr. 26, 2016]

§ 944.550 Kiwifruit import regulation.

(a) Pursuant to section 8e of the Agricultural Marketing Agreement Act of 1937, as amended, the importation into the United States of any kiwifruit is prohibited unless such kiwifruit meets all the requirements of a U.S. No. 1 grade as defined in the United States Standards for Grades of Kiwifruit (7 CFR 51.2335 through 51.2340), except that the kiwifruit shall be "not badly misshapen," and an additional tolerance of 16 percent is provided for kiwifruit that is "badly misshapen," and except that such kiwifruit shall have a minimum of 6.2 percent soluble solids. Such fruit shall be at least Size 45, which means there shall be a maximum of 55 pieces of fruit and the average weight of all samples in a specific lot must weigh at least 8 pounds (3.632 kilograms), provided that no individual sample may be less than 7 pounds 12 ounces (3.472 kilograms).

(b) The Federal or Federal-State Inspection Service, Fruit and Vegetable Division, Agricultural Marketing Service, United States Department of Agriculture, is designated as the governmental inspection service for certifying the quality and size of kiwifruit imported into the United States. Inspection by the Federal or Federal-State Inspection Service with evidence thereof in the form of an official inspection certificate, issued by the respective service, applicable to a particular shipment of kiwifruit, is required on all imports. The inspection and certification services will be available upon application in accordance with the rules and regulations governing the inspection and certification of fresh fruits, vegetables, and other products (7 CFR part 51) and in accordance with the procedure for requesting inspection and designating the agencies to perform required inspection and certification (7 CFR 944.400).

(c) The term *importation* means release from custody of the United States Customs Service. The term *commercial processing into products* means that the kiwifruit is physically altered in form or chemical composition through freezing, canning, dehydrating, pulping, juicing, or heating of the product. The act of slicing, dicing, or peeling shall not be considered commercial processing into products.

(d) Any lot or portion thereof which fails to meet the import requirements and is not being imported for purposes of consumption by charitable institutions, distribution by relief agencies, or commercial processing into products may be reconditioned or exported. Any failed lot which is not reconditioned or exported shall be disposed of under supervision of the Federal or Federal-State Inspection Service with the costs of certifying the disposal of said lot borne by the importer.

(e) Any person may import up to 200 pounds of kiwifruit in any one shipment exempt from the requirements of this section.

(f) The grade, size, quality, and maturity requirements of this section shall not be applicable to kiwifruit imported for consumption by charitable institutions, distribution by relief agencies, or commercial processing into products, but shall be subject to the safeguard provisions contained in §944.350.

[56 FR 10504, Mar. 13, 1991, as amended at 57 FR 42688, Sept. 16, 1992; 58 FR 69186, Dec. 30, 1993; 59 FR 45620, Sept. 2, 1994; 61 FR 13059, Mar. 26, 1996; 65 FR 54948, Sept. 12, 2000; 78 FR 43760, July 22, 2013]

§944.700 Fresh prune import regulation.

(a) Pursuant to section 8e of the Agricultural Marketing Agreement Act of 1937, as amended, the importation into the United States of any fresh prunes, other than the Brooks variety, during the period July 15 through September 30 of each year is prohibited unless such fresh prunes meet the following requirements:

(1) Such fresh prunes grade at least U.S. No. 1, except that at least two-thirds of the surface of the fresh prune is required to be purplish in color, and such fresh prunes measure not less than 1¼ inches in diameter as measured by a rigid ring: Provided, That the following tolerances, by count, of the fresh prunes in any lot shall apply in lieu of the tolerance for defects provided in the United States Standards for Grades of Fresh Plums and Prunes (7 CFR 51.1520 through 51.1538): A total of not more than 15 percent for defects, including therein not more than the following percentage for the defect listed:

(i) 10 percent for fresh prunes which fail to meet the color requirement;

(ii) 10 percent for fresh prunes which fail to meet the minimum diameter requirement;

(iii) 10 percent for fresh prunes which fail to meet the remaining requirements of the grade: Provided, That not more than one-half of this amount, or 5 percent, shall be allowed for defects causing serious damage, including in the latter amount not more than 1 percent for decay.

(2) [Reserved]

(b) The importation of any individual shipment which, in the aggregate, does not exceed 500 pounds net weight, of fresh prunes of the Stanley or Merton varieties, or 350 pounds net weight, of fresh prunes of any variety other than the Stanley or Merton varieties, is exempt from the requirements specified in this section.

(c) The grade, size and quality requirements of this section shall not be applicable to fresh prunes imported for consumption by charitable institutions, distribution by relief agencies, or commercial processing into products, but such prunes shall be subject to the safeguard provisions in §944.350.

(d) The term *U.S. No. 1* shall have the same meaning as when used in the United States Standards for Grades of Fresh Plums and Prunes (7 CFR 51.1520 through 51.1538); the term *purplish color* shall have the same meaning as when used in the Washington State Department of Agriculture Standards for Italian Prunes (April 28, 1978), and the Oregon State Department of Agriculture Standards for Italian Prunes (October 5, 1977); the term *diameter* means the greatest dimension measured at right angles to a line from the stem to the blossom end of the fruit.

(e) The term *Prunes* means all varieties of plums, classified botanically as Prunus domestica, except those of the President variety.

(f) The term *importation* means release from custody of the United States Customs Service.

(g) Inspection and certification service is required for imports and will be available in accordance with the regulation designating inspection services and procedure for obtaining inspection and certification (7 CFR 944.400).

(h) Any lot or portion thereof which fails to meet the import requirements, and is not being imported for purposes of consumption by charitable institutions, distribution by relief agencies, or commercial processing into products, prior to or after reconditioning may be exported or disposed of under the supervision of the Federal or Federal-State Inspection Service with the costs of certifying the disposal of such fresh prunes borne by the importer.

(i) It is determined that fresh prunes imported into the United States shall meet the same minimum grade, size and quality requirements as those established for fresh prunes under Marketing Order No. 924 (7 CFR part 924).

[61 FR 40959, Aug. 7, 1996]

EFFECTIVE DATE NOTE: At 71 FR 26821, May 9, 2006, § 944.700 was suspended indefinitely, effective May 10, 2006.

PART 945—IRISH POTATOES GROWN IN CERTAIN DESIGNATED COUNTIES IN IDAHO, AND MALHEUR COUNTY, OREGON

Subpart A—Order Regulating Handling

Subpart B—Administrative Requirements

Subpart A—Order Regulating Handling

SOURCE: 23 FR 5709, July 30, 1958, unless otherwise noted. Redesignated at 26 FR 12751, Dec. 30, 1961.

DEFINITIONS

§ 945.1   Secretary.

Secretary means the Secretary of Agriculture of the United States, or any officer or employee of the United States Department of Agriculture to whom authority has heretofore been delegated, or to whom authority may hereafter be delegated, to act in his stead.

§ 945.2   Act.

Act means Public Act No. 10, 73d Congress, as amended and as reenacted and amended by the Agricultural Marketing Agreement Act of 1937, as amended (48 Stat. 31, as amended; 7 U.S.C. 601 *et seq.;* 68 Stat. 906, 1047).

§ 945.3   Person.

Person means an individual, partnership, corporation, association, or any other business unit.

§ 945.4   Production area.

Production area means all territory included within Malheur County, Oregon, and the counties of Adams, Valley, Lemhi, Clark, and Fremont in the State of Idaho, and all of the counties in Idaho lying south thereof.

§ 945.5   Potatoes.

Potatoes means all varieties of Irish potatoes grown within the aforesaid production area.

§ 945.6   Varieties.

Varieties means and includes all classifications or subdivisions of Irish potatoes according to those definitive characteristics now or hereafter recognized by the United States Department of Agriculture.

§ 945.7   Certified seed potatoes.

Certified seed potatoes means and includes all potatoes officially certified and tagged, marked, or otherwise appropriately identified, under the supervision of the official seed potato certifying agency of the State in which the potatoes are grown, or other seed certification agencies which the Secretary may designate.

§ 945.8   Handler.

Handler is synonymous with shipper and means any person (except a common or contract carrier of potatoes owned by another person) who ships potatoes.

§ 945.9   Ship or handle.

Ship or *handle* means to pack, sell, consign, transport or in any other way to place potatoes grown in the production area, or cause such potatoes to be placed, in the current of commerce within the production area or between the production area and any point outside thereof, so as to directly burden, obstruct, or affect any such commerce: *Provided*, That the definition of *ship* or *handle* shall not include the transportation of ungraded potatoes within the production area for the purpose of having such potatoes stored or prepared for market, except that the committee may impose safeguards pursuant to § 945.53 with respect to such potatoes.

[60 FR 29726, June 5, 1995]

§ 945.10   Producer.

Producer means any person engaged in the production of potatoes for market.

293

§ 945.11 Committee.

Committee means the administrative committee, called the Idaho-Eastern Oregon Potato Committee, established pursuant to § 945.20.

§ 945.12 Fiscal period.

Fiscal period means the period beginning and ending on the dates approved by the Secretary pursuant to recommendations by the committee.

§ 945.13 Grade and size.

Grade means any one of the officially established grades of potatoes, and *size* means any one of the officially established sizes of potatoes, as defined and set forth in:

(a) The United States Standards for Potatoes issued by the United States Department of Agriculture (§§ 51.1540 to 51.1556 of this title), or amendments thereto, or modifications thereof, or variations based thereon;

(b) The United States Consumer Standards for Potatoes as issued by the United States Department of Agriculture (§§ 51.1575 to 51.1587 of this title), or amendments thereto, or modifications thereof, or variations based thereon; or

(c) Standards for potatoes issued by the State from which the potatoes are shipped, or amendments thereto, or modifications thereof, or variations based thereon.

§ 945.14 Export.

Export means shipment of potatoes beyond the boundaries of continental United States.

§ 945.15 Pack.

Pack means a quantity of potatoes in any type of container and which falls within specific weight limits or within specific grade and/or size limits, or any combination thereof, recommended by the committee and approved by the Secretary.

§ 945.16 Container.

Container means a sack, box, bag, crate, hamper, basket, carton, package, barrel, or any other type of receptacle used in the packaging, transportation, sale, or other handling of potatoes.

§ 945.17 District.

District means each of the geographical divisions of the production area established pursuant to § 945.22 or as reestablished pursuant to § 945.23.

ADMINISTRATIVE COMMITTEE

§ 945.20 Establishment and membership.

(a) The Idaho-Eastern Oregon Potato Committee is hereby established consisting of eight members, of whom four shall currently be producers of potatoes for the fresh market who produced such potatoes during at least three of the last five years; at least one member shall be a producer predominately of potatoes for seed during a similar period; and three shall be handlers. For each member of the committee, there shall be an alternate who shall have the same qualifications as the member. The number of producer and/or handler members and alternates on the committee may be increased and the composition of the committee between producers and handlers may be changed as provided in § 945.23.

(b) Each person selected as a committee member or alternate to represent producers shall be an individual who is a producer in the district for which selected or an officer or employee of a producer in such district, and shall be a resident thereof. A producer who handles potatoes other than of his own production shall qualify as a producer under this section, and §§ 945.24, 945.25, 945.27, and 945.29, only if the potatoes of his own production constituted 51 percent or more of the total quantity of potatoes handled by him during the portion of the then current season preceding his nomination.

(c) Each person selected as a committee member or alternate to represent handlers shall be an individual who is a handler or an officer or employee of a handler, and shall be a resident of the production area.

(d) At least every six years, the committee shall review committee size, composition, and representation and recommend to the Secretary whether changes should be made, as provided in §945.23.

[23 FR 5709, July 30, 1958. Redesignated at 26 FR 12751, Dec. 30, 1961, as amended at 60 FR 29726, June 5, 1995]

§945.21 Term of office.

(a) Except as otherwise provided in this section, the term of office of committee members and alternates shall be for two years beginning June 1 or such other date as recommended by the committee and approved by the Secretary. The term of office of members and alternates shall be so determined that approximately one-half of the total producer and handler committee membership shall terminate each year.

(b) Committee members and alternates shall serve during the term of office for which they are selected and have qualified and continue until their successors are selected and have qualified. Beginning with the 1987 term of office, no member or alternate shall serve more than three full consecutive terms: *Provided,* That an alternate member may serve up to three consecutive terms and then serve as a member for up to three consecutive terms without a break in service. Members serving three consecutive terms could again become eligible to serve on the committee by not serving for one full term as either member or alternate member: *Provided,* That in the event a position would otherwise remain vacant for lack of eligible nominees or eligible persons willing to serve, the Secretary may authorize a member or alternate member to serve more than three full consecutive terms.

[53 FR 3188, Feb. 4, 1988]

§945.22 Districts.

For the purpose of selecting committee members and alternate members, the following districts of the production area are hereby established: *Provided,* That these districts may be changed as provided in §945.23.

(a) *District No. 1:* The counties of Bonneville, Butte, Clark, Fremont, Jefferson, Madison, and Teton;

(b) *District No. 2:* The counties of Bannock, Bear Lake, Bingham, Caribou, Franklin, Oneida, and Power; and

(c) *District No. 3:* Malheur County, Oregon, and the remaining designated counties in Idaho included in the production area, and not included in District No. 1 or District No. 2.

[60 FR 29726, June 5, 1995]

§945.23 Redistricting and reapportionment.

(a) The Secretary, upon recommendation of the committee, may reestablish districts within the production area, may reapportion committee membership among the various districts, may increase the number of producer and/or handler members and alternates on the committee, and may change the composition of the committee by changing the ratio between producer and handler members, including their alternates. At least every six years, the committee shall review committee size, composition and representation and recommend to the Secretary whether changes should be made. In recommending any such changes, the committee shall give consideration to:

(1) Shifts in potato acreage within districts and within the production area during recent years;

(2) the importance of new potato production in its relation to existing districts;

(3) the equitable relationship between committee membership and districts;

(4) economies to result for producers in promoting efficient administration due to redistricting or reapportionment of members within districts; and

(5) other relevant factors.

(b) Membership of the committee shall be apportioned among the districts of the production area so as to provide the following representation or such other representation as recommended by the committee and approved by the Secretary:

(1) Three producer members, including at least one who predominately produces seed potatoes, and one handler member, with their respective alternates, from District No. 1;

(2) One producer member and one handler member, with their respective alternates, from District No. 2; and

295

(3) One producer member and one handler member, with their respective alternates, from District No. 3.

[60 FR 29727, June 5, 1995]

§ 945.24 Selection.

Members and alternates of the committee shall be selected by the Secretary on the basis specified in § 945.23 (b) from nominations made pursuant to § 945.25 or from other eligible persons.

[60 FR 29727, June 5, 1995]

§ 945.25 Nominations.

For the selection by the Secretary of the members and alternates of the Idaho-Eastern Oregon Potato Committee, nominations may be made in the manner indicated in this section. Nominations for members and alternates may be submitted by producers or handlers, as the case may be, or groups of either thereof, on an elective basis or otherwise.

(a) In order to provide nominations for producer and handler committee members and alternates, the committee shall hold, or cause to be held, prior to April 1 of each year, or such other date as the Secretary may designate, one or more meetings of producers and of handlers in each district to nominate such members and alternates; or the committee may conduct nominations by mail in a manner recommended by the committee and approved by the Secretary.

(b) In arranging for such meetings, the committee may, if it deems it to be desirable, utilize the services and facilities of existing organizations and agencies, and may combine its meetings with others.

(c) At least one nominee shall be designated for each position as member and for each position as alternate member on the committee.

(d) Only producers may participate in designating nominees for producer members and alternates, and only handlers may participate in designating nominees for handler members and alternates.

(e) Each person who is both a handler and a producer may vote either as a handler or as a producer and may elect the group in which he will vote.

(f) Regardless of the number of districts in which a person produces or handles potatoes, each such person is entitled to cast only one vote on behalf of himself, his agents, subsidiaries, affiliates, and representatives, in designating nominees for committee members and alternates. In the event a person is engaged in producing or handling potatoes in more than one district, such person shall elect the district within which he may participate, as aforesaid, in designating nominees. An eligible voter's privilege of casting only one vote, as aforesaid, shall be construed to permit a voter to cast one vote for each position to be filled in the district in which he elects to vote.

(g) Nominations shall be supplied to the Secretary in such manner and form as the Secretary may prescribe, not later than May 1 of each year, or such other date as the Secretary may specify.

[23 FR 5709, July 30, 1958. Redesignated at 26 FR 12751, Dec. 30, 1961, as amended at 53 FR 3188, Feb. 4, 1988]

§ 945.26 Failure to nominate.

If nominations are not made within the time and in the manner specified by the Secretary pursuant to § 945.25, the Secretary may, without regard to nominations, select the committee members and alternates on the basis of the representation prescribed in this subpart.

§ 945.27 Acceptance.

Any person nominated to serve on the committee as a member or as an alternate shall qualify by filing a statement of willingness to serve with the Secretary.

[53 FR 3189, Feb. 4, 1988]

§ 945.28 Vacancies.

To fill any vacancy occasioned by the failure of any person selected as a committee member or as an alternate to qualify, or in the event of the death, removal, resignation, or disqualification of any qualified member or alternate, a successor for his unexpired term may be selected by the Secretary from nominations made in the manner specified in § 945.25 or the Secretary may select such committee member or alternate from previously unselected nominees on the current nominee list from the district involved. If the names of nominees to fill any such vacancy are not made available to the Secretary within 30 days after such vacancy occurs, the Secretary may fill such vacancy without regard to nominations, which selection shall be made on the basis of the representation provided for in §945.24.

§945.29 Alternate members.

An alternate member of the committee shall act in the place and stead of the member for whom he is an alternate during such member's absence and may perform such other duties as may be assigned or requested by the committee. In the event of the death, removal, resignation, or disqualification of a member his alternate shall act for him until a successor to such member is selected and has qualified. The committee may request the attendance of one or more alternates at any or all meetings, notwithstanding the expected or actual presence of the respective members.

§945.30 Procedure.

(a) A simple majority of all members of the committee, including alternates acting for members, shall be necessary to constitute a quorum or to pass any motion or approve any committee action, except any motion regarding a change in committee size shall require a unanimous vote. At any assembled meeting, all votes shall be cast in person.

(b) The committee may provide for meetings by telephone, telegraph or other means of communication and any vote cast at such meeting shall be confirmed promptly in writing.

[23 FR 5709, July 30, 1958. Redesignated at 26 FR 12751, Dec. 30, 1961, as amended at 60 FR 29727, June 5, 1995]

§945.31 Expenses.

Committee members and alternates shall be reimbursed for reasonable expenses necessarily incurred by them in the performance of their duties and in the exercise of their powers under this subpart, and may receive compensation at a rate determined by the committee, and approved by the Secretary, for each day or portion thereof, spent in conducting committee business.

[53 FR 3189, Feb. 4, 1988]

§945.32 Powers.

The committee shall have the following powers:

(a) To administer the provisions of this subpart in accordance with its terms;

(b) To make rules and regulations to effectuate the terms and provisions of this subpart;

(c) To receive, investigate, and report to the Secretary complaints of violation of the provisions of this subpart; and

(d) To recommend to the Secretary amendments to this subpart.

§945.33 Duties.

It shall be the duty of the committee:

(a) To act as intermediary between the Secretary and any producer or handler;

(b) To select a chairman and such other officers as may be necessary, to select subcommittees of committee members, and to adopt such rules and regulations for the conduct of its business as it may deem advisable;

(c) To appoint such employees, agents, and representatives as it may deem necessary and to determine the salaries and define the duties of each such person;

(d) To investigate, from time to time, and to assemble data on the growing, harvesting, shipping and marketing conditions with respect to potatoes, and to engage in such research and service activities which relate to the handling or marketing of potatoes as may be approved by the Secretary;

(e) To furnish to the Secretary such available information as he may request;

(f) To keep minutes, books, and records which clearly reflect all of the acts and transactions of the committee and such minutes, books, and records shall be subject to examination at any time by the Secretary or his authorized agent or representative;

(g) To make available to producers and handlers the committee voting record on recommended regulations and on other matters of policy;

(h) At the beginning of each fiscal period to submit to the Secretary a budget of its expenses for such fiscal period, together with a report thereon;

(i) To cause the books of the committee to be audited by a competent accountant at least once each fiscal period, and at such other time as the committee may deem necessary or as the Secretary may request. The report of such audit shall show the receipt and expenditure of funds collected pursuant to this subpart; a copy of each such report shall be furnished to the Secretary and a copy of each such report shall be made available at the principal office of the committee for inspection by producers and handlers; and

(j) To consult, cooperate and exchange information when deemed desirable by the committee with other potato marketing committees and other individuals or agencies in connection with all proper committee activities and objectives under this subpart.

BUDGET, EXPENSES AND ASSESSMENTS

§ 945.40 Expenses.

The committee is authorized to incur such expenses as the Secretary may find are reasonable and likely to be incurred during each fiscal period for its maintenance and functioning, and for such purposes as the Secretary, pursuant to this subpart, determines to be appropriate. Handlers shall share such expenses upon the basis of a fiscal period. Each handler's share of such expense shall be proportionate to the ratio between the total quantity of potatoes handled by him as the first handler thereof during a fiscal period and the total quantity of potatoes handled by all handlers as first handlers thereof during the same period.

§ 945.41 Budget.

At the beginning of each fiscal period, and as may be necessary thereafter, the committee shall prepare an estimated budget of income and expenditures necessary for the administration of this part. The committee may recommend a rate of assessment calculated to provide adequate funds to defray its proposed expenses as authorized in § 945.40. The committee shall present such budget promptly to the Secretary with an accompanying report showing the basis for its calculations.

§ 945.42 Assessments.

(a) The funds to cover the committee's expenses pursuant to § 945.40 shall be acquired by the levying of assessments upon handlers as provided in this subpart. Each handler who ships potatoes as the first handler thereof shall pay assessments to the committee upon demand, which assessments shall be in payment of such handler's pro rata share of such expenses.

(b) Assessments shall be levied upon handlers at a rate per hundredweight of potatoes or equivalent established by the Secretary. Such rate may be established upon the basis of the committee's budget recommendations, and other available information.

(c) At any time during or subsequent to a given fiscal period, the committee may recommend the approval of an amended budget and an increase in the rate of assessment. Upon the basis of such recommendation, or other available information, the Secretary may approve an amended budget and increase the rate of assessment. Such increase shall be applicable to all potatoes assessable under this part and handled by the first handler thereof during such fiscal period.

(d) The committee may impose a late payment charge or an interest charge, or both, on any handler who fails to pay, on or before the due date established by the Secretary, the total assessment for which such handler is liable. Such due date and the late payment fee and interest rate shall be recommended by the committee and approved by the Secretary.

(e) In order to provide funds to carry out its function, after the effective date of this subpart the committee may accept advance assessments from handlers. Advance assessments received from a handler shall be credited toward assessments levied against that handler during that fiscal period. In the case of an extreme emergency, the committee may also borrow money on a short term basis to provide funds for the administration of this part. Any such borrowed money shall only be used to meet the committee's current financial obligations, and the committee shall repay all borrowed money by the end of the next fiscal period from assessment income.

[23 FR 5709, July 30, 1958. Redesignated at 26 FR 12751, Dec. 30, 1961, as amended at 60 FR 29727, June 5, 1995]

§ 945.43 Accounting.

(a) All funds received by the committee pursuant to the provisions of this part shall be used solely for the purposes specified in this part.

(b) The Secretary may at any time require the committee, its members and alternates, employees, agents, and all other persons to account for all receipts and disbursements, funds, property, and records for which they are responsible. Whenever any person ceases to be a member or alternate of the committee, he shall account for all receipts, disbursements, funds, and property (including but not limited to books and other records) pertaining to the committee's activities for which he is responsible, and deliver all such property and funds in his hands to such successor, agency, or person as may be designated by the Secretary, and shall execute such assignments and other instruments as may be necessary or appropriate to vest in such successor, agency, or designated person, the right to all of such property and funds and all claims vested in such person.

(c) The committee may make recommendations to the Secretary for one or more of the members thereof, or any other person or persons to act as trustee or trustees for holding records, funds, or any other committee property during periods of suspension of this part, or during any period or periods when regulations are not in effect and, if the Secretary determines such action appropriate, he may direct that such person or persons shall act as trustee or trustees for the committee.

§ 945.44 Excess funds.

(a) The funds remaining at the end of a fiscal period which are in excess of the expenses necessary for committee operations during such period may be carried over, with the approval of the Secretary, into following periods as a reserve. Such reserve shall be established at an amount not to exceed approximately one fiscal period's budgeted expenses. Funds in such reserve shall be available for use by the committee for expenses authorized under § 945.40.

(b) Funds in excess of those placed in the operating reserve shall be credited proportionately against a handler's operations of the following fiscal period, except that if the handler demands payment, such proportionate refund shall be paid to such handler.

(c) Upon termination of this part, any funds not required to defray the necessary expenses of liquidation shall be disposed of in such manner as the Secretary may determine to be appropriate. To the extent practical, such funds shall be returned pro rata to the persons from whom such funds were collected.

[23 FR 5709, July 30, 1958. Redesignated at 26 FR 12751, Dec. 30, 1961, as amended at 53 FR 3189, Feb. 4, 1988]

REGULATIONS

§ 945.50 Marketing policy.

(a) *Preparation.* Prior to or at the same time as recommendations are made pursuant to § 945.51, the committee shall consider, and prepare, a proposed policy for the marketing of potatoes. In developing its marketing policy the committee shall investigate relevant supply and demand conditions for potatoes. In such investigations the committee shall give appropriate consideration to the following:

(1) Market prices for potatoes, including prices by grade, size, and quality, in different packs, and in different containers;

(2) Supplies of potatoes by grade, size, and quality in the production area and in other potato producing areas;

(3) The trend and level of consumer income;

(4) Establishing and maintaining orderly marketing conditions for potatoes;

(5) Orderly marketing of potatoes as will be in the public interest; and

(6) Other relevant factors.

(b) *Reports.* (1) The committee shall promptly submit a report to the Secretary setting forth the aforesaid marketing policy and shall notify producers and handlers of the contents of such report.

(2) In the event it becomes advisable to deviate from such marketing policy because of changed supply and demand conditions, the committee shall formulate a new or revised marketing policy in the manner set forth in this section. The committee shall promptly submit a report thereon to the Secretary and notify producers and handlers of the contents of such report on the new or revised marketing policy.

§ 945.51 Recommendation for regulations.

Whenever the committee deems it advisable that the handling of potatoes be regulated pursuant to § 945.52, or § 945.53, or both, it shall recommend to the Secretary grade, size, quality, or maturity regulation, or any combination thereof, or amendment thereto, or modification, suspension, or termination thereof, whenever it finds that such regulation, as provided in such sections, will tend to effectuate the declared policy of the act.

§ 945.52 Issuance of regulations.

(a) The Secretary shall limit the handling of potatoes whenever he finds from the recommendations and information submitted by the committee, or from other available information, that such regulation will tend to effectuate the declared policy of the act. Such limitation may:

(1) Regulate in any or all portions of the production area, the handling of particular grades, sizes, qualities, or maturities, or any combination thereof, of any or all varieties of potatoes during any period; or

(2) Regulate the shipment of particular grades, sizes, qualities, or maturities of potatoes differently, for different varieties, for different portions of the production area, for different packs, for different containers, or for any combination of the foregoing, during any period; or

(3) Fix the size, capacity, weight, dimensions, pack, labeling or marking of the container, or containers, which may be used in the packaging or handling of potatoes, or both; or

(4) Regulate the shipment of potatoes by establishing, in terms of grades, sizes, or both, minimum standards of quality and maturity.

(b) [Reserved]

[23 FR 5709, July 30, 1958. Redesignated at 26 FR 12751, Dec. 30, 1961, as amended at 60 FR 29727, June 5, 1995]

§ 945.53 Shipments for specified purposes.

Whenever the Secretary finds, upon the basis of the recommendations and information submitted by the committee, or from other available information, that it will tend to effectuate the declared policy of the act, he shall modify, suspend, or terminate regulations under or pursuant to § 945.42, § 945.52, or § 945.65, or any combination thereof in order to facilitate shipments of potatoes for the following purposes:

(a) Export;

(b) Relief or charity;

(c) Livestock feed;

(d) Certified seed potatoes;

(e) Processing into specified products; and

(f) Such other purposes which may be specified by the Committee, with the approval of the Secretary.

§ 945.54 Minimum quantity exemption.

The committee, with the approval of the Secretary, may establish, for any or all portions of the production area, minimum quantities below which shipments will be free from regulations issued or in effect pursuant to §§ 945.40 to 945.65, inclusive, or any combination thereof.

§ 945.55 Notification of regulation.

The Secretary shall notify the committee of any regulations issued or of any modifications, suspension, or termination thereof. The committee shall give reasonable notice thereof to handlers.

§ 945.56 Safeguards.

(a) The committee, with the approval of the Secretary, may prescribe adequate safeguards to prevent shipments pursuant to § 945.53 from entering channels of trade and other outlets for other than the specific purpose authorized therefor.

(b) Safeguards, provided by this section, may include, but shall not be limited to, requirements that handlers:

(1) Shall obtain the inspection required by §945.65 or pay the assessment provided by §945.42, or both, in connection with the potato shipments effected in accordance with §945.53; and

(2) Shall obtain Certificates of Privilege from the committee for shipments of potatoes effected or to be effected under provisions of §945.53.

(c) The committee, with the approval of the Secretary, shall prescribe rules governing the issuance and the contents of Certificates of Privilege.

(d) The committee may rescind, or deny to any handler, Certificates of Privilege if proof satisfactory to the committee is obtained that potatoes shipped by him for the purposes stated in §945.53 were handled contrary to the provisions of this section.

(e) The committee shall make reports to the Secretary, as requested, showing the number of applications for such certificates, the quantity of potatoes covered by such applications for such certificates, the number of such applications denied and certificates granted, the quantity of potatoes shipped under duly issued certificates, and such other information as may be requested by the Secretary.

INSPECTION AND CERTIFICATION

§945.65 Inspection and certification.

(a) During any period in which regulations are in effect pursuant to §945.42, §945.52, or §945.53, or any combination thereof, no handler shall handle potatoes unless such potatoes are inspected by an authorized representative of the Federal-State Inspection Service, and are covered by a valid inspection certificate, except when relieved from such requirements pursuant to recommendations by the committee and approved by the Secretary.

(b) Regrading, resorting, or repacking any lot of potatoes shall invalidate any prior inspection certificates covering such potatoes insofar as the requirements of this section are concerned. During any period in which shipments of potatoes are regulated, as

aforesaid, no handler shall handle potatoes after they have been regraded, resorted, repacked, or in any way further prepared for market, unless such potatoes are inspected and covered by a valid inspection certificate as required in paragraph (a) of this section.

(c) Insofar as the requirements of this section are concerned, the length of time for which an inspection certificate shall be valid may be established by the committee with the approval of the Secretary; and such length of time may be different for shipments for different purposes.

(d) When potatoes are inspected in accordance with the requirements of this section, a copy of each inspection certificate issued shall be made available promptly to the committee by the inspection service.

COMPLIANCE

§945.70 Compliance.

Except as provided in this part, no handler shall ship potatoes, the shipment of which has been prohibited by the Secretary in accordance with provisions of this subpart, and no handler shall ship potatoes except in conformity to the provisions of this subpart.

MISCELLANEOUS PROVISIONS

§945.80 Reports.

(a) Upon the request of the committee, with approval of the Secretary, every handler shall furnish to the committee, in such manner and at such time as may be prescribed, such information as will enable the committee to exercise its powers and perform its duties under this subpart. The Secretary shall have the right to modify, change, or rescind any requests for reports pursuant to this section.

(b) All data or other information constituting a trade secret, or disclosing a trade position or business condition of a particular handler shall be treated as confidential and shall at all times be received by and kept in the custody and under the control of one or more designated employees of the committee. Information which would reveal the circumstances of a single handler shall be disclosed to no person other than the Secretary.

(c) Each handler shall maintain for at least two succeeding fiscal periods such records of potatoes received and of potatoes disposed of by such handler as may be necessary to verify reports required pursuant to this section. The committee, with the approval of the Secretary, may prescribe rules and regulations issued pursuant to this section specifying handler records and reports which the committee may need to perform its functions.

(d) For the purpose of assuring compliance and checking and verifying reports filed by handlers, the Secretary and the committee, through its duly authorized agents, shall have access to any premises where applicable records are maintained, where potatoes are held, and, at any time during reasonable business hours, shall be permitted to inspect such handlers' premises and any and all records of such handlers with respect to matters within the purview of this part.

[23 FR 5709, July 30, 1958. Redesignated at 26 FR 12751, Dec. 30, 1961, as amended at 60 FR 29727, June 5, 1995]

§ 945.81   Right of the Secretary.

The members of the committee (including successors and alternates), and any agent or employee appointed or employed by the committee, shall be subject to removal or suspension by the Secretary at any time. Each and every order, regulation, decision, determination or other act of the committee shall be subject to the continuing right of the Secretary to disapprove of the same at any time. Upon such disapproval, the disapproved action of the said committee shall be deemed null and void, except as to acts done in reliance thereon or in compliance therewith prior to such disapproval by the Secretary.

§ 945.82   Effective time.

The provisions of this subpart shall become effective at such time as the Secretary may declare above his signature attached to this subpart, and shall continue in force until terminated in one of the ways specified in this subpart.

§ 945.83   Termination.

(a) The Secretary may, at any time, terminate the provision of this subpart by giving at least one day's notice by means of a press release or in any other manner which he may determine.

(b) The Secretary may terminate or suspend the operation of any or all of the provisions of this subpart whenever he finds that such provisions do not tend to effectuate the declared policy of the act.

(c) The Secretary shall terminate the provisions of this subpart at the end of any fiscal period whenever he finds that such termination is favored by a majority of producers who, during the preceding fiscal period, have been engaged in the production for market of potatoes: *Provided*, That such majority has, during such period, produced for market more than fifty percent of the volume of such potatoes produced for market; but such termination shall be effective only if announced on or before April 30 of the then current fiscal period.

(d) The Secretary shall conduct a referendum as soon as practicable after July 31, 1992, and at such time every sixth year thereafter, to ascertain whether continuance of this order is favored by potato producers. The Secretary may terminate the provisions of this order at the end of any fiscal period in which the Secretary has found that continuance of this order is not favored by producers who, during a representative period determined by the Secretary, have been engaged in the production for market of potatoes in the production area. Termination of the order shall be effective only if announced on or before July 1 of the then current fiscal period.

(e) The provisions of this subpart shall, in any event, terminate whenever the provisions of the act authorizing them cease to be in effect.

[23 FR 5709, July 30, 1958. Redesignated at 26 FR 12751, Dec. 30, 1961, as amended at 53 FR 3189, Feb. 4, 1988]

§945.84 Proceedings after termination.

(a) Upon the termination of the provisions of this subpart the then functioning members of the committee shall continue as trustees, for the purpose of liquidating the affairs of the committee, of all the funds and property then in the possession of or under control of the committee, including claims for any funds unpaid or property not delivered at the time of such termination. Action by said trusteeship shall require the concurrence of a majority of the said trustees.

(b) The said trustees shall continue in such capacity until discharged by the Secretary; shall, from time to time account for all receipts and disbursements and deliver all property on hand, together with all books and records of the committee and of the trustees, to such person as the Secretary may direct; and shall upon request of the Secretary execute such assignments or other instruments necessary or appropriate to vest in such person full title and right to all of the funds, property, and claims vested in the committee or the trustees pursuant thereto.

(c) Any person to whom funds, property, or claims have been transferred or delivered by the committee or its members, pursuant to this section, shall be subject to the same obligations imposed upon the members of the committee and upon the said trustees.

§945.85 Effect of termination or amendments.

(a) Unless otherwise expressly provided by the Secretary, the termination of this subpart or of any regulation issued pursuant to this subpart, or the issuance of any amendments to either thereof, shall not (1) affect or waive any right, duty, obligation, or liability which shall have arisen or which may thereafter arise in connection with any provision of this subpart or any regulation issued under this subpart, or (2) release or extinguish any violation of this subpart or of any regulation issued under this subpart, or (3) affect or impair any rights or remedies of the Secretary or of any other person with respect to any such violation.

(b) The persons who are committee members and alternates on the effective date of this subpart shall continue in office until their successors have been selected and have qualified. All rules and regulations issued or approved by the Secretary pursuant to this part (Order No. 945, as amended) and not in conflict herewith, which are in effect immediately prior to the date of this amendment shall continue in effect under this subpart as originally issued, or subsequently modified, until such rules and regulations are changed, modified, or suspended in accordance with this subpart.

§945.86 Duration of immunities.

The benefits, privileges, and immunities conferred upon any person by virtue of this subpart shall cease upon the termination of this subpart, except with respect to acts done under and during the existence of this subpart.

§945.87 Agents.

The Secretary may, by designation in writing, name any person, including any officer or employee of the Government, or name any bureau or division in the United States Department of Agriculture, to act as his agent or representative in connection with any of the provisions of this subpart.

§945.88 Derogation.

Nothing contained in this subpart is, or shall be construed to be, in derogation or in modification of the rights of the Secretary or of the United States to exercise any powers, granted by the act or otherwise, or, in accordance with such powers, to act in the premises whenever such action is deemed advisable.

§945.89 Personal liability.

No member or alternate of the committee, nor any employee or agent thereof, shall be held personally responsible, either individually or jointly with others, in any way whatsoever, to any handler or to any person for errors in judgment, mistakes, or other acts, either of commission or omission, as such member, alternate, or employee, except for acts of dishonesty.

§945.90 Separability.

If any provision of this subpart is declared invalid, or the applicability

303

thereof to any person, circumstance, or thing is held invalid, the validity of the remainder of this subpart, or the applicability thereof to any other person, circumstance, or thing, shall not be affected thereby.

§ 945.91 Amendments.

Amendments to this subpart may be proposed, from time to time, by the committee or by the Secretary.

Subpart B—Administrative Requirements

Source: 24 FR 8688, Oct. 27, 1959, unless otherwise noted. Redesignated at 26 FR 12751, Dec. 30, 1961.

§ 945.100 Communications.

Unless otherwise provided by specific direction of the committee, all reports, applications, submittals, requests, and communications in connection with the marketing agreement and order, both as amended, shall be addressed to the committee at its principal office.

DEFINITIONS

§ 945.110 Order.

Order means Order No. 945, as amended, effective September 1, 1958 (§§ 945.1 through 945.91) regulating the handling of Irish potatoes grown in Malheur County, Oregon, and the counties of Adams, Valley, Lemhi, Clark, and Fremont in the State of Idaho, and all of the counties in Idaho lying south thereof.

§ 945.111 Fiscal period.

The fiscal period that began June 1, 1981, shall end July 31, 1982. Each year thereafter *fiscal period* shall mean the period beginning August 1 and ending the following July 31.

[47 FR 17272, Apr. 22, 1982]

§ 945.112 Terms.

Terms used in this subpart shall have the same meaning as when used in the marketing agreement and order, both as amended.

CERTIFICATES OF PRIVILEGE

§ 945.120 General.

Whenever shipments of potatoes for special purposes pursuant to § 945.53 are relieved in whole or in part from grade and size regulations issued under § 945.52 the committee shall require information and evidence as to the manner, methods, and timing of such shipments as safeguards against the entry of any such potatoes into trade channels other than those for which intended. Such information and evidence shall include the requirements set forth below with respect to Certificates of Privilege.

§ 945.121 Qualification.

Before handling potatoes for special purposes which do not meet regulations issued pursuant to § 945.52 a handler must qualify with the committee to handle shipments for special purposes. To qualify he must (a) apply for and receive a Certificate of Privilege indicating his intent to so handle potatoes; (b) agree to comply with reporting and other requirements set forth in §§ 945.121 to 945.125, inclusive, with respect to such shipments; and (c) receive approval of the committee, or its duly authorized agents, to so handle potatoes. Such approval will be based upon evidence furnished in his application for a Certificate of Privilege, and other information available to the committee.

§ 945.122 Application.

(a) Application for a Certificate of Privilege shall be made on forms furnished by the committee. Each application may contain, but need not be limited to, the name and address of the handler; the quantity by grade, size, quality and variety of the potatoes to be shipped; the mode of transportation; the consignee; the destination; the purpose for which the potatoes are to be used; a certification to the United States Department of Agriculture and to the committee as to the truthfulness of the information shown thereon; and any other appropriate information or documents deemed necessary by the committee or its duly authorized agents for the purpose stated in § 945.120.

(b) The committee may require each handler making shipments of potatoes for export to include with his application a copy of the Department of Commerce Shipper's Export Declaration Form No. 7525–V applicable to such shipment.

§945.123 Approval.

The committee or its duly authorized agents shall give prompt consideration to each application for a Certificate of Privilege. Approval of an application, based upon a determination as to whether the information contained therein and other information available to the committee supports approval, shall be evidenced by the issuance of a Certificate of Privilege to the applicant. Each certificate shall cover a specified period, and specified qualities and quantities of potatoes to be sold or transported to the designated consignee for the purposes declared.

§945.124 Reports.

Each handler of potatoes shipping under Certificates of Privilege shall supply the committee with reports as requested by the committee or its duly authorized agents showing the name and address of the shipper; the car or truck identification; the loading point; destination; consignee; the inspection certificate number when inspection is required; and any other information deemed necessary by the committee.

§945.125 Disqualification.

The committee from time to time may conduct surveys of handling of potatoes for special purposes requiring Certificates of Privilege to determine whether handlers are complying with the requirements and regulations applicable to such certificates. Whenever the committee finds that a handler or consignee is failing to comply with requirements and regulations applicable to handling of potatoes in special outlets, and requiring such certificates, a Certificate or Certificates of Privilege issued such handler may be rescinded and further certificates denied. Such disqualification shall apply to, and not exceed, a reasonable period of time as determined by the committee but in no event shall it extend beyond the end of

the succeeding fiscal period. Any handler who has a certificate rescinded or denied may appeal to the committee in writing for reconsideration of his disqualification.

Subpart C—Assessment Rates

§945.249 Assessment rate.

On and after August 1, 2017, an assessment rate of $0.002 per hundredweight is established for Idaho-Eastern Oregon potatoes.

[82 FR 28552, June 23, 2017]

Subpart D—Handling Requirements

§945.341 Handling regulation.

No person shall handle any lot of potatoes unless such potatoes meet the requirements of paragraphs (a) through (d) of this section, or unless such potatoes are handled in accordance with paragraphs (e) and (f), or (g) of this section.

(a) *Minimum quality requirements*—(1) *Grade—All varieties.* U.S. No. 2 or better grade.

(2) *Size*—(i) *All varieties, except Russet types.* 1⅞ inches minimum diameter, unless otherwise specified on the container in connection with the grade.

(ii) *Russet types.* 2 inches minimum diameter, or 4 ounces minimum weight: *Provided,* That at least 40 percent of the potatoes in each lot shall be 5 ounces or heavier.

(iii) *All varieties, U.S. No. 1 grade or better.* (A) Size B (1½ to 2¼ inches diameter).

(B) Creamer (¾ to 1⅝ inches diameter).

(3) *Cleanness—All varieties.* "Fairly clean."

(b) *Minimum maturity requirements*—(1) *White Rose and red skin varieties.* Each year from August 1 through December 31, "moderately skinned"; during other periods no maturity requirements.

(2) *All other varieties.* "Slightly skinned."

(3) *Exceptions.* (i) Subject to complaince with paragraph (b)(3)(iii) of this section, any lot of potatoes not exceeding a total of 50 hundredweight of such variety may be handled for any

producer without regard to the foregoing maturity requirements.

(ii) If an officially inspected lot of potatoes meets the foregoing maturity requirements, but fails to meet the grade and size requirements, the lot may be regraded. If, after regrading, such lot then meets the grade and size requirements but fails to meet the maturity requirements, as indicated by the applicable Federal-State inspection certificate, such lot if not exceeding 100 hundredweight shall be exempt from the foregoing maturity requirements if the handler complies with paragraph (b)(3)(iii) of this section.

(iii) Prior to each shipment of potatoes exempt from the foregoing maturity requirements, the handler thereof shall report to the committee the name and address of the producer of such potatoes, and each such shipment shall be handled as an identifiable entity.

(c) *Pack and marking.* (1) When 50-pound containers (except master containers) of potatoes are marked with a count, size or similar designation, they must meet the count, average count and weight ranges for the count designation listed below.

| Size | Range | | |
	Count	Average count [1]	Weight
Larger than 50	(2)	(3)	(4)
50	45–55	48–53	12–19
60	54–66	57–63	10–16
70	63–77	67–74	9–15
80	72–88	76–84	8–13
90	81–99	86–95	7–12
100	90–110	95–105	6–10
110	99–121	105–116	5–9
120	108–132	114–126	4–8
130	117–143	124–137	4–8
140	126–154	133–147	4–8
Smaller than 140	(2)	(3)	4–8

[1] Applicable to lots.
[2] 10 percent over or under.
[3] 5 percent over or under.
[4] 15 ounces or larger.

The following tolerances by weight, are provided for potatoes in any lot which fail to meet the weight range for the designated count:

(i) Not to exceed 5 percent for undersize; and

(ii) Not to exceed 10 percent for oversize.

(2) Potatoes packed in cartons (except when used as a master container) shall be either:

(i) U.S. No. 1 grade or better, except potatoes of U.S. Extra No. 1 shall be no smaller than 110 size nor larger than 60 size; or

(ii) U.S. No. 2 grade in 50-pound fiberboard cartons of natural kraft color, provided the cartons are permanently and conspicuously marked as to grade.

(3) Size shall be conspicuously marked on all cartons (except when used as a master container) consistent with § 51.1545 of the United States Standards for Grades of Potatoes (7 CFR 51.1540–51.1566).

(d) *Inspection.* Except when relieved of such requirement pursuant to paragraphs (e) and (f), or (g) of this section:

(1) No handler shall handle potatoes unless such potatoes are inspected by either the Idaho Federal-State Inspection Service or Oregon Federal-State Inspection Service and are covered and accompanied by a valid inspection certificate, numbered notesheet, or shipping clearance report: *Provided,* That a valid inspection certificate, numbered notesheet, or shipping clearance report is not required to accompany positive lot identified potatoes.

(2) Each lot shipped shall be accompanied by a copy of a valid inspection certificate, a numbered notesheet, shipping clearance report, or the lot must meet PLI requirements established by the Fresh Products Branch, Fruit and Vegetable Division, Agricultural Marketing Service of the U.S. Department of Agriculture.

(3) Inspection certificates, numbered notesheets or shipping clearance reports for potatoes to be shipped must be issued within four days of such shipment. Otherwise, such potatoes, including lots that are positive lot identified, can only be shipped if a new inspection is performed to verify that the potatoes meet the requirements specified in paragraphs (a), (b), and (c) of this section. If the subsequent inspection verifies that the lot meets the requirements of paragraphs (a), (b), and (c) of this section, a new certificate, a new numbered notesheet, or a new shipping clearance report shall be issued and, if positive lot identified, shall reference the original PLI number, and a new PLI number need not be applied to the

lot. However, if upon subsequent inspection, the lot does not meet the requirements specified in either paragraphs (a), (b), or (c) of this section, the lot shall be reconditioned in the presence of an authorized representative of the Idaho Federal-State Inspection Service or Oregon Federal-State Inspection Service prior to the close of the business day. If the lot is reconditioned prior to the close of the business day, a new certificate, a new numbered notesheet, or a new shipping clearance report must be issued, and either a new PLI number must be applied to the lot or the original PLI number must be modified. If the PLI numbered lot is not reconditioned prior to the close of the business day, all PLI numbers must be obliterated. Any inspection certificate, numbered notesheet, or shipping clearance report issued upon a subsequent inspection, including when a lot is reconditioned, must be issued within four days of shipment of the potatoes.

(4) Handlers shall provide the Committee with the destination zip codes of all potatoes handled by permitting the Idaho Federal-State Inspection Service or Oregon Federal-State Inspection Service to review the bills of lading upon inspection to determine the destination zip codes. The destination zip codes shall be included on the inspection certificates. The destination zip codes and the quantity shall be provided by the handler to the Committee on lots which are positive lot identified, either orally or in writing. Whenever potatoes are diverted to a different destination, the handler shall notify the Committee of the new destination zip code and quantity orally or in writing as soon as practicable.

(e) *Special purpose shipments.* (1) The minimum grade, size, cleanness, maturity, and pack requirements set forth in paragraphs (a), (b), and (c) of this section shall not be applicable to shipments of prepeeled potatoes as defined in paragraph (h) of this section or potatoes for any of the following purposes:

(i) Charity;

(ii) Certified seed;

(iii) Experimentation; and

(iv) Canning, freezing and "other processing" as hereinafter defined. Also, shipments of potatoes for the purpose specified in this subdivision

(iv) shall be exempt from inspection requirements specified in §945.65 and paragraph (d) of this section and from assessment requirements specified in §945.42.

(2) The minimum grade, size, cleanness, maturity and pack requirements set forth in paragraphs (a), (b), (c) and (d) of this section shall be applicable to shipment of potatoes for each of the following purposes:

(i) *Export:* Except potatoes of a size not smaller than 1½ inches in diameter may be shipped if the potatoes grade not less than U.S. No. 2; and

(ii) *Prepeeling:* Except potatoes of a size not smaller than 1½ inches in diameter may be shipped if the potatoes grade not less than Idaho Utility or Oregon Utility grade.

(f) *Safeguards.* (1) Each handler making shipments of potatoes for charity, experimentation, or export pursuant to paragraph (e) of this section shall:

(i) First, apply to the committee for and obtain a Certificate of Privilege to make shipments for each purpose;

(ii) Upon request by the committee, furnish reports of each shipment pursuant to the applicable Certificate of Privilege;

(iii) At the time of applying to the committee for a Certificate of Privilege, or promptly thereafter, furnish the committee with a receiver's or buyer's certification that the potatoes so handled are to be used only for the purpose stated in the application and that such receiver will complete and return to the committee such periodic receiver's reports that the committee may require.

(iv) Mail to the office of the committee a copy of the bill of lading for each Certificate of Privilege shipment promptly after the date of shipment, unless other arrangements are made with the committee office;

(v) Bill each shipment directly to the applicable receiver.

(2) Each handler making shipments of potatoes for canning, freezing, or "other processing" pursuant to paragraph (e) of this section shall:

(i) First apply to the committee for and obtain a Certificate of Privilege to make shipments for processing;

(ii) Make shipments only to those firms whose names appear on the committee's current list of manufacturers of potato products;

(iii) Upon request by the committee, furnish reports of each shipment pursuant to the applicable Certificate of Privilege;

(iv) Mail to the committee's office a copy of the bill of lading for each Certificate of Privilege shipment promptly after the date of shipment, unless other arrangements are made with the committee office;

(v) Bill each shipment directly to the applicable processor.

(3) Each receiver of potatoes for processing pursuant to paragraph (e) of this section shall:

(i) Complete and return an application form for listing as a manufacturer of potato products;

(ii) Certify to the committee and to the Secretary that potatoes received from the production area for processing will be used for such purposes and will not be placed in fresh market channels;

(iii) Report on shipments received as the committee may require and the Secretary approve.

(4) Each handler making shipments of certified seed potatoes pursuant to paragraph (e) of this section shall furnish, at the request of the committee, reports on the total volume of seed potatoes handled.

(g) *Minimum quantity exemption.* Each handler may ship up to, but not to exceed, five hundredweight of potatoes any day without regard to the inspection and assessment requirements of this part, but this exception shall not apply to any shipment that exceeds five hundredweight of potatoes.

(h) *Definitions.* The terms *U.S. Extra No. 1, U.S. No. 1, U.S. No. 2, Size B, fairly clean, moderately skinned,* and *slightly skinned* shall have the same meaning as when used in the United States Standards for Potatoes (7 CFR 51.1540–51.1566), including the tolerances set forth therein. The term *prepeeling* means the commercial preparation in a prepeeling plant of clean, sound, fresh potatoes by washing, peeling, or otherwise removing the outer skin, trimming, sorting, and properly treating to prevent discoloration preparatory to sale in one or more of the styles of peeled potatoes described in §52.2422 of the United States Standards for Peeled Potatoes (7 CFR 52.2421–52.2433). The term *other processing* has the same meaning as the term appearing in the act and includes, but is not restricted to, potatoes for dehydration, chips, shoestrings, starch, and flour. It includes only that preparation of potatoes for market which involves the application of heat or cold to such an extent that the natural form or stability of the commodity undergoes a substantial change. The act of peeling, cooling, slicing, dicing, or applying material to prevent oxidation does not constitute "other processing." The terms *Idaho Utility* grade and *Oregon Utility* grade shall have the same meaning as when used in the standards for potatoes for the respective State. Other terms used in this section shall have the same meaning as when used in Marketing Agreement No. 98 and Order No. 945, both as amended.

(Secs. 1–19, 48 Stat. 31, as amended; 7 U.S.C. 601–674)

[47 FR 34355, Aug. 9, 1982, as amended at 52 FR 5530, Feb. 25, 1987; 52 FR 41695, Oct. 30, 1987; 53 FR 48634, Dec. 2, 1988; 57 FR 62167, Dec. 30, 1992; 59 FR 46723, Sept. 12, 1994; 60 FR 57905, Nov. 24, 1995; 65 FR 25627, May 3, 2000; 65 FR 48144, Aug. 7, 2000; 67 FR 66531, Nov. 1, 2002; 70 FR 21330, Apr. 26, 2005; 71 FR 29567, May 23, 2006; 74 FR 45734, Sept. 4, 2009; 79 FR 45675, Aug. 6, 2014; 83 FR 49779, Oct. 3, 2018]

PART 946—IRISH POTATOES GROWN IN WASHINGTON

Subpart A—Order Regulating Handling

DEFINITIONS

Subpart B—Administrative Requirements

Subpart C—Handling Requirements

AUTHORITY: 7 U.S.C. 601–674.

Subpart A—Order Regulating Handling

SOURCE: 17 FR 2912, Apr. 4, 1952, unless otherwise noted. Redesignated at 26 FR 12751, Dec. 30, 1961.

DEFINITIONS

§ 946.1 Secretary.

Secretary means the Secretary of Agriculture of the United States, or any other officer, or member of the United States Department of Agriculture, who is, or may hereafter be authorized to exercise the powers and to perform the duties of the Secretary of Agriculture.

§ 946.2 Act.

Act means Public Act No. 10, 73d Congress, as amended and reenacted and amended by the Agricultural Marketing Agreement Act of 1937, as amended (48 Stat. 31, as amended; 7 U.S.C. 601 *et seq.*).

§ 946.3 Person.

Person means an individual, partnership, corporation, association, legal representative, or any organized group or business unit.

§ 946.4 Production area.

Production area means all territory included within the boundaries of the State of Washington.

§ 946.5 Potatoes.

Potatoes means all varieties of Irish potatoes grown within the State of Washington.

§ 946.6 Handler.

Handler is synonymous with *shipper* and means any person (except a common or contract carrier of potatoes owned by another person) who handles potatoes or causes potatoes to be handled.

[37 FR 10916, June 1, 1972]

§ 946.7 Handle.

Handle is synonymous with *ship* and means to transport, sell, or in any other way to place potatoes grown in the State of Washington, or cause such potatoes to be placed, in the current of commerce within the production area or between the production area and any point outside thereof, or from any point in the adjoining States of Oregon and Idaho to any other point: *Provided,* That, the definition of "handle" shall not include the transportation of ungraded potatoes within the production area for the purpose of having such potatoes prepared for market, or stored, except that the committee may impose safeguards pursuant to § 946.55 with respect to such potatoes.

[37 FR 10916, June 1, 1972]

§ 946.8 Producer.

Producer means any person engaged in the production of potatoes for market.

§ 946.9 Fiscal period.

Fiscal period means the period beginning on July 1 of each year and ending June 30 of the following year, or such other period as the Secretary may establish pursuant to recommendation of the committee.

[37 FR 10916, June 1, 1972]

§ 946.10 Committee.

Committee means the administrative committee, called the State of Washington Potato Committee, established pursuant to § 946.22.

§ 946.11 Varieties.

Varieties means and includes all classifications or subdivisions of Irish potatoes according to those definitive characteristics now or hereafter recognized by the United States Department of Agriculture.

§ 946.12 Seed potatoes.

Seed potatoes means and includes all potatoes officially certified and tagged, marked or otherwise appropriately identified under the supervision of the official seed potato certifying agency of the State of Washington or other seed certification agencies which the Secretary may recognize.

§ 946.13 Grade and size.

Grade means any one of the officially established grades of potatoes, and *size* means any one of the officially established sizes of potatoes as defined and set forth in:

(a) The U.S. Standards for Potatoes issued by the U.S. Department of Agriculture (§§ 51.1540 to 51.1566 of this title), or amendments thereto or modifications thereof, or variations based thereon;

(b) U.S. Standards for Grades of Potatoes for Processing as issued by the U.S. Department of Agriculture (§§ 51.3410 to 51.3424 of this title), or amendments thereto, or modifications thereof, or variations based thereon;

(c) U.S. Standards for Grades of Peeled Potatoes (§§ 52.2421 to 52.2433 of this title), or amendments thereto or modifications thereof, or variations based thereon; and

(d) State of Washington Standards for Potatoes issued by the State of Washington Director of Agriculture, or amendments thereto, or modifications thereof, or variations based thereon.

[37 FR 10916, June 1, 1972]

§ 946.14 Grading.

Grading is synonymous with *preparing for market* which means the sorting or separating of potatoes into grades and sizes for market purposes.

[37 FR 10916, June 1, 1972]

§ 946.15 Export.

Export means shipment of potatoes beyond the boundaries of the 48 contiguous States of the United States, or the District of Columbia.

[37 FR 10916, June 1, 1972]

§946.16 District.

District means each one of the geographical divisions of the production area established pursuant to §946.31.

[17 FR 2912, Apr. 4, 1952. Redesignated at 26 FR 12751, Dec. 30, 1961, and further redesignated at 37 FR 10916, June 1, 1972]

§946.17 Pack.

Pack means a quantity of potatoes in any type of container and which falls within the specific weight limits or within specific grade and/or size limits, or any combination thereof, recommended by the committee and approved by the Secretary.

[70 FR 41133, July 18, 2005]

§946.18 Container.

Container means a sack, box, bag, crate, hamper, basket, carton, package, barrel, or any other type of receptacle used in the packing, transportation, sale or other handling of potatoes.

[70 FR 41133, July 18, 2005]

ADMINISTRATIVE COMMITTEE

§946.22 Establishment and membership.

(a) The State of Washington Potato Committee consisting of fifteen members, of whom ten shall be producers and five shall be handlers, is hereby established. For each member of the committee there shall be an alternate who shall have the same qualifications as the member.

(b) The Secretary, upon recommendation of the committee, may reestablish districts, may reapportion members among districts, may change the number of members and alternate members, and may change the composition by changing the ratio of members, including their alternates. In recommending any such changes, the following shall be considered:

(1) Shifts in acreage within districts and within the production area during recent years;

(2) The importance of new production in its relation to existing districts;

(3) The equitable relationship between committee apportionment and districts; and

(4) Other relevant factors.

[17 FR 2912, Apr. 4, 1952. Redesignated at 26 FR 12751, Dec. 30, 1961, as amended at 70 FR 41133, July 18, 2005]

§946.23 Alternate members.

(a) An alternate member of the committee shall act in the place and stead of the member for whom he is an alternate, during such member's absence. In the event of the death, removal, resignation, or disqualification of a member, his alternate shall act for him until a successor of such member is selected and has qualified.

(b) In the event that both a member and his or her alternate are unable to attend a Committee meeting, the member, the alternate member, or the Committee members present, in that order, may designate another alternate of the same classification (handler or producer) to serve in such member's place and stead.

[17 FR 2912, Apr. 4, 1952. Redesignated at 26 FR 12751, Dec. 30, 1961, as amended at 70 FR 41133, July 18, 2005]

§946.24 Procedure.

(a) Sixty percent of the committee members shall constitute a quorum and a concurring vote of 60 percent of the committee members will be required to pass any motion or approve any committee action.

(b) The quorum and voting requirements of paragraph (a) of this section shall not apply to the designation of temporary alternates as provided in §946.23.

(c) The committee may provide for meetings by telephone, telegraph, or other means of communication and any vote cast at such a meeting shall be confirmed promptly in writing: *Provided,* That if any assembled meeting is held, all votes shall be cast in person.

[17 FR 2912, Apr. 4, 1952. Redesignated at 26 FR 12751, Dec. 30, 1961, as amended at 70 FR 41133, July 18, 2005]

§946.25 Selection.

(a) Persons selected as committee members or alternates to represent producers shall be individuals who are producers of fresh potatoes in the respective district for which selected, or officers or employees of a corporate

311

producer in such district. Such individuals must also have produced potatoes for the fresh market for at least three out of the five years prior to nomination.

(b) Persons selected as committee members or alternates to represent handlers shall be individuals who are handlers in the State of Washington, or officers or employees of a corporate handler in the aforesaid State, and such persons shall be residents of the State of Washington.

(c) The Secretary shall select committee membership so that, during each fiscal period, each district, as designated in § 946.31, will be represented as follows:

(1) District No. 1—Three producer members and one handler member;

(2) District No. 2—Two producer members and one handler member;

(3) District No. 3—Two producer members and one handler member;

(4) District No. 4—Two producer members and one handler member;

(5) District No. 5—One producer member and one handler member.

[17 FR 2912, Apr. 4, 1952. Redesignated at 26 FR 12751, Dec. 30, 1961, as amended at 37 FR 10916, June 1, 1972; 52 FR 13070, Apr. 21, 1987; 70 FR 41133, July 18, 2005]

§ 946.26 Acceptance.

Any person selected by the Secretary as a committee member or as an alternate shall qualify by filing a written acceptance with the Secretary within ten days after being notified of such selection.

§ 946.27 Term of office.

(a) The term of office of committee members and alternates shall be for 3 years beginning on the 1st day of July and continuing until their successors are selected and have qualified: *Provided, however,* That the terms of office of the initial committee under the amended order shall be determined by the Secretary so that the terms of office of one-third of the initial members and alternates shall be for 1 year, one-third for 2 years, and one-third for 3 years.

(b) Committee members and alternates shall serve during the term of office for which they are selected and have qualified, or during that portion

thereof beginning on the date on which they qualify during the term of office and continuing until the end thereof, and until their successors are selected and have qualified.

[17 FR 2912, Apr. 4, 1952. Redesignated at 26 FR 12751, Dec. 30, 1961, as amended at 37 FR 10916, June 1, 1972]

§ 946.28 Powers.

The committee shall have the following powers:

(a) To administer the provisions of this subpart in accordance with its terms;

(b) To make rules and regulations to effectuate the terms and provisions of this subpart;

(c) To receive, investigate, and report to the Secretary complaints of violation of the provisions of this subpart; and

(d) To recommend to the Secretary amendments to this subpart.

§ 946.29 Duties.

It shall be the duty of the committee:

(a) At the beginning of each fiscal year, to meet and organize, to select a chairman and such other officers as may be necessary, to select subcommittees of committee members, and to adopt such rules and regulations for the conduct of its business as it may deem advisable;

(b) To act as intermediary between the Secretary and any producer or handler;

(c) To furnish to the Secretary such available information as he may request;

(d) To appoint such employees, agents, and representatives as it may deem necessary and to determine the salaries and define the duties of each such person;

(e) To investigate, from time to time, and to assemble data on the growing, harvesting, shipping, and marketing conditions with respect to potatoes, and to engage in such research and service activities which relate to the handling or marketing of potatoes as may be approved by the Secretary;

(f) To keep minutes, books, and records which clearly reflect all of the acts and transactions of the committee and such minutes, books, and records shall be subject to examination at any time by the Secretary or his authorized agent or representative;

(g) To make available to producers and handlers the committee voting record on recommended regulations and on other matters of policy;

(h) At the beginning of each fiscal year, to submit to the Secretary a budget of its expenses for such fiscal year, together with a report thereon;

(i) To cause the books of the committee to be audited by a competent accountant at least once each fiscal year, and at such other time as the committee may deem necessary or as the Secretary may request. The report of such audit shall show the receipt and expenditure of funds collected pursuant to this subpart; a copy of each such report shall be furnished to the Secretary and a copy of each such report shall be made available at the principal office of the committee for inspection by producers and handlers; and

(j) To consult, cooperate, and exchange information with the other potato marketing committees and other individuals or agencies in connection with all proper committee activities and objectives under this subpart.

§946.30 Expenses and compensation.

Committee members and their respective alternates when acting on committee business shall be reimbursed for reasonable expenses necessarily incurred by them in the performance of their duties and in the exercise of their powers under this subpart. In addition, they may receive reasonable compensation at a rate recommended by the committee and approved by the Secretary.

[37 FR 10916, June 1, 1972]

§946.31 Districts.

For the purpose of determining the basis for selecting committee members, the following districts of the production area are hereby established:

(a) District No. 1—The counties of Ferry, Stevens, Pend Oreille, Spokane, Whitman, and Lincoln, plus the East Irrigation District of the Columbia Basin Project, plus the area of Grant County not included in either the Quincy or South Irrigation Districts which lies east of township vertical line R27E,

plus the area of Adams County not included in either of the South or Quincy Irrigation Districts.

(b) District No. 2—The counties of Kittitas, Douglas, Chelan, and Okanogan, plus the Quincy Irrigation District of the Columbia Basin Project, plus the area of Grant County not included in the East or South Irrigation Districts which lies west of township line R28E.

(c) District No. 3—The counties of Benton, Klickitat, and Yakima.

(d) District No. 4—The counties of Walla Walla, Columbia, Garfield, and Asotin, plus the South Irrigation District of the Columbia Basin Project, plus the area of Franklin County not included in the South District.

(e) District No. 5—All of the remaining counties in the State of Washington not included in Districts No. 1, 2, 3, and 4 of this section.

[70 FR 41134, July 18, 2005]

§946.32 Nomination.

The Secretary may select the members of the State of Washington Potato Committee and their respective alternates from nominations which may be made in the following manner, or from among such other qualified persons:

(a) Nominations for Committee members and alternate members shall be made at a meeting or meetings of producers and handlers held by the Committee or at other industry meetings or events not later than May 1 of each year; or the Committee may conduct nominations by mail not later than May 1 of each year in a manner recommended by the Committee and approved by the Secretary.

(b) At least one nominee shall be designated for each position as member and for each position as alternate member on the committee which is vacant, or which is to become vacant the following July 1;

(c) The names of nominees shall be supplied to the Secretary in such manner and form as he may prescribe, not later than June 1 of each year, or by such other date as may be specified by the Secretary;

(d) Only producers may participate in designating producer nominees, and only handlers may participate in designating handler nominees. Any person who operates in more than one district or is engaged in producing and handling potatoes, shall elect the classification (i.e., producer or handler), and the district within which he desires to participate in designating nominees;

(e) Regardless of the number of districts in which a person produces or handles potatoes, each such person is entitled to cast only one vote on behalf of himself, his agents, subsidiaries, affiliates, and representatives in designating nominees for committee members and alternates. An eligible voter's privilege of casting only one vote as aforesaid shall be construed to permit a voter to cast one vote for each position to be filled in the district in which he elects to vote; and

(f) If nominations are not made within the time and in the manner specified in this section, the Secretary may, without regard to nominations, select the committee members and alternates on the basis of the representation provided for in this subpart.

[37 FR 10916, June 1, 1972, as amended at 70 FR 41134, July 18, 2005]

§ 946.33 Vacancies.

To fill any vacancy occasioned by the failure of any person selected as a committee member or as an alternate to qualify, or in the event of the death, removal, resignation, or disqualification of any qualified member or alternate, a successor for his unexpired term may be selected by the Secretary from nominations made in the manner specified in § 946.32, or the Secretary may select such committee member or alternate from previously unselected nominees on the current nominee list from the district involved. If the names of nominees to fill any such vacancy are not made available to the Secretary within 30 days after such vacancy occurs, the Secretary may fill such vacancy without regard to nominations, which selection shall be made on the basis of the representation provided for in this subpart.

[17 FR 2912, Apr. 4, 1952. Redesignated at 26 FR 12751, Dec. 30, 1961, and further redesignated at 37 FR 10917, June 1, 1972]

EXPENSES AND ASSESSMENTS

§ 946.40 Expenses.

The committee is authorized to incur such expenses as the Secretary finds are reasonable and likely to be incurred by it during each fiscal period for its maintenance and functioning, and for such other purposes as the Secretary, pursuant to this subpart, determines to be appropriate. The committee shall submit to the Secretary a budget for each fiscal period, including an explanation of the items appearing therein, and a recommendation as to the rate of assessment for such fiscal period.

[37 FR 10917, June 1, 1972]

§ 946.41 Assessments.

Each handler shall pay to the committee upon demand, his pro rata share of the expenses authorized by the Secretary for each fiscal period. Each handler's pro rata share shall be the rate of assessment per hundredweight fixed by the Secretary times the quantity of potatoes which he handles as the first handler thereof. At any time during or after a fiscal period, the Secretary may increase the rate of assessment as necessary to cover authorized expenses. Such increase shall be applicable to all potatoes handled during the given fiscal period. The payment of expenses for the maintenance and functioning of the committee may be required during periods when no regulations are in effect. If a handler does not pay his assessment within the time prescribed by the committee, the assessment may be increased by a late payment charge or an interest charge, or both, at rates prescribed by the committee with the approval of the Secretary.

[37 FR 10917, June 1, 1972]

§ 946.42 Accounting.

(a) *Excess funds.* At the end of a fiscal period, funds in excess of the year's expenses shall be placed in an operating reserve not to exceed approximately two fiscal periods' operational expenses or such lower limits as the committee, with the approval of the Secretary, may establish. Funds in such reserve shall be available for use by the committee for expenses authorized pursuant to §946.40. Funds in excess of those placed in the operating reserve shall be refunded to handlers. Each handler's share of such excess shall be the amount of assessments he paid in excess of his pro rata share of the actual expenses of the committee and the addition, if any, to the operating reserve.

(b) *Accounting of funds upon termination of order.* Any money collected as assessments pursuant to this subpart and remaining unexpended in the possession of the committee after termination of this part shall be distributed in such manner as the Secretary may direct: *Provided,* That to the extent practical, such funds shall be returned pro rata to the persons from whom such funds were collected.

[37 FR 10917, June 1, 1972]

§946.43 Funds.

All funds received by the committee pursuant to any provisions of this subpart shall be used solely for the purposes specified in this subpart and shall be accounted for in the following manner:

(a) The Secretary may at any time require the committee and its members to account for all receipts and disbursements; and

(b) Whenever any person ceases to be a committee member or alternate, he shall account for all receipts and disbursements and deliver all property and funds in his hands, together with all books and records in his possession, to his successor in office or to such person as the Secretary may designate, and shall execute such assignments and other instruments as may be necessary or appropriate to vest in such successor or in such designated person the right to all the property, funds, or claims vested in such member or alternate.

REGULATION

§946.50 Marketing policy.

(a) Prior to each marketing season, the committee shall consider and prepare a policy statement for the marketing of potatoes. In developing its marketing policy, the committee shall investigate relevant supply and demand conditions for potatoes. In such investigations, the committee shall give appropriate considerations to the following:

(1) Market prices of potatoes, including prices by grade, size, quality, and maturity in different packs of fresh potatoes and of the various forms of processed potatoes;

(2) Supplies of potatoes by grade, size, quality, and maturity in the production area and in other production areas, of fresh potatoes, and the supplies of various forms of processed potatoes;

(3) The trend and level of consumer income;

(4) Establishing and maintaining orderly marketing conditions for potatoes;

(5) Orderly marketing of potatoes as will be in the public interest; and

(6) Other relevant factors.

(b) In the event it becomes advisable to deviate from such marketing policy because of changed supply and demand conditions, the committee shall formulate a revised marketing policy statement in accordance with the appropriate considerations in paragraph (a) of this section.

(c) The committee shall submit a report to the Secretary setting forth such marketing policy. Notice of each such marketing policy and any revision thereof shall be given to producers, handlers, and other interested parties by bulletins, newspapers, or other appropriate media, and copies thereof shall be available for examination at the committee office to all interested parties.

[37 FR 10917, June 1, 1972]

§946.51 Recommendation for regulations.

The committee shall recommend to the Secretary regulations, or amendments, modifications, suspension, or termination thereof, whenever it finds that such regulations as provided in §946.52 are in accordance with the marketing policy established pursuant to §946.50 and that such regulations will tend to effectuate the declared policy of the act.

[37 FR 10917, June 1, 1972]

§ 946.52 Issuance of regulations.

(a) The Secretary shall limit the shipment of potatoes as set forth in this subpart whenever he finds from the recommendation and information submitted by the committee, or from other available information, that it would tend to effectuate the declared policy of the act:

(1) To regulate, in any or all portions of the production area the handling of particular grades, sizes, qualities or maturity of any or all varieties of potatoes during any period;

(2) To regulate the handling of particular grades, sizes, qualities or maturities of any or all varieties of potatoes, or for any combination of the foregoing during any period in the States of Oregon and Idaho which have been shipped from the production area to specified locations therein for grading or storage pursuant to § 946.54;

(3) To regulate the handling of particular grades, sizes, qualities or maturities of any or all varieties differently for: Different portions of the production area, different uses or outlets, different packs or for any combination of the foregoing, during any period;

(4) To regulate the handling of potatoes by establishing in terms of grades, sizes, or both, minimum standards of quality and maturity.

(5) To regulate the size, capacity, weight, dimensions, pack, and marking or labeling of the container, or containers, which may be used in the packing or handling of potatoes, or both.

(b) The Secretary may amend any regulation issued under this subpart whenever he finds that such amendment would tend to effectuate the declared policy of the act. The Secretary may also terminate or suspend any regulation whenever he finds that such regulation obstructs or no longer tends to effectuate the declared policy of the act.

(c) The Secretary shall notify the committee of any such regulation issued pursuant to this section and the committee shall give reasonable notice thereof to handlers.

[37 FR 10917, June 1, 1972, as amended at 70 FR 41134, July 18, 2005]

§ 946.53 Minimum quantities.

The committee, with the approval of the Secretary, may establish, for any or all portions of the production area, minimum quantities below which shipments will be free from regulations issued pursuant to this part.

[37 FR 10918, June 1, 1972]

§ 946.54 Shipments for specified purposes.

(a) Whenever the Secretary finds, upon the basis of the recommendations and information submitted by the committee, or from other available information, that it will tend to effectuate the declared policy of the act, he shall modify, suspend, or terminate any or all regulations issued pursuant to this part in order to facilitate shipments of potatoes for the following purposes:

(1) Livestock feed;

(2) Charity;

(3) Export;

(4) Seed;

(5) Prepeeling;

(6) Such other purposes as may be specified by the committee with the approval of the Secretary; and

(7) Grading or storing between the districts within the production area or to and within specified locations in the adjoining States of Idaho and Oregon.

(b) The Secretary shall give prompt notice to the committee of any modification, suspension, or termination of regulations pursuant to this section, or of any approval issued by him under the provisions of this section.

[37 FR 10918, June 1, 1972]

§ 946.55 Safeguards.

(a) The committee, with the approval of the Secretary, may prescribe adequate safeguards to prevent shipments pursuant to § 946.54 from entering channels of trade and other outlets for other than the specific purposes authorized therefor, and the transportation of potatoes for grading and storing to points outside the production area.

(b) Safeguards provided by this section may include, but shall not be limited to, requirements that handlers:

(1) Shall obtain the inspection required by § 946.60 or pay the assessment

provided by § 946.41, or both, in connection with the potato shipments effected in accordance with § 946.54, and

(2) Shall obtain a special purpose certificate from the committee for shipments of potatoes effected or to be effected under provisions of § 946.54.

(c) The committee, with the approval of the Secretary, shall prescribe rules governing the issuance and the contents of the special purpose certificate.

(d) The committee may rescind, or deny to any handler the special purpose certificate if proof satisfactory to the committee is obtained that potatoes shipped by him for the purpose stated in § 946.54 were handled contrary to the provisions of this section.

(e) The committee shall make reports to the Secretary, as requested, showing the number of applications for such certificates, the quantity of potatoes covered by such applications for such certificates, the number of such applications denied, and certificates granted, the quantity of potatoes shipped under duly issued certificates, and such other information as may be requested by the Secretary.

[37 FR 10918, June 1, 1972]

INSPECTION AND CERTIFICATION

§ 946.60 Inspection and certification.

(a) During any period in which the Secretary regulates the shipment of potatoes pursuant to the provisions of this subpart, each handler who first ships potatoes shall, prior to making shipment, cause each shipment to be inspected by an authorized representative of the Federal-State inspection service or such other inspection service as the Secretary shall designate. The committee may, with the approval of the Secretary, prescribe rules and regulations modifying the inspection requirements of this section in circumstances under which such requirements would create an undue hardship on growers or shippers: *Provided,* That all such shipments shall comply with all regulations in effect: *And provided further,* That proper safeguards to assure compliance are adopted.

(b) Each such handler shall make arrangements with the inspecting agency to forward promptly to the committee a copy of such inspection certificate:

Provided, however, That (1) each handler making shipments of potatoes during such period shall prior to making such shipment, determine if such shipment has been inspected and if such shipment has not been so inspected and is not covered by an inspection certificate, each handler making such determinations shall have such potatoes inspected and shall arrange for a copy of the inspection certificate to be forwarded to the committee as aforesaid, and (2) each handler who first ships potatoes after such potatoes are regraded, resorted, or repacked, or in any other way further prepared for market shall have each shipment of such potatoes inspected as provided in this section.

[17 FR 2912, Apr. 14, 1952. Redesignated at 26 FR 12751. Dec. 30, 1961, and further redesignated and amended at 37 FR 10918, June 1, 1972]

EFFECTIVE TIME AND TERMINATION

§ 946.62 Effective time.

The provisions of this subpart shall become effective at such time as the Secretary may declare above his signature attached to this subpart, and shall continue in force until terminated in one of the ways specified in this subpart.

§ 946.63 Termination.

(a) The Secretary may, at any time, terminate the provisions of this subpart by giving at least one day's notice by means of a press release or in any other manner which he may determine.

(b) The Secretary may terminate or suspend the operation of any or all of the provisions of this subpart whenever he finds that such provisions do not tend to effectuate the declared policy of the act.

(c) The Secretary shall terminate the provisions of this subpart at the end of any fiscal year whenever he finds that such termination is favored by a majority of producers who, during the preceding fiscal year, have been engaged in the production for market of potatoes: *Provided,* That such majority has during such year, produced for market more than fifty percent of the volume of such potatoes produced for market; but such termination shall be effective only if announced on or before May 31 of the then current fiscal year.

(d) The Secretary shall conduct a referendum six years after the effective date of this paragraph and every sixth thereafter to ascertain whether producers favor continuance of this part.

(e) The provisions of this subpart shall, in any event, terminate whenever the provisions of the act authorizing them cease to be in effect.

[37 FR 10917, June 1, 1972, as amended at 70 FR 41134, July 18, 2005]

§ 946.64 Proceedings after termination.

(a) Upon the termination of the provisions of this subpart, the then functioning members of the committee shall continue as trustees, for the purpose of liquidating the affairs of the committee, of all the funds and property then in the possession of or under control of the committee, including claims for any funds unpaid or property not delivered at the time of such termination. Action by said trusteeship shall require the concurrence of a majority of the said trustees.

(b) The said trustee shall continue in such capacity until discharged by the Secretary; shall from time to time, account for all receipts and disbursements and deliver all property on hand, together with all books and records of the committee and of the trustees, to such person as the Secretary may direct; and shall upon request of the Secretary, execute such assignments or other instruments necessary or appropriate to vest in such person full title and right to all of the funds, property, and claims vested in the committee or the trustees pursuant thereto.

(c) Any person to whom funds, property, or claims have been transferred or delivered by the committee or its members, pursuant to this section, shall be subject to the same obligations imposed upon the members of the committee and upon the said trustees.

§ 946.65 Effect of termination or amendment.

Unless otherwise expressly provided by the Secretary the termination of this subpart or of any regulation issued pursuant to this subpart, or the issuance of any amendments to either thereof, shall not (a) affect or waive any right, duty, obligation, or liability which shall have arisen or which may thereafter arise in connection with any provisions of this subpart or any regulation issued under this subpart, or (b) release or extinguish any violation of this subpart or of any regulation issued under this subpart, or (c) affect or impair any rights or remedies of the Secretary or of any other person with respect to any such violation.

MISCELLANEOUS PROVISIONS

§ 946.70 Reports and records.

(a) Upon the request of the committee, with the approval of the Secretary, every handler shall furnish to the committee in such manner and at such time as may be prescribed, such information as will enable the committee to exercise its duties under this subpart.

(b) Each handler shall establish and maintain for at least 2 succeeding years such records and documents with respect to potatoes received and potatoes disposed of by him as will substantiate the required reports.

(c) For the purpose of assuring compliance with the recordkeeping requirements and verifying reports filed by handlers, the Secretary and the committee through its duly authorized employees, shall have access to such records.

(d) All reports and records furnished or submitted by handlers to, or obtained by the employees of, the committee which contain data for information constituting a trade secret or disclosing the trade position, financial condition, or business operations of the particular handler from whom received, shall be treated as confidential, and the reports and all information obtained from records shall at all times be kept in the custody and under the control of one or more employees of the committee who shall disclose such information to no person other than the Secretary, or his authorized agents. Compilations of general reports from data and information submitted by handlers is authorized subject to the prohibition of disclosure of individual handlers' identity or operations.

[37 FR 10918, June 1, 1972]

§946.71 Compliance.

Except as provided in this subpart, no handler shall ship potatoes, the shipment of which has been prohibited by the Secretary in accordance with provisions of this subpart, and no handler shall ship potatoes except in conformity to the provisions of this subpart.

§946.72 Right of the Secretary.

The members of the committee (including successors and alternates), and any agent or employee appointed or employed by the committee, shall be subject to removal or suspension by the Secretary at any time. Each and every order, regulation, decision, determination or other act of the committee shall be subject to the continuing right of the Secretary to disapprove of the same at any time. Upon such disapproval the disapproved action of the said committee shall be deemed null and void, except as to acts done in reliance thereon or in compliance therewith prior to such disapproval by the Secretary.

§946.73 Duration of immunities.

The benefits, privileges, and immunities conferred upon any person by virtue of this subpart shall cease upon the termination of this subpart, except, with respect to acts done under and during the existence of this subpart.

§946.74 Agents.

The Secretary may, by designation in writing, name any person, including any officer or employee of the Government or name any bureau or division in the United States Department of Agriculture, to act as his agent or representative in connection with any of the provisions of this subpart.

§946.75 Derogation.

Nothing contained in this subpart is, or shall be construed to be, in derogation or in modification of the rights of the Secretary or of the United States to exercise any powers granted by the act or otherwise, or, in accordance with such powers, to act in the premises whenever such action is deemed advisable.

§946.76 Personal liability.

No member or alternate of the committee, nor any employee or agent thereof, shall be held personally responsible, either individually or jointly with others, in any way whatsoever, to any handler or to any person for errors in judgment, mistakes, or other acts, either of commission or omission, as such member, alternate, or employee, except for acts of dishonesty.

§946.77 Separability.

If any provision of this subpart is declared invalid, or the applicability thereof to any person, circumstance, or thing is held invalid, the validity of the remainder of this subpart, or the applicability thereof, to any other person, circumstance, or thing, shall not be affected thereby.

§946.78 Amendments.

Amendments to this subpart may be proposed, from time to time, by the committee or by the Secretary.

Subpart B—Administrative Requirements

DEFINITIONS

§946.100 Order.

Order means Order No. 946 (§§946.1 to 946.78), as amended, regulating the handling of Irish potatoes grown in the State of Washington.

[39 FR 1971, Jan. 16, 1974]

§946.101 Marketing agreement.

Marketing agreement means Marketing Agreement No. 113, as amended.

[39 FR 1972, Jan. 16, 1974]

§946.102 Terms.

Terms used in this subpart shall have the same meaning as set forth in said marketing agreement and order.

[22 FR 8177, Oct. 16, 1957. Redesignated at 26 FR 12751, Dec. 30, 1961]

§946.103 Reestablishment of districts.

Pursuant to §946.22, on and after July 1, 2007, the following districts are reestablished:

(a) District No. 1—the counties of Douglas, Chelan, Okanogan, Grant,

Adams, Ferry, Stevens, Pend Oreille, Spokane, Whitman, and Lincoln.

(b) District No. 2—the counties of Kittitas, Yakima, Klickitat, Benton, Franklin, Walla Walla, Columbia, Garfield, and Asotin.

(c) District No. 3—all of the remaining counties in the State of Washington, not included in Districts No. 1 and No. 2 of this paragraph.

[72 FR 17795, Apr. 10, 2007]

§ 946.104 Reestablishment and reapportionment of committee.

(a) Pursuant to § 946.22, on and after July 1, 2007, the State of Washington Potato Committee consisting of nine members, of whom six shall be producers and three shall be handlers, is hereby reestablished. For each member of the committee there shall be an alternate who shall have the same qualifications as the member.

(b) Pursuant to § 946.22, on and after July 1, 2007, membership representation of the State of Washington Potato Committee shall be reapportioned among the districts of the production area so as to provide that each of the three districts as defined in § 946.103 are represented by two producer members and one handler member and their respective alternates.

[72 FR 17795, Apr. 10, 2007]

SPECIAL PURPOSE CERTIFICATES

§ 946.120 Application.

(a) Whenever shipments for special purposes pursuant to § 946.54 are relieved in whole or in part from regulations issued under § 946.52, each handler desiring to make shipments of potatoes for the following purposes shall submit an application to the committee, prior to initiating such shipments, for a special purpose certificate permitting such shipments:

(1) Charity: *Provided*, That handlers making shipments for charity of 1,000 pounds or less are exempt from these application requirements;

(2) Prepeeling;

(3) Canning, freezing, and "other processing";

(4) Grading or storing at any specified location in Morrow or Umatilla Counties in the State of Oregon; and

(5) Experimentation.

(b) Applications for special purpose shipment certificates shall be made on forms furnished by the committee. Such application shall contain the name and address of the handler, and such other information that the committee may require such as the estimated amount of potatoes to be shipped, the grades and sizes of potatoes to be shipped (when applicable), expected consignees and destinations, certification by applicant that statements are correct and that he will comply with disposition stated therein, and other information or documents as the committee may require in safeguarding against entry of such potatoes into trade channels other than those for which the special purpose certificate was granted.

[39 FR 1972, Jan. 16, 1974, as amended at 65 FR 70463, Nov. 24, 2000; 70 FR 44256, Aug. 2, 2005]

§ 946.121 Issuance.

The committee, or its duly authorized agents, shall give prompt consideration to each applicant for a special purpose certificate. Upon approval of the application, a special purpose certificate shall be issued authorizing the applicant named therein to ship potatoes for a specified purpose for a specified period of time.

[39 FR 1972, Jan. 16, 1974]

§ 946.122 Reports.

Each handler shipping potatoes under and pursuant to a special purpose certificate shall supply to the committee, upon request, a report thereon showing the name and address of the shipper, car or truck number, Federal-State Inspection Certificate number (if such inspection is required by regulations in effect at the time of such shipment), loading point, destination and consignee.

[39 FR 1972, Jan. 16, 1974]

§ 946.123 Denial and appeals.

The committee may rescind a special purpose certificate issued to a handler for the purpose specified in § 946.120(a), or deny such special purpose certificates to a handler, upon proof satisfactory to the committee that such handler has shipped potatoes contrary to those provisions. Such committee action denying or rescinding a special purpose certificate shall apply to and not exceed a reasonable period of time as determined by the committee. Any handler who has been denied a special purpose certificate or who has had a special purpose certificate rescinded may appeal to the committee for reconsideration. Such appeal shall be in writing.

[39 FR 1972, Jan. 16, 1974]

MODIFICATION OF INSPECTION
REQUIREMENTS

§946.130 Application.

Any handler whose packing facilities are located in an area where a Washington State Department of Agriculture, Plant Industry Division Office or Federal-State Inspector is not readily available to perform the inspection required by this part may, pursuant to §946.60(a), apply to the committee for a permit authorizing modification of inspection requirements. Applications shall be made on forms furnished by the committee and shall contain such information as the committee, with approval of the Secretary, may find necessary in making a determination regarding the issuance of such permit.

[39 FR 1972, Jan. 16, 1974]

§946.131 Issuance.

The committee, or its duly authorized agents, shall give prompt consideration to each application for an inspection modification permit. Approval of an application shall be evidenced by the issuance of an applicable permit.

[39 FR 1972, Jan. 16, 1974]

§946.132 Reports.

Each handler shipping potatoes pursuant to an inspection modification permit shall report periodically as specified by the committee on forms furnished by the committee the following information on each shipment: quantity of potatoes, variety or varieties, grade, minimum size, type of container(s), date of shipment, carrier, destination, and name and address of receiver.

[39 FR 1972, Jan. 16, 1974]

§946.133 Cancellation.

Whenever the committee finds that shipments of potatoes pursuant to an inspection modification permit are not in accordance with the application provisions of the order, such inspection modification permit may be cancelled.

[39 FR 1972, Jan. 16, 1974]

§946.140 Handling potatoes for commercial processing into products.

Pursuant to §946.54(a)(6), shipments of potatoes for commercial processing into products may be made only in accordance with paragraphs (a) or (b) of this section.

(a) Shipments may be made to persons whose names are on the State of Washington Potato Committee's list of manufacturers of potato products. Such list may consist of firms actively engaged in the business of canning, freezing, or "other processing" as defined in the act.

(1) Persons desiring to have their name placed on the committee's list shall apply to the committee. Such application shall contain the following:

(i) Name and address of applicant;

(ii) Location and description of facilities for commercial processing into products;

(iii) Expected source of potatoes for commercial processing into products;

(iv) Such other information as the committee, with approval of the Secretary, may deem necessary.

(2) Upon receipt of an application for such listing, the State of Washington Potato Committee shall make such investigation as it deems necessary, and if it appears that the applicant may reasonably be expected to use potatoes covered by the application in accordance with the requirements of this section, it shall place the applicant's name on the State of Washington Potato Committee's list of manufacturers of potato products.

(b) For each shipment to a person whose name is not on the committee's list, the handler must provide evidence to the committee prior to shipment that the potatoes will be used only for processing into products. Further, he shall submit reports as prescribed by the committee and approved by the Secretary.

[39 FR 1972, Jan. 16, 1974]

§ 946.141 Late payment and interest charge.

The Committee shall impose an interest charge on any handler who fails to pay his or her assessment within sixty (60) days of the billing date shown on the handler's assessment statement received from the Committee. The interest charge shall, after 60 days, be one percent of the unpaid assessment balance. In the event the handler fails to pay the delinquent assessment, the one percent interest charge shall be applied monthly thereafter to the unpaid balance, including any accumulated unpaid interest. Any amount paid by a handler as an assessment, including any charges imposed pursuant to this paragraph, shall be credited when the payment is received in the Committee office.

[73 FR 74348, Dec. 8, 2008]

§ 946.142 Operating reserve.

(a) The Committee, with the approval of the Secretary, may carry over excess funds into subsequent fiscal periods as an operating reserve: *Provided,* That funds in the operating reserve may not exceed approximately two fiscal periods' expenses.

(b) The funds in said operating reserve may be used:

(1) To defray expenses incurred during any fiscal period prior to the time assessment income is sufficient to cover such expenses,

(2) To cover deficits incurred during any fiscal period when assessment income is less than expenses,

(3) To defray expenses incurred during any period when assessments are suspended or are inoperative and

(4) To cover necessary expenses of liquidation in the event of termination of this part.

(c) Upon termination of this part any funds not required to defray the necessary expenses of liquidation shall be disposed of in such manner as the Secretary may determine to be appropriate. To the extent practical, such funds shall be returned pro rata to the handlers from whom they were collected.

(d) Terms used in this section shall have the same meaning as when used in said marketing agreement and this part.

[32 FR 16199, Nov. 28, 1967. Redesignated at 44 FR 73012, Dec. 17, 1979; 60 FR 27683, May 25, 1995]

§ 946.143 Assessment reports.

During the period that russet, red, yellow fleshed, and white types of potatoes are exempt from handling requirements under § 946.336, each person handling russet, red, yellow fleshed, and white types of potatoes shall submit a monthly report to the Committee by the 10th day of the month following the month such potatoes are handled. Each assessment report shall contain the following information:

(a) The name and address of the handler;

(b) The date and quantity of russet, red, yellow fleshed, and white types of potatoes handled;

(c) The assessment payment due; and

(d) Other information as may be requested by the Committee.

[79 FR 8256, Feb. 12, 2014]

§ 946.248 Assessment rate.

On and after July 1, 2013, an assessment rate of $0.0025 per hundredweight is established for Washington potatoes.

[78 FR 24983, Apr. 29, 2013]

Subpart C—Handling Requirements

§ 946.336 Handling regulation.

No person shall handle any lot of potatoes unless such potatoes meet the requirements of paragraphs (a), (b), (c), and (g) of this section or unless such potatoes are handled in accordance with paragraphs (d) and (e), or (f) of this section, except that shipments of the blue or purple flesh varieties of potatoes shall be exempt from both this handling regulation and the assessment requirements specified in § 946.41: *Provided,* That yellow fleshed, white, red, and russet type potatoes shall be exempt from the requirements of paragraphs (a), (b), (c), (e), and (g) of this section.

(a) *Minimum quality requirements*—(1) *Grade: All varieties*—U.S. No. 2 or better grade.

(2) *Size:* (i) At least 1⅞ inches in diameter, except that all red, yellow fleshed, and white types may be ¾ inch (19.1 mm) minimum diameter, if they otherwise meet the requirements of U.S. No. 1.

(ii) All Russet types, 2 inches (54.0 mm) minimum diameter, or 4 ounces minimum weight.

(iii) Any type of any size may be packed in a 3-pound or less container if the potatoes otherwise meet the requirements of U.S. No. 1 grade or better at the time of packing.

(iv) *Tolerances*—The tolerance for size contained in the U.S. Standards for Grades of Potatoes shall apply.

(3) *Cleanness:* All varieties and grades—as required in the United States Standards for Grades of Potatoes. For example: U.S. No. 2—"not seriously damaged by dirt," and U.S. No. 1—"fairly clean."

(b) *Minimum maturity requirements*—(1) *Red, yellow fleshed and white types:* Not more than "moderately skinned."

(2) *Russet types:* Not more than "slightly skinned."

(c) *Pack and marking:*

(1) *Domestic:* Potatoes packed in cartons shall be either:

(i) U.S. No. 1 grade or better, except that potatoes which fail to meet the U.S. No. 1 grade only because of internal defects may be shipped without regard to this requirement provided the lot contains no more than 10 percent damage by any internal defect or combination of internal defects but not more than 5 percent serious damage by any internal defect or combination of internal defects.

(ii) U.S. No. 2 grade, provided the cartons are permanently and conspicuously marked as to grade. This marking requirement does not apply to cartons containing potatoes meeting the requirements of (c)(1)(i).

(2) *Export:* Potatoes packed in cartons shall be U.S. No. 1 grade or better.

(d) *Special purpose shipments.* (1) The minimum grade, size, cleanness, maturity, and pack requirements set forth in paragraphs (a), (b), and (c) of this section shall not apply to shipments of potatoes for any of the following purposes:

(i) Livestock feed;

(ii) Charity;

(iii) Seed;

(iv) Prepeeling;

(v) Canning, freezing, and "other processing" as hereinafter defined;

(vi) Grading or storing at any specified location in Morrow or Umatilla Counties in the State of Oregon;

(vii) Experimentation.

(2) Shipments of potatoes for the purposes specified in paragraphs (d)(1)(i) through (vii) of this section shall be exempt from the inspection requirements specified in paragraph (g) of this section, except that shipments pursuant to paragraph (d)(1)(vi) of this section shall comply with the inspection requirements of paragraph (e)(2) of this section. Shipments specified in paragraphs (d)(1)(i), (ii), (iii), (v) and (vii) of this section shall be exempt from assessment requirements as specified in § 946.248 and established pursuant to § 946.41.

(e) *Safeguards.* (1) Handlers desiring to make shipments of potatoes for prepeeling shall:

(i) Notify the committee of intent to ship potatoes by applying on forms furnished by the committee for a certificate applicable to such special purpose shipments;

(ii) Prepare on forms furnished by the committee a special purpose shipment report on each such shipment, a copy of which must also accompany each shipment. The handler shall forward copies of each such special purpose shipment report to the committee office and to the receiver with instructions to the receiver to sign and return a copy to the committee office. Failure of the handler or receiver to report such shipments by promptly signing and returning the applicable special purpose shipment report to the committee office shall be cause for cancellation of such handler's certificate applicable to such special purpose shipments and/or the receiver's eligibility to receive further shipments pursuant to such certificate. Upon cancellation of such certificate, the handler may appeal to the committee for reconsideration; such appeal shall be in writing;

(iii) Before diverting any such special purpose shipment from the receiver of record as previously furnished to the committee by the handler such handler shall submit to the committee a revised special purpose shipment report.

(2) Handlers desiring to ship potatoes for grading or storing to any specified location in Morrow or Umatilla Counties in the State of Oregon shall:

(i) Notify the committee of intent to ship potatoes by applying on forms furnished by the committee for a certificate applicable to such special purpose shipment. Upon receiving such application, the committee shall supply to the handler the appropriate certificate after it has determined that adequate facilities exist to accommodate such shipments and that such potatoes will be used only for authorized purposes;

(ii) If reshipment is for any purpose other than as specified in paragraph (d) of this section, each handler desiring to make reshipment of potatoes which have been graded or stored shall, prior to reshipment, cause each such shipment to be inspected by an authorized representative of the Federal-State Inspection Service. Such shipments must comply with the minimum grade, size, cleanliness, maturity, and pack requirements specified in paragraphs (a), (b), and (c) of this section;

(iii) If reshipment is for any of the purposes specified in paragraph (d) of this section, each handler making reshipment of potatoes which have been graded or stored shall do so in accordance with the applicable safeguard requirements specified in paragraph (e) of this section.

(3) Each handler making shipments of potatoes for canning, freezing, or "other processing" pursuant to paragraph (d) of this section shall:

(i) First apply to the committee for and obtain a Special Purpose Certificate to make shipments for processing;

(ii) Make shipments only to those firms whose names appear on the committee's list of canners, freezers, or other processors of potato products maintained by the committee, or to persons not on the list provided the handler furnishes the committee, prior

to such shipment, evidence that the receiver may reasonably be expected to use the potatoes only for canning, freezing, or other processing;

(iii) Upon request by the committee, furnish reports, or cause reports to be furnished, for each shipment pursuant to the applicable Special Purpose Certificate;

(iv) Mail to the office of the committee a copy of the bill of lading for each Special Purpose Certificate shipment promptly after the date of shipment unless other arrangements are made;

(v) Bill each shipment directly to the applicable processor.

(4) Each receiver of potatoes for processing pursuant to paragraph (d) of this section shall:

(i) Complete and return an application form for consideration of approval as a canner, freezer, or other processor of potato products;

(ii) Certify to the committee and to the Secretary that potatoes received from the production area for processing will be used for such purpose and will not be placed in fresh market channels;

(iii) Report on shipments received as the committee may require and the Secretary approve.

(5) Each handler desiring to make shipments of potatoes for experimentation shall:

(i) First apply to the committee for and obtain a Special Purpose Certificate to make shipments for experimentation;

(ii) Upon request by the committee, furnish reports of each shipment pursuant to the applicable Special Purpose Certificate.

(6) Handlers diverting potatoes to livestock feed are not required to apply for a Special Purpose Certificate nor report such shipments to the committee.

(7) Each handler desiring to make shipments of potatoes for charity shall:

(i) First apply to the committee for, and obtain, a Special Purpose Certificate for the purpose of making shipments for charity: *Provided,* That shipments for charity of 1,000 pounds or less are exempt from the application and reporting requirements: *And provided further,* That potatoes previously graded, assessed, and inspected in preparation for shipment to the fresh market are exempt from the application and reporting requirements.

(ii) Each handler shipping potatoes to charity must inform the recipient that the potatoes cannot be resold or otherwise placed in commercial market channels.

(8) Each handler making shipments of seed potatoes shall furnish, at the request of the committee, reports on the total volume of seed potatoes handled.

(f) *Minimum quantity exemption.* Each handler may ship up to, but not to exceed 5 hundredweight of potatoes per day without regard to the inspection and assessment requirements of this part, but this exception shall not apply to any shipment over 5 hundredweight of potatoes.

(g) *Inspection.* (1) Except when relieved by paragraphs (d) or (f) of this section, no person may handle any potatoes unless a Federal-State Inspection Notesheet or certificate covering them has been issued by an authorized representative of the Federal-State Inspection Service and the document is valid at the time of shipment.

(2) U.S. No. 1 grade or better potatoes in the State of Washington which are resorted or repacked within 72 hours of being inspected and certified are exempt from reinspection.

(h) *Definitions.* The terms *U.S. No. 1, U.S. No. 2, not seriously damaged by dirt, fairly clean, slightly skinned,* and *moderately skinned* shall have the same meaning as when used in the United States Standards for Grades of Potatoes (7 CFR 51.1540–51.1566), including the tolerances set forth in it. The term *prepeeling* means the commercial preparation in the prepeeling plant of clean, sound, fresh tubers by washing, peeling or otherwise removing the outer skin, trimming, sorting, and properly treating to prevent discoloration preparatory to sale in one or more of the styles of peeled potatoes described in § 52.2422 United States Standards for Grades of Peeled Potatoes (7 CFR 52.2421–52.2433). The term *other processing* has the same meaning as the term appearing in the Act and includes, but is not restricted to, potatoes for dehydration, chips, shoe-strings, starch, and flour. It includes the application of heat or cold to such an extent that the natural form or stability of the commodity undergoes a substantial change. The act of peeling, cooling, slicing, dicing, or applying material to prevent oxidation does not constitute "other processing." Other terms used in this section have the same meaning as when used in the marketing agreement, as amended, and this part.

[75 FR 77752, Dec. 14, 2010, as amended at 76 FR 27852, May 13, 2011; 78 FR 62969, Oct. 23, 2013; 79 FR 8256, Feb. 12, 2014; 79 FR 26111, May 7, 2014]

PART 948—IRISH POTATOES GROWN IN COLORADO

Subpart A—Order Regulating Handling

AUTHORITY: 7 U.S.C. 601–674.

Subpart A—Order Regulating Handling

SOURCE: 25 FR 7092, July 27, 1960, unless otherwise noted. Redesignated at 26 FR 12751, Dec. 30, 1961.

DEFINITIONS

§ 948.1   Secretary.

Secretary means the Secretary of Agriculture of the United States, or any officer or employee of the Department of Agriculture to whom authority has heretofore been delegated, or to whom authority hereafter may be delegated, to act in his stead.

§ 948.2   Act.

Act means Public Act No. 10 73d Congress, as amended and as reenacted and amended by the Agricultural Marketing Agreement Act of 1937, as amended (sections 1–19, 48 Stat. 31, as amended; 7 U.S.C. 601–674).

§ 948.3   Person.

Person means an individual, partnership, corporation, association, legal representative, or any organized group or business unit of individuals.

§ 948.4   Area.

Area means any of the subdivisions of the State of Colorado as set forth in this section or as reestablished pursuant to § 948.53.

(a) *Area No. 1*, commonly known as the Western Slope, includes and consists of the counties of Routt, Eagle, Pitkin, Gunnison, Hinsdale, La Plata, in the State of Colorado, and all counties in said State west of the aforesaid counties.

(b) *Area No. 2*, commonly known as the San Luis Valley, includes and consists of the counties of Saguache, Huerfano, Las Animas, Mineral, Archuleta, in the State of Colorado, and all counties in said State, south of the counties enumerated in this definition of Area No. 2.

(c) *Area No. 3* includes and consists of all the remaining counties in the State of Colorado which are not included in Area No. 1 or Area No. 2.

§948.5 Potatoes.

Potatoes means and includes all varieties of Irish potatoes grown within any of the aforesaid areas.

§948.6 Seed potatoes.

Seed potatoes or *seed* means any potatoes which have been certified by the official seed certification agency of the State of Colorado and bear the official tags, seals, or other appropriate identification indicating such certification.

§948.7 Handler.

Handler is synonymous with *shipper* and means any person, except a common or contract carrier of potatoes owned by another person, who handles potatoes.

§948.8 Handle or ship.

Handle or *ship* means to transport, sell, or in any way to place potatoes in the current of the commerce between the State of Colorado and any point outside thereof.

§948.9 Producer.

Producer means any person engaged in the production of potatoes for market.

§948.10 Fiscal period.

Fiscal period means the period beginning and ending on the dates approved by the Secretary pursuant to recommendations by an area committee.

§948.11 Grade, size and maturity.

Grade, means any of the officially established grades of potatoes, *Size* means any of the officially established sizes of potatoes, and *Maturity* means any of the stages of development or condition of the outer skin (epidermis) of potatoes, as defined in the United States Standards for Potatoes issued by the United States Department of Agriculture (§§51.1540 to 51.1556, inclusive of this title) or Colorado grades established by the Commissioner, or amendments thereto, or modifications thereof, or variations based on any of the foregoing.

§948.12 Varieties.

Varieties means all classifications or subdivisions of Irish potatoes according to those definitive characteristics now or hereafter recognized by the United States Department of Agriculture.

§948.13 Pack.

Pack means a quantity of potatoes in any type of container, which falls within specific weight limits, numerical limits, grade limits, or any combination of these recommended by the committee and approved by the Secretary.

§948.14 Container.

Container means a sack, bag, crate, box, basket, barrel, or bulk load or any other receptacle used in the packaging, transportation, or sale of potatoes.

§948.15 Culls.

Culls means potatoes which do not meet the requirements set forth in §948.20.

§948.16 Committee.

Committee means any of the area committees established pursuant to §948.50 or the Colorado Potato Committee established pursuant to §948.51.

§948.17 Export.

Export means the shipment of potatoes to any destination which is not within the 48 contiguous States, or the District of Columbia, of the United States.

REGULATION

§ 948.20 Marketing policy.

(a) *General cull regulation.* (1) It shall be the marketing policy for the production area to maintain a general cull regulation in effect prohibiting the handling of potatoes for fresh market, except as otherwise provided in this subpart, which do not meet the requirements of the U.S. No. 2, or better, grade, 1½ inches minimum diameter and larger.

(2) Upon recommendation of the Colorado Potato Committee, or on other available information, the general cull regulation may be suspended or modified by the Secretary during a specified period with respect to any or all varieties of potatoes.

(b) *Area marketing policies.* Each season prior to or at the same time as initial recommendations are made pursuant to § 948.21, each area committee shall submit to the Secretary a report setting forth the marketing policy it deems desirable for the industry to follow in handling the respective area's potatoes during the ensuing season. Additional reports shall be submitted from time to time if it is deemed advisable by an area committee to adopt a new marketing policy because of changes in the demand and supply situation with respect to potatoes. The committee shall publicly announce the submission of each such marketing policy report and copies thereof shall be available at the committee's office for inspection by any producer or any handler. In determining each such marketing policy the committee shall give due consideration to the following:

(1) Supply of potatoes by grade, size, quality, and maturity in the respective area, in the production area, and in other areas;

(2) Market prices for fresh potatoes, including grower, shipping point, and terminal market prices by grade, size, and quality in different packs or in different containers;

(3) Market prices for potatoes in other outlets, including growers' and other market price levels by grade, size, and quality;

(4) The trend and level of consumer income;

(5) Establishing and maintaining such orderly marketing conditions for potatoes as will be in the public interest; and

(6) Other relevant factors.

§ 948.21 Recommendations for regulations.

An area committee upon complying with the requirements of § 948.20 may recommend regulations, or modifications, suspension or termination thereof, to the Secretary whenever it finds that such regulations as provided for in this subpart will tend to effectuate the declared policies of the act.

§ 948.22 Issuance of regulations.

(a) The Secretary shall limit by regulation the handling of potatoes whenever he finds from recommendations and information submitted by an area committee, or from other available information, that such regulation would tend to effectuate the declared policy of the act. Such regulation may:

(1) Limit the handling of particular grades, sizes, qualities, or maturities of any or all varieties of potatoes, or any combination of the foregoing during any period.

(2) Limit the handling of particular grades, sizes, qualities, or maturities of potatoes differently, for different varieties, for different containers, for different packs, for different portions of the production area, for different purposes under § 948.23, or for any combination of the foregoing, during any period.

(3) Provide a method through rules and regulations issued pursuant to this subpart for fixing the size, capacity, weight, dimensions, or pack of the container, or containers, which may be used in the packaging or handling of potatoes, or both.

(4) Establish in terms of grades, sizes, or both, minimum standards of quality and maturity.

(b) Any regulation issued hereunder may be amended, modified, suspended, or terminated by the Secretary on recommendations by an area committee, or on other available information, to provide for

(1) Such changes in regulations found necessary by changes in supplies, demand, or prices;

(2) Minimum quantities which should be relieved of regulatory or administrative obligations; or

(3) Relief from regulations no longer tending to effectuate the declared policies of the Act.

(c) The Secretary shall notify each committee of each regulation recommended by it and issued pursuant to this section. The respective committee shall give reasonable notice thereof to handlers. No regulation, except when relieving limitations, shall become effective less than two days after issuance thereof.

§948.23 Handling for special purposes.

Upon the basis of recommendations and information submitted by an area committee, or other available information, the Secretary, whenever he finds that it will tend to effectuate the declared purposes of the Act, shall modify, suspend, or terminate requirements in effect pursuant to §§948.20 to 948.22, inclusive, or §§948.40 or 948.77, or any combination thereof, to facilitate handling of potatoes for

(a) Relief or charity;

(b) Livestock feed;

(c) Export;

(d) Seed;

(e) Potatoes, other than certified seed, sold to a producer exclusively for planting within specific geographic limits;

(f) Manufacture or conversion into specified products;

(g) Other purposes recommended by the committees and approved by the Secretary.

§948.24 Safeguards.

(a) Each area committee, with the approval of the Secretary, shall prescribe adequate safeguards for potatoes handled pursuant to §948.23 from entering trade channels other than those authorized by regulations and by such rules as may be necessary and incidental thereto.

(b) Such safeguards may include requirements that handlers or processors desiring to handle potatoes pursuant to §948.23 shall:

(1) Apply for and obtain Certificates of Privilege from the area committee for handling potatoes affected or to be affected under the provisions of §948.23;

(2) Obtain inspection as required by §948.40, or pay the assessment levied pursuant to §948.77, or both, except as modified pursuant to §948.23 in connection with shipments made under any such certificate; and

(3) Furnish the committee such information, and execute or obtain execution of such documents, as the committee may require.

(c) An area committee may rescind or deny to any handler permission to handle potatoes pursuant to §948.23 of this subpart if proof satisfactory to the committee is obtained that potatoes handled by him for a purpose stated in §948.23 were handled contrary to the provisions of this subpart.

(d) The committee shall make reports to the Secretary, as requested, showing the number of applications for such certificates, the quantity of potatoes covered by such applications, the number of such applications denied and certificates granted, the quantity of potatoes handled under duly issued certificates, and such other information as may be requested.

EXEMPTIONS

§948.28 Policy.

Any producer whose potatoes have been adversely affected by acts beyond the control or reasonable expectation of a prudent grower and who, by reason of any regulation issued pursuant to this part, is or will be prevented from shipping or having shipped during the then current marketing season, or a specific portion thereof, as large a proportion of his potato crop as the average proportion shipped or to be shipped during comparable portions of the season by all producers in his immediate area of production, may apply to the committee for exemptions from such regulations for the purpose of obtaining equitable treatment under such regulations.

§948.29 Procedure.

Rules and procedures for granting exemptions may be issued by the Secretary, upon recommendation of area committees. Such rules and procedures may provide for methods of determinations by area committees of average proportions of crops shipped or being

329

shipped in respective areas or subdivisions thereof during any or all portions of a season, for processing applications for exemption, for issuing or denying certificates of exemption, for administrative compliance with certificates issued, for reports by handlers thereon, and for such other procedures as may be necessary to administration hereof.

§ 948.30 Granting exemptions.

An area committee may issue certificates of exemption to any qualified applicant who furnishes adequate evidence to such committee:

(a) That the grade, size, or quality of the applicant's potatoes have been adversely affected by acts beyond his control or reasonable expectations;

(b) That by reason of regulations issued pursuant to § 948.20 or § 948.22, the applicant will be prevented as a producer from shipping or having shipped as large a proportion of his production as the average proportion of production shipped by all producers in said applicant's immediate area of production during the season, or a specific portion thereof.

(c) Each such certificate issued shall permit the person identified therein to ship or have shipped the potatoes described thereon, and evidence of such certificates shall be made available to subsequent handlers thereof.

§ 948.31 Investigation.

An area committee shall be permitted at any time to make a thorough investigation of any applicant's claim pertaining to exemptions.

§ 948.32 Appeal.

If any applicant for exemption certificates is dissatisfied with the determination by an area committee with respect to his application, he may file an appeal with the committee. Any applicant filing an appeal shall furnish evidence satisfactory to the committee for a determination on the appeal.

RESEARCH AND DEVELOPMENT

§ 948.35 Research and development.

The committee, with the approval of the Secretary, may provide for the establishment of marketing research and development projects designed to as-

sist, improve, or promote the marketing, distribution, and consumption of potatoes and may make available committee information and data to any person, or to any employee of an agency or its agent, authorized by the committee as its agent with the approval of the Secretary, to conduct such projects.

INSPECTION

§ 948.40 Inspection and certification.

(a) During any period in which the handling of potatoes is regulated pursuant to § 948.20 through § 948.24, inclusive, no handler shall handle potatoes unless such potatoes are inspected by an authorized representative of the Federal or a Federal-State Inspection Service and are covered by a valid inspection certificate, except when relieved of such requirements by § 948.22(b), § 948.23, or § 948.40(b).

(b) Rules may be issued by the Secretary, upon recommendation of the Colorado Potato Committee requiring inspection on regraded, resorted or repacked lots, or providing for special inspection requirements or relief therefrom. Such rules may provide distinctions, insofar as practical, between handling at shipping point and handling in receiving markets within the production area.

(c) Upon recommendation of an area committee and approval by the Secretary, any or all potatoes so inspected and certified shall be identified by appropriate seals, stamps, or tags to be affixed to the containers by the handler under the direction and supervision of a Federal or Federal-State Inspector or the committee. Master containers may bear the identification instead of the individual containers within said master container.

(d) Insofar as the requirements of this section are concerned, the length of time for which an inspection certificate is valid may be established by the committee with the approval of the Secretary.

(e) When potatoes are inspected in accordance with the requirements of this section, a copy of each inspection certificate issued shall be made available to the committee by the inspection service.

(f) Area committees with the approval of the Colorado Potato Committee may recommend and the Secretary may require that no handler shall transport or cause the transportation of potatoes by motor vehicle or by other means unless such shipment is accompanied by a copy of the inspection certificate issued thereon, or other document authorized by the committee to indicate that such inspection has been performed. Such certificate or document shall be surrendered to such authority as may be designated.

COMMITTEES

§948.50 Area committees.

A committee is hereby established as an administrative agency for each area. Each area committee shall be comprised of members and alternates as set forth in this section or as reestablished by §948.53.

(a) Area No. 1 (Western Slope): Four producers and three handlers selected as follows:

Two (2) producers and one (1) handler from the counties of Eagle, Garfield, Pitkin, Moffat, and Routt, in the State of Colorado;

Two (2) producers and one (1) handler from the remaining counties of Area No. 1;

One (1) handler representing all producers' cooperative marketing associations in Area No. 1.

(b) Area No. 2 (San Luis Valley): Seven producers and five handlers selected as follows:

Three (3) producers from Rio Grande County;

One (1) producer from Saguache County;

One (1) producer from Conejos County;

One (1) producer from Alamosa County;

One (1) producer from all other counties in Area No. 2;

Two (2) handlers representing all producers' cooperative marketing associations in Area No. 2;

Three (3) handlers representing handlers in Area No. 2 other than producers' cooperative marketing associations.

(c) Area No. 3: Five Producers and four handlers selected as follows:

Three (3) producers from Weld County;

One (1) producer from Morgan County;

One (1) producer from the remaining counties of Area No. 3;

Four (4) handlers from Area No. 3.

EFFECTIVE DATE NOTE: At 57 FR 61774, Dec. 29, 1992, in §948.50, paragraph (a) was suspended indefinitely.

§948.51 Colorado Potato Committee.

The Colorado Potato Committee is hereby established consisting of six members, with alternates. Two members and alternates shall be selected from each area committee. Committeemen shall be selected by the Secretary from nominations of area committee members or alternates.

EFFECTIVE DATE NOTE: At 57 FR 61774, Dec. 29, 1992, §948.51 was amended by suspending indefinitely the second sentence.

§948.52 Alternates.

(a) For each committee member there shall be an alternate who shall have the same qualifications. During a member's absence, or when called upon to do so in accordance with the terms hereof, or in the event of a member's death, removal, resignation, or disqualification, an alternate shall act in his place and stead until the member's successor is selected and has qualified.

(b) Area committees, with the Secretary's approval, may provide through rules for members or for alternates to recommend regulations for early crop potatoes or for late crop potatoes and to specify the particular crop for which each group shall be responsible.

§948.53 Reestablishment.

Areas, subdivisions of areas, the distribution of representation among the subdivision of areas, or among marketing organizations within respective areas may be reestablished by the Secretary upon area committee recommendations. Upon approval therefor of respective committees affected thereby, areas may be reestablished. In recommending any such changes, the committee shall consider (a) the relative importance of new producing sections, (b) relative production, (c) changes in marketing organizations and their relative status in the industry, (d) the geographic locations of producing sections as they would affect the efficiency of administration of this part, and (e) other relevant factors.

§ 948.54 Eligibility.

Area committee members and alternates shall be individuals who shall be residents of, and producers or handlers, as the case may be, in the respective area. Also, each member or alternate to qualify as a representative (a) for producers shall be a producer, or an officer or employee of a producer; (b) for producer's cooperative marketing associations shall be members or employees of such associations; or (c) for handlers other than cooperative marketing associations shall be a handler, or an officer or employee of a handler.

§ 948.55 Term of office.

The term of office of each area committee member and alternate shall be for two years. The term of office for Colorado Potato Committee members and alternates shall be for one year. The dates on which terms of office for each committee shall begin and end shall be established by the Secretary pursuant to respective committee recommendation. Terms of office of area committee members shall be arranged so that approximately one-half shall terminate each year. Determination of which initial members and alternates shall serve for one year or two years shall be by lot.

§ 948.56 Nomination and selection.

(a) Each area committee shall hold or cause to be held, not less than 15 days prior to the expiration date of respective terms of office, meetings of producers and handlers for each subdivision in which terms expire or in which vacancies otherwise occur.

(b) At each such meeting one or more nominees shall be designated for each impending vacancy as member or alternate. Such designation may be by ballot or by motion at the option of those present in voting capacity.

(c) Only producers may participate in designating producer nominees; only handlers may participate in designating handler nominees; and only duly authorized representatives of producers' cooperative marketing associations may participate in designating nominees to represent such associations. If no separate representation is provided for producers' cooperative marketing associations, duly author-ized representatives of such associations may participate in designating handler nominees.

(d) Each producers' cooperative marketing association shall be entitled to cast only one vote in designating nominees to represent such associations. Each producer and each handler shall be entitled to cast only one vote on behalf of himself, his agents, subsidiaries, affiliates, and representatives.

(e) If a producer, handler, or producers' cooperative marketing association is engaged in producing or handling potatoes in more than one area, or in more than one subdivision of an area, such producer, handler, or producers' cooperative marketing association shall elect the area or subdivision in which he may participate in designating nominees. In no event shall there be participation in more than one area or subdivision.

§ 948.57 Failure to nominate.

If nominations are not made pursuant to the provisions of § 948.56 by the date provided therein, the Secretary may, without regard to nominations, select members and alternates on the basis of the representation provided for in this part.

§ 948.58 Vacancies.

To fill any vacancy occasioned by the failure of any person selected as a member or as an alternate to qualify, or in the event of the death, removal, resignation, or disqualification of a member or alternate, a successor for his unexpired term may be selected by the Secretary from nominations made pursuant to § 948.56, from previously unselected nominees on the current nominee list, or from other eligible persons.

§ 948.59 Qualification.

Each person selected as a member or as an alternate shall qualify by promptly filing a written acceptance with the Secretary.

§ 948.60 Compensation and expenses.

(a) Members of each area committee and their alternates shall serve without salary, but may be compensated at a rate not in excess of $10 per day while engaged on committee business, and may be reimbursed for necessary expenses actually incurred while so engaged. At the discretion of an area committee, alternates may be requested to attend any or all committee meetings and receive compensation and expenses therefor regardless of attendance by the respective members.

(b) The compensation and expenses of members and alternates of the Colorado Potato Committee shall be paid by the respective area committee they represent.

(c) Such other expenses as may be incurred by the Colorado Potato Committee pursuant to a budget of expenses approved by the Secretary shall be allotted to, and paid by, one or more of the area committees, as may be specified in an order issued by the Secretary pursuant to the provisions of this subpart.

§ 948.61 Procedure.

(a) A majority of all members of a committee shall be necessary to constitute a quorum or to pass and motion or approve any committee action.

(b) Each committee may provide for the members thereof, including the alternate members when acting as members, to vote by mail, telegraph, telephone, or other means of communication, provided that any such vote cast orally shall be confirmed promptly in writing. If any assembled meeting is held all votes shall be cast in person.

§ 948.62 Powers.

Each committee shall have the following powers:

(a) To administer the provisions of this subpart as specified herein;

(b) To make rules and regulations to effectuate the terms and provisions of this subpart;

(c) To receive, investigate, and report to the Secretary complaints of violation of the provisions of this part; and

(d) To recommend to the Secretary amendments to this part.

§ 948.63 Duties.

(a) Each committee shall:

(1) Meet and organize as soon as practical after the beginning of each term of office, select a chairman and such other officers' as may be necessary, select subcommittees and adopt such rules and procedures for the conduct of its business as it may deem advisable;

(2) Act as intermediary between the Secretary and any producer or handler;

(3) Appoint such employees, agents and representatives as it may deem necessary and determine the salaries and define the duties of each;

(4) Keep minutes, books, and records which clearly reflect all its acts and transactions. Such minutes, books and records shall be subject to examination at any time by the Secretary;

(5) Furnish promptly notices of meetings, copies of the minutes of each committee meeting, and such other reports or information as may be requested by the Secretary, including annual reports of each area committee's operations for the preceding marketing season or fiscal period;

(6) Make available to producers, and to other area committees and the Colorado Potato Committee the committee's voting record on recommended regulations and other matters of policy;

(7) Meet jointly with other area committees when requested to do so by the Colorado Potato Committee;

(8) Consult, cooperate, and exchange information with other area committees, with other marketing agreement committees and other agencies or individuals in connection with proper committee activities and objectives;

(9) Take any proper action necessary to carry out the provisions of this subpart; and

(10) Cause the books of the committee to be audited by a competent accountant at least once each fiscal period.

(b) The Colorado Potato Committee shall also:

(1) Supervise the regulation of shipments pursuant to the provisions of the general cull regulation in the absence of more restrictive regulations, and shall cooperate with any area committee in administering any regulation issued pursuant to this subpart;

(2) Make recommendations to the Secretary with respect to suspending or modifying the provisions of the general cull regulation;

(3) Make available to area committees its voting record on recommendations for modification of the cull regulation and other matters of policy;

(4) Submit to each area committee such available information as may be requested; and

(5) Call joint meetings of area committees on matters requiring consideration of statewide marketing policies when requested to do so by an area committee.

EXPENSES AND ASSESSMENTS

§ 948.75 Expenses.

Each area committee is authorized to incur such expenses as the Secretary may find are reasonable and likely to be incurred during each fiscal period for its maintenance and functioning, and for purposes determined to be appropriate for administration of this part. Handlers shall share expenses upon the basis of a fiscal period. Each handler's share of such expenses shall be proportionate to the ratio between the total quantity of potatoes handled by him as the first handler thereof during a fiscal period and the total quantity of potatoes handled by all handlers as first handlers thereof during such fiscal period.

§ 948.76 Budget.

As soon as practicable after the beginning of each fiscal period and as may be necessary thereafter, each area committee shall prepare an estimated budget of income and expenditures necessary for its administration of this part. Each area committee may recommend a rate of assessment calculated to provide adequate funds to defray its proposed expenditures. Each area committee shall present such budget to the Secretary with an accompanying report showing the basis for its calculations.

§ 948.77 Assessments.

(a) The funds to cover each area committee's expenses shall be acquired by the levying of assessments upon handlers as provided in this subpart. Each handler who first handles potatoes under this part, shall pay assessments to his respective area committee upon demand, which assessments shall be in payment of such handler's pro rata share of the area committee's expenses.

(b) Assessments shall be levied upon handlers at rates established by the Secretary. Such rates may be established upon the basis of each area committee's budget, recommendations, and other available information. Such rates may be applied to specified containers used in the production area.

(c) At any time during, or subsequent to, a given fiscal period each area committee may recommend the approval of an amended budget and an increase in the rate of assessment. Upon the basis of such recommendations, or other available information, the Secretary may approve an amended budget and increase the rate of assessment. Such increase shall be applicable to all potatoes grown within the particular area where an area committee recommends such increase and which were handled by the first handler thereof during such fiscal period.

(d) The payment of assessments for the maintenance and functioning of each area committee may be required under this part throughout the period it is in effect irrespective to whether particular provisions thereof are suspended or become inoperative.

(e) In order to provide funds to enable each area committee to perform its functions under this part, handlers may make advance payment of assessments.

§ 948.78 Accounting.

(a) If, at the end of a fiscal period, the assessments collected are in excess of expenses incurred, such excess shall be accounted for in accordance with one of the following:

(1) If such excess is not retained in a reserve, as provided in paragraph (a)(2) of this section, it shall be refunded proportionately to the persons from whom it was collected.

(2) An area committee, with the approval of the Secretary, may carry over such excess into subsequent fiscal periods as a reserve: *Provided,* That funds already in the reserve are less than approximately two fiscal period's expenses. Such reserve funds may be used (i) to defray expenses, during any fiscal period, prior to the time assessment income is sufficient to cover such

expenses; (ii) to cover deficits incurred during any fiscal period when assessment income is less than expenses; (iii) to defray expenses incurred during any period when any or all provisions of this subpart are suspended or are inoperative; (iv) to cover necessary expenses of liquidation in the event of termination of this subpart. Upon such termination, any funds not required to defray the necessary expenses of liquidation shall be disposed of in such manner as the Secretary may determine to be appropriate. To the extent practical, such funds shall be returned pro rata to the persons from whom such funds were collected.

(b) All funds received by an area committee pursuant to the provisions of this part shall be used solely for the purposes specified herein. The Secretary may at any time require an area committee and its members to account for all receipts and disbursements.

(c) Upon the removal or expiration of the term of office of any member of an area committee, such member shall account for all receipts and disbursements and deliver all property and funds in his possession to such committee, and shall execute such assignments and other instruments as may be necessary or appropriate to vest in such committee full title to all of the property funds and claims vested in such member pursuant to this part.

(d) Each area committee may make recommendations to the Secretary for one or more of the members thereof, or any other person, to act as a trustee for holding records, funds, or any other committee property during periods of suspension of this subpart, or during any period or periods when regulations are not in effect and if the Secretary determines such action appropriate, he may direct that such person or persons shall act as trustee or trustees for such committee.

REPORTS

§948.80 Reports.

Upon request of an area committee or of the Colorado Potato Committee through an area committee, each handler within the respective area of such area committee shall furnish to the area committee in such manner and at such time as it may prescribe, reports and other information as may be necessary for the committee to perform its duties under this part.

(a) Such reports may include, but are not necessarily limited to the following examples:

(1) The quantities of potatoes received by a handler during any or all periods of a season;

(2) The quantities disposed of by him, segregated as to quantities subject to regulation, and where necessary segregated as to types of outlets and special or modified regulations applicable to alternative outlets, and including quantities not subject to grade, inspection, assessment, or other similar regulations;

(3) The date of each such disposition and the identification of the carrier transporting such potatoes;

(4) Information essential to identification of any or all specific quantities, lots, and disposition of potatoes handled under §§948.23 to 948.30, inclusive, which may include identification of inspection certificates, exemption certificates, certificates of privilege, or other appropriate identification, including the destination of each special shipment, where necessary.

(b) All such reports shall be held under appropriate protective classification and custody by the committee, or duly appointed employees thereof, so that the information contained therein which may adversely affect the competitive position of any handler in relation to other handlers will not be disclosed. Compilations of general reports from data submitted by handlers is authorized, subject to prohibition of disclosure of individual handlers' identities or operations.

(c) Each handler shall maintain for at least two succeeding years such records of the potatoes received and disposed of by such handler as may be necessary to verify the reports he submits to the committee pursuant to this section.

COMPLIANCE

§948.81 Compliance.

Except as provided in this subpart, no handler shall handle potatoes, the handling of which has been prohibited by the Secretary in accordance with provisions of this subpart, and no handler shall handle potatoes except in conformity to the provisions of this subpart.

MISCELLANEOUS PROVISIONS

§ 948.82 Right of the Secretary.

The members of each area committee (including successors and alternates) and any agent or employee appointed or employed by any committee shall be subject to removal or suspension by the Secretary at any time. Each and every order, regulation, decision, determination or other act of each committee shall be subject to the continuing right of the Secretary to disapprove of the same at any time. Upon such disapproval, the disapproved action of the said committee shall be deemed null and void, except as to acts done in reliance thereon or in compliance therewith prior to such disapproval by the Secretary.

§ 948.83 Effective time.

The provisions of this subpart or any amendments thereto shall become effective at such time as the Secretary may declare and shall continue in force until terminated in one of the ways specified in this subpart.

[25 FR 7092, July 27, 1960, as amended at 26 FR 11483, Dec. 5, 1961. Redesignated at 26 FR 12751, Dec. 30, 1961]

§ 948.84 Termination.

(a) The Secretary may at any time terminate any or all provisions of this subpart by giving at least one day's notice by means of a press release or in any other manner which he may determine.

(b) The Secretary may at any time terminate or suspend the operations of any or all of the provisions of this subpart whenever he finds that such provisions do not tend to effectuate the declared policy of the Act.

(c) The Secretary shall terminate the provisions of this subpart at the end of any fiscal period whenever he finds that such termination is favored by a majority of producers, who during a representative period, as determined by the Secretary have been engaged in the production of potatoes for market: *Pro-*

vided, That such majority has, during such representative period, produced for market more than fifty percent of the volume of such potatoes produced for market.

(d) The provisions of this subpart shall in any event terminate whenever the provisions of the Act authorizing them cease to be in effect.

[25 FR 7092, July 27, 1960, as amended at 26 FR 11483, Dec. 5, 1961. Redesignated at 26 FR 12751, Dec. 30, 1961]

§ 948.85 Proceedings after termination.

(a) Upon the termination of the provisions of this subpart the then functioning members of each area committee shall continue as joint trustees for the purpose of liquidating the affairs of their respective area committee of all funds and property then in the possession of or under control of the committee, including claims for any funds unpaid or property not delivered at the time of such termination. Action by said trusteeship shall require the concurrence of a majority of the said trustees.

(b) The said trustees shall continue in such capacity until discharged by the Secretary; shall from time to time account for all receipts and disbursements and deliver all property on hand, together with all books and records of said committees and of the trustees, to such person as the Secretary may direct; and shall upon the request of the Secretary, execute such assignments or other instruments necessary or appropriate to vest in such person full title and right to all of the funds, property, and claims vested in said committee or the trustees pursuant to this subpart.

(c) Any person to whom funds, property, or claims have been transferred or delivered by an area committee or its members pursuant to this section shall be subject to the same obligations imposed upon the members of such committees and upon the said trustees.

§ 948.86 Effect of termination or amendment.

Unless otherwise expressly provided by the Secretary, the termination of this subpart or of any regulation issued pursuant to this subpart or the issuance of any amendments to either thereof, shall not (a) effect or waive any right, duty, obligation, or liability which shall have arisen or which may thereafter arise in connection with any provision of this subpart or any regulation issued under this subpart; or (b) release or extinguish any violation of this subpart or of any regulations issued under this subpart; or (c) affect or impair any rights or remedies of the Secretary or of any other person with respect to any such violations.

§948.87 Duration of immunities.

The benefits, privileges and immunities conferred upon any person by virtue of this subpart shall cease upon the termination of this subpart, except with respect to acts done under and during the existence of this subpart.

§948.88 Agents.

The Secretary may, by designation in writing, name any person, including any officer or employee of the United States or name any agency in the United States Department of Agriculture, to act as his agent or representative in connection with any of the provisions of this subpart.

§948.89 Derogation.

Nothing contained in this subpart is, or shall be construed to be, in derogation or in modification of the rights of the Secretary or of the United States to exercise any powers granted by the Act or otherwise, or in accordance with such powers, to act in the premises whenever such action is deemed advisable.

§948.90 Personal liability.

No member or alternate of any committee or any employee or agent thereof, shall be held personally responsible, either individually or jointly with others, in any way whatsoever, to any handler or to any person for errors in judgment, mistakes, or other acts, either of commission or omission, as such member, alternate, agent, or employee, except for acts of dishonesty, willful misconduct or gross negligence.

§948.91 Separability.

If any provision of this subpart is declared invalid or the applicability thereof to any person, circumstance or thing is held invalid, the validity of the remainder of this subpart, or the applicability thereof to any other person, circumstance or thing shall not be affected thereby.

§948.92 Amendments.

Amendments to this subpart may be proposed from time to time by a committee or by the Secretary.

Subpart B—Administrative Requirements

GENERAL

SOURCE: 26 FR 5219, June 10, 1961, unless otherwise noted. Redesignated at 26 FR 12751, Dec. 30, 1961.

§948.100 Order.

Order means §§948.1 to 948.92 (Order No. 948 as amended) regulating the handling of Irish potatoes grown in the State of Colorado.

§948.101 Terms.

The terms used in this subpart shall have the same meaning as when used in §§948.1 to 948.92.

§948.102 Communications.

Unless otherwise provided in §§948.1 to 948.92, or by specific direction of an area committee, all reports, applications, submittals, requests and communications in connection with the order shall be addressed to the office of the committee for the area in which the potatoes involved are grown.

§948.103 Fiscal period.

Pursuant to §948.10, the fiscal periods for each area shall be as follows:

(a) Area No. 1 and Area No. 3 shall begin July 1 and end June 30, of the following year, both dates inclusive;

(b) Area No. 2 shall begin September 1 and end August 31, of the following year, both dates inclusive. The 1986–87 fiscal period which began July 1, 1986, will be extended two months to August 31, 1987.

[52 FR 12515, Apr. 17, 1987]

EFFECTIVE DATE NOTE: At 57 FR 61774, Dec. 29, 1992, in §948.103, in paragraph (a), the words "Area No. 1 and" were suspended indefinitely.

§ 948.104   Term of office.

(a) Pursuant to § 948.55, the two-year term of office for area committee members and alternates shall be as follows:

(1) Area No. 1 and Area No. 2 shall begin June 1 and end May 31 of the second year following;

(2) Area No. 3 shall begin May 1 and end April 30 of the second year following.

(b) The one-year term of office of Colorado Potato Committee members shall begin as of June 1 of each year.

[52 FR 12515, Apr. 17, 1987]

Effective Date Note: At 57 FR 61774, Dec. 29, 1992, in § 948.104, in paragraph (a)(1), the words "Area No. 1 and" were suspended indefinitely.

Safeguards

Source: Sections 948.120 through 948.126 appear at 26 FR 10792, Nov. 18, 1961, unless otherwise noted. Redesignated at 26 FR 12751, Dec. 30, 1961.

§ 948.120   General.

Whenever shipments of potatoes for special purposes under § 948.23 are relieved in whole or in part from grade and size regulations issued under § 948.22 the committee shall require information and evidence as to the manner, methods, and timing of such shipments as safeguards against the entry of any such potatoes into trade channels other than those for which intended. Such information and evidence shall include the requirements set forth below with respect to Certificates of Privilege.

§ 948.121   Qualification.

Before handling potatoes for special purposes which do not meet regulations issued under § 948.22 a handler must qualify with the committee to handle shipments for special purposes. To qualify he must (a) apply for and receive a Certificate of Privilege indicating his intent to so handle potatoes; (b) agree to comply with reporting and other requirements set forth in §§ 948.121 to 948.125, inclusive, with respect to such shipments; and (c) receive approval of the committee to so handle potatoes. Such approval will be based upon evidence furnished in his application for a Certificate of Privilege, and other information available to the committee.

§ 948.122   Application.

(a) Application for Certificate of Privilege shall be made in person, by telephone, or on forms furnished by the committee. Each application may contain, but need not be limited to, the name and address of the handler; the quantity by grade, size, quality and variety of the potatoes to be shipped; the mode of transportation; the consignee; the destination; the purpose for which the potatoes are to be used; a certification to the United States Department of Agriculture and to the committee as to the truthfulness of the information shown thereon; and any other appropriate information or documents deemed necessary by the committee for the purposes stated in § 948.120.

(b) [Reserved]

§ 948.123   Approval.

The committee or its duly authorized agents shall give prompt consideration to each application for a Certificate of Privilege. Approval of an application based upon a determination as to whether the information contained therein and other information available to the committee supports approval, shall be evidenced by the issuance of a Certificate of Privilege to the applicant. Each certificate shall cover a specified period, and specified qualities and quantities of potatoes to be sold or transported to the designated consignee for the purposes declared.

§ 948.124   Reports.

Each handler of potatoes shipping under Certificates of Privilege shall supply the committee with reports as requested by the committee or its duly authorized agents showing the name and address of the shipper; the car or truck identification; the loading point; destination; consignee; the inspection certificate number when inspection is required; and any other information deemed necessary by the committee.

§ 948.125 Disqualification.

The committee from time to time may conduct surveys of handling of potatoes for special purposes requiring Certificates of Privilege to determine whether handlers are complying with the requirements and regulations applicable to such certificates. Whenever the committee finds that a handler or consignee is failing to comply with requirements and regulations applicable to handling of potatoes in special outlets, and requiring such certificates, a Certificate or Certificates of Privilege issued such handler may be rescinded and further certificates denied. Such disqualification shall apply to, and not exceed, a reasonable period of time as determined by the committee but in no event shall it extend beyond the end of the succeeding fiscal period. Any handler who has a certificate rescinded or denied may appeal to the committee in writing for reconsideration of his disqualification.

§ 948.126 General cull regulation.

(a) No handler shall handle potatoes grown in the State of Colorado which do not meet the requirements of U.S. No. 2 or better grade, or are less than ¾-inch in diameter.

(b) This General Cull Regulation shall remain in effect until suspended or modified pursuant to § 948.20(a)(2).

(c) The term U.S. No. 2 grade has the same meaning as when used in the U.S. Standards for Potatoes (§§ 51.1540 to 51.1556 of this title), or amendments thereto or modifications thereof.

(d) Applicability to imports: Pursuant to section 608e–1 of the act and § 980.1 *Import Regulations; Irish potatoes* (part 980 of this chapter), in the absence of more restrictive regulations in effect for potatoes grown in Areas Nos. 2 and 3 in Colorado, this cull regulation shall be used in a basis for import regulations for the red skinned, round type and for other round type potatoes, during the periods specified and as designated in said § 980.1 of this chapter.

[35 FR 11988, July 25, 1970, as amended at 78 FR 35745, June 14, 2013]

EXEMPTIONS

Source: Sections 948.130 through 948.132 appear at 26 FR 10793, Nov. 18, 1961, unless otherwise noted. Redesignated at 26 FR 12751, Dec. 30, 1961.

§ 948.130 Application for exemption certificates.

Any producer applying for exemption from any grade and size regulation issued under § 948.22 shall make application to the respective area committee for the area in which the applicant's potatoes were grown or are stored, on forms to be furnished by the area committee. The application shall include:

(a) The name and address of the applicant;

(b) The location, or locations, of the potatoes with respect to which exemption is requested;

(c) The total estimated quantity of potatoes (excluding culls) produced by the applicant during the current season, stated in hundredweights, by varieties, grades, and sizes;

(d) The estimated percentage of the applicant's potato crop (excluding culls) which cannot be shipped because of grade and size regulations then in effect and the acts beyond his control or reasonable expectation adversely affecting his potatoes;

(e) The quantity of potatoes of each variety (excluding culls) which has already been sold or otherwise shipped during the current season;

(f) The signature of the applicant and certification that the statements given in the application are true and correct; and

(g) Such additional information as the area committee may find necessary in making a determination regarding the granting of an exemption certificate.

§ 948.131 Federal-State inspection reports.

Each application for exemption shall be accompanied by a written report of a Federal-State Inspector, which shall contain the following:

(a) A statement by the inspector that he personally inspected the potatoes with respect to which exemption is requested, and that he took a representative sample of such potatoes;

(b) A statement of the percentage of the potatoes (excluding culls) which fail to meet the requirements of the

339

grade and size regulations then in effect;

(c) A statement of the defects or damage causing the potatoes to fail to meet grade and size requirements then in effect.

In the event that more than one variety of potatoes is being regulated the above percentage shall be determined separately for each variety of the applicant's potatoes. The cost of Federal-State inspection and report shall be borne by the applicant for exemption.

§ 948.132 Issuance of exemption certificates.

(a) The respective area committee receiving an application for exemption shall give prompt consideration thereto and determine on the basis of the statements and facts therein contained and the factors set forth in § 948.30 whether the application may be approved. The determination, if favorable, shall be evidenced by the issuance of a certificate of exemption pursuant to §§ 948.28 through 948.32. If the applicant's request for exemption is denied, he shall be so notified in writing.

(b) Each certificate of exemption issued as provided in this subpart, shall contain the name and address of the applicant, the location of his farm or ranch, the location, or locations, of all potatoes remaining to be shipped, the total quantity of potatoes which may be shipped under the certificate of exemption, and such other information as the area committee may deem desirable.

(c) The committee may furnish each applicant receiving a certificate of exemption with appropriate subcertificates of exemption to identify each lot of exempted potatoes and a subcertificate shall be transferable with the lot of potatoes to which it applies. Each applicant receiving a certificate of exemption shall report each shipment of potatoes made under such certificate to the respective area committee issuing the certificate. The report shall state the name and address of the person to whom the potatoes were sold, the quantity sold, the date of transfer, and such other information as the committee may request.

MODIFICATION OF INSPECTION
REQUIREMENTS

§ 948.140 Application.

Any handler whose packing facilities are located in an area where inspection is not readily available or the actual cost for inspection would otherwise exceed 1⅓ times the current per hundredweight inspection fee, may apply to the respective area committee for a waiver from the reinspection requirements. Applications shall be made on forms furnished by the respective area committee and shall contain such information as the respective area committee, with the approval of the Secretary, may find necessary in making a determination regarding the issuance of such waiver.

[55 FR 41181, Oct. 10, 1990]

§ 948.141 Issuance.

Each respective area committee shall give prompt consideration to each application for a waiver from reinspection. In granting a waiver, the handler shall agree to comply with all marketing order requirements. Approval of an application shall be evidenced by the issuance of an applicable waiver by the respective area committee to the handler.

[55 FR 41181, Oct. 10, 1990]

§ 948.142 Reports.

Each handler shipping potatoes pursuant to a waiver from reinspection shall report periodically as specified by the respective area committee on forms furnished by the respective committee the following information one ach shipment: quantity of potatoes, variety or varieties, grade, size, type of container(s), date of shipment, carrier, destination, and name and address of receiver.

[55 FR 41181, Oct. 10, 1990]

§ 948.143 Cancellation.

Whenever the respective area committee finds that shipments of potatoes pursuant to a reinspection waiver are not in accordance with the established application and safeguard provisions, such waiver may be cancelled.

[55 FR 41181, Oct. 10, 1990]

§ 948.150 Reestablishment of committee membership.

Pursuant to § 948.53, membership on each area committee shall be reestablished as follows:

(a) Area No. 2 (San Luis Valley): Nine producers and five handlers selected as follows:

(1) Two (2) producers from Rio Grande County;

(2) Two (2) producers from either Saguache County or Chaffee County;

(3) One (1) producer from either Conejos or Costilla County.

(4) Two (2) producers from Alamosa County;

(5) One (1) producer from all other counties in Area No. 2;

(6) One (1) producer representing certified seed producers in Area No. 2;

(7) Two (2) handlers representing bulk handlers in Area No. 2;

(8) Three (3) handlers representing handlers in Area No. 2 other than bulk handlers.

(b) *Area No. 3:* Three producers and two handlers selected as follows: Three (3) producers and two (2) handlers from any county in Area No. 3.

[52 FR 12515, Apr. 17, 1987, as amended at 58 FR 8541, Feb. 16, 1993; 60 FR 16566, Mar. 31, 1995; 67 FR 68021, Nov. 8, 2002; 68 FR 40119, July 7, 2003; 78 FR 30745, May 23, 2013]

§ 948.151 Colorado Potato Committee membership.

The Colorado Potato Committee shall be comprised of six members and alternates selected by the Secretary. Three members and three alternates shall be selected from nominations of Area 2 committee members or alternates, and three members and three alternates shall be selected from nominations of Area 3 committee members or alternates.

[57 FR 61774, Dec. 29, 1992]

§ 948.153 Reestablishment of area.

Pursuant to § 948.53, Area No. 2 is reestablished as follows:

Area No. 2 (San Luis Valley) includes and consists of the counties of Chaffee, Saguache, Huerfano, Las Animas, Mineral, Archuleta, Rio Grande, Conejos, Costilla, and Alamosa, in the State of Colorado.

[60 FR 16566, Mar. 31, 1995]

Subpart C—Accounting and Collections

§ 948.200 Accounting and collections.

(a) Each handler's assessment account with Area No. 2 (San Luis Valley) Committee shall become due and payable upon presentation of a statement thereof to such handler.

(b) If settlement of such an assessment account is not completed on or before the 20th day following presentation of a statement of such account, each handler failing to so complete settlement of his account may be declared delinquent by said area committee.

(c) The name of each person who is declared delinquent may be forwarded to the Secretary and, in addition, the names of persons declared delinquent pursuant to paragraph (b) of this section may be publicized by said area committee.

(d) Terms used in this section shall have the same meaning as when used in Marketing Agreement No. 97 and Order No. 948 (§§ 948.1 to 948.92).

[19 FR 8647, Dec. 17, 1954. Redesignated at 26 FR 12751, Dec. 30, 1961]

§ 948.215 Assessment rate.

On or after July 1, 2005, an assessment rate of $0.02 per hundredweight is established for Colorado Area No. 3 potatoes.

[70 FR 36816, June 27, 2005]

§ 948.216 Assessment rate.

On and after September 1, 2018, an assessment rate of $0.006 per hundredweight is established for Colorado Area No. 2 potatoes.

[83 FR 43503, Aug. 27, 2018]

Subpart D—Handling Requirements

§ 948.386 Handling regulation.

No person shall handle any lot of potatoes grown in Area No. 2 unless such potatoes meet the requirements of paragraphs (a), (b), and (c) of this section, or unless such potatoes are handled in accordance with paragraphs (d) and (e), or (f) of this section.

(a) *Minimum grade and size requirements*—(1) *All varieties.* U.S. No. 2 or better grade, 2 inches minimum diameter or 4 ounces minimum weight.

(2) *1½-inch minimum to 2¼-inch maximum diameter (Size B).* U.S. Commercial grade or better, except that red varieties may be U.S. No. 2 grade or better.

(3) *¾-inch minimum to 1⅞-inch maximum diameter.* U.S. Commercial grade or better.

(4) None of the above categories of potatoes identified in paragraphs (a)(1) through (a)(4) of this section may be commingled in the same bag or other container.

(b) *Maturity (skinning) requirements.* From August 1 through October 31 shall be:

(1) *For U.S. No. 2 grade.* Not more than "moderately skinned."

(2) *All other grades.* Not more than "slightly skinned."

(c) *Inspection.* (1) No handler shall handle any potatoes for which inspection is required unless an appropriate inspection certificate has been issued with respect thereto and the certificate is valid at the time of shipment. For purposes of operation under this part it is hereby determined pursuant to § 948.40(d) that each inspection certificate shall be valid for a period not to exceed five days following the date of inspection as shown on the inspection certificate.

(2) No handler may transport or cause the transportation by motor vehicle of any shipment of potatoes for which an inspection certificate is required unless each shipment is accompanied by a copy of the inspection certificate applicable thereto and the copy is made available for examination at any time upon request.

(3) Each handler who handles potatoes after such potatoes are regraded, resorted, or repacked shall have such potatoes reinspected, unless such handler has received a waiver from reinspection pursuant to rules established by the Secretary upon the recommendation of the committee.

(d) *Special purpose shipments.* (1) The grade, size, maturity, and inspection requirements of paragraphs (a), and (b), and (c) of this section and the assessment requirements of this part shall not be applicable to shipments of potatoes for:

(i) Livestock feed;

(ii) Relief or charity; or

(iii) Canning, freezing, and "other processing" as hereinafter defined.

(2) The grade, size, maturity and inspection requirements of paragraphs (a), (b), and (c) of this section shall not be applicable to shipments of potatoes for experimentation, the manufacture or conversion into specified products, or for seed pursuant to section 948.6, but such shipments shall be subject to assessments.

(e) *Safeguards.* Each handler of potatoes which do not meet the grade, size, and maturity requirements of paragraphs (a) and (b) of this section and which are handled pursuant to paragraph (d) of this section for any of the special purposes set forth therein shall:

(1) Prior to handling, apply for and obtain a Certificate of Privilege from the committee.

(2) Furnish the committee such reports and documents as requested, including certification by the buyer or receiver as to the use of such potatoes; and

(3) Bill each shipment directly to the applicable processor or receiver.

(f) *Minimum quantity.* For purposes of regulation under this part, each person may handle up to but not to exceed 2,000 pounds of potatoes without regard to the requirements of paragraphs (a), (b), and (c) of this section, but this exception shall not apply to any shipment which exceeds 2,000 pounds of potatoes.

(g) *Definitions.* The terms *U.S. No. 1, U.S. Commercial, U.S. No. 2, Size B, slightly skinned,* and *moderately skinned* shall have the same meaning as when used in the U.S. Standards for Potatoes (7 CFR 2851.1540–2851.1566), including the tolerances set forth therein. The term *other processing* has the same meaning as the term appearing in the act and includes, but is not restricted to, potatoes for dehydration, chips, shoestrings, starch, and flour. It includes only that preparation of potatoes for market which involves the application of heat or cold to such an extend that the natural form or stability of the commodity undergoes a substantial change. The act of peeling, cooling, slicing, dicing, or applying material to prevent oxidation does not constitute "other processing." The term *manufacture or conversion into specified products* means the preparation of potatoes for market into products by peeling, slicing, dicing, applying material to prevent oxidation, or other means approved by the committee, but not including other processing. Other terms used in this section shall have the same meaning as when used in Marketing Agreement No. 97, as amended, and this part.

(Secs. 1–19, 48 Stat. 31, as amended; 7 U.S.C. 601–674)

[46 FR 52324, Oct. 27, 1981]

EDITORIAL NOTE: For FEDERAL REGISTER citations affecting §948.386, see the List of CFR Sections Affected, which appears in the Finding Aids section of the printed volume and at *www.govinfo.gov*.

§948.387   Handling regulation.

On and after August 1, 1982, no person shall handle any lot of potatoes grown in Area No. 3 unless such potatoes meet the requirements of paragraphs (a), (b), and (c) of this section, or unless such potatoes are handled in accordance with paragraphs (d) and (e), or (f) of this section.

(a) *Minimum grade and size requirements—All varieties.* (1) U.S. No. 2 or better grade, 1⅞ inches minimum diameter or 4 ounces minimum weight.

(2) U.S. No. 1 grade, Size B (1½ inches minimum to 2¼ inches maximum diameter).

(3) U.S. No. 1 grade, ¾ inch minimum to 1⅞ inches maximum diameter.

(b) *Maturity (skinning) requirements—All Varieties.* During the period beginning July 1 and ending December 31 each season for U.S. No. 2 grade, not more than "moderately skinned," and for all other grades, not more than "slightly skinned"; thereafter no maturity requirements.

(c) *Inspection.* (1) No handler shall handle any potatoes for which inspection is required unless an appropriate inspection certificate has been issued with respect thereto and the certificate is valid at the time of shipment. For purpose of operation under this part it is hereby determined pursuant to paragraph (d) of §948.40, that each inspection certificate shall be valid for a period not to exceed five days following the date of inspection as shown on the inspection certificate.

(2) No handler may transport or cause the transportation by motor vehicle of any shipment of potatoes for which an inspection certificate is required unless each shipment is accompanied by a copy of the inspection certificate applicable thereto and the copy is made available for examination at any time upon request.

(3) Each handler who handles potatoes after such potatoes are regraded, resorted, or repacked shall have such potatoes reinspected, unless such handler has received a waiver from reinspection pursuant to rules established by the Secretary upon the recommendation of the committee.

(d) *Special purpose shipments.* (1) The grade, size, maturity and inspection requirements of paragraphs (a), (b), and (c) of this section and the assessment requirements of this part shall not be applicable to shipments of potatoes for:

(i) Livestock feed;

(ii) Charity;

(iii) Canning, freezing, and "other processing" as hereinafter defined; and

(iv) Certified seed potatoes (§948.6).

(v) Experimentation and the manufacture or conversion into specified products.

(2) The maturity requirements set forth in paragraph (b) of this section shall not be applicable to shipments of potatoes for prepeeling.

(e) *Safeguards.* Each handler making shipments of potatoes pursuant to paragraph (d) of this section shall:

(1) Prior to shipment, apply for and obtain a Certificate of Privilege from the committee;

(2) Furnish the committee such reports and documents as required, including certification by the buyer or receiver on the use of such potatoes; and

(3) Bill each shipment directly to the applicable buyer or receiver.

(f) *Minimum quantity.* For purpose of regulation under this part, each person may handle up to but not to exceed 2,000 pounds of potatoes per shipment without regard to the requirements of paragraphs (a) and (b) of this section, but this exception shall not apply to any shipment of over 2,000 pounds of potatoes.

(g) *Definitions.* The terms *U.S. No. 1, U.S. No. 2, Size B, moderately skinned* and *slightly skinned* shall have the same meaning as when used in the United States Standards for Grades of Potatoes (7 CFR 51.1540–51.1566) including the tolerances set forth therein. The term *prepeeling* means the commercial preparation in a prepeeling plant of clean, sound, fresh potatoes by washing, peeling or otherwise removing the outer skin, trimming, sorting, and properly treating to prevent discoloration preparatory to sale in one or more of the styles of peeled potatoes described in § 52.2422 United States Standards for Grades of Peeled Potatoes (7 CFR 52.2421–52.2433). The term *other processing* has the same meaning as the term appearing in the act and includes, but is not restricted to, potatoes for dehydration, chips, shoestrings, starch, and flour. It includes only that preparation of potatoes for market which involves the application of heat or cold to such an extent that the natural form or stability of the commodity undergoes a substantial change. The act of peeling, cooling, slicing, dicing, or applying material to prevent oxidation does not constitute "other processing." The term *manufacture or conversion into specified products* means the preparation of potatoes for market into products by peeling, slicing, dicing, applying material to prevent oxidation, or other means approved by the committee, but not including other processing. All other terms used in this section shall have the same meaning as when used in Marketing Agreement No. 97, as amended, and this part.

(Secs. 1–19, 48 Stat. 31, as amended; 7 U.S.C. 601–674)

[47 FR 32911, July 30, 1982, as amended at 52 FR 7269, Mar. 10, 1987; 55 FR 41181, Oct. 10, 1990; 66 FR 49513, Sept. 28, 2001; 74 FR 65393, Dec. 10, 2009; 75 FR 17036, Apr. 5, 2010; 76 FR 80214, Dec. 23, 2011; 80 FR 3142, Jan. 22, 2015]

PART 955—VIDALIA ONIONS GROWN IN GEORGIA

Subpart A—Order Regulating Handling

AUTHORITY: 7 U.S.C. 601–674.

SOURCE: 55 FR 717, Jan. 9, 1990, unless otherwise noted.

Subpart A—Order Regulating Handling

DEFINITIONS

§955.1 Secretary.

Secretary means the Secretary of Agriculture of the United States, or any officer or employee of the Department of Agriculture who has been delegated, or who may hereafter be delegated, the authority to act for the Secretary.

§955.2 Act.

Act means Public Act No. 10, 73d Congress (May 12, 1933), as amended and as reenacted and amended by the Agricultural Marketing Agreement Act of 1937, as amended (Sec. 1–19, 48 Stat. 31, as amended; 7 U.S.C. 601 *et seq.*).

§955.3 Person.

Person means an individual, partnership, corporation, association, or any other business unit.

§955.4 Production area.

Production area means that part of the State of Georgia enclosed by the following boundaries:

Beginning at a point in Laurens County where U.S. Highway 441 intersects Highway 16; thence continue southerly along U.S. Highway 441 to a point where it intersects the southern boundary of Laurens County; thence southwesterly along the border of Laurens County to a point where it intersects the county road known as Jay Bird Springs Road; thence southeasterly along Jay Bird Springs Road to a point where it intersects U.S. Highway 23; thence easterly to a point where U.S. Highway 23 intersects the western border of Telfair County; thence southwesterly following the western and southern border of Telfair County to a point where it intersects with Jeff Davis County; thence following the southern border of Jeff Davis County to a point where it intersects with the western border of Bacon County; thence southerly and easterly along the border of Bacon County to a point where it intersects Georgia State Road 32; thence easterly along Georgia State Road 32 to Seaboard Coastline Railroad; thence northeasterly along the tracks of Seaboard Coastline Railroad to a point where they intersect Long County and Liberty County; thence northwesterly and northerly along the southwestern border of Liberty County to a point where the border of Liberty County intersects the southern border of Evans County; thence northeasterly along the eastern border of Evans County to the intersection of the Bulloch County border; thence northeasterly along the Bulloch County border to a point where it intersects with the Ogeechee River; thence northerly along the main channel of the Ogeechee River to a point where it intersects with the southeastern border of Screven County; thence northeasterly along the southeasterly border of Screven County to the main channel of the Savannah River; thence northerly along the main channel of the Savannah River to a point where the northwestern boundary of Hampton County, South Carolina intersects the Savannah River; thence due west to a point where State Road 24 intersects Brannen Bridge Road; thence westerly along Brannen Bridge Road to a point where it intersects with State Road 21; thence westerly along State Road 21 to the intersection of State Road 17; thence westerly along State Road 17 to the intersection of State Road 56 and southerly to the northern border of Emanuel County; thence westerly and southerly along the border of Emanuel County to a point where it intersects the Treutlen County border; thence southerly to a point where the Truetlen County border intersects Interstate Highway 16; thence westerly to the point of beginning in Laurens County.

§955.5 Vidalia onion.

Vidalia onion means all varieties of *Allium cepa* of the hybrid yellow granex, granex parentage or any other similar variety recommended by the committee and approved by the Secretary, that are grown in the production area.

§955.6 Handler.

Handler is synonymous with *shipper* and means any person (except a common or contract carrier of Vidalia onions owned by another person) who handles Vidalia onions, or causes Vidalia onions to be handled.

§955.7 Handle.

Handle or *ship* means to package, load, sell, transport, or in any other way to place Vidalia onions, or cause Vidalia onions to be placed, in the current of commerce within the production area or between the production area and any point outside thereof. Such term shall not include the transportation, sale, or delivery of field-run Vidalia onions to a person within the production area for the purpose of having such Vidalia onions prepared for market.

§ 955.9 Producer.

Producer is synonymous with *grower* and means any person engaged in a proprietary capacity in the production of Vidalia onions for market.

§ 955.10 Producer-handler.

Producer-Handler means a producer who handles Vidalia onions.

§ 955.12 Committee.

Committee means the Vidalia Onion Committee, established pursuant to § 955.20.

§ 955.13 Fiscal period.

Fiscal period means the 12-month period beginning on September 16 and ending on September 15 of the next year or such other period that may be recommended by the committee and approved by the Secretary.

COMMITTEE

§ 955.20 Establishment and membership.

(a) There is hereby established a Vidalia Onion Committee, consisting of nine members, to administer the terms and provisions of this part. Eight members shall be producers, and one shall be a public member. At least four of the producer members shall be producer-handlers. Each member shall have an alternate who shall have the same qualifications as the member.

(b) Each member, other than the public member, shall be an individual who is, prior to selection and during such member's term of office, a resident of the production area and a grower or an officer or employee of a grower.

(c) The public member shall be a resident of the production area and shall have no direct financial interest in the commercial production, financing, buying, packing or marketing of Vidalia onions, except as a consumer, nor shall such person be a director, officer or employee of any firm so engaged.

§ 955.21 Term of office.

(a) Except as otherwise provided in paragraph (b) of this section, the term

of office of committee members and their respective alternates shall be for two years and shall begin as of September 16 or for such other period as the committee may recommend and the Secretary approve. The terms shall be determined so that approximately one-half of the total committee membership shall terminate each year. Members and alternates shall serve in such capacity during the term of office or portion thereof for which they are selected and until their respective successors are selected.

(b) The term of office of the initial members and alternates shall begin as soon as possible after effective date of this part. As determined by lot drawn at the initial nomination meeting, one-fourth of the initial grower members and alternates shall serve for a one-year term, one-fourth shall serve for a two-year term, one-fourth shall serve for a three-year term, and one-fourth shall serve for a four-year term. The term of office for the initial public member and alternate shall be for two years.

(c) The consecutive terms of office of members shall be limited to three 2-year terms.

§ 955.22 Nominations.

(a) *Initial members.* For nominations to the initial committee, a meeting of producers shall be held by the Secretary.

(b) *Successor members.* (1) The committee shall hold or cause to be held not later than August 1 of each year, or such other date as may be specified by the Secretary, a meeting or meetings of growers for the purpose of designating one nominee for each position as member and for each position as alternate member of the committee which is vacant, or which is about to become vacant.

(2) Nominations for members and alternates shall be supplied to the Secretary in such manner and form as the Secretary may prescribe, not later than August 15 of each year, or by such other date as may be specified by the Secretary.

(3) The Secretary may, upon recommendation of the committee, divide the production area into districts for

346

the purpose of nominating committee members and their alternates.

(c) Only producers may participate in designating nominees to serve as committee members. Each producer is entitled to cast only one vote on behalf of such producer and such producer's agents, subsidiaries, affiliates, and representatives in designating nominees for committee members and alternates. An eligible voter's privilege of casting only one vote shall be construed to permit a voter to cast one vote for each position to be filled.

(d) The producer members shall nominate the public member and alternate member at the first meeting following the selection of members for a new term of office. Nominations for the public member and alternate member shall be supplied to the Secretary in such manner and form as the Secretary may prescribe, not later than November 1, or such other date as may be specified by the Secretary.

§955.23 Selection.

From the nominations made pursuant to §955.22 or from other qualified persons, the Secretary shall select members and alternate members of the committee.

§955.24 Acceptance.

Any person nominated to serve as a member or alternate member of the committee shall, prior to selection by the Secretary, qualify by filing a written acceptance indicating such person's willingness to serve in the position for which nominated.

§955.25 Alternates.

An alternate member of the committee shall act in the place and stead of the member for whom such person is an alternate during such member's absence or when designated to do so by such member. In the event both a member of the committee and that member's alternate are unable to attend a committee meeting, the member, the alternate, or the committee, in that order, may designate another alternate from the same district (if applicable) and the same group (producer or producer-handler) to serve in such member's stead. Only the public member's alternate is authorized to serve in

the place and stead of the public member. In the event of the death, removal, resignation or disqualification of a member, that member's alternate shall serve until a successor to such member is selected.

§955.26 Vacancies.

To fill any vacancy occasioned by the failure of any person nominated as a member or as an alternate to qualify, or in the event of the death, removal, resignation, or disqualification of a member or alternate, a successor for the unexpired term may be selected by the Secretary from nominations made pursuant to §955.22, or from other eligible persons.

§955.27 Failure to nominate.

If nominations are not made within the time and manner prescribed in §955.22, the Secretary may, without regard to nominations, select members and alternates on the basis of the representation provided for in §955.20.

§955.28 Procedure.

(a) Five members of the committee shall constitute a quorum, and five concurring votes shall be required to pass any motion or approve any committee action.

(b) The committee may provide for meetings by telephone, telegraph, or other means of communication, and any vote cast orally at such meetings shall be confirmed promptly in writing: *Provided,* That if an assembled meeting is held, all votes shall be cast in person.

§955.29 Expenses.

Members and alternates shall serve without compensation but shall be reimbursed for such expenses authorized by the committee and necessarily incurred by them in attending committee meetings and in the performance of their duties under this part.

§955.30 Powers.

The committee shall have the following powers:

(a) To administer the provisions of this part in accordance with its terms;

(b) To make rules and regulations to effectuate the terms and provisions of this part;

(c) To receive, investigate, and report to the Secretary complaints of violation of the provisions of this part; and

(d) To recommend to the Secretary amendments to this part.

§ 955.31   Duties.

The committee shall have, among others, the following duties:

(a) As soon as practicable after the beginning of each term of office, to meet and organize, to select a chairman and such other officers as may be necessary, to select subcommittees of committee members or alternates, and to adopt such rules and regulations for the conduct of its business as it deems necessary;

(b) To act as intermediary between the Secretary and any producer or handler;

(c) To furnish to the Secretary such available information as may be requested;

(d) To appoint such employees, agents, and representatives as it may deem necessary, to determine the compensation and define the duties of each such person, and to protect the handling of committee funds;

(e) To investigate from time to time and to assemble data on the growing, harvesting, shipping, and marketing conditions with respect to Vidalia onions;

(f) To keep minutes, books, and records which clearly reflect all of the acts and transactions of the committee. Such minutes, books, and records shall be subject to examination at any time by the Secretary or the Secretary's authorized agent or representative. Minutes of each committee meeting shall be furnished promptly to the Secretary;

(g) Prior to the beginning of each fiscal period, to prepare and submit to the Secretary a budget of its projected income and expenses for such fiscal period, together with a report thereon and a recommendation as to the rate of assessment for such period;

(h) To cause its books to be audited by a Certified Public Accountant at least once each fiscal period, and at such other time as the committee may deem necessary or as the Secretary may request. The report of such audit shall show the receipt and expenditure of funds collected pursuant to this part. A copy of each report shall be furnished to the Secretary. A copy shall also be made available at the principal office of the committee for inspection by producers and handlers provided that confidential information shall be removed;

(i) To give the Secretary the same notice of meetings of the committee and its subcommittees as is given to its members.

EXPENSES AND ASSESSMENTS

§ 955.40   Expenses.

The committee is authorized to incur such expenses as the Secretary may find are reasonable and likely to be incurred by the committee for its maintenance and functioning, and to enable it to exercise its powers and perform its duties in accordance with the provisions of this part. The funds to cover such expenses shall be acquired in the manner prescribed in §§ 955.42 and 955.45.

§ 955.41   Budget.

At least 60 days prior to each fiscal period, or such other date as may be specified by the Secretary, and as may be necessary thereafter, the committee shall prepare an estimated budget of income and expenditures necessary for the administration of this part. The committee may recommend a rate of assessment calculated to provide adequate funds to defray its proposed expenditures. The committee shall present such budget to the Secretary with an accompanying report showing the basis for its calculations.

§ 955.42   Assessments.

(a) The funds to cover the committee's expenses shall be acquired by the levying of assessments upon handlers as provided in this subpart. Each person who first handles Vidalia onions shall pay assessments to the committee upon demand, which assessments shall be in payment of such handler's pro rata share of the committee's expenses.

(b) Assessments shall be levied upon handlers at rates established by the

Secretary. Such rates may be established upon the basis of the committee's recommendations or other available information.

(c) At any time during, or subsequent to, a given fiscal period the committee may recommend the approval of an amended budget and an increase in the rate of assessment. Upon the basis of such recommendations, or other available information, the Secretary may approve an amended budget and increase the assessment rate. Such increase shall be applicable to all Vidalia onions which were handled during such fiscal period.

(d) The payment of assessments for the maintenance and functioning of the committee may be required under this part throughout the period it is in effect irrespective of whether particular provisions of this part are suspended or become inoperative.

(e) To provide funds for the administration of the provisions of this part during the initial fiscal period or the first part of a fiscal period when neither sufficient operating reserve funds nor sufficient revenue from assessments on the current seasons's shipments are available, the committee may accept payment of assessments in advance or may borrow money for such purposes.

(f) The committee may impose a late payment charge or an interest charge or both, on any handler who fails to pay any assessment in a timely manner. Such time and the rates shall be recommended by the committee and approved by the Secretary.

§955.43 Accounting.

(a) All funds received by the committee pursuant to the provisions of this part shall be used solely for the purposes specified in this part.

(b) The Secretary may at any time require the committee, its members and alternates, employees, agents and all other persons to account for all receipts and disbursements, funds, property, or records for which they are responsible. Whenever any person ceases to be a member or alternate of the committee, such person shall account for all receipts and disbursements and deliver all property and funds in such member's possession to the committee,

pertaining to the committee's activities for which such person was responsible, and shall execute such assignments and other instruments as may be necessary or appropriate to vest in the committee full title to all of the property, funds, and claims vested in such person.

(c) The committee may make recommendations to the Secretary for one or more of the members thereof, or any other person, to act as a trustee for holding records, funds, or any other committee property during periods of suspension of this part, or during any period or periods when regulations are not in effect and, upon determining such action is appropriate, the Secretary may direct that such person or persons shall act as trustee or trustees for the committee.

§955.44 Excess funds.

If, at the end of a fiscal period, the assessments collected are in excess of expenses incurred, such excess shall be accounted for as follows:

(a) The committee, with the approval of the Secretary, may establish an operating reserve and may carry over to subsequent fiscal periods excess funds in a reserve so established, except funds in the reserve shall not exceed the equivalent of approximately three fiscal periods' budgeted expenses. Such reserve funds may be used:

(1) To defray any expenses authorized under this part;

(2) To defray expenses during any fiscal period prior to the time assessment income is sufficient to cover such expenses;

(3) To cover deficits incurred during any fiscal period when assessment income is less than expenses;

(4) To defray expenses incurred during any period when any or all provisions of this part are suspended or are inoperative; and

(5) To cover necessary expenses of liquidation in the event of termination of this part.

Upon termination of this part, any funds not required to defray the necessary expenses of liquidation shall be disposed of in such manner as the Secretary may determine to be appropriate except that to the extent practicable, such funds shall be returned pro rata to the persons from whom such funds were collected.

(b) If such excess is not retained in a reserve as provided in paragraph (a) of this section, each handler entitled to a proportionate refund of the excess assessments collected shall be credited at the end of a fiscal period with such refund against the operations of the following fiscal period unless such handler demands payment thereof, in which event such proportionate refund shall be paid.

§ 955.45 Contributions.

The committee may accept voluntary contributions but these shall only be used to pay expenses incurred pursuant to § 955.50. Such contributions shall be free from any encumbrances by the donor, and the committee shall retain complete control of their use.

RESEARCH AND DEVELOPMENT

§ 955.50 Research and development.

(a) The committee, with the approval of the Secretary, may establish or provide for the establishment of production research, marketing research and development and marketing promotion projects, including paid advertising, designed to assist, improve, or promote the marketing, distribution, consumption, or efficient production of Vidalia onions. Any such project for the promotion and advertising of Vidalia onions may utilize an identifying mark which shall be made available for use by all handlers in accordance with such terms and conditions as the committee, with the approval of the Secretary, may prescribe. The expense of such projects shall be paid from funds collected pursuant to § 955.42 or § 955.45.

(b) In recommending projects pursuant to this section, the committee shall give consideration to the following:

(1) The expected supply of Vidalia onions in relation to market requirements;

(2) The supply situation among competing areas and commodities;

(3) The anticipated benefits from such projects in relation to their costs;

(4) The need for marketing research with respect to any market development activity; and

(5) Other relevant factors.

(c) If the committee should conclude that a program of research and development should be undertaken, or continued, in any fiscal period, it shall submit the following for the approval of the Secretary;

(1) Its recommendations as to the funds to be obtained pursuant to § 955.42 or § 955.45;

(2) Its recommendation as to any research projects; and

(3) Its recommendations as to promotion activity and paid advertising.

(d) Upon conclusion of each activity, but at least annually, the committee shall summarize and report the results of such activity to the Secretary.

(e) All marketing promotion activity engaged in by the committee, including paid advertising, shall be subject to the following terms and conditions:

(1) No marketing promotion, including paid advertising, shall refer to any private brand, private trademark or private trade name;

(2) No promotion or advertising shall disparage the quality, use, value or sale of like or any other agricultural commodity or product, and no false or unwarranted claims shall be made in connection with the product; and

(3) No promotion or advertising shall be undertaken without reason to believe that returns to producers will be improved by such activity.

REPORTS AND RECORDKEEPING

§ 955.60 Reports and recordkeeping.

Upon request of the committee, made with the approval of the Secretary, each handler shall furnish to the committee, in such manner and at such time as it may prescribe, such reports and other information as may be necessary for the committee to perform its duties under this part.

(a) Such reports may include, but are not limited to, the following:

(1) The quantities of Vidalia onions received by a handler;

(2) The quantities disposed of by the handler;

(3) The date of each such disposition; and

(4) The identification of the carrier transporting such Vidalia onions.

(b) All such reports shall be held under appropriate protective classification and custody by duly appointed employees of the committee, so that the information contained therein which may adversely affect the competitive position of any handler in relation to other handlers will not be disclosed. Compilations of general reports from data submitted by handlers is authorized, subject to the prohibition of disclosure of an individual handler's identity or operations.

(c) Each handler shall maintain for at least two succeeding years such records of the Vidalia onions received and disposed of by such handler as may be necessary to verify reports submitted to the committee pursuant to this section.

MISCELLANEOUS PROVISIONS

§ 955.71 **Termination or suspension.**

(a) The Secretary may at any time terminate the provisions of this part by giving at least one day's notice by means of a press release or in any other manner which the Secretary may determine.

(b) The Secretary shall terminate or suspend the operations of any or all of the provisions of this part whenever it is found that such provisions do not tend to effectuate the declared policy of the Act.

(c) The Secretary shall terminate the provisions of this part at the end of any fiscal period whenever it is found that such termination is favored by a majority of producers who, during a representative period, have been engaged in the production of Vidalia onions:

Provided, That such majority has, during such representative period, produced for market more than fifty percent of the volume of such Vidalia onions produced for market, but such termination shall be effective only if announced on or before June 15 of the then current fiscal period.

(d) Within six years of the effective date of this part, the Secretary shall conduct a continuance referendum to ascertain whether continuance of this part is favored by producers. Subsequent referenda to ascertain continu-

ance shall be conducted every six years thereafter.

(e) The provisions of this part shall, in any event, terminate whenever the provisions of the Act authorizing them cease to be in effect.

§ 955.72 **Proceedings after termination.**

(a) Upon the termination of the provisions of this subpart, the then functioning members of the committee shall continue as joint trustees, for the purpose of liquidating the affairs of the committee, of all funds and property then in the possession, or under control, of the committee, including claims for any funds unpaid or property not delivered at the time of such termination. Action by said trusteeship shall require the concurrence of a majority of the said trustees.

(b) The said trustees shall continue in such capacity until discharged by the Secretary; shall, from time to time, account for all receipts and disbursements and deliver all property on hand, together with all books and records of said committee and of the trustees, to such person as the Secretary may direct; and shall upon the request of the Secretary, execute such assignments or other instruments necessary or appropriate to vest in such person full title and right to all of the funds, property, and claims vested in said committee or the trustees pursuant to this subpart.

(c) Any person to whom funds, property, or claims have been transferred or delivered by the committee or its members pursuant to this section shall be subject to the same obligations imposed upon the members of the committee and upon the said trustees.

§ 955.73 **Effect of termination or amendment.**

Unless otherwise expressly provided by the Secretary, the termination of this subpart or of any regulation issued pursuant to this subpart, or the issuance of any amendments to either thereof, shall not:

(a) Affect or waive any right, duty, obligation, or liability which shall have arisen or which may thereafter arise in connection with any provision of this subpart or any regulation issued under this subpart;

(b) Release or extinguish any violation of this subpart or of any regulations issued under this subpart; or

(c) Affect or impair any rights or remedies of the Secretary or of any other person with respect to any such violations.

§ 955.80 Compliance.

No handler shall handle Vidalia onions except in conformity with the provisions of this part.

§ 955.81 Right of the Secretary.

The members of the committee (including successors and alternates) and any agent or employee appointed or employed by the committee shall be subject to removal or suspension by the Secretary at any time. Each and every order, regulation, decision, determination, or other act of the committee shall be subject to the continuing right of the Secretary to disapprove of the same at any time. Upon such disapproval, the disapproved action of the committee shall be deemed null and void except as to acts done in reliance thereon or in compliance therewith prior to such disapproval by the Secretary.

§ 955.82 Duration of immunities.

The benefits, privileges, and immunities conferred upon any person by virtue of this part shall cease upon the termination of this part, except with respect to acts done under and during the existence of this part.

§ 955.83 Agents.

The Secretary may, by designation in writing, name any person, including any officer or employee of the Government, or name any agency in the United States Department of Agriculture, to act as the Secretary's agent or representative in connection with any of the provisions of this part.

§ 955.84 Derogation.

Nothing contained in this part is, or shall be construed to be, in derogation or in modification of the rights of the Secretary or of the United States to exercise any powers granted by the Act or otherwise, or, in accordance with such powers, to act in the premises whenever such action is deemed advisable.

§ 955.85 Personal liability.

No member or alternate of the committee or any employee or agent thereof, shall be held personally responsible, either individually or jointly with others, in any way whatsoever, to any handler or to any person for errors in judgment, mistakes, or other acts, either of commission or omission, as such member, alternate, employee, or agent, except for acts of dishonesty, willful misconduct, or gross negligence.

§ 955.86 Separability.

If any provision of this part is declared invalid, or the applicability thereof to any person, circumstance, or thing is held invalid, the validity of the remainder of this part, or the applicability thereof to any other person, circumstance, or thing shall not be affected thereby.

§ 955.87 Amendments.

Amendments to this part may be proposed, from time to time, by the committee or by the Secretary.

MARKETING AGREEMENT

§ 955.90 Counterparts.

This agreement may be executed in multiple counterparts and when one counterpart is signed by the Secretary, all such counterparts shall constitute, when taken together, one and the same instrument as if all signatures were contained in one original.

§ 955.91 Additional parties.

After the effective date thereof, any handler may become a party to this agreement if a counterpart is executed by such handler and delivered to the Secretary. This agreement shall take effect as to such new contracting part at the time such counterpart is delivered to the Secretary, and the benefits, privileges, and immunities conferred by this agreement shall then be effective as to such new contracting party.

§ 955.92 Order with marketing agreement.

Each signatory hereby requests the Secretary to issue, pursuant to the Act, an order providing for regulating the handling of Vidalia onions in the same manner as is provided for in this agreement.

Subpart B—Administrative Requirements

§ 955.101 Vidalia Onion Handler Report.

(a) Each handler shall furnish shipping reports with the Vidalia Onion Committee on a monthly basis. Such reports shall be made on forms provided by the Committee and shall include:

(1) The name and address of the handler;

(2) Monthly period covered by the report;

(3) Total quantity of Vidalia onions received;

(4) Total fresh market shipments of Vidalia onions;

(5) Shipment volume coming from acreage owned by the handler;

(6) Total assessments owed;

(7) Volume of onions packed under contract for another handler and those handler names;

(8) Onions sold to another handler; and

(9) Information on onions placed in Controlled Atmosphere storage.

(b) Handlers shall file reports each fiscal period beginning the first month they make shipments and shall continue filing reports until they submit a final report for the season. Each such report shall be filed with the Committee not later than 5 p.m. on the tenth day of each month following the month in which any shipments were made. Should the tenth day of the month fall on a weekend or holiday, reports are due by the first business day following the tenth day of the month.

[67 FR 41816, June 20, 2002, as amended at 71 FR 34509, June 15, 2006; 78 FR 28120, May 14, 2013]

§ 955.113 Fiscal period.

Pursuant to § 955.13, *fiscal period* shall mean the period beginning January 1 and ending December 31 of each year, except that the fiscal period that began on September 16, 1998, shall end on December 31, 1999.

[64 FR 48245, Sept. 3, 1999]

§ 955.121 Change in term of office.

Pursuant to § 955.21, the term of office for the Committee shall be for two years beginning January 1 and ending December 31, except that, the term of office for members and alternates whose terms expired on September 15, 1999, shall end on December 31, 1999, or until qualified successors are selected.

[64 FR 72269, Dec. 27, 1999]

§ 955.122 Change in nomination deadlines.

Pursuant to § 955.22, the Committee shall hold or cause to be held not later than October 1 of each year a meeting or meetings of growers for the purpose of designating one nominee for each position as member and for each position as alternate of the Committee which is vacant, or about to become vacant. Such nominations shall be supplied to the Secretary in such manner and form as the Secretary may prescribe, not later than October 15 of each year. The grower members shall nominate the public member and alternate public member at the first meeting following the selection of members for a new term of office. Nominations for the public member and alternate public member shall be supplied to the Secretary in such manner and form as the Secretary may prescribe, not later than February 15.

[64 FR 72269, Dec. 27, 1999]

§ 955.142 Delinquent assessments.

(a) Each handler shall submit assessments to the Vidalia Onion Committee on a monthly basis for each month during the fiscal period in which they made shipments. Each such assessment shall be paid to the Committee not later than 5 p.m. on the tenth day of each month following the month in which any shipments were made. Should the tenth day of the month fall on a weekend or holiday, assessments are due by the first business day following the tenth day of the month.

(b) Each handler shall pay interest of 1.5 percent per month on any assessments levied pursuant to § 955.42 and on any accrued unpaid interest beginning the day immediately after the date the monthly assessments were due, until the delinquent handler's assessments, plus applicable interest, have been paid in full. In addition to the interest charge, the Committee shall impose a late payment charge on any handler whose assessment payment has not been received within 10 days of the due date. The late payment charge shall be 10 percent of the late assessments.

[71 FR 34509, June 15, 2006, as amended at 76 FR 37620, June 28, 2011; 78 FR 28120, May 14, 2013]

ASSESSMENT RATES

§ 955.209 Assessment rate.

On and after January 1, 2008, an assessment rate of $0.13 per 40-pound carton or equivalent is established for Vidalia onions.

[73 FR 31607, June 3, 2008]

PART 956—SWEET ONIONS GROWN IN THE WALLA WALLA VALLEY OF SOUTHEAST WASHINGTON AND NORTHEAST OREGON

Subpart A—Order Regulating Handling

DEFINITIONS

Subpart B—Administrative Requirements

AUTHORITY: 7 U.S.C. 601–674.

SOURCE: 60 FR 27626, May 24, 1995, unless otherwise noted.

Subpart A—Order Regulating Handling

DEFINITIONS

§ 956.1 Secretary.

Secretary means the Secretary of Agriculture of the United States or any officer or employee of the Department of Agriculture who has been delegated, or to whom authority may hereafter be delegated, the authority to act for the Secretary.

§ 956.2 Act.

Act means Public Act No. 10, 73d Congress (May 12, 1933), as amended and as reenacted and amended by the Agricultural Marketing Agreement Act of 1937, as amended (Sec. 1–19, 48 Stat. 31, as amended; 7 U.S.C. 601 *et seq.*).

§ 956.3 Person.

Person means an individual, partnership, corporation, association, or any other business unit.

§ 956.4 Production area.

Production area means a tract of land in Umatilla County, Oregon, and Walla Walla County, Washington, based on surveyors' maps, enclosed by the following boundaries:

Commencing at the Southeast corner of Section 13, Township (Twp.) 5 North, Range (Rge.) 36 East, W.M.; thence Westerly along the South line of Sections 13, 14, 15, 16, 17, and 18 in Twp. 5 North, Rge. 36 East, Sections 13, 14, 15, 16, 17, and 18 in Twp. 5 North, Rge. 35 East, Sections 13, 14, 15, 16, 17, and 18 in Twp. 5 North, Rge. 34 East, Sections 13, 14, and 15 in Twp. 5 North, Rge. 33 East, W.M. to the East right of way line of the Northern Pacific Railway, as it runs Northwesterly through Vansyckle Canyon; thence Northwesterly along said Easterly right of way line to a point in the Northwest ¼ of Section 20, Twp. 7 North, Rge. 32 East, W.M. where said line intersects the South right of way of the Union Pacific Railway, said intersection being commonly known as Zangar Junction; thence Easterly along said South right of way line of the Union Pacific Railway to a point in the Southwest ¼ of Section 23, Twp. 7 North, Rge. 32 East where said line intersects the South right of way line of Washington State Highway No. 12; thence Easterly along said South right of way line to the intersection with the West line of Section 34, Twp. 7 North, Rge. 33 East, W.M.; thence North, along the West line of Sec-

tions 34, 27, 22, 15, 10, and 3 in Twp. 7 North, Rge. 33 East, W.M., and the West line of Sections 34, 27, and 22 in Twp. 8 North, Rge. 33 East, W.M. to the Northwest corner of said Section 22; thence East along the North line of said Section 22 to the Northeast corner thereof; thence North along the West line of Sections 14, 11, and 2 in Twp. 8 North, Rge. 33 East, W.M. to the Northwest corner of said Section 2; thence East along North lines of Sections 2 and 1 in Twp. 8 North, Rge. 33 East, W.M. and the North line of Section 6, Twp. 8 North, Rge. 34 East, W.M. to the centerline of the Touchet River; thence northerly and Easterly along said centerline of the Touchet River as it runs through Twp. 9 North, Rge. 34 East, Twp. 9 North, Rge. 35 East, Twp. 10 North, Rge. 35 East, Twp. 10 North, Rge. 36 East, Twp. 9 North, Rge. 36 East, and Twp. 9 North, Rge. 37 East to a point on the East line of Section 11 in Twp. 9 North, Rge. 37 East, W.M., thence South along the East line of Sections 11, 14, 23, 26, and 35 in Twp. 9 North, Rge. 37 East, W.M., the East lines of Sections 2, 11, 14, 23, 26, and 35 in Twp. 8 North, Rge. 37 East, W.M., the East lines of Sections 2, 11, 14, 23, 26, and 35 in Twp. 7 North, Rge. 37 East, W.M., and the East lines of Sections 2, 11, and fractional Section 14 in Twp. 6 North, Rge. 37 East, W.M., to a point on the Washington-Oregon State line; thence West along said State Line to the closing corner on the West side of Section 18 in Twp. 6 North, Rge. 37 East, W.M.; thence South along the West line of Sections 18, 19, 30, and 31 in Twp. 6 North, Rge. 37 East, W.M. and the West line of Sections 6, 7, and 18 in Twp. 5 North, Rge. 37 East to corner common to Sections 18 and 19 in Twp. 5 North, Rge. 37 East, W.M. and 13 and 24 in Twp. 5 North, Rge. 36 East, W.M., Being the True Point of Beginning of this Legal Description.

§ 956.5 Walla Walla Sweet Onions.

Walla Walla Sweet Onions means all varieties of *Allium cepa* grown within the production area, except Spanish hybrid varieties. The committee may, with the approval of the Secretary, exempt individual varieties from any or all regulations issued under this part.

§ 956.6 Handler.

Handler is synonymous with *shipper* and means any person (except a common or contract carrier of Walla Walla Sweet Onions owned by another person) who handles Walla Walla Sweet Onions or causes Walla Walla Sweet Onions to be handled.

§ 956.7   Registered handler.

Registered handler means any person with adequate facilities for preparing Walla Walla Sweet Onions for commercial market, who has requested such registration and is so recorded by the committee, or any person who has access to such facilities and has recorded with the committee the ability and willingness to assume customary obligations of preparing Walla Walla Sweet Onions for commercial market. The committee may recommend, for approval of the Secretary, procedures with respect to handler registration.

§ 956.8   Handle.

Handle is synonymous with *ship* and means to package, load, sell, transport, or in any way place Walla Walla Sweet Onions or cause Walla Walla Sweet Onions to be placed in the current of commerce within the production area or between the production area and any point outside thereof. Such term shall not include the transportation, sale, or delivery of harvested Walla Walla Sweet Onions to a handler within the production area for the purpose of having such Walla Walla Sweet Onions prepared for market.

§ 956.9   Container.

Container means a box, bag, crate, hamper, basket, package, or any other receptacle used in the packaging, transporting, sale, shipment, or other handling of Walla Walla Sweet Onions.

§ 956.10   Producer.

Producer is synonymous with *grower* and means any person engaged in a proprietary capacity in the production of Walla Walla Sweet Onions for market.

§ 956.11   Varieties.

Varieties means and includes all classifications, subdivisions, or types of Walla Walla Sweet Onions according to those definitive characteristics now or hereafter recognized by the United States Department of Agriculture or recommended by the committee and approved by the Secretary.

§ 956.12   Committee.

Committee means the Walla Walla Sweet Onion Committee established pursuant to § 956.20.

§ 956.13   Fiscal period.

Fiscal period means the period beginning on June 1 and ending on May 31 of each year, or other such period as may be recommended by the committee and approved by the Secretary.

§ 956.14   [Reserved]

§ 956.15   Grade and size.

Grade means any of the officially established grades of onions, including maturity requirements and *size* means any of the officially established sizes of onions as set forth in the United States standards for grades of onions or amendments thereto, or modifications thereof, or variations based thereon, or States of Washington or Oregon standards of onions or amendments thereto or modifications thereof or variations based thereon, recommended by the committee and approved by the Secretary.

[64 FR 4933, Feb. 1, 1999]

§ 956.16   Pack.

Pack means a quantity of Walla Walla Sweet Onions specified by grade, size, weight, or count, or by type or condition of container, or any combination of these recommended by the committee and approved by the Secretary.

[64 FR 4933, Feb. 1, 1999]

ADMINISTRATIVE COMMITTEE

§ 956.20   Establishment and membership.

(a) The Walla Walla Sweet Onion Marketing Committee, consisting of seven members, is hereby established. The Committee shall consist of four producer members, two handler members, and one public member. Each member shall have an alternate who shall have the same qualifications as the member.

(b) A producer shall have three years of experience in producing onions in order to qualify for committee membership. At the time of selection, no more than two producer members may be affiliated with the same handler.

[60 FR 27626, May 24, 1995, as amended at 64 FR 4933, Feb. 1, 1999; 84 FR 13515, May 6, 2019]

§ 956.21 Term of office.

(a) Except as otherwise provided in paragraph (b) of this section, the term of office of grower and handler Committee members and their respective alternates shall be two fiscal periods beginning on June 1 or such other date as recommended by the Committee and approved by the Secretary. The terms shall be determined so that one-half of the grower membership and one-half of the handler membership shall terminate each year. Members and alternates shall serve during the term of office for which they are selected and have been qualified, or during that portion thereof beginning on the date on which they qualify during such term of office and continuing until the end thereof, or until their successors are selected and have qualified.

(b) The term of office of the initial members and alternates shall begin as soon as possible after May 6, 2019. One-half of the initial industry grower and handler members and alternates shall serve for a one-year term and one-half shall serve for a two-year term. The initial as well as all successive terms of office of the public member and alternate member shall be for three years.

(c) The consecutive terms of office for all members shall be limited to two two-year terms. There shall be no such limitation for alternate members.

[84 FR 13515, May 6, 2019]

§ 956.22 Nominations.

Nominations from which the Secretary may select the members of the committee and their respective alternates may be made in the following manner:

(a) The committee shall hold or cause to be held, within the production area and prior to April 1 of each year or by such other date as may be specified by the Secretary, one or more meetings of producers and handlers for the purpose of designating one nominee for each of the member and alternate member positions which are vacant or will be vacant at the end of the fiscal period;

(b) In arranging for such meetings the committee may, if it deems such desirable, cooperate with existing organizations and agencies;

(c) Nominations for committee members and alternate members shall be provided to the Secretary, in such manner and form as the Secretary may prescribe, not later than 30 days prior to the end of the fiscal period within which the current term of office expires;

(d) Only producers may participate in designating nominees for producer committee members and their alternates and only handlers may participate in designating nominees for handler committee members and their alternates;

(e) Each person who is both a handler and a producer may vote either as a handler or as a producer, but not both;

(f) Each person is entitled to cast only one vote on behalf of him or herself, his or her partners, agents, subsidiaries, affiliates and representatives, in designating nominees for committee members and alternates. An eligible producer's or handler's privilege of casting only one vote, as aforesaid, shall be construed to permit such voter to cast one vote for each producer member and alternate member position to be filled or each handler member and alternate member position to be filled, but not both.

(g) Every three years, at the first meeting following selection, the committee shall nominate the public member and alternate for a three-year term of office.

(h) The committee shall prescribe such additional qualifications, administrative rules and procedures for selection and voting for each candidate as it deems necessary and as the Secretary approves.

§ 956.23 Selection.

The Secretary shall select members and alternate members of the committee from the nominations made pursuant to § 956.22 or from other qualified persons.

§ 956.24 Qualification and acceptance.

Any person nominated to serve as a member or alternate member of the committee shall, prior to selection by the Secretary, qualify by filing a written background and acceptance statement indicating such person's willingness to serve in the position for which nominated.

§ 956.25 Alternates.

An alternate member of the committee shall act in the place and stead of the member for whom such person is an alternate, during such member's absence. In the event of the death, removal, resignation, or disqualification of a member, that member's alternate shall serve until a successor to such member has qualified and is selected.

§ 956.26 Vacancies.

To fill any vacancy occasioned by the failure of any person nominated as a member or as an alternate to qualify, or in the event of the death, removal, resignation, or disqualification of a member or alternate, a successor for the unexpired term may be selected by the Secretary from nominations made pursuant to § 956.22 from previously unselected nominees on the current nominee list, or from other eligible persons.

§ 956.27 Failure to nominate.

If nominations are not made within the time and manner prescribed in § 956.22 the Secretary may, without regard to nominations, select the members and alternates on the basis of the representation provided for in § 956.20.

§ 956.28 Procedure.

(a) Four members of the Committee shall constitute a quorum, and four concurring votes shall be required to pass any motion or approve any Committee action, except that recommendations made pursuant to § 956.61 shall require five concurring votes.

(b) The committee may provide for meetings by telephone, telegraph, facsimile, or other means of communication, and any vote cast orally at such meetings shall be confirmed promptly in writing: *Provided,* That if an assembled meeting is held, all votes shall be cast in person.

[60 FR 27626, May 24, 1995, as amended at 84 FR 13516]

§ 956.29 Expenses.

Members and alternates shall serve without compensation but shall be reimbursed for such expenses authorized by the committee and necessarily incurred by them in attending committee meetings and in the performance of their duties under this part.

§ 956.30 Powers.

The committee shall have the following powers:

(a) To administer the provisions of this part in accordance with its terms;

(b) To make rules and regulations to effectuate the terms and provisions of this part;

(c) To receive, investigate, and report to the Secretary complaints of violations of the provisions of this part; and

(d) To recommend to the Secretary amendments to this part.

§ 956.31 Duties.

It shall be among the duties of the committee:

(a) At the beginning of each fiscal period, or as soon thereafter as practicable, to meet and organize, to select a chairperson and such other officers as may be necessary, to select subcommittees, and to adopt such rules and regulations for the conduct of its business as it may deem advisable;

(b) To act as intermediary between the Secretary and any producer or handler;

(c) To furnish to the Secretary such available information as the Secretary may request;

(d) To appoint such employees, agents, and representatives as it may deem necessary and to determine the salaries and define the duties of each such person;

(e) To investigate from time to time and to assemble data on the growing, harvesting, shipping, and marketing conditions with respect to Walla Walla Sweet Onions and to engage in such research and service activities which relate to the production, handling, or marketing of Walla Walla Sweet Onions as may be approved by the Secretary;

(f) To keep minutes, books, and records which clearly reflect all of the acts and transactions of the committee. Such minutes, books, and records shall be subject to examination at any time by the Secretary or the Secretary's authorized agent or representative;

(g) To make available to producers and handlers the committee voting record on recommended regulations and on other matters of policy;

(h) Prior to each fiscal period, to submit to the Secretary a budget of its proposed expenses for such fiscal period, together with a report thereon, and a recommendation as to the rate of assessment for such period;

(i) To cause its books to be audited by a competent accountant at least once each fiscal period, and at such other time as the committee may deem necessary or as the Secretary may require; the report of such audit shall show the receipt and expenditure of funds collected pursuant to this part; a copy of each such report shall be furnished to the Secretary, and a copy of each such report shall be made available at the principal office of the committee for inspection by producers and handlers: *Provided*, that confidential information shall be removed from all copies made available to the public; and

(j) To consult, cooperate, and exchange information with other onion marketing committees and other individuals or agencies in connection with all proper committee activities and objectives under this subpart.

EXPENSES AND ASSESSMENTS

§956.40 Expenses.

The committee is authorized to incur such expenses as the Secretary may find are reasonable and likely to be incurred by the committee for its maintenance and functioning, and to enable it to exercise its powers and perform its duties in accordance with the provisions of this part. The funds to cover such expenses shall be acquired in the manner prescribed in §§956.42 and 956.45.

§956.41 Budget.

Prior to each fiscal period and as may be necessary thereafter, the committee shall prepare an estimated budget of income and expenditures necessary for the administration of this part. The committee shall recommend a rate of assessment calculated to provide adequate funds to defray its proposed expenditures. The committee shall present such budget to the Secretary with an accompanying report showing the basis for its calculations.

§956.42 Assessments.

(a) The funds to cover the committee's expenses shall be acquired by the levying of assessments upon handlers as provided in this subpart. Each person who first handles Walla Walla Sweet Onions shall pay assessments to the committee upon demand, which assessments shall be in payment of such handler's pro rata share of the committee's expenses.

(b) Assessments shall be levied upon handlers, at rates established by the Secretary. Such rates may be established upon the basis of the committee's recommendations or other available information.

(c) At any time during, or subsequent to, a given fiscal period, the committee may recommend the approval of an amended budget and an increase in the rate of assessment. Upon the basis of such recommendations, or other available information, the Secretary may approve an amended budget and increase the assessment rate. Such increase in the assessment rate shall be applicable to all Walla Walla Sweet Onions which were handled by each handler thereof during such fiscal period.

(d) The payment of assessments for the maintenance and functioning of the committee may be required under this part throughout the period it is in effect, irrespective of whether particular provisions of this part are suspended or become inoperative.

(e) To provide funds for the administration of the provisions of this part during the initial fiscal period or the first part of a fiscal period when neither sufficient operating reserve funds nor sufficient revenue from assessments on the current season's shipments are available, the committee

359

may accept payment of assessments in advance or may borrow money for such purposes.

(f) The committee may impose a late payment charge or an interest charge, or both, on any handler who fails to pay any assessment in a timely manner. Such time and the rates shall be recommended by the committee and approved by the Secretary.

§ 956.43 Accounting.

(a) All funds received by the committee pursuant to the provisions of this part shall be used solely for the purposes specified in this part.

(b) The Secretary may at any time require the committee, its members and alternate members, employees, agents, and all other such persons associated with the committee to account for all receipts, disbursements, funds, property, or records for which they are responsible. Whenever any person ceases to be a member, alternate member, employee, or agent of the committee, such person shall account for all receipts, disbursements, funds, property, and records pertaining to the committee's activities for which such person was responsible, deliver all property and funds in such person's possession to the committee, and execute such assignments and other instruments as may be necessary or appropriate to vest in the committee full title to all of the property, funds, and claims vested in such person pursuant to this part.

(c) The committee may make recommendations to the Secretary for one or more of the members thereof, or any other person, to act as a trustee for holding records, funds, or any other committee property during periods of suspension of this part, or during any period or periods when regulations are not in effect and, upon determining such action is appropriate, the Secretary may direct that such person or persons shall act as trustee or trustees for the committee.

§ 956.44 Excess funds.

If, at the end of a fiscal period, the assessments collected are in excess of expenses incurred, such excess shall be accounted for as follows:

(a) The committee, with approval of the Secretary, may establish an operating reserve and may carry over to subsequent fiscal periods excess funds in a reserve so established, except funds in the reserve shall not exceed the equivalent of approximately two fiscal period's budgeted expenses. Such reserve funds may be used:

(1) To defray any expenses authorized under this part;

(2) To defray expenses during any fiscal period prior to the time assessment income is sufficient to cover such expenses;

(3) To cover deficits incurred during any fiscal period when assessment income is less than expenses;

(4) To defray expenses incurred during any period when any or all provisions of this part are suspended or are inoperative; and

(5) To cover necessary expenses of liquidation in the event of termination of this part.

(b) Upon termination of this part, any funds not required to defray the necessary expenses of liquidation shall be disposed of in such manner as the Secretary may determine to be appropriate except that to the extent practicable, such funds shall be returned pro rata to the persons from whom such funds were collected.

(c) If such excess is not retained in a reserve as provided in paragraph (a) of this section, each handler entitled to a proportionate refund of the excess assessments collected shall be credited at the end of a fiscal period with such refund against the operations of the following fiscal period unless such handler demands payment thereof, in which event such proportionate refund shall be paid as soon as practicable.

§ 956.45 Contributions.

The committee may accept voluntary contributions but these shall be used only to pay expenses incurred pursuant to § 956.50. Such contributions shall be free from any encumbrances by the donor, and the committee shall retain complete control of their use.

§956.50 Research and development.

(a) The committee, with the approval of the Secretary, may establish or provide for the establishment of production research, marketing research and development, and marketing promotion projects, including paid advertising, designed to assist, improve, or promote the marketing, distribution, consumption, or efficient production of Walla Walla Sweet Onions. Any such project for the promotion and advertising of Walla Walla Sweet Onions may utilize an identifying mark, including but not limited to registered trademarks and logos, which shall be made available for use by all handlers in accordance with such terms and conditions as the committee, with the approval of the Secretary, may prescribe. The committee may register such logos with the Commissioner of Patents and Trademarks, U.S. Patent and Trademark Office. The expense of such projects shall be paid from funds collected pursuant to §§956.42 and 956.45.

(b) In recommending projects pursuant to this section, the committee shall give consideration to the following:

(1) The expected supply of Walla Walla Sweet Onions in relation to market requirements;

(2) The supply situation among competing onion areas and communities;

(3) The anticipated benefits from such projects in relation to their costs;

(4) The need for marketing research with respect to any market development activity; and

(5) Other relevant factors.

(c) If the committee concludes that a program of research and development should be undertaken, or continued, in any fiscal period, it shall submit the following for the approval of the Secretary:

(1) Its recommendations as to the funds to be obtained pursuant to §§956.42 and 956.45;

(2) Its recommendations as to any research projects; and

(3) Its recommendations as to promotion activity and paid advertising.

(d) Upon conclusion of each activity, but at least annually, the committee shall summarize and report the results of such activity to the Secretary.

(e) All marketing promotion activity engaged in by the committee, including paid advertising, shall be subject to the following terms and conditions:

(1) No marketing promotion, including paid advertising, shall refer to any private brand, private trademark, or private trade name;

(2) No promotion or advertising shall disparage the quality, use, value, or sale of like or any other agricultural commodity or product, and no false or unwarranted claims shall be made in connection with the product; and

(3) No promotion or advertising shall be undertaken without reason to believe that returns to producers will be improved by such activity.

§956.60 Marketing policy.

(a) *Preparation.* Prior to each marketing season, the committee shall consider and prepare a proposed policy for the marketing of Walla Walla Sweet Onions. In developing its marketing policy, the committee shall investigate relevant supply and demand conditions for Walla Walla Sweet Onions. In such investigations, the committee shall give appropriate consideration to the following:

(1) Market prices for sweet onions, including prices by variety, grade, size, quality, and maturity, and by different packs;

(2) Supply of sweet onions by grade, size, quality, maturity, and variety in the production area and in other sweet onion producing sections;

(3) The trend and level of consumer income;

(4) Establishing and maintaining orderly marketing conditions for Walla Walla Sweet Onions;

(5) Orderly marketing of Walla Walla Sweet Onions as will be in the public interest; and

(6) Other relevant factors.

(b) *Reports.* (1) The committee shall submit a report to the Secretary setting forth the aforesaid marketing policy, and the committee shall notify producers and handlers of the contents of such report.

(2) In the event it becomes advisable to shift from such marketing policy because of changed supply and demand conditions, the committee shall prepare an amended or revised marketing policy in accordance with the manner previously outlined. The committee shall submit a report thereon to the Secretary and notify producers and handlers of the contents of such report on the revised or amended marketing policy.

[64 FR 4933, Feb. 1, 1999]

§ 956.61 Recommendation for regulations.

The committee shall recommend regulations to the Secretary whenever it deems it advisable, as provided in § 956.62. The committee also may recommend modification, suspension, or termination of any regulation, or amendments thereto, in order to facilitate the handling of Walla Walla Sweet Onions for the purposes authorized in § 956.63. The committee may also recommend amendment, modification, termination, or suspension of any regulation issued under this part.

§ 956.62 Issuance of regulations.

(a) Except as otherwise provided in this part, the Secretary shall limit the shipment of Walla Walla Sweet Onions by any one or more of the methods hereinafter set forth whenever the Secretary finds from the recommendations and information submitted by the committee, or from other available information, that such regulation would tend to effectuate the declared policy of the Act. Such limitation may:

(1) Regulate in any or all portions of the production area, the handling of particular grades, sizes, qualities, or maturities of any or all varieties of Walla Walla Sweet Onions, or combinations thereof, during any period or periods;

(2) Regulate the handling of particular grades, sizes, qualities, or maturities of Walla Walla Sweet Onions differently, for different varieties or packs, or for any combination of the foregoing, during any period or periods;

(3) Provide a method, through rules and regulations issued pursuant to this part, for fixing the size, capacity, weight, dimensions, markings or pack of the container or containers, which may be used in the packaging or handling of Walla Walla Sweet Onions, including appropriate logo or other container markings to identify the contents thereof;

(4) Regulate the handling of Walla Walla Sweet Onions by establishing, in terms of grades, sizes, or both, minimum standards of quality and maturity.

(b) The Secretary may amend any regulation issued under this part whenever the Secretary finds that such amendment would tend to effectuate the declared policy of the Act. The Secretary may also terminate or suspend any regulation or amendment thereof whenever the Secretary finds that such regulation or amendment obstructs or no longer tends to effectuate the declared policy of the Act.

[64 FR 4933, Feb. 1, 1999]

§ 956.63 Handling for specified purposes.

Upon the basis of recommendations and information submitted by the committee, or other available information, the Secretary may issue special regulations, or modify, suspend, or terminate requirements in effect pursuant to §§ 956.42 and 956.62 or any combination thereof, in order to facilitate the handling of onions for the following purposes:

(a) Shipments of Walla Walla Sweet Onions for relief or to charitable institutions;

(b) Shipments of Walla Walla Sweet Onions for livestock feed;

(c) Shipments of Walla Walla Sweet Onions for planting and for plants;

(d) Shipments of Walla Walla Sweet Onions as salad onions;

(e) Shipments of Walla Walla Sweet Onions for all processing uses including, pickling, peeling, dehydration, juicing, or other processing;

(f) Shipments of Walla Walla Sweet Onions for disposal;

(g) Shipments of Walla Walla Sweet Onions for seed;

(h) Shipments of Walla Walla Sweet Onions for packing or storing within the production area or outside the production area, but within specified locations in the States of Oregon and Washington; and

(i) Shipments of Walla Walla Sweet Onions for other purposes which may be specified.

§956.64 **Minimum quantities.**

During any period in which shipments of Walla Walla Sweet Onions are regulated pursuant to this part, each handler may handle up to, but not to exceed, 2,000 pounds of Walla Walla Sweet Onions per shipment without regard to the inspection requirements of this part: *Provided,* That such Walla Walla Sweet Onion shipments meet the minimum requirements in effect at the time of the shipment pursuant to §956.62. The committee, with the approval of the Secretary, may recommend modifications to this section and the establishment of such other minimum quantities below which Walla Walla Sweet Onion shipments will be free from the requirements in, or pursuant to, §§956.42, 956.62, 956.63, and 956.70, or any combination thereof.

[64 FR 4934, Feb. 1, 1999]

§956.65 **Notification of regulations.**

The Secretary shall notify the committee of each regulation issued and of each amendment, modification, suspension, or termination thereof. The committee shall give reasonable notice thereof to handlers.

§956.66 **Safeguards.**

(a) The committee, with the approval of the Secretary, may prescribe adequate safeguards to prevent Walla Walla Sweet Onions shipped, pursuant to §§956.63 and 956.64, from entering channels of trade for other than the purpose authorized therefor.

(b) The committee, with the approval of the Secretary, may also prescribe rules and regulations governing the issuance, and the contents, of Certificates of Privilege, if such certificates are prescribed as safeguards by the committee. Such safeguards may include requirements that:

(1) Handlers shall first file applications with the committee to ship such Walla Walla Sweet Onions.

(2) Handlers shall pay the pro rata share of expenses provided by §956.42 in connection with such Walla Walla Sweet Onions.

(3) Handlers shall obtain Certificates of Privilege from the committee prior to effecting the particular onion shipment.

(c) The committee may rescind any Certificate of Privilege, or refuse to issue any Certificate of Privilege, to any handler if proof is obtained that Walla Walla Sweet Onions shipped by the handler for the purposes stated in the Certificate of Privilege were handled contrary to the provisions of this part.

(d) The Secretary shall have the right to modify, change, alter, or rescind any safeguards prescribed and any certificates issued by the committee pursuant to the provisions of this section.

(e) The committee shall make reports to the Secretary as requested, showing the number of applications for such certificates, the quantity of Walla Walla Sweet Onions covered by such applications, the number of such applications denied and certificates granted, the quantity of Walla Walla Sweet Onions handled under duly issued certificates, and such other information as may be requested.

INSPECTION

§956.70 **Inspection and certification.**

(a) During any period in which shipments of Walla Walla Sweet Onions are regulated pursuant to this subpart, no handler shall handle Walla Walla Sweet Onions unless such onions are inspected by an authorized representative of the Federal-State Inspection Service, or such other inspection service as the Secretary shall designate and are covered by a valid inspection certificate, except when relieved from such requirements pursuant to §956.63 or §956.64, or both. Upon recommendation of the committee, with approval of the Secretary, inspection providers and certification requirements may be modified to facilitate the handling of Walla Walla Sweet Onions.

(b) Regrading, resorting, or repacking any lot of Walla Walla Sweet Onions shall invalidate prior inspection certificates insofar as the requirements of this section are concerned. No handler shall ship Walla Walla Sweet Onions after they have been regraded, resorted, repacked, or in any other way further prepared for market, unless such onions are inspected by an authorized representative of the Federal-State Inspection Service, or such other inspection service as the Secretary shall designate: *Provided*, That such inspection requirements on regraded, resorted, or repacked Walla Walla Sweet Onions may be modified, suspended, or terminated under rules and regulations recommended by the committee, and approved by the Secretary.

(c) Upon recommendation of the committee, and approval of the Secretary, all Walla Walla Sweet Onions that are required to be inspected and certified in accordance with this section shall be identified by appropriate seals, stamps, tags, or other identification to be furnished by the committee and affixed to the containers by the handler under the direction and supervision of the Federal-State or Federal inspector, or the committee. Master containers may bear the identification instead of the individual containers within said master container.

(d) Insofar as the requirements of this section are concerned, the length of time for which an inspection certificate is valid may be established by the committee with the approval of the Secretary.

(e) When Walla Walla Sweet Onions are inspected in accordance with the requirements of this section, a copy of each inspection certificate issued shall be made available to the committee by the inspection service.

(f) The committee may enter into an agreement with an inspection service with respect to the costs of the inspection as provided by paragraph (a) of this section, and may collect from handlers their respective pro rata shares of such costs.

[64 FR 4934, Feb. 1, 1999]

REPORTS

§ 956.80 Reports and recordkeeping.

Upon request of the committee, made with the approval of the Secretary, each handler shall furnish to the committee, in such manner and at such time as it may prescribe, such reports and other information as may be necessary for the committee to perform its duties under this part.

(a) Such reports may include, but are not necessarily limited to, the following:

(1) The acreage of Walla Walla Sweet Onions grown;

(2) The quantities of Walla Walla Sweet Onions received by such handler;

(3) The quantities of Walla Walla Sweet Onions disposed of by such handler;

(4) The disposition date of such Walla Walla Sweet Onions;

(5) The manner of disposition of such Walla Walla Sweet Onions; and

(6) The identification of the carrier transporting such Walla Walla Sweet Onions.

(b) All such reports shall be held under appropriate protective classification and custody by the committee, or duly appointed employees thereof, so that any information contained therein which may adversely affect the competitive position of any handler in relation to other handlers will not be disclosed. Compilations of general reports from data submitted by handlers is authorized, subject to the prohibition of disclosure of individual handler's identity or operations.

(c) Each handler shall maintain for at least two succeeding years such records of the Walla Walla Sweet Onions received and disposed of by such handler as may be necessary to verify reports submitted to the committee pursuant to this section.

MISCELLANEOUS PROVISIONS

§ 956.85 Termination or suspension.

(a) The Secretary may at any time terminate the provisions of this subpart by giving at least one day's notice by means of a press release or in any other manner which the Secretary may determine.

(b) The Secretary shall terminate or suspend the operations of any or all of the provisions of this subpart whenever it is found that such provisions do not tend to effectuate the declared policy of the act.

(c) The Secretary shall terminate the provisions of this subpart at the end of any fiscal period whenever it is found that such termination is favored by a majority of producers who, during a representative period, have been engaged in the production of Walla Walla Sweet Onions: Provided, That such majority has, during such representative period, produced for market more than fifty percent of the volume of such Walla Walla Sweet Onions produced for market, but such termination shall be announced at least 90 days before the end of the current fiscal period.

(d) Within six years of the effective date of this subpart the Secretary shall conduct a continuance referendum to ascertain whether continuance of this subpart is favored by producers. Subsequent referenda to ascertain continuance shall be conducted every six years thereafter. The Secretary may terminate the provisions of this part at the end of any fiscal period in which the Secretary has found that continuance of this subpart is not favored by a majority of producers who, during a representative period determined by the Secretary, have been engaged in the production for market of Walla Walla Sweet Onions in the production area. Such termination shall be announced on or before the end of the fiscal period.

(e) The provisions of this subpart shall, in any event, terminate whenever the provisions of the Act authorizing them cease to be in effect.

§956.87 Proceedings after termination.

(a) Upon the termination of the provisions of this subpart, the then functioning members of the committee shall continue as joint trustees, for the purpose of liquidating the affairs of the committee, of all funds and property then in the possession, or under control, of the committee, including claims for any funds unpaid or property not delivered at the time of such termination. Action by said trusteeship shall require the concurrence of a majority of the said trustees.

(b) The said trustees shall continue in such capacity until discharged by the Secretary; shall, from time to time, account for all receipts and disbursements and deliver all property on hand, together with all books and records of said committee and of the

trustees, to such person as the Secretary may direct; and shall upon the request of the Secretary, execute such assignments or other instruments necessary or appropriate to vest in such person full title and right to all of the funds, property, and claims vested in said committee or the trustees pursuant to this subpart.

(c) Any person to whom funds, property, or claims have been transferred or delivered by the committee or its members pursuant to this section shall be subject to the same obligations imposed upon the members of the committee and upon the said trustees.

§956.88 Effect of termination or amendment.

Unless otherwise expressly provided by the Secretary, the termination of this subpart or of any regulation issued pursuant to this subpart, or the issuance of any amendments to either thereof, shall not:

(a) Affect or waive any right, duty, obligation, or liability which shall have arisen or which may thereafter arise in connection with any provision of this subpart;

(b) Release or extinguish any violation of this subpart or of any regulations issued under this subpart; and

(c) Affect or impair any rights or remedies of the Secretary or of any other person with respect to any such violations.

§956.89 Compliance.

No handler shall handle Walla Walla Sweet Onions except in conformity to the provisions of this part.

§956.90 Right of the Secretary.

The members of the committee, including successors and alternates, and any agent or employee appointed or employed by the committee shall be subject to removal or suspension by the Secretary at any time. Each and every order, regulation, decision, determination, or other act of the committee shall be subject to the continuing right of the Secretary to disapprove of the same at any time. Upon such disapproval, the disapproved action of the committee shall be deemed null and void except as to acts done in reliance thereon or in compliance therewith prior to such disapproval by the Secretary.

§ 956.91 Duration of immunities.

The benefits, privileges, and immunities conferred upon any person by virtue of this subpart shall cease upon the termination of this subpart, except with respect to acts done under and during the existence of this subpart.

§ 956.92 Agents.

The Secretary may, by designation in writing, name any person, including any officer or employee of the Government, or name any agency in the United States Department of Agriculture, to act as the Secretary's agent or representative in connection with any of the provisions of this part.

§ 956.93 Derogation.

Nothing contained in this part is, or shall be construed to be, in derogation or in modification of the rights of the Secretary or of the United States to exercise any powers granted by the Act or otherwise, or, in accordance with such powers, to act in the premises whenever such action is deemed advisable.

§ 956.94 Personal liability.

No member or alternate of the committee or any employee or agent thereof, shall be held personally responsible, either individually or jointly with others, in any way whatsoever, to any handler or to any person for errors in judgment, mistakes, or other acts, either of commission or omission, as such member, alternate, employee, or agent, except for acts of dishonesty, willful misconduct, or gross negligence.

§ 956.95 Separability.

If any provision of this subpart is declared invalid, or the applicability thereof to any person, circumstance, or thing is held invalid, the validity of the remainder of this subpart, or the applicability thereof to any other person, circumstance, or thing shall not be affected thereby.

§ 956.96 Amendments.

Amendments to this subpart may be proposed, from time to time, by the committee or by the Secretary.

Subpart B—Administrative Requirements

SOURCE: 61 FR 44151, Aug. 28, 1996, unless otherwise noted.

§ 956.113 Fiscal period.

Pursuant to § 956.13, *fiscal period* shall mean the period beginning January 1 and ending December 31 of each year.

[68 FR 57326, Oct. 3, 2003]

§ 956.142 Interest charges.

For Walla Walla Sweet Onions handled prior to September 1, the Committee shall impose an interest charge on any handler who fails to pay his or her annual assessments within thirty (30) days of the due date of September 30. For Walla Walla Sweet Onions handled during the period September 1 through May 31, the Committee shall impose an interest charge on any handler who fails to pay his or her assessments within thirty (30) days of the last day of the month in which such shipments are made. The interest charge shall be 1½ percent of the unpaid assessment balance. In the event the handler fails to pay the delinquent assessment amount within 60 days following the due date, the 1½ percent interest charge shall be applied monthly thereafter to the unpaid balance, including any accumulated interest. Any amount paid by a handler as an assessment, including any charges imposed pursuant to this paragraph, shall be credited when the payment is received in the Committee office.

[75 FR 34347, June 17, 2010]

§ 956.162 Container markings.

Effective April 15, 1997, no handler shall ship any container of Walla Walla Sweet Onions except in accordance with the following terms and provisions:

(a) Each container of Walla Walla Sweet Onions shall be conspicuously marked with the "Genuine Walla Walla Sweet Onion" logo. The marking may be in the form of a decal or a stamped imprint of any color and size: *Provided,* That the decal or stamped imprint must be placed in plain sight and easy to read.

(b) Walla Walla Sweet Onions may be handled not subject to the marking requirements of this section when handlers ship such onions pursuant to §956.163, or ship such onions in field packed bulk bins containing more than 500 pounds net weight for sale to roadside stands and farmers' market operators for repacking and direct consumer sale: *Provided,* That subject to Committee verification of handler container inventories, handlers may use their existing inventories of unmarked containers until April 15, 1999.

[62 FR 18026, Apr. 14, 1997]

§956.163 Handling for specified purposes.

(a) Assessment and container marking requirements specified in this part shall not be applicable to shipments of onions for any of the following purposes:

(1) Shipments of Walla Walla Sweet Onions for relief or to charitable institutions: *Provided,* That such shipments must be donated and not sold in order for this exemption to apply;

(2) Shipments of Walla Walla Sweet Onions for livestock feed;

(3) Shipments of Walla Walla Sweet Onions for planting and for plants;

(4) Shipments of Walla Walla Sweet Onions as salad onions;

(5) Shipments of Walla Walla Sweet Onions for all processing uses including, pickling, peeling, dehydration, juicing, or other processing;

(6) Shipments of Walla Walla Sweet Onions for disposal;

(7) Shipments of Walla Walla Sweet Onions for seed.

(b) *Market preparation outside the production area.* (1) Persons desiring to ship or receive Walla Walla sweet onions for grading, packing, or storing outside the production area, but within Oregon and Washington, shall apply to the Committee on a *"Shippers/Receivers Application for Certificate of Privilege"* form. Such application shall contain the following:

(i) Company name, contact name, address, contact telephone numbers, date, and signature of the applicant;

(ii) Whether the applicant is the shipper or receiver;

(iii) Agreement to provide a *Special Purpose Shipment Report* to the Committee as required after shipping or receiving Walla Walla sweet onions for grading, packing, or storing out of the production area under a Certificate of Privilege.

(iv) Certification by the applicant that all provisions of the rules and regulations of this part will be adhered to including, but not limited to, any grade, size, quality, maturity, pack, or container requirements that may be currently in effect;

(v) Certification by the applicant, if a receiver under the Certificate of Privilege, that they will forward to the Committee office all assessments due on Walla Walla sweet onions handled.

(vi) Such other information as the Committee may require.

(2) Each approved applicant shall furnish to the Committee a *Special Purpose Shipment Report* form no later than thirty (30) days after the final shipment of sweet onions are shipped or received pursuant to the Certificate of Privilege. That report shall contain the following information:

(i) Company name, contact name, address, contact telephone numbers, signature, and date;

(ii) Names of shippers or receivers who have either shipped Walla Walla sweet onions out of the production area or received the same;

(iii) The total quantity of Walla Walla sweet onions shipped or received under this section during the period covered;

(iv) Certification by the receiver that all assessments due on Walla Walla sweet onions handled under the respective Certificate of Privilege are being forwarded to the Committee; and

(v) Such other information as the Committee may require.

(3) The Committee may cancel any Certificate of Privilege if proof satisfactory to the Committee is obtained that any Walla Walla sweet onions shipped or received were done so contrary to the provisions of this section. Upon cancellation of such Certificate

of Privilege the shipper or receiver may appeal to the Committee for reconsideration.

[62 FR 18026, Apr. 14, 1997, as amended at 69 FR 22382, Apr. 26, 2004]

§ 956.180 Reports.

(a) Each handler shall furnish to the Committee, no later than May 31 each year, a preseason *Walla Walla Sweet Onion Handler Registration Form*. Such form shall include:

(1) Company name, contact name, mailing and physical addresses, contact telephone numbers, and signature of handler;

(2) Season covered by registration;

(3) Brand names or labels to be used; and

(4) Estimated number of acres of fall planted and spring planted Walla Walla Sweet Onions to be packed during the season.

(b) Each handler shall furnish to the Committee a *Handler's Statement of Walla Walla Sweet Onion Shipments* containing the information in paragraphs (a)(1), (a)(2), and (a)(3) of this section, except that gift box and roadside stand sales shall be exempt from paragraph (a)(2) of this section: *Provided*, That for Walla Walla Sweet Onions handled prior to September 1, such report shall be furnished to the Committee by September 30, and that for Walla Walla Sweet Onions handled during the period September 1 through May 31, such report shall be furnished to the Committee no later than thirty (30) days after the end of the month in which such onions were handled:

(1) The number of 50 lb. equivalents of Walla Walla Sweet Onions shipped by each handler during each week of the shipping season and the total for the season;

(2) The geographical regions as defined by the Committee to which each shipment is made;

(3) The name, address, and signature of each handler; and

(4) The name of each producer and the number of 50 lb. equivalents of Walla Walla Sweet Onions that were handled on behalf of or acquired from that producer.

[69 FR 22382, Apr. 26, 2004, as amended at 75 FR 34347, June 17, 2010]

§ 956.202 Assessment rate.

On and after January 1, 2017, an assessment rate of $0.10 per 50-pound bag or equivalent is established for Walla Walla sweet onions.

[82 FR 11791, Feb. 27, 2017]

PART 958—ONIONS GROWN IN CERTAIN DESIGNATED COUNTIES IN IDAHO, AND MALHEUR COUNTY, OREGON

Subpart A—Order Regulating Handling

DEFINITIONS

AUTHORITY: 7 U.S.C. 601–674.

SOURCE: 22 FR 26, Jan. 3, 1957, unless otherwise noted. Redesignated at 26 FR 12751, Dec. 30, 1961.

Subpart A—Order Regulating Handling

DEFINITIONS

§ 958.1 Secretary.

Secretary means the Secretary of Agriculture of the United States or any officer or employee of the United States Department of Agriculture to whom authority has heretofore been delegated, or to whom authority may hereafter be delegated, to act in his stead.

§ 958.2 Act.

Act means Public Act No. 10, 73d Congress, as amended and as reenacted and amended by the Agricultural Marketing Agreement Act of 1937, as amended (48 Stat. 31, as amended; 7 U.S.C. 601 *et seq.;* 68 Stat. 906, 1047).

§ 958.3 Person.

Person means an individual, partnership, corporation, association, or any other business unit.

§ 958.4 Production area.

Production area means all territory included within the boundaries of the County of Malheur in Oregon, and all counties south and southeast of the southern boundary of Idaho County in the State of Idaho.

§ 958.5 Onions.

Onions means all varieties of *Allium cepa,* commonly known as onions, grown, or which may be grown in the production area.

[41 FR 36196, Aug. 27, 1976]

§ 958.6 Handler.

Handler is synonymous with *shipper* and means any person (except a common or contract carrier of onions owned by another person) who handles onions.

§ 958.7 Handle.

Handle is synonymous with *ship* and means to sell or transport onions, or cause onions to be sold or transported, within the production area or between the production area and any point outside thereof. Except as otherwise provided in §§ 958.56 and 958.65, this definition of "handle" shall not be applicable to onions that are transported within the production area for grading or storing therein, or to onions that are transported or sold to commercial dehydrators for processing by such dehydrators into dehydrated onion products.

§ 958.8 Grading.

Grading is synonymous with *prepare for market* and means the sorting or separation of onions into grades and sizes for market purposes.

369

§ 958.9 Grade and size.

Grade means any of the officially established grades of onions, and *size* means any of the officially established sizes of onions, as set forth in:

(a) The United States Standards for grades of onions (other than Bermuda-Granex and Creole Types) (§§ 51.2830 to 51.2850 of this title), or amendments thereto, or modifications thereof, or variations based thereon; and

(b) Any other United States Standards, or State of Idaho or Oregon Standards for onions, or amendments thereto, or modifications thereof, or variations based thereon.

The term *size* also includes any of the sizes recognized by the onion trade in the production area.

§ 958.10 Producer.

Producer means any person engaged in the production of onions for market.

§ 958.11 Committee.

Committee means the Idaho-Eastern Oregon Onion Committee established pursuant to § 958.20.

§ 958.12 Fiscal period.

Fiscal period means the period beginning and ending on the dates approved by the Secretary pursuant to recommendations by the committee.

§ 958.13 Variety or varieties.

Variety or *varieties* means and includes all classifications of onions according to those definitive characteristics now or hereafter recognized by the United States Department of Agriculture.

§ 958.14 Export.

Export means shipment of onions beyond the boundaries of continental United States.

§ 958.15 District.

District means each of the geographical divisions of the production area initially established or as reestablished pursuant to § 958.27.

§ 958.16 Pack.

Pack means a quantity of onions in any type of container and which falls within specific weight limits or within specific grade or size limits, or both, as may be recommended by the committee and approved by the Secretary.

§ 958.17 Container.

Container means a sack, box, bag, crate, hamper, basket, carton, package, or any other type of receptacle used in the packaging, transportation, sale, shipment or other handling of onions.

ADMINISTRATIVE COMMITTEE

§ 958.20 Establishment and membership.

(a) The Idaho-Eastern Oregon Onion Committee, consisting of six producer members, four handler members, and one public member is hereby established. Each shall have an alternate who shall have the same qualifications as the member.

(b) An alternate member of the committee shall act in the place and stead of the member for whom he is an alternate, during such member's absence or inability to act, and shall perform other duties as assigned. In the event of the death, removal, resignation or disqualification of a member, his alternate shall act for him until a successor for such member is selected and has qualified.

[22 FR 26, Jan. 3, 1957. Redesignated at 26 FR 12751, Dec. 30, 1961, as amended at 47 FR 8000, Feb. 24, 1982]

§ 958.21 Procedure.

(a) Seven members of the committee shall be necessary to constitute a quorum and seven concurring votes shall be required to pass any motion or approve any committee action.

(b) The committee may provide for voting by telephone, telegraph, or other means of communication and any such vote shall be confirmed promptly in writing: *Provided,* That if an assembled meeting is held, all votes shall be cast in person.

§ 958.22 Selection.

The Secretary shall select committee members and alternates from the nominee lists submitted pursuant to this part or from among other eligible persons.

(a) Each person selected as a committee member or alternate to represent producers shall be an individual who is a producer, or an officer or employee of a producer, in the district for which selected.

(b) Each person selected as a committee member or alternate to represent handlers shall be an individual who is a handler, or an officer or employee of a handler in the portion of the production area for which selected.

(c) The Secretary shall select one producer member of the committee, and alternate, from each of the districts established, or reestablished, pursuant to §958.27. The Secretary shall also select one handler member of the committee, and his alternate, from the Idaho portion of the production area and one member and his alternate from Malheur County, Oregon, and two handler members, and their respective alternates, from the production area-at-large.

(d) Each person selected by the Secretary as a committee member or alternate shall qualify by filing a written acceptance promptly with the Secretary.

(e) The public member shall be a resident of the production area and have no direct financial interest in the commercial production, financing, buying, packing or marketing of onions except as a consumer nor be a director, officer or employee of any firm so engaged.

[22 FR 26, Jan. 3, 1957. Redesignated at 26 FR 12751, Dec. 30, 1961, as amended at 47 FR 8000, Feb. 24, 1982]

§958.23 Term of office.

(a) The term of office of committee members and alternates shall be for two years beginning on the first day of June and continuing through May 31. The terms of office of members and alternates shall be so determined that one-half of the total committee membership shall terminate each May 31.

(b) Committee members and alternates shall serve during the term of office for which they are selected and have qualified, or during that portion thereof beginning on the date on which they qualify during the current term of office and continuing until the end thereof, and until their successors are selected and have qualified.

§958.24 Powers.

The committee shall have the following powers:

(a) To administer the provisions of this part in accordance with its terms;

(b) To make rules and regulations to effectuate the terms and provisions of this part;

(c) To receive, investigate, and report to the Secretary complaints of violations of the provisions of this part; and

(d) To recommend to the Secretary amendments to this part.

§958.25 Duties.

It shall be the duty of the committee:

(a) At the beginning of each fiscal period, or as soon thereafter as practicable, to meet and organize, to select a chairman and such other officers as may be necessary, to select subcommittees of committee members, and to adopt such rules and regulations for the conduct of its business as it may deem advisable;

(b) To act as intermediary between the Secretary and any producer or handler;

(c) To furnish to the Secretary such available information as he may request;

(d) To appoint such employees, agents, and representatives as it may deem necessary and to determine the salaries and define the duties of each such person;

(e) To investigate from time to time and to assemble data on the growing, harvesting, shipping and marketing conditions with respect to onions and to engage in such research and service activity which relate to the production, handling or marketing of onions as may be approved by the Secretary;

(f) To keep minutes, books, and records which clearly reflect all of the acts and transactions of the committee and such minutes, books, and records shall be subject to examination at any time by the Secretary or his authorized agent or representative;

(g) To make available to producers and handlers the committee voting record on recommended regulations and on other matters of policy;

(h) Prior to each fiscal period, to submit to the Secretary a budget of its proposed expenses for such fiscal period, together with a report thereon;

(i) To cause the books of the committee to be audited by a competent accountant at least once each fiscal period, and at such other time as the committee may deem necessary or as the Secretary may request; and the report of such audit shall show the receipt and expenditure of funds collected pursuant to this part; a copy of each such report shall be furnished to the Secretary and a copy of each such report shall be made available at the principal office of the committee for inspection by producers and handlers; and

(j) To consult, cooperate, and exchange information, with other onion marketing committees and other individuals or agencies in connection with all proper committee activities and objectives under this subpart; and

(k) To recommend nominees for the public member and alternate.

[22 FR 26, Jan. 3, 1957. Redesignated at 26 FR 12751, Dec. 30, 1961, as amended at 41 FR 36196, Aug. 27, 1976; 47 FR 8000, Feb. 24, 1982]

§ 958.26 Expenses.

Committee members and alternates when acting on committee business shall be reimbursed for reasonable expenses necessarily incurred by them in the performance of their duties and in the exercise of their powers under this part. However, at its discretion the committee may request the attendance of alternates at any or all meetings notwithstanding the expected or actual presence of the respective members.

[41 FR 36196, Aug. 27, 1976]

§ 958.27 Districts.

(a) For the purpose of selecting committee members, the following districts of the production area are hereby initially established:

District No. 1 (Emmett, Payette, Weiser Area): All territory within the boundaries of Washington, Payette and Gem Counties, in Idaho.

District No. 2 (Oregon Slope): All territory within a boundary following the Snake River northwesterly from its junction with the Malheur River, to the west line of Range 46E; thence south along said west line to the south line of Township 17S, and thence east along said south line to its junction with the Malheur River, and thence northeasterly along the Malheur

River to the junction with the Snake River, the point of beginning.

District No. 3 (Ontario, Vale, Jamieson, Brogan): All territory within a boundary starting at the junction of the Malheur River with the Snake River and extending southwestward along the Malheur River to its junction with the south line of Township 17S, E. W. M.; thence westward along this line to its junction with the west line of Range 46E; thence north along this line to its junction with the Snake River; thence northwest along the Snake River to its junction with the north boundary of Malheur County; thence west along the north boundary of Malheur County to the west boundary of the county; thence south along the west boundary of Malheur County to its intersection with the south line of Township 20S; thence east along this line to its junction with the Hyline Canal and Siphon; thence northeast along the Hyline Canal to its intersection with Highway 20; thence east along Highway 20 to Cairo Junction; thence south ⅛ mile to the junction of Highway 20 to Oregon Avenue; thence east along Oregon Avenue to its termination at the Snake River; thence north along the Snake River to its junction with the Malheur River, the point of beginning.

District No. 4 (Nyssa-Adrian): All the area of Malheur County, Oregon, south of District No. 3.

District No. 5 (Parma, Wilder, Nampa, and Notus Area): Canyon County, Idaho.

District No. 6 (Homedale, Marsing, Meredian, Melba, Mountain Home, Glenns Ferry and Twin Falls Area): All counties in the Idaho portion of the production area not included within Districts Nos. 1 and 5.

(b) The Secretary, upon the recommendation of the committee, may reestablish districts within the production area and may reapportion committee membership among the various districts: *Provided*, That in recommending any such changes in districts or representation, the committee shall give consideration to: (1) The relative importance of new producing sections; (2) changes in the relative position of existing districts with respect to onion production; (3) the geographic location of areas of production as they would affect the efficiency of administering this part; (4) other relevant factors: *Provided, further*, That there shall be no change in the total number of committee members or in the total number of districts.

§958.28 Nominations.

Nominations from which the Secretary may select the members of the Idaho-Eastern Oregon Onion Committee and their respective alternates may be made in the following manner:

(a) The committee shall hold or cause to be held prior to April 1 of each year, after the effective date of this subpart, one or more meetings of producers and of handlers in each of the districts, or portions of the production area, in which the then current terms of office will expire the following May 31;

(b) In arranging for such meetings the committee may, if it deems desirable, cooperate with existing organizations and agencies and may combine its meetings with others;

(c) Nominations for committee members and alternate members shall be supplied to the Secretary, in such manner and form as he may prescribe, not later than 30 days prior to the end of each fiscal period;

(d) Only producers may participate in designating nominees for producer committee members and their alternates and only handlers may participate in designating nominees for handler committee members and their alternates;

(e) Each person who is both a handler and a producer may vote either as a handler or as a producer and may select the group in which he will vote;

(f) Regardless of the number of districts in which a person produces or handles onions, each such person is entitled to cast only one vote on behalf of himself, his partners, agents, subsidiaries, affiliates and representatives, in designating nominees for committee members and alternates. In the event a person is a producer engaged in producing onions in more than one district, such person shall select the district within which he may participate as aforesaid in designating nominees. Similarly, a person who is a handler both in Malheur County, Oregon, and in the Idaho portion of the production area, may select either Malheur County or the Idaho portion of the production area in which to cast his vote for the applicable committee handler member and alternate. Each such handler shall also be entitled to cast his vote for the committee member and alternate to represent the production area-at-large. An eligible voter's privilege of casting only one vote, as aforesaid, shall be construed to permit such voter to cast one vote for each member and alternate position to be filled in the respective district or portion of the production area, as the case may be, in which he elects to vote; and

(g) The producer and handler members of the committee shall nominate the public member and alternate. The committee shall prescribe such additional qualifications, administrative rules and procedures for selection and voting for each candidate as it deems necessary and as the Secretary approves.

[22 FR 26, Jan. 3, 1957. Redesignated at 26 FR 12751, Dec. 30, 1961, as amended at 47 FR 8000, Feb. 24, 1982]

§958.29 Failure to nominate.

If nominations are not made within the time and in the manner specified by the Secretary pursuant to §958.28, the Secretary may, without regard to nominations, select the committee members and alternates on the basis of the representation provided for in this subpart.

§958.30 Vacancies.

To fill any vacancy occasioned by the failure of any person, selected as a committee member or alternate, to qualify, or in the event of the death, removal, resignation, or disqualification of any qualified member or alternate, a successor for his unexpired term may be selected by the Secretary from nominations made in the manner specified in §958.28, or the Secretary may select such committee member or alternate from previously unselected nominees on the current nominee list from the district or portion of the production area, as the case may be, that is involved, or from other eligible persons. If the names of nominees to fill any such vacancy are not made available to the Secretary within 30 days after such vacancy occurs, the Secretary may fill such vacancy without regard to nominations, which selection shall be made on the basis of the representation provided for in this subpart.

EXPENSES AND ASSESSMENTS

§ 958.40 Expenses.

The committee is authorized to incur such expenses as the Secretary may find are reasonable and likely to be incurred by it during each fiscal period for its maintenance and functioning, and for such purposes as the Secretary, pursuant to this subpart, determines to be appropriate. Handlers shall share expenses upon the basis of a fiscal period. Each handler's share of such expenses shall be proportionate to the ratio between the total quantity of such handler's onion shipments inspected pursuant to this part that are handled by him as the first handler thereof during a fiscal period, and the total quantity of such onions handled by all handlers as first handlers thereof during the same period.

§ 958.41 Budget.

Prior to each fiscal period, and as may be necessary thereafter the committee shall prepare a budget of estimated income and expenditures necessary for the administration of this part. The committee shall recommend to the Secretary a rate of assessment calculated to provide adequate funds to defray its proposed expenditures. The committee shall present such budget promptly to the Secretary with an accompanying report thereon showing the basis for its calculations and recommended rate.

[22 FR 26, Jan. 3, 1957. Redesignated at 26 FR 12751, Dec. 30, 1961, as amended at 41 FR 36196, Aug. 27, 1976]

§ 958.42 Assessments.

(a) The funds to cover the committee's expenses pursuant to § 958.40 shall be acquired by the levying of assessments upon handlers as provided in this subpart. Each handler who handles onions as the first handler thereof which are inspected pursuant to this part shall pay assessments to the committee upon demand, which assessments shall be in payment of such handler's pro rata share of such expenses.

(b) Assessments shall be levied upon handlers at rates established by the Secretary. Such rates may be established upon the basis of the commit-

tee's recommendations or other available information.

(c) At any time during or subsequent to a given fiscal period, the committee may recommend the approval of an amended budget and an increase in the rate of assessment. Upon the basis of such recommendation, or other available information, the Secretary may approve an amended budget and increase the rate of assessment. Such increase shall be applicable to all onion shipments inspected pursuant to this part during such fiscal period.

§ 958.43 Accounting.

(a) All funds received by the committee pursuant to the provisions of this part shall be used solely for the purposes specified in this part.

(b) The Secretary may at any time require the committee, its members and alternates, employees, agents, and all other persons to account for all receipts and disbursements, funds, property, or records for which they are responsible. Whenever any person ceases to be a member or alternate of the committee, he shall account for all receipts, disbursements, funds, and property (including, but not being limited to, books and other records) pertaining to the committee's activities for which he is responsible, and deliver all such property and funds in his hands to such successor, agency, or person as may be designated by the Secretary, and shall execute such assignments and other instruments as may be necessary or appropriate to vest in each such successor, agency, or person as may be designated by the Secretary the right to all of such property and funds and all claims vested in such person.

(c) The committee may make recommendations to the Secretary for one or more of the members thereof, or any other person, to act as a trustee for holding records, funds, and any other committee property during periods of suspension of this part, or during any periods when regulations are not in effect; and, if the Secretary determines such action appropriate, he may direct that such person or persons shall so act as trustee or trustees.

§958.44 Reserve fund.

At the end of each fiscal period, funds in excess of the committee's expenses may be placed in an operating reserve not to exceed approximately 1 fiscal year's operational expenses or such lower limits as the committee, with the approval of the Secretary, may establish. Also, the committee, with the approval of the Secretary, may include in its budget an item for such reserve. Funds in the reserve shall be available for use by the committee for expenses authorized pursuant to §958.40. Funds in excess of those placed in the operating reserve shall be refunded to handlers. Each handler's share of such excess shall be the amount he paid in excess of his pro rata share of the expenses of the committee.

[32 FR 11261, Aug. 3, 1967]

§958.45 Accounting of funds upon termination of the order.

Any funds collected as assessments pursuant to this subpart and remaining unexpended in the possession of the committee after termination of this part shall be distributed in such manner as the Secretary may direct: *Provided*, That to the extent practical, such funds shall be returned pro rata to the persons from whom such funds were collected.

[32 FR 11262, Aug. 3, 1967]

§958.46 Contributions.

The committee may accept voluntary contributions but these shall only be used to pay expenses incurred pursuant to §958.47. Furthermore, such contributions shall be free from any encumbrances by the donor and the committee shall retain complete control of their use.

[41 FR 36196, Aug. 27, 1976]

RESEARCH AND DEVELOPMENT

§958.47 Research and development.

(a) The committee with the approval of the Secretary, may establish or provide for the establishment of projects involving production research, marketing research and development projects, and marketing promotion including paid advertising, designed to assist, improve, or promote the marketing, distribution, consumption or efficient production of onions. Any such project for the promotion and advertising of onions may utilize an identifying mark which shall be made available for use by all handlers in accordance with such terms and conditions as the committee, with the approval of the Secretary, may prescribe. The expenses of such projects shall be paid from funds collected pursuant to §958.42 or §958.46.

(b) In recommending projects pursuant to this section the committee shall give consideration to the following:

(1) The expected supply of onions in relation to market requirements;

(2) The supply situation among competing areas and commodities;

(3) The anticipated benefits from such projects in relation to their costs;

(4) The need for marketing research with respect to any market development activity; and

(5) The need for a coordinated effort with USDA's Food Marketing Alert or other similar programs.

(c) If the committee should conclude that a program of research or development should be undertaken, or continued, in any crop year, it shall submit the following for the approval of the Secretary:

(1) Its recommendations as to the funds to be obtained pursuant to §958.42 or §958.46;

(2) Its recommendation as to any research projects; and

(3) Its recommendation as to promotion activity and paid advertising.

(d) Upon conclusion of each activity, but at least annually, the committee shall summarize and report the results of such activity to its members and to the Secretary.

[41 FR 36196, Aug. 27, 1976]

REGULATION

§958.50 Marketing policy.

(a) *Preparation.* Prior to each marketing season the committee shall consider and prepare a proposed policy for the marketing of onions. In developing its marketing policy the committee shall investigate relevant supply and demand conditions for onions. In such investigations the committee shall give appropriate consideration to the following:

(1) Market prices for onions, including prices by variety, grade, size, and quality, and by different packs;

(2) Supply of onions by grade, size, quality, and variety in the production area and in other onion producing sections;

(3) The trend and level of consumer income;

(4) Establishing and maintaining orderly marketing conditions for onions;

(5) Orderly marketing of onions as will be in the public interest; and

(6) Other relevant factors.

(b) *Reports.* (1) The committee shall submit a report to the Secretary setting forth the aforesaid marketing policy; and the committee shall notify producers and handlers of the contents of such report.

(2) In the event it becomes advisable to shift from such marketing policy because of changed supply and demand conditions, the committee shall prepare an amended or revised marketing policy in accordance with the manner previously outlined. The committee shall submit a report thereon to the Secretary and notify producers and handlers of the contents of such report on the revised or amended marketing policy.

§ 958.51 Recommendations for regulations.

The committee shall recommend regulations to the Secretary whenever it finds that such regulations as provided in § 958.52 will tend to effectuate the declared policy of the act. The committee also may recommend modification, suspension, or termination of any regulation, or amendments thereto, in order to facilitate the handling of onions for the purposes authorized in § 958.53. The committee may also recommend amendment, modification, termination, or suspension of any regulation issued under this part.

§ 958.52 Issuance of regulations.

(a) Except as otherwise provided in this part, the Secretary shall limit the shipment of onions by any one or more of the methods hereinafter set forth whenever he finds from the recommendations and information submitted by the committee, or from other available information, that such regulation would tend to effectuate the declared policy of the act. Such limitation may:

(1) Regulate in any or all portions of the production area, the handling of particular grades, sizes, or qualities of any or all varieties of onions, or combinations thereof, during any period or periods;

(2) Regulate the handling of particular grades, sizes, or qualities, of onions differently, for different varieties, for different portions of the production area, for different packs, or for any combination of the foregoing, during any period or periods;

(3) Provide a method, through rules and regulations issued pursuant to this part, for fixing the size, capacity, weight, dimensions, or pack of the container, or containers, which may be used in the packaging or handling of onions, including appropriate container markings to identify the contents thereof;

(4) Regulate the handling of onions by establishing, in terms of grades, sizes, or both, minimum standards of quality and maturity; or

(5) Limit the shipment of the total quantity of onions by prohibiting the handling thereof during a specified period or periods. No regulation issued pursuant to this subparagraph shall be effective for more than 96 consecutive hours: *Provided,* That not less than 72 consecutive hours shall elapse between the termination of any such period of prohibition and the beginning of the next such period.

(6) Regulate the handling of onions by establishing, in terms of total weight or total number of layers of containers of onions, the maximum load in railcars, taking into account types of containers and sizes of railcars used, potential resulting damage, and other relevant factors.

(b) In the event the handling of onions is regulated pursuant to paragraph (a)(5) of this section, no handler shall handle any onions which were prepared for market or loaded during the effective period of such regulation. However, during any such period, no such regulation shall be deemed to limit the right of any person to sell or contract to sell onions for future shipment or delivery.

(c) The Secretary may amend any regulation issued under this part whenever he finds that such amendment would tend to effectuate the declared policy of the act. The Secretary may also terminate or suspend any regulation or amendment thereof whenever he finds that such regulation or amendment obstructs or no longer tends to effectuate the declared policy of the act.

[22 FR 26, Jan. 3, 1957. Redesignated at 26 FR 12751, Dec. 30, 1961, as amended at 32 FR 11262, Aug. 3, 1967; 47 FR 8000, Feb. 24, 1982]

§958.53 Handling for specified purposes.

Upon the basis of recommendations and information submitted by the committee, or other available information, the Secretary shall issue special regulations, or modify, suspend, or terminate requirements in effect pursuant to §§958.42, 958.52, 958.60, or any combination thereof, in order to facilitate the handling of onions for the following purposes whenever he finds that to do so will tend to effectuate the declared policy of the act:

(a) Shipments of onions for export;

(b) Shipments of onions for relief or to charitable institutions;

(c) Shipments of onions for livestock feed;

(d) Shipments of onions for planting; and

(e) Shipments of onions for other purposes which may be specified.

§958.54 Minimum quantities.

The committee, with the approval of the Secretary, may establish minimum quantities below which onion shipments will be free from the requirements in, or pursuant to, §§958.42, 958.52, 958.53, 958.60, or any combination thereof.

§958.55 Notification of regulations.

The Secretary shall notify the committee of each regulation issued, and of each amendment, modification, suspension, or termination thereof. The committee shall give reasonable notice thereof to handlers.

§958.56 Safeguards.

(a) The committee, with the approval of the Secretary, may prescribe adequate safeguards to prevent onions shipped,

(1) Pursuant to §958.53 or §958.54; or

(2) To commercial dehydrators for processing by such dehydrators into dehydrated onion products,

from entering channels of trade for other than the purpose authorized therefor.

(b) The committee, with the approval of the Secretary, may also prescribe rules and regulations governing the issuance, and the contents, of Certificates of Privilege if such certificates are prescribed as safeguards by the committee. Such safeguards may include requirements that:

(1) Handlers shall first file applications with the committee to ship such onions;

(2) Handlers shall obtain inspection provided by §958.60, or pay the pro rata share of expenses provided by §958.42, or both, in connection with such onions; and

(3) Handlers shall obtain Certificates of Privilege from the committee prior to effecting the particular onion shipment.

(c) The committee may rescind any Certificate of Privilege, or refuse to issue any Certificate of Privilege to any handler if proof is obtained that onions shipped by him for the purposes stated in the Certificate of Privilege were handled contrary to the provisions of this part.

(d) The Secretary shall have the right to modify, change, alter, or rescind any safeguards prescribed and any certificates issued by the committee pursuant to the provisions of this section.

(e) The committee shall make reports to the Secretary, as requested, showing the number of applications for such certificates, the quantity of onions covered by such applications, the number of such applications denied and certificates granted, the quantity of onions handled under duly issued certificates, and such other information as may be requested.

INSPECTION

§ 958.60 Inspection and certification.

(a) During any period in which shipments of onions are regulated pursuant to this subpart, no handler shall handle onions unless such onions are inspected by an authorized representative of the Federal-State Inspection Service, or such other inspection service as the Secretary shall designate and are covered by a valid inspection certificate, except when relieved from such requirements pursuant to § 958.53, § 958.54, or both.

(b) Regarding, resorting, or repacking any lot of onions shall invalidate prior inspection certificates insofar as the requirements of this section are concerned. No handler shall ship onions after they have been regarded, resorted, repacked or in any other way further prepared for market, unless such onions are inspected by an authorized representative of the Federal-State Inspection Service, or such other inspection service as the Secretary shall designate.

(c) Upon recommendation of the committee, and approval of the Secretary, all onions that are required to be inspected and certified in accordance with this section, shall be identified by appropriate seals, stamps, tags, or other identification to be furnished by the committee and affixed to the containers by the handler under the direction and supervision of the Federal-State, or Federal inspector, or the committee. Master containers may bear the identification instead of the individual containers within said master container.

(d) Insofar as the requirements of this section are concerned, the length of time for which an inspection certificate is valid may be established by the committee with the approval of the Secretary.

(e) When onions are inspected in accordance with the requirements of this section, a copy of each inspection certificate issued shall be made available to the committee by the inspection service.

REPORTS

§ 958.65 Reports.

Upon request of the committee, made with the approval of the Secretary, each handler shall furnish to the committee, in such manner and at such time as it may prescribe, such reports and other information as may be necessary for the committee to perform its duties under this part.

(a) Such reports may include, but are not necessarily limited to, the following:

(1) The quantities of onions received by a handler;

(2) The quantities disposed of by him, segregated as to the respective quantities subject to regulation and not subject to regulation;

(3) The date of each such disposition and the identification of the carrier transporting such onions; and

(4) identification of the inspection certificates relating to the onions which were handled pursuant to §§ 958.53 and 958.54.

(b) All such reports shall be held under appropriate protective classification and custody by the committee, or duly appointed employees thereof, so that the information contained therein which may adversely affect the competitive position of any handler in relation to other handlers will not be disclosed. Compilations of general reports from data submitted by handlers is authorized, subject to the prohibition of disclosure of individual handler's identities or operations.

(c) Each handler shall maintain for at least two succeeding years such records of the onions received, and of onions disposed of, by such handler as may be necessary to verify the reports he submits to the committee pursuant to this section.

EFFECTIVE TIME AND TERMINATION

§ 958.70 Effective time.

The provisions of this subpart, or any amendment thereto, shall become effective at such time as the Secretary may declare and shall continue in force until terminated in one of the ways specified in this subpart.

§958.71 Termination.

(a) The Secretary may at any time terminate the provisions of this subpart by giving at least one day's notice by means of a press release or in any other manner which he may determine.

(b) The Secretary may terminate or suspend the operations of any or all of the provisions of this subpart whenever he finds that such provisions do not tend to effectuate the declared policy of the act.

(c) The Secretary shall terminate the provisions of this subpart at the end of any fiscal period whenever he finds that such termination is favored by a majority of producers who, during a representative period, have been engaged in the production for market of onions: *Provided,* That such majority has, during such representative period, produced for market more than fifty percent of the volume of such onions produced for market, but such termination shall be effective only if announced on or before May 31 of the then current fiscal period.

(d) The provisions of this subpart shall, in any event, terminate whenever the provisions of the act authorizing them cease to be in effect.

§958.72 Proceeding after termination.

(a) Upon the termination of the provisions of this subpart, the then functioning members of the committee shall continue as joint trustees, for the purpose of liquidating the affairs of the committee, of all the funds and property then in the possession, or under control, of the committee, including claims for any funds unpaid and property not delivered at the time of such termination. Action by said trusteeship shall require the concurrence of a majority of the said trustees.

(b) The said trustees shall continue in such capacity until discharged by the Secretary; shall, from time to time, account for all receipts and disbursements and deliver all property on hand, together with all books and records of the committee and of the trustees, to such person as the Secretary may direct; and shall, upon request of the Secretary, execute such assignments or other instruments necessary or appropriate to vest in such person full title and right to all of the

funds, property, and claims vested in the committee or the trustees pursuant to this subpart.

(c) Any person to whom funds, property, or claims have been transferred or delivered by the committee or its members pursuant to this section, shall be subject to the same obligations imposed upon the members of the committee and upon the said trustees.

§958.73 Effect of termination or amendment.

Unless otherwise expressly provided by the Secretary, the termination of this subpart or of any regulation issued pursuant to this subpart, or the issuance of any amendment to either thereof, shall not (a) affect or waive any right, duty, obligation, or liability which shall have arisen or which may thereafter arise in connection with any provision of this subpart or of any regulation issued under this subpart; (b) release or extinguish any violation of this subpart or of any regulations issued under this subpart; or (c) affect or impair any rights or remedies of the Secretary or of any other person with respect to any such violations.

MISCELLANEOUS PROVISIONS

§958.81 Compliance.

No handler shall handle onions the handling of which has been prohibited or otherwise limited by the Secretary in accordance with provisions of this part; and no handler shall handle onions except in conformity to the provisions of this part.

§958.82 Right of the Secretary.

The members of the committee (including successors and alternates) and any agent or employee appointed or employed by the committee shall be subject to removal or suspension by the Secretary at any time. Each and every order, regulation, decision, determination, or other act of the committee shall be subject to the continuing right of the Secretary to disapprove of the same at any time. Upon such disapproval, the disapproved action of the said committee shall be deemed null and void except as to acts done in reliance thereon or in compliance therewith prior to such disapproval by the Secretary.

§ 958.83 Duration of immunities.

The benefits, privileges, and immunities conferred upon any person by virtue of this subpart shall cease upon the termination of this subpart, except with respect to acts done under and during the existence of this subpart.

§ 958.84 Agents.

The Secretary may, by designation in writing, name any person, including any officer or employee of the Government, or name any agency in the United States Department of Agriculture, to act as his agent or representative in connection with any of the provisions of this part.

§ 958.85 Derogation.

Nothing contained in this subpart is, or shall be construed to be, in derogation or in modification of the rights of the Secretary or of the United States to exercise any powers granted by the act or otherwise, or, in accordance with such powers, to act in the premises whenever such action is deemed advisable.

§ 958.86 Personal liability.

No member or alternate of the committee nor any employee or agent thereof, shall be held personally responsible, either individually or jointly with others, in any way whatsoever, to any handler or to any person for errors in judgment, mistakes, or other acts, either of commission or omission, as such member, alternate, employee, or agent, except for acts of dishonesty, wilful misconduct, or gross negligence.

§ 958.87 Separability.

If any provision of this subpart is declared invalid, or the applicability thereof to any person, circumstance, or thing is held invalid, the validity of the remainder of this subpart, or the applicability thereof to any other person, circumstance, or thing, shall not be affected thereby.

§ 958.88 Amendments.

Amendments to this subpart may be proposed, from time to time, by the committee or by the Secretary.

§ 958.89 Counterparts.

This agreement may be executed in multiple counterparts and when one counterpart is signed by the Secretary, all such counterparts shall constitute, when taken together, one and the same instrument as if all signatures were contained in one original.

[41 FR 29135, July 15, 1976]

§ 958.90 Additional parties.

After the effective date hereof, any handler may become a party to this agreement if a counterpart is executed by him and delivered to the Secretary. This agreement shall take effect as to such new contracting party at the time such counterpart is delivered to the Secretary, and the benefits, privileges, and immunities conferred by this agreement shall then be effective as to such new contracting party.

[41 FR 29135, July 15, 1976]

§ 958.91 Order with marketing agreement.

Each signatory handler requests the Secretary to issue, pursuant to the act, an order providing for regulating the handling of onions in the same manner as is provided for in this agreement.

The undersigned hereby authorizes the Director, or Acting Director, Fruit and Vegetable Division, Agricultural Marketing Service, United States Department of Agriculture, to correct any typographical errors which may have been made in this marketing agreement.

In witness whereof, the contracting parties, acting under the provisions of the act, for the purpose and subject to the limitations therein contained, and not otherwise, have hereto set their respective signatures and seals.

```
                              (Firm name)
By:   _____
                              (Signature)[1]
                              (Mailing address)
                              (Title)
```

[1] If one of the contracting parties to this agreement is a corporation my signature

(Corporate Seal; if none, so state)

(Date of execution)

[41 FR 29136, July 15, 1976]

Subpart B—Administrative Requirements

§958.112 Fiscal period.

The fiscal period shall begin July 1 of each year and end June 30 of the following year, both dates inclusive.

[68 FR 48531, Aug. 14, 2003]

§958.160 Reestablishment of Districts.

(a) Pursuant to §958.27(b) the following districts are reestablished:

(1) District No. 5 (Parma-Wilder area): That portion of Canyon County lying west and north of a line commencing at the junction of the north boundary of Canyon County and Range 4, Township 12 east, thence south along this line to Soeck Road, thence west along Soeck Road one-fourth mile to Notus Road, thence south along Notus Road to Highway 19, thence west one mile along Highway 19 to Friends Road, thence south along Friends Road to Boundary Road, thence east one-half mile along Boundary Road to Plum Road, thence south along Plum Road to Homedale Road, thence west along Homedale Road to the western boundary of Canyon County.

(2) District No. 6 (Caldwell-Nampa-Homedale and southern Idaho area): That portion of Canyon County not included in District No. 5 plus all of the counties in the Idaho portion of the production area not included within District No. 1.

(b) Terms used in this section have the same meaning as when used in said marketing agreement and this part.

[39 FR 1601, Jan. 11, 1974]

§958.240 Assessment rate.

On and after July 1, 2015, an assessment rate of $0.05 per hundredweight is established for Idaho-Eastern Oregon onions.

[80 FR 50195, Aug. 19, 2015]

---

constitutes certification that I have the power granted to me by the Board of Directors to bind this corporation to the marketing agreement.

§958.250 Assessment Credit Report.

Each handler may receive a credit for assessments on onions that have been levied in accordance with §§958.42 and 958.240 and are subsequently regraded, resorted, or repacked within the production area, or shipped in accordance with §958.328(e) by furnishing the "Assessment Credit Report" and such other information as required to the committee.

[71 FR 65040, Nov. 7, 2006]

Subpart C—Handling Requirements

§958.328 Handling regulation.

No person shall handle any lot of onions, except braided red onions, unless such onions are at least "moderately cured," as defined in paragraph (h) of this section, and meet the requirements of paragraphs (a), (b), and (c) of this section, or unless such onions are handled in accordance with paragraphs (d), (e) and (f) or (g) of this section.

(a) *Grade and size requirements—(1) White varieties (except cipolline (Borettana) varieties).* Shall be either:

(i) U.S. No. 1, 1 inch minimum to 2 inches maximum diameter; or

(ii) U.S. No. 1, at least 1½ inches minimum diameter. However, neither of these two categories of onions may be commingled in the same bag or other container.

(2) *Cipolline (Borettana) varieties and red varieties.* U.S. No. 2 or better grade, at least 1½ inches minimum diameter.

(3) *All other varieties.* Shall be either:

(i) U.S. No. 2 or U.S. Commercial grade, at least 3 inches minimum diameter, but not more than 30 percent of the lot shall be comprised of onions of U.S. No. 1 quality when packed in containers weighing less than 60 pounds; or

(ii) U.S. No. 1, 1¾ inches minimum to 2¾ maximum diameter; or

(iii) U.S. No. 1, at least 2¼ inches minimum diameter.

However, none of these three categories of onions may be commingled in the same bag or other container.

(b) *Pack.* Onions packed as U.S. Commercial grade in containers weighing less than 60 pounds shall have the grade marked permanently and conspicuously on the container.

(c) *Inspection.* No handler may handle any onions regulated hereunder unless such onions are inspected by the Federal-State Inspection Service and are covered by a valid applicable inspection certificate, except when relieved of such requirement pursuant to paragraph (d), (e) or (g) of this section.

(d) *Onions for peeling, chopping, or slicing.* Onions that have been inspected and certified as meeting the requirements of paragraphs (a) and (b) of this section and that are subsequently peeled, chopped, or sliced for fresh market within the production area may be handled without reinspection: *Provided the following:*

(1) Each handler making shipments of onions for alteration or performing alteration by peeling, chopping, or slicing must furnish the committee the following information on the "Fresh Cut Report" and such other documents as required:

(i) Business name, address, telephone number, signature, and the date the form was signed;

(ii) The date of peeling, chopping, or slicing;

(iii) Inspection certificate number;

(iv) The quantity of onions; and

(v) Such other information as may be required by the committee.

(2) Handlers who peel, chop, or slice onions produced outside the production area must provide the committee with documentation showing that the onions so prepared were produced outside the production area.

(e) *Special purpose shipments.* (1) The minimum grade, size, maturity, pack, assessment, and inspection requirements of this section shall not be applicable to shipments of onions for any of the following purposes:

(i) Planting,

(ii) Livestock feed,

(iii) Charity,

(iv) Dehydration,

(v) Canning,

(vi) Freezing,

(vii) Extraction,

(viii) Pickling, and

(ix) Disposal.

(2) Shipments of onions for the purpose of experimentation, as approved by the Committee, may be made without regard to the minimum grade, size, maturity, pack, and inspection requirements of this section. Assessment requirements shall be applicable to such shipments.

(3) The minimum grade, size, and maturity requirements set forth in paragraph (a) of this section shall not be applicable to shipments of pearl onions, but the maximum size requirement in paragraph (h) of this section and the assessment and inspection requirements shall be applicable to shipments of pearl onions.

(f) *Safeguards.* Each handler making shipments of onions outside the production area for dehydration, canning, freezing, extraction, pickling, or experimentation pursuant to paragraph (e) of this section shall:

(1) Furnish "Application to Make Special Purpose Shipments—Certificate of Privilege" and such other information to the committee as required. The committee will review and verify each "Application to Make Special Purpose Shipments—Certificate of Privilege" and notify the handler of approval or disapproval. The committee may contact the receiver or receiver's agent of the special purpose shipment for verification and request the receiver or receiver's agent to complete a "Special Purpose Shipment Receiver Certification"

(2) Bill or consign each shipment directly to the applicable receiver or receiver's agent of the special purpose shipment;

(3) Furnish "Onion Diversion Report" and such other information to the committee as required. Failure of the handler to furnish such report and information as required to the committee may be cause for cancellation of such handlers' Certificate of Privilege. Upon cancellation of any such Certificate of Privilege the handler may appeal to the committee for reconsideration. The committee may audit a receiver or receiver's agent of the special purpose shipment to verify reports and information submitted by handlers. Failure of a receiver or receiver's agent of a special purpose shipment to comply with the committee may be cause for cancellation of the receiver's or receiver agent's eligibility to receive further special purpose shipments from

the production area. Upon cancellation of any such Certificate of Privilege the receiver or the receiver's agent may appeal to the committee for reconsideration.

(g) *Minimum quantity exemption.* Each handler may ship up to, but not to exceed, one ton of onions each day without regard to the inspection and assessment requirements of this part, if such onions meet minimum grade, size and maturity requirements of this section. This exception shall not apply to any portion of a shipment that exceeds one ton of onions.

(h) *Definitions.* The terms "U.S. No. 1", "U.S. Commercial," and "U.S. No. 2" have the same meaning as defined in the United States Standards for Grades of Onions (Other than Bermuda Granex-Grano and Creole Types), as amended (7 CFR 51.2830 through 51.2854), or the United States Standards for Grades of Bermuda-Granex-Grano Type Onions (7 CFR 51.3195 through 51.3209), as amended, whichever is applicable to the particular variety, or variations thereof specified in this section. The term "braided red onions" means onions of red varieties with tops braided (interlaced). "Pearl onions" means onions produced using specific cultural practices that limit growth to the same general size as boilers and picklers (defined in the United States Standards specified in this paragraph), and that have been inspected and certified as measuring 2 inches in diameter or less. The term "moderately cured" means the onions are mature and are more nearly well cured than fairly well cured. Other terms used in this section have the same meaning as when used in Marketing Agreement No. 130 and this part.

[47 FR 32913, July 30, 1982, as amended at 49 FR 31257, Aug. 6, 1984; 50 FR 50157, Dec. 9, 1985; 53 FR 32597, Aug. 26, 1988; 55 FR 31036, July 31, 1990; 55 FR 36601, Sept. 6, 1990; 58 FR 60369, Nov. 16, 1993; 61 FR 35593, July 8, 1996; 61 FR 39841, July 31, 1996; 63 FR 55783, Oct. 19, 1998; 69 FR 56671, Sept. 22, 2004; 71 FR 65040, Nov. 7, 2006; 76 FR 67319, Nov. 1, 2011; 76 FR 67319, Nov. 1, 2011]

PART 959—ONIONS GROWN IN SOUTH TEXAS

Subpart A—Order Regulating Handling

AUTHORITY: 7 U.S.C. 601–674.

Subpart A—Order Regulating Handling

SOURCE: 26 FR 704, Jan. 25, 1961, unless otherwise noted. Redesignated at 26 FR 12751, Dec. 30, 1961.

DEFINITIONS

§ 959.1 Secretary.

Secretary means the Secretary of Agriculture of the United States, or any officer or employee of the Department to whom authority has heretofore been delegated, or to whom authority may be hereafter delegated, to act in his stead.

§ 959.2 Act.

Act means Public Act No. 10, 73d Congress, as amended and as reenacted and amended by the Agricultural Marketing Agreement Act of 1937, as amended (sections 1–19, 48 Stat. 31, as amended; 7 U.S.C. 601–674).

§ 959.3 Person.

Person means an individual, partnership, corporation, association or any other business unit.

§ 959.4 Production area.

Production area means the counties of Val Verde, Kinney, Uvalde, Medina, Wilson, Karnes, Goliad, Victoria, Calhoun, Maverick, Zavala, Frio, Atascosa, Dimmit, La Salle, McMullen, Live Oak, Bee, Refugio, Webb, Duval, Jim Wells, San Patricio, Nueces, Zapata, Jim Hogg, Brooks, Kleberg, Kenedy, Starr, De Witt, Aransas, Hidalgo, Willacy, and Cameron, in the State of Texas.

§ 959.5 Onions.

Onions means all varieties of Allium cepa commonly known as onions grown within the production area and marketed dry.

§ 959.6 Handler.

Handler is synonymous with *shipper* and means any person (except a common or contract carrier of onions owned by another person) who handles onions or causes onions to be handled.

§ 959.7 Handle.

Handle or *ship* means to package, load, sell, transport, or in any way to place onions in the current of the commerce within the production area or between the production area and any point outside thereof. Such term shall not include the transportation, sale, or delivery of field-run onions to a person in the production area who is a registered handler.

[27 FR 227, Mar. 9, 1962, as amended at 34 FR 6440, Apr. 12, 1969]

§ 959.8 Registered handler.

Registered handler means any person with adequate facilities within the production area for preparing onions for commercial market, who customarily does so, and who is so recorded by the committee, or any person who has access to such facilities within the production area, and has recorded with the committee his ability and willingness to assume customary obligations of preparing onions for commercial market.

§959.9 Producer.

Producer means any person engaged in a proprietary capacity in the production of onions for market.

§959.10 Grading.

Grading is synonymous with *preparation for market* and means the sorting or separation of onions into grades, sizes, and packs for market purposes.

§959.11 Grade and size.

Grade means any of the established grades of onions, and *size* means any of the established sizes of onions as defined and set forth in the United States Standards for Bermuda-Granex Type Onions (§§ 51.3195 to 51.3209 of this title) or any other United States Standards for onions, or amendments thereto or modifications thereof, or variations based thereon, recommended by the committee and approved by the Secretary.

§959.12 Pack.

Pack means a quantity of onions specified by grade, size, weight, or count, or by type or condition of container, or any combination of these recommended by the committee and approved by the Secretary.

[27 FR 2278, Mar. 9, 1962]

§959.13 Container.

Container means a box, bag, crate, hamper, basket, package, or any other receptacle used in the packaging, transportation, sale, shipment or other handling of onions.

§959.14 Varieties.

Varieties means and includes all classifications, subdivisions, or types of onions according to those definitive characteristics now or hereafter recognized by the United States Department of Agriculture or recommended by the committee and approved by the Secretary.

§959.15 Committee.

Committee means the South Texas Onion Committee, established pursuant to § 959.22.

§959.16 Fiscal period.

Fiscal period means the annual period beginning and ending on such dates as may be approved by the Secretary pursuant to recommendations of the committee.

§959.17 District.

District means each of the geographic divisions of the production area initially established pursuant to § 959.24 or as reestablished pursuant to § 959.25.

§959.18 Export.

Export means to ship onions to any destination which is not within the 48 contiguous States, or the District of Columbia, of the United States.

COMMITTEE

§959.22 Establishment and membership.

The South Texas Onion Committee, consisting of thirteen members, eight of whom shall be producers and five of whom shall be handlers, is hereby established. For each member of the Committee there shall be an alternate. Producer members and alternates shall not have a proprietary interest in or be employees of a handler organization.

[84 FR 10667, Mar. 22, 2019]

§959.23 Term of office.

(a) The term of office of committee members and their respective alternates shall be for two years and shall begin as of August 1 and end as of July 31. The terms shall be so determined that about one-half of the total committee membership shall terminate each year.

(b) Committee members and alternates shall serve during the term of office for which they are selected and have qualified, or during that portion thereof beginning on the date on which they qualify during such term of office and continuing until the end thereof, and until their successors are selected and have qualified.

§ 959.24　Districts.

To determine a basis for selecting Committee members, the following districts of the production area are hereby established:

(a) *District No. 1.* (Coastal Bend-Lower Valley) The Counties of Victoria, Calhoun, Goliad, Refugio, Bee, Live Oak, San Patricio, Aransas, Jim Wells, Nueces, Kleberg, Brooks, Kenedy, Duval, McMullen, Cameron, Hidalgo, Starr, and Willacy in the State of Texas.

(b) *District No. 2.* (Laredo-Winter Garden) The Counties of Zapata, Webb, Jim Hogg De Witt, Wilson, Atascosa, Karnes Val Verde, Frio, Kinney, Uvalde, Medina, Maverick, Zavala, Dimmit, and La Salle in the State of Texas.

[84 FR 10667, Mar. 22, 2019]

§ 959.25　Redistricting.

The committee may recommend, and pursuant thereto, the Secretary may approve, the reapportionment of members among districts, and the reestablishment of districts within the production area. In recommending any such changes, the committee shall give consideration to:

(a) Shifts in onion acreage within the districts and within the production area during recent years;

(b) The importance of new production in its relation to existing districts;

(c) The equitable relationship of committee membership and districts;

(d) Economies to result for producers in promoting efficient administration due to redistricting or reapportionment of members within districts; and

(e) Other relevant factors. No change in districting or in apportionment of members within districts may become effective less than 30 days prior to the date on which terms of office begin each year and no recommendations for such redistricting or reapportionment may be made less than six months prior to such date.

§ 959.26　Selection.

The Secretary shall select members and respective alternates from districts established pursuant to § 959.24 or § 959.25. Selections shall be as follows:

(a) *District No. 1.* Five producer members and alternates; three handler members and alternates.

(b) *District No. 2.* Three producer members and alternates; two handler members and alternates.

[84 FR 10667, Mar. 22, 2019]

§ 959.27　Nomination.

The Secretary may select the members of the committee and alternates from nominations which may be made in the following manner:

(a) A meeting or meetings of producers and handlers shall be held for each district to nominate members and alternates for the committee. For nominations to the initial committee, the meetings may be sponsored by the United States Department of Agriculture or by any agency or group requested to do so by such department. For nominations for succeeding members and alternates on the committee, the committee shall hold such meetings or cause them to be held prior to June 15 of each year, after the effective date of this subpart, or by such other date as may be specified by the Secretary;

(b) At each such meeting at least one nominee shall be designated for each position as member and for each position as alternate member on the committee;

(c) Nominations for committee members and alternates shall be supplied to the Secretary in such manner and form as he may prescribe, not later than July 1 of each year, or by such other date as may be specified by the Secretary;

(d) Only producers may participate in designating producer nominees, and only handlers may participate in naming handler nominees. In the event a person is engaged in producing or handling onions in more than one district, such person shall elect the district within which he may participate as aforesaid in designating nominees;

(e) Regardless of the number of districts in which a person produces or handles onions, each such person is entitled to cast only one vote on behalf of himself, his agents, subsidiaries, affiliates, and representatives in designating nominees for committee members and alternates. An eligible voter's privilege of casting only one vote as aforesaid shall be construed to permit a voter to cast one vote for each position to be filled in the respective district in which he elects to vote.

[26 FR 12751, Dec. 30, 1961, as amended at 34 FR 6440, Apr. 12, 1969]

§959.28 Failure to nominate.

If nominations are not made within the time and in the manner specified in §959.27, the Secretary may, without regard to nominations, select the committee members and alternates, which selection shall be on the basis of the representation provided for in §§959.22 through 959.26.

§959.29 Acceptance.

Any person selected as a committee member or alternate shall qualify by filing a written acceptance within ten days after being notified of such selection.

§959.30 Vacancies.

To fill committee vacancies, the Secretary may select such members or alternates from unselected nominees on the current nominee list from the district involved, or from nominations made in the manner specified in §959.27. If the names of nominees to fill any such vacancy are not made available to the Secretary within 30 days after such vacancy occurs, such vacancy may be filled without regard to nominations, which selection shall be made on the basis of the representation provided for in §§959.24 to 959.26.

§959.31 Alternate members.

An alternate member of the committee shall act in the place and stead of the member for whom he is an alternate, during such member's absence or when designated to do so by the member for whom he is an alternate. In the event both a member of the committee and his alternate are unable to attend a committee meeting, the member or his alternate or the committee (in that order) may designate another alternate from the same district and the same group (handler or grower) to serve in such member's place and stead. In the event of the death, removal, resignation, or disqualification of a member,

his alternate shall act for him until a successor of such member is selected and has qualified. The committee may request the attendance of alternates at any or all meetings, notwithstanding the expected or actual presence of the respective members.

[27 FR 2278, Mar. 9, 1962]

§959.32 Procedure.

(a) Nine members of the Committee shall be necessary to constitute a quorum. Seven concurring votes, or two-thirds of the votes cast, whichever is greater, shall be required to pass any motion or approve any Committee action. At assembled meetings all votes shall be cast in person.

(b) The committee may meet by telephone, telegraph, or other means of communication and any vote at such a meeting shall be promptly confirmed in writing. On such occasions unanimous vote of committee members voting will be required to approve any action.

[26 FR 704, Jan. 25, 1961, as amended at 84 FR 10667, Mar. 22, 2019]

§959.33 Expenses and compensation.

Committee members and alternates when acting on committee business shall be reimbursed for reasonable expenses necessarily incurred by them in the performance of their duties and in the exercise of their powers under this part. In addition they may receive compensation at a rate to be determined by the committee and approved by the Secretary, not to exceed $10 for each day, or portion thereof, spent in attending to committee business.

§959.34 Powers.

The committee shall have the following powers:

(a) To administer the provisions of this part in accordance with its terms and provisions;

(b) To make rules and regulations to effectuate the terms and provisions of this part;

(c) To receive, investigate, and report to the Secretary complaints of violation of the provisions of this part; and

(d) To recommend to the Secretary amendments to this part.

§ 959.35   Duties.

It shall be, among other things, the duty of the committee:

(a) As soon as practicable after the beginning of each term of office, to meet and organize, to select a chairman and such other officers as may be necessary, to select subcommittees of committee members and alternates, and to adopt such rules and regulations for the conduct of its business as it may deem advisable;

(b) To act as intermediary between the Secretary and any producer or handler;

(c) To furnish to the Secretary such available information as he may request;

(d) To appoint such employees, agents, and representatives as it may deem necessary and to determine the salaries and define the duties of each such person, and to protect the handling of committee funds through fidelity bonds for employees;

(e) To investigate from time to time and to assemble data on the growing, harvesting, shipping, and marketing conditions with respect to onions;

(f) To prepare a marketing policy;

(g) To recommend marketing regulations to the Secretary;

(h) To recommend rules and procedures for, and to make determinations in connection with, issuance of certificates of privilege;

(i) To keep minutes, books, and records which clearly reflect all of the acts and transactions of the committee, and such minutes, books and records shall be subject to examination at any time by the Secretary or by his authorized agent or representative. Minutes of each committee meeting shall be reported promptly to the Secretary;

(j) At the beginning of each fiscal period, to prepare a budget of its expenses for such fiscal period, together with a report thereon;

(k) To cause the books of the committee to be audited by a competent accountant at least once each fiscal period, and at such other time as the committee may deem necessary or as the Secretary may request. The report of such audit shall show the receipt and expenditure of funds collected pursuant to this part. A copy of each such report shall be made available at the principal office of the committee for inspection by producers and handlers, and a copy of each such report shall be furnished the Secretary;

(l) To consult, cooperate, and exchange information with other marketing agreement committees and other individuals or agencies in connection with all proper committee activities and objectives under this part.

EXPENSES AND ASSESSMENTS

§ 959.40   Expenses.

The committee is authorized to incur such expenses as the Secretary may find are reasonable and likely to be incurred during each fiscal period for its maintenance and functioning, and for such purposes as the Secretary, pursuant to this subpart, determines to be appropriate. Handlers shall share expenses on the basis of a fiscal period. Each handler's share of such expenses shall be proportionate to the ratio between the total quantity of onions handled by him as the first handler thereof during a fiscal period and the quantity of onions handled by all handlers as first handlers thereof during such fiscal period.

§ 959.41   Budget.

As soon as practicable after the beginning of each fiscal period and as may be necessary thereafter, the committee shall prepare an estimated budget of income and expenditures necessary for the administration of this part. The committee may recommend a rate of assessment calculated to provide adequate funds to defray its proposed expenditures. The committee shall present such budget to the Secretary with an accompanying report showing the basis for its calculations.

§ 959.42   Assessments.

(a) The funds to cover the committee's expenses shall be acquired by the levying of assessments upon handlers as provided in this subpart. Each handler who first handles onions, which are regulated under this part, shall pay assessments to the committee upon demand, which assessments shall be in payment of such handler's pro rata share of the committee's expenses.

(b) Assessments shall be levied upon handlers at rates established by the Secretary. Such rates may be established upon the basis of the committee's recommendations and other available information. Such rates may be applied to specified containers used in the production area.

(c) At any time during, or subsequent to, a given fiscal period the committee may recommend the approval of an amended budget and an increase in the rate of assessment. Upon the basis of such recommendations, or other available information, the Secretary may approve an amended budget and increase the rate of assessment. Such increase shall be applicable to all onions which were regulated under this part and which were handled by the first handlers thereof during such fiscal period.

(d) The payment of assessments for the maintenance and functioning of the committee may be required under this part throughout the period it is in effect irrespective of whether particular provisions thereof are suspended or become inoperative.

(e) If a handler does not pay assessments within the time prescribed by the committee, the assessment may be increased by a late payment charge and/or an interest rate charge at amounts prescribed by the committee with approval of the Secretary.

[26 FR 704, Jan. 25, 1961, as amended at 73 FR 10976, Feb. 29, 2008]

§959.43 Accounting.

(a) Assessments collected in excess of expenses incurred shall be accounted for in accordance with one of the following:

(1) Excess funds not retained in a reserve, as provided in paragraph (a)(2) of this section shall be refunded proportionately to the persons from whom they were collected.

(2) The committee, with the approval of the Secretary, may carry over excess funds into subsequent fiscal periods as reserves: *Provided,* That funds already in reserves do not equal approximately two fiscal periods' expenses. Such reserve funds may be used (i) to defray expenses during any fiscal period prior to the time assessment income is sufficient to cover such expenses, (ii) to cover deficits incurred during any fiscal period when assessment income is less than expenses, (iii) to defray expenses incurred during any period when any or all provisions of this part are suspended or are inoperative, (iv) to cover necessary expenses of liquidation in the event of termination of this part. Upon such termination, any funds not required to defray the necessary expenses of liquidation shall be disposed of in such manner as the Secretary may determine to be appropriate. To the extent practical, such funds shall be returned pro rata to the persons from whom such funds were collected.

(b) All funds received by the committee pursuant to the provisions of this part shall be used solely for the purpose specified in this part and shall be accounted for in the manner provided for in this part. The Secretary may at any time require the committee and its members to account for all receipts and disbursements.

(c) Upon the removal or expiration of the term of office of any member of the committee, such member shall account for all receipts and disbursements and deliver all property and funds in his possession to the committee, and shall execute such assignments and other instruments as may be necessary or appropriate to vest in the committee full title to all of the property, funds, and claims vested in such member pursuant to this part.

(d) The committee may make recommendations to the Secretary for one or more of the members thereof, or any other person, to act as a trustee for holding records, funds, or any other committee property during periods of suspension of this subpart, or during any period or periods when regulations are not in effect and if the Secretary determines such action appropriate, he may direct that such person or persons shall act as trustee or trustees for the committee.

[26 FR 12751, Dec. 30, 1961, as amended at 34 FR 6440, Apr. 12, 1969]

RESEARCH AND DEVELOPMENT

§959.48 Research and development.

The committee, with the approval of the Secretary, may establish or provide

389

for the establishment of production research, marketing research, and development projects designed to assist, improve, or promote the marketing, distribution, consumption or efficient production of onions. The expenses of such projects shall be paid from funds collected pursuant to § 959.42.

[38 FR 31516, Nov. 15, 1973]

REGULATIONS

§ 959.50 Marketing policy.

(a) At the beginning of each season, and as the Secretary may require, the committee shall prepare a marketing policy. Such policy shall indicate the data on onion supplies and demand on which the committee bases its judgments and recommendations. It shall indicate also the kind or types of regulations contemplated during the ensuing season, and, to the extent practical, shall include recommendations for specific regulations. Notice of such marketing policy shall be given to producers, handlers, and other interested parties by bulletins, newspapers, or other appropriate media, and copies thereof shall be submitted to the Secretary and shall be available generally.

(b) Marketing policy statements relating to recommendations for regulations shall give appropriate consideration to onion supplies for the season, with special consideration to:

(1) Estimates of total supplies, including grade, size, and quality thereof, in the production area;

(2) Estimates of supplies in the competing areas;

(3) Market prices by grades, sizes, containers, and packs;

(4) Estimates of supplies of competing commodities;

(5) Anticipated marketing problems;

(6) Level and trend of consumer income; and

(7) Other relevant factors.

§ 959.51 Recommendations for regulations.

Upon complying with the requirements of § 959.50 the committee may recommend regulations to the Secretary whenever it finds that such regulations as are provided for in this subpart will tend to effectuate the declared policy of the act.

§ 959.52 Issuance of regulations.

(a) The Secretary shall limit the handling of onions by regulations specified in this section whenever he finds from the recommendations and information submitted by the committee, or from other available information, that such regulations would tend to effectuate the declared policy of the act.

(b) Such regulations may:

(1) Limit in any or all portions of the production area the handling of particular grades, sizes, qualities or packs, or any combination thereof, of any or all varieties of onions during any period;

(2) Limit the handling of particular grades, sizes, qualities, or packs of onions differently for different varieties, for different containers, for different portions of the production area, or any combination of the foregoing, during any period;

(3) Limit the handling of onions by establishing, in terms of grades, sizes, or both, minimum standards of quality and maturity;

(4) Fix the size, capacity, weight, dimensions, or pack of the container or containers which may be used in the packaging, transportation, sale, preparation for market, shipment, or other handling of onions;

(5) Establish holidays by prohibiting throughout the entire production area, the packaging or loading, or both, of onions on Sundays;

(6) Prohibit the packaging or loading, or both, of onions except during specified consecutive hours of any calendar day or days: *Provided,* That, any handler may, upon such notice to the committee as it may prescribe with approval of the Secretary, package or load onions during a different period in such day consisting of the same number of consecutive hours: *Provided further,* That any handler who, due to conditions specified in regulations established by the committee with the approval of the Secretary as being beyond a handler's reasonable control, is prevented for more than one of such consecutive hours from so packaging or loading onions may, in accordance with such regulations, obtain permission from the committee to package or load onions, or both, during a comparable number of additional hours in the same

day or a later day as specified by the committee.

(c) Regulations issued hereunder may be amended, modified, suspended, or terminated whenever it is determined:

(1) That such action is warranted upon recommendation of the committee or other available information;

(2) That such action is essential to provide relief from inspection, assessment, or regulations under paragraph (b) of this section for minimum quantities less than customary commercial transactions; or

(3) That regulations issued hereunder no longer tend to effectuate the declared policy of the act.

(d) No handler may handle onions that were packaged or loaded or both during any period when such packaging or loading or both was prohibited by any regulation issued pursuant to paragraphs (b)(5) or (6) of this section, except such onions as were exempted thereunder.

[26 FR 704, Jan. 25, 1961. Redesignated at 26 FR 12751, Dec. 30, 1961, as amended at 34 FR 6440, Apr. 12, 1969]

§959.53 Handling for special purposes.

Regulations in effect pursuant to §§959.42, 959.52, or 959.60 may be modified, suspended, or terminated to facilitate handling of onions for:

(a) Relief or charity;

(b) Experimental purposes;

(c) Export; and

(d) Other purposes which may be recommended by the committee and approved by the Secretary.

§959.54 Safeguards.

The committee, with the approval of the Secretary, may establish through rules such requirements as may be necessary to establish that shipments made pursuant to §959.53 were handled and used for the purpose stated.

§959.55 Notification of regulation.

The Secretary shall promptly notify the committee of regulations issued or of any modification, suspension, or termination thereof. The committee shall give reasonable notice thereof to handlers.

INSPECTION

§959.60 Inspection and certification.

(a) Whenever the handling of onions is regulated pursuant to §959.52, or at other times when recommended by the committee and approved by the Secretary, no handlers shall handle onions unless they are inspected by an authorized representative of the Federal or Federal-State Inspection Service and are covered by a valid inspection certificate, except when relieved from such requirements pursuant to §959.52(c) or §959.54, or paragraph (b) of this section.

(b) Regarding, resorting, or repacking any lot of onions shall invalidate any prior inspection certificate insofar as the requirements of this section are concerned. No handler shall handle onions after they have been regraded, resorted, or repacked unless such onions are inspected by an authorized representative of the Federal or Federal-State Inspection Service. Such inspection requirements on regraded, resorted, or repacked onions may be modified, suspended, or terminated upon recommendation by the committee and approval of the Secretary.

(c) Upon recommendation of the committee and approval by the Secretary, any or all onions so inspected and certified shall be identified by appropriate seals, stamps, or tags to be affixed to the containers by the handler under the direction and supervision of a Federal or Federal-State Inspector or the Committee. Master containers may bear the identification instead of the individual containers within said master container.

(d) At any time this marketing order is inoperative, compulsory inspection is not required.

(e) Insofar as the requirements of this section are concerned, the length of time for which an inspection certificate is valid may be established by the committee with the approval of the Secretary.

(f) When onions are inspected in accordance with the requirements of this section, a copy of each inspection certificate issued shall be made available to the committee by the Inspection Service.

(g) The committee may recommend and the Secretary may require that no handler shall transport or cause the transportation of onions by motor vehicle or by other means unless such shipment is accompanied by a copy of the inspection certificate issued thereon, or other document authorized by the committee to indicate that such inspection has been performed. Such certificate or document shall be surrendered to such authority as may be designated.

REPORTS

§ 959.80   Reports.

Upon request of the committee, made with the approval of the Secretary, each handler shall furnish to the committee, in such manner or form and at such time as it may prescribe, such reports and other information as may be necessary for the committee to perform its duties under this part.

(a) Such reports may include, but are not necessarily limited to, the following:

(1) The quantities of onions received by a handler;

(2) The quantities disposed of by him segregated as to the respective quantities subject to regulation and not subject to regulation;

(3) The date of each such disposition and the identification of the carrier transporting such onions; and

(4) Identification of the inspection certificates relating to the onions which were handled pursuant to § 959.52 or § 959.53, or both.

(b) All such reports shall be held under appropriate protective classification and custody by the committee, or duly appointed employees thereof, so that the information contained therein which may adversely affect the competitive position of any handler in relation to other handlers will not be disclosed. Compilations of general reports from data submitted by handlers is authorized, subject to the prohibition of disclosure of individual handlers' identities or operations.

(c) Each handler shall maintain for at least two succeeding years such records and documents on onions received and onions disposed of by him as may be necessary to verify reports he submits to the committee pursuant to this section.

COMPLIANCE

§ 959.81   Compliance.

Except as provided in this subpart, no handler shall handle onions, the handling of which has been prohibited by the Secretary in accordance with provisions of this subpart, or the rules and regulations thereunder, and no handler shall handle onions except in conformity to the provisions of this subpart.

MISCELLANEOUS PROVISIONS

§ 959.82   Right of the Secretary.

The members of the committee (including successors and alternates), and any agent or employee appointed or employed by the committee, shall be subject to removal or suspension by the Secretary at any time. Each and every order, regulation, decision, determination or other act of the committee shall be subject to the continuing right of the Secretary to disapprove of the same at any time. Upon such disapproval, the disapproved action of the said committee shall be deemed null and void, except as to acts done in reliance thereon or in compliance therewith prior to such disapproval by the Secretary.

§ 959.83   Effective time.

The provisions of this subpart, or any amendment thereto, shall become effective at such time as the Secretary may declare and shall continue in force until terminated in one of the ways specified in this subpart.

§ 959.84   Termination.

(a) The Secretary may, at any time, terminate the provisions of this subpart by giving at least one day's notice by means of a press release or in any other manner which he may determine.

(b) The Secretary shall terminate or suspend the operation of any or all of the provisions of this subpart whenever he finds that such provisions do not tend to effectuate the declared policy of the act.

(c) The Secretary shall terminate the provisions of this subpart at the end of any fiscal period whenever he finds that such termination is favored by a majority of producers who, during a representative period, have been engaged in the production of onions for market: *Provided,* That such majority has, during such representative period, produced for market more than fifty percent of the volume of such onions produced for market.

(d) The Secretary shall conduct a referendum within six years after the effective date of this paragraph and every sixth year thereafter to ascertain whether continuance is favored by producers. The Secretary would consider termination of this part if less than two-thirds of the growers voting in the referendum and growers of less than two-thirds of the volume of onions represented in the referendum favor continuance.

(e) The provisions of this subpart shall, in any event, terminate whenever the provisions of the act authorizing them cease to be in effect.

[26 FR 704, Jan. 25, 1961, as amended at 73 FR 10976, Feb. 29, 2008]

§959.85 Proceeding after termination.

(a) Upon the termination of the provisions of this subpart the then functioning members of the committee shall continue as joint trustees for the purpose of settling the affairs of the committee by liquidating all of the funds and property then in the possession of or under control of the committee, including claims for any funds unpaid or property not delivered at the time of such termination. Action by said trusteeship shall require the concurrence of a majority of the said trustees.

(b) The said trustees shall continue in such capacity until discharged by the Secretary; shall, from time to time, account for all receipts and disbursements and deliver all property on hand, together with all books and records of the committee and of the trustees, to such person as the Secretary may direct; and shall, upon request of the Secretary, execute such assignments or other instruments necessary or appropriate to vest in such persons full title and right to all of the funds, property, and claims vested in

the committee or the trustees pursuant to this subpart.

(c) Any person to whom funds, property, or claims have been transferred or delivered by the committee or its members, pursuant to this section, shall be subject to the same obligations imposed upon the members of the committee and upon the said trustees.

§959.86 Effect of termination or amendments.

Unless otherwise expressly provided by the Secretary, the termination of this subpart or of any regulation issued pursuant to this subpart, or the issuance of any amendments to either thereof, shall not (a) affect or waive any right, duty, obligation, or liability which shall have arisen or which may thereafter arise in connection with any provisions of this subpart or any regulation issued under this subpart, or (b) release or extinguish any violation of this subpart or of any regulation issued under this subpart, or (c) affect or impair any rights or remedies of the Secretary or of any other person with respect to any such violation.

§959.87 Duration of immunities.

The benefits, privileges, and immunities conferred upon any person by virtue of this subpart shall cease upon the termination of this subpart, except with respect to acts done under and during the existence of this subpart.

§959.88 Agents.

The Secretary may, by designation in writing, name any person, including any officer or employee of the United States Department of Agriculture, to act as his agent or representative in connection with any of the provisions of this subpart.

§959.89 Derogation.

Nothing contained in this subpart is, or shall be construed to be, in derogation or in modification of the rights of the Secretary or of the United States to exercise any powers granted by the act or otherwise, or, in accordance with such powers, to act in the premises whenever such action is deemed advisable.

§ 959.90   Personal liability.

No member or alternate of the committee nor any employee or agent thereof, shall be held personally responsible, either individually or jointly with others, in any way whatsoever, to any handler or to any person for errors in judgment, mistakes, or other acts, either of commission or omission, as such member, alternate, agent, or employee, except for acts of dishonesty, willful misconduct, or gross negligence.

§ 959.91   Separability.

If any provision of this subpart is declared invalid, or the applicability thereof to any person, circumstance, or thing is held invalid, the validity of the remainder of this subpart, or the applicability thereof to any other person, circumstance, or thing, shall not be affected thereby.

§ 959.92   Amendments.

Amendments to this subpart may be proposed, from time to time, by the committee or by the Secretary.

Subpart B—Administrative Provisions

Source: 26 FR 2560, Mar. 25, 1961, unless otherwise noted. Redesignated at 26 FR 12751, Dec. 30, 1961.

GENERAL

§ 959.100   Order.

Order means Order No. 959 (§§ 959.1 to 959.92; 26 FR 704) regulating the handling of onions grown in South Texas.

§ 959.101   Terms.

The terms used in this subpart shall have the same meaning as when used in the order.

§ 959.102   Communications.

Unless otherwise provided in the order, or by specific direction of the committee, all reports, applications, submittals, requests and communications in connection with the order shall be addressed to the South Texas Onion Committee, at its principal office.

§ 959.103   Registered handler.

For purposes of this part any person who operates an established packing house within the production area with commonly accepted adequate facilities for grading and packing onions for market, and who customarily buys onions from producers for grading, packing, and marketing shall be recorded by the committee as a registered handler. Any other person who wishes to be listed as a registered handler may make application for registration on forms furnished by the committee. If such applicant has facilities available to him that are determined by the committee to be adequate for grading and packing onions for market, and he assumes responsibility for inspection of onions handled by him, and for assessments thereon, he may be approved and recorded as a registered handler. If the committee determines from the available information that the applicant is not entitled to be registered with the committee, he shall be so informed by written notice stating the reason for denial of his application. Any registration of a handler pursuant to this section may be canceled by the committee under circumstances which would have justified denial of his application. Any handler whose registration has been canceled shall be so informed by written notice thereof stating the reason therefor. The committee shall also notify producers of each such cancellation of handler registration through committee bulletins or published notice in local newspapers of general distribution, or both.

§ 959.104   Fiscal period.

The fiscal period shall begin August 1 of each year and end July 31 of the following year, both dates inclusive.

[68 FR 11466, Mar. 11, 2003]

§§ 959.110—959.111   [Reserved]

§ 959.115   Planting reports.

Each handler shall furnish every two weeks during the planting season to the committee, on a form provided by the committee, the number of acres of onions planted by the handler or growers for whom the handler packs onions during such period and the location of such plantings.

[53 FR 7330, Mar. 8, 1988]

SAFEGUARDS

§959.120 Policy.

Whenever shipments of onions for special purposes pursuant to §959.53 are relieved in whole or in part from regulations issued under §959.52, the committee may require information and evidence on the manner, methods, and timing of such shipments as safeguards against the entry of any such onions in trade channels other than those for which intended. Such information and evidence shall include requirements set forth below with respect to Certificates of Privilege.

§959.121 Qualification.

Before handling onions for special purposes which do not meet regulations issued pursuant to §959.52, a handler, when required by such regulations, must qualify with the committee to handle shipments for special purposes. To qualify he must (a) apply for and receive a Certificate of Privilege indicating his intent to so handle onions, (b) agree to comply with reporting and other requirements set forth in §§959.120 to 959.125, inclusive, with respect to such shipments, and (c) receive approval of the committee, or its duly authorized agents, to so handle onions. Such approval will be based upon evidence furnished in his application for Certificate of Privilege and other information available to the committee.

§959.122 Application.

(a) Applications for a Certificate of Privilege shall be made on forms furnished by the committee. Each application may contain, but need not be limited to, the name and address of the handler; the quantity by grade, size, and quality of the onions to be shipped; the mode of transportation; the consignee; the destination; the purpose for which the onions are to be used; and certification to the United States Department of Agriculture and to the committee as to the truthfulness of the information shown thereon, and any other appropriate information or documents deemed necessary by the com-

mittee or its duly authorized agents for the purposes stated in §959.120.

(b) The committee may require each handler making shipments of onions for export to include with his application a copy of the Department of Commerce Shippers Export Declaration Form No. 7525–V applicable to such shipment.

§959.123 Approval.

The committee or its duly authorized agents shall give prompt consideration to each application for a Certificate of Privilege. Approval of an application, based upon the determination as to whether the information contained therein and other information available to the committee supports approval, shall be evidenced by the issuance of a Certificate of Privilege to the applicant. Each certificate shall cover a specified period and specified qualities and quantities of onions to be sold or transported to a designated consignee for the purpose declared.

§959.124 Reports.

Each handler of onions shipping under Certificates of Privilege shall supply the committee with reports as requested by the committee, or its duly authorized agents, showing the name and address of the shipper; the car or truck identification; the loading point; destination; consignee; the inspection certificate number when inspection is required; and any other information deemed necessary by the committee.

§959.125 Disqualification.

The committee from time to time may conduct surveys of handling of onions for special purposes requiring Certificates of Privilege to determine whether handlers are complying with the requirements and regulations applicable to such certificates. Whenever the committee finds that the handler or consignee is failing to comply with requirements and regulations applicable to handling of onions in special outlets and requiring such certificates, a Certificate or Certificates of Privilege issued such handler may be rescinded and subsequent certificates denied. Such disqualification shall apply to, and not exceed, a reasonable period of time as determined by the committee, but in no event shall it extend beyond the date of the succeeding fiscal period. Any handler who has a Certificate rescinded or denied may appeal to the committee in writing for reconsideration of his disqualification.

§ 959.126 Handling of culls.

(a) The handling of culls, i.e., onions which fail to meet the grade, size and quality requirements established under § 959.52(b) of this part, is prohibited, unless such onions are:

(1) Mechanically mutilated at the packing shed rendering them unsuitable for fresh market;

(2) Handled for special purpose outlets approved under § 959.53 of this part; or

(3) Handled for canning or freezing.

(b) As a safeguard against culls entering fresh market channels each handler of culls under paragraphs (a) (2) or (3) of this section shall apply for and obtain a certificate from the committee which shall require the handler to furnish such reports or other information as the committee may request.

[28 FR 60, Jan. 3, 1963]

§ 959.237 Assessment rate.

On and after August 1, 2017, an assessment rate of $0.065 per 50-pound equivalent is established for South Texas onions.

[83 FR 594, Jan. 5, 2018]

§ 959.322 Handling regulation.

During the period beginning March 1 and ending June 4, no handler shall handle any onions, including onions for peeling, chopping, and slicing, unless they comply with paragraphs (a) through (c) or (d) or (e) of this section; except that onions handled during the period June 5 through July 15 shall comply with paragraphs (c) or (d) or (e) of this section.

(a) *Grade requirements.* Not to exceed 20 percent defects of U.S. No. 1 grade. In percentage grade lots, tolerances for serious damage shall not exceed 10 percent including not more than 2 percent decay. Double the lot tolerance shall be permitted in individual packages in percentage grade lots. Application of tolerances in U.S. onion standards shall apply to in-grade lots.

(b) *Size requirements.* (1) "Small"—1 to 2¼ inches in diameter, and limited to whites only;

(2) "Repacker"—1¾ to 3 inches in diameter, with 60 percent or more 2 inches in diameter or larger;

(3) "Medium"—2 to 3½ inches in diameter; or

(4) "Jumbo" or "Large"—3 inches or larger in diameter; or

(5) "Colossal"—3¾ inches or larger in diameter.

(6) Tolerances for size in the U.S. onion standards shall apply except that for "repacker" and "medium" sizes not more than 20 percent, by weight, of onions in any lot may be larger than the maximum diameter specified. Application of tolerances in the U.S. onion standards shall apply.

(c) *Inspection.* (1) No handler may handle any onions regulated hereunder, except pursuant to paragraphs (d), (e)(1), or (e)(2) of this section unless an inspection certificate has been issued by the Federal or Federal-State Inspection Service, Texas Cooperative Inspection Program, covering them and the certificate is valid at the time of shipment. City destinations shall be listed on inspection certificates and release forms.

(2) No handler may transport by motor vehicle or cause such transportation of any shipment of onions for which an inspection certificate is required unless each such shipment is accompanied by a copy of the inspection certificate applicable thereto or the shipment release form furnished by the inspection service identifying truck lots to which a valid inspection certificate is applicable. A copy of such inspection certificate or shipment release form shall be surrendered upon request to Texas Department of Agriculture personnel designated by the committee.

(3) For purposes of operation under this part, each inspection certificate, shipment release form, or committee form required as evidence of inspection is hereby determined to be valid for a period not to exceed 72 hours following completion of inspection as shown on the certificate.

(4) Handlers shall pay assessment on all assessable onions according to the provisions of § 959.42.

(d) *Minimum quantity exemption.* Any handler may handle, other than for resale, up to, but not to exceed 110 pounds of onions per day without regard to the requirements of this section, but this exemption shall not apply to any shipment or any portion thereof of over 110 pounds of onions.

(e) *Special purpose shipments.* (1) The minimum grade, size, quality, and inspection requirements set forth in paragraphs (a) through (c) of this section shall not be applicable to shipments of onions for charity, relief, export, and processing if handled in accordance with paragraph (f) of this section.

(2) *Experimental shipments.* Upon approval by the committee, onions may be shipped for experimental purposes exempt from regulations issued pursuant to §§959.42, 959.52 and 959.60, provided they are handled in accordance with the safeguard provisions of paragraph (f) of this section.

(3) *Onions failing to meet requirements.* Onions failing to meet the grade and size requirements of this section, and not exempt under paragraphs (d) or (e) of this section, may be handled only pursuant to §959.126. Such onions not handled in accordance with paragraph (f) of this section shall be mechanically mutilated at the packing shed rendering them unsuitable for fresh market.

(f) *Safeguards.* Each handler making shipments of onions for charity, relief, export, processing, or experimental purposes shall:

(1) Apply to the committee for and obtain a Certificate of Privilege to make such shipments;

(2) Furnish reports of each shipment made under the applicable Certificate of Privilege;

(3) Such reports, in accordance with §959.80, shall be furnished to the committee in such manner, on such forms and at such times as it may prescribe. Each handler shall maintain records of such shipments pursuant to §959.80(c), and the records shall be subject to review and audit by the committee to verify reports thereon.

(4) In addition to provisions in the preceding paragraphs, each handler making shipments for processing shall:

(i) Weigh or cause to be weighed each shipment prior to, or upon arrival at, the processor.

(ii) Attach a copy of the weight ticket to a completed copy of the Report of Special Purpose Onion Shipment and return both promptly to the committee office.

(iii) Make each shipment directly to the processor or the processor's subcontractor and attach a copy of the Report of Special Purpose Onion Shipment.

(iv) Each processor or processor's subcontractor who receives cull onions shall weigh the onions upon receipt, complete the Report of Special Purpose Onion Shipment which accompanies each load and mail it immediately to the committee office.

(v) Each processor who receives cull onions shall make available at its business office at any reasonable time during business hours, copies of all applicable purchase orders, sales contracts, or disposition documents for examination by the Department or by the committee, together with any other information which the committee or the Department may deem necessary to enable it to determine the disposition of the onions.

(vi) If a processor employs a subcontractor for any stage of processing, such processor shall be responsible for ensuring that the subcontractor accounts for all quantities of onions received and processed or otherwise disposed of, and that the subcontractor reports to the committee in the same manner and frequently as the processor.

(5) Cull onions transported in bags shall be transported in unlabelled bags, or shall have labelled bags reversed so that the label is not visible.

(g) *Definitions. U.S. onion standards* means the United States Standards for Grades of Bermuda-Granex-Grano Type Onions (7 CFR 51.3195–51.3209), or the United States Standards for Grades of Onions (Other Than Bermuda-Granex-Grano and Creole Types) (7 CFR 51.2830–51.2854), whichever is applicable to the particular variety, or variations thereof specified in this section. The term *U.S. No. 1* shall have the same

397

meaning as set forth in these standards. *Processing* means cooking or freezing the onions in such a way, or with such other food components, that the consistency of the product is changed. Canning and freezing shall be considered forms of processing. All other terms used in this section shall have the same meaning as when used in Marketing Agreement No. 143, as amended, and this part.

[47 FR 8552, Mar. 1, 1982]

EDITORIAL NOTE: For FEDERAL REGISTER citations affecting § 959.322, see the List of CFR Sections Affected, which appears in the Finding Aids section of the printed volume and at *www.govinfo.gov.*

PART 966—TOMATOES GROWN IN FLORIDA

Subpart A—Order Regulating Handling

Subpart B—Administrative Requirements

966.131   Investigations.
966.132   Issuance.
966.133   Disposition of certificates.
966.134   Reports.
966.135   Appeals.

INSPECTION

966.140   Truck shipments.

INTERPRETATIVE RULES

966.150   Meaning of "producer".

REESTABLISHMENT OF DISTRICTS

966.160   Reestablishment of districts.
966.161   Reapportionment of committee membership.

Subpart C—Assessment Rates

966.234   Assessment rate.

Subpart D—Handling Requirements

966.323   Handling regulation.

AUTHORITY: 7 U.S.C. 601–674.

SOURCE: 20 FR 7357, Oct. 4, 1955, unless otherwise noted. Redesignated at 26 FR 12751, Dec. 30, 1961.

Subpart A—Order Regulating Handling

DEFINITIONS

§ 966.1   Secretary.

Secretary means the Secretary of Agriculture of the United States, or any officer or employee of the Department to whom authority has heretofore been delegated, or to whom authority may hereafter be delegated, to act in his stead.

§ 966.2   Act.

Act means Public Act No. 10, 73d Congress, as amended and as reenacted and amended by the Agricultural Marketing Agreement Act of 1937, as amended (48 Stat. 31, as amended; 7 U.S.C. 601 *et seq.;* 68 Stat. 906, 1047).

§ 966.3   Person.

Person means an individual, partnership, corporation, association, or any other business unit.

§ 966.4   Production area and regulated area.

(a) *Production area* means the counties of Pinellas, Hillsborough, Polk, Osceola, and Brevard in the State of Florida, and all the counties of that State situated south of such counties.

(b) *Regulated area* means that portion of the State of Florida which is bounded by the Suwannee River, the Georgia border, the Atlantic Ocean, and the Gulf of Mexico.

[33 FR 8585, June 12, 1968, as amended at 34 FR 19186, Dec. 4, 1969]

§ 966.5   Tomatoes.

Tomatoes means all varieties of the edible fruit (Lycopersicon esculentum) commonly known as tomatoes and grown within the production area.

§ 966.6   Handler.

Handler is synonymous with *shipper* and means any person (except a common or contract carrier transporting tomatoes for another person) who, as owner, agent, or otherwise, handles fresh tomatoes or causes fresh tomatoes to be handled.

[33 FR 8585, June 12, 1968]

§ 966.7   Handle.

Handle or *ship* means to sell, transport, deliver, or in any other way to place fresh tomatoes, produced in the production area, in the current of commerce within the regulated area or between any point in the regulated area and any point outside thereof. Such term shall not include the transportation, sale or delivery of field-run tomatoes within the production area by the producer thereof to a registered handler for the purpose of having such tomatoes prepared for market. A registered handler is a handler who has adequate facilities in the production area for grading and packing tomatoes and who is registered with the committee pursuant to rules established with the approval of the Secretary.

[34 FR 19186, Dec. 4, 1969]

§ 966.8   Producer.

Producer means any person engaged in a proprietary capacity in the production of tomatoes for market.

§ 966.9   Grading.

Grading is synonymous with *preparation for market* and means the sorting or separation of tomatoes into grades, sizes, maturities, and packs for market purposes.

§ 966.10 Grade and size.

Grade means any one of the established grades of tomatoes and *size* means any one of the established sizes of tomatoes as defined and set forth in U.S. Standards for Fresh Tomatoes (§§ 51.1855 to 51.1877 of this title or U.S. Consumer Standards for Fresh Tomatoes (§§ 51.1900 to 51.1913 of this title), both issued by the United States Department of Agriculture, or amendments thereto, or modifications thereof, or variations based thereon recommended by the committee and approved by the Secretary.

§ 966.11 Pack.

Pack means any of the packs of tomatoes as defined and set forth in the United States Standards for Fresh Tomatoes issued by the United States Department of Agriculture (§§ 51.1855 to 51.1877 of this title), or any pack of tomatoes recommended by the committee and approved by the Secretary.

§ 966.12 Maturity.

Maturity means any of the various degrees of ripeness of tomatoes as established by the committee with approval of the Secretary as determined at the time of the inspection, pursuant to § 966.60(a).

[34 FR 19186, Dec. 4, 1969]

§ 966.13 Container.

Container means a box, bag, crate, hamper, basket, package, tube, bulk load or any other type of unit used in the packaging, transportation, sale, shipment, or handling of tomatoes.

§ 966.14 Varieties.

Varieties means and includes all classifications or subdivisions of tomatoes according to those definitive characteristics now or hereafter recognized by the United States Department of Agriculture.

§ 966.15 Committee.

Committee means the Florida Tomato Committee, established pursuant to § 966.22.

§ 966.16 Fiscal period.

Fiscal period means the period beginning August 1 and ending July 31 following.

§ 966.17 District.

District means each one of the geographic divisions of the production area initially established pursuant to § 966.24, or as reestablished pursuant to § 966.25.

§ 966.18 Export.

Export means shipment of tomatoes beyond the boundaries of the 48 contiguous States (including the District of Columbia) of the United States.

[34 FR 19186, Dec. 4, 1969]

COMMITTEE

§ 966.22 Establishment and membership.

(a) The Florida Tomato Committee, consisting of 12 producer members, is hereby established. For each member of the committee there shall be an alternate who shall have the same qualifications as the member.

(b) Each person selected as a committee member or alternate shall be an individual who is a producer, or an officer or an employee of a corporate producer, in the district for which selected and a resident of the production area.

[33 FR 8586, June 12, 1968]

§ 966.23 Term of office.

(a) The term of office of committee members, and their respective alternates, shall be for 1 year and shall begin as of August 1 and end as of July 31.

(b) Committee members and alternates shall serve during the term of office for which they are selected and have qualified, or during that portion thereof beginning on the date on which they qualify during such term of office and continuing until the end thereof, and until their successors are selected and have qualified.

§ 966.24 Districts.

For the purpose of determining the basis for selecting committee members the following districts of the production area are hereby initially established.

District No. 1. The counties of Broward and Dade in the State of Florida;

District No. 2. The counties of Brevard, Glades, Indian River, Martin, Osceola, Okeechobee, Palm Beach, and St. Lucie in the State of Florida;

District No. 3. The counties of Charlotte, Collier, Hendry, Lee, and Monroe in the State of Florida; and

District No. 4. The counties of De Soto, Hardee, Highlands, Hillsborough, Manatee, Pinellas, Polk, and Sarasota in the State of Florida.

[33 FR 8586, June 12, 1968]

§ 966.25 Redistricting.

The committee may recommend, and pursuant thereto, the Secretary may approve, the reapportionment of members among districts, and the reestablishment of districts within the production area. In recommending any such changes, the committee shall give consideration to: (a) Shifts in tomato acreage within districts and within the production area during recent years; (b) the importance of new production in its relation to existing districts; (c) the equitable relationship of committee membership and districts; (d) economies to result for producers in promoting efficient administration due to redistricting or reapportionment of members within districts; and (e) other relevant factors. No change in districting or in apportionment of members within districts may become effective within less than 30 days prior to the date on which terms of office begin each year and no recommendations for such redistricting or reapportionment may be made less than six months prior to such date.

§ 966.26 Selection.

The Secretary shall select initially 3 members of the committee with their respective alternates, from each district.

§ 966.27 Nomination.

The Secretary may select the members of the committee and alternates from nominations which may be made in the following manner:

(a) A meeting or meetings of producers shall be held in each district to nominate members and alternates for the committee. The committee shall hold such meetings or cause them to be held prior to June 15 of each year or by such other date as may be approved by the Secretary pursuant to recommendation of the committee.

(b) At each such meeting at least one nominee shall be designated for each position as member and for each position as alternate on the committee.

(c) Nominations for committee members and alternates shall be supplied to the Secretary in such manner and form as he may prescribe, not later than July 15 of each year, or by such other date as may be approved by the Secretary pursuant to recommendation of the committee.

(d) Only producers may participate in designating nominees for members and alternates on the committee. In the event a person is engaged in producing tomatoes in more than one district, such person shall elect the district within which he may participate as aforesaid in designating nominees; and

(e) Regardless of the number of districts in which a person produces tomatoes, each such person is entitled to cast only one vote on behalf of himself, his agents, subsidiaries, affiliates, and representatives in designating nominees for committee members and alternates. An eligible voter's privilege of casting only one vote as aforesaid shall be construed to permit a voter to cast one vote for each position to be filled in the respective district in which he elects to vote.

[20 FR 7357, Oct. 4, 1955. Redesignated at 26 FR 12751, Dec. 30, 1961, as amended at 33 FR 8586, June 12, 1968]

§ 966.28 Failure to nominate.

If nominations are not made within the time and in the manner specified in § 966.27, the Secretary may, without regard to nominations, select the committee members and alternates, which selection shall be on the basis of the representation provided for in §§ 966.24 through 966.26 inclusive.

§ 966.29 Acceptance.

Any person selected as a committee member or alternate shall qualify by filing a written acceptance with the Secretary within ten days after being notified of such selection.

§ 966.30 Vacancies.

To fill committee vacancies, the Secretary may select such members or alternates from unselected nominees on the current nominee list from the district involved, or from nominations made in the manner specified in § 966.27. If the names of nominees to fill any such vacancy are not made available to the Secretary within 30 days after such vacancy occurs, such vacancy may be filled without regard to nominations, which selection shall be made on the basis of the representation provided for in §§ 966.24 through 966.26 inclusive.

§ 966.31 Alternate members.

An alternate member of the committee shall act in the place and stead of the member for whom he is an alternate, during such member's absence. In the event of the death, removal, resignation, or disqualification of a member, his alternate shall act for him until a successor of such member is selected and has qualified.

§ 966.32 Procedure.

(a) Eight members of the committee shall be necessary to constitute a quorum and the same number of concurring votes shall be required to pass any motion or approve any committee action.

(b) If both a member and respective alternate are unable to attend a committee meeting, the committee may designate any other alternate present from the same district to serve in place of the absent member.

(c) The committee may provide for meeting by telephone, telegraph, or other means of communication, and any vote cast at such a meeting shall be promptly confirmed in writing: *Provided*, That if any assembled meeting is held, all votes shall be cast in person.

[20 FR 7357, Oct. 4, 1955. Redesignated at 26 FR 12751, Dec. 30, 1961, as amended at 33 FR 8586, June 12, 1968; 51 FR 30474, Aug. 27, 1986]

§ 966.33 Expenses and compensation.

Committee members and alternates may be reimbursed for expenses nec-essarily incurred by them in the performance of duties and in the exercise of powers under this part.

§ 966.34 Powers.

The committee shall have the following powers:

(a) To administer the provisions of this part in accordance with its terms;

(b) To make rules and regulations to effectuate the terms and provisions of this part;

(c) To receive, investigate, and report to the Secretary complaints of violation of the provisions of this part; and

(d) To recommend to the Secretary amendments to this part.

§ 966.35 Duties.

It shall be, among other things, the duty of the committee:

(a) At the beginning of each term of office, to meet and organize, to select a chairman and such other officers as may be necessary, to select subcommittees of committee members, and to adopt such rules and regulations for the conduct of its business as it may deem advisable;

(b) To act as intermediary between the Secretary and any producer or handler;

(c) To furnish to the Secretary such available information as he may request;

(d) To appoint such employees, agents, and representatives as it may deem necessary and to determine the salaries and define the duties of each such person;

(e) To investigate from time to time and to assemble data on the growing, harvesting, shipping, and marketing conditions with respect to tomatoes;

(f) To prepare a marketing policy;

(g) To recommend marketing regulations to the Secretary;

(h) To recommend rules and procedures for, and to make determinations in connection with, issuance of certificates of privilege or exemptions, or both;

(i) To investigate an applicant's claim for exemptions;

(j) To keep minutes, books, and records which clearly reflect all of the acts and transactions of the committee and such minutes, books and records shall be subject to examination at any time by the Secretary or his authorized agent or representative. Minutes of each committee meeting shall be reported promptly to the Secretary;

(k) At the beginning of each fiscal period, to prepare a budget of its expenses for such fiscal period, together with a report thereon;

(l) To cause the books of the committee to be audited by a competent accountant at least once each fiscal period, and at such other time as the committee may deem necessary or as the Secretary may request. The report of such audit shall show the receipt and expenditure of funds collected pursuant to this part; a copy of each such report shall be furnished to the Secretary and a copy of each such report shall be made available at the principal office of the committee for inspection by producers and handlers; and

(m) To consult, cooperate, and exchange information with other marketing agreement committees and other individuals or agencies in connection with all proper committee activities and objectives under this part.

EXPENSES AND ASSESSMENTS

§966.40   Expenses.

The committee is authorized to incur such expenses as the Secretary may find are reasonable and likely to be incurred during each fiscal period for its maintenance and functioning, and for such purposes as the Secretary, pursuant to this subpart, determines to be appropriate. Handlers shall share expenses upon the basis of a fiscal period. Each handler's share of such expense shall be proportionate to the ratio between the total quantity of tomatoes handled by him as the first handler thereof during a fiscal period and the total quantity of tomatoes handled by all handlers as first handlers thereof during such fiscal period.

§966.41   Budget.

At the beginning of each fiscal period and as may be necessary thereafter, the committee shall prepare an estimated budget of income and expenditures necessary for the administration of this part. The committee may recommend a rate of assessment calculated to provide adequate funds to defray its proposed expenditures. The committee shall present such budget to the Secretary with an accompanying report showing the basis for its calculations.

§966.42   Assessments.

(a) The funds to cover the committee's expenses shall be acquired by the levying of assessments upon handlers as provided in this subpart. Each handler who first handles tomatoes shall pay assessments to the committee upon demand, which assessments shall be in payment of such handler's pro rata share of the committee's expenses.

(b) Assessments shall be levied upon handlers at rates established by the Secretary. Such rates may be established upon the basis of the committee's recommendations and other available information. Such rates may be applied to specified containers used in the production area.

(c) At any time during, or subsequent to, a given fiscal period the committee may recommend the approval of an amended budget and an increase in the rate of assessment. Upon the basis of such recommendations, or other available information, the Secretary may approve an amended budget and increase the rate of assessment. Such increase shall be applicable to all tomatoes which were regulated under this part and which were shipped by the first handler thereof during such fiscal period.

(d) The payment of assessments for the maintenance and functioning of the committee may be required under this part throughout the period it is in effect irrespective whether particular provisions thereof are suspended or become inoperative.

(e) In order to provide funds for the administration of the provisions of this part, the committee may accept the payment of assessments in advance, or may borrow money on a short-term basis not to exceed one full-year coinciding with the existing committee's term of office. The authority of the committee to borrow money may be used only to meet financial obligations as they occur and to allow the committee a season to adjust its reserve funds to meet any additional obligations.

[20 FR 7357, Oct. 4, 1955. Redesignated at 26 FR 12751, Dec. 30, 1961, as amended at 51 FR 30474, Aug. 27, 1986]

§ 966.43 Accounting.

(a) All funds received by the committee pursuant to the provisions of this subpart shall be used solely for the purposes specified in this part.

(b) The Secretary may at any time require the committee, its members and alternates, employees, agents and all other persons to account for all receipts and disbursements, funds, property, or records for which they are responsible. Whenever any person ceases to be a member of the committee or alternate, he shall account to his successor, the committee, or to the person designated by the Secretary, for all receipts, disbursements, funds and property (including but not being limited to books and other records) pertaining to the committee's activities for which he is responsible, and shall execute such assignments and other instruments as may be necessary or appropriate to vest in such successor, committee, or designated person, the right to all of such property and funds and all claims vested in such person.

(c) The committee may make recommendations to the Secretary for one or more of the members thereof, or any other person, to act as a trustee for holding records, funds, or any other committee property during periods of suspension of this subpart, or during any period or periods when regulations are not in effect and, if the Secretary determines such action appropriate, he may direct that such person or persons shall act as trustee or trustees for the committee.

§ 966.44 Excess funds.

(a) If, at the end of a fiscal period, the assessments collected are in excess of expenses incurred, such excess shall be accounted for in accordance with one of the following:

(1) If such excess is not retained in a reserve, as provided in paragraph (a)(2) of this section, to the extent practical it shall be refunded proportionately to the persons from whom it was collected.

(2) The committee, with the approval of the Secretary, may establish an operating monetary reserve and may carry over to subsequent fiscal periods excess funds in a reserve so established: *Provided,* That funds in the reserve shall not exceed approximately one fiscal period's expenses. Such reserve funds may be used (i) to defray any expenses authorized under this part, (ii) to defray expenses during any fiscal period prior to the time assessment income is sufficient to cover such expenses, (iii) to cover deficits incurred during any fiscal period when assessment income is less than expenses, (iv) to defray expenses incurred during any period when any or all provisions of this part are suspended or are inoperative, and (v) to cover necessary expenses of liquidation in the event of termination of this part. Upon such termination any funds not required to defray the necessary expenses of liquidation, and after reasonable effort by the committee it is found impracticable to return such remaining funds to handlers, such funds shall be disposed of in such manner as the Secretary may determine to be appropriate.

(b) [Reserved]

[33 FR 8586, June 12, 1968]

§ 966.45 Contributions.

The committee may accept voluntary contributions but these shall only be used for production research, market research and development and marketing and promotion including paid advertising pursuant to § 966.48. Furthermore, such contributions shall be free from any encumbrances by the donor and the committee shall retain complete control of their use. The committee is prohibited from accepting contributions from handlers subject to the order, or any person whose contributions would constitute a conflict of interest.

[51 FR 30474, Aug. 27, 1986]

RESEARCH AND DEVELOPMENT

§ 966.48 Research and promotion.

The committee may, with the approval of the Secretary, establish, or provide for the establishment of projects including production research, marketing research and development projects, and marketing promotion including paid advertising, designed to assist, improve or promote the marketing, distribution and consumption or efficient production of tomatoes. The expenses of such projects shall be paid by funds collected pursuant to §§966.42 and 966.45. Upon conclusion of each project, but at least annually, the committee shall summarize the program status and accomplishments, to its members and the Secretary. A similar report to the committee shall be required of any contracting party on any project carried out under this section. Also, for each project the contracting party shall be required to maintain records of money received and expenditures and such shall be available to the committee and the Secretary.

[51 FR 30474, Aug. 27, 1986]

REGULATION

§966.50 Marketing policy.

Prior to or at the same time as initial recommendations are made pursuant to §966.51, the committee shall submit to the Secretary a report setting forth the marketing policy it deems desirable for the industry to follow in shipping tomatoes from the production area during the ensuring season. Additional reports shall be submitted from time to time if it is deemed advisable by the committee to adopt a new or modified marketing policy because of changes in the demand and supply situation with respect to tomatoes. The committee shall publicly announce the submission of each such marketing policy report and copies thereof shall be available at the committee's office for inspection by any producer or any handler. In determining each such marketing policy the committee shall give due consideration to the following:

(a) Market prices of tomatoes, including prices by grades, sizes, and quality in different packs, and such prices by foreign competing areas;

(b) Supply of tomatoes, by grade, size, and quality in the production area, and in other production areas, including foreign competing production areas;

(c) Trend and level of consumer income;

(d) Marketing conditions affecting tomato prices; and

(e) Other relevant factors.

§966.51 Recommendations for regulations.

The committee, upon complying with the requirements of §966.50, may recommend regulations to the Secretary whenever it finds that such regulations, as are provided for in this subpart, will tend to effectuate the declared policies of the act.

§966.52 Issuance of regulations.

The Secretary shall limit the handling of tomatoes whenever he finds from the recommendation and information submitted by the Committee, or from other available information, that such regulation would tend to effectuate the declared policy of the act. Such regulation may:

(a) Limit, in any or all portions of the production area, the handling of particular grades, sizes, qualities (including maturity as a factor of grade or quality), or packs of any or all varieties of tomatoes, during any period; or

(b) Limit the handling of particular grades, sizes, qualities, or packs of tomatoes differently, from different varieties, for different stages of maturity, for different portions of the production area, for different containers, for different markets, for different purposes specified in §966.54, or any combination of the foregoing, during any period; or

(c) Limit the handling of tomatoes by establishing, in terms of grades, sizes, or both, minimum standards of quality and maturity; or

(d) Fix the size, weight, capacity, dimensions, markings (including labels and stamps), or pack of the container or containers which may be used in the packaging, transportation, sale, shipment, or other handling of tomatoes.

[20 FR 7357, Oct. 4, 1955. Redesignated at 26 FR 12751, Dec. 30, 1961, as amended at 33 FR 8586, June 12, 1968; 34 FR 19186, Dec. 4, 1969]

§966.53 Minimum quantities.

The committee, with the approval of the Secretary, may establish, for any or all portions of the production area, minimum quantities below which handling will be free from regulations issued or effective pursuant to §§ 966.42, 966.52, 966.54, 966.60, or any combination thereof.

§ 966.54 Shipments for special purposes.

Upon the basis of recommendations and information submitted by the committee, or other available information, the Secretary, whenever he finds that it will tend to effectuate the declared policy of the act, shall modify, suspend, or terminate regulations issued pursuant to §§ 966.42, 966.52, 966.53, 966.60, or any combination thereof, in order to facilitate handling of tomatoes for the following purposes:

(a) For export;

(b) For relief or for charity;

(c) For processing; or

(d) For other purposes which may be specified by the committee, with the approval of the Secretary.

§ 966.55 Notification of regulation.

The Secretary shall notify the committee of any regulations issued or of any modification, suspension, or termination thereof. The committee shall give reasonable notice thereof to handlers.

§ 966.56 Safeguards.

(a) The committee, with the approval of the Secretary, may prescribe adequate safeguards to prevent handling of tomatoes pursuant to § 966.53 or § 966.54 from entering channels of trade for other than the specific purpose authorized therefor, and rules governing the issuance and the contents of Certificates of Privilege if such certificates are prescribed as safeguards by the committee. Such safeguards may include requirements that:

(1) Handlers shall file applications with the committee to ship tomatoes pursuant to §§ 966.53 and 966.54; or

(2) Handlers shall obtain inspection provided by § 966.60, or pay the assessment levied pursuant to § 966.42, or both, in connection with shipments made under § 966.54; or

(3) Handlers shall obtain Certificates of Privilege from the committee to handle tomatoes effected or to be ef-

fected under the provisions of §§ 966.53 and 966.54.

(b) The committee may rescind or deny Certificates of Privilege to any handler if proof is obtained that tomatoes handled by him for the purposes stated in §§ 966.53 and 966.54 were handled contrary to the provisions of this part.

(c) The Secretary shall have the right to modify, change, alter, or rescind any safeguards prescribed and any certificates issued by the committee pursuant to the provisions of this section.

(d) The committee shall make reports to the Secretary, as requested, showing the number of applications for such certificates, the quantity of tomatoes covered by such applications, the number of such applications denied and certificates granted, the quantity of tomatoes handled under duly issued certificates, and such other information as may be requested.

INSPECTION

§ 966.60 Inspection and certification.

(a) During any period in which the handling of tomatoes is regulated pursuant to this subpart no handler shall handle tomatoes unless such tomatoes have been inspected and certified as meeting the requirements of this subpart by an authorized representative of the Federal or Federal-State Inspection Service, or such other inspection service as the Secretary shall designate, and such tomatoes are covered by a valid inspection certificate except when relieved from such requirements pursuant to § 966.53 or § 966.54 or both.

(b) Insofar as the requirements of this section are concerned, the length of time for which an inspection certificate is valid may be established by the Secretary upon the recommendation of the committee.

(c) When tomatoes are inspected in accordance with the requirements of this section a copy of each inspection certificate issued shall be made available to the committee by the inspection service.

[34 FR 19186, Dec. 4, 1969]

§966.70 Procedure.

The committee may adopt, with approval of the Secretary, the procedures pursuant to which certificates of exemption will be issued to producers or handlers.

§966.71 Granting exemptions.

The committee shall issue certificates of exemption to any producer who applies for such exemption and furnishes adequate evidence to the committee, that by reason of a regulation issued pursuant to §966.52 he will be prevented from handling as large a proportion of his production as the average proportion of production handled during the entire season, or such portion thereof as may be determined by the committee, by all producers in said applicant's immediate production area and that the grade, size, or quality of the applicant's tomatoes have been adversely affected by acts beyond the applicant's control and by acts beyond reasonable expectation. Each certificate shall permit the producer to handle the amount of tomatoes specified thereon. Such certificate shall be transferred with such tomatoes at time of transportation or sale.

§966.72 Investigation.

The committee shall be permitted at any time to make a thorough investigation of any producer's or handler's claim pertaining to exemptions.

§966.73 Appeal.

If any applicant for exemption certificates is dissatisfied with the determination by the committee with respect to his application, said applicant may file an appeal with the committee. Such an appeal must be taken promptly after the determination by the committee from which the appeal is taken. Any applicant filing an appeal shall furnish evidence satisfactory to the committee for a determination on the appeal. The committee shall thereupon reconsider the application, examine all available evidence, and make a final determination concerning the application. The committee shall notify the appellant of the final determination, and shall furnish the Secretary with a copy of the appeal and a statement of considerations involved in making the final determination.

§966.74 Records.

(a) The committee shall maintain a record of all applications submitted for exemption certificates, a record of all exemption certificates issued and denied, the quantity of tomatoes covered by such exemption certificates, a record of the amount of tomatoes handled under exemption certificates, a record of appeals for reconsideration of applications, and such information as may be requested by the Secretary. Periodic reports on such records shall be compiled and issued by the committee upon request of the Secretary.

(b) The Secretary shall have the right, to modify, change, alter, or rescind any procedure and any exemptions granted pursuant to §§966.70, 966.71, 966.72, 966.73, or any combination thereof.

§966.80 Reports.

Upon request of the committee, made with approval of the Secretary, each handler shall furnish to the committee, in such manner and at such time as it may prescribe, such reports and other information as may be necessary for the committee to perform its duties under this part.

(a) Such reports may include, but are not necessarily limited to, the following: (1) The quantities of tomatoes received by a handler; (2) the quantities disposed of by him, segregated as to the respective quantities subject to regulation and not subject to regulation; (3) the date of each such disposition and the identification of the carrier transporting such tomatoes; and (4) identification of the inspection certificates and the exemption certificates, if any, pursuant to which the tomatoes were handled, together with the destination of each exempted disposition, and of all tomatoes handled pursuant to §§966.53 and 966.54.

(b) All such reports shall be held under appropriate protective classification and custody by the committee, or duly appointed employees thereof, so that the information contained therein which may adversely affect the competitive position of any handler in relation to other handlers will not be disclosed. Compilations of general reports from data submitted by handlers is authorized, subject to prohibition of disclosure of individual handlers identities or operations.

(c) Each handler shall maintain for at least two succeeding years such records of the tomatoes received and disposed of by such handler as may be necessary to verify the reports he submits to the committee pursuant to this section.

MISCELLANEOUS PROVISIONS

§ 966.81 Compliance.

Except as provided in this subpart, no handler shall handle tomatoes, the handling of which has been prohibited by the Secretary in accordance with provisions of this subpart, and no handler shall handle tomatoes except in conformity to the provisions of this subpart.

§ 966.82 Right of the Secretary.

The members of the committee (including successors and alternates), and any agent or employee appointed or employed by the committee, shall be subject to removal or suspension by the Secretary at any time. Each and every order, regulation, decision, determination or other act of the committee shall be subject to the continuing right of the Secretary to disapprove of the same at any time. Upon such disapproval, the disapproved action of the said committee shall be deemed null and void, except as to acts done in reliance thereon or in compliance therewith prior to such disapproval by the Secretary.

§ 966.83 Effective time.

The provisions of this subpart, or any amendment thereto, shall become effective at such time as the Secretary may declare and shall continue in force until terminated in one of the ways specified in this subpart.

§ 966.84 Termination.

(a) The Secretary may, at any time, terminate the provisions of this subpart by giving at least one day's notice by means of a press release or in any other manner which he may determine.

(b) The Secretary may terminate or suspend the operations of any or all of the provisions of this subpart whenever he finds that such provisions do not tend to effectuate the declared policy of the act.

(c) The Secretary shall terminate the provisions of this subpart at the end of any fiscal period whenever he finds that such termination is favored by a majority of producers, who during a representative period, have been engaged in the production for market of tomatoes: *Provided*, That such majority has, during such representative period, produced for market more than fifty percent of the volume of such tomatoes produced for market.

(d) The provisions of this subpart shall, in any event, terminate whenever the provisions of the act authorizing them cease to be in effect.

§ 966.85 Proceedings after termination.

(a) Upon the termination of the provisions of this subpart the then functioning members of the committee shall continue as joint trustees for the purpose of liquidating the affairs of the committee of all the funds and property then in the possession of or under control of the committee, including claims for any funds unpaid or property not delivered at the time of such termination. Action by said trusteeship shall require the concurrence of a majority of the said trustees.

(b) The said trustees shall continue in such capacity until discharged by the Secretary; shall, from time to time, account for all receipts and disbursements and deliver all property on hand, together with all books and records of the committee and of the trustees, to such person as the Secretary may direct; and shall, upon request of the Secretary, execute such assignments or other instruments necessary or appropriate to vest in such person full title and right to all of the funds, property and claims vested in the committee or the trustees pursuant to this subpart.

(c) Any person to whom funds, property, or claims have been transferred or delivered by the committee or its members pursuant to this section, shall be subject to the same obligations imposed upon the members of the committee and upon the said trustees.

§966.86 Effect of termination or amendment.

Unless otherwise expressly provided by the Secretary, the termination of this subpart or of any regulation issued pursuant to this subpart, or the issuance of any amendments to either thereof, shall not (a) affect or waive any right, duty, obligation, or liability which shall have arisen or which may thereafter arise in connection with any provision of this subpart or any regulation issued under this subpart, or (b) release or extinguish any violation of this subpart or of any regulations issued under this subpart, or (c) affect or impair any rights or remedies of the Secretary or of any other person with respect to any such violations.

§966.87 Duration of immunities.

The benefits, privileges, and immunities conferred upon any person by virtue of this subpart shall cease upon the termination of this subpart, except with respect to acts done under and during the existence of this subpart.

§966.88 Agents.

The Secretary may, by designation in writing, name any person, including any officer or employee of the United States, or name any agency in the United States Department of Agriculture, to act as his agent or representative in connection with any of the provisions of this subpart.

§966.89 Derogation.

Nothing contained in this subpart is, or shall be construed to be, in derogation or in modification of the rights of the Secretary or of the United States to exercise any powers granted by the act or otherwise, or, in accordance with such powers, to act in the premises whenever such action is deemed advisable.

§966.90 Personal liability.

No member or alternate of the committee nor any employee or agent thereof, shall be held personally responsible, either individually or jointly with others, in any way whatsoever, to any handler or to any person for errors in judgment, mistakes, or other acts, either of commission or omission, as such member, alternate, agent, or employee except for acts of dishonesty, willful misconduct, or gross negligence.

§966.91 Separability.

If any provision of this subpart is declared invalid, or the applicability thereof to any person, circumstance, or thing is held invalid, the validity of the remainder of this subpart, or the applicability thereof to any other person, circumstance, or thing, shall not be affected thereby.

§966.92 Amendments.

Amendments to this subpart may be proposed, from time to time, by the committee or by the Secretary.

Subpart B—Administrative Requirements

GENERAL

§966.100 Communications.

Unless otherwise provided in the marketing agreement and order, or by specific direction of the committee, all reports, applications, submittals, requests, and communications in connection with the marketing agreement and order shall be addressed to the Florida Tomato Committee at its principal office.

DEFINITIONS

§966.110 Order.

Order means Order No. 966 (§§966.1 through 966.92) regulating the handling of tomatoes grown in Florida, also referenced in this part as *marketing order and agreement.*

[84 FR 59292, Dec. 5, 2019]

§966.111 Marketing Agreement.

The Marketing Agreement associated with Order No. 966 is Marketing Agreement No. 125.

[84 FR 59292, Dec. 5, 2019]

§ 966.112 Terms.

Terms used in this subpart shall have the same meaning as when used in the marketing agreement and order.

§ 966.113 Registered handler certification.

Each handler who handles tomatoes grown in the production area must be certified as a registered handler by the committee in order to ship such tomatoes outside of the regulated area. A handler who is certified as a registered handler is a handler who has adequate facilities to meet the requirements for preparing tomatoes for market, obtains inspection on tomatoes handled, agrees to handle tomatoes in compliance with the order's grade, size and container requirements, pays applicable assessments on a timely basis, submits reports required by the committee, and agrees to comply with other regulatory requirements on the handling of tomatoes grown in the production area.

(a) Based on the criteria specified in this section, the committee shall determine eligibility for certification as a registered handler. The committee or its authorized agent shall inspect a handler's facilities to determine if the facilities are adequate for preparing tomatoes for market. In order to be adequate for such purposes, the facilities must be permanent, nonportable buildings located in the production area with equipment that is nonportable for the proper washing, grading, sizing and packing of tomatoes grown in the production area.

(b) Application for certification shall be executed by the handler and filed with the committee on a form, prescribed by and available at the principal office of the committee, containing the following information:

(1) Business name,

(2) Address of handling facilities (including telephone and facsimile number),

(3) Mailing address (if different from handling facility),

(4) Number of years in tomato business in Florida,

(5) Type of business, and

(6) Names of senior officers, partners, or principal owners with financial interest in the business.

(c) If the committee determines from available information that an applicant meets the criteria specified in this section, such applicant shall be certified as a registered handler and shall be so informed by written notice from the committee. If certification is denied, such denial shall be made by the committee in writing, stating the reasons for denial.

(d) A registered handler's certification shall be cancelled by the committee, with the approval of the Secretary, if the handler fails to pay assessments within 45 days of the end of the assessment billing period, fails to provide reports, or no longer has adequate facilities as described in this section. Cancellation of a handler's registration shall be made in writing to the handler and shall specify the reason(s) for and effective date of such cancellation. The committee shall recertify the handler's registration at such time as the handler corrects the deficiencies which resulted in the cancellation. Certification is permanent until the committee determines, based on criteria herein, that cancellation is warranted. Persons who make deliveries of ungraded tomatoes to such certified registered handlers are hereby determined to be exempt from otherwise applicable regulations pursuant to this part.

(e) During any period in which the handling of tomatoes is regulated pursuant to this part, no handler shall obtain an inspection certifying that said handler's tomatoes meet the requirements of the marketing order unless said handler has been certified as a registered handler. Any person who is not certified as a registered handler may receive inspection on tomatoes from the Federal-State Inspection Service. Such inspection certificate shall state "Fails to meet the requirements of Marketing Order No. 966 because the handler is not a registered handler."

[59 FR 51090, Oct. 7, 1994]

SAFEGUARDS

§ 966.120 Application for Certificate of Privilege.

(a) Whenever handling is regulated pursuant to § 966.54, each handler desiring to make shipments of tomatoes for any of the following purposes shall, prior thereto, apply to the committee for and obtain a Certificate of Privilege permitting such shipment:

(1) For pickling, or

(2) For processing, or

(3) For experimental purposes, or

(4) For relief or charity, or

(5) For export, or

(6) For other purposes which may be specified by the committee, with the approval of the Secretary.

(b) Applications for Certificates of Privilege shall be made on forms furnished by the committee. Each application shall contain the name and address of the handler, and such other information as such committee may require, such as, but not limited, to the quantity (by grade, size, quality, and variety) of tomatoes to be shipped, the mode of transportation, consignee, destination, and other appropriate information or documents necessary to safeguard against the entry of such tomatoes into trade channels other than those for which the Certificate of Privilege is granted.

[21 FR 353, Jan. 19, 1956. Redesignated at 26 FR 12751, Dec. 30, 1961, as amended at 59 FR 51091, Oct. 7, 1994]

§966.121 Issuance.

The committee, or its duly authorized agents, shall give prompt consideration to each application for a Certificate of Privilege and shall determine whether the application is approved. Approval of an application shall be evidenced by the issuance of a Certificate of Privilege authorizing the applicant named therein to ship tomatoes for a specified purpose for a specified period of time.

§966.122 Reports.

Each handler handling tomatoes under and pursuant to a Certificate of Privilege shall supply the committee with a report thereon within the time specified on the application for such certificate showing the name and address of the shipper, car or truck identification, loading point, destination, consignee, and, when inspection is required, the Federal-State Inspection Certificate number.

§966.123 Denial and appeal.

The committee may rescind a Certificate of Privilege issued to a handler, or deny a Certificate of Privilege to a handler, upon proof satisfactory to such committee, that such handler has shipped tomatoes contrary to the provisions of this part. Such committee action denying a Certificate of Privilege shall apply to and not exceed a reasonable period of time as determined by such committee. Any handler who has been denied a Certificate of Privilege, or who has had a Certificate of Privilege rescinded, may appeal to the committee for reconsideration. Such appeal shall be in writing.

§966.124 Approved receiver.

(a) *Approved receiver.* Any person who desires to acquire, as an approved receiver, tomatoes for purposes as set forth in §966.120(a), shall annually, prior thereto, file an application with the committee on a form approved by it, which shall contain, but not be limited to, the following information:

(1) Name, address, contact person, telephone number, and e-mail address of applicant;

(2) Purpose of shipment;

(3) Physical address of where manufacturing or other specified purpose is to occur;

(4) Whether or not the receiver packs, repacks or sells fresh tomatoes;

(5) A statement that the tomatoes obtained exempt from the fresh tomato regulations will not be resold or transferred for resale, directly or indirectly, but will be used only for the purpose specified in the corresponding certificate of privilege;

(6) A statement agreeing to undergo random inspection by the committee;

(7) A statement agreeing to submit such reports as is required by the committee.

(b) The committee, or its duly authorized agents, shall give prompt consideration to each application for an approved receiver and shall determine whether the application is approved or disapproved and notify the applicant accordingly.

(c) The committee, or its duly authorized agents, may rescind a person's approved receiver status upon proof satisfactory that such a receiver has

411

handled tomatoes contrary to the provisions established under the Certificate of Privilege. Such action rescinding approved receiver status shall apply to and not exceed a reasonable period of time as determined by the committee or its duly authorized agents. Any person who has been denied as an approved receiver or who has had their approved receiver status rescinded, may appeal to the committee for reconsideration. Such an appeal shall be made in writing.

[70 FR 53540, Sept. 9, 2005]

EXEMPTION PROCEDURES

§ 966.130   Application.

Any person applying for exemption from regulations issued pursuant to § 966.52 shall file such application with the committee, or its duly authorized agent for such purpose, on forms to be furnished by such committee. Each application shall state the name and address of the applicant, the grade, size, and quality regulations from which exemption is requested; and facts demonstrating that the tomatoes, for which exemption is requested, were adversely affected by acts beyond his control or by acts beyond the applicant's reasonable expectation. Applications shall set forth such additional information as the committee may find necessary in making determinations with respect thereto, including, without limitation thereto, the information required on producers' applications by paragraphs (a) and (b) of this section.

(a) The location and acreage of the farm on which tomatoes for which exemption is requested, the location where such tomatoes are to be prepared for market, and the loading point from which such tomatoes are to be shipped if exemption is granted;

(b) Quantity (by grade, size, quality, and variety) of tomatoes harvested during the current season or any specific portion thereof prior to the date of application and to be harvested, subsequent to such date, during the remainder of the current season or any specific portion thereof (as may be determined pursuant to this part); an estimate of the portion of such tomatoes which can be handled under regulation issued pursuant to § 966.52, during the remainder of the season; and the reasons why all of such tomatoes cannot be handled under such regulations.

[22 FR 9132, Nov. 16, 1957. Redesignated at 26 FR 12751, Dec. 30, 1961]

§ 966.131   Investigations.

The committee may authorize investigations of applications by its employees, and such other persons as may be necessary to procure adequate information to pass upon the merits of such applications.

[22 FR 9132, Nov. 16, 1957. Redesignated at 26 FR 12751, Dec. 30, 1961]

§ 966.132   Issuance.

(a) The committee, or its duly authorized agents, shall give prompt consideration to all statements and facts relating to each application for exemption, and, pursuant to applicable provisions of this part, a determination shall be made as to whether or not the application is approved. The determination, if approving the application, shall be evidenced by the issuance of a certificate of exemption pursuant to § 966.71: *Provided*, That a separate certificate may be issued, at the request of an applicant, for each affected field.

(b) The applicant shall be notified in writing if his request for exemption is denied.

(c) Each exemption certificate issued pursuant to this subpart shall be on a form duly approved by the committee and signed by an authorized representative of such committee. At least one copy of each exemption certificate issued shall be retained in the committee records. Each such certificate shall contain the name and address of the recipient, the location of all tomatoes authorized to be shipped thereunder, the quantity (by grade, size, quality and variety) of tomatoes which will be permitted in the exempted shipments and such other information as may be deemed necessary by the committee to provide such committee, the recipient, or both, with adequate and specific information regarding such exempted tomatoes.

[22 FR 9132, Nov. 16, 1957. Redesignated at 26 FR 12751, Dec. 30, 1961]

§966.133 Disposition of certificates.

(a) Each lot of tomatoes handled under an exemption certificate shall be accompanied by such certificate, or such appropriate identifying information with respect to such certificate, as the committee may require, to facilitate the administration of regulatory provisions applicable thereto.

(b) Each shipment of a lot or portion thereof, of tomatoes covered by an exemption certificate shall be accompanied by a Federal-State Inspection Certificate which shall show the exemption certificate number covering the lot.

[22 FR 9132, Nov. 16, 1957. Redesignated at 26 FR 12751, Dec. 30, 1961]

§966.134 Reports.

Persons handling tomatoes under exemption certificates shall, at such times as may be specified in such certificates, report thereon to the committee the names and addresses of the receivers of such tomatoes, the quantity shipped (by grade, size, quality, and variety), the inspection certificates issued with respect thereto, the dates of such shipments, and such other information as may be requested by such committee in order to administer the regulatory provisions applicable thereto.

[22 FR 9132, Nov. 16, 1957. Redesignated at 26 FR 12751, Dec. 30, 1961]

§966.135 Appeals.

If any applicant is dissatisfied with the determination of the committee regarding an application for an exemption certificate, or any duly issued exemption certificate an appeal by such applicant may be taken to such committee in accordance with §966.73.

[22 FR 9132, Nov. 16, 1957. Redesignated at 26 FR 12751, Dec. 30, 1961]

INSPECTION

§966.140 Truck shipments.

In case of the transportation by truck outside of the production area of any tomatoes which are required to be inspected and certified as complying with any applicable requirements under this part, such tomatoes shall be accompanied by, and made available for examination at any time upon request, a copy of the appropriate inspection certificate or a copy of the appropriate transfer clearance receipt issued by the Federal-State Inspection Service, the official inspection agency for this program, showing that such tomatoes have been so inspected and certified.

[21 FR 3000, May 5, 1956. Redesignated at 26 FR 12751, Dec. 30, 1961, as amended at 65 FR 8253, Feb. 18, 2000]

INTERPRETATIVE RULES

§966.150 Meaning of "producer".

The term "producer" is defined in §966.8 as being any person engaged in a proprietary capacity in the production of tomatoes for market. Under the definition of "tomatoes" in §966.5, such production must have been in the production area. Section 966.22 provides that each person selected as a committee member or alternate must be a producer, or an officer or an employee of a corporate producer. Section 966.27 provides that producers may vote for nominees for members and alternates on the Florida Tomato Committee, the administrative agency established pursuant to said marketing agreement and order. Section 966.3 defines a person as an individual, partnership, corporation, association, or other business unit. The term "person" is construed to mean the business unit which produces the tomatoes for market.

(a) The prevailing principle which shall apply to the determination of "producer" is who or which interest as a unit, whether an individual, partnership, corporation, association, or any other business unit, has the authority to pass title to the tomatoes grown and made a part of the marketable supply of tomatoes. In other words, the terms shall be limited to those who have an ownership in tomatoes produced in the production area.

(b) *Producer* means any person, as defined in this section:

(1) Who or which owns and farms land resulting in his or its ownership of the tomatoes produced thereon;

(2) Who or which rents or farms land, resulting in his or its ownership of all or a portion of the tomatoes produced thereon; or

(3) Who or which owns land which he or it does not farm and, as rental for such land, obtains the ownership of a portion of the tomatoes produced thereon.

(c) The term "partnership" shall be deemed to include a husband and wife with respect to land, the title to which, or leasehold interest in which, is vested in them as tenants in common, joint tenants, tenants by entirety, or, under community property laws, as community property. The term "partnership" shall also be deemed to include individuals, partnerships or corporations which join together by agreement, informal or otherwise, for the purpose of growing tomatoes and which, as a unit, have authority to transfer title to such tomatoes at the time they are harvested or subsequent thereto. The term "partnership" shall also include so-called "joint ventures," wherein one or more parties to the arrangement contributes capital and others contribute labor, management, equipment, or other services, or any variation of such contributions by two or more parties, so that it results in the growing of tomatoes and the authority to transfer title to the tomatoes so produced from that business unit to some other parties in the marketing chain.

(d) Each legal entity, whether an individual, a partnership, a "joint venture," or a corporation, so engaged in the production of tomatoes for market shall have one vote for each position which is to be filled for the district for which he or it is eligible to vote. In the case of a partnership or a "joint venture," such vote shall not be accepted in the absence of unanimous agreement of the respective members. In the case of a corporation, such vote shall be cast pursuant to the authorization of its board of directors. In the case of a person who owns land which he or it does not farm but, as rental for such land, obtains the ownership of a portion of the tomatoes produced thereon, such person shall be regarded as the producer of that portion and entitled to one vote, and the tenant on such land shall be regarded as the producer of the remaining portion produced on such land and also entitled to one vote.

(e) A producer eligible to vote is a person who produced tomatoes for market in a proprietary capacity in the production area during the then current fiscal period, i.e., between August 1, of the previous year and July 31 of the then current year. If a person who would otherwise qualify as a producer in a proprietary capacity in the production area planted tomatoes for market as fresh tomatoes during the current fiscal period, but (1) did not market any tomatoes in the fresh market during the current fiscal period due to adverse weather conditions, or (2) has tomatoes in production for fresh market during the current fiscal period, although still unharvested, he shall, nevertheless, be eligible as a producer to vote for committee nominees, if he produced and marketed tomatoes grown in the production area in the next preceding fiscal period.

[23 FR 2588, Apr. 19, 1958. Redesignated at 26 FR 12751, Dec. 30, 1961]

REESTABLISHMENT OF DISTRICTS

§ 966.160   Reestablishment of districts.

(a) District No. 1: The counties of Charlotte, Glades, Palm Beach, Lee, Hendry, Collier, Broward, Monroe, and Dade in the State of Florida.

(b) District No. 2: The counties of Pinellas, Hillsborough, Polk, Osceola, Brevard, Manatee, Hardee, Highlands, Okeechobee, Indian River, St. Lucie, Sarasota, De Soto, and Martin in the State of Florida.

(c) Terms used in this section have the same meaning as when used in said marketing agreement and this part.

[35 FR 19633, Dec. 25, 1970, as amended at 84 FR 50713, Sept. 26, 2019]

§ 966.161   Reapportionment of committee membership.

Pursuant to § 966.25, industry membership on the Florida Tomato Committee shall be reapportioned as follows:

(a) District 1—six members and their alternates.

(b) District 2—six members and their alternates.

[84 FR 50713, Sept. 26, 2019]

Subpart C—Assessment Rates

§966.234 Assessment rate.

On and after August 1, 2017, an assessment rate of $0.025 per 25-pound container is established for Florida tomatoes.

[83 FR 14359, Apr. 4, 2018]

Subpart D—Handling Requirements

§966.323 Handling regulation.

From October 10 through June 15 of each season, except as provided in paragraphs (b) and (d) of this section, no person shall handle any lot of tomatoes produced in the production area for shipment outside the regulated area unless it meets the requirements of paragraph (a) of this section.

(a) *Grade, size, container, and inspection requirements*—(1) *Grade.* Tomatoes shall be graded and meet the requirements specified for U.S. No. 1, U.S. Combination, or U.S. No. 2 of the U.S. Standards for Grades of Fresh Tomatoes. When not more than 15 percent of the tomatoes in any lot fail to meet the requirements of U.S. No. 1 grade and not more than one-third of this 15 percent (or 5 percent) are comprised of defects causing very serious damage including not more than 1 percent of tomatoes which are soft or affected by decay, such tomatoes may be shipped and designated as at least 85 percent U.S. No. 1 grade.

(2) *Size.* (i) All tomatoes packed by a registered handler shall be at least 2⁹/₃₂ inches in diameter and shall be sized with proper equipment in one or more of the following ranges of diameters. Tomatoes shipped outside the regulated area shall also be sized with proper equipment in one or more of the following ranges of diameters. Measurements of diameters shall be in accordance with the methods prescribed in §51.1859 of the U.S. Standards for Grades of Fresh Tomatoes.

Size designation	Inches minimum diameter	Inches maximum diameter
6 × 7	2⁹/₃₂	2¹⁹/₃₂
6 × 6	2¹⁷/₃₂	2²⁹/₃₂
5 × 6	2²⁵/₃₂	

(ii) Tomatoes of designated sizes may not be commingled, and each container or lid shall be marked to indicate the designated size.

(iii) Only 6 × 7, 6 × 6, or 5 × 6, may be used to indicate the above listed size designations or containers of tomatoes.

(iv) To allow for variations incident to proper sizing, not more than a total of ten (10) percent, by count, of the tomatoes in any lot may be smaller than the specified minimum diameter or larger than the maximum diameter.

(3) *Containers.* (i) All tomatoes packed by a registered handler shall be packed in containers of 10, 20, and 25 pounds designated net weights. The net weight of the contents shall not be less than the designated net weight and shall not exceed the designated net weight by more than two pounds.

(ii) Each container or lid shall be marked to indicate the designated net weight and must show the name and address of the registered handler (as defined in 966.7) in letters at least one-fourth (¼) inch high, and such containers must be packed at the registered handler's facilities. The use of inverted, previously printed container lids is limited to the registered handler identified by the labels or marks that originally appeared on the lid.

(iii) The container in which the tomatoes are packed must be clean and bright in appearance without marks, stains, or other evidence of previous use.

(4) *Inspection.* Tomatoes shall be inspected and certified pursuant to the provisions of §966.60. Each handler who applies for inspection shall register with the committee pursuant to §966.113. Persons not certified by the committee as a registered handler shall be issued inspection certificates on shipments handled by such persons stating "Fails to meet the requirements of Marketing Order No. 966 because the handler is not a registered handler." Evidence of inspection must accompany truck shipments.

(b) *Special purpose shipments.* The requirements of paragraph (a) of this section shall not be applicable to shipments of tomatoes for pickling, processing, experimental purposes, relief, charity, export, or other outlets recommended by the committee and approved by the Secretary, if the handler thereof complies with the safeguard requirements of paragraph (c) of this section. Shipments for processing are also exempt from the assessment requirements of this part.

(c) *Safeguards.* Each handler making shipments of tomatoes for pickling, canning, experimental purposes, relief, charity, or export in accordance with paragraph (b) of this section shall:

(1) Apply to the committee and obtain a Certificate of Privilege to make such shipments.

(2) Prepare on forms furnished by the committee a report in quadruplicate on such shipments authorized in paragraph (b) of this section.

(3) Bill or consign each shipment directly to the designated applicable receiver.

(4) Forward one copy of such report to the committee office and two copies to the receiver for signing and returning one copy to the committee office. Failure of the handler or receiver to report such shipments by signing and returning the applicable report to the committee office within ten days after shipment may be cause for cancellation of such handler's certificate and/or receiver's eligibility to receive further shipments pursuant to such certificate. Upon cancellation of any such certificate, the handler may appeal to the committee for reconsideration.

(5) Make shipments only to those who have qualified with the committee as approved receivers.

(d) *Exemption*—(1) *For types.* The following types of tomatoes are exempt from these regulations: Elongated types commonly referred to as pear shaped or paste tomatoes and including but not limited to San Marzano, Red Top, and Roma varieties; cerasiform type tomatoes commonly referred to as cherry tomatoes; hydroponic tomatoes; and greenhouse tomatoes. Specialty packed red ripe tomatoes, yellow meated tomatoes, and single layer and two layer place packed tomatoes are exempt from the container net weight requirements specified in paragraph (a)(3)(i) of this section, and the requirement that each container or lid shall be marked to indicate the designated

net weight as specified in paragraph (a)(3)(ii) of this section, but must meet the other requirements of this section. Producer field-packed tomatoes must meet all of the requirements of this section except for the requirement that all containers must be packed at registered handler facilities as specified in paragraph (a)(3)(ii) of this section, and the requirement that such tomatoes designated as size 6 × 6 must meet the maximum diameter requirement specified in paragraph (a)(2)(i) of this section: *Provided,* That 6 × 6 and larger is used to indicate the listed size designation on containers.

(2) *For minimum quantity.* For purposes of this regulation each person subject thereto may handle up to but not exceed 50 pounds of tomatoes per day without regard to the requirements of this regulation, but this exemption shall not apply to any shipment or any portion thereof of over 50 pounds of tomatoes.

(3) *For special packed tomatoes.* Tomatoes which met the inspection requirements of paragraph (a)(4) of this section which are resorted, regraded, and repacked by a handler who has been designated as a "Certified Tomato Repacker" by the committee are exempt from:

(i) The tomato grade classifications of paragraph (a)(1) of this section;

(ii) The size classifications of paragraph (a)(2) of this section, except that the tomatoes shall be at least 2–9/32 inches in diameter; and

(iii) The container weight requirements of paragraph (a)(3) of this section.

(4) *For varieties.* Upon recommendation of the committee, varieties of tomatoes that are elongated or otherwise misshapen due to adverse growing conditions may be exempted by the Secretary from the provisions of paragraph (a)(2) of this section.

(5) *For UglyRipe™ and Vintage Ripes™ tomatoes.* UglyRipe™ and Vintage Ripes™ tomatoes must meet all the requirements of this section: *Provided,* That UglyRipe™ and Vintage Ripes™ tomatoes shall be graded and at least meet the requirements specified for U.S. No. 2 under the U.S. Standards for Grades of Fresh Tomatoes, except they are exempt from the

requirements that they be reasonably well formed and not more than slightly rough, and *Provided,* Further that the UglyRipe™ and Vintage Ripes™ tomatoes meet the requirements of the Identity Preservation program, Fresh Products Branch, Fruit and Vegetable Programs, AMS, USDA.

(e) *Report of packouts.* Each registered handler shall, at the end of each day during which handling activities have been conducted, or the following morning as the committee may prescribe, provide to the committee or its designated agent a complete and accurate accounting of the number of containers of tomatoes packed that day. The report shall include an accounting of the grade, size, maturity, and net weight of the containers packed in each such category. The total packout report shall be provided to the committee or its authorized agent in a timely fashion that allows the committee to compile a daily, industry-wide packout report.

(f) *Assessments.* Handlers shall pay assessments as provided in §966.42. Assessment will be based on inspection certificates supplied to the committee by the Federal-State Inspection Service.

(g) *Definitions. Hydroponic tomatoes* means tomatoes grown in solution without soil; *greenhouse tomatoes* means tomatoes grown indoors; *specialty packed red ripe tomatoes* means tomatoes which at the time of inspection are #5 or #6 color (according to color classification requirements in the U.S. tomato standards) with their calyx ends and stems attached and cell packed in a single layer container; and *producer field-packed tomatoes* means tomatoes which at the time of inspection are #3 color or higher (according to color classification requirements in the U.S. tomato standards), that are picked and place packed in new containers in the field by a producer as defined in §966.150 and transferred to a registered handler's facilities for final preparation for market. A *Certified Tomato Repacker* is a repacker of tomatoes in the regulated area who has the facilities for handling, regrading, resorting, and repacking tomatoes into consumer sized packages and has been certified as such by the committee. *Processing* as used in §§966.120 and

966.323 means the manufacture of any tomato product which has been converted into juice, or preserved by any commercial process, including canning, dehydrating, drying, and the addition of chemical substances. Further, all processing procedures must result in a product that does not require refrigeration until opened. *Pickling* as used in §§966.120 and 966.323 means to preserve tomatoes in a brine or vinegar solution. *U.S. tomato standards* means the revised United States Standards for Fresh Tomatoes (7 CFR 51.1855 through 51.1877) effective October 1, 1991, as amended, or variations thereof specified in this section, provided that §51.1863 shall not apply to tomatoes covered by this part. Other terms in this section shall have the same meaning as when used in this part and the U.S. tomato standards.

[52 FR 46347, Dec. 7, 1987]

EDITORIAL NOTE: For FEDERAL REGISTER citations affecting §966.323, see the List of CFR Sections Affected, which appears in the Finding Aids section of the printed volume and at *www.govinfo.gov.*

PART 980—VEGETABLES; IMPORT REGULATIONS

AUTHORITY: 7 U.S.C. 601–674.

§980.1 Import regulations; Irish potatoes.

(a) *Findings and determinations with respect to imports of Irish potatoes.* (1) Pursuant to section 8e of the Agricultural Marketing Agreement Act of 1937, as amended (7 U.S.C. 601–674), it is hereby found that:

(i) Grade, size, quality, and maturity regulations have been issued from time to time pursuant to the following marketing orders: No. 945 (part 945 of this chapter), No. 948 (part 948 of this chapter), No. 947 (part 947 of this chapter), No. 946 (part 946 of this chapter), and No. 953 (part 953 of this chapter).

(ii) During the past several years, grade, size, quality, and maturity regulations have been in effect pursuant to two or more of such orders during each month of the year;

(iii) The marketing of Irish potatoes can be reasonably distinguished by the several seasonal categories, i.e., winter, early spring, late spring, early summer, late summer, and fall. The bulk of the fall crop is harvested and placed in storage in the fall and marketed over a period of several months extending into the following summer. But potatoes harvested from the other seasonal crops are generally marketed as the potatoes are harvested. The marketing seasons for these crops overlap.

(iv) Concurrent grade, size, quality, and maturity regulations under two or more of the aforesaid marketing orders are expected in the ensuing and future seasons, as in the past.

(2) Therefore it is hereby determined that:

(i) Imports of red-skinned, round type potatoes during each month of the marketing year are in most direct competition with potatoes of the same type produced in the area covered by Marketing Order No. 946 (part 946 of this chapter).

(ii) Imports of all other round type potatoes during each month of the marketing year are in most direct competition with potatoes of the same type produced in Area 2, Colorado (San Luis Valley) covered by Marketing Order No. 948, as amended (part 948 of this chapter).

(iii) Imports of long type potatoes during each month of the marketing year are in most direct competition with potatoes of the same type produced in the area covered by Order No. 945 (part 945 of this chapter).

(b) *Grade, size, quality, and maturity requirements.* On and after the effective date hereof importation of Irish potatoes, except certified seed potatoes and red skinned, round types of potatoes, shall be prohibited unless they comply with the following requirements.

(1) Through the entire year, the grade, size, quality, and maturity requirements of Area II, Colorado (San Luis Valley) covered by Marketing Order No. 948, as amended (part 948 of

this chapter), applicable to potatoes of the round type, other than red-skinned varieties, shall be the respective grade, size, quality, and maturity requirements for imports of all other round type potatoes.

(2) Through the entire year the grade, size, quality, and maturity requirements of Marketing Order 945, as amended (part 945 of this chapter) applicable to potatoes of all long types shall be the respective grade, size, quality, and maturity requirements for imported potatoes of all long types.

(3) The grade, size, quality, and maturity requirements as provided for in this paragraph shall apply to imports of similar types of potatoes, unless otherwise ordered, on and after the effective date of the applicable domestic regulation or amendment thereto, as provided in this paragraph or 3 days following publication of such regulation or amendment in the FEDERAL REGISTER, whichever is later.

(c) *Minimum quantities.* Any importation which, in the aggregate, does not exceed 500 pounds of red skinned, round type or long type potatoes, or 2,000 pounds for all other round type potatoes, may be imported without regard to the provisions of this section.

(d) *Plant quarantine.* No provisions of this section shall supersede the restrictions or prohibitions of potatoes under the Plant Quarantine Act of 1912.

(e) *Certified seed.* Certified seed potatoes shall include only those potatoes which are officially certified and tagged as seed potatoes by the Plant Health and Production Division, Plant Products Directorate, Canadian Food Inspection Agency, and which are subsequently used as seed.

(f) *Designation of governmental inspection services.* The Federal or Federal-State Inspection Service, Fruit and Vegetable Programs, Agricultural Marketing Service, U.S. Department of Agriculture and the Food of Plant Origin Division, Plant Products Directorate, Canadian Food Inspection Agency, are hereby designated as governmental inspection services for the purpose of certifying the grade, size, quality, and maturity of Irish potatoes that are imported, or to be imported, into the United States under the provisions of § 608e of the Act.

(g) *Inspection and official inspection certificates.* An official inspection certificate certifying the potatoes meet the United States import requirements for Irish potatoes under section 8e (7 U.S.C. 608e) issued by a designated governmental inspection service applicable to a particular shipment of potatoes is required on all imports of potatoes other than certified seed.

(1)(i) Inspection and certification by the Federal or Federal-State Inspection Service will be available and performed in accordance with the rules and regulations governing certification of fresh fruits, vegetables, and other products (part 51 of this title), and each lot shall be made available and accessible for inspection as provided therein. Cost of inspection and certification shall be borne by the applicant.

(ii) Since inspectors may not be stationed in the immediate vicinity of a port, or point of entry, an importer of uninspected and uncertified Irish potatoes should make advance arrangements for inspection. Each importer should give at least the specified advance notice to one of the following applicable inspection offices prior to the time the Irish potatoes will be imported.

Ports and points	Inspection offices	Advance notice (days)
All Maine ports and points of entry ...	In-Charge, Post Office Box 1058, Presque Isle, ME 04767 (PH 207–764–2100).	1
Port of Boston, MA	In-Charge, Boston Market Terminal Building, Room 1, 34 Market Street, Everett, MA 02149 (PH 617–389–2480).	1
Port of New York, NY	In-Charge, 465B New York City Terminal Market, Bronx, NY 10474 (PH 718–991–7665).	1
Port of Philadelphia, PA	In-Charge, 210 Produce Building, 3301 South Galloway Street, Philadelphia, PA 19148 (PH 215–336–0845).	1
All other ports and points of entry	Head, Field Operations Section, Fresh Products Branch, Fruit and Vegetable Programs, AMS, USDA, Washington, DC 20250–0240 (PH 1–800–811–2373).	3

(2) In the event the required inspection is performed prior to the arrival of the potatoes at the port of entry, the inspection certificate that is issued must show that the inspection was performed at the time of loading such potatoes for direct transportation to the United States; and if transportation is by water, the certificate must show that the inspection was performed at the time of loading onto the vessel.

(3) Inspection certificates shall cover only the quantity of potatoes that is being imported at a particular port of entry by particular importers.

(4) Each inspection certificate issued with respect to any Irish potatoes to be imported into the United States shall set forth, among other things:

(i) The date and place of inspection;

(ii) The name of the shipper, or applicant;

(iii) The Customs entry number pertaining to the lot or shipment covered by the certificate;

(iv) The commodity inspected;

(v) The quantity of the commodity covered by the Certificate;

(vi) The principal identifying marks of the containers;

(vii) The railroad car initials and number, the truck and trailer number, the name of the vessel, or other identification of the shipment; and

(viii) The following statement if the facts warrant: Meets U.S. Import requirements under section 8e of the Agricultural Marketing Agreement Act of 1937.

(h) *Reconditioning prior to importation.* Nothing contained in this part shall be deemed to preclude any importer from reconditioning prior to importation any shipment of Irish potatoes for the purpose of making it eligible for importation under the Act.

(i) *Definitions.* (1) For the purpose of this part potatoes meeting the requirements of Canada No. 1 grade and Canada No. 2 grade shall be deemed to comply with the requirements of the U.S. No. 1 grade and U.S. No. 2 grade, respectively, and the tolerances for size, as set forth in the U.S. Standards for Grades of Potatoes (§§51.1540 to

51.1556, inclusive of this title) may be used.

(2) *Importation* means release from the custody of the U.S. Customs Service.

(j) *Exemptions.* (1) The grade, size, quality and maturity requirements of this section shall not be applicable to potatoes imported for canning, freezing, other processing, livestock feed, charity, or relief, but such potatoes shall be subject to the safeguard provisions contained in § 980.501. Processing includes canning, freezing, dehydration, chips, shoestrings, starch and flour. Processing does not include potatoes that are only peeled, or cooled, sliced, diced, or treated to prevent oxidation, or made into fresh potato salad.

(2) There shall be no size requirements for potatoes that are imported in containers with a net weight of 3 pounds or less, if the potatoes are otherwise U.S. No. 1 grade or better.

[34 FR 8044, May 22, 1969, as amended at 35 FR 8204, May 26, 1970; 36 FR 9634, May 27, 1971; 37 FR 8059, Apr. 25, 1972; 54 FR 22577, May 25, 1989; 57 FR 30382, July 9, 1992; 58 FR 69189, Dec. 30, 1993; 61 FR 13060, Mar. 26, 1996; 67 FR 66531, Nov. 1, 2002; 74 FR 2808, Jan. 16, 2009; 74 FR 65394, Dec. 10, 2009; 79 FR 8256, Feb. 12, 2014; 80 FR 22361, Apr. 22, 2015]

§ 980.117 Import regulations; onions.

(a) *Findings and determinations with respect to onions.* (1) Under section 8e of the Agricultural Marketing Agreement Act of 1937, as amended (7 U.S.C. 601–674), it is hereby found that:

(i) Grade, size, quality, and maturity regulations have been issued regularly under Marketing Orders No. 958 and 959, both as amended;

(ii) Since December 9, 1985, grade, size, quality, and maturity regulations have been in effect pursuant to these orders during the period August through July;

(iii) The marketing of onions can be reasonably distinguished by the seasonal categories, i.e., late summer and early spring. The bulk of the late summer crop is harvested and placed in storage in late summer and early fall and marketed over a period of several months extending into the following spring. But the onions harvested from the early spring crop are generally marketed as soon as the onions are harvested. The marketing seasons for these crops overlap;

(iv) Concurrent grade, size, quality, and maturity regulations under the two marketing orders are expected in future seasons, as in the past.

(2) Therefore, it is hereby determined that: Imports of onions during the June 5 through March 9 period, and the entire year for imports of pearl and cipolline varieties of onions, are in most direct competition with the marketing of onions produced in designated counties of Idaho and Malheur County, Oregon, covered by Marketing Order No. 958, as amended (7 CFR Part 958) and during the March 10 through June 4 period the marketing of imported onions, not including pearl or cipolline varieties of onions, is in most direct competition with onions produced in designated counties in South Texas covered by Marketing Order No. 959, as amended (7 CFR part 959).

(b) *Grade, size, quality, and maturity requirements.* On and after the effective date hereof no person may import onions as defined herein unless they are inspected and meet the following requirements:

(1) During the period June 5 through March 9 of each marketing year, and the entire year for pearl and cipolline onions, whenever onions grown in designated counties in Idaho and Malheur County, Oregon, are regulated under Marketing Order No. 958, imported onions shall comply with the grade, size, quality, and maturity requirements imposed under that order.

(2) During the period March 10 through June 4 of each marketing year, whenever onions grown in designated counties in South Texas are regulated under Marketing Order No. 959, imported onions, not including pearl and cipolline onions, shall comply with the grade, size, quality, and maturity requirements imposed under that order.

(c) *Minimum quantity exemption.* Any importation which in the aggregate does not exceed 110 pounds (50 kilograms) may be imported without regard to the provisions of this section.

(d) *Plant quarantine.* Provisions of this section shall not supercede the restrictions or prohibitions on onions under the Plant Quarantine Act of 1912.

(e) *Designation of governmental inspection service.* The Federal or Federal-State Inspection Service, Fruit and Vegetable Programs, Agricultural Marketing Service, U.S. Department of Agriculture and the Food of Plant Origin Division, Plant Products Directorate, Canadian Food Inspection Agency, are hereby designated as governmental inspection services for the purpose of certifying the grade, size, quality, and maturity of onions that are imported, or to be imported, into the United States under the provisions of section 8e of the Act.

(f) *Inspection and official inspection certificates.* (1) An official inspection certificate certifying the onions meet the U.S. import requirements for onions under section 8e (7 U.S.C. 608e–1), issued by a designated governmental inspection service and applicable to a specified lot is required on all imports of onions.

(2) Inspection and certification by the Federal or Federal-State Inspection Service will be available and performed in accordance with the rules and regulations governing certification of fresh fruits, vegetables and other products (7 CFR part 51). Each lot shall be made available and accessible for inspection as provided therein. Cost of inspection and certification shall be borne by the applicant.

(3) Since inspectors may not be stationed in the immediate vicinity of some smaller ports of entry, importers should make advance arrangements for inspection by ascertaining whether or not there is an inspector located at their particular port of entry. For all ports of entry where an inspection office is not located, each importer must give the specified advance notice to the applicable office listed below prior to the time the onions will be imported.

Ports	Office	Advance notice (days)
All Texas points	Officer-in-charge, 1301 West Expressway, Alamo, Tex. 78516. Phone 512–787–4091 or 512–787–6881.	1
All Arizona points	Officer-in-charge, P.O. Box 1614, Nogales, Ariz. 85621. Phone 602–287–4783.	1

Ports	Office	Advance notice (days)
All California points	Officer-in-charge, 784 South Central Ave., room 266, Los Angeles, Calif. 90021. Phone 213–688–2489.	3
All Hawaii points	Officer-in-charge, P.O. Box 22159, Pawaa Sub-station, Honolulu, Hawaii 96822. Phone 808–941–3071.	1
All Puerto Rico points	Officer-in-charge, P.O. Box 9112, Santurce, P.R. 00908. Phone 809–783–2230 or 809–783–4116.	2
New York City, N.Y	Officer-in-charge, room 28A, Hunts Point Market, Bronx, N.Y. 10474. Phone 212–991–7669 or 212–991–7668.	1
New Orleans, La	Officer-in-charge, 5027 U.S. Postal Service Bldg., 701 Loyola Ave., New Orleans, La. 70113. Phone 504–589–6741 or 504–589–6742.	1
Miami, Fla	Officer-in-charge, 1350 Northwest 12th Ave., room 530, Miami, Fla. 33136. Phone 305–324–6116 or 305–324–6117.	1
All other Florida points.	Officer-in-charge, P.O. Box 1232, Winter Haven, Fla. 33880. Phone 813–294–3511, extension 33.	1
All other points	Chief, Fresh Products Branch, Fruit and Vegetable Quality Division, Food Safety and Quality Service, Washington, D.C. 20250. Phone 202–447–5870.	3

(4) Inspection certificates shall cover only the quantity of onions that is being imported at a particular port of entry by a particular importer.

(5) Each inspection certificate issued with respect to any onions to be imported into the United States shall set forth, among other things:

(i) The date and place of inspection;

(ii) The name of the shipper, or applicant;

(iii) The Customs entry number pertaining to the lot or shipment covered by the certificate;

(iv) The commodity inspected;

(v) The quantity of the commodity covered by the certificate;

(vi) The principal identifying marks on the containers;

(vii) The railroad car initials and number, the truck and trailer license number, the name of the vessel, or other identification of the shipment; and

(viii) The following statement, if the facts warrant: Meets import requirements of 7 U.S.C. 608e–1.

(g) *Reconditioning prior to importation.* Nothing contained in this part shall be deemed to preclude any importer from reconditioning prior to importation any shipment of onions for the purpose of making it eligible for importation.

(h) *Definitions.* For the purpose of this section, *Onions* means all varieties of *Allium cepa* marketed dry, except dehydrated, canned, or frozen onions, pickling onions in brine, onion sets, green onions, or braided red onions. The term *U.S. No. 2* has the same meaning as set forth in the United States Standards for Grades of Bermuda-Granex-Grano Type Onions (7 CFR 51.3195 through 51.3209), the United States Standards for Grades of Creole Onions (7 CFR 51.3955 through 51.3970), or the United States Standards for Grades of Onions Other Than Bermuda-Granex-Grano and Creole Types (7 CFR 51.2830 through 51.2854), whichever is applicable to the particular variety, and variations thereof specified in this section. The term *moderately cured* means the onions are mature and are more nearly well cured than fairly well cured. *Importation* means release from the custody of U.S. Customs and Border Protection. The term *pearl onions* means onions produced using specific cultural practices that limit growth to 2 inches in diameter or less.

(i) *Exemptions.* The grade, size, quality and maturity requirements of this section shall not be applicable to onions imported for processing, livestock feed, charity, or relief, and pearl onions, onion sets (plantings), braided red onions, and minimum quantity shipments of 110 pounds, but such onions shall be subject to the safeguard provisions in § 980.501. Processing includes canning, freezing, dehydration, extraction (juice) and pickling in brine. Processing does not include fresh chop, fresh cut, convenience food or other pre-packaged salad operations. Pearl onions must be inspected for size prior to entry into the United States.

[43 FR 5500, Feb. 9, 1978, as amended at 52 FR 8872, Mar. 20, 1987; 52 FR 19281, May 22, 1987; 54 FR 8520, Mar. 1, 1989; 58 FR 69189, Dec. 30, 1993; 59 FR 46912, Sept. 13, 1994; 61 FR 13060, Mar. 26, 1996; 61 FR 25557, May 22, 1996; 69 FR 56671, Sept. 22, 2004; 74 FR 2808, Jan. 16, 2009; 74 FR 65394, Dec. 10, 2009; 75 FR 1269, Jan. 11, 2010]

§ 980.212 Import regulations; tomatoes.

(a) *Findings and determinations with respect to fresh tomatoes.* (1) Under Section 8e of the Agricultural Marketing Agreement Act of 1937, as amended (7 U.S.C. 601–674), it is hereby found that:

(i) Grade, size, quality and maturity regulations have been issued from time to time under Marketing Order No. 966, as amended;

(ii) The marketing of fresh tomatoes from Florida covered by Marketing Order No. 966, as amended, can reasonably be expected to occur during the months of October through June;

(2) Therefore, it is hereby determined that imports of fresh tomatoes during the months of October through June are in most direct competition with the marketing of fresh tomatoes produced in Florida covered by Marketing Order No. 966, as amended.

(b) *Grade, size, quality and maturity requirements.* On and after the effective date hereof no person may import fresh tomatoes except pear shaped, cherry, hydroponic and greenhouse tomatoes as defined herein, unless they are inspected and meet the following requirements:

(1) From October 10 through June 15 of each season, tomatoes offered for importation shall be at least 2⁹⁄₃₂ inches in diameter. Not more than 10 percent, by count, in any lot may be smaller than the minimum specified diameter. All lots of tomatoes shall be at least U.S. No. 2 grade. *Provided,* That UglyRipe™ and Vintage Ripes™ tomatoes shall be graded and at least meet the requirements specified for U.S. No. 2 under the U.S. Standards for Grades of Fresh Tomatoes, except they are exempt from the requirements that they be reasonably well formed and not more than slightly rough, and *Provided,*

Further that the UglyRipe™ and Vintage Ripes™ tomatoes meet the requirements of the Identity Preservation program, Fresh Products Branch, Fruit and Vegetable Programs, AMS, USDA.

(2)—(3) [Reserved]

(c) *Minimum quantity exemption.* Any importation which in the aggregate does not exceed 60 pounds may be imported without regard to the provisions of this section.

(d) *Plant quarantine.* Provisions of this section shall not supersede the restrictions or prohibitions on tomatoes under the Plant Quarantine Act of 1912.

(e) *Designation of governmental inspection service.* The Federal or Federal-State Inspection Service, Fruit and Vegetable Programs, Agricultural Marketing Service, U.S. Department of Agriculture and the Food of Plant Origin Division, Plant Products Directorate, Canadian Food Inspection Agency, are hereby designated as governmental inspection services for the purpose of certifying the grade, size, quality, and maturity of tomatoes that are imported, or to be imported, into the United States under the provisions of section 8e of the Act.

(f) *Inspection and official inspection certificates.* (1) An official inspection certificate certifying the tomatoes meet the United States import requirements for tomatoes under Section 8e (7 U.S.C. 608e–1), issued by a designated governmental inspection service and applicable to a specified lot is required on all imports of fresh tomatoes.

(2) Inspection and certification by the Federal or Federal-State Inspection Service will be available and performed in accordance with the rules and regulations governing certification of fresh fruits, vegetables and other products (7 CFR part 51). Each lot shall be made available and accessible for inspection as provided therein. Cost of inspection and certification shall be borne by the applicant.

(3) Since the inspectors may not be stationed in the immediate vicinity of some smaller ports of entry, importers should make advance arrangements for inspection by ascertaining whether or not there is an inspector located at their particular port of entry. For all ports of entry where an inspection office is not located, each importer must give the specified advance notice to the applicable office listed below prior to the time the tomatoes will be imported.

Ports	Office	Advance notice (days)
All Texas points	Officer-in-charge, 1301 West Expressway, Alamo, Tex. 78516, phone 512–787–4091 or 6881.	1
All Arizona points	Officer-in-charge, P.O. Box 1614, Nogales, Ariz. 85621, phone 602–287–2902.	1
All California points	Officer-in-charge, 784 South Central Ave., room 266, Los Angeles, Calif. 90021, phone 213–688–2489.	1
All Hawaii points	Officer-in-charge, P.O. Box 22159, Pawaa substation, Honolulu, Hawaii 96822, phone 808–941–3071.	1
All Puerto Rico points	Officer-in-charge, P.O. Box 9112, Santurce, P.R. 00908, phone 809–783–2230 or 4116.	2
New York, N.Y	Officer-in-charge, room 28A, Hunts Point Market, Bronx, N.Y. 10474, phone 212–991–7669 or 7668.	1
New Orleans, La	Officer-in-charge, 5027 U.S. Postal Service Bldg., 701 Loyola Ave., New Orleans La. 70113, phone 504–589–6741 or 6742.	1
Miami, Fla	Officer-in-charge, 1350 Northwest 12th Ave., room 530, Miami, Fla. 33136, phone 305–324–6116 or 6117.	1
All other Florida points.	Officer-in-charge, P.O. Box 1232, Winter Haven, Fla. 33880, phone 813–294–3511, ext. 33.	1
All other points	Chief, Fresh Products Branch, Fruit and Vegetable Quality Division, Food Safety and Quality Service, Washington, DC 20250, phone 202–447–5870.	3

(4) Inspection certificates shall cover only the quantity of tomatoes that is being imported at a particular port of entry by a particular importer.

(5) Each inspection certificate issued with respect to any tomatoes to be imported into the United States shall set forth, among other things:

(i) The date and place of inspection;

(ii) The name of the shipper, or applicant;

423

(iii) The Customs entry number pertaining to the lot or shipment covered by the certificate;

(iv) The commodity inspected;

(v) The quantity of the commodity covered by the certificate;

(vi) The principal identifying marks on the containers;

(vii) The railroad car initials and number, the truck and trailer license number, the name of the vessel, or other identification of the shipment; and

(viii) The following statement, if the facts warrant: Meets import requirements of 7 U.S.C. 608e–1.

(g) *Reconditioning prior to importation.* Nothing contained in this part shall be deemed to preclude any importer from reconditioning prior to importation any shipment of tomatoes for the purpose of making it eligible for importation.

(h) *Definitions.* For the purpose of this section, *Importation* means release from custody of the United States Bureau of Customs. *Cherry tomatoes* means cerasiform types commonly referred to as "cherry tomatoes." *Pear shaped tomatoes* means elongated types, commonly referred to as pear shaped or paste tomatoes and include San Marzano, Red Top and Roma varieties. *Hydroponic tomatoes* means tomatoes grown in solution without soil. *Greenhouse tomatoes* means tomatoes grown indoors. The terms relating to grade and size, as used herein, shall have the same meaning as when used in the U.S. Standards for Grades of Fresh Tomatoes (7 CFR 51.1855 through 51.1877).

(i) *Exemptions.* The grade, size, quality and maturity requirements of this section shall not apply to tomatoes for charity, relief, canning or pickling, but such tomatoes shall be subject to the safeguard provisions contained in § 980.501. Processing includes canning and pickling.

[42 FR 55192, Oct. 14, 1977, as amended at 43 FR 3349, Jan. 25, 1978; 57 FR 27352, June 19, 1992; 58 FR 69189, Dec. 30, 1993; 61 FR 13060, Mar. 26, 1996; 63 FR 12401, Mar. 13, 1998; 72 FR 2172, Jan. 18, 2007; 74 FR 2808, Jan. 16, 2009; 74 FR 45736, Sept. 4, 2009; 74 FR 65394, Dec. 10, 2009; 81 FR 87412, Dec. 5, 2016]

§ 980.501 **Safeguard procedures for potatoes, onions, and tomatoes exempt from grade, size, quality, and maturity requirements.**

(a) Each person who imports or receives any of the commodities listed in paragraphs (a)(1) through (5) of this section shall file (electronically or paper) an "Importer's Exempt Commodity Form" (FV–6) with the Marketing Order and Agreement Division, Specialty Crops Program, AMS, USDA. A "person who imports" may include a customs broker, acting as an importer's representative (hereinafter referred to as "importer"). A copy of the completed form (electronic or paper) shall be provided to the U.S. Customs and Border Protection. If a paper form is used, a copy of the form shall accompany the lot to the exempt outlet specified on the form. Any lot of any commodity offered for inspection and, all or a portion thereof, subsequently imported as exempt under this provision shall also be reported on an FV–6 form. Such form (electronic or paper) shall be provided to the Marketing Order and Agreement Division in accordance with paragraph (d) of this section. The applicable commodities are:

(1) Potatoes, onions or tomatoes for consumption by charitable institutions or distribution by relief agencies;

(2) Potatoes, onions, or tomatoes for processing;

(3) Potatoes or onions for livestock feed; or

(4) Pearl onions; or

(5) Tomatoes to be used in noncommercial outlets for experimental

(b) *Certification of exempt use.* (1) Each importer of an exempt commodity as specified in paragraph (a) of this section shall certify on the FV–6 form (electronic or paper) as to the intended exempt outlet (*e.g.*, processing, charity, livestock feed). If certification is made using a paper FV–6 form, the importer shall provide a handwritten signature on the form.

(2) Each receiver of an exempt commodity as specified in paragraph (a) of this section shall also receive a copy of the associated FV–6 form (electronic or paper) filed by the importer. Within two days of receipt of the exempt lot, the receiver shall certify on the form (electronic or paper) to the Marketing

Order and Agreement Division that such lot has been received and will be utilized in the exempt outlet as certified by the importer. If certification is made using a paper FV–6 form, the receiver shall provide a handwritten signature on the form.

(c) It is the responsibility of the importer to notify the Marketing Order and Agreement Division of any lot of exempt commodity rejected by a receiver, shipped to an alternative exempt receiver, returned to the country of origin, or otherwise disposed of. In such cases, a second FV–6 form must be filed by the importer, providing sufficient information to determine ultimate disposition of the exempt lot, and such disposition shall be so certified by the final receiver.

(d) All FV–6 forms and other correspondence regarding entry of exempt commodities must be submitted electronically, by mail, or by fax to the Marketing Order and Agreement Division, Fruit and Vegetable Program, AMS, USDA, 1400 Independence Avenue SW, STOP 0237, Washington, DC 20250–0237; telephone (202) 720–2491; email *ComplianceInfo@ams.usda.gov;* or fax (202) 720–5698.

[80 FR 15677, Mar. 25, 2015, as amended at 81 FR 87412, Dec. 5, 2016]

PART 981—ALMONDS GROWN IN CALIFORNIA

Subpart A—Order Regulating Handling

DEFINITIONS

AUTHORITY: 7 U.S.C. 601–674.

SOURCE: 35 FR 11372, July 16, 1970, unless otherwise noted.

Subpart A—Order Regulating Handling

DEFINITIONS

§ 981.1   Secretary.

Secretary means the Secretary of Agriculture of the United States, or any other officer or employee of the United States Department of Agriculture who is, or who may be, authorized to perform the duties under this part of the Secretary of Agriculture of the United States.

§ 981.2   Act.

Act means Public Act No. 10, 73d Congress, as amended and as reenacted and amended by the Agricultural Marketing Agreement Act of 1937, as amended (48 Stat. 31, as amended; 62 Stat. 1247; 63 Stat. 282, 1051; 7 U.S.C. 601 *et seq.*).

§ 981.3   Person.

Person means an individual, partnership, corporation, association, or any other business unit.

§ 981.4   Almonds.

Almonds means (unless otherwise specified) all varieties of almonds (except bitter almonds), either shelled or unshelled, grown in the State of California, and for the purposes of research includes almond shells and hulls.

[41 FR 26852, June 30, 1976]

§ 981.5   Unshelled almonds.

Unshelled almonds means almonds the kernels of which are contained in the shell.

§ 981.6   Shelled almonds.

Shelled almonds mean raw or roasted almonds after the shells are removed and includes blanched, diced, sliced, slivered, cut, halved, or broken almonds, or any combination thereof. Additional almond products may be included by the Secretary from time to time upon consideration of a recommendation from the Board or other pertinent information.

§ 981.7   Edible kernel.

Edible kernel means a kernel, piece, or particle of almond kernel that is not inedible.

[41 FR 26852, June 30, 1976]

§ 981.8   Inedible kernel.

Inedible kernel means a kernel, piece, or particle of almond kernel with any defect scored as serious damage, or damage due to mold, gum, shrivel, or brown spot, as defined in the United States Standards for Shelled Almonds, or which has embedded dirt not easily removed by washing. This definition may be modified by the Board with the approval of the Secretary: *Provided,* That the Board shall submit any recommendation for modification to the Secretary not later than August 1.

[41 FR 26852, June 30, 1976]

§981.9 Kernel weight.

Kernel weight means the weight of kernels, including pieces and particles, regardless of whether edible or inedible, contained in any lot of almonds, unshelled or shelled.

§981.10 Almonds received for his own account.

Almonds received for his own account means all almonds which are received by a handler (including all almonds of his own production), except those which are received by him for storage or processing for the account of any other person and with respect to which such handler performs no handling function.

§981.11 Area of production.

Area of production means the State of California.

§981.12 Grower.

Grower is synonymous with *producer* and means any person engaging, in a proprietary capacity, in the commercial production of almonds.

§981.13 Handler.

Handler means any person handling almonds during any crop year, except that such term shall not include either a grower who sells only almonds of his own production at retail at a roadside stand operated by him, or a person receiving almonds from growers and other persons and delivering these almonds to a handler.

[41 FR 26852, June 30, 1976]

§981.14 Cooperative handler.

Cooperative handler means any handler as defined in §981.13 of this subpart which qualifies for treatment as a nonprofit cooperative association as defined in Section 54001, *et seq.* of the California Food and Agricultural Code. The Board, with the approval of the Secretary, may modify this definition, if necessary.

[61 FR 32920, June 26, 1996]

§981.15 Almond product.

Almond product means any edible preparation other than those included under the definition of "shelled almonds," manufactured entirely or partially from raw shelled almonds, and nut mixtures containing shelled or unshelled almonds.

§981.16 To handle.

To handle means to use almonds commercially of own production or to sell, consign, transport, ship (except as a common carrier of almonds owned by another) or in any other way to put almonds grown in the area of production into any channel of trade for human consumption worldwide, either within the area of production or by transfer from the area of production to points outside or by receipt as first receiver at any point of entry in the United States or Puerto Rico of almonds grown in the area of production, exported therefrom and submitted for reentry or which are reentered free of duty. However, sales or deliveries by a grower to handlers, hullers or other processors within the area of production shall not, in itself, be considered as handling by a grower.

[61 FR 32920, June 26, 1996]

§981.17 Inspection agency.

Inspection agency means the Federal-State Inspection Service or, when specifically designated, the Federal Inspection Service.

§981.18 Settlement weight.

Settlement weight means the actual gross weight of any lot of almonds received for his own account by any handler, less adjustments as follows:

  (a) For weight of containers,

  (b) For excess moisture,

  (c) For trash or other foreign material of any kind, and

  (d) For inedible kernels as defined in §981.8.

[35 FR 11372, July 16, 1970, as amended at 61 FR 32920, June 26, 1996]

§981.19 Crop year.

Crop year means the twelve month period from August 1 to the following July 31, inclusive. Any new crop almonds harvested or received prior to August 1 will be applied to the next crop year for marketing order purposes. The first crop year after the implementation of this amendment shall be a 13-month period.

[61 FR 32920, June 26, 1996]

§ 981.20 Handler carryover.

Handler carryover as of any given date means all almonds, wherever located, then held by handlers for their own accounts (whether or not sold) but not including any almond products.

[41 FR 26852, June 30, 1976]

§ 981.21 Trade demand.

Trade demand means the quantity of almonds (kernelweight basis) which commercial distributors and users such as the wholesale, chain store, confectionery, bakery, ice cream, and nut salting trades will acquire from all handlers during a crop year for distribution worldwide.

[61 FR 32920, June 26, 1996]

§ 981.21a Salable almonds.

Salable almonds means those almonds which are free to be handled pursuant to any salable percentage established by the Secretary pursuant to § 981.47 or § 981.48 and, in the absence of a reserve percentage being established for a crop year, all almonds received by handlers for their own accounts during that crop year.

§ 981.21b Reserve almonds.

Reserve almonds means those almonds which must be withheld from handling in satisfaction of a reserve obligation arising from application of a reserve percentage established by the Secretary pursuant to § 981.47 or § 981.48.

§ 981.22 Board.

Board means the Almond Board of California which is the administrative agency established by this subpart.

[41 FR 26852, June 30, 1976]

§ 981.23 Part and subpart.

Part means the order regulating the handling of almonds grown in the State of California, and all rules, regulations, and supplementary orders issued thereunder, and the aforesaid order shall be a *subpart* of such part.

ALMOND BOARD OF CALIFORNIA

§ 981.30 Establishment.

A Board of ten members, with an alternate member for each such member, is hereby established.

§ 981.31 Membership representation.

Membership of the Board will be determined in the following manner:

(a) Two members and an alternate for each member shall be selected from nominees submitted by each of the following groups designated in paragraphs (a) (1) and (2) of this section, or from among other qualified persons belonging to such groups:

(1) Those growers who market their almonds through cooperative handlers; and

(2) Those growers who market their almonds through other than cooperative handlers.

(b) Two members and an alternate for each member shall be selected from nominees submitted by each of the following groups designated in paragraphs (b) (1) and (2) of this section, or from among other qualified persons belonging to such groups:

(1) Cooperative handlers; and

(2) All handlers, other than cooperative handlers.

(c) One member and an alternate shall be selected from nominees submitted by each of the following groups designated in paragraphs (c) (1) and (2) of this section, or from among other qualified persons belonging to such groups:

(1) The group of cooperative handlers or the group of handlers other than cooperative handlers, whichever received for their account more than 50 percent of the almonds delivered by all growers as determined by December 31 of the then current crop year; and

(2) Those growers whose almonds were marketed through the handler group identified in paragraph (c)(1) of this section.

[61 FR 32920, June 26, 1996]

§ 981.32 Nominations.

(a) *Method.* (1) Each year the terms of office of three of the members elected pursuant to § 981.31(a) and (b) shall expire, except every third year when the term of office for two of those members shall expire. Nominees for each respective member and alternate member shall be chosen by ballot delivered to the Board. Nominees chosen by the Board in this manner shall be submitted by the Board to the Secretary on or before June 1 of each year together with such information as the Secretary may require. If a nomination for any Board member or alternate is not received by the Secretary on or before June 1, the Secretary may select such member or alternate from persons belonging to the group to be represented without nomination. The Board shall mail to all handlers and growers, other than the cooperative(s) of record, the required ballots with all necessary voting information including the names of incumbents willing to accept renomination, and, to such growers, the name of any person proposed for nomination in a petition signed by at least 15 such growers and filed with the Board on or before April 1. Distribution of ballots shall be announced by press release, furnishing pertinent information on balloting, issued by the Board through newspapers and other publications having general circulation in the almond producing areas.

(2) Nominees for the positions described in §981.31(c) shall be handled in the same manner as described in paragraph (a)(1) of this section except that those terms of office shall expire annually.

(3) The Board may recommend, subject to the approval of the Secretary, a change to the nomination method, should the Board determine that a revision is necessary.

(b) *Voting.* (1) Nominees for each member and alternate member position shall be voted upon separately by the group proposing them. The handler or grower group which is determined to be eligible for additional representation pursuant to §981.31 (e) and (f), respectively, shall nominate such representatives in the same manner prescribed for choosing other nominees.

(2) Each handler may vote for a nominee for each position representing the group to which he belongs. Each handler vote shall be weighted by the quantity of almonds (kernel weight basis computed to the nearest whole ton) handled for his own account

through December 31 of the crop year in which nominations are made. The nominee for each position shall be the person receiving the highest weighted vote for the position.

(3) Growers who market their almonds through cooperative handlers shall vote through their respective organizations. Each cooperative shall cast a vote for nominees for each position representing the cooperative grower group and such ballots shall be weighted by the number of growers who are members of, or under contract with, such cooperative. The nominee for each position shall be the person receiving the highest weighted vote for that position.

(4) Growers who market their almonds through other than cooperative handlers shall each have one equal vote. The nominees for each position representing such grower group shall be the person receiving the highest number of votes for that position.

[35 FR 11372, July 16, 1970, as amended at 61 FR 32920, June 26, 1996; 84 FR 50716, Sept. 26, 2019]

§981.33 Selection and term of office.

(a) Members and their respective alternates for positions open on the Board shall be selected by the Secretary from persons nominated pursuant to §981.32, or, at the discretion of the Secretary, from other qualified persons, for a term of office beginning August 1. Members and alternates shall continue to serve until their respective successors are selected and qualified.

(b) The term of office of members of the Board shall be for a period of three years beginning on August 1 of the years selected except where otherwise provided. However, for the initial eight members of the Board selected pursuant to this section and to paragraphs (a) and (b) of §981.31, two members shall serve for a term of one year; three members shall serve for a term of two years; and three members shall serve for a term of three years. For the initial terms of office, at the time of nomination under §981.32, the Board shall make this designation by lot. The term of office for the two members selected under paragraph (c) of §981.31 shall always be for a period of one year.

(c) Board members may serve for a total of six consecutive years. Members who have served for six consecutive years must leave the Board for at least one year before becoming eligible to serve again. A person who has served less than six consecutive years on the Board may not be nominated to a new three year term if his or her total consecutive years on the Board at the end of that new term would exceed six years. This limitation on tenure shall not apply to alternate members.

(d) The Board may recommend, subject to approval of the Secretary, revisions to the start date for the term of office of members of the Board.

[61 FR 32920, June 26, 1996, as amended at 84 FR 50716, Sept. 26, 2019]

§ 981.34 Qualification and acceptance.

(a) Any person to be selected as a member or alternate of the Board shall, prior to such selection, qualify by providing such background information as necessary and by advising the Secretary that he/she agrees to serve in the position for which nominated. Grower members and alternates shall be growers or employees of growers, and handler members and alternates shall be handlers or employees of handlers. In the event any member or alternate ceases to be qualified for the position for which selected, that position shall be deemed vacant.

(b) The Board, with approval of the Secretary, may establish additional eligibility requirements for grower members on the Board.

[61 FR 32921, June 26, 1996]

§ 981.35 Alternates.

An alternate for a member for the Board shall act in the place and stead of such member (a) in his absence, or (b) in the event of his death, removal, resignation or disqualification, until a successor for his unexpired term has been selected and has qualified.

§ 981.36 Vacancy.

To fill any vacancy occasioned by the death, removal, resignation, or disqualification of any member or alternate of the Board, a successor for his unexpired term shall be selected by the Secretary after consideration of recommendations which may be submitted by members of the group for which such vacancy exists, unless such selection is deemed unnecessary by the Secretary.

§ 981.37 Expenses.

The members of the Board shall serve without compensation, but shall be allowed their necessary expenses.

§ 981.38 Powers.

The Board shall have the following powers:

(a) To administer the provisions of this part in accordance with its terms;

(b) To make rules and regulations to effectuate the terms and provisions of this part;

(c) To receive, investigate and report to the Secretary complaints of violations of this part; and

(d) To recommend to the Secretary amendments to this part.

§ 981.39 Duties.

The Board shall have, among other things, the following duties:

(a) To act as intermediary between the Secretary and any handler or grower;

(b) To keep minute books and records which will clearly reflect all of its acts and transactions, and such minute books and records shall be subject to examination by the Secretary at any time;

(c) To investigate the growing, shipping, and marketing conditions with respect to almonds and to assemble data in connection therewith;

(d) To furnish to the Secretary such available information as may be deemed pertinent or as he may request;

(e) To appoint such employees as it may deem necessary and to determine the salaries, define the duties and fix the bonds of such employees; and

(f) To cause the books of the Board to be audited by one or more competent certified public accountants at least once for each crop year, and at such other times as the Board may deem necessary or as the Secretary may request; and the report of each such audit shall show, among other things, the receipt and expenditure of funds pursuant hereto: and to file with the

Secretary three copies of all audit reports made.

§981.40 Procedure.

(a) *Organization and rules.* The members of the Board shall select a chairman from their membership. The Board shall select such other officers and adopt such rules for the conduct of its business as it may deem advisable. The Board shall give to the Secretary or his designated agent and representatives the same notice of meetings of the Board as is given to members of the Board.

(b) *Quorum.* All decisions of the Board, except where otherwise specifically provided, shall be by a majority vote of the members present. The presence of six members shall be required to constitute a quorum.

(c) *Voting by mail, telegram, fax or other electronic means.* The Board may vote by mail, telegram, fax or other electronic means upon written notice to all members, or alternates acting in their place, including in the notice a statement of a reasonable time, not to exceed 10 days, in which a vote by mail, telegram, fax or other electronic means must be received by the Board for counting. Voting by mail, telegram, fax or other electronic means shall not be permitted at any assembled meeting of the Board. When a proposition is submitted for vote by mail, telegram, fax or other electronic means, at least eight members of the Board must vote in favor of its passage or the proposition shall be defeated.

(d) *Right of the Secretary.* The members of the Board (including successors or alternates), and any agent or employee appointed or employed by the Board, shall be subject to removal or suspension by the Secretary at any time. Each and every order, regulation, decision, determination, or other act of the Board shall be subject to the continuing right of the Secretary to disapprove of the same at any time, and, upon such disapproval, shall be deemed null and void except as to acts done in reliance thereon or in compliance therewith.

(e) *Additional voting requirements.* Adoption of recommendations by the Board with respect to projects pursuant to §981.41 involving production re-search, marketing research and development projects, and marketing promotion including paid advertising and crediting the pro rata expense assessment obligation of handlers with such portion of their direct expenditures for marketing promotion including paid advertising, shall require at least seven affirmative votes.

[35 FR 11372, July 16, 1970, as amended at 37 FR 3984, Feb. 15, 1972; 61 FR 32921, June 26, 1996]

RESEARCH

§981.41 Research and development.

(a) *General.* The Board, with the approval of the Secretary, may establish or provide for the establishment of projects involving production research, marketing research and development projects, and marketing promotion including paid advertising, designed to assist, improve, or promote the marketing, distribution, consumption or efficient production of almonds. The Board may also provide for crediting the pro rata expense assessment obligations of a handler with such portion of his direct expenditure for such marketing promotion including paid advertising as may be authorized. The expenses of such projects shall be paid from funds collected pursuant to §981.81(a) or credited pursuant to paragraph (c) of this section.

(b) *Authorization.* If, on the basis of a Control Board recommendation pursuant to §981.40(e) with respect to projects pursuant to this section, and appertaining rules and regulations established by the Secretary on recommendation of the Board, and other available information, the Secretary concurs that such activities should be permitted, he shall authorize such activities.

(c) *Creditable expenditures.* The Board, with the approval of the Secretary, may provide for crediting all or any portion of a handler's direct expenditures for marketing promotion including paid advertising, that promotes the sale of almonds, almond products or their uses. No handler shall receive credit for any allowable direct expenditures that would exceed the total of his

431

assessment obligation which is attributable to that portion of his assessment designated for marketing promotion including paid advertising. Such expenditures may include, but are not limited to, money spent for advertising space or time in newspaper, magazines, radio, television, transit, and outdoor media, including the actual standard agency commission costs not to exceed 15 percent.

(d) *Promotion guidelines.* All marketing promotion activity engaged in by the Board, including paid advertising, shall be subject to the following terms and conditions:

(1) No marketing promotion, including paid advertising shall refer to any private brand, private trademark or private trade name;

(2) No promotion or advertising shall disparage the quality, use, value, or sale of like or any other agricultural commodity or product, and no false or unwarranted claims shall be made in connection with the product;

(3) No promotion or advertising shall be undertaken without reason to believe that returns to producers will be improved by such activity; and

(4) Upon conclusion of each activity, but at least annually, the Board shall summarize and report the results of such activity to its members and to the Secretary.

(e) *Rules and regulations.* Before any project involving marketing promotion, including paid advertising and the crediting of the pro rata expense assessment obligation of handlers is undertaken pursuant to this section, the Secretary, after recommendation by the Board, shall prescribe appropriate rules and regulations as are necessary to effectively regulate such activity.

[37 FR 3984, Feb. 25, 1972, as amended at 61 FR 32921, June 26, 1996]

QUALITY CONTROL

§ 981.42   Quality control.

(a) *Incoming.* Except as provided in this paragraph, each handler shall cause to be determined, through the inspection agency, and at handler expense, the percent of inedible kernels in each variety received by him and shall report the determination to the Board. The quantity of inedible kernels in each variety in excess of two percent of the kernel weight received, shall constitute a weight obligation to be accumulated in the course of processing and shall be delivered to the Board, or Board accepted crushers, feed manufacturers, or feeders. The Board, with the approval of the Secretary, may change this percentage for any crop year, may authorize additional outlets, may exempt bleaching stock from inedible kernel determination or obligation and may establish rules and regulations necessary and incidental to the administration of this provision, including the method of determining inedible kernel content and satisfaction of the disposition obligation. The Board for good cause may waive portions of obligations for those handlers not generating inedible material from such sources as blanching or manufacturing.

(b) *Outgoing.* For any crop year the Board may establish, with the approval of the Secretary, such minimum quality and inspection requirements applicable to almonds to be handled or to be processed into manufactured products, as will contribute to orderly marketing or be in the public interest. In such crop year, no handler shall handle or process almonds into manufactured items or products unless they meet the applicable requirements as evidenced by certification acceptable to the Board. The Board may, with the approval of the Secretary, establish different outgoing quality requirements for different markets. The Board, with the approval of the Secretary, may establish rules and regulations necessary and incidental to the administration of this provision.

[41 FR 26853, June 30, 1976, as amended at 41 FR 53651, Dec. 8, 1976; 73 FR 45156, Aug. 4, 2008]

§ 981.43   Marking or labeling of containers.

The Board may, with the approval of the Secretary, establish regulations to require handlers to mark or label their containers that are used in packaging or handling of bulk almonds. For purposes of this section, *container* means a box, bin, bag, carton, or any other type of receptacle used in the packaging or handling of bulk almonds.

[73 FR 45156, Aug. 4, 2008]

VOLUME REGULATION

§981.45 General.

In order to effectuate the declared policy of the act, no handler shall handle almonds except in accordance with the terms and conditions of this part.

§981.46 Withholding reserve.

When a reserve percentage has been fixed for any crop year, as hereinafter provided, no handler shall handle almonds except on condition that he comply with the requirements in respect to withholding reserve almonds and the prescribed disposition thereof.

§981.47 Method of establishing salable and reserve percentages.

Whenever the Secretary finds, from the recommendations and supporting information supplied by the Board or from any other available information, that to designate the percentages of almonds during any crop year which shall be salable almonds and reserve almonds would tend to effectuate the declared policy of the act, he shall designate such percentages. Except as provided in §981.50 the salable and reserve percentages shall each be applied to the kernel weight of almonds received by a handler for his own account during the crop year. In establishing such salable and reserve percentages, the Secretary shall give consideration to the ratio of estimated trade demand (domestic plus export, less the handler carryover available to satisfy trade demand plus the desirable handler carryover at the end of the crop year) to the estimated production of marketable almonds (all expressed in terms of kernel weight) or the allocation quantity (marketable production plus almonds diverted to oil or feed when eligible for reserve satisfaction) whichever is applicable; the recommendation submitted to him by the Board; and such other information as he deems appropriate. The total of the salable and re-

serve percentages established each crop year shall equal 100 percent.

[41 FR 26853, June 30, 1976, as amended at 61 FR 32921, June 26, 1996]

§981.48 Increase of salable percentage.

Upon request filed prior to May 15 by the Board or, if the Board should fail to request, by two or more handlers who have handled at least 15 percent of all almonds handled in the preceding crop year, and after findings of fact (based upon a revision of the estimates required under §981.49 and other pertinent information) that the quantity of salable almonds is not sufficient to satisfy trade demand and desirable carryover requirements for the crop year, the Secretary may increase the salable percentage. Such findings shall be made in the manner specified in §981.47.

§981.49 Board estimates and recommendations.

To aid the Secretary in fixing the salable and reserve percentages, the Board shall furnish to the Secretary, not later than August 1, the following estimates (kernel weight basis) and recommendations for the crop year, each of which, or any later revisions thereof, shall be adopted by the affirmative vote of at least six members:

(a) The quantity of marketable almonds to be produced;

(b) The estimated handler carryover and the estimated reserve inventory as of July 31;

(c) The desirable handler carryover and the probable reserve inventory at the end of the crop year;

(d) The trade demand, taking into consideration anticipated imports, economic conditions and the anticipated market price (within the limitations of the act); and

(e) The recommended salable and reserve percentages to be established.

The Board shall also furnish to the Secretary a complete report of the proceedings of the Board meeting at which the recommended salable and reserve percentages were considered. If, for any reason, the Board fails to make these estimates or to recommend to the Secretary salable and reserve percentages as required hereby, reports representing the views of members with respect to such matters may be submitted to the Secretary who may act on the basis of such reports or other information available to him.

[35 FR 11372, July 16, 1970, as amended at 41 FR 26853, June 30, 1976; 61 FR 32921, June 26, 1996]

§ 981.50 Reserve obligation.

Whenever salable and reserve percentages are in effect for a crop year, each handler shall withhold from handling a quantity of almonds having a kernel weight equal to the reserve percentage of the kernel weight of all almonds such handler receives for his own account during the crop year: *Provided*, That, any quantity of almonds delivered to outlets such as poultry or animal feed or crushing into oil, in a manner permitting accountability to the Board, shall not be included in such receipts. The quantity of almonds hereby required to be withheld from handling shall constitute, and may be referred to as the "reserve" or "reserve obligation" of a handler. The almonds handled as salable almonds by any handler, in accordance with the provisions of this part, shall be deemed to be that handler's quota fixed by the Secretary within the meaning of section 8a(5) of the act.

[41 FR 26853, June 30, 1976]

§ 981.51 Requirements for reserve.

Each handler may satisfy his reserve obligation with such almonds specified in the terms of the agency agreement authorized in § 981.67, including all applicable inspection and certification requirements. Any handler who does not become an agent may receive credit by similarly delivering almonds to the Board or its designees. These requirements may be established by the Board, with the approval of the Secretary, and from time to time so modified, and may include grade requirements for reserve almonds delivered to human consumption outlets.

[41 FR 26853, June 30, 1976]

§ 981.52 Holding requirement and delivery.

Each handler shall, at all times, hold in his possession or under his control, in proper storage for the account of the Board, the quantity of almonds necessary to meet his reserve obligation less: (a) Any quantity which was disposed of by him pursuant to § 981.67; and (b) any quantity for which he is otherwise relieved by the Board of responsibility to so hold almonds. Upon demand of the Board reserve almonds shall be delivered to the Board f.o.b. handler's warehouse or point of storage, except that the Board shall not make such demand upon a handler with respect to reserve almonds for which he has agreed to undertake disposition pursuant to § 981.67. Any handler who does not act as agent for the Board in the disposition of reserve almonds shall be subject to the applicable inspection and certification requirements prescribed by the Board pursuant to § 981.67.

[41 FR 26853, June 30, 1976]

§ 981.54 Payment to handlers for services rendered.

The Board may pay handlers for necessary services rendered by them in connection with almonds eventually disposed of directly by the Board as reserve including but not limited to storing, shelling, sorting, bleaching, grading, packaging, fumigating, and other services in accordance with such schedule of payments and under such conditions as may be established by the Secretary after recommendation of the Board.

§ 981.55 Interhandler transfers.

(a) Any handler may, upon notice to and under the supervision and direction of the Board, transfer almonds or reserve credits to another handler. Any such transfers shall be accounted for in such manner that the reserve obligation and assessments on the combined transactions of the participating handlers shall be fully met and such reserve withholding obligation and assessments may be divided between such handlers in accordance with their arrangements subject to approval of the Board.

(b) When salable and reserve percentages are in effect, any handler may transfer reserve withholding obligation to other handlers. Terms and conditions implementing this provision must be recommended by the Board and approved by the Secretary.

[35 FR 11372, July 16, 1970, as amended at 61 FR 32921, June 26, 1996]

§981.56 Assistance of Board in accounting for reserve.

The Board, on written request, may assist handlers in accounting for their reserve obligations and may aid any handler in acquiring almonds to meet any deficiency in his reserve.

§981.57 Application of salable and reserve percentages after end of crop year.

The salable and reserve percentages established for any crop year shall continue in effect with respect to all almonds for which the reserve obligation has not been previously met, which are received for his own account or handled by any handler after the end of such crop year and before salable and reserve percentages are established for the succeeding crop year. After such percentages are established for the new crop year, the withholding requirements for all such almonds theretofore received for his own account or handled during that crop year shall be adjusted to the newly established percentages.

§981.59 Adjustment upon increase of salable percentage.

(a) Upon any increase in the salable percentage and corresponding decrease in the reserve percentage, the reserve obligation of each handler for the entire crop year to the effective date of such action shall be computed in accordance with such revised salable and reserve percentages. From the reserve almonds that may have been withheld by him and not yet disposed of, any handler authorized to act and acting as agent of the Board in disposing of reserve pursuant to §981.66 shall be permitted to select, under the supervision and direction of the Control Board, the particular reserve almonds to be restored to his salable percentage, and such restoration shall be deemed to ful-

fill the obligation of the Board with respect to such increase.

(b) In the case of handlers who have not been authorized to dispose of their own reserves, and handlers who have terminated their agencies to dispose of their own reserves, prior to an increase in the salable percentage, insofar as practicable each such handler shall be permitted to select almonds from his own reserve to be restored to his salable quantity. In the event there are not sufficient reserve almonds held by the Board at the time the salable percentage is increased, to make full restoration, as represented by the increase in the salable percentage, to all such handlers, the restoration to the salable quantities of the respective handlers shall be pro rata on the basis of certified kernel weight poundage of reserve contributed by said handlers during the crop year to the date of increase of the salable percentage: *Provided*, That restoration shall be made in a manner that will result, to the extent practicable, in a comparable percentage of reserve disposition for each such handler and that no handler shall receive almonds in excess of his contribution. Such restoration to the salable quantity shall be deemed to fulfill the obligation of the Board with respect to the increase in the salable percentage.

§981.60 Determination of kernel weight.

(a) *Almonds for which settlement is made on kernel weight.* All lots of almonds, whether shelled or unshelled, for which settlement is made on the basis of kernel weight shall be included in the total kernel weight for any handler at the settlement weight.

(b) *Almonds for which settlement is made on unshelled weight.* The settlement weight for unshelled almonds shall be determined on the basis of representative samples of unshelled almonds reduced to shelled weight.

[35 FR 11372, July 16, 1970, as amended at 61 FR 32921, June 26, 1996]

§981.61 Redetermination of kernel weight.

The Board, on the basis of reports by handlers, shall redetermine the kernel weight of almonds received by each handler for his own account during each crop year through each of the following dates: December 31, March 31, and June 30. Such redetermined kernel weight for each handler shall be the basis for computing his reserve obligation for the crop year through such dates, except that adjustment shall be made for almonds on which the obligation has been assumed by another handler. The redetermined kernel weight of each handler's receipts, as of any date during the crop year, shall be his carryover as of that date plus the weight of almonds delivered or used in products minus the carryover at the beginning of the crop year, the weight on which another handler has assumed the obligations, and the weight delivered to exempt outlets. Weights used in such computations for various classifications of almonds shall be:

(a) For unshelled almonds, the kernelweight based on representative samples reduced to shelled weight;

(b) For shelled almonds, the net weight; and

(c) For shelled almonds used in production of almond products, the net weight of such almonds.

[41 FR 26853, June 30, 1976, as amended at 61 FR 32921, June 26, 1996]

DISPOSITION OF RESERVE

§ 981.65 Prohibition on the use or disposition of reserve almonds.

Except as provided in §§ 981.66 and 981.67, almonds that are withheld as reserve pursuant to the requirements of § 981.50 or are creditable in satisfaction of a reserve withholding obligation thereunder, shall not be used or disposed of by any handler or any other person.

§ 981.66 Conditions governing disposition of reserve.

(a) *General.* The Board shall have power and authority to sell or dispose of any and all reserve almonds withheld upon the best terms and at the highest return obtainable consistent with the ultimate complete disposition of reserve, subject to all conditions of this section.

(b) *Exclusion from salable normal trade channels.* No reserve almonds shall be sold in the United States, Puerto Rico, and the Canal Zone other than to governmental agencies or to charitable institutions for charitable purposes, except for diversion into almond oil, almond butter, poultry or animal feed, or into other channels which the Board finds are noncompetitive with existing normal markets for almonds, and with proper safeguards in each case to prevent such almonds thereafter entering the channels of trade in such normal markets.

(c) *Disposition after December 31.* Any reserve almonds remaining unsold as of December 31 shall be disposed of by the Board as soon as practicable through the most readily available reserve outlets. The date of December 31 herein specified may be extended to a later date by the Secretary, upon recommendation of the Board or other information.

(d) *Expenses.* Direct expenses incurred by the Board in the maintenance and disposition of reserve almonds shall be charged against the proceeds of sales of such almonds.

(e) *Distribution of proceeds.* Net proceeds from the disposition of reserve almonds by the Board shall be distributed to each handler in proportion to his relative share of such disposition in terms of creditable reserve kernel weight pursuant to § 981.51 or such other basis as the Board may adopt with the approval of the Secretary.

[35 FR 11372, July 16, 1970, as amended at 37 FR 3984, Feb. 25, 1972; 41 FR 26854, June 30, 1976; 61 FR 32921, June 26, 1996]

§ 981.67 Disposition by handler.

Upon request of a handler, made prior to the delivery by him of any reserve to the Board in any crop year, the Board shall authorize such handler to act as agent of the Board, upon such reasonable terms and conditions, including inspection and certification requirements, as the Board may specify and subject to the conditions of § 981.66 in disposing of the reserve withheld from handling by such handler for that crop year. Any handler who is authorized to dispose of his reserve may, through arrangement with another handler dispose of such reserve through such other handler or, in lieu of disposition, may acquire credits for reserve disposition from another handler.

In the first instance, the second handler shall also be subject to the conditions of §981.66. It shall be the obligation of any handler authorized to dispose of such reserve to effect disposition thereof in accordance with all applicable requirements and conditions. The proceeds of such disposition shall be retained by the handler making the disposition, except that, in case he disposes of the reserve of another handler, the proceeds from that disposition shall be divided between the two handlers on the basis of a mutual agreement. Such authorization shall expire as of December 31 of the next crop year, and any reserve then remaining undisposed of by the handler shall be returned to the Board. If the date of December 31 specified in §981.66(e) is extended, the date of December 31 shall be extended correspondingly. Any handler who has been authorized to act as agent of the Board in disposing of his reserve may terminate such agency as of April 1 of the particular crop year by giving written notice to the Board to that effect not later than the previous March 20, in which event such handler shall return to the Board, for disposition by it, all reserve almonds remaining in his possession. In case a handler does not terminate his agency as of April 1, he shall be required to continue to serve as such agent until December 31 of the next crop year. The Board shall not terminate such an agency prior to December 31 unless the agent violates the terms and conditions specified by the Board or other provisions of the order. During the period of such agency the Board, as principal, shall not dispose of the reserve withheld from handling by said agent. The Board, with the approval of the Secretary may prescribe such rules and regulations as are necessary to regulate disposition of reserve almonds including methods for crediting as reserve any salable almonds sold and delivered to reserve outlets.

[35 FR 11372, July 16, 1970, as amended at 61 FR 32921, June 26, 1996]

RECORDS AND REPORTS

§981.70 Records and verification.

Each handler shall keep records which will clearly show the details of his or her receipts of almonds, withholdings, sales, shipments, inventories, reserve disposition, advertising and promotion activities, as well as other pertinent information regarding his or her operation pursuant to the provisions of this part: *Provided*, that, such records shall be kept in the State of California. Such records shall be retained by the handler for 2 years after the end of the crop year to which they apply. Each handler's premises shall be accessible to authorized representatives of the Board and the Secretary for examination and audit of the aforesaid records and for inspection and observation of almonds. The Board shall make such checks of almonds or audits of each handler's records as it deems appropriate or are requested by the Secretary to insure that accurate information as required in this part is being furnished by handlers.

[35 FR 11372, July 16, 1970, as amended at 37 FR 3984, Feb. 25, 1972; 61 FR 32921, June 26, 1996]

§981.71 Record of receipts.

For the purpose of establishing the reserve obligation and furnishing statistical information to the Board necessary for the conduct of its operations, each handler, on receiving almonds for his own account, shall issue to the person from whom so received a receipt therefor. At least two duplicates thereof shall be made at the time of issuance, one of which shall be retained by the handler as a part of his records and the other submitted to the Board as hereinafter provided. Such receipts shall be serially numbered and shall accurately show for each lot received, the identity of the handler, the name and address of the person from whom received, the number of containers in the lot, the variety, whether shelled or unshelled, and the settlement weight for each such variety. The character and amount of all adjustments deducted from the gross weight shall be shown with the gross weight on the receipt issued by the handler.

EFFECTIVE DATE NOTE: At 40 FR 4416, Jan. 30, 1975, §981.71 was suspended indefinitely.

§ 981.72 Reports of receipts.

Each handler receiving almonds for his own account shall tabulate such receipts by varieties and shall submit reports thereof to the Board in such form and at such intervals as the Board may prescribe for all receipts issued by him. Such reports shall be accompanied by duplicate copies of the receipts issued pursuant to the provisions of § 981.71 for all almonds included in such report. The Board, after checking such reports in such manner as it deems desirable, shall determine in the manner specified in § 981.60 the kernel weight of the almonds so received.

EFFECTIVE DATE NOTE: At 40 FR 4416, Jan. 30, 1975, in § 981.72, the second sentence was suspended indefinitely.

§ 981.73 Periodic reports.

On or before January 15, and April 15, and August 15 of each crop year, each handler shall file with the Board a written report, certified to the Board and to the Secretary by such handler as to its completeness and correctness, showing as of the close of business on December 31, March 31, and July 31, respectively, such information as may be prescribed by the Board for use in redetermination of kernel weight and marketing policy considerations.

[35 FR 11372, July 16, 1970, as amended at 61 FR 32922, June 26, 1996]

§ 981.74 Other reports.

Upon the request of the Board, made with the approval of the Secretary, every handler shall furnish to the Board in such manner and at such times as it prescribes (in addition to such other reports as are specifically provided for in this part) such other information as will enable the Board to perform its duties and exercise its powers hereunder.

§ 981.75 Confidential nature of records and reports.

All information contained in handler records made available to the Board or the Secretary, or in reports to the Board, constituting a trade secret or disclosing the trade position, financial condition, or business operations of any handler shall be considered as confidential information. Such informa-

tion received by the Board, shall be kept in the custody and under the control of one or more employees of the Board, who shall disclose such information to no person except the Secretary.

§ 981.76 Handler list of growers.

No later than December 31 of each crop year, each handler other than a cooperative handler (hereinafter, referred to as independent handler) governed by this subpart shall, upon request, submit to the Board a complete list of growers who have delivered almonds to such independent handler during that crop year.

[61 FR 32921, June 26, 1996]

EXPENSES AND ASSESSMENTS

§ 981.80 Expenses.

The Board is authorized to incur such expenses as the Secretary may find are reasonable and likely to be incurred by it during each crop year, for the maintenance and functioning of the Board, including the accumulation and maintenance of an operating reserve fund, and for such purposes as the Secretary may, pursuant to the provisions of this subpart, determine to be appropriate. The recommendation of the Board as to the expenses for each such year, together with all data supporting such recommendation, shall be submitted to the Secretary on or before August 1 of the crop year in connection with which such recommendation is made.

[35 FR 11372, July 16, 1970, as amended at 37 FR 3984, Feb. 25, 1972]

§ 981.81 Assessment.

(a) *Requirement for payment.* Each handler shall pay to the Board on demand by the Board, from time to time, such sum less any amounts credited pursuant to § 981.41, based on such rate per pound of almonds, kernel weight basis, received by him for his own account (except as to receipts from other handlers on which assessments have been paid) as the Secretary finds is necessary to provide funds to meet the authorized board expenses and the operating reserve requirements, and establishes for the crop year. Upon redetermination of the kernel weight of almonds received by handlers for their own account as provided in §981.61, such redetermined kernel weight for each handler, adjusted for receipts on which assessments have been paid, shall be the basis upon which he shall pay assessments. At any time during or after a crop year, the Secretary may increase the rate of assessments to apply to all such almonds during such crop year to secure sufficient funds to cover the expenses authorized by §981.80 or by any later finding by the Secretary relative to the expenses of the Board, and such additional assessments shall be paid to the Board by each handler on demand. The payment of assessments for the maintenance and functioning of the Board may be required under this part throughout the period it is in effect irrespective of whether particular provisions thereof are suspended or become inoperative.

(b) *Refunds.* Any money collected as assessments for either the administrative (maintenance and functioning) or research activities of the Board and not used for the expenses of the applicable crop year, may be used in paying the Board's expenses of the first four months of the succeeding crop year. No later than the fifth month the amount not expended from assessments collected for administrative-research in the previous crop year shall be retained in the operating reserve fund. Any amounts, not credited pursuant to §981.41 for a crop year may be used by the Board for its marketing promotion expenses of the succeeding crop year, and any unexpended portion of those amounts at the end of that crop year shall be retained in the marketing promotion portion of the operating reserve fund. Any funds in each portion of the operating reserve fund in excess of the level authorized pursuant to paragraph (c) of this section shall be refunded to handlers or used to reduce the assessment rate of the subsequent crop year, as the Board may determine. Each handler's share of a refund shall be the amount by which his payment of assessments exceeds his pro rata share of the two major classifications of Board expenses. For the purpose of computing any refund from the marketing promotion portion, each handler's payment of assessments shall include any amount credited to the handler pursu-

ant to §981.41. In lieu of a refund, each handler may have the amount due him credited to his assessment obligation of the crop year in which the amount would be refunded.

(c) *Reserves.* The Board may maintain an operating reserve fund consisting of an administrative-research portion and a marketing promotion portion. The amount in each portion shall not exceed approximately six-months' budget for the activity area or such lower amount as the Board may establish with the approval of the Secretary: *Provided,* That this limitation shall not restrict the temporary retention of excess funds for the purpose of stabilizing or reducing the assessment rate of a crop year. To the extent that funds from current crop year assessments are inadequate, funds in the operating reserve may be used for the authorized activities of the crop year. Funds so used, and not exceeding the six-month limitation, shall be replaced to the extent practicable from assessments subsequently collected for the crop year.

(d) *Disposition of funds upon termination.* Any money collected from assessments hereunder and remaining unexpended in possession of the Board upon the termination of this part shall be distributed in such manner as the Secretary may direct.

(e) Any assessment not paid by a handler within a period of time prescribed by the Board may be subject to an interest or late payment charge or both. The period of time, rate of interest and late payment charge shall be as recommended by the Board and approved by the Secretary. Subsequent to such approval, all assessments not paid within the prescribed period of time shall be subject to an interest or late payment charge or both.

[35 FR 11372, July 16, 1970, as amended at 37 FR 3984, Feb. 25, 1972; 41 FR 26854, June 30, 1976; 61 FR 32921, June 26, 1996]

MISCELLANEOUS PROVISIONS

§981.85 Personal liability.

No member or alternate member of the Board, or any employee or agent thereof, shall be held personally responsible, either individually or jointly with others, in any way whatsoever, to any handler or any other person for errors in judgment, mistakes, or other acts either of commission or omission, as such member, alternate member, agent, or employee, except for acts of dishonesty.

§ 981.86　Separability.

If any provision of this subpart is declared invalid, or the applicability thereof to any person, circumstance, or thing is held invalid, the validity of the remainder hereof or the applicability thereof to any other person, circumstance, or thing shall not be affected thereby.

§ 981.87　Derogation.

Nothing contained in this subpart is, or shall be construed to be, in derogation or in modification of the rights of the Secretary or of the United States to exercise any powers granted by the act or otherwise, or, in accordance with such powers, to act in the premises whenever such action is deemed advisable.

§ 981.88　Duration of immunities.

The benefits, privileges, and immunities conferred upon any person by virtue of this subpart shall cease upon its termination except with respect to acts done under and during its existence.

§ 981.89　Agents.

The Secretary may, by a designation in writing, name any person, including any officer or employee of the United States Government, or name any bureau or division of the United States Department of Agriculture, to act as his agent or representative in connection with any of the provisions of this subpart.

§ 981.90　Effective time, suspension, or termination.

(a) *Effective time.* The provisions of this subpart, as well as any amendments to this subpart, shall become effective at such time as the Secretary may declare, and shall continue in force until terminated or suspended in one of the ways hereinafter specified in this section.

(b) *Suspension or termination*—(1) *Failure to effectuate policy of act.* The Secretary shall terminate or suspend the operation of any or all of the provisions of this subpart, whenever he finds that such provisions do not tend to effectuate the declared policy of the act.

(2) The Secretary shall conduct a referendum as soon as practical after the end of the fiscal year ending two years after implementation of this amendment, and at such time every fifth year thereafter, to ascertain whether continuation of the order is favored by growers who have been engaged in the production of almonds for market within the State of California during the current crop year.

(3) *When favored by growers.* The Secretary shall terminate the provisions of this subpart at the end of any crop year whenever he finds that such termination is favored by a majority of the growers of almonds who during the crop year have been engaged in the production for market of almonds in the State of California: *Provided,* That such majority have during such period produced for market more than 50 percent of the volume of such almonds produced for market within said State; but such termination shall be effected only if announced on or before July 1 of the then current crop year.

(4) *If enabling legislation is terminated.* The provisions of this subpart shall, in any event, terminate whenever the provisions of the act authorizing them cease to be in effect.

(c) *Proceedings after termination*—(1) *Designation of trustees.* Upon the termination of the provisions of this subpart, the members of the Board then functioning shall continue as joint trustees, for the purpose of liquidating the affairs of the Board, of all funds and property then in the possession or under the control of the Board, including claims for any funds unpaid or property not delivered at the time of such termination. Action by said trusteeship shall require the concurrence of a majority of the said trustees.

(2) *Duties of trustees.* Said trustees shall continue in such capacity until discharged by the Secretary; shall, from time to time, account for all receipts and disbursements and deliver all property on hand, together with all books and records of the Board and the joint trustees, to such person as the Secretary may direct; and shall, upon

request of the Secretary, execute such assignments or other instruments necessary or appropriate to vest in such person full title and right to all of the funds, property, and claims vested in the Board or the joint trustees pursuant thereto.

(3) *Obligations of persons other than board members and trustees.* Any person to whom funds, property, or claims have been transferred or delivered by the Board or its members, pursuant to this section, shall be subject to the same obligations imposed upon the members of the said Board and upon the said joint trustees.

[35 FR 11372, July 16, 1970, as amended at 61 FR 32921, June 26, 1996]

§981.91 Effect of termination or amendment.

Unless otherwise expressly provided by the Secretary, the termination of this subpart or of any regulation issued pursuant to this subpart, or the issuance of any amendment to either thereof, shall not (a) affect or waive any right, duty, obligation, or liability which shall have arisen or which may thereafter arise in connection with any provision of this subpart or any regulation issued under this subpart, or (b) release or extinguish any violation of this subpart or of any regulation issued under this subpart, or (c) affect or impair any rights or remedies of the Secretary or of any other person, with respect to any such violation.

§981.92 Amendments.

Amendments to this subpart may be proposed, from time to time, by any person or by the Board.

Subpart B—Assessment Rates

§981.343 Assessment rate.

For the period August 1, 2016, through July 31, 2019, the assessment rate shall be $0.04 per pound for California almonds. Of the $0.04 assessment rate, 60 percent per assessable pound is available for handler credit-back. On and after August 1, 2019, an assessment rate of $0.03 per pound is established for California almonds. Of the $0.03 assessment rate, 60 percent per assessable pound is available for handler credit-back.

[81 FR 92564, Dec. 20, 2016]

Subpart C—Administrative Requirements

§981.401 Adjusted kernel weight.

(a) *Definition. Adjusted kernel weight* shall mean the actual gross weight of any lot of almonds: Less weight of containers; less moisture of kernels in excess of five percent; less shells, if applicable; less processing loss of one percent for deliveries with less than 95 percent kernels; less trash or other foreign material. The adjusted kernel weight shall be determined by sampling certified by the inspection agency.

(b) *Computation.* The computation of adjusted kernel weight shall be in the manner shown in the following examples. The examples are based on the analysis of a 1,000 gram sample taken from a lot of almonds weighing 10,000 pounds with less than 95 percent kernels, and a 1,000 gram sample taken from a lot of almonds weighing 10,000 pounds with 95 percent or more kernels. The first computation example is for the lot with less than 95 percent kernels containing the following: Edible kernels, 530 grams; inedible kernels, 120 grams; foreign material, 350 grams, and moisture content of kernels, seven percent. Excess moisture is two percent. The second computation example is for the lot with 95 percent or more kernels containing the following: Edible kernels, 840 grams; inedible kernels, 120 grams; foreign material, 40 grams; and moisture content of kernels, seven percent. Excess moisture is two percent. The example computations are as follows:

| | Computation number 1 | | Computation number 2 | |
| | Deliveries with less than 95 percent kernels | | Deliveries with 95 percent or more kernels | |
	Percent of sample	Weight (pounds)	Percent of sample	Weight (pounds)
1. Actual gross weight of delivery		10,000		10,000
2. Percent of edible kernel weight	53.000		84.000	
3. Less weight loss in processing [1]	1.000		0.000	
4. Less excess moisture of edible kernels (excess moisture × line 2)	1.060		1.680	
5. Net percent shell out (line 2 − lines 3 and 4)	50.940		82.320	
6. Net edible kernels (line 5 × line 1)		5,094		8,232
7. Percent of inedible kernels (from sample)	12.000		12.000	
8. Less excess moisture of inedible kernels (excess moisture from sample × line 7)	0.240		0.240	
9. Net percent inedible kernels (line 7 − line 8)	11.760		11.760	
10. Total inedible kernels (line 9 × line 1)		1,176		1,176
11. Adjusted kernel weight (line 6 + line 10)		6,270		9,408

[1] Only applies to deliveries with less than 95 percent kernels.

(c) *Computation adjustments.* If applicable, adjustments shall be made by rounding such that the sample computation percentages total equals 100 percent. Rounding adjustments shall be made as follows: First adjust the foreign material percentage; if there is no foreign material in the sample, then adjust the excess moisture percentage; or if there is no foreign material or excess moisture in the sample, adjust the inedible kernels percentage.

[45 FR 68630, Oct. 16, 1980, as amended at 61 FR 42991, Aug. 20, 1996; 83 FR 28525, June 20, 2018]

§ 981.408 Inedible kernel.

Pursuant to § 981.8, the definition of inedible kernel is modified to mean a kernel, piece, or particle of almond kernel with any defect scored as serious damage, or damage due to mold, gum, shrivel, or brown spot, as defined in the United States Standards for Shelled Almonds, or which has embedded dirt or other foreign material not easily removed by washing: Provided, That the presence of web or frass shall not be considered serious damage for the purposes of determining inedible kernels, pieces, or particles of almond kernels.

[59 FR 39419, Aug. 3, 1994]

§ 981.413 Roadside stand exemption.

The term *at retail at a roadside stand* as used in § 981.13 shall be defined to mean sales for home use and not for resale which are not in excess of 100 pounds net kernel weight to any one customer per day. Sales of almonds at certified farmers' markets in compliance with section 1392 of the regulations of the California Department of Food and Agriculture shall be construed as "roadside" sales for the purpose of § 981.13 where these conditions are met.

[50 FR 30264, July 25, 1985]

§ 981.441 Credit for market promotion activities, including paid advertising.

(a) In order for a handler to receive credit for his/her own promotional activities from his/her pro rata portion of advertising assessment payments, pursuant to § 981.41(c), the Board must determine that such expenditures meet the applicable requirements of this section. Credit will be granted either in the form of a payment from the Board, or as an offset to that portion of the assessment if activities are conducted and documented to the satisfaction of the Board at least 2 weeks prior to the Board's first and second assessment billings, and at least 3 weeks prior to the Board's third and fourth assessment billings in a crop year. Credit, hereinafter termed "Credit-Back", will be granted in an amount not to exceed 66⅔ percent of a handler's proven expenditures for qualified activities.

(b) The portion of the handler assessment for which credit may be received under this section will be billed, and is due and payable, at the same time as the portion of the handler assessment used for the Board's administrative expenses, unless the handler(s) conduct and document activities at least 2 weeks prior to the first and second assessment billings and 3 weeks prior to the third and fourth assessment billings. If the handler(s) conduct activities and submit documentation according to applicable provisions in this section, their advertising assessment obligation will be reduced according to the amount of proven activities approved by the Board.

(c) The Board shall grant Credit-Back for qualifying activities only to the handler who performed such activities and who filed a claim for Credit-Back in accordance with this section.

(d) Credit-Back shall be granted only for qualified promotional activities which are conducted and completed during the crop year for which Credit-Back is requested.

(e) The following requirements shall apply to Credit-Back for all promotional activities:

(1) Credit-Back granted by the Board shall be that which is appropriate when compared to accepted professional practices and rates for the type of activity conducted. In the case of claims for Credit-Back activities not covered by specific and established criteria, the Board shall grant the claim if it is consistent with practices and rates for similar activities. To this end, the Board may issue guidelines for qualifying activities from time to time as warranted. For activities in markets other than the United States and Canada, paragraph (e)(5) of this section shall also apply.

(2) The clear and evident purpose of each activity shall be to promote the sale, consumption or use of California almonds, and nothing therein shall detract from this purpose.

(3) No Credit-Back will be given for advertising placed in publications that target the farming or grower trade. No Credit-Back shall be given for any outdoor advertising in California almond growing counties with more than 1,000 bearing acres: *Provided,* That outdoor advertising in these counties which specifically directs consumers to a handler-operated outlet offering direct purchase of almonds will be eligible for Credit-Back.

(4) Credit-Back shall be granted for those qualified activities specified below, except that Credit-Back will not be allowed in any case for travel expenses, or for any promotional activities that result in price discounting.

(i) *Paid advertising directed to end-users, trade or industrial users.* Credit-Back shall be granted for money spent on paid advertising space or time including, but not limited to, newspapers, magazines, radio, television, transit and outdoor media, and including the standard agency commission costs not to exceed 15 percent of gross.

(ii) *Other market promotion activities.* Credit-Back shall be granted for market promotion other than paid advertising, for the following activities:

(A) Marketing research (except pretesting and test-marketing of paid advertising);

(B) Trade and consumer product publicity: *Provided,* That no Credit-Back shall be given for related fees charged by an advertising or public relations agency;

(C) Printing costs for promotional material;

(D) Direct mail printing and distribution;

(E) Retail in-store demonstrations;

(F) Point-of-sale materials (not including packaging);

(G) Sales and marketing presentation kits;

(H) Trade fairs and exhibits;

(I) 50/50 advertising with retailers;

(J) Couponing (printing, distribution, and handling costs only); and

(K) Development and use of web-site on the Internet for advertising and public relations purposes, including E-commerce (mail ordering through the Internet): *Provided,* That Credit-Back shall be limited to $20,000 per year for such activities, and no credit shall be given for costs for E-commerce administration, Extranet (restricted Web sites within the Internet), Intranet (inter-office communication network), or portions of a web-site that target the farming or grower trade.

(iii) For any qualified activity involving joint participation by a handler and a manufacturer or seller of a complementary product(s), or a handler selling multiple complementary products, including other nuts, with such activity including the handler's name or brand, or the words "California Almonds", the amount allowed for Credit-Back claim shall reflect that portion of the activity represented by almonds, or the handler's actual payment, whichever is less.

(iv) Except as otherwise provided in paragraph (e)(4)(v) of this section, when products containing almonds are promoted, the amount allowed for Credit-Back shall reflect that portion of the product weight represented by almonds, or the handler's actual payment, whichever is less: *Provided,* That, except for mixed nut products, the amount of Credit-Back for qualified promotional activities for products containing almonds shall be granted at 66⅔ percent of proven expenditures, if the product is owned or distributed by the handler and such ownership or distributorship is stated on the package: *Provided Further,* That to receive any level of credit, the product must display the handler's name, the handler's brand, or the words "California Almonds" on the primary, face label.

(5) Credit-Back for promotional activities in a foreign market shall be granted at 66⅔ percent of a handler's unreimbursed expenditures for qualified activities in any foreign market, if the handler is promoting pursuant to a contract with the Foreign Agricultural Service, USDA (FAS) and/or the California Department of Food and Agriculture (CDFA). Such activities must also meet the requirements of paragraphs (e)(1), (2), (3), (4), and (6) of this section. Unless the Board is administering the foreign marketing program, such activities shall not be eligible for Credit-Back unless the handler certifies that he/she was not and will not be reimbursed by either FAS or the CDFA for the amount claimed for Credit-Back, and has on record with the Board all claims for reimbursement made to FAS and/or the CDFA. Foreign market expenses paid by third parties as part of a handler's contract with FAS or CDFA will not be eligible for Credit-Back.

(6) A handler must file claims with the Board to obtain Credit-Back for promotional expenditures, as follows:

(i) All claims submitted to the Board for any qualified activity must include:

(A) A description of the activity and when and where it was conducted;

(B) Copies of all invoices from suppliers or agencies;

(C) Copies of all canceled checks issued by the handler in payment of these invoices; and

(D) An actual sample, picture or other physical evidence of the activity.

(ii) Handlers may receive credit against their assessment obligation up to the advertising amount of the assessment installment due: *Provided,* That handlers submit the required documentation for a qualified activity at least 2 weeks prior to the mailing of the Board's first and second assessment notices, and at least 3 weeks prior to the mailing of the Board's third and fourth assessment notices in a crop year. In all other instances, handlers must remit the advertising assessment to the Board when billed, and a refund will be issued to the extent of proven, qualified activities.

(iii) Checks from the Board in payment of approved Credit-Back claims will be mailed to handlers on February 15, April 15, June 15, and 30 days after submission of final claims for the crop year pursuant to paragraph (e)(6)(iv) of this section. To receive payment on these dates, handler claims must be submitted, with all required elements, at least one month prior to the payment date. A handler can receive Credit-Back for his/her allowable direct expenditures only up to the amount of that portion of the handler's assessment designated for marketing promotion, including paid advertising.

(iv) A statement of the Credit-Back commitments outstanding as of the close of a crop year must be submitted in full to the Board within 15 days after the close of that crop year. Final claims pertaining to such commitments outstanding must be submitted with all required elements within 76 days after the close of that crop year. All other final claims for which no statement of Credit-Back commitments outstanding has been filed must be submitted by August 15 of that calendar year.

(f) *Appeals.* If a determination is made by the Board staff that a particular promotional activity is not eligible for Credit-Back because it does not meet the criteria specified herein, or for any other reason, the affected handler may request the Public Relations and Advertising Committee to review the Board staff's decision. If the affected handler disagrees with the decision of the Public Relations and Advertising Committee, the handler may request that the Board review the Committee decision. If the handler disagrees with the decision of the Board, the handler, through the Board, may request that the Secretary review the Board's decision. Handlers have the right to request anonymity in the review of their appeal. The Secretary maintains the right to review any decisions made by the aforementioned bodies at his/her discretion.

[59 FR 35233, July 11, 1994, as amended at 64 FR 41028, July 29, 1999; 64 FR 58766, Nov. 1, 1999; 70 FR 36818, June 27, 2005]

§981.442   Quality control.

(a) *Incoming.* Pursuant to §981.42(a), the quantity of inedible kernels in each variety of almonds received by a handler, including almonds of his own production, shall be determined and disposed of in accordance with the provisions of this paragraph.

(1) *Sampling.* Each handler shall cause a representative sample of almonds to be drawn from each lot of any variety received. The sample shall be drawn before inedible kernels are removed from the lot, or the lot is processed or stored by the handler. For receipts at premises with mechanical sampling equipment and under contracts providing for payment by the handler to the producer for sound meat content, samples shall be drawn by the handler in a manner acceptable to the Board and the inspection agency. The inspection agency shall make periodic checks of the mechanical sampling procedures. For all other receipts, including but not limited to field examination and purchase receipts, accumulations purchased for cash at the handler's door or from an accumulator, or almonds of the handler's own production, sampling shall be conducted or monitored by the inspection agency in a manner acceptable to the Board. All samples shall be bagged and identified in a manner acceptable to the Board and the inspection agency.

(2) *Variety.* For the purpose of classifying receipts by variety to determine a handler's disposition obligation, "variety" shall mean that variety of almonds which constitutes at least 90 percent of the lot: *Provided,* That lots containing a combination of Butte and Padre varieties only, shall be classified as "Butte-Padre", regardless of the percentage of each variety in the lot. If no variety constitutes at least 90 percent of the almonds in a lot, the lot shall be classified as "mixed": *Provided further,* That if the variety or varieties of almonds in a lot are not identified, the lot shall be classified as "mixed", regardless of the percentage of each variety in a lot.

(3) *Analysis of sample.* Each sample shall be analyzed by or under the surveillance of the inspection agency to determine the kernel content and the proportion of inedible kernels in the sample. The inspection agency shall prepare a report for each handler showing, by variety, the total adjusted kernel weight received by handler, the inedible kernel weight and any other information as the Board may prescribe. The report shall cover the handler's daily receipt or the handler's total receipts during a period not exceeding one week, and shall be submitted by the inspection agency to the Board and the handler.

(4) *Disposition obligation.* (i) Beginning August 1, 2016, the weight of inedible kernels in excess of 2 percent of kernel weight reported to the Board of any variety received by a handler shall constitute that handler's disposition obligation. For any almonds sold inshell, the weight may be reported to the Board and the disposition obligation for that variety reduced proportionately.

(ii) If a sufficient sample is not available for any lot of almonds, the handler may establish and substantiate, to the satisfaction of the Board, the received weight, the edible and inedible kernel weights, and the adjusted kernel weight by providing sufficient information as the Board may prescribe. If the handler is only able to establish and substantiate the approximate received weight, an inedible disposition obligation of 10 percent of such received weight may be applied, upon agreement between the Board and the handler.

(5) *Meeting the disposition obligation.* Each handler shall meet its disposition obligation by delivering packer pickouts, kernels rejected in blanching, pieces of kernels, meal accumulated in manufacturing, or other material, to crushers, feed manufacturers, feeders, or dealers in nut wastes on record with the Board as accepted users. Handlers shall notify the Board at least 72 hours prior to delivery: *Provided,* That the Board or its employees may lessen this notification time whenever it determines that the 72 hour requirement is impracticable. The Board may supervise deliveries at its option. In the case of a handler having an annual total obligation of less than 1,000 pounds, delivery may be to the Board in lieu of an accepted user, in which case the Board would certify the disposition lot and report the results to the USDA. For dispositions by handlers with mechanical sampling equipment, samples may be drawn by the handler in a manner acceptable to the Board and the inspection agency. For all other dispositions, samples shall be drawn by or under supervision of the inspection agency. Upon approval by the Board and the inspection agency, sampling may be accomplished at the accepted user's destination. The edible and inedible almond meat content of each delivery shall be determined by the inspection agency and reported by the inspection agency to the Board and the handler. The handler's disposition obligation will be credited upon satisfactory completion of ABC Form 8. ABC Form 8, Part A, is filled out by the handler, and Part B by the accepted user. Beginning August 1, 2006, at least 50 percent of a handler's total crop year inedible disposition obligation shall be satisfied with dispositions consisting of inedible kernels as defined in § 981.408: *Provided,* That this 50 percent requirement shall not apply to handlers with total annual obligations of less than 1,000 pounds. Each handler's disposition obligation shall be satisfied when the almond meat content of the material delivered to accepted users equals the disposition obligation, but no later than September 30 succeeding the crop year in which the obligation was incurred.

(6) *Inedible almonds unfit for processing.* All lots received from growers as "inedible almonds unfit for processing," shall be exempt from the requirements of paragraphs (a) (1) and (3) of this section, but shall be disposed of in their entirety (other than as pickouts), as provided in paragraph (a)(5) of this section. Disposition of these lots shall not be credited toward the disposition obligation of paragraph (a)(4) of this section. If a grower sells or ships inedible almonds to a person other than a handler, the grower thereby becomes a handler and subject to all the requirements of this paragraph.

(7) *Accepted users.* An accepted user's eligibility shall be subject to the following criteria:

(i) Annual completion of an application with the Board for accepted user status;

(ii) Annual submission of a business data sheet to the Board;

(iii) Annual submission of an Accepted User Plan (Form ABC 30) to the Board by July 31 of each year;

(iv) The accurate and prompt submission of Form ABC 8, Part B, to the Board for each lot of almonds received. Each lot of inedible almonds received must be documented by a public weighmaster weight certificate issued at the request of the accepted user at the time of receipt of the lot. Weighmaster weight certificates must be submitted to the Board within 10 business days of issuance;

(v) Disposal of inedible almond material within 6 months of receipt; and

(vi) Disposal of inedible almond material received with no transfer of the material between accepted users.

(vii) The Board may deny or revoke accepted user status at any time if the applicant or accepted user fails to meet the terms and conditions of § 981.442, or if the applicant or accepted user fails to meet the terms and conditions set forth in the accepted user application (Form ABC 34).

(viii) The eligibility of accepted users shall be reviewed annually by the Board. Handlers will not receive credit towards their disposition obligations

446

pursuant to paragraph (a)(4) of this section for inedible lots where the difference between the weight of the lot reported by the inspection agency on Form ABC 8 and the weight of the lot reported on the public weighmaster weight certificate exceeds 2.0 percent.

(b) *Outgoing*. Pursuant to §981.42(b), beginning September 1, 2007, and except as provided in §981.13 and in paragraph (b)(6) of this section, handlers shall subject their almonds to a treatment process or processes prior to shipment to reduce potential *Salmonella* bacteria contamination in accordance with the provisions of this section.

(1) *Treatment process*. Treatment processes shall utilize technologies that have been determined to achieve in total a minimum 4-log reduction of *Salmonella* bacteria in almonds, pursuant to a letter of determination issued by the Food and Drug Administration (FDA), or acceptance by a scientific review panel as identified by the Board (Technical Expert Review Panel or "TERP"). Such panel shall be approved at least annually by the Board prior to the beginning of each crop year, or as needed during the crop year.

(2) *On-site versus off-site treatment*. Handlers shall subject almonds to a treatment process or processes prior to shipment either at their handling facility (on-site), or at an off-site treatment facility located within the production area. Transportation of almonds by a handler to an off-site treatment facility shall not be deemed a shipment.

(3) *Validation by process authorities*. Handlers shall only use, or transport their almonds to off-site treatment facilities that use treatment processes that have been validated by a Board-approved process authority. Treatment technology and equipment that have been modified to a point where operating parameters such as time, temperature, or volume change, shall be revalidated.

(i) Validation means that the treatment technology and equipment have been demonstrated to achieve in total a minimum 4-log reduction of *Salmonella* bacteria in almonds. Validation data prepared by a Board-approved process authority must be submitted to and accepted by the TERP for each piece of equipment used to treat almonds prior to its use under the program.

(ii) A process authority is a person that has expert knowledge of appropriate processes for the treatment of almonds as defined in paragraph (b)(1) of this section, and meets the following criteria:

(A) Knowledge about the equipment used for the treatment process;

(B) Experience in conducting appropriate studies to determine the ability of the equipment to deliver the appropriate treatment (such as heat penetration or heat distribution); and

(C) Able to determine that sufficient data has been gathered to identify the critical factors needed to ensure the quality of the final product.

(iii) Process authorities may be employees of the entity for which they are conducting validation. The Board shall provide process authorities specific protocols and parameters for treatment processes that are FDA determined or TERP accepted.

(iv) Process authorities must submit an initial application to the Board on ABC Form No. 51, "Application for Process Authority for Almonds," and be approved by the TERP. Should the applicant disagree with the TERP's decision concerning approval, the applicant may appeal the decision in writing to the Board, and ultimately to USDA. For subsequent crop years, approved applicants with no changes to their initial application must send the Board a letter, signed and dated, indicating that there are no changes to the application the Board has on file.

(v) The TERP may revoke any approval for cause. The TERP shall notify the process authority in writing of the reasons for revoking the approval. Should the process authority disagree with the TERP's decision, he/she may appeal the decision in writing to the Board, and ultimately to USDA. A process authority whose approval has been revoked must submit a new application to the TERP and await approval.

(4) *Compliance and verification*. In accordance with the requirements of this paragraph, handlers shall utilize either an on-site verification program (traditional), or an audit-based verification program to ensure that their almonds

have been subjected to a treatment process to reduce *Salmonella* bacteria prior to shipment. Each handler may decide which verification program would be the most cost-effective for his or her operation.

(i) By May 31, each handler shall submit to the Board a Treatment Plan for the upcoming crop year. A Treatment Plan shall describe how a handler plans to treat his or her almonds, and must address specific parameters as outlined by the Board for the handler to ship almonds. Such plan shall be reviewed by the Board, in conjunction with the inspection agency, to ensure it is complete and can be verified, and be approved by the Board. Almonds sent by a handler for treatment to an off-site facility affiliated with another handler shall be subject to the approved Treatment Plan utilized at that facility. Handlers shall follow their own approved Treatment Plans for almonds sent to an off-site facility that is not affiliated with another handler.

(ii) Handlers utilizing an on-site verification program shall cause the inspection agency to verify that their Treatment Plans have been followed, and that their almonds have been subjected to a treatment process that has been validated by a Board-approved process authority. Such handlers shall submit, or cause to be submitted, a verification report to the Board. The inspection agency must physically observe the treatment process to issue such report.

(iii) Handlers utilizing an audit-based verification program shall be subject to periodic audits conducted by the inspection agency. The inspection agency shall provide copies of the audit report to the Board. Handlers who do not comply with an audit-based verification program shall be required to revert to an on-site verification program.

(iv) Interhandler transfers of almonds may or may not be treated prior to transfer. Handlers receiving untreated almonds from another handler shall be responsible for treating the product. Handlers receiving treated almonds from another handler must have procedures outlined in their Treatment Plan addressing how the integrity of the treated almonds will be maintained. In all instances involving interhandler transfers, the receiving handler shall be responsible for ensuring that the almonds are treated prior to shipment and maintaining documentation to that effect.

(v) An off-site treatment facility that does not handle almonds, pursuant to § 981.16, shall provide access to the inspection agency and Board staff for verification of treatment and review of treatment records. A treatment process at an off-site treatment facility that has been validated by a Board approved process authority is deemed to be approved by the Board for handler use. The Board may revoke any such approval for cause. The Board shall notify the off-site treatment facility of the reasons for revoking the approval. Should the off-site facility disagree with the Board's decision, it may appeal the decision in writing to USDA. Handlers may treat their almonds only at off-site treatment facilities that have been deemed to be approved by the Board.

(5) *Records.* Handlers shall maintain records and documentation that will be subject to audit by the Board for the purpose of verifying compliance with this section. Records must be maintained for two full years following the end of the crop year, and must identify lots from the point of treatment forward to the point of shipment by the handler. Lot identification shall also provide the ability to differentiate treated from untreated product. Off-site treatment facilities that do not handle almonds pursuant to § 981.16, shall maintain treatment records for 2 full years following the end of a crop year and make such records available to the Board.

(6) *Exemptions.* Handlers may ship untreated almonds under the following conditions. For purposes of this section, container means a box, bin, bag, carton, or any other type of receptacle used in the packaging of bulk almonds.

(i) Handlers may ship untreated almonds for further processing directly to manufacturers located within the U.S., Canada or Mexico. This program shall be termed the Direct Verifiable (DV) program. Handlers may only ship untreated almonds to manufacturers who have submitted ABC Form No. 52,

"Application for Direct Verifiable (DV) Program for Further Processing of Untreated Almonds," and have been approved by the TERP. Such almonds must be shipped directly to approved manufacturing locations, as specified on Form No. 52. Such manufacturers DV users must submit an initial Form No. 52 to the Board and be approved by the TERP. Should the applicant disagree with the TERP's decision concerning approval, it may appeal the decision in writing to the Board, and ultimately to USDA. For subsequent crop years, approved applicants with no changes to their initial application must send the Board a letter, signed and dated, indicating that there are no changes to the application the Board has on file. The TERP may revoke any approval for cause. The TERP shall notify the manufacturer in writing of the reasons for revoking the approval. Should the manufacturer disagree with the TERP's decision, it may appeal the decision in writing to the Board, and ultimately to USDA. A manufacturer whose approval has been revoked must submit a new application to the TERP and await approval. The Board shall issue a DV User code to an approved manufacturer. Handlers must reference such code in all documentation accompanying the lot and identify each container of such almonds with the term "unpasteurized." Such lettering shall be on one outside principal display panel, at least ½ inch in height, clear and legible. If a third party is involved in the transaction, the handler must provide sufficient documentation to the Board to track the shipment from the handler's facility to the approved DV user. While a third party may be involved in such transactions, shipments to a third party and then to a manufacturing location are not permitted under the DV program. Approved DV Users shall:

(A) Subject such almonds to a treatment process or processes using technologies that achieve in total a minimum 4-log reduction of *Salmonella* bacteria as determined by the FDA, accepted by the TERP, or established by a process authority approved in accordance with and subject to the provisions and procedures of paragraph (b)(3) of this section. Establish means that the treatment process and protocol have been evaluated to ensure the technology's ability to deliver a lethal treatment for *Salmonella* bacteria in almonds to achieve a minimum 4-log reduction;

(B) Identify the manufacturing locations where treatment will occur;

(C) Have their treatment technology and equipment validated by a Board-approved process authority, and provide documentation with their DV application to verify that their treatment technology and equipment have been validated by a Board-approved process authority. Such documentation may include, but not be limited to, a letter from such process authority certifying the validation. Such documentation shall be sufficient to demonstrate that the treatment processes and equipment achieve a 4-log reduction in *Salmonella* bacteria. Treatment technology and equipment that have been modified to a point where operating parameters such as time, temperature, or volume change, shall be revalidated;

(D) Have their technology and procedures verified by a Board-approved DV auditor to ensure they are being applied appropriately. A DV auditor may not be an employee of the manufacturer that he/she is auditing. DV auditors must submit a report to the Board after conducting each audit. DV auditors must submit an initial application to the Board on ABC Form No. 53, "Application for Direct Verifiable (DV) Program Auditors," and be approved by the TERP. Should the applicant disagree with the TERP's decision concerning approval, it may appeal the decision in writing to the Board, and ultimately to USDA. For subsequent crop years, approved DV auditors with no changes to their initial application must send the Board a letter, signed and dated, indicating that there are no changes to the application the Board has on file. The TERP may revoke any approval for cause. The TERP shall notify the DV auditor in writing of the reasons for revoking the approval. Should the DV auditor disagree with the TERP's decision, it may appeal the decision in writing to the Board, and ultimately to USDA. A DV auditor whose approval has been revoked must submit a new application to the TERP and await approval;

(E) Maintain all records regarding validation and verification of treatment methods, processing, and product traceability. Such records shall be retained for two years and shall be made available for review by the Board; and,

(F) Ship any almonds which will not be treated to a handler, to another approved DV user, to locations outside the U.S., Canada, and Mexico (containers must remain identified with the term "unpasteurized"), as specified in § 981.442(b)(6)(i), or dispose of such almonds in non-edible channels.

(ii) Handlers may ship untreated almonds directly or through a third party to locations outside the U.S., Canada, and Mexico, provided that each container of such almonds is identified with the term "unpasteurized." Such lettering shall be on one outside principal display panel, at least ½ inch in height, clear and legible. If a third party is involved in the transaction, the handler must provide sufficient documentation to the Board to track the shipment from the handler's facility to the importer in the foreign country.

(7) *Other restrictions.* The provisions of this section do not supersede any restrictions or prohibitions regarding almonds grown in California under the Federal Food, Drug and Cosmetic Act, or any other applicable laws or regulations or the need to comply with applicable food and sanitary regulations of city, county, State or Federal agencies.

[42 FR 3160, Jan. 17, 1977]

EDITORIAL NOTE: For FEDERAL REGISTER citations affecting § 981.442, see the List of CFR Sections Affected, which appears in the Finding Aids section of the printed volume and at *www.govinfo.gov.*

§ 981.450   Exempt dispositions.

As provided in § 981.50 any handler disposing of almonds for crushing into oil, or for poultry or animal feed, may have the kernel weight of these almonds excluded from his receipts, and exempt from program obligations so long as the handler qualifies as, or delivers such almonds to, a crusher, feeder, or dealer in nut waste; the crusher, feeder, or dealer are acceptable to the Board; each delivery is made directly to the crusher, feeder, or dealer, by June 30 of the crop year; and each delivery is certified to the Board by the handler on ABC Form 8.

[42 FR 19322, Apr. 13, 1977]

§ 981.455   Interhandler transfers.

(a) *Transfers of almonds.* Interhandler transfers of almonds pursuant to § 981.55 shall be reported to the Board on ABC Form 7. The report shall contain the following information:

(1) Date of transfer;

(2) The names, and plant locations of both the transferring and receiving handlers;

(3) The variety of almonds transferred;

(4) Whether the almonds are shelled or unshelled;

(5) The name of the handler assuming reserve and assessment obligations on the almonds transferred;

(6) Whether the almonds had been treated to achieve a 4-log reduction in *Salmonella* bacteria, pursuant to § 981.442(b); and

(7) A unique handler identification number for each lot.

(b) *Transfers of reserve credits.* A handler may transfer reserve credits to another handler after having filed with the Board, in accordance with § 981.474, a completed ABC Form 13/14 covering the almonds to be diverted to a noncompetitive outlet and all the documentation applicable thereto. Such a transfer does not relieve the transferring handler of any reserve obligations for the applicable crop year. The transferred credit shall not exceed the quantity needed by the receiving handler to cover that handler's reserve obligation. The Board shall complete the transfer upon receipt of an ABC Form 11 executed by both handlers. No transfer of reserve credits shall be made to satisfy a handler's inedible disposition obligation incurred pursuant to § 981.42(a).

(c) *Transfers of reserve withholding obligation.* A handler may transfer reserve withholding obligation to other handlers pursuant to § 981.55 after having filed with the Board an ABC Form 11 executed by both handlers. The Board shall approve the transfer upon receipt of the properly completed form.

(d) Transfer of inedible obligation may be made, with the approval of the

Board, only when the inedible kernels are physically transferred with the entire lot of almonds. The transfer of the lot shall be reported on ABC Form 9, showing date of transfer and, for the transferring handler, the (1) original inspection certificate number, (2) total weight shown on the certificate, and (3) weight of inedible kernels shown on the certificate. For the receiving handler, ABC Form 9 shall show the (1) new inspection certificate number, (2) total weight shown on the certificate, and (3) weight of inedible kernels shown on the certificate. ABC Form 9 shall be signed by both, the transferring handler and the receiving handler, and submitted by the receiving handler to the Board for approval.

[42 FR 19322, Apr. 13, 1977, as amended at 44 FR 30076, May 24, 1979; 56 FR 19794, Apr. 30, 1991; 62 FR 56051, Oct. 29, 1997; 72 FR 51992, Sept. 12, 2007]

§ 981.466 Almond butter.

Almond butter as used in § 981.66(c) is hereby defined as a comminuted food product prepared by grinding shelled or blanched almonds into a homogeneous plastic or semiplastic mass or liquid having very few particles larger than ¹⁄₁₆ inch in any dimension. To produce chunky style almond butter, almond chunks or pieces may be added up to a maximum of 25 percent by weight of the finished product. The size of the almond pieces used to make chunky style almond butter may not exceed ⁵⁄₁₆ inch in any dimension.

[48 FR 11250, Mar. 17, 1983]

§ 981.467 Disposition in reserve outlets by handlers.

(a) *Agents of Board.* Beginning with August 1 of any crop year, a handler may become an agent of the Board pursuant to § 981.67 for the purpose of disposing of reserve almonds of such crop year, in the authorized outlets. The agency shall be established upon a handler executing a reserve agreement (ABC Form 12) ABC, applicable to diversion, containing terms and conditions specified by the Board.

(b) *Reserve credit.* Credit in satisfaction of a reserve obligation shall not exceed the accrued reserve obligation derived by applying the reserve percentage to the quanity of almonds re-

ceived by a handler for his own account during the crop year. Disposition by an agent of the Board in eligible reserve outlets within a crop year in excess of his reserve obligation shall be held to be a disposition of salable almonds. Whenever such disposition has been inspected and certified, if required, and has complied with the terms, conditions, and documentation applicable to disposition of reserve almonds as determined by the Board, the disposition may be credited against any reserve obligation subsequently incurred by the handler during that crop year, or the disposition may be credited pursuant to § 981.455(b) against the reserve obligation of another handler.

(c) *Minimum prices.* Minimum prices shall apply to 1990–91 crop year reserve almonds diverted to almond butter, natural almond paste, foil packets for sales to airlines, and sales to government agencies, including federal and state school lunch programs. Prices are F.O.B. handlers plant. The prices may contain a maximum of two percent brokerage commission. No cash discounts are allowed. The prices are as follows for various grades or categories of almonds:

Grade or category	Price per pound
U.S. Select Sheller Run or better, unblanched.	75 cents.
U.S. Standard Sheller Run, unblanched	74 cents.
U.S. No. 1 Whole and Broken, unblanched	73 cents.
U.S. No. 1 Pieces, unblanched	73 cents.
U.S. No. 1 Pieces or better, unblanched, to be used for almond butter manufactured in the 48 contiguous states and shipped to EEC countries.	60 cents.
Blanched made from U.S. No. 1 Pieces or better.	95 cents.
Blanched made from U.S. No. 1 Pieces or better to be used for almond butter manufactured in the 48 contiguous states and shipped to EEC countries.	82 cents.

(d) For the 1990–91 crop year only, the reserve disposition obligation date is extended until September 1, 1992, and the date for submitting documentation verifying reserve dispositions is extended to December 1, 1992.

[42 FR 19322, Apr. 13, 1977, as amended at 56 FR 10508, Mar. 13, 1991; 56 FR 51150, Oct. 10, 1991; 57 FR 27353, June 19, 1992; 61 FR 32922, June 26, 1996]

§ 981.472 Report of almonds received.

(a) Each handler shall report to the Board, on or before the 5th calendar day of each month, on ABC Form 1, the total adjusted kernel weight of almonds, by variety, received by it for its own account for the preceding month.

(b) [Reserved]

[58 FR 34696, June 29, 1993, as amended at 61 FR 32922, June 26, 1996; 62 FR 37488, July 14, 1997; 64 FR 18802, Apr. 16, 1999]

§ 981.473 Redetermination reports.

Each handler shall furnish for use by the Board in redetermination of the kernel weight of almonds received for his own account and for marketing policy considerations, the information listed and described in this section. Such information shall be reported within the applicable times specified in § 981.73 on forms provided by the Board.

(a) *Handler carryover.* Report the weight of all almonds, whether unshelled or shelled, wherever located, held by the handler for the handler's own account, whether or not sold.

(b) *Delivered sales.* Report the weight of salable almonds sold and delivered (shipments), showing the weight, and whether unshelled or shelled, including those disposed of pursuant to the requirements for reserve disposition, or used in almond products.

(c) *Transfers.* A report of almonds transferred to another handler showing the weight of each lot transferred, whether unshelled or shelled.

(d) *Remaining inedible obligation.* Report the quantity of almonds the handler intends to deliver to Board approved outlets to meet the disposition obligation pursuant to § 981.42(a).

[42 FR 19322, Apr. 13, 1977, as amended at 42 FR 56488, Oct. 26, 1977; 58 FR 34696, June 29, 1993]

§ 981.474 Other reports.

(a) *Report of shipments and commitments.* Each handler shall report on ABC Form 25–1 all shipments of almonds, inshell, shelled, and products by classification (domestic and export by countries of destination); and on ABC Form 25–2 all commitments (almonds not shipped, but sold or otherwise obligated) whether domestic contract, export contract, or non-contract.

If the destination of any export is unknown to the handler, such handler shall have the broker/exporter furnish this information to the Board. In support of this report, the handler shall keep invoices on the shipments, or such other documentation as may be acceptable to the Board. The reports shall be received by the Board within five calendar days after the close of each month of the crop year.

(b) *Reserve reports.* In any crop year when reserve almonds are diverted to noncompetitive outlets, such handler shall report such handler's intentions to divert on ABC Form 13 and the completion of diversion on ABC Form 14. Upon notice to all handlers, the Board may waive the requirements to file ABC Form 13 for diversion of almonds to noncompetitive outlets which are acceptable to the Board.

(c) *Handler information reports.* Each handler shall file no later than September 1 of each year ABC Form 42, a Handler Information Sheet, listing the handler's name, address, phone number, ownership or corporate information and acknowledging receipt of marketing order program information.

[50 FR 47709, Nov. 20, 1985, as amended at 51 FR 9763, Mar. 21, 1986; 54 FR 5409, Feb. 3, 1989; 58 FR 34696, June 29, 1993]

§ 981.481 Interest and late payment charges.

(a) Pursuant to § 981.481, the Board shall impose an interest charge on any handler whose assessment payment has not been received in the Board's office, or the envelope containing the payment legibly postmarked by the U.S. Postal Service, within 30 days of the invoice date shown on the handler's statement. The interest charge shall be a rate of one and one half percent per month and shall be applied to the unpaid assessment balance for the number of days all or any part of the unpaid balance is delinquent beyond the 30 day payment period.

(b) In addition to the interest charge specified in paragraph (a) of this section, the Board shall impose a late payment charge on any handler whose payment has not been received in the Board's office, or the envelope containing the payment legibly postmarked by the U.S. Postal Service, within 60 days of the invoice date. The late payment charge shall be 10 percent of the unpaid balance.

[61 FR 64603, Dec. 6, 1996]

PART 982—HAZELNUTS GROWN IN OREGON AND WASHINGTON

Subpart A—Order Regulating Handling

982.466 Reports of inshell hazelnuts handled, shelled and withheld.
982.467 Report of receipts and dispositions of hazelnuts grown outside the United States.
982.468 Report of hazelnut receipts, disposition, and inventory.
982.471 Records.

AUTHORITY: 7 U.S.C. 601–674.

SOURCE: 24 FR 6185, Aug. 1, 1959, unless otherwise noted. Redesignated at 26 FR 12751, Dec. 30, 1961.

EDITORIAL NOTE: Nomenclature changes to part 982 appear at 61 FR 17559, Apr. 22, 1996.

Subpart A—Order Regulating Handling

DEFINITIONS

§ 982.1 Secretary.

Secretary means the Secretary of Agriculture of the United States, or any other officer or employee of the United States Department of Agriculture who is, or who may be, authorized to perform the duties of the Secretary of Agriculture of the United States.

§ 982.2 Act.

Act means Public Act No. 10, 73d Congress, as amended and as reenacted and amended by the Agricultural Marketing Agreement Act of 1937, as amended (7 U.S.C. 601 *et seq.;* 48 Stat. 31, as amended).

§ 982.3 Person.

Person means an individual, partnership, corporation, association, or any other business unit.

§ 982.4 Hazelnuts.

Hazelnuts means hazelnuts or filberts produced in the States of Oregon and Washington from trees of the genus Corylus.

[61 FR 17559, Apr. 22, 1996]

§ 982.5 Area of production.

Area of production means the States of Oregon and Washington.

§ 982.6 Grower.

Grower is synonymous with *producer* and means any person engaged, in a proprietary capacity, in the commercial production of hazelnuts.

§ 982.7 To handle.

To handle means to sell, consign, transport or ship (except as a common carrier of hazelnuts owned by another person), or in any other way to put hazelnuts, inshell or shelled, into the channels of trade either within the area of production or from such area to points outside thereof: *Provided,* That sales or deliveries by growers to handlers within the area of production or authorized disposition of restricted hazelnuts and substandard hazelnuts shall not be considered as handling.

§ 982.8 Handler.

Handler means any person who handles hazelnuts.

§ 982.11 Pack.

Pack means a specific commercial classification according to size, internal quality, and external appearance and condition of hazelnuts packed in accordance with any of the pack specifications prescribed pursuant to § 982.45.

§ 982.12 Merchantable hazelnuts.

Merchantable hazelnuts means inshell hazelnuts that meet the grade, size, and quality regulations in effect pursuant to § 982.45 and are likely to be available for handling as inshell hazelnuts.

[83 FR 52949, Oct. 19, 2018]

§ 982.13 Substandard hazelnuts.

Substandard hazelnuts means hazelnuts, inshell or shelled, that do not meet the minimum standards effective pursuant to § 982.45.

§ 982.14 Restricted hazelnuts.

Restricted hazelnuts means inshell hazelnuts withheld in satisfaction of a restricted obligation.

§ 982.15 Inshell handler carryover.

Inshell handler carryover as of any given date means all inshell hazelnuts (except restricted hazelnuts) wherever located then held by handlers or for their accounts, whether or not sold, including certified merchantable hazelnuts and the estimated merchantable content of those uncertified hazelnuts then held by handlers which are intended for handling as inshell hazelnuts.

§ 982.16 Inshell trade acquisitions.

Inshell trade acquisitions means the quantity of inshell hazelnuts acquired by the trade from all handlers during a marketing year for distribution in the continental United States and such other distribution areas as may be recommended by the Board and established by the Secretary.

[61 FR 17559, Apr. 22, 1996]

§ 982.17 Marketing year.

Marketing year means the 12 months from July 1 to the following June 30, both inclusive, or such other period of time as may be recommended by the Board and established by the Secretary.

[51 FR 29546, Aug. 19, 1986]

§ 982.18 Board.

Board means the Hazelnut Marketing Board established pursuant to § 982.30.

[46 FR 26038, May 11, 1981]

§ 982.19 Disappearance.

Disappearance means the difference between orchard-run production and the available supply of merchantable hazelnuts and merchantable equivalent of shelled hazelnuts.

[46 FR 26038, May 11, 1981]

§ 982.20 Part and subpart.

Part means the order, as amended, regulating the handling of hazelnuts grown in Oregon and Washington, and all rules, regulations, and supplementary orders issued thereunder. This order, as amended, regulating the handling of hazelnuts grown in Oregon and Washington shall be a *subpart* of such part.

[24 FR 6185, Aug. 1, 1959. Redesignated at 26 FR 12751, Dec. 30, 1961, and further redesignated at 46 FR 26038, May 11, 1981]

HAZELNUT CONTROL BOARD

§ 982.30 Establishment and membership.

(a) There is hereby established a Hazelnut Marketing Board consisting of 10 members, each of whom shall have an alternate member, to administer the terms and provisions of this part. Each member and alternate shall meet the same eligibility qualifications. The 10 member positions shall be allocated as follows:

(b) Four of the members shall represent handlers, as follows:

(1) One member shall be nominated by the handler who handled the largest volume of hazelnuts during the two marketing years preceding the marketing year in which nominations are made;

(2) One member shall be nominated by the handler who handled the second largest volume of hazelnuts during the two marketing years preceding the marketing year in which nominations are made;

(3) One member shall be nominated by the handler who handled the third largest volume of hazelnuts during the two marketing years preceding the marketing year in which nominations are made;

(4) The fourth handler member shall be nominated by and represent all other handlers.

(c) Five members shall represent growers and shall be nominated for the districts designated in or established pursuant to § 982.31. One grower member shall represent each of the five grower districts unless changes are made pursuant to § 982.31(b).

(d) One member shall be a public member who is neither a grower nor a handler.

(e) The Secretary, or the Board with the approval of the Secretary, may revise the handler representation on the Board if the Board ceases to be representative of the industry.

[51 FR 29546, Aug. 19, 1986, as amended at 61 FR 17559, Apr. 22, 1996]

§ 982.31 Grower districts.

(a) For the purpose of nominating grower members and alternate members, the following districts within the production area are hereby established:

(1) District 1—The State of Washington, and Clackamas and Multnomah Counties in Oregon.

(2) District 2—Marion and Polk Counties in Oregon.

(3) District 3—Linn, Lane, and Benton Counties in Oregon.

(4) District 4—Yamhill County in Oregon.

(5) District 5—All other Oregon counties within the production area.

(b) The Secretary, upon the recommendation of the Board, may reestablish districts within the production area and may reapportion grower membership among the various districts: *Provided,* That in recommending any such changes, the Board shall give consideration to (1) the relative importance of production in each district and the number of growers in each district; (2) the geographic location of districts as they would affect the efficiency of administering this part; and (3) other relevant factors.

[51 FR 29547, Aug. 19, 1986]

§ 982.32 Initial members and nomination of successor members.

(a) Members and alternate members of the Board serving immediately prior to the effective date of this amended subpart shall continue to serve on the Board until their respective successors have been selected.

(b) Nominations for successor handler members and alternate members specified in § 982.30(b) (1) through (3) shall be made by the largest, second largest, and third largest handler determined according to the tonnage of certified merchantable hazelnuts and, when shelled hazelnut grade and size regulations are in effect, the inshell equivalent of certified shelled hazelnuts (computed to the nearest whole ton) recorded by the Board as handled by each such handler during the two marketing years preceding the marketing year in which nominations are made.

(c) Nominations for successor handler member and alternate handler member positions specified in § 982.30(b)(4) shall be made by the handlers in that category by mail ballot. All votes cast shall be weighted according to the tonnage of certified merchantable hazelnuts and, when shelled hazelnut grade and size regulations are in effect, the inshell equivalent of certified shelled hazelnuts (computed to the nearest whole ton) recorded by the Board as handled by each handler during the two

marketing years preceding the marketing year in which nominations are made. If less than one ton is recorded for any such handler, the vote shall be weighted as one ton. Voting will be by position, and each eligible handler can vote for a member and an alternate member. The person receiving the highest number of weighted votes for each position shall be the nominee for that respective position.

(d) For the purposes of nominating and voting for handler members and alternates, the tonnage of hazelnuts shall be credited to the handler responsible under the order for the payment of assessments of those hazelnuts.

(e) Nominees to successor grower member and alternate member positions shall be submitted to the Secretary after the Board conducts balloting of growers, or officers or employees of growers, in the grower districts according to the following procedure: Names of the candidates to be shown on the ballot for a particular district may be submitted to the Board on petitions signed by not less than 10 growers on record with the Board as growers being in that district; each grower may sign only as many petitions as there are persons to be nominated within that district. If such petitions fail to result in submission of at least two names for a district, the Board shall request County Agricultural Extension Agents in that district to recommend one or more eligible growers to be included on the ballot. Ballots, accompanied by the names of all such candidates, with spaces to indicate voters' choices and spaces for write-in candidates, together with voting instructions, shall be mailed to all growers who are on record with the Board. The person receiving the highest number of votes shall be the member nominee for that district, and the person receiving the second highest number of votes shall be the alternate member. The Board shall recommend one candidate in case of a tie vote.

(f) Nominations received in the foregoing manner by the Board for all handler and grower member and alternate member positions shall be certified and sent to the Secretary at least 60 days prior to the beginning of each two-year term of office, together with all necessary data and other information deemed by the Board to be pertinent or requested by the Secretary. If nominations are not made within the time and manner specified in this subpart, the Secretary may, without regard to nominations, select the Board members and alternates on the basis of the representation provided for in this subpart.

(g) The members of the Board shall nominate the public member and alternate public member at the first meeting following the selection of members for a new term of office.

(h) The Board with the approval of the Secretary shall issue rules and regulations necessary to carry out the provisions of this section or to change the procedures in this section in the event they are no longer practical.

[51 FR 29547, Aug. 19, 1986, as amended at 61 FR 17559, Apr. 22, 1996]

§982.33 Selection and term of office.

(a) *Selection.* Members and their respective alternates shall be selected by the Secretary from nominees submitted by the Board or from among other qualified persons.

(b) *Term of office.* The term of office of Board members and their alternates shall be for two years beginning on July 1 and ending on June 30, but they shall serve until their respective successors are selected and have qualified: *Provided,* That beginning with the 1996–97 marketing year, no member shall serve more than three consecutive two-year terms as member and no alternate member shall serve more than three consecutive two-year terms as alternate unless specifically exempted by the Secretary. Nomination elections for all Board grower and handler member and alternate positions shall be held every two years.

(c) The members on the Board shall continue to serve until the new members and alternates have been selected and have qualified.

[51 FR 29547, Aug. 19, 1986, as amended at 61 FR 17559, Apr. 22, 1996]

§982.34 Qualification.

(a) Any person prior to selection as a member or an alternate member of the Board shall qualify by filing with the Secretary a written acceptance of willingness to serve on the Board.

(b) Each grower member and alternate shall be, at the time of selection and during the term of office, a grower or an officer, employee, or agent of a grower in the district for which nominated.

(c) Each handler member and alternate shall be, at the time of selection and during the term of office, a handler or an officer, employee, or agent of a handler.

(d) Any member or alternate member who at the time of selection was a member (or employed by or an agent of a member) of the group which nominated that person shall, upon ceasing to be such, become disqualified to serve further and that position shall be deemed vacant. In the event any grower member or alternate member of the Board handles hazelnuts produced by other growers or becomes an employee or agent of a handler, that person shall be disqualified to continue to serve on the Board in that capacity.

(e) No person nominated to serve as a public member or alternate member shall have a financial interest in any hazelnut growing or handling operation.

(f) The Board, with the approval of the Secretary, may issue rules and regulations covering matters of qualifications for members or alternate members.

[51 FR 29547, Aug. 19, 1986]

§982.35 Vacancy.

To fill any vacancy occasioned by the death, removal, resignation, or disqualification of any member or alternate of the Board, a successor for his unexpired term shall be nominated and selected in the manner provided in §§982.32 and 982.33, so far as applicable, unless selection is deemed unnecessary by the Secretary.

§982.36 Alternates.

An alternate for a member of the Board shall act in the place of the member during such member's absence or, upon the member's death, removal, resignation, or disqualification, until a successor for that member's term has been selected and has qualified.

[51 FR 29548, Aug. 19, 1986]

§ 982.37 Procedure.

(a) Seven members of the Board shall constitute a quorum at an assembled meeting of the Board, and any action of the Board shall require the concurring vote of at least six members. At any assembled meeting, all votes shall be cast in person.

(b) The Board may vote by mail, telephone, telegraph, or other means of communication: *Provided,* That any votes (except mail votes) so cast shall be confirmed at the next regularly scheduled meeting. When any proposition is submitted for voting by any such method, its adoption shall require 10 concurring votes.

(c) The members of the Board and their alternates shall serve without compensation, but members and alternates acting as members shall be allowed their necessary expenses: *Provided,* That the Board may request the attendance of one or more alternates not acting as members at any meeting of the Board, and such alternates may be allowed their necessary expenses.

[26 FR 6185, Aug. 1, 1959. Redesignated at 26 FR 12751, Dec. 30, 1961, as amended at 51 FR 29548, Aug. 19, 1986; 61 FR 17559, Apr. 22, 1996]

§ 982.38 Powers.

The Board shall have the following powers:

(a) To administer the provisions of this subpart in accordance with its terms;

(b) To make rules and regulations to effectuate the terms and provisions of this subpart;

(c) To receive, investigate, and report to the Secretary complaints of violations of this subpart;

(d) To recommend to the Secretary amendments to this subpart.

§ 982.39 Duties.

The Board shall have among others the following duties:

(a) To select from among its members such officers and adopt rules or bylaws for the conduct of its meetings as it deems advisable;

(b) To act as intermediary between the Secretary and any handler or grower;

(c) To keep minute books and records which will clearly reflect all of its acts and transactions, and such books and records shall be available for examination by the Secretary at any time;

(d) To furnish to the Secretary such available information as he may request;

(e) To appoint such employees as it deems necessary and determine the salaries, define the duties and fix the bonds of such employees;

(f) To cause the books of the Board to be audited by one or more public accountants approved by the Board at least once for each marketing year and at such other times as the Board deems necessary or as the Secretary may request, and to file with the Secretary reports of all audits made;

(g) To investigate the growing, shipping and marketing conditions with respect to hazelnuts, and assemble data in connection therewith;

(h) To give the Secretary the same notice of the meetings of the Board as is given to its members; and

(i) To furnish to the Secretary a report of the proceedings of each meeting of the Board held for the purpose of making marketing policy recommendations.

[24 FR 6185, Aug. 1, 1959, as amended at 46 FR 26039, May 11, 1981; 61 FR 17559, Apr. 22, 1996]

MARKETING POLICY

§ 982.40 Marketing policy and volume regulation.

(a) *General.* As provided in this section, prior to September 20 of each marketing year, the Board may hold meetings for the purpose of computing its marketing policy for that year and shall do so for the purpose of submitting any recommendations on its policy to the Secretary. The Board may designate one of its employees to compute and announce the preliminary computed free and restricted percentages.

(b) *Inshell trade demand.* If the Board determines that volume regulation would tend to effectuate the declared policy of the act, it shall compute and announce an inshell trade demand for that year prior to September 20. The inshell trade demand shall equal the average of the preceding three years' trade acquisitions of inshell hazelnuts: *Provided,* That the Board may increase such average by no more than 25 percent if market conditions justify such an increase. If the trade acquisitions during any or all of these years were abnormal because of crop or marketing conditions, the Board may use a prior year or years in determining the three-year average.

(c) *Inshell allocation—*(1) *Preliminary computed percentages.* Prior to September 20 of a marketing year, the Board shall compute and announce preliminary computed free and restricted percentages for that year, to release 80 percent of the inshell trade demand for that year. The preliminary computed free percentage shall be computed by multiplying that trade demand, adjusted by the declared carryin, by 80 percent, and by dividing that amount by the Board's estimate of orchard-run production less the average disappearance during the preceding three years, plus the undeclared carryin. The difference between 100 percent and the preliminary free percentage shall be the preliminary computed restricted percentage. At the same time, the Board may announce the portion of the restricted supply that may be shelled or exported, and the remainder of that supply to be disposed of in outlets approved by the Board pursuant to §982.52.

(2) *Interim final and final percentages.* On or before November 15, the Board shall meet to recommend to the Secretary the interim final and final free and restricted percentages, including the portion of the restricted supply that may be shelled or exported. The interim final percentages shall release 100 percent of the inshell trade demand previously computed by the Board for the marketing year. The final free and restricted percentages may release an additional 15 percent of the average of the preceding three years' trade acquisitions of inshell hazelnuts for desirable carryout. If the trade acquisitions during any or all of these years were abnormal, the Board may use a prior year or years in determining this three-year average. The final free and

restricted percentages shall become effective 30 days prior to the end of the marketing year, or earlier as may be recommended by the Board and approved by the Secretary. The recommendations to the Secretary shall include the following:

(i) The estimated tonnage of merchantable hazelnuts expected to be produced during the marketing year.

(ii) The estimated tonnage of inshell hazelnuts held by handlers on the first day of the marketing year which may be available for handling as inshell hazelnuts thereafter.

(iii) Any other pertinent factors bearing on the marketing of hazelnuts during the marketing year.

Whenever the Secretary finds, on the basis of the recommendation of the Board or other available information that, to establish the interim final and final free and restricted percentages would tend to effectuate the declared policy of the act, the Secretary shall establish such percentages.

(d) *Grade, size, and quality regulations.* Prior to September 20, the Board may consider grade, size, and quality regulations in effect and may recommend modifications thereof to the Secretary.

(e) *Revision of marketing policy.* At any time prior to February 15 of the marketing year, the Board may recommend to the Secretary revisions in the marketing policy for that year: *Provided,* That in no event shall any such recommendation provide for free and restricted percentages based on an inshell trade demand which is more than 125 percent of the average of the preceding three years' trade acquisitions computed pursuant to paragraph (b) of this section for that marketing year. At any time during the period December 1 through February 10 at the request of two or more handlers, who during the preceding marketing year handled at least 10 percent of all hazelnuts handled, the Board shall meet to determine whether the marketing policy should be revised.

[51 FR 29548, Aug. 19, 1986, as amended at 61 FR 17560, Apr. 22, 1996; 83 FR 52949, Oct. 19, 2018]

§ 982.41 Free and restricted percentages.

The free and restricted percentages computed by the Board or established by the Secretary pursuant to § 982.40 shall apply to all merchantable hazelnuts handled during the current marketing year. Until the preliminary computed free and restricted percentages are computed by the Board for the current marketing year, the percentages in effect at the end of the previous marketing year shall be applicable.

[51 FR 29548, Aug. 19, 1986]

GRADE, SIZE, AND QUALITY REGULATION

§ 982.45 Establishment of grade, size, and quality regulations.

(a) *Minimum standards.* No handler shall handle any inshell or shelled hazelnuts unless such inshell hazelnuts meet requirements of Oregon No. 1 grade and medium size (as defined in the Oregon Grade Standards Hazelnuts In Shell), and such shelled hazelnuts meet such requirements as are established by the Secretary on the basis of a recommendation of the Board, except as may be otherwise provided in § 982.57. These minimum standards may be modified by the Secretary on the basis of a recommendation of the Board or other information whenever he finds that such modification would tend to effectuate the declared policy of the act. Such minimum standards and the provisions of this part relating to the administration thereof shall continue in effect irrespective of whether the season average price of hazelnuts is above the parity level specified in section 2(1) of the act.

(b) *Additional grade and size regulations.* When the season average price of hazelnuts is not determined to be above parity, the Secretary may establish additional grade and size regulations for inshell hazelnuts in the form of a more restrictive minimum standard than that specified in paragraph (a) of this section, or pack specifications as to grades and sizes that may be handled, if he finds, on the basis of a recommendation of the Board or other information, that such regulations would tend to effectuate the declared policy of the act.

(c) *Quality regulations.* For any marketing year, the Board may establish, with the approval of the Secretary, such minimum quality and inspection requirements applicable to hazelnuts to facilitate the reduction of pathogens as will contribute to orderly marketing or will be in the public interest. In such marketing year, no handler shall handle hazelnuts unless they meet applicable minimum quality and inspection requirements as evidenced by certification acceptable to the Board.

(d) *Different regulations for different markets.* The Board may, with the approval of the Secretary, recommend different outgoing quality requirements for different markets. The Board, with the approval of the Secretary, may establish rules and regulations necessary and incidental to the administration of this provision.

[24 FR 6185, Aug. 1, 1959. Redesignated at 26 FR 12751, Dec. 30, 1961, as amended at 37 FR 589, Jan. 14, 1972; 83 FR 52949, Oct. 19, 2018]

§ 982.46 Inspection and certification.

(a) Before or upon handling any hazelnuts, or before any inshell or shelled hazelnuts are credited (under § 982.50 or § 982.51) in satisfaction of a restricted obligation, each handler shall, at his own expense, cause such hazelnuts to be inspected and certified by the Federal-State Inspection Service as meeting the then effective grade and size regulations or, if inshell or shelled hazelnuts are withheld under § 982.51, the applicable requirements specified in that section. The handler obtaining such inspection of hazelnuts shall cause a copy of the certificate issued by such inspection service applicable to such hazelnuts to be furnished to the Board.

(b) All hazelnuts so inspected and certified shall be identified as prescribed by the Board. Such identification shall be affixed to the hazelnut containers by the handler under direction and supervision of the Board or the Federal-State Inspection Service, and shall not be removed or altered by any person except as directed by the Board.

(c) Whenever the Board determines that the length of time in storage and conditions of storage of any lot of certified merchantable hazelnuts have been or are such as to normally cause deterioration, it may require that such lot of hazelnuts be reinspected at the handler's expense prior to handling.

(d) Whenever quality regulations are in effect pursuant to §982.45, each handler shall certify that all product to be handled or credited in satisfaction of a restricted obligation meets the quality regulations as prescribed.

[40 FR 53227, Nov. 17, 1975, as amended at 61 FR 17560, Apr. 22, 1996; 83 FR 52949, Oct. 19, 2018]

CONTROL OF DISTRIBUTION

§982.50 Restricted obligation.

(a) No handler shall handle inshell hazelnuts unless prior to or upon shipment thereof, he: (1) Has withheld from handling a quantity, by weight, of certified merchantable hazelnuts determined by dividing the quantity handled, or to be handled, by the applicable free percentage and multiplying the quotient by the restricted percentage; (2) has withheld from handling an equivalent quantity of creditable ungraded inshell hazelnuts under §982.51(a); or (3) has under §982.51(b), declared in lieu of a quantity of certified merchantable hazelnuts, under paragraph (a)(1) of this section, the equivalent quantity, by weight as determined under that section, of shelled hazelnuts certified as meeting the standards in effect for Oregon No. 1 grade for shelled hazelnuts as contained in Oregon Grade Standards for Hazelnut Kernels or such other standards as may be recommended by the Board and established by the Secretary. Any handler who intends to withhold shelled hazelnuts in satisfaction of a restricted obligation must make such declaration to the Board prior to shelling any such hazelnuts. Withholding may be temporarily deferred under the bonding provisions in §982.54. The quantity of hazelnuts required to be withheld shall be the restricted obligation. Certified merchantable hazelnuts handled in accordance with this subpart shall be deemed to be the handler's quota fixed by the Secretary within the meaning of section 8a(5) of the Act.

(b) Inshell hazelnuts withheld by a handler in satisfaction of his restricted obligation shall not be handled and shall be held by him subject to examination by and accounting control of, the Board until disposed of pursuant to this part.

(c) A handler having certified merchantable hazelnuts which have not been handled at the end of a marketing year may elect to have those hazelnuts bear the restricted and assessment obligations of that year or of the marketing year in which handled. The Board shall establish such procedures as are necessary to facilitate the administration of this option among handlers.

(d) Whenever the restricted percentage for a marketing year is reduced, each handler's restricted obligation shall be reduced to conform with the new restricted percentage. Any handler who, upon such reduction, is withholding restricted hazelnuts in excess of his new restricted obligation may have the excess freed from withholding by complying with such procedures as the Board may require to insure identification of the remaining hazelnuts withheld.

[40 FR 53227, Nov. 17, 1975, as amended at 46 FR 26039, May 11, 1981]

§982.51 Restricted credit for ungraded inshell hazelnuts and for shelled hazelnuts.

(a) A handler may withhold ungraded inshell hazelnuts in lieu of certified merchantable hazelnuts in satisfaction of that handler's restricted obligations, and the weight on which credit may be received shall be the shelled hazelnut equivalent weight as inspected by the Federal-State Inspection Service multiplied by 2.5. Any lot of ungraded hazelnuts not meeting the moisture requirements for certified merchantable hazelnuts shall not be eligible for credit. All determinations as to the shelled hazelnut equivalent weight shall be made by the Federal-State Inspection Service at the handler's expense. Hazelnuts so withheld shall be subject to the applicable requirements of §982.50. The weight of all such lots for which a handler has received credit shall be adjusted by the Board when the lots are handled or disposed of so that the creditable weight is equal to the amount of certified merchantable inshell hazelnuts or certified shelled hazelnuts that are subsequently handled or disposed of from those lots. If this adjustment causes the handler to no longer be in satisfaction of that handler's restricted obligation as required by § 982.50, the deficiency shall be satisfied in the subsequent marketing year. If this adjustment results in a handler disposing of, in restricted outlets, a quantity in excess of that handler's restricted obligation, such excess shall not be credited to such handler's restricted obligation during the subsequent marketing year.

(b) A handler may withhold, in accordance with § 982.50(a), certified shelled hazelnuts in lieu of merchantable hazelnuts in satisfaction of such handler's restricted obligation, subject to such terms and conditions as are recommended by the Board and established by the Secretary. The inshell equivalent of such hazelnuts shall be determined by multiplying the weight of the shelled hazelnuts by 2.5.

(c) The Secretary upon recommendation of the Board and other available data may modify these procedures, change the conversion factors, and specify factors for conversion for different varieties of hazelnuts.

[51 FR 29548, Aug. 19, 1986, as amended at 61 FR 17560, Apr. 22, 1996]

§ 982.52 Disposition of restricted hazelnuts.

Hazelnuts withheld from handling as inshell hazelnuts pursuant to §§ 982.50 and 982.51 may be disposed of as follows:

(a) *Shelling.* Any handler may dispose of such hazelnuts by shelling them under the direction or supervision of the Board or by delivering them to an authorized sheller. Any person who desires to become an authorized sheller in any marketing year may submit written application during such year to the Board. Such application shall be granted only upon condition that the applicant agrees:

(1) To use such restricted hazelnuts as he may receive for no purpose other than shelling;

(2) To dispose of or deliver such restricted hazelnuts, as inshell hazelnuts, to no one other than another authorized sheller;

(3) To comply fully with all laws and regulations applicable to shelling of hazelnuts; and

(4) To make such reports, certified to the Board and to the Secretary as to their correctness, as the Board may require.

(b) *Export.* Sales of certified merchantable restricted hazelnuts for shipment to destinations outside the continental United States and such other distribution areas as may be recommended by the Board and established by the Secretary shall be made only by the Board. Any handler desiring to export any part or all of that handler's certified merchantable restricted hazelnuts shall deliver to the Board the certified merchantable restricted hazelnuts to be exported, but the Board shall be obligated to sell in export only such quantities for which it may be able to find satisfactory export outlets. Any hazelnuts so delivered for export which the Board is unable to export shall be returned to the handler delivering them. Sales for export shall be made by the Board only on execution of an agreement to prevent exportation into the area designated in § 982.16. A handler may be permitted to act as an agent of the Board, upon such terms and conditions as the Board may specify, in negotiating export sales, and when so acting shall be entitled to receive a selling commission as authorized by the Board. The proceeds of all export sales, after deducting all expenses actually and necessarily incurred, shall be paid to the handler whose certified merchantable restricted hazelnuts are so sold by the Board.

(c) *Other outlets.* In addition to the dispositions authorized in paragraphs (a) and (b) of this section, the Board may designate such other outlets into which such hazelnuts may be disposed which it determines are noncompetitive with normal market outlets for inshell hazelnuts. Such dispositions shall be made under the direction or supervision of the Board.

(d) *Restricted credits.* During any marketing year, handlers who dispose of a quantity of eligible hazelnuts in restricted outlets in excess of their restricted obligations, may transfer such excess credits to another handler or handlers. Upon a handler's written request to the Board during a marketing year, the Board shall transfer any or all of such excess restricted credits to such other handler or handlers that the handler may designate. The Board, with the approval of the Secretary, shall establish rules and regulations for the transfer of excess restricted credits.

[40 FR 53227, Nov. 17, 1975, as amended at 51 FR 29549, Aug. 19, 1986; 61 FR 17560, Apr. 22, 1996]

§ 982.53 Substandard hazelnuts.

The Board shall, with the approval of the Secretary, establish such reporting and disposition procedures as it deems necessary to insure that hazelnuts which do not meet the effective inshell or shelled hazelnut minimum standards do not enter normal market outlets for certified hazelnuts.

§ 982.54 Deferment of restricted obligation.

(a) *Bonding.* Compliance by any handler with the requirements of § 982.50 when restricted hazelnuts may be withheld shall be temporarily deferred to any date requested by the handler, but not later than 60 days prior to the end of the marketing year. Such deferment shall be conditioned upon the voluntary execution and delivery by the handler to the Board of a written undertaking before beginning to handle merchantable hazelnuts during the marketing year. Such written undertaking shall be secured by a bond or bonds with a surety or sureties acceptable to the Board that on or prior to such date the handler will have fully satisfied the restricted obligation required by § 982.50, subject to any adjustment pursuant to § 982.51.

(b) *Bonding requirement.* Such bond or bonds shall, at all times during their effective period, be in such amounts that the aggregate thereof shall be no less than the total bonding value of the handler's deferred restricted obligation. The bonding value shall be the deferred restricted obligation poundage multiplied by the applicable bonding rate. The cost of such bond or bonds shall be borne by the handler filing same.

(c) *Bonding rate.* Said bonding rate shall be an amount per pound as established by the Board. Such bonding rate shall be based on the estimated value of restricted credits for the current marketing year. Until bonding rates for a marketing year are fixed, the rates in effect for the preceding marketing year shall continue in effect. The Board should make any necessary adjustments once such new rates are fixed.

(d) *Restricted credit purchases.* Any sums collected through default of a handler on the handler's bond shall be used by the Board to purchase restricted credits from handlers, who have such restricted credits in excess of their needs, and are willing to part with them. The Board shall at all times purchase the lowest priced restricted credits offered, and the purchases shall be made from the various handlers as nearly as practicable in proportion to the quantity of their respective offerings of the restricted credits to be purchased.

(e) *Unexpended sums.* Any unexpended sums which have been collected by the Board through default of a handler on the handler's bond, remaining in the possession of the Board at the end of a marketing year, shall be used to reimburse the Board for its expenses, including administrative and other costs incurred in the collection of such sums, and in the purchase of restricted credits as provided in paragraph (d) of this section.

(f) *Transfer of restricted credit purchases.* Restricted credits purchased as provided for in this section shall be turned over to those handlers who have defaulted on their bonds for liquidation of their restricted obligation. The quantity delivered to each handler shall be that quantity represented by sums collected through default.

(g) *Collection upon bonds.* Collection upon any defaulted bond shall be deemed a satisfaction of the restricted obligation represented by the collection.

[40 FR 53228, Nov. 17, 1975, as amended at 46 FR 26039, May 11, 1981; 51 FR 29549, Aug. 19, 1986; 61 FR 17560, Apr. 22, 1996]

§ 982.55 Exchange of certified merchantable hazelnuts withheld.

Any handler who has withheld from handling certified merchantable hazelnuts pursuant to the requirements of § 982.50 may exchange therefor an equal quantity, by weight, of other certified merchantable hazelnuts. Any such exchange shall be made under the direction or supervision of the Board.

§ 982.56 Interhandler transfers.

Within the area of production, interhandler transfers of hazelnuts may be made as follows:

(a) Uncertified inshell hazelnuts may be sold or delivered by one handler to another for packing or shelling, and the receiving handler shall be responsible for compliance with the regulations effective pursuant to this part with respect to such hazelnuts.

(b) Restricted hazelnuts withheld by a handler may be sold or delivered to another handler for shelling, export, or other authorized outlet subject to the disposition requirements set forth in § 982.52.

(c) Certified hazelnuts other than restricted hazelnuts may be sold or delivered by one handler to another and the transferring handler shall be responsible for compliance with the requirements effective pursuant to this part, unless specified and agreed upon in writing by both handlers that the receiving handler shall be responsible for such compliance and a copy of such agreement is furnished to the Board.

(d) The Board, with the approval of the Secretary, shall establish procedures, including necessary reports, for such transfers.

§ 982.57 Exemptions.

(a) *General.* The Board, with the approval of the Secretary, may establish such rules, regulations, and safeguards that exempt from any or all requirements pursuant to this part such quantities of hazelnuts or types of shipments as do not interfere with the volume and quality control objectives of this part, and shall require such reports, certifications, or other conditions as are necessary to ensure that such hazelnuts are handled or used only as authorized.

(b) *Sales by growers direct to consumers.* Any hazelnut grower may sell hazelnuts of such grower's own production free of the regulatory and assessment provisions of this part if such grower sells such hazelnuts in the area of production directly to end users at such grower's ranch or orchard or at roadside stands and farmers' markets. The Board, with the approval of the Secretary, may establish such rules, regulations, and safeguards and require such reports, certifications, and other conditions, as are necessary to ensure that such hazelnuts are disposed of only as authorized. Mail order sales are not exempt sales under this part.

[51 FR 29549, Aug. 19, 1986, as amended at 61 FR 17560, Apr. 22, 1996]

MARKET DEVELOPMENT

§ 982.58 Research, promotion, and market development.

(a) *General.* The Board, with the approval of the Secretary, may establish or provide for the establishment of projects involving production research, marketing research and development, and marketing promotion, including paid advertising, designed to assist, improve, or promote the marketing, distribution, consumption, or efficient production of hazelnuts. The Board may also provide for crediting the pro rata expense assessment obligations of a handler with such portion of such handler's direct expenditures for such marketing promotion including paid advertising as may be authorized. The expenses of such projects shall be paid from funds collected pursuant to § 982.61, § 982.63, or credited pursuant to paragraph (b) of this section.

(b) *Creditable expenditures.* The Board, with the approval of the Secretary, may provide for crediting all or any portion of a handler's direct expenditures for marketing promotion including paid advertising, that promotes the sale of hazelnuts, hazelnut products, or their uses. No handler shall receive credit for any allowable direct expenditures that would exceed the total of the handler's assessment obligation which is attributable to that portion of the handler's assessment designated for marketing promotion including paid advertising.

(c) *Rules and regulations.* Before any projects involving marketing promotion, including paid advertising and the crediting of the pro rata expense assessment obligation of handlers is undertaken pursuant to this section, the Secretary, after recommendation by the Board, shall prescribe appropriate rules and regulations as are necessary to effectively administer such projects.

[51 FR 29549, Aug. 19, 1986, as amended at 61 FR 17560, Apr. 22, 1996]

EXPENSES AND ASSESSMENTS

§ 982.60 Expenses.

The Board is authorized to incur such expenses including maintenance of an operating reserve fund as the Secretary may find are reasonable and likely to be incurred by it during each marketing year, for the maintenance and functioning of the Board and for such purposes as the Secretary may, pursuant to the provisions of this subpart, determine to be appropriate. The recommendation of the Board as to the expenses and size of the operating reserve for each such marketing year, together with all data supporting such recommendations, shall be submitted to the Secretary at the beginning of the fiscal year in connection with which such recommendation is made. The funds to cover such expenses shall be acquired by levying assessments as provided in § 982.61.

§ 982.61 Assessments.

(a) For each marketing year, the Secretary shall fix an assessment rate per pound of hazelnuts handled and withheld, including the creditable weight of ungraded restricted hazelnuts withheld pursuant to § 982.51 and, when subject to regulation pursuant to § 982.45, the inshell equivalent of shelled hazelnuts certified which are produced from other than restricted hazelnuts that will provide sufficient funds to meet the authorized expenses and reserve requirements of the Board. At any time during or after a marketing year when he determines, on the basis of a Board recommendation or other information, that a different rate is necessary, the Secretary may modify the assessment rate and the new rate shall be applicable to all such hazelnuts. Each handler shall pay to the Board on demand, assessments on all such assessable hazelnuts at the rate fixed by the Secretary, less any amounts credited pursuant to § 982.58. The Board shall impose a late payment charge on any handler who fails to pay his assessment within the time prescribed by the Board. In the event the handler thereafter fails to pay the amount outstanding, including the late payment charge, within the prescribed time, the Board shall impose an additional charge in the form of interest on such outstanding amount. The rate of such charges shall be prescribed by the Board, with the approval of the Secretary.

(b) In order to provide funds for the administration of the provisions of this part during the first part of a fiscal period before sufficient operating income is available from assessments on the current year's shipments, the Board may accept the payment of assessments in advance, and may also borrow money for such purpose. Further, payment discounts may be authorized by the Board upon the approval of the Secretary to handlers making such advance assessment payments.

[24 FR 6185, Aug. 1, 1959. Redesignated at 26 FR 12751, Dec. 30, 1961, as amended at 37 FR 589, Jan. 14, 1972; 51 FR 29550, Aug. 19, 1986; 61 FR 17560, Apr. 22, 1996]

§ 982.62 Accounting.

(a) *Operating reserve.* The Board with the approval of the Secretary may establish and maintain an operating monetary reserve in an amount not to exceed approximately one marketing year's operational expenses or such lower limits as the Board with the approval of the Secretary may establish.

(b) *Refunds.* At the end of a marketing year funds in excess of the marketing year's expenses and reserve requirements shall be refunded to handlers from whom collected and each handler's share of such excess funds shall be the amount of assessments the handler paid in excess of the handler's pro rata share of expenses of the Board. However, excess funds may be maintained and used by the Board until December 1 following the end of any such marketing year: *Provided,* That the Board shall refund to each handler upon request, or credit to the handler's account with the Board, the handler's share of such excess prior to January 1.

(c) *Termination.* Upon termination of this subpart any money remaining unexpended in possession of the Board shall be distributed in such manner as the Secretary may direct: *Provided,* That to the extent practical, such funds shall be returned pro rata to the persons from whom such funds were collected.

[24 FR 6185, Aug. 1, 1959, as amended at 46 FR 26040, May 11, 1981]

§ 982.63 Contributions.

The Board may accept voluntary contributions but these shall only be used to pay expenses incurred pursuant to § 982.58. Furthermore, such contributions shall be free from any encumbrances by the donor and the Board shall retain complete control of their use.

[61 FR 17560, Apr. 22, 1996]

RECORDS AND REPORTS

§ 982.64 Creditable promotion and advertising reports.

Each handler shall file such reports of creditable promotion including paid advertising conducted pursuant to § 982.58 as recommended by the Board and approved by the Secretary.

[51 FR 29550, Aug. 19, 1986]

§ 982.65 Carryover reports.

As of January 1, May 1, and August 1, or such other dates as the Board may recommend and the Secretary approve, each handler shall report within 10 days to the Board the handler's inventory of inshell and shelled hazelnuts. Such reports shall be certified to the Board and the Secretary as to their accuracy and completeness and shall show, among other items, the following: (a) Certified merchantable hazelnuts on which the restricted obligation has been met; (b) merchantable hazelnuts on which the restricted obligation has not been met; (c) the merchantable equivalent of any hazelnuts intended for handling as inshell hazelnuts; and (d) restricted hazelnuts withheld.

[46 FR 26040, May 11, 1981]

§ 982.66 Shipment reports.

Each handler shall report to the Board the respective quantities of inshell and shelled hazelnuts handled by him during such periods and in such manner as are prescribed by the Board with the approval of the Secretary.

§ 982.67 Reports of disposition of restricted hazelnuts.

(a) Each handler, before he disposes of any quantity of restricted hazelnuts held by him, shall file with the Board a report of his intention to dispose of such quantity of restricted hazelnuts. This report shall be filed not less than five days prior to the date on which the restricted hazelnuts are disposed of, unless the five-day period is expressly waived by the Board.

(b) Each handler, within 15 days after the disposition of any quantity of restricted hazelnuts, shall file with the Board a report of the actual disposition of such quantity of restricted hazelnuts. Such reports shall be certified to the Board and to the Secretary as to their correctness and accuracy.

(c) All reports required by this section shall show the quantity, pack, and location of the hazelnuts covered by such reports; the applicable handler's storage lot and inspection certificate numbers; and the disposition of the restricted hazelnuts which is intended or which has been accomplished.

§ 982.68 Other reports.

Each handler shall furnish to the Board such other reports as the Board, with the approval of the Secretary, may require to enable it to exercise its powers and to perform its duties.

§ 982.69 Verification of reports.

For the purpose of checking and verifying reports submitted by handlers, the Secretary and the Board, through its duly authorized agents, shall have access to each handler's premises at any time during reasonable business hours and shall be permitted to inspect any hazelnuts held by such handler and all records of the handler with respect to hazelnuts held or disposed of by such handler and all records of the handler with respect to promotion and advertising activities conducted pursuant to § 982.58. Each handler shall furnish all labor necessary to facilitate such inspections as the Secretary or the Board may make of such handler's holdings of any hazelnuts. Each handler shall store hazelnuts in such manner as to facilitate inspection, and shall maintain adequate storage records which will permit accurate identification of all such hazelnuts held.

[24 FR 6185, Aug. 1, 1959. Redesignated at 26 FR 12751, Dec. 30, 1961, as amended at 37 FR 589, Jan. 14, 1972; 51 FR 29550, Aug. 19, 1986]

§982.70 Confidential information.

All reports and records furnished or submitted by handlers to the Board, which include data or information constituting a trade secret or disclosing of the trade position, financial condition, or business operations of the particular handler from whom received, shall be kept in the custody and under the control of one or more employees of the Board, and shall be disclosed to no person except the Secretary.

§982.71 Records.

Each handler shall maintain such records of hazelnuts received, held, and disposed of by the handler, and such records detailing such handler's promotion and advertising activities, as may be prescribed by the Board in order to perform its function under this part. Such records shall be retained and be available for examination by authorized representatives of the Board or the Secretary for a period of two years after the end of the marketing year in which the transactions occurred.

[40 FR 53228, Nov. 17, 1975, as amended at 51 FR 29550, Aug. 19, 1986]

MISCELLANEOUS PROVISIONS

§982.80 Right of the Secretary.

The members of the Board (including successors, alternates, or other persons selected by the Secretary), and any agent or employee appointed or employed by the Board, shall be subject to removal or suspension by the Secretary, in his discretion, at any time. Each and every order, regulation, decision, determination, or other act of the Board shall be subject to the continuing right of the Secretary to dis-

approve of the same at any time, and, upon such disapproval, shall be deemed null and void except as to acts done in reliance thereon or in compliance therewith.

§982.81 Personal liability.

No member or alternate member of the Board, or any employee or agent thereof, shall be held personally responsible, either individually or jointly with others, in any way whatsoever, to any handler or any other person for errors in judgment, mistakes, or other acts either of commission or omission, as such member, alternate member, agent or employee, except for acts of dishonesty.

§982.82 Separability.

If any provision of this subpart is declared invalid, or the applicability thereof to any person, circumstance, or thing is held invalid, the validity of the remainder of this subpart or the applicability thereof to any other person, circumstance, or thing shall not be affected thereby.

§982.83 Derogation.

Nothing contained in this subpart is, or shall be construed to be, in derogation or in modification of the rights of the Secretary or of the United States to exercise any powers granted by the act or otherwise, or, in accordance with such powers, to act in the premises whenever such action is deemed advisable.

§982.84 Duration of immunities.

The benefits, privileges, and immunities conferred upon any person by virtue of this subpart shall cease upon the termination of this subpart, except with respect to acts done under and during the existence of this subpart.

§982.85 Agents.

The Secretary may, by a designation in writing, name any person, including any officer or employee of the United States Department of Agriculture, to act as his agent or representative in connection with any of the provisions of this subpart.

§ 982.86 Effective time, termination or suspension.

(a) *Effective time.* The provisions of this subpart, as well as any amendments to this subpart, shall become effective at such time as the Secretary may declare, and shall continue in force until terminated or suspended in one of the ways specified in this section.

(b) *Suspension or termination.* (1) The Secretary may, at any time, terminate the provisions of this subpart by giving at least one day's notice by means of a press release or in any other manner which he may determine.

(2) The Secretary shall terminate or suspend the operation of any or all of the provisions of this subpart whenever he finds that such provisions do not tend to effectuate the declared policy of the act.

(3) *Referendum.* The Board shall recommend to the Secretary during the first half of every 10-year period starting January 1, 1990, that a referendum be conducted to ascertain whether continuance of this subpart is favored by the producers.

(4) The Secretary shall terminate the provisions of this subpart at the end of any marketing year whenever the Secretary finds that such termination is favored by a majority of the producers of hazelnuts who during the preceding marketing year have been engaged in the production for marketing of hazelnuts in the States of Oregon and Washington: *Provided,* That such majority have during such period produced for market more than 50 percent of the volume of such hazelnuts produced for market within said States; but such termination shall be effected only if announced 30 days or more before the end of the then current marketing year.

(5) The provisions of this subpart shall, in any event, terminate whenever the provisions of the act authorizing them cease to be in effect.

(c) *Proceedings after termination.* (1) Upon the termination of the provisions of this subpart, the members of the Board then functioning shall continue as joint trustees, for the purpose of liquidating the affairs of the Board, of all funds and property then in the possession or under the control of the Board, including claims for any funds unpaid or property not delivered at the time of such termination. Action by said trusteeship shall require the concurrence of a majority of the said trustees.

(2) Said trustees shall continue in such capacity until discharged by the Secretary; shall, from time to time, account for all receipts and disbursements and deliver all property on hand, together with all books and records of the Board and the joint trustees, to such person as the Secretary may direct; and shall, upon the request of the Secretary, execute such assignments or other instruments necessary or appropriate to vest in such person full title and right to all of the funds, property, and claims vested in the Board or the joint trustees pursuant to this subpart.

(3) Any person to whom funds, property, or claims have been transferred or delivered by the Board or its members, pursuant to this section shall be subject to the same obligations imposed upon the members of the said Board and upon said joint trustees.

[24 FR 6185, Aug. 1, 1959. Redesignated at 26 FR 12751, Dec. 30, 1961, as amended at 46 FR 26040, May 11, 1981; 51 FR 29550, Aug. 19, 1986]

§ 982.87 Effect of termination or amendment.

(a) Unless otherwise expressly provided by the Secretary, the termination of this subpart or of any regulation issued pursuant to this subpart, or the issuance of any amendment to either thereof, shall not (1) affect or waive any right, duty, obligation, or liability which shall have arisen or which may thereafter arise in connection with any provision of this subpart or any regulation issued under this subpart, or (2) release or extinguish any violation of this subpart or of any regulation issued under this subpart, or (3) affect or impair any right or remedies of the Secretary or of any other person, with respect to any such violation.

(b) All rules and regulations in this part which are in effect immediately prior to this amendment of this subpart and not inconsistent with such amendment shall continue in effect until otherwise prescribed pursuant to this subpart.

§982.88 Amendments.

Amendments to this subpart may be proposed, from time to time, by any person or by the Board.

Subpart B—Grade and Size Requirements

§982.101 Grade requirements for shelled hazelnuts.

(a) Pursuant to §982.45(a), no handler shall handle any shelled hazelnuts unless such hazelnuts meet the grade requirements for shelled hazelnuts as contained in exhibit A of this section.

(b) Pursuant to §§982.50(a) and 982.51(b), a handler may declare and withhold shelled hazelnuts in lieu of merchantable hazelnuts in satisfaction of the handler's restricted obligation. Shelled hazelnuts so declared and withheld shall, in lieu of the standards prescribed in §982.50(a)(3), meet the grade requirements contained in exhibit A of this section.

EXHIBIT A

Grade Requirements for Shelled Hazelnuts

Hazelnut kernels or portions of hazelnut kernels shall meet the following requirements:

(1) Well dried and clean;
(2) Free from foreign material, mold, rancidity, decay or insect injury; and
(3) Free from serious damage caused by serious shriveling, or other means.

Tolerances

In order to allow for variations incident to proper grading and handling the following tolerances, by weight, are permitted as specified:

(1) For Foreign Material: 0.02 of one percent, for foreign material.
(2) For Defects: Five percent for kernels or portions of kernels which are below the requirements of this grade, including not more than the following: Two percent for mold, rancidity, decay or insect injury: *Provided, That* not more than one percent shall be for mold, rancidity, or insect injury.

Definitions

(1) *Well dried* means that the kernels are firm and crisp, not containing more than 6 percent moisture.
(2) *Clean* means practically free from plainly visible adhering dirt or other foreign material.
(3) *Foreign material* means any substance other than the hazelnut kernels, or portions of kernels. (Loose skins, pellicles or corky tissue which have become separated from the kernels shall not be considered as foreign material, provided that this material does not exceed .02 of one percent by weight.)

(4) *Serious damage* means any specific defect described in this section, or any equally objectionable variation of any one of these defects, or any other defects, or any combination of defects, which seriously detracts from the appearance or the edible or marketing quality of the individual portion of the kernel or of the lot as a whole. The following defects shall be considered as serious damage.

(i) *Serious shriveling* means when the kernel is seriously shrunken, wrinkled and tough.
(ii) *Mold* means that there is a visible growth of mold either on the outside or inside of the kernel.
(iii) *Rancidity* means that the kernel is noticeably rancid to the taste. An oily appearance of the flesh does not necessarily indicate a rancid condition.
(iv) *Decay* means that any portion of the kernel is decomposed.
(v) *Insect injury* means that the insect, frass or web is present, or the kernel or portion of kernel show definite evidence of insect feeding.

[47 FR 12611, Mar. 24, 1982, as amended at 48 FR 34015, July 27, 1983]

Subpart C—Free and Restricted Percentages

§982.254 Free and restricted percentages—2006–2007 marketing year.

The final free and restricted percentages for merchantable hazelnuts for the 2006–2007 marketing year shall be 8.2840 percent and 91.7160 percent, respectively.

[72 FR 2603, Jan. 22, 2007]

§982.255 Free and restricted percentages—2007–2008 marketing year.

(a) The interim final free and restricted percentages for merchantable hazelnuts for the 2007–2008 marketing year shall be 8.1863 and 91.8137 percent, respectively.

(b) On May 1, 2008, the final free and restricted percentages for merchantable hazelnuts for the 2007–2008 marketing year shall be 9.2671 and 90.7329 percent, respectively.

[73 FR 9005, Feb. 19, 2008]

Subpart D—Assessment Rates

§ 982.340 Assessment rate.

On and after July 1, 2017, an assessment rate of $0.006 per pound is established for Oregon and Washington hazelnuts.

[82 FR 61675, Dec. 29, 2017]

Subpart E—Administrative Requirements

SOURCE: 26 FR 4191, May 16, 1961, unless otherwise noted. Redesignated at 26 FR 12751, Dec. 30, 1961.

§ 982.446 Inspection documentation.

Pursuant to § 982.46(b), handlers are required to use the following identification on bags and cartons of 25 pounds or larger capacity which contain certified hazelnuts:

(a) The words "This Produce Inspected and Certified Per Federal Marketing Order No. 982" shall be contained within an outline of the combined States of Oregon and Washington; and

(b) This identification shall be printed on the upper right quarter of the printed side of a bag; or

(c) This identification shall be printed on the upper right quarter of one of the side panels of a carton.

[54 FR 46720, Nov. 7, 1989]

§ 982.450 Application of restricted obligation.

(a) Each handler required to withhold restricted hazelnuts pursuant to § 982.50 or § 982.51 shall hold such hazelnuts separate from all other hazelnuts and shall maintain the identity of each lot so withheld. The restricted product withheld must be reported to the Board on F/H Form 1d, Restricted Inshell Certified.

(b) Each handler making the election pursuant to § 982.50(c) in connection with certified merchantable hazelnuts which have not been handled, shall thereupon give written notification to the Board on F/H Form 4 of the particular election and of the weight and identity of the hazelnuts involved.

(c) Pursuant to § 982.50(d), a handler may withdraw from withholding restricted hazelnuts in excess of such handler's restricted obligation upon advising the Board of the weight and lot identity of the hazelnuts to be withdrawn. When the quantity of restricted hazelnuts to be withdrawn from withholding consists of a part of a lot of ungraded hazelnuts, no part of such lot shall be withdrawn unless the remainder of such lot is reinspected and meets the requirements of § 982.51. Handlers will use F/H Form 1d prior to the end of the marketing year or F/H Form 7 after the end of the marketing year, when reporting the withdrawal of restricted hazelnuts from withholding status.

[54 FR 46720, Nov. 7, 1989]

§ 982.452 Disposition of restricted hazelnuts.

(a) *Shelling.* (1) Any person desiring to shell restricted hazelnuts during a fiscal year may do so upon being designated by the Board as an authorized sheller for such year. Application for such designation shall be made in duplicate on F/H Form B and include, in addition to the conditions specified in § 982.52(a), the following: (i) The location of the applicant's shelling operation; (ii) the number of years such person has operated a hazelnut shelling plant; and (iii) the daily (8-hour) shelling capacity of the plant. Designation of an authorized sheller shall be effected by the board manager signing the application form and returning a signed copy of the form to the applicant. Each such designation shall continue in effect during the particular fiscal year so long as the authorized sheller is in compliance with the requirements and conditions pursuant to § 982.52 applicable to authorized shellers.

(2) When an authorized sheller completes the shelling of a lot of restricted hazelnuts, the sheller shall submit a report thereon to the Board on F/H Form 7 showing:

(i) The date shelling was completed;

(ii) The inspection certificate or lot number;

(iii) The quantity shelled;

(iv) The weight of the kernels produced; and

(v) The location where restricted hazelnuts were held immediately prior to shelling.

(b) *Exports.* Any handler who desires to act as agent of the Board in negotiating export sales of certified merchantable restricted hazelnuts may do so upon the execution of an "Export Agreement", F/H Form A, wherein the handler agrees, among other things, to negotiate such export sales at not less than such price as the Board may prescribe, and in conformity to and compliance with the other terms and conditions of the Export Agreement including those set forth in §982.52(b).

(c) *Other authorized outlets.* Under the direction or supervision of the Board, a handler may dispose of restricted hazelnuts for charitable purposes and for promoting the consumption of hazelnuts on behalf of the hazelnut industry in general. The report required under §982.67(b) following each such disposition shall be accompanied by a certification by the person receiving such hazelnuts from the handler that they will be used for charitable or promotional purposes, as authorized.

[26 FR 4191, May 16, 1961. Redesignated at 26 FR 12751, Dec. 30, 1960, as amended at 54 FR 46721, Nov. 7, 1989]

§982.453 Disposition of substandard hazelnuts.

The Board shall maintain a list of approved users who are crushers, livestock feed manufacturers, or livestock feeders, and of the locations of the facilities to which substandard hazelnuts may be shipped. Users interested in purchasing substandard hazelnuts or hazelnut waste must make prior application to the Board on F/H Form D to be included on the approved list of such users. Each handler who disposes of substandard hazelnuts to an approved user shall, upon shipment, report to the Board on F/H Form D1 the quantities disposed of or shipped. Substandard hazelnuts disposed of to an approved user may only be shipped directly to an approved location where the crushing, feed manufacture, or feeding is to take place. The Board may deny approval to any user application, or may remove any user from the approved list when such denial or removal is deemed necessary to ensure control over disposition of substandard hazelnuts. This may occur if the Board determines that substandard hazelnuts are not properly shipped to, or utilized at, approved facilities, in compliance with this requirement. F/H Form D includes the location and description of the disposal facilities to be used as well as a certification to the Board and the Secretary of Agriculture that the applicant will:

(a) Crush, manufacture feed, or feed to livestock such hazelnuts at the location;

(b) Use such hazelnuts for no other purpose than for crushing into oil, manufacturing into livestock feed, or livestock feeding;

(c) Permit such inspection of premises and of hazelnuts received and held, and such examination of books and records covering hazelnut transactions as the Board may require;

(d) Keep a record of receipts, holdings, and use of substandard hazelnuts available for examination by authorized representatives of the Board and the U.S. Department of Agriculture for a period of two years after the end of the marketing year in which the recorded transactions are completed; and

(e) Make such reports, certified to the Board and the Secretary of Agriculture as to their correctness, as the Board with the approval of the Secretary may require.

[54 FR 24328, June 7, 1989]

§982.454 Sureties acceptable to the Board.

Bonds secured by cash, cashier's or certified checks, or by assets that are entirely separate and apart from the handler named in the bond may be accepted by the Board pursuant to §982.54(a). As a condition of accepting any surety, the Board may require such financial statements or other information relating to the ability of such surety to guarantee a handler's bond as it deems necessary. Handlers are also required to submit F/H Form C to the Board to document the handler's execution of a bond.

[54 FR 46721, Nov. 7, 1989]

§982.455 Exchange of certified merchantable hazelnuts withheld.

Each handler desiring to exchange hazelnuts pursuant to §982.55 shall prior thereto file a written notification with the Board setting forth for the respective quantities of hazelnuts involved in the exchange, the inspection certificate numbers, quantities, locations, and applicable lot numbers.

[54 FR 46721, Nov. 7, 1989]

§ 982.456   Interhandler transfers.

Each interhandler transfer of hazelnuts pursuant to § 982.56 (a) and (c) may be made upon notification to the Board in triplicate by the receiving handler on F/H Form 2 signed by both the transferring handler and the receiving handler which shall include the following information:

(a) Date of transfer;

(b) Names of the transferring and receiving handlers;

(c) Locations between which the hazelnuts were transferred;

(d) Whether uncertified inshell or certified merchantable;

(e) Net weight of the hazelnuts transferred, by size and variety;

(f) The inspection certificate, or lot number covering the hazelnuts; and

(g) If certified merchantable, the name of the handler responsible for compliance with the applicable requirements pursuant to this part relating to such hazelnuts.

[54 FR 46721, Nov. 7, 1989]

§ 982.460   Transfer of excess restricted credits.

(a) *Notification.* Each handler having excess restricted credits who wants to transfer all or a portion thereof to another handler or handlers, may notify the Board accordingly. The Board shall make available to all handlers such information on a weekly basis.

(b) *Application.* Each handler who has excess restricted credits and desires to transfer them to another handler, may submit such request to the Board on F/H Form 3. This form shall include:

(1) The name and signature of the handler requesting the transfer;

(2) The name and signature of the designated handler to whom the transfer is to be made;

(3) The amount of excess restricted credits to be transferred; and

(4) Such other information as may be needed by the Board to enable the Board to effect the requested transfer of the excess restricted credits.

(c) *Transfer.* The Board shall transfer the requested amount of the excess restricted credits from one handler to a designated handler upon receipt of a completed F/H Form 3 together with such information as may be required by this section.

[37 FR 3630, Feb. 18, 1972, as amended at 54 FR 46721, Nov. 7, 1989]

§ 982.461   Late payment and interest charges.

The Board shall impose a late payment charge on any handler failing to pay his assessment within 30 days of the billing date shown on the handler's assessment statement received from the Board. Such amount shall be shown on the statement as the "Assessment Due". The late payment charge shall be 5 percent of the unpaid balance of that amount. In the event the handler fails to pay the delinquent amount, including the late payment charge, within 60 days following the billing date, an additional 1 percent interest charge shall be applied monthly thereafter to the unpaid balance, including any accumulated interest. Any amount paid by a handler as assessments, including any charges imposed pursuant to this paragraph, shall be credited when the payment is received in the Board's office.

[38 FR 5151, Feb. 26, 1973]

§ 982.466   Reports of inshell hazelnuts handled, shelled and withheld.

Each handler shall report to the Board monthly on F/H Form 1 and F/H Forms 1a through 1e, as applicable, the quantities of inshell hazelnuts handled or withheld for restricted use and all product shelled and certified since the last report. All reports shall be submitted to include transactions through the end of each month, or other reporting periods established by the Board, and are due in the Board office on the tenth day following the end of the reporting period. The quantities of inshell hazelnuts handled shall be reported by size. The respective quantities of merchantable or ungraded hazelnuts withheld as restricted product shall be reported separately, and with respect to hazelnuts certified for shelling, or certified kernels withheld, the kernel weight and inshell equivalent weight shall be reported separately by size.

[54 FR 46721, Nov. 7, 1989]

§ 982.467 Report of receipts and dispositions of hazelnuts grown outside the United States.

Each handler who receives hazelnuts grown outside the United States shall report to the Board monthly on *F/H Form 1f* the receipt and disposition of such hazelnuts. All reports submitted shall include transactions through the end of each month, or other reporting periods established by the Board, and are due in the Board office on the tenth day following the end of the reporting period. The report shall include the quantity of such hazelnuts received, the country of origin for such hazelnuts, inspection certificate number, whether such hazelnuts are inshell or kernels, the disposition outlet, and shipment date of such hazelnuts. With each report, the handler shall submit copies of the applicable inspection certificates.

[67 FR 5445, Feb. 6, 2002]

§ 982.468 Report of hazelnut receipts, disposition, and inventory.

On or before January 15 and July 15, or any other date requested by the Board with the approval of the Secretary, each handler shall:

(a) Report to the Board on F/H Form 6 such handler's receipts and disposition of inshell hazelnuts and production of hazelnut kernels during the respective preceding six-month period of July 1 to December 31, and the preceding 12-month period of July 1 to June 30; and

(b) Report to the Board on F/H Form 5 such handler's inventory of hazelnuts as of January 1 and July 1, respectively, showing the quantities of inshell hazelnuts separately in terms of certified merchantable, graded uncertified merchantable, restricted, and ungraded. The certified merchantable hazelnuts shall be reported on the basis of whether located within or outside the production area and whether

or not the restricted obligation has been met.

[54 FR 46721, Nov. 7, 1989]

§ 982.471 Records.

Each handler shall maintain complete and accurate records showing the receipt, shipment and sale of all hazelnuts handled, used or otherwise disposed of and shall retain such records for the two-year period prescribed in § 982.71. Handlers shall also maintain a current record of all hazelnuts held in inventory.

[54 FR 46721, Nov. 7, 1989]

PART 983—PISTACHIOS GROWN IN CALIFORNIA, ARIZONA, AND NEW MEXICO

Subpart A—Order Regulating Handling

DEFINITIONS

AUTHORITY: 7 U.S.C. 601–674.

SOURCE: 69 FR 17850, Apr. 4, 2004, unless otherwise noted.

Subpart A—Order Regulating Handling

DEFINITIONS

§ 983.1 Accredited laboratory.

An *accredited laboratory* is a laboratory that has been approved or accredited by the U.S. Department of Agriculture.

[74 FR 56539, Nov. 2, 2009]

§ 983.2 Act.

Act means Public Act No. 10, 73rd Congress (May 12, 1933), as amended and as re-enacted and amended by the Agricultural Marketing Order Act of 1937, as amended (48 Stat. 31, as amended; 7 U.S.C. 601 *et seq.*).

§ 983.3 Affiliation.

Affiliation. This term normally appears as "affiliate of", or "affiliated with", and means a person such as a producer or handler who is: A producer or handler that directly, or indirectly through one or more intermediaries, owns or controls, or is controlled by, or is under common control with the producer or handler specified; or a producer or handler that directly, or indirectly through one or more intermediaries, is connected in a proprietary capacity, or shares the ownership or control of the specified producer or handler with one or more other producers or handlers. As used in this part, the term "control" (including the terms "controlling", "controlled by", and "under the common control with") means the possession, direct or indirect, of the power to direct or cause the direction of the management and policies of a handler or a producer, whether through voting securities, membership in a cooperative, by contract or otherwise.

§ 983.4 Aflatoxin.

Aflatoxin is one of a group of mycotoxins produced by the molds *Aspergillus flavus* and *Aspergillus parasiticus*. Aflatoxins are naturally occurring compounds produced by molds, which can be spread in improperly processed and stored nuts, dried fruits and grains.

§983.5 Aflatoxin inspection certificate.

Aflatoxin inspection certificate is a certificate issued by an accredited laboratory or by a USDA laboratory.

§983.6 Assessed weight.

Assessed weight means pounds of inshell pistachios, with the weight computed at 5 percent moisture, received for processing by a handler within each production year: *Provided,* That for loose kernels, the actual weight shall be multiplied by two to obtain an inshell weight; *Provided further,* That the assessed weight may be based upon quality requirements for inshell pistachios that may be recommended by the Committee and approved by the Secretary.

[74 FR 56539, Nov. 2, 2009]

§983.7 Certified pistachios.

Certified pistachios are those that meet the inspection and certification requirements under this part.

[74 FR 56539, Nov. 2, 2009]

§983.8 Committee.

Committee means the Administrative Committee for Pistachios established pursuant to §983.41.

[74 FR 56539, Nov. 2, 2009]

§983.9 Confidential data or information.

Confidential data or information submitted to the committee consists of data or information constituting a trade secret or disclosure of the trade position, financial condition, or business operations of a particular entity or its customers.

§983.10 Department or USDA.

Department or USDA means the United States Department of Agriculture.

§983.11 Districts.

(a) *Districts* shall consist of the following:

(1) *District 1* consists of Tulare, Kern, San Bernardino, San Luis Obispo, Santa Barbara, Ventura, Los Angeles, Orange, Riverside, San Diego, and Imperial Counties of California.

(2) *District 2* consists of Kings, Fresno, Madera, and Merced Counties of California.

(3) *District 3* consists of all counties in California where pistachios are produced that are not included in Districts 1 and 2.

(4) *District 4* consists of the States of Arizona and New Mexico.

(b) With the approval of the Secretary, the boundaries of any district may be changed by the committee to ensure proper representation. The boundaries need not coincide with county lines.

[69 FR 17850, Apr. 4, 2004, as amended at 72 FR 69141, Dec. 7, 2007; 74 FR 56539, Nov. 2, 2009]

§983.12 Domestic shipments.

Domestic shipments means shipments to the fifty states of the United States or to territories of the United States and the District of Columbia.

§983.14 Handle.

Handle means to engage in:

(a) Receiving pistachios;

(b) Hulling and drying pistachios;

(c) Further preparing pistachios by sorting, sizing, shelling, roasting, cleaning, salting, and/or packaging for marketing in or transporting to any and all markets in the current of interstate or foreign commerce; and/or

(d) Placing pistachios into the current of commerce from within the production area to points outside thereof: *Provided,* however, that transportation within the production area between handlers and from the orchard to the processing facility is not handling.

§983.15 Handler.

Handler means any person who handles pistachios.

§983.16 Inshell pistachios.

Inshell pistachios means pistachios that have a shell that has not been removed.

§983.17 Inspector.

Inspector means any inspector authorized by the USDA to inspect pistachios.

§ 983.18 Lot.

Lot means any quantity of pistachios that is submitted for testing purposes under this part.

§ 983.20 Part and subpart.

Part means the order regulating the handling of pistachios grown in the States of California, Arizona and New Mexico, and all the rules, regulations and supplementary orders issued thereunder. The aforesaid order regulating the handling of pistachios grown in California, Arizona and New Mexico shall be a subpart of such part.

[74 FR 56539, Nov. 2, 2009]

§ 983.21 Person.

Person means an individual, partnership, limited liability corporation, corporation, trust, association, or any other business unit.

[69 FR 17850, Apr. 4, 2004. Redesignated at 74 FR 56539, Nov. 2, 2009]

§ 983.22 Pistachios.

Pistachios means the nuts of the pistachio tree of the genus and species *Pistacia vera* grown in the production area, whether inshell or shelled.

[74 FR 56539, Nov. 2, 2009]

§ 983.23 Processing.

Processing means hulling and drying pistachios in preparation for market.

[69 FR 17850, Apr. 4, 2004. Redesignated at 74 FR 56539, Nov. 2, 2009]

§ 983.24 Producer.

Producer means any person engaged within the production area in a proprietary capacity in the production of pistachios for sale.

[69 FR 17850, Apr. 4, 2004. Redesignated at 74 FR 56539, Nov. 2, 2009]

§ 983.25 Production area.

Production Area means the States of California, Arizona, and New Mexico.

[74 FR 56539, Nov. 2, 2009]

§ 983.26 Production year.

Production year is synonymous with "fiscal period" and means the period beginning on September 1 and ending on August 31 of each year or such other period as may be recommended by the committee and approved by the Secretary. Pistachios harvested and received in August of any year shall be applied to the subsequent production year for marketing order purposes.

[69 FR 17850, Apr. 4, 2004. Redesignated at 74 FR 56539, Nov. 2, 2009]

§ 983.27 Proprietary capacity.

Proprietary capacity means the capacity or interest of a producer or handler that, either directly or through one or more intermediaries, is a property owner together with all the appurtenant rights of an owner including the right to vote the interest in that capacity as an individual, a shareholder, member of a cooperative, partner, trustee or in any other capacity with respect to any other business unit.

[69 FR 17850, Apr. 4, 2004. Redesignated at 74 FR 56539, Nov. 2, 2009]

§ 983.28 Secretary.

Secretary means the Secretary of Agriculture of the United States or any officer or employee of the United States Department of Agriculture who is, or who may hereafter be, authorized to act in his/her stead.

[69 FR 17850, Apr. 4, 2004. Redesignated at 74 FR 56539, Nov. 2, 2009]

§ 983.29 Shelled pistachios.

Shelled pistachios means pistachio kernels, or portions of kernels, after the pistachio shells have been removed.

[69 FR 17850, Apr. 4, 2004. Redesignated at 74 FR 56539, Nov. 2, 2009]

§ 983.30 Substandard pistachios.

Substandard pistachios means pistachios, inshell or shelled, which do not meet regulations established pursuant to §§ 983.50 and 983.51.

[74 FR 56540, Nov. 2, 2009]

ADMINISTRATIVE COMMITTEE

§ 983.41 Establishment and membership.

There is hereby established an administrative committee for pistachios to administer the terms and provisions of this part. This committee, consisting of twelve (12) member positions, each of whom shall have an alternate, shall be allocated as follows:

(a) *Handlers.* Two of the members shall represent handlers, as follows:

(1) One handler member nominated by one vote for each handler; and

(2) One handler member nominated by voting based on each handler casting one vote for each ton (or portion thereof) of the assessed weight of pistachios processed by such handler during the two production years preceding the production year in which the nominations are made.

(b) *Producers.* Nine members shall represent producers. Producers within the respective districts shall nominate four producers from District 1, three producers from District 2, one producer from District 3, and one producer from District 4. The Secretary, upon recommendation of the committee, may reapportion producer representation among the districts to ensure proper representation.

(c) *Public member.* One member shall be a public member who is neither a producer nor a handler and shall have all the powers, rights and privileges of any other member of the committee. The public member and alternate public member shall be nominated by the committee and selected by the Secretary.

[69 FR 17850, Apr. 4, 2004. Redesignated and amended at 74 FR 56541, Nov. 2, 2009]

§983.42 Initial members and nomination of successor members.

Nomination of committee members and alternates shall follow the procedure set forth in this section or as may be changed as recommended by the committee and approved by the Secretary.

(a) *Initial members.* Nominations for initial producer and handler members shall be conducted by the Secretary by either holding meetings of handlers and producers, or by mail.

(b) *Successor members.* Subsequent to the first nomination of committee members under this part, persons to be nominated to serve on the committee as producer or handler members shall be selected pursuant to nomination procedures that shall be established by the committee with the approval of the Secretary: *Provided,* That:

(1) Any qualified individuals who seek nomination as a producer member shall submit to the committee an intent to seek office in one designated district on such form and with such information as the committee shall designate; ballots, accompanied by the names of all such candidates, with spaces to indicate voters' choices and spaces for write-in candidates, together with voting instructions, shall be mailed to all producers who are on record with the committee within the respective districts; the person(s) receiving the highest number of votes shall be the member nominee(s) for that district, and the person(s) receiving the second highest number of votes shall be the alternate member nominee(s). In case of a tie vote, the nominee shall be selected by a drawing.

(2) Any qualified individuals who seek nomination as a handler member shall submit to the committee an intent to seek office with such information as the committee shall designate; ballots, accompanied by the names of all such candidates, with spaces to indicate voters' choices and spaces for write-in candidates, together with voting instructions, shall be mailed to all handlers who are on record with the committee. For the first handler member seat, the person receiving the highest number of votes shall be the handler member nominee for that seat, and the person receiving the second highest number of votes shall be the alternate member nominee. For the second handler member seat, the person receiving the highest number of votes representing handler volume shall be the handler member nominee for that seat, and the person receiving the second highest number of votes representing handler volume shall be the alternate member nominee. In case of a tie vote, the nominee shall be selected by a drawing.

(c) *Handlers.* Only handlers, including duly authorized officers or employees of handlers, may participate in the nomination of the two handler member nominees and their alternates. Nomination of the two handler members and their alternates shall be as follows:

(1) For one handler member nomination, each handler entity shall be entitled to one vote;

477

(2) For the second handler member nomination, each handler entity shall be entitled to cast one vote respectively for each ton of assessed weight of pistachios processed by that handler during the two production years preceding the production year in which the nominations are made. For the purposes of nominating handler members and alternates by volume, the assessed weight of pistachios shall be credited to the handler responsible under the order for the payment of assessments of those pistachios. The committee with the approval of the Secretary, may revise the handler representation on the committee if the committee ceases to be representative of the industry.

(d) *Producers.* Only producers, including duly authorized officers or employees of producers, may participate in the nomination of nominees for producer members and their alternates. Each producer shall be entitled to cast only one vote, whether directly or through an authorized officer or employee, for each position to be filled in the district in which the producer produces pistachios. If a producer is engaged in producing pistachios in more than one district, such producer shall select the district in which to participate in the nomination. If a person is both a producer and a handler of pistachios, such person may participate in both producer and handler nominations, provided, however, that a single member may not hold concurrent seats as both a producer and handler.

(e) *Member's affiliation.* Not more than two members and not more than two alternate members shall be persons employed by or affiliated with producers or handlers that are affiliated with the same handler and/or producer. Additionally, only one member and one alternate in any one district representing producers and only one member and one alternate representing handlers shall be employed by, or affiliated with the same handler and/or producer. No handler, and all of its affiliated handlers, can be represented by more than one handler member.

(f) *Cooperative affiliation.* In the case of a producer cooperative, a producer shall not be deemed to be connected in a proprietary capacity with the cooperative notwithstanding any outstanding retains, contributions or financial indebtedness owed by the cooperative to a producer if the producer has not marketed pistachios through the cooperative during the current and one preceding production year. A cooperative that has as its members one or more other cooperatives that are handlers shall not be considered as a handler for the purpose of nominating or voting under this part.

(g) *Alternate members.* Each member of the committee shall have an alternate member to be nominated in the same manner as the member. Any alternate serving in the same district as a member where both are employed by, or connected in a proprietary capacity with the same corporation, firm, partnership, association, or business organization, shall serve as the alternate to that member. An alternate member, in the absence of the member for whom that alternate is selected shall serve in place of that member on the committee, and shall have and be able to exercise all the rights, privileges, and powers of the member when serving on the committee. In the event of death, removal, resignation, or the disqualification of a member, the alternate shall act as a member on the committee until a successor member is selected and has been qualified.

(h) *Selection by Secretary.* Nominations under paragraph (g) of this section received by the committee for all handler and producer members and alternate member positions shall be certified and sent to the Secretary at least 60 days prior to the beginning of each two-year term of office, together with all necessary data and other information deemed by the committee to be pertinent or requested by the Secretary. From those nominations, the Secretary shall select the ten producer and handler members of the committee and an alternate for each member.

(i) *Acceptance.* Each person to be selected by the Secretary as a member or as an alternate member of the committee shall, prior to such selection, qualify by advising the Secretary that if selected, such person agrees to serve in the position for which that nomination has been made.

(j) *Failure to nominate.* If nominations are not made within the time and manner specified in this part, the Secretary may, without regard to nominations, select the committee members and alternates qualified to serve on the basis of the representation provided for in §983.41.

(k) *Term of office.* Selected members and alternate members of the committee shall serve for terms of two years: *Provided,* That four of the initially selected producer members and one handler member and their alternates shall, by a drawing, be seated for terms of one year so that approximately half of the memberships' terms expire each year. Each member and alternate member shall continue to serve until a successor is selected and has qualified. The term of office shall begin on July 1st of each year. Committee members and alternates may serve up to four consecutive, two-year terms of office. In no event shall any member or alternate serve more than eight consecutive years on the committee. For purposes of determining when a member or alternate has served four consecutive terms, the accrual of terms shall begin following any period of at least twelve consecutive months out of office.

(l) *Qualifications.* (1) Each producer member and alternate shall be, at the time of selection and during the term of office, a producer or an officer, or employee, of a producer in the district for which nominated.

(2) Each handler member and alternate shall be, at the time of selection and during the term of office, a handler or an officer or employee of a handler.

(3) Any member or alternate member who at the time of selection was employed by or affiliated with the person who is nominated, that member shall, upon termination of that relationship, become disqualified to serve further as a member and that position shall be deemed vacant.

(4) No person nominated to serve as a public member or alternate public member shall have a financial interest in any pistachio growing or handling operation.

(m) *Vacancy.* Any vacancy on the committee occurring by the failure of any person selected to the committee to qualify as a member or alternate member due to a change in status making the member ineligible to serve, or due to death, removal, or resignation, shall be filled, by a majority vote of the committee for the unexpired portion of the term. However, that person shall fulfill all the qualifications set forth in this part as required for the member whose office that person is to fill. The qualifications of any person to fill a vacancy on the committee shall be certified in writing to the Secretary. The Secretary shall notify the committee if the Secretary determines that any such person is not qualified.

(n) The committee, with the approval of the Secretary, may issue rules and regulations implementing §§983.41, 983.42 and 983.43.

[69 FR 17850, Apr. 4, 2004. Redesignated and amended at 74 FR 56541, Nov. 2, 2009]

§983.43 Procedure.

(a) *Quorum.* A quorum of the committee shall be any seven voting committee members. The vote of a majority of members present at a meeting at which there is a quorum shall constitute the act of the committee: *Provided,* That:

(1) Actions of the committee with respect to the following issues shall require twelve (12) concurring votes of the voting members regarding any recommendation to the Secretary for adoption or change in:

(i) Quality regulation;

(ii) Aflatoxin regulation;

(iii) Research under §983.46; and

(2) Actions of the committee with respect to the following issues shall require eight (8) concurring votes of the voting members regarding recommendation to the Secretary for adoption or change in:

(i) Inspection programs;

(ii) The establishment of the committee.

(b) *Voting.* Members of the committee may participate in a meeting by attendance in person or through the use of a conference telephone or similar communication equipment, as long as all members participating in such a meeting can communicate with one another. An action required or permitted to be taken by the committee may be taken without a meeting, if all members of the committee shall consent in writing to that action.

(c) *Compensation.* The members of the committee and their alternates shall serve without compensation, but members and alternates acting as members shall be allowed their necessary expenses: *Provided,* That the committee may request the attendance of one or more alternates not acting as members at any meeting of the committee, and such alternates may be allowed their necessary expenses; and, *Provided further,* That the public member and the alternate for the public member may be paid reasonable compensation in addition to necessary expenses.

[69 FR 17850, Apr. 4, 2004. Redesignated and amended at 74 FR 56541, Nov. 2, 2009]

§ 983.44   Powers.

The committee shall have the following powers:

(a) To administer the provisions of this part in accordance with its terms;

(b) To make and adopt bylaws, rules and regulations to effectuate the terms and provisions of this part with the approval of the Secretary;

(c) To receive, investigate, and report to the Secretary complaints of violations of this part; and

(d) To recommend to the Secretary amendments to this part.

[69 FR 17850, Apr. 4, 2004. Redesignated at 74 FR 56541, Nov. 2, 2009]

§ 983.45   Duties.

The committee shall have, among others, the following duties:

(a) To adopt bylaws and rules for the conduct of its meetings and the selection of such officers from among its membership, including a chairperson and vice-chairperson, as may be necessary, and define the duties of such officers; and adopt such other bylaws, regulations and rules as may be necessary to accomplish the purposes of the Act and the efficient administration of this part;

(b) To employ or contract with such persons or agents as the committee deems necessary and to determine the duties and compensation of such persons or agents;

(c) To select such subcommittees as may be necessary;

(d) To submit to the Secretary a budget for each fiscal period, prior to the beginning of such period, including a report explaining the items appearing therein and a recommendation as to the rate of assessments for such period;

(e) To keep minutes, books, and records which will reflect all of the acts and transactions of the committee and which shall be subject to examination by the Secretary;

(f) To prepare periodic statements of the financial operations of the committee and to make copies of each statement available to producers and handlers for examination at the office of the committee;

(g) To cause its financial statements to be audited by a certified public accountant at least once each fiscal year and at such times as the Secretary may request. Such audit shall include an examination of the receipt of assessments and the disbursement of all funds. The committee shall provide the Secretary with a copy of all audits and shall make copies of such audits, after the removal of any confidential individual or handler information that may be contained in them, available for examination at the offices of the committee;

(h) To act as intermediary between the Secretary and any producer or handler with respect to the operations of this part;

(i) To investigate and assemble data on the growing, handling, shipping and marketing conditions with respect to pistachios;

(j) To apprise the Secretary of all committee meetings in a timely manner;

(k) To submit to the Secretary such available information as the Secretary may request;

(l) To investigate compliance with the provisions of this part;

(m) To provide, through communication to producers and handlers, information regarding the activities of the committee and to respond to industry inquiries about committee activities;

(n) To oversee the collection of assessments levied under this part;

(o) To borrow such funds, subject to the approval of the Secretary and not to exceed the expected expenses of one fiscal year, as are necessary for administering its responsibilities and obligations under this part.

[69 FR 17850, Apr. 4, 2004. Redesignated at 74 FR 56541, Nov. 2, 2009]

RESEARCH

§ 983.46 Research.

The committee, with the approval of the Secretary, may establish or provide for the establishment of projects involving research designed to assist or improve the efficient production and postharvest handling of quality pistachios. The committee, with the approval of the Secretary, may also establish or provide for the establishment of projects designed to determine the effects of pistachio consumption on human health and nutrition. Pursuant to § 983.43(a), such research projects may only be established with 12 concurring votes of the voting members of the committee. The expenses of such projects shall be paid from funds collected pursuant to §§ 983.71 and 983.72.

[74 FR 56542, Nov. 2, 2009]

MARKETING POLICY

§ 983.47 Marketing policy.

Prior to August 1st each year, the committee shall prepare and submit to the Secretary a report setting forth its recommended marketing policy covering quality regulations for the pending crop. In the event it becomes advisable to modify such policy, because of changed crop conditions, the committee shall formulate a new policy and shall submit a report thereon to the Secretary. In developing the marketing policy, the committee shall give consideration to the production, harvesting, processing and storage conditions of that crop. The committee may also give consideration to current prices being received and the probable general level of prices to be received for pistachios by producers and handlers. Notice of the committee's marketing policy, and of any modifications thereof, shall be given promptly by rea-sonable publicity, to producers and handlers.

[69 FR 17850, Apr. 4, 2004. Redesignated at 74 FR 56541, Nov. 2, 2009]

REGULATIONS

§ 983.50 Aflatoxin regulations.

The committee shall establish, with the approval of the Secretary, such aflatoxin sampling, analysis, and inspection requirements applicable to pistachios to be shipped for domestic human consumption as will contribute to orderly marketing or be in the public interest. The committee may also establish, with the approval of the Secretary, such requirements for pistachios to be shipped for human consumption in export markets. No handler shall ship, for human consumption in domestic, or if applicable, export markets, pistachios that exceed an aflatoxin level established by the committee and approved by the Secretary. All shipments to markets for which requirements have been established must be covered by an aflatoxin inspection certificate. The committee may, with the approval of the Secretary, establish different sampling, analysis, and inspection requirements, and different aflatoxin level requirements, for different markets.

[77 FR 36123, June 18, 2012]

§ 983.51 Quality regulations.

For any production year, the committee may establish, with the approval of the Secretary, such quality and inspection requirements applicable to pistachios shipped for human consumption in domestic or export markets as will contribute to orderly marketing or be in the public interest. In such production year, no handler shall ship pistachios for human consumption in domestic, or if applicable, export markets unless they meet the applicable requirements as evidenced by certification acceptable to the committee. The committee may, with the approval of the Secretary, establish different quality and inspection requirements for different markets.

[77 FR 36123, June 18, 2012]

§ 983.52   Failed lots/rework procedure.

(a) *Substandard pistachios.* Each lot of substandard pistachios may be reworked to meet aflatoxin or quality requirements. The committee may establish, with the Secretary's approval, appropriate rework procedures.

(b) *Failed lot reporting.* If a lot fails to meet the aflatoxin and/or the quality requirements of this part, a failed lot notification report shall be completed and sent to the committee within 10 working days of the test failure. This form must be completed and submitted to the committee each time a lot fails either aflatoxin or quality testing. The accredited laboratories shall send the failed lot notification reports for aflatoxin tests to the committee, and the handler, under the supervision of an inspector, shall send the failed lot notification reports for the lots that do not meet the quality requirements to the committee.

[74 FR 56542, Nov. 2, 2009]

§ 983.53   Testing of minimal quantities.

(a) *Aflatoxin.* Handlers who handle less than 1 million pounds of assessed weight per year have the option of utilizing both of the following methods for testing for aflatoxin:

(1) The handler may have an inspector sample and test his or her entire inventory of hulled and dried pistachios for the aflatoxin certification before further processing.

(2) The handler may segregate receipts into various lots at the handler's discretion and have an inspector sample and test each specific lot. Any lots that are found to have less aflatoxin than the level established by the committee and approved by the Secretary can be certified by an inspector to be negative as to aflatoxin. Any lots that are found to have aflatoxin exceeding the level established by the committee and approved by the Secretary may be tested after reworking in the same manner as specified in § 983.52.

(b) *Quality.* The committee may, with the approval of the Secretary, establish regulations regarding the testing of minimal quantities of pistachios for quality.

[74 FR 56540, Nov. 2, 2009, as amended at 77 FR 36123, June 18, 2012]

§ 983.54   Commingling.

Certified lots may be commingled with other certified lots, but the commingling of certified and uncertified lots shall cause the loss of certification for the commingled lots.

[74 FR 56540, Nov. 2, 2009]

§ 983.55   Reinspection.

The Secretary, upon recommendation of the committee, may establish rules and regulations to establish conditions under which pistachios would be subject to reinspection.

[69 FR 17850, Apr. 4, 2004. Redesignated at 74 FR 56540, Nov. 2, 2009]

§ 983.56   Inspection, certification and identification.

Upon recommendation of the committee and approval of the Secretary, all pistachios that are required to be inspected and certified in accordance with this part shall be identified by appropriate seals, stamps, tags, or other identification to be affixed to the containers by the handler. All inspections shall be at the expense of the handler, *Provided,* That for handlers making shipments from facilities located in an area where inspection costs for inspector travel and shipment of samples for aflatoxin testing would otherwise exceed the average of those same inspection costs for comparable handling operations located in Districts 1 and 2, such handlers may be reimbursed by the committee for the difference between their respective inspection costs and such average, or as otherwise recommended by the committee and approved by the Secretary.

[74 FR 56540, Nov. 2, 2009]

§ 983.57   Substandard pistachios.

The committee shall, with the approval of the Secretary, establish such reporting and disposition procedures as it deems necessary to ensure that pistachios which do not meet aflatoxin and quality requirements are not shipped for human consumption in those markets for which such requirements exist pursuant to § 983.50 and § 983.51.

[77 FR 36123, June 18, 2012]

§ 983.58 Interhandler transfers.

Within the production area, any handler may transfer pistachios to another handler for additional handling, and any assessments, inspection requirements, aflatoxin testing requirements, and any other marketing order requirements with respect to pistachios so transferred may be assumed by the receiving handler. The committee, with the approval of the Secretary, may establish methods and procedures, including necessary reports, to maintain accurate records for such transfers.

[74 FR 56541, Nov. 2, 2009]

§ 983.59 Modification or suspension of regulations.

(a) In the event that the committee, at any time, finds that by reason of changed conditions, any regulations issued pursuant to §§ 983.50 through 983.58 should be modified or suspended, it shall, pursuant to § 983.43, so recommend to the Secretary.

(b) Whenever the Secretary finds from the recommendations and information submitted by the committee or from other available information, that a regulation should be modified, suspended, or terminated with respect to any or all shipments of pistachios in order to effectuate the declared policy of the Act, the Secretary shall modify or suspend such provisions. If the Secretary finds that a regulation obstructs or does not tend to effectuate the declared policy of the Act, the Secretary shall suspend or terminate such regulation.

(c) The Secretary, upon recommendation of committee, may issue rules and regulations implementing §§ 983.50 through 983.58.

[74 FR 56541, Nov. 2, 2009]

REPORTS, BOOKS AND RECORDS

§ 983.64 Reports.

Upon the request of the committee, with the approval of the Secretary, each handler shall furnish such reports and information on such forms as are needed to enable the Secretary and the committee to perform their functions and enforce the regulations under this part. The committee shall provide a uniform report format for the handlers.

[69 FR 17850, Apr. 4, 2004. Redesignated at 74 FR 56541, Nov. 2, 2009]

§ 983.65 Confidential information.

All reports and records furnished or submitted by handlers to the committee which include confidential data or information constituting a trade secret or disclosing the trade position, financial condition, or business operations of the particular handler or their customers shall be received by, and at all times kept in the custody and under the control of, one or more employees of the committee, who shall disclose such data and information to no person except the Secretary. However, such data or information may be disclosed only with the approval of the Secretary, to the committee when reasonably necessary to enable the committee to carry out its functions under this part.

[69 FR 17850, Apr. 4, 2004. Redesignated at 74 FR 56541, Nov. 2, 2009]

§ 983.66 Records.

Records of pistachios received, held and shipped by him, as will substantiate any required reports and will show performance under this part will be maintained by each handler for at least three years beyond the crop year of their applicability.

[69 FR 17850, Apr. 4, 2004. Redesignated at 74 FR 56541, Nov. 2, 2009]

§ 983.67 Random verification audits.

(a) All handlers' pistachio inventory shall be subject to random verification audits by the committee to ensure compliance with the terms of the order, and regulations adopted pursuant thereto.

(b) Committee staff or agents of the committee, based on information from the industry or knowledge of possible violations, may make buys of handler product in retail locations. If it is determined that violations of the order have occurred as a result of the buys, the matter will be referred to the Secretary for appropriate action.

[69 FR 17850, Apr. 4, 2004. Redesignated at 74 FR 56541, Nov. 2, 2009]

§ 983.68 Verification of reports.

For the purpose of checking and verifying reports filed by handlers or the operation of handlers under the provisions of this part, the Secretary and the committee, through their duly authorized agents, shall have access to any premises where pistachios and records relating thereto may be held by any handler and at any time during reasonable business hours, shall be permitted to inspect any pistachios so held by such handler and any and all records of such handler with respect to the acquisition, holding, or disposition of all pistachios which may be held or which may have been shipped by him/her.

[69 FR 17850, Apr. 4, 2004. Redesignated at 74 FR 56541, Nov. 2, 2009]

EXPENSES AND ASSESSMENTS

§ 983.70 Expenses.

The committee is authorized to incur such expenses as the Secretary finds are reasonable and likely to be incurred by it during each production year for the maintenance and functioning of the committee and for such other purposes as the Secretary may, pursuant to the provisions of this part, determine to be appropriate.

[69 FR 17850, Apr. 4, 2004. Redesignated at 74 FR 56541, Nov. 2, 2009]

§ 983.71 Assessments.

(a) Each handler who receives pistachios for processing in each production year, except as provided in § 983.58, shall pay the committee on demand, an assessment based on the *pro rata* share of the expenses authorized by the Secretary for that year attributable to the assessed weight of pistachios received by that handler in that year.

(b) The committee, prior to the beginning of each production year, shall recommend and the Secretary shall set the assessment for the following production year, which shall not exceed one-half of one percent of the average price received by producers in the preceding production year. The committee, with the approval of the Secretary, may revise the assessment if it determines, based on information including crop size and value, that the

action is necessary, and if the revision does not exceed the assessment limitation specified in this section and is made prior to the final billing of the assessment.

[69 FR 17850, Apr. 4, 2004. Redesignated and amended at 74 FR 56540, Nov. 2, 2009]

§ 983.72 Contributions.

The committee may accept voluntary contributions but these shall only be used to pay for committee expenses unless specified in support of research under § 983.46. Furthermore, research contributions shall be free of additional encumbrances by the donor and the committee shall retain complete control of their use.

[74 FR 56540, Nov. 2, 2009]

§ 983.73 Delinquent assessments.

Any handler who fails to pay any assessment within the time required by the committee, shall pay to the committee a late payment charge of 10 percent of the amount of the assessment determined to be past due and, in addition, interest on the unpaid balance at the rate of one and one-half percent per month. The late payment and interest charges may be modified by the Secretary upon recommendation of the committee.

[69 FR 17850, Apr. 4, 2004. Redesignated at 74 FR 56540, Nov. 2, 2009]

§ 983.74 Accounting.

(a) If, at the end of a production year, the assessments collected are in excess of expenses incurred, such excess shall be accounted for in accordance with one of the following:

(1) If such excess is not retained in a reserve, as provided in paragraph (a)(2) of this section, it shall be refunded proportionately to the persons from whom it was collected in accordance with § 983.71: *Provided,* That any sum paid by a person in excess of his/her *pro rata* share of the expenses during any production year may be applied by the committee at the end of such production year as credit for such person, toward the committee's fiscal operations of the following production year;

(2) The committee, with the approval of the Secretary, may carry over such excess into subsequent production

years as a reserve: *Provided,* That funds already in the reserve do not exceed approximately two production years' budgeted expenses. In the event that funds exceed two production years' budgeted expenses, future assessments will be reduced to bring the reserves to an amount that is less than or equal to two production years' budgeted expenses. Such reserve funds may be used:

(i) To defray expenses, during any production year, prior to the time assessment income is sufficient to cover such expenses;

(ii) To cover deficits incurred during any production year when assessment income is less than expenses;

(iii) To defray expenses incurred during any period when any or all provisions of this part are suspended; and

(iv) To cover necessary expenses of liquidation in the event of termination of this part. Upon such termination, any funds not required to defray the necessary expenses of liquidation shall be disposed of in such manner as the Secretary may determine to be appropriate: *Provided,* That to the extent practical, such funds shall be returned *pro rata* to the persons from whom such funds were collected.

(b) All funds received by the committee pursuant to the provisions of this part shall be used solely for the purpose specified in this part and shall be accounted for in the manner provided in this part. The Secretary may at any time require the committee and its members to account for all receipts and disbursements.

(c) Upon the removal or expiration of the term of office of any member of the committee, such member shall account for all receipts and disbursements for which that member was personally responsible, deliver all committee property and funds in the possession of such member to the committee, and execute such assignments and other instruments as may be necessary or appropriate to vest in the committee full title to all of the committee property, funds, and claims vested in such member pursuant to this part.

[69 FR 17850, Apr. 4, 2004. Redesignated and amended at 74 FR 56540, Nov. 2, 2009]

§ 983.75 Implementation and amendments.

The Secretary, upon the recommendation of a majority of the committee, may issue rules and regulations implementing or modifying §§ 983.64 through 983.74 inclusive.

[74 FR 56540, Nov. 2, 2009]

MISCELLANEOUS PROVISIONS

§ 983.80 Compliance.

Except as provided in this part, no handler shall handle pistachios, the handling of which has been prohibited or otherwise limited by the Secretary in accordance with provisions of this part; and no handler shall handle pistachios except in conformity to the provision of this part.

[69 FR 17850, Apr. 4, 2004. Redesignated at 74 FR 56540, Nov. 2, 2009]

§ 983.81 Right of the Secretary.

The members of the committee (including successors or alternates) and any agent or employee appointed or employed by the committee, shall be subject to removal or suspension at the discretion of the Secretary, at any time. Each and every decision, determination, or other act of the committee shall be subject to the continuing right of the Secretary to disapprove of the same at any time, and upon such disapproval, shall be deemed null and void.

[69 FR 17850, Apr. 4, 2004. Redesignated at 74 FR 56540, Nov. 2, 2009]

§ 983.82 Personal liability.

No member or alternate member of the committee, nor any employee, representative, or agent of the committee shall be held personally responsible to any handler, either individually, or jointly with others, in any way whatsoever, to any person, for errors in judgment, mistakes, or other acts, either of commission or omission, as such member, alternate member, employee, representative, or agent, except for acts of dishonesty, willful misconduct, or gross negligence.

[69 FR 17850, Apr. 4, 2004. Redesignated at 74 FR 56540, Nov. 2, 2009]

§ 983.83 Separability.

If any provision of this part is declared invalid, or the applicability thereof to any person, circumstance, or thing is held invalid, the validity of the remainder, or the applicability thereof to any other person, circumstance, or thing, shall not be affected thereby.

[69 FR 17850, Apr. 4, 2004. Redesignated at 74 FR 56540, Nov. 2, 2009]

§ 983.84 Derogation.

Nothing contained in this part is, or shall be construed to be, in derogation or in modification of the rights of the Secretary or of the United States to exercise any powers granted by the Act or otherwise, or, in accordance with such powers, to act in the premises whenever such action is deemed advisable.

[69 FR 17850, Apr. 4, 2004. Redesignated at 74 FR 56540, Nov. 2, 2009]

§ 983.85 Duration of immunities.

The benefits, privileges, and immunities conferred upon any person by virtue of this part shall cease upon its termination, except with respect to acts done under and during the existence thereof.

[69 FR 17850, Apr. 4, 2004. Redesignated at 74 FR 56540, Nov. 2, 2009]

§ 983.86 Agents.

The Secretary may, by a designation in writing, name any person, including any officer or employee of the United States Government, or name any service, division or branch in the United States Department of Agriculture, to act as agent or representative of the Secretary in connection with any of the provisions of this part.

[69 FR 17850, Apr. 4, 2004. Redesignated at 74 FR 56540, Nov. 2, 2009]

§ 983.87 Effective time.

The provisions of this part, as well as any amendments, shall become effective at such time as the Secretary may declare, and shall continue in force until terminated or suspended in one of the ways specified in § 983.88 or § 983.89.

[74 FR 56540, Nov. 2, 2009]

§ 983.88 Suspension or termination.

The Secretary shall terminate or suspend the operation of any or all of the provisions of this part, whenever he/she finds that such provisions do not tend to effectuate the declared policy of the Act.

[69 FR 17850, Apr. 4, 2004. Redesignated at 74 FR 56540, Nov. 2, 2009]

§ 983.89 Termination.

(a) The Secretary may at any time terminate the provisions of this part.

(b) The Secretary shall terminate or suspend the operations of any or all of the provisions of this part whenever it is found that such provisions do not tend to effectuate the declared policy of the Act.

(c) The Secretary shall terminate the provisions of this part at the end of any fiscal period whenever it is found that such termination is favored by a majority of producers who, during a representative period, have been engaged in the production of pistachios: *Provided,* That such majority has, during such representative period, produced for market more than fifty percent of the volume of such pistachios produced for market, but such termination shall be announced at least 90 days before the end of the current fiscal period.

(d) Within six years of the effective date of this part the Secretary shall conduct a referendum to ascertain whether continuance of this part is favored by producers. Subsequent referenda to ascertain continuance shall be conducted every six years thereafter. The Secretary may terminate the provisions of this part at the end of any fiscal period in which the Secretary has found that continuance of this part is not favored by a two thirds (⅔) majority of voting producers, or a two thirds (⅔) majority of volume represented thereby, who, during a representative period determined by the Secretary, have been engaged in the production for market of pistachios in the production area. Such termination shall be announced on or before the end of the production year.

(e) The provisions of this part shall, in any event, terminate whenever the provisions of the Act authorizing them cease.

[69 FR 17850, Apr. 4, 2004. Redesignated at 74 FR 56540, Nov. 2, 2009]

§983.90 Procedure upon termination.

Upon the termination of this part, the members of the committee then functioning shall continue as joint trustees, for the purpose of liquidating the affairs of the committee. Action by such trustees shall require the concurrence of a majority of said trustees. Such trustees shall continue in such capacity until discharged by the Secretary, and shall account for all receipts and disbursements and deliver all property on hand, together with all books and records of the committee and the joint trustees, to such persons as the Secretary may direct; and shall upon the request of the Secretary, execute such assignments or other instruments necessary or appropriate to vest in such person full title and right to all the funds, properties, and claims vested in the committee or the joint trustees, pursuant to this part. Any person to whom funds, property, or claims have been transferred or delivered by the committee or the joint trustees, pursuant to this section, shall be subject to the same obligations imposed upon the members of said committee and upon said joint trustees.

[69 FR 17850, Apr. 4, 2002. Redesignated at 74 FR 56540, Nov. 2, 2009]

§983.91 Effect of termination or amendment.

Unless otherwise expressly provided by the Secretary, the termination of this part or of any regulation issued pursuant thereto, or the issuance of any amendment to either thereof, shall not:

(a) Affect or waive any right, duty, obligation, or liability which shall have arisen or which may thereafter arise, in connection with any provisions of this part or any regulation issued there under,

(b) Release or extinguish any violation of this part or any regulation issued there under, or

(c) Affect or impair any rights or remedies of the Secretary, or of any

other persons, with respect to such violation.

[69 FR 17850, Apr. 4, 2004. Redesignated at 74 FR 56540, Nov. 2, 2009]

§983.92 Exemption.

Any handler may handle pistachios within the production area free of the requirements in §§983.50 through 983.58 and §983.71 if such pistachios are handled in quantities not exceeding 5,000 dried pounds during any production year. The Secretary, upon recommendation of the committee, may issue rules and regulations changing the 5,000 pound quantity applicable to this exemption.

[74 FR 56540, Nov. 2, 2009]

Subpart B—Administrative Requirements

§983.150 Aflatoxin regulations.

(a) *Maximum level.* No handler shall ship for domestic human consumption, pistachios that exceed an aflatoxin level of 15 ppb. All shipments must also be covered by an aflatoxin inspection certificate. Pistachios that fail to meet the aflatoxin requirements shall be disposed in such manner as described in Failed lots/rework procedure of this part.

(b) *Change in level.* The committee may recommend to the Secretary changes in the aflatoxin level specified in this section. If the Secretary finds, on the basis of such recommendation or other information, that such an adjustment of the aflatoxin level would tend to effectuate the declared policy of the Act, such change shall be made accordingly.

(c) *Transfers between handlers.* Transfers between handlers within the production area are exempt from the aflatoxin regulation of this section.

(d) *Aflatoxin testing procedures.* To obtain an aflatoxin inspection certificate, each lot to be certified shall be uniquely identified, be traceable from testing through shipment by the handler, and be subjected to the following:

(1) *Samples for testing.* Prior to testing, each handler shall cause a representative sample to be drawn from each lot ("lot samples") of sufficient

487

weight to comply with Tables 1 and 2 of this section.

(i) At premises with mechanical sampling equipment (auto-samplers) approved by the USDA Federal-State Inspection Service, samples shall be drawn by the handler in a manner acceptable to the Committee and the USDA Federal-State Inspection Service.

(ii) At premises without mechanical sampling equipment, sampling shall be conducted by or under the supervision of an inspector, or as approved under an alternative USDA-recognized inspection program.

(2) *Test samples for aflatoxin.* Prior to submission of samples to an accredited laboratory for aflatoxin analysis, one sample ("test sample") shall be created from the pistachios designated for aflatoxin testing in compliance with Tables 1 and 2 of this paragraph for inshell and kernel pistachio lots that weigh up to and including 4,400 pounds.

For lot sizes larger than 4,400 pounds, two samples ("test samples") shall be created equally from the pistachios designated for aflatoxin testing in compliance with the requirements of Tables 1 and 2 of this paragraph. The test samples shall be prepared by, or under the supervision of an inspector, or as approved under an alternative USDA-recognized inspection program. The test samples shall be designated by an inspector as Test Sample #1 and Test Sample #2. Each sample shall be placed in a suitable container, with the lot number clearly identified, and then submitted to an accredited laboratory. The gross weight of the inshell lot sample for aflatoxin testing and the minimum number of incremental samples required are shown in Table 1 of this paragraph. The gross weight of the kernel lot sample for aflatoxin testing and the minimum number of incremental samples required is shown in the Table 2 of this paragraph.

TABLE 1 TO § 983.150(d)(2)—INSHELL PISTACHIO LOT SAMPLING INCREMENTS FOR AFLATOXIN CERTIFICATION

Lot weight (lbs.)	Minimum number of incremental samples for the lot sample	Total weight of lot sample (kilograms)	Weight of test sample (kilograms)
220 or less	10	2.0	2.0
221–440	15	3.0	3.0
441–1,100	20	4.0	4.0
1,101–2,200	30	6.0	6.0
2,201–4,400	40	8.0	8.0
4,401–11,000	60	12.0	6.0
11,001–22,000	80	16.0	8.0
22,001–150,000	100	20.0	10.0

TABLE 2 TO § 983.150(d)(2)—SHELLED PISTACHIO KERNEL LOT SAMPLING INCREMENTS FOR AFLATOXIN CERTIFICATION

Lot weight (lbs.)	Minimum number of incremental samples for the lot sample	Total weight of lot sample (kilograms)	Weight of test sample (kilograms)
220 or less	10	1.0	1.0
221–440	15	1.5	1.5
441–1,100	20	2.0	2.0
1,101–2,200	30	3.0	3.0
2,201–4,400	40	4.0	4.0
4,401–11,000	60	6.0	3.0
11,001–22,000	80	8.0	4.0
22,001–150,000	100	10.0	5.0

(3) *Testing of pistachios.* Test samples shall be received and logged by an accredited laboratory and each test sample shall be prepared and analyzed using High Pressure Liquid Chromatography (HPLC), Vicam Method (Aflatest), or other methods as recommended by not fewer than eight

members of the committee and approved by the Secretary. The aflatoxin level shall be calculated on a kernel weight basis.

(4) *Certification of lots "negative" as to aflatoxin.* (i) Lots which require a single test sample will be certified as "negative" on the aflatoxin certificate if the sample has an aflatoxin level at or below 15 ppb. If the aflatoxin level is above 15 ppb, the lot fails and the accredited laboratory shall fill out a failed lot notification report as specified in §§983.52 and 983.152.

(ii) Lots which require two test samples will be certified as "negative" on the aflatoxin inspection certificate if Test Sample #1 has an aflatoxin level at or below 10 ppb. If the aflatoxin level of Test Sample #1 is above 20 ppb, the lot fails and the accredited laboratory shall fill out a failed lot notification report as specified in §§983.52 and 983.152. If the aflatoxin level of Test Sample #1 is above 10 ppb and at or below 20 ppb, the accredited laboratory may at the handler's discretion analyze Test Sample #2 and the test results of Test Samples #1 and #2 will be averaged. Alternately, the handler may elect to withdraw the lot from testing, rework the lot, and resubmit it for testing after reworking. If the handler directs the laboratory to proceed with the analysis of Test Sample #2, a lot will be certified as negative to aflatoxin and the laboratory shall issue an aflatoxin inspection certificate if the averaged results of Test Sample #1 and Test Sample #2 is at or below 15 ppb. If the averaged aflatoxin level of Test Samples #1 and #2 is above 15 ppb, the lot fails and the accredited laboratory shall fill out a failed lot notification report as specified in §§983.52 and 983.152.

(iii) The accredited laboratory shall send a copy of the failed lot notification report to the Committee and to the failed lot's owner within 10 working days of any failure described in this section. If the lot is certified as negative as described in this section, the aflatoxin inspection certificate shall certify the lot using a certification form identifying each lot by weight, grade, and date. The certification expires for the lot or remainder of the lot after 12 months.

(5) *Certification of aflatoxin levels.* Each accredited laboratory shall complete aflatoxin testing and reporting and shall certify that every lot of pistachios shipped domestically does not exceed the aflatoxin levels as required in paragraph (a) of this section or as provided under §983.50. Each handler shall keep a record of each test, along with a record of final shipping disposition. These records must be maintained for three years beyond the production year of their applicability, and are subject to audit by the Secretary or the committee at any time.

(6) *Test samples that are not used for analysis.* If a handler does not elect to use Test Sample #2 for certification purposes, the handler may request that the laboratory return it to the handler.

[74 FR 56530, Nov. 2, 2009, as amended at 75 FR 43048, July 23, 2010; 79 FR 37932, July 3, 2014]

§983.152 **Failed lots/rework procedure.**

(a) *Inshell rework procedure for aflatoxin.* If inshell rework is selected as a remedy to meet the aflatoxin regulations of this part, then 100% of the product within that lot shall be removed from the bulk and/or retail packaging containers and reworked to remove the portion of the lot that caused the failure. Reworking shall consist of mechanical, electronic, or manual procedures normally used in the handling of pistachios. After the rework procedure has been completed, the total weight of the accepted product and the total weight of the rejected product shall be reported to the committee. The reworked lot shall be sampled and tested for aflatoxin as specified in §983.150, except that the lot sample size and the test sample size shall be doubled. If, after the lot has been reworked and tested, it fails the aflatoxin test for a second time, the lot may be shelled and the kernels reworked, sampled, and tested in the manner specified for an original lot of kernels, or the failed lot may be used for non-human consumption or otherwise disposed of.

(b) *Kernel rework procedure for aflatoxin.* If pistachio kernel rework is selected as a remedy to meet the aflatoxin regulations in §983.150, then 100% of the product within that lot shall be removed from the bulk and/or retail packaging containers and re-worked to remove the portion of the lot that caused the failure. Reworking shall consist of mechanical, electronic, or manual procedures normally used in the handling of pistachios. After the rework procedure has been completed, the total weight of the accepted prod-uct and the total weight of the rejected product shall be reported to the com-mittee. The reworked lot shall be sam-pled and tested for aflatoxin as speci-fied in § 983.150.

[74 FR 56531, Nov. 2, 2009]

§ 983.155 Reinspection.

(a) Any lot of inshell pistachios that is pin-picked, hand-sorted, color-sort-ed, and/or resized is considered to be "materially changed." Pistachios which are roasted, salted, flavored, air-legged, dyed, color-coated, cleaned, and otherwise subjected to similar proc-esses are not considered to be materi-ally changed.

(b) Each handler who handles pis-tachios shall cause any lot or portion of a lot initially certified for aflatoxin, and subsequently materially changed, to be reinspected for aflatoxin and cer-tified as a new lot or new lots: *Pro-vided*, That, handlers exempted from order requirements under § 983.92 are exempt from all reinspection require-ments.

[70 FR 61226, Oct. 21, 2005, as amended at 71 FR 51987, Sept. 1, 2006; 72 FR 69141, Dec. 7, 2007. Redesignated and amended at 74 FR 56530, Nov. 2, 2009]

§ 983.164 Reports.

(a) *ACP–2, Failed Lot Notification.* Each handler shall notify the Adminis-trative Committee for Pistachios (com-mittee) of all lots that fail to meet the order's maximum aflatoxin require-ments by completing section A of this form. Handlers shall furnish this report to the committee no later than 10 days after completion of the aflatoxin test. Each USDA-approved aflatoxin testing laboratory shall complete section C of this report, and forward this report and the failing aflatoxin test results to the committee and to the handler within 10 days of the test failure.

(b) *ACP–3, Failed Lot Disposition and Rework Report.* Each handler who re-works a failing lot of pistachios shall complete this report and shall forward it to the committee no later than 10 days after the rework is completed. If rework is not selected as a remedy, the handler shall submit the form to the committee office within 10 days of dis-position of the lot.

(c) *ACP–4, Federal Marketing Order Ex-empt Handler Notification.* Each handler who handles less than 5,000 pounds of assessed weight pistachios in a produc-tion year shall complete and furnish this report to the committee no later than November 15 of each production year.

(d) *ACP–5, Minimal Testing Form.* Each handler who handles less than 1,000,000 pounds of dried weight pistachios in a production year and who wishes to re-quest permission to handle under the minimal quantities provisions (§ 983.53) of the order shall furnish this report to the committee office no later than Au-gust 1 of each production year.

(e) *ACP–6, Inter-handler Transfer.* Each handler who transfers uninspected pistachios to another han-dler within the production area shall complete the ACP–6 and sign Part A. The transferring handler shall forward the original ACP–6 and one copy to the handler who receives the uninspected pistachios. The transferring handler shall furnish one copy of ACP–6 to the committee within 30 days of the trans-fer. The handler receiving the uninspected pistachios (receiving han-dler) shall sign Part B of the original ACP–6 and shall file it with the com-mittee within 30 days of the transfer.

(f) *ACP–7 Monthly Report of Inventory/ Shipments.* Each handler of pistachios shall file this report with the com-mittee by the 10th day of each month for the previous month's inventory and shipment information.

(g) *ACP–8, Producer Delivery Report.* Each handler of pistachios shall file this report with the committee by the 15th day of December of each produc-tion year: *Provided*, That for the 2007–08 production year, handlers must file this report with the committee by April 17, 2008, to report his/her receipts of pistachios during the current production year, the names of the handlers' producing entities, business type, and the following information concerning each producing entity: Federal Tax Identification number; mailing and e-mail address; telephone and fax number; total bearing acres; county of production; and for the current production year, the total receipts of open inshell, closed shell, shelling stock of each producing entity; and total pounds of processed pistachios produced by each producing entity.

(h) *Exemptions.* Handlers who handle less than 5,000 pounds of assessed weight pistachios during any production year are exempt from filing all forms, with the exception of the ACP–4.

(i) *Records.* Each handler shall maintain all records of pistachios received, held, shipped, and disposed of for at least 3 years following each crop year to show compliance with the marketing order provisions.

[70 FR 39907, July 12, 2005, as amended at 71 FR 51987, Sept. 1, 2006; 72 FR 69141, Dec. 7, 2007; 73 FR 18705, Apr. 7, 2008. Redesignated and amended at 74 FR 56530, Nov. 2, 2009]

Subpart C—Assessment Rate

§ 983.253   Assessment rate.

On and after September 1, 2017, an assessment rate of $0.0001 per pound is established for California, Arizona, and New Mexico pistachios.

[82 FR 49089, Oct. 24, 2017]

PART 984—WALNUTS GROWN IN CALIFORNIA

Subpart A—Order Regulating Handling

AUTHORITY: 7 U.S.C. 601–674.

Subpart A—Order Regulating Handling

SOURCE: 27 FR 9094, Sept. 13, 1962, unless otherwise noted.

DEFINITIONS

§ 984.1 Secretary.

Secretary means the Secretary of Agriculture of the United States, or any other officer or employee of the United States Department of Agriculture who is, or who may be, authorized to perform the duties of the Secretary of Agriculture of the United States.

§ 984.2 Act.

Act means Public Act No. 10, 73d Congress, as amended and as reenacted and amended by the Agricultural Marketing Agreement Act of 1937, as amended (7 U.S.C. 601 *et seq.*).

§ 984.3 Person.

Person means an individual, partnership, corporation, association, or any other business unit.

§ 984.4 Area of production.

Area of production means the State of California.

[41 FR 31542, July 29, 1976]

§ 984.5 Grower.

Grower is synonymous with *producer* and means any person engaged in a proprietary capacity in the commercial production of walnuts.

§ 984.6 Board.

Board means the California Walnut Board established pursuant to § 934.35.

[73 FR 11336, Mar. 3, 2008]

§ 984.7 Marketing year.

Marketing year means the twelve months from September 1 to the following August 31, both inclusive, or any other such period deemed appropriate and recommended by the Board for approval by the Secretary.

[73 FR 11337, Mar. 3, 2008]

§ 984.8 Walnuts.

Walnuts means only walnuts of the "English" (Juglans regia) varieties grown in California.

[41 FR 31542, July 29, 1976]

§ 984.9 Inshell walnuts.

Inshell walnuts means walnuts the kernels of which are contained in the shell.

§ 984.10 Shelled walnuts.

Shelled walnuts means walnut kernels after the shells are removed.

§ 984.11 Merchantable walnuts.

(a) *Inshell.* Merchantable inshell walnuts means all inshell walnuts meeting the minimum grade and size regulations effective pursuant to § 984.50.

(b) *Shelled.* Merchantable shelled walnuts means all shelled walnuts meeting the minimum grade and size regulations effective pursuant to §984.50.

[27 FR 9094, Sept. 13, 1962, as amended at 39 FR 35328, Oct. 1, 1974]

§984.12 Substandard walnuts.

Substandard walnuts means all walnuts (whether inshell or shelled) the kernels of which do not meet the minimum standard prescribed for merchantable shelled walnuts.

§984.13 To handle.

To handle means to pack, sell, consign, transport, or ship (except as a common or contract carrier of walnuts owned by another person), or in any other way to put walnuts, inshell or shelled, into the current of commerce either within the area of production or from such area to any point outside thereof, or for a manufacturer or retailer within the area of production to purchase directly from a grower: The term "to handle" shall not include sales and deliveries within the area of production by growers to handlers, or between handlers.

[73 FR 11337, Mar. 3, 2008]

§984.14 Handler.

Handler means any person who handles inshell or shelled walnuts.

[73 FR 11337, Mar. 3, 2008]

§984.15 Pack.

Pack means to bleach, clean, grade, shell or otherwise prepare walnuts for market as inshell or shelled walnuts.

[73 FR 11337, Mar. 3, 2008]

§984.19 Manufacturer.

Manufacturer means any person who uses walnuts in the production of bakery goods, ice cream, candy, or other food products, except walnut oil.

§984.20 Kernelweight.

Kernelweight means the determined weight of the kernels in a quantity of walnuts regardless of their quality.

[39 FR 35328, Oct. 1, 1974]

§984.21 Handler inventory.

Handler inventory as of any date means all walnuts, inshell or shelled

(except those held in satisfaction of a reserve obligation), wherever located, then held by a handler or for his or her account.

[73 FR 11337, Mar. 3, 2008]

§984.22 Trade demand.

(a) *Inshell.* The quantity of merchantable inshell walnuts that the trade will acquire from all handlers during a marketing year for distribution in the United States and its territories.

(b) *Shelled.* The quantity of merchantable shelled walnuts that the trade will acquire from all handlers during a marketing year for distribution in the United States and its territories.

[73 FR 11337, Mar. 3, 2008]

§984.23 Free walnuts.

Free walnuts means walnuts which are included in the free percentage established by the Secretary pursuant to §984.49.

[39 FR 35328, Oct. 1, 1974]

§984.26 Reserve walnuts.

Reserve walnuts means those walnuts which are held to meet a reserve obligation.

[41 FR 31542, July 29, 1976]

§984.31 Part and subpart.

Part means the order regulating the handling of walnuts grown in California, and all rules, regulations, and supplementary orders issued thereunder. This order regulating the handling of walnuts grown in California shall be a *subpart* of such part.

[41 FR 31542, July 29, 1976]

§984.32 To certify.

To certify means the issuance of a certification of inspection of walnuts by the inspection service.

[41 FR 31542, July 29, 1976]

§984.33 Hold.

Hold means to maintain possession or keep control of, in proper storage at all times, the kernelweight of certified merchantable walnuts necessary to meet a reserve obligation.

[41 FR 31542, July 29, 1976]

ADMINISTRATIVE BODY

§ 984.35 California Walnut Board.

(a) A California Walnut Board is hereby established consisting of 10 members selected by the Secretary, each of whom shall have an alternate nominated and selected in the same way and with the same qualifications as the member. The members and their alternates shall be selected by the Secretary from nominees submitted by each of the following groups or from other eligible persons belonging to such groups:

(1) Two handler members from District 1;

(2) Two handler members from District 2;

(3) Two grower members from District 1;

(4) Two grower members from District 2;

(5) One grower member nominated at-large from the production area; and,

(6) One member and alternate who shall be selected after the selection of the nine handler and grower members and after the opportunity for such members to nominate the tenth member and alternate. The tenth member and his or her alternate shall be neither a walnut grower nor a handler.

(b) In the event that one handler handles 35% or more of the crop the membership of the Board shall be as follows:

(1) Two handler members to represent the handler that handles 35% or more of the crop;

(2) Two members to represent growers who market their walnuts through the handler that handles 35% or more of the crop;

(3) Two handler members to represent handlers that do not handle 35% or more of the crop;

(4) One member to represent growers from District 1 who market their walnuts through handlers that do not handle 35% or more of the crop;

(5) One member to represent growers from District 2 who market their walnuts through handlers that do not handle 35% or more of the crop;

(6) One member to represent growers who market their walnuts through handlers that do not handle 35% or more of the crop shall be nominated at large from the production area; and,

(7) One member and alternate who shall be selected after the selection of the nine handler and grower members and after the opportunity for such members to nominate the tenth member and alternate. The tenth member and his or her alternate shall be neither a walnut grower nor a handler.

(c) Grower Districts:

(1) *District 1.* District 1 encompasses the counties in the State of California that lie north of a line drawn on the south boundaries of San Mateo, Alameda, San Joaquin, Calaveras, and Alpine Counties.

(2) *District 2.* District 2 shall consist of all other walnut producing counties in the State of California south of the boundary line set forth in paragraph (c)(1) of this section.

(d) The Secretary, upon recommendation of the Board, may reestablish districts, may reapportion members among districts, and may revise the groups eligible for representation on the Board as specified in paragraphs (a) and (b) of this section: Provided, That any such recommendation shall require at least six concurring votes of the voting members of the Board. In recommending any such changes, the following shall be considered:

(1) Shifts in acreage within districts and within the production area during recent years;

(2) The importance of new production in its relation to existing districts;

(3) The equitable relationship between Board apportionment and districts;

(4) Changes in industry structure and/or the percentage of crop represented by various industry entities resulting in the existence of two or more major handlers;

(5) Other relevant factors.

[73 FR 11337, Mar. 3, 2008]

§ 984.36 Term of office.

The term of office for Board members and their alternates shall be for a period of two years ending on August 31 of odd-numbered years, but they shall serve until their respective successors are selected and have qualified.

[74 FR 18464, Apr. 23, 2009]

§984.37 Nominations.

(a) Nominations for all grower members shall be submitted by ballot pursuant to an announcement by press releases of the Board to the news media in the walnut producing areas. Such releases shall provide pertinent voting information, including the names of candidates and the location where ballots may be obtained. Ballots shall be accompanied by full instructions as to their markings and mailing and shall include the names of incumbents who are willing to continue serving on the Board and such other candidates as may be proposed pursuant to methods established by the Board with the approval of the Secretary. Each grower, regardless of the number and location of his or her walnut orchard(s), shall be entitled to cast only one ballot in the nomination and each vote shall be given equal weight. If the grower has orchards in both grower districts, he or she shall advise the Board of the district in which he/she desires to vote. The person receiving the highest number of votes for each grower position shall be the nominee.

(b) Nominations for handler members shall be submitted on ballots mailed by the Board to all handlers in their respective Districts. All handlers' votes shall be weighted by the kernelweight of walnuts certified as merchantable by each handler during the preceding marketing year. Each handler in the production area may vote for handler member nominees and their alternates. However, no handler with less than 35% of the crop shall have more than one member and one alternate member. The person receiving the highest number of votes for each handler member position shall be the nominee for that position.

(c) A calculation to determine whether or not a handler who handles 35 percent or more of the crop shall be made prior to nominations. For the first nominations held upon implementation of this language, the 35 percent threshold shall be calculated using an average of crop handled for the year in which nominations are made and one

year's handling prior. For all future nominations, the 35 percent handling calculation shall be based in the average of the two years prior to the year in which nominations are made. In the event that one handler handles 35% or more of the crop the membership of the Board, nominations shall be as follows:

(1) Nominations of growers who market their walnuts to the handler that handles 35% or more of the crop shall be conducted by that handler and the names of the nominees shall be forwarded to the Board for approval and appointment by the Secretary.

(2) Nominations for the two handler members representing the major handler shall be conducted by the major handler and the names of the nominees shall be forwarded to the Board for approval and appointment by the Secretary.

(3) Nominations on behalf of all other grower members (Groups (b)(4), (5) and (6) of §984.35) shall be submitted after ballot by such growers pursuant to an announcement by press releases of the Board to the news media in the walnut producing areas. Such releases shall provide pertinent voting information, including the names of candidates and the location where ballots may be obtained. Ballots shall be accompanied by full instructions as to their markings and mailing and shall include the names of incumbents who are willing to continue serving on the Board and such other candidates as may be proposed pursuant to methods established by the Board with the approval of the Secretary. Each grower in Groups (Groups (b)(4), (5) and (6) of §984.35), regardless of the number and location of his or her walnut orchard(s), shall be entitled to cast only one ballot in the nomination and each vote shall be given equal weight. If the grower has orchard(s) in both grower districts he or she shall advise the Board of the district in which he or she desires to vote. The person receiving the highest number of votes for grower position shall be the nominee.

(4) Nominations for handler members representing handlers that do not handle 35% or more of the crop shall be submitted on ballots mailed by the Board to those handlers. The votes of these handlers shall be weighted by the

kernelweight of walnuts certified as merchantable by each handler during the preceding marketing year. Each handler in the production area may vote for handler member nominees and their alternates of this subsection. However, no handler shall have more than one person on the Board either as member or alternate member. The person receiving the highest number of votes for a handler member position of this subsection shall be the nominee for that position.

(d) Each grower is entitled to participate in only one nomination process, regardless of the number of handler entities to whom he or she delivers walnuts. If a grower delivers walnuts to more than one handler entity, the grower must choose which nomination process he or she participates in.

(e) The nine members shall nominate one person as member and one person as alternate for the tenth member position. The tenth member and alternate shall be nominated by not less than 6 votes cast by the nine members of the Board.

(f) Nominations in the foregoing manner received by the Board shall be reported to the Secretary on or before June 15 of each odd-numbered year, together with a certified summary of the results of the nominations. If the Board fails to report nominations to the Secretary in the manner herein specified by June 15 of each odd-numbered year, the Secretary may select the members without nomination. If nominations for the tenth member are not submitted by September 1 of any such year, the Secretary may select such member without nomination.

(g) The Board may recommend, subject to the approval of the Secretary, a change to these nomination procedures should the Board determine that a revision is necessary.

[73 FR 11337, Mar. 3, 2008]

§ 984.38 Eligibility.

No person shall be selected or continue to serve as a member or alternate to represent one of the groups specified in § 984.35(a)(1) through (6) or § 984.38(b)(1) through (6), unless he or she is engaged in the business he or she is to represent, or represents, either in his or her own behalf or as an officer or employee if the business unit engaged in such business. Also, each member or alternate member representing growers in District 1 or District 2 shall be a grower, or officer or employee of the group he or she is to represent.

[73 FR 11338, Mar. 3, 2008]

§ 984.39 Qualify by acceptance.

Any person nominated to serve as a member or alternate member of the Board shall, prior to selection by USDA, qualify by filing a written qualification and acceptance statement indicating such person's willingness to serve in the position for which nominated.

[73 FR 11338, Mar. 3, 2008]

§ 984.40 Alternate.

(a) An alternate for a member of the Board shall act in the place and stead of such member in his or her absence or in the event of his or her death, removal, resignation, or disqualification, until a successor for his or her unexpired term has been selected and has qualified.

(b) In the event any member of the Board and his or her alternate are both unable to attend a meeting of the Board, any alternate for any other member representing the same group as the absent member may serve in the place of the absent member, or in the event such other alternate cannot attend, or there is no such other alternate, such member, or in the event of his disability or a vacancy, his or her alternate may designate, subject to the disapproval of the Secretary, a temporary substitute to attend such meeting. At such meeting such temporary substitute may act in the place of such member.

[73 FR 11338, Mar. 3, 2008]

§ 984.41 Vacancy.

Any vacancy occasioned by the removal, resignation, disqualification, or death of any member of alternate, or any need to select a successor through failure of any person selected as a member or alternate to qualify, shall be recognized by the Board causing a nomination to be made by the appropriate group and certifying to the Secretary a new nominee within 60 calendar days.

[39 FR 35330, Oct. 1, 1974, as amended at 41 FR 31543, July 29, 1976]

§ 984.42 Expenses.

The members and their alternates of the Board shall serve without compensation, but shall be allowed their necessary expenses incurred by them in the performance of their duties under this part.

[73 FR 11338, Mar. 3, 2008]

§ 984.43 Powers.

The Board shall have the following powers:

(a) To administer the provisions of this part in accordance with its terms;

(b) To make rules and regulations to effectuate the terms and provisions of this part;

(c) To receive, investigate, and report to the Secretary complaints of violations of this part; and

(d) To recommend to the Secretary amendments to this part.

§ 984.44 Duties.

The duties of the Board shall be as follows:

(a) To act as intermediary between the Secretary and any handler or grower;

(b) To keep minute books and records which will clearly reflect all of its acts and transactions, and such minute books and records shall at any time be subject to the examination of the Secretary;

(c) To furnish to the Secretary a complete report of all meetings and such other available information as he may request;

(d) To appoint such employees as it may deem necessary and to determine the salaries, define the duties, and fix the bonds of such employees;

(e) To cause the books of the Board to be audited by one or more competent public accountants at least once for each marketing year and at such other times as the Board deems necessary or as the Secretary may request, and to file with the Secretary three copies of all audit reports made;

(f) To investigate the growing, shipping and marketing conditions with respect to walnuts and to assemble data in connection therewith;

(g) To investigate compliance with the provisions of this part; and

(h) To recommend rules and regulations for the purpose of administering this subpart.

§ 984.45 Procedure.

(a) The members of the Board shall select a chairman from their membership, and shall select such other officers and adopt such rules for the conduct of Board business as they deem advisable. The Board shall give the Secretary the same notice of its meetings as is given to members of the Board.

(b) All decisions of the Board, except where otherwise specifically provided (see § 984.35(d)), shall be by a sixty-percent (60%) super-majority vote of the members present. A quorum of six members, or the equivalent of sixty percent (60%) of the Board, shall be required for the conduct of Board business.

(c) The Board may vote by mail or telegram, or by any other means of communication, upon due notice to all members. The Board, with the approval of the Secretary, shall prescribe the minimum number of votes that must be cast when voting is by any of these methods, and any other procedures necessary to carry out the objectives of this paragraph.

(d) The Board may provide for meetings by telephone, or other means of communication and any vote cast at such a meeting shall be confirmed promptly in writing: Provided, That if any assembled meeting is held, all votes shall be cast in person.

[27 FR 9094, Sept. 13, 1962, as amended at 39 FR 35330, Oct. 1, 1974; 73 FR 11338, Mar. 3, 2008]

§ 984.46 Research and development.

The Board, with the approval of the Secretary, may establish or provide for the establishment of production research, marketing research and development projects, and marketing promotion, including paid advertising, designed to assist, improve, or promote

497

the marketing, distribution, and consumption or efficient production of walnuts. The expenses of such projects shall be paid from funds collected pursuant to § 984.69 and § 984.70.

[73 FR 11339, Mar. 3, 2008]

MARKETING POLICY

§ 984.48 Marketing estimates and recommendations.

(a) Each marketing year the Board shall hold a meeting, prior to October 20, for the purpose of recommending to the Secretary a marketing policy for such year. Each year such recommendation shall be adopted by the affirmative vote of at least 60% of the Board and shall include the following, and where applicable, on a kernelweight basis:

(1) Its estimate of the orchard-run production in the area of production for the marketing year;

(2) The Board's estimate of the handler inventory on September 1 of inshell and shelled walnuts;

(3) Its estimate of the merchantable and substandard walnuts in the production;

(4) The Board's estimate of the trade demand for such marketing year for shelled and inshell walnuts, taking into consideration trade inventory, imports, prices, competing nut supplies, and other factors;

(5) The Board's recommendation for desirable handler inventory of inshell and shelled walnuts on August 31 of each marketing year;

(6) Its recommendation as to the free and reserve percentages to be established for walnuts;

(7) Its recommendation of the percentage of reserve walnuts that may be exported pursuant to § 984.56, when it determines that the quantity of reserve walnuts that may be exported should be limited;

(8) Its opinion as to whether grower prices are likely to exceed parity; and

(9) Its recommendation for change, if any, in grade and size regulations.

(b) [Reserved]

[39 FR 35330, Oct. 1, 1974; 39 FR 35999, Oct. 7, 1974, as amended at 41 FR 31543, July 29, 1976; 73 FR 11339, Mar. 3, 2008]

§ 984.49 Volume regulation.

(a) *Free, reserve, and export percentages.* Whenever the Secretary finds, on the basis of the Board's recommendation or other information, that limiting the quantity of walnuts that may be handled in domestic markets for merchantable free walnuts during a marketing year will tend to effectuate the declared policy of the act, he shall establish a free percentage to prescribe the portion of such walnuts which may be handled as free walnuts, and a reserve percentage to prescribe the portion that must be withheld as reserve walnuts. Whenever the Board recommends an export percentage pursuant to § 984.48(a)(7), the Secretary shall establish a percentage if he finds it would tend to effectuate the declared policy of the act.

(b) *Revision of percentages.* (1) On or before February 15 of the marketing year, the Board may recommend that the free percentage be increased and the reserve percentage be decreased. On the basis of the Board's recommendation or other information the Secretary may establish such revisions. If the reserve percentage is reduced when an export percentage is in effect, an increase shall be made in the export percentage so that the quantity previously authorized for export will not be reduced. If the revised reserve quantity is less than the quantity previously authorized for export the export percentage shall be 100 percent. Upon revision, all reserve obligations that are theretofore accrued on merchantable walnuts certified during such year on the basis of the previously effective percentages shall be adjusted accordingly.

(2) Any time prior to July 1, the Board may recommend an increase in the export percentage, if it finds that there is an insufficient volume of reserve walnuts available for export and additional demand exists, which would not adversely affect the disposition of the oncoming crop. On the basis of the Board's recommendation or other information, the Secretary may establish such revision.

[41 FR 31543, July 29, 1976]

EFFECTIVE DATE NOTE: At 60 FR 40064, Aug. 7, 1995, in § 989.49, in paragraph (b)(1), the words "On or before February 15 of the marketing year," were suspended.

QUALITY CONTROL

§984.50 Grade, quality and size regulations.

(a) *Minimum standard for inshell walnuts.* Except as provided in §984.64, no handler shall handle inshell walnuts unless such walnuts are equal to or better than the requirements of U.S. No. 2 grade and baby size as defined in the then effective United States Standards for Walnuts (Juglans regia) in the Shell. This minimum standard may be modified by the Secretary on the basis of a Board recommendation or other information.

(b) *Minimum standard for shelled walnuts.* Except as provided in §984.64, no handler shall handle shelled walnuts unless such walnuts are equal to or better than the requirements of the U.S. Commercial grade as defined in the then effective United States Standards for Shelled Walnuts (Juglans regia) and the minimum size shall be pieces not more than 5 percent of which will pass through a round opening ⁹⁄₆₄ inch in diameter. This minimum standard may be modified by the Secretary on the basis of a Board recommendation or other information.

(c) *Effective period.* The minimum standards established pursuant to paragraphs (a) and (b) of this section and the provisions of this part relating to the administration thereof, shall continue in effect irrespective of whether the season average price for walnuts is above the parity level specified in section 2(1) of the Act.

(d) *Additional grade, size or other quality regulation.* The Board may recommend to the Secretary additional grade, size or other quality regulations, and may also recommend different regulations for different market destinations. If the Secretary finds on the basis of such recommendation or other information that such additional regulations would tend to effectuate the declared policy of the Act, he or she shall establish such regulations.

(e) *Minimum requirements for reserve.* The Board, with the approval of the Secretary, may specify the minimum kernel content and related requirements for any lot of walnuts acceptable for disposition for credit against a reserve obligation: *Provided,* That reserve walnuts exported must meet the requirements of paragraph (a) of this section if inshell, or paragraph (b) of this section if shelled.

[27 FR 9094, Sept. 13, 1962, as amended at 39 FR 35330, Oct. 1, 1974; 41 FR 31543, July 29, 1976; 73 FR 11339, Mar. 3, 2008]

§984.51 Inspection and certification of inshell and shelled walnuts.

(a) Before or upon handling of any walnuts for use as free or reserve walnuts, each handler at his or her own expense shall cause such walnuts to be inspected to determine whether they meet the then applicable grade and size regulations. Such inspection shall be performed by the inspection service or services designated by the Board with the approval of the Secretary; Provided, That if more than one inspection service is designated, the functions performed by each service shall be separate, and shall not duplicate each other. Handlers shall obtain a certificate for each inspection and cause a copy of each certificate issued by the inspection service to be furnished to the Board. Each certificate shall show the identity of the handler, quantity of walnuts, the date of inspection, and for inshell walnuts the grade and size of such walnuts as set forth in the United States Standards for Walnuts (Juglans regia) in the Shell. Certificates covering reserve shelled walnuts for export shall also show the grade, size, and color of such walnuts as set forth in the United States Standards for Shelled Walnuts (Juglans regia). The Board, with the approval of the Secretary, may prescribe procedures for the administration of this provision.

(b) Inshell merchantable walnuts certified shall be converted to the kernelweight equivalent at 45 percent of their inshell weight. This conversion percentage may be changed by the Board with the approval of the Secretary.

(c) Upon inspection, all walnuts for use as free or reserve walnuts shall be identified by tags, stamps, or other means of identification prescribed by the Board and affixed to the container by the handler under the supervision of the Board or of a designated inspector

499

and such identification shall not be altered or removed except as directed by the Board. The assessment requirements in § 984.69 shall be incurred at the time of certification.

(d) Whenever the Board determines that the length of time in storage or conditions of storage of any lot of merchantable walnuts which has been previously inspected have been or are such as normally to cause deterioration, such lot of walnuts shall be reinspected at the handler's expense and recertified as merchantable prior to shipment.

[39 FR 35330, Oct. 1, 1974, as amended at 41 FR 31543, July 29, 1976; 73 FR 11339, Mar. 3, 2008]

§ 984.52 Processing of shelled walnuts.

(a) No handler shall slice, chop, grind, or in any manner change the form of shelled walnuts unless such walnuts have been certified as merchantable or unless such walnuts meet quality regulations established under § 984.50(d) if such regulations are in effect.

(b) Any lot of shelled walnuts which, upon inspection, fails to meet the minimum standard effective pursuant to § 984.50 solely due to excess shriveling may be certified for processing provided that the total amount of shrivel does not exceed 20 percent, by weight, of the lot. All such walnuts must be reinspected after processing and shall be certified as merchantable if the processed material meets the effective minimum standard. The provisions of this paragraph may be modified by the Secretary, upon recommendation of the Board or other information.

(c) The Board shall establish such procedures as are necessary to insure that all such walnuts are inspected prior to being placed into the current of commerce.

[39 FR 35330, Oct. 1, 1974, as amended at 73 FR 11339, Mar. 3, 2008]

RESERVE WALNUTS

§ 984.54 Establishment of obligation.

(a) *Reserve obligation.* Whenever free and reserve percentages are in effect for a marketing year, each handler shall withhold a kernelweight of certified merchantable walnuts equal to a quantity derived by the application of the reserve percentage to the kernelweight of merchantable walnuts certified. The kernelweight of certified merchantable walnuts which handlers are required to withhold shall be the "reserve obligation." The walnuts handled for use as free walnuts by any handler in accordance with the provisions of this part shall be deemed to be that handler's quota fixed by the Secretary within the meaning of section 8(a)(5) of the act.

(b) *Holding requirements.* Each handler shall at all times hold in his possession or under his control in proper storage the kernelweight of certified merchantable walnuts necessary to meet his reserve obligation less:

(1) Any quantity which was disposed of by him pursuant to § 984.56; and

(2) Any quantity for which he is otherwise relieved by the Board of responsibility to so hold walnuts.

[41 FR 31543, July 29, 1976]

§ 984.56 Disposition of reserve walnuts.

(a) *General.* The Board shall have power and authority to sell or dispose of any and all reserve walnuts withheld upon the best terms and at the highest returns obtainable consistent with the ultimate complete disposition of reserve, subject to all conditions of this section. The Board may dispose of reserve walnuts through handlers acting as agents of the Board under the terms and conditions specified by the Board.

(b) *Export.* The Board may export or authorize the disposition in export to the destinations outside the United States, Puerto Rico, and the Canal Zone, the quantity of reserve walnuts permitted to be exported by the export percentage establishment pursuant to § 984.49. Reserve walnuts may be exported by any handler as an agent of the Board under the terms and conditions specified by the Board.

(c) *Pooling.* At any time during the marketing year a handler may deliver reserve walnuts and any substandard walnuts meeting the minimum kernel content requirements effective pursuant to § 984.50(e) to the Board for pooling and crediting against his reserve obligation. Any reserve walnuts that the handler as agent of the Board has not disposed of by the end of the marketing year shall thereafter be delivered to the Board for pooling on demand. The Board shall dispose of these walnuts for use in the following outlets: Government agencies, charitable institutions, poultry or animal feed, walnut oil or other markets noncompetitive with markets for merchantable free walnuts. The Board may rent and operate or arrange the use of facilities for storage and disposition of reserve walnuts delivered to it.

(d) *Crediting.* The kernelweight of walnuts disposed of in accordance with this section shall be credited to the handler's reserve obligation. At any time during the marketing year, upon a handler's written request, the Board shall transfer part or all of the handler's credit in excess of his reserve obligation to any handler he designates.

(e) *Pool proceeds.* The proceeds remaining after the payment of all expenses incurred by the Board in receiving, holding, and disposing of pooled walnuts shall be distributed pro rata by the Board to each handler in proportion to his contribution thereto, measured in kernelweight, or such other basis as the Board may adopt with the approval of the Secretary.

(f) *Rules and regulations.* The Board, with the approval of the Secretary, may prescribe such rules and regulations as are necessary to carry out the provisions of this section.

[41 FR 31544, July 29, 1976]

§984.59 Interhandler transfers.

For the purposes of this part, transfer means the sale of inshell and shelled walnuts within the area of production by one handler to another. The Board, with the approval of the Secretary, may establish methods and procedures, including necessary reports, for such transfers.

[73 FR 11339, Mar. 3, 2008]

§984.64 Disposition of substandard walnuts.

Substandard walnuts may be disposed of only for manufacture into oil, livestock feed, or such other uses as the Board determines to be noncompetitive with existing domestic and export markets for merchantable wal-

nuts and with proper safeguards to prevent such walnuts from thereafter entering channels of trade in such markets. Wherever free and reserve percentages are in effect, the kernelweight of any walnuts meeting the minimum kernel content requirements effective pursuant to §984.50(e), may be pooled and the disposition credited to the handler's reserve obligation pursuant to §984.56. Each handler shall submit, in such form and at such intervals as the Board may determine, reports of (a) his production and holdings of substandard walnuts and (b) the disposition of all substandard walnuts to any other person, showing the quantity, lot, date, name and address of the person to whom delivered, the approved use and such other information pertaining thereto as the Board may specify.

[41 FR 31544, July 29, 1976]

§984.65 Compliance.

Except as provided in this subpart, no person shall handle walnuts, inshell or shelled, during any marketing year in which this subpart and any regulations issued by the Secretary hereunder are in effect, unless such person has previously met the obligations imposed by each such regulation and the provisions of this subpart.

§984.66 Assistance of the Board in meeting reserve obligation.

The Board may assist any handler in accounting for his reserve obligation and may aid any handler in acquiring walnuts to meet any deficiency in his reserve obligation, or in accounting for, or disposing of reserve walnuts.

[41 FR 31544, July 29, 1976]

§984.67 Exemptions.

(a) Exemption from volume regulation. Reserve percentages shall not apply to lots of merchantable inshell walnuts which are of jumbo size or larger as defined in the then effective United States Standards for Walnuts in the Shell, or to such quantities as the Board may, with the approval of the Secretary, prescribe.

(b) *Exemptions from assessments, quality, and volume regulations*—(1) *Sales by growers direct to consumers.* Any walnut

501

grower may handle walnuts of his own production free of the regulatory and assessment provisions of this part if he sells such walnuts in the area of production directly to consumers under the following types of exemptions.

(i) At roadside stands and farmers' markets;

(ii) In quantities not exceeding an aggregate of 500 pounds of inshell walnuts or 200 pounds of shelled walnuts during any marketing year (at locations other than those specified in (b)(i) of this section); and

(iii) If shipped by parcel post or express in quantities not exceeding 10 pounds of inshell walnuts or 4 pounds of shelled walnuts to any one consumer in any one calendar day.

(2) *Green walnuts.* Walnuts which are green and which are so immature that they cannot be used for drying and sale as dried walnuts may be handled without regard to the provisions of this part.

(3) *Noncompetitive outlets.* Any person may handle walnuts, free of the provisions of this part, for use by charitable institutions, relief agencies, governmental agencies for school lunch programs, and diversion to animal feed or oil manufacture pursuant to an authorized governmental diversion program.

(c) *Rules and modifications.* The Board may establish, with the approval of the Secretary, such rules, regulations and safeguards and such modifications as will promote the objectives of this subpart.

[27 FR 9094, Sept. 13, 1962, as amended at 41 FR 31544, July 29, 1976; 73 FR 11339, Mar. 3, 2008]

EXPENSES AND ASSESSMENTS

§ 984.68   Expenses.

The Board is authorized to incur such expenses as the Secretary finds are reasonable and likely to be incurred by it during each marketing year for the maintenance and functioning of the Board, and for such other purposes as the Secretary may, pursuant to this part, determine to be appropriate. The Board shall file a proposed budget of expenses and a rate of assessment with the Secretary as soon as practicable

after the beginning of each marketing year.

[41 FR 31544, July 29, 1976]

§ 984.69   Assessments.

(a) *Requirement for payment.* Each handler shall pay the Board, on demand, his pro rata share of the expenses authorized by the Secretary for each marketing year. Each handler's pro rata share shall be the rate of assessment per kernelweight pound of walnuts fixed by the Secretary times the kernelweight of merchantable walnuts he has certified. At any time during or after the marketing year the Secretary may increase the assessment rate as necessary to cover authorized expenses and each handler's pro rata share shall be adjusted accordingly.

(b) *Reserve walnut pool expenses.* The Board is authorized temporary use of funds derived from assessments collected pursuant to paragraph (a) of this section to defray expenses incurred in disposing of reserve walnuts pooled. All such expenses shall be deducted from the proceeds obtained by the Board from the sale or other disposal of pooled reserve walnuts.

(c) *Accounting.* If at the end of a marketing year the assessments collected are in excess of expenses incurred, such excess shall be accounted for in accordance with one of the following:

(1) If such excess is not retained in a reserve, as provided in paragraph (c)(2) or (c)(3) of this section, it shall be refunded to handlers from whom collected and each handler's share of such excess funds shall be the amount of assessments he or she has paid in excess of his or her pro rata share of the actual expenses of the Board.

(2) Excess funds may be used temporarily by the Board to defray expenses of the subsequent marketing year: Provided, That each handler's share of such excess shall be made available to him or her by the Board within five months after the end of the year.

(3) The Board may carry over such excess into subsequent marketing years as a reserve: Provided, That funds already in reserve do not exceed approximately two years' budgeted expenses. In the event that funds exceed

two marketing years' budgeted expenses, future assessments will be reduced to bring the reserves to an amount that is less than or equal to two marketing years' budgeted expenses. Such reserve funds may be used:

(i) To defray expenses, during any marketing year, prior to the time assessment income is sufficient to cover such expenses;

(ii) To cover deficits incurred during any year when assessment income is less than expenses;

(iii) To defray expenses incurred during any period when any or all provisions of this part are suspended;

(iv) To meet any other such costs recommended by the Board and approved by the Secretary.

(d) *Advanced assessments and commercial loans.* To provide funds for the administration of the provisions of this part during the part of a fiscal period when neither sufficient operating reserve funds nor sufficient revenue from assessments on the current season's certifications are available, the Board may accept payment of assessments in advance or may borrow money from a commercial lending institution for such purposes.

(e) *Termination.* Any money collected from assessments hereunder and remaining unexpended in the possession of the Board upon termination of this part shall be distributed in such manner as the Secretary may direct.

[27 FR 9094, Sept. 13, 1962, as amended at 41 FR 31544, July 29, 1976; 73 FR 11339, Mar. 3, 2008; 83 FR 21843, June 11, 2018]

§984.70 Contributions.

The Board may accept voluntary contributions but these shall only be used to pay expenses incurred pursuant to §984.46, Research and development. Furthermore, such contributions shall be free from any encumbrances by the donor and the Board shall retain complete control of their use.

[73 FR 11339, Mar. 3, 2008]

REPORTS, BOOKS, AND OTHER RECORDS

§984.71 Reports of handler inventory.

Each handler shall submit to the Board in such form and on such dates as the Board may prescribe, reports showing his or her inventory of inshell and shelled walnuts.

[73 FR 11339, Mar. 3, 2008]

§984.72 Reports of merchantable walnuts handled.

Each handler who handles merchantable walnuts, inshell or shelled, at any time during a marketing year shall submit to the Board in such form and at such intervals as the Board may prescribe, reports showing the quantity so handled and such other information pertinent thereto as the Board may specify.

§984.73 Reports of walnut receipts.

Each handler shall file such reports of his or her walnut receipts from growers, handlers, or others in such form and at such times as may be requested by the Board with the approval of the Secretary.

[73 FR 11339, Mar. 3, 2008]

§984.76 Other reports.

Upon request of the Board made with the approval of the Secretary each handler shall furnish such other reports and information as are needed to enable the Board to perform its duties and exercise its powers under this subpart.

§984.77 Verification of reports.

For the purpose of verifying and checking reports filed by handlers or the operations of handlers, the Secretary and the Board through its duly authorized representatives shall have access to any premises where walnuts and walnut records are held. Such access shall be available at any time during reasonable business hours. Authorized representatives shall be permitted to inspect any walnuts held and any and all records of the handler with respect to matters within the purview of this part. Each handler shall maintain complete records on the receiving, holding, and disposition of both inshell and shelled walnuts. Each handler shall furnish all labor necessary to facilitate such inspections at no expense to the Board or the Secretary. Each handler shall store all walnuts held by him in such manner as to facilitate inspection and shall maintain adequate storage records which will permit accurate identification with respect to inspection certificates of respective lots and of all such walnuts held or disposed of theretofore. The Board, with the approval of the Secretary, may establish any methods and procedures needed to verify reports.

[41 FR 31544, July 29, 1976]

§ 984.78 Certification of reports.

All reports submitted to the Board as required in this part shall be certified to the Secretary and the Board as to the completeness and correctness of the information contained therein.

§ 984.79 Confidential information.

All reports and records submitted by handlers to the Board, which include data or information constituting a trade secret or disclosing the trade position, or financial condition or business operations of the handler shall be kept in custody of one or more employees of the Board and shall be disclosed to no person except the Secretary.

§ 984.80 Books and other records.

Each handler shall maintain such records of walnuts received, held and disposed of by him as may be prescribed by the Board for the purpose of performing its functions under this subpart. Such books and records shall be retained and be available for examination by authorized representatives of the Board and the Secretary for a period of two years after the end of the marketing year in which the recorded transactions are completed.

MISCELLANEOUS PROVISIONS

§ 984.83 Rights of the Secretary.

The members and alternates of the Board and any agent or employee appointed or employed by the Board, shall be subject to removal or suspension by the Secretary, at his discretion, at any time. Each and every decision, determination, or other act of the Board shall be subject to the continuing right of the Secretary to disapprove of the same at any time, and upon such disapproval, shall be deemed null and void.

§ 984.84 Personal liability.

No member or alternate of the Board, nor any employee or agent thereof shall be held personally responsible either individually or jointly with others, in any way whatsoever, to any handler or any person for errors in judgment, mistakes, or other acts either of commission or omission, as such member, alternate employee or agent, except for acts of dishonesty.

[39 FR 35332, Oct. 1, 1974, as amended at 41 FR 31545, July 29, 1976]

§ 984.85 Separability.

If any provision of this subpart is declared invalid, or the applicability thereof to any person, circumstance, or thing is held invalid, the validity of the remainder hereof or the applicability thereof to any other person, circumstance, or thing shall not be affected thereby.

§ 984.86 Derogation.

Nothing contained in this subpart is, or shall be construed to be, in derogation or in modification of the rights of the Secretary or of the United States to exercise any powers granted by the act or otherwise, or in accordance with such powers, to act in the premises whenever such action is deemed advisable.

§ 984.87 Duration of immunities.

The benefits, privileges, and immunities conferred upon any person by virtue of this subpart shall cease upon the termination hereof except with respect to acts done under and during the existence hereof.

§ 984.88 Agents.

The Secretary may, by a designation in writing, name any person, including any officer or employee of the Government, or name any subdivision of the United States Department of Agriculture, to act as his agent or representative in connection with any of the provisions of this subpart.

§ 984.89 Effective time and termination.

(a) *Effective time.* The provisions of this subpart shall become effective at such time as the Secretary may declare above his signature attached to this subpart, and shall continue in force until terminated in one of the ways hereinafter specified.

(b) *Termination.* (1) The Secretary may, at any time, terminate the provisions of this subpart by giving at least one day's notice by means of a press release or in any other manner which he may determine.

(2) The Secretary may terminate or suspend the operation of any or all of the provisions of this subpart, whenever he finds that such provisions do not tend to effectuate the declared policy of the act.

(3) The Secretary shall terminate the provisions of this subpart at the end of any marketing year whenever he finds that such termination is favored by a majority of the producers of walnuts who during the preceding marketing year have been engaged in the production for market of walnuts in the State of California: *Provided,* That such majority have during such period produced for market more than 50 percent of the volume of such walnuts produced for market within said States, but such termination shall be effected only if announced on or before July 1 of the then current marketing year.

(4) Within six years of the effective date of this amendment the Secretary shall conduct a referendum to ascertain whether continuance of this part is favored by producers. Subsequent referenda to ascertain continuance shall be conducted every six years thereafter. The Secretary may terminate the provisions of this part at the end of any fiscal period in which the Secretary has found that continuance of this part is not favored by a two-thirds (⅔) majority of voting producers, or a two-thirds (⅔) majority of volume represented thereby, who, during a representative period determined by the Secretary, have been engaged in the production for market of walnuts in the production area. Such termination shall be announced on or before the end of the production year.

(5) The provisions of this subpart shall, in any event, terminate whenever the provisions of the act authorizing them cease to be in effect.

(c) *Proceedings after termination.* (1) Upon the termination of the provisions of this subpart, the members of the Board then functioning shall continue as joint trustees, for the purpose of liquidating the affairs of the Board, of all funds and property then in the possession or under the control of the Board, including claims for any funds unpaid or property not delivered at the time of such termination. Action by said trusteeship shall require the concurrence of a majority of the said trustees.

(2) Said trustees shall continue in such capacity until discharged by the Secretary; shall from time to time, account for all receipts and disbursements and deliver all property on hand, together with all books and records of the Board and the joint trustees to such person as the Secretary may direct; and shall, upon the request of the Secretary, execute such assignments or other instruments necessary or appropriate to vest in such person full title and right to all of the funds, property, and claims vested in the Board or the joint trustees pursuant hereto.

(3) Any person to whom funds, property or claims have been transferred or delivered by the Board or its members, pursuant to this section, shall be subject to the same obligation imposed upon the members of the said Board and upon said joint trustees.

[27 FR 9094, Sept. 13, 1962, as amended at 41 FR 31545, July 29, 1976; 73 FR 11340, Mar. 3, 2008]

§984.90 Effect of termination or amendment.

Unless otherwise expressly provided by the Secretary, the termination of this subpart or of any regulation issued pursuant to this subpart, or the issuance of any amendment to either thereof, shall not (a) affect or waive any right, duty, obligation, or liability which shall have arisen or which may thereafter arise in connection with any provision of this subpart or any regulation issued under this subpart, or (b) release or extinguish any violation of this subpart or of any regulation issued under this subpart, or (c) affect or impair any rights or remedies of the Secretary or of any other person, with respect to any such violation.

§ 984.91 Relationship with the California Walnut Commission.

In conducting Board activities and other objectives under this part, the Board may deliberate, consult, cooperate and exchange information with the California Walnut Commission, whose activities compliment those of the Board. Any sharing of information gathered under this subpart shall be kept confidential in accordance with provisions under section 10(i) of the Act.

[73 FR 11340, Mar. 3, 2008]

Subpart B—Assessment Rates

§ 984.347　Assessment rate.

On and after September 1, 2017, an assessment rate of $0.0400 per kernelweight pound is established for California merchantable walnuts.

[82 FR 33778, July 21, 2017]

Subpart C—Administrative Requirements

§ 984.437　Methods for proposing names of additional candidates to be included on walnut growers' nomination ballots.

(a) With regard to Board grower member positions specified in § 984.35(a)(5) and (b)(6), any ten or more such growers who marketed an aggregate of 500 or more tons of walnuts through handlers who did not handle 35% or more of the crop during the marketing year preceding the year in which Board nominations are held, may petition the Board to include on the nomination ballot the name of an eligible candidate for this position, and the name of an eligible candidate to serve as his or her alternate. The names of the eligible candidates proposed pursuant to this paragraph shall be included on the ballot together with the names of any incumbents who are willing to continue serving on the Board.

(b) Any ten or more growers eligible to serve in the grower member positions specified in § 984.35(a)(3) and (4) or § 984.35(b)(4) and (5) and who marketed an aggregate of 500 or more tons of walnuts through handlers who did not handle 35% or more of the crop during the

marketing year preceding the year in which Board nominations are held, may petition the Board to include on the nomination ballot for a district the name of an eligible candidate for the applicable position, and the name of an eligible candidate to serve as his or her alternate. The names of the eligible candidates proposed pursuant to this paragraph shall be included on the ballot together with the names of any incumbents who are willing to continue serving on the Board.

(c) Petitions made pursuant to paragraphs (a) and (b) of this section shall be on forms supplied by the Board and filed no later than April 1 of the nomination year.

[41 FR 54476, Dec. 14, 1976, as amended at 73 FR 73997, Dec. 5, 2008; 74 FR 9047, Mar. 2, 2009]

§ 984.445　Procedures for voting by mail, e-mail, telephone, videoconference, facsimile, or any other means of communication.

(a) Whenever the Board votes upon any proposition by mail, e-mail, or facsimile, at least six members or alternates acting as members must vote and one dissenting vote shall prevent its adoption. Each proposition to be voted upon by mail, e-mail, or facsimile shall specify a time limit for members to vote, after which the alternates shall be given the opportunity to vote.

(b) Whenever the Board conducts meetings by telephone, videoconference, or any technology that enables member interaction, the vote shall be conducted by roll call.

[75 FR 1527, Jan. 12, 2010]

§ 984.450　Grade and size regulations.

(a) *Minimum kernel content requirements for inshell walnuts for reserve disposition credit.* For purposes of §§ 984.54 and 984.56, no lot of inshell walnuts may be held, exported, or disposed of for use by governmental agencies or charitable institutions unless it meets the minimum requirements for merchantable inshell walnuts effective pursuant to § 984.50(a). The disposition of any lot of inshell walnuts pursuant to § 984.64 having at least a certified kernelweight of not less than 10 percent of the inshell weight of the lot may be credited against a handler's reserve obligation.

(b) *Minimum kernel content requirements for shelled walnuts for reserve disposition credit.* For the purposes of §§984.54 and 984.56, no lot of shelled walnuts may be held, exported, or disposed of for use by government agencies or charitable institutions unless it meets the minimum requirements for merchantable shelled walnuts effective pursuant to §984.50(b). The disposition of any lot of shelled walnuts pursuant to §984.64 having at least a certified kernelweight of kernels six sixty-fourths of an inch or larger of not less than 10 percent of the total weight of the lot may be credited against a handler's reserve obligation: *Provided,* That such minimum kernel content requirements shall not apply to any lot of walnut meal certified by the designated inspection service as having been derived from chopping, slicing, or dicing merchantable shelled walnuts.

(c) *Inspection and certification of shelled walnuts that are manufactured into products.* For purposes of §§984.50(d) and 984.52(c), shelled walnuts may be cut or diced without prior inspection and certification: *Provided,* That the end product, except for walnut meal, is inspected and certified. For purposes of this section, *end product* shall be defined as walnut pieces equal to or larger than eight sixty-fourths of an inch in diameter. *Walnut meal* shall be defined as walnut pieces smaller than eight sixty-fourths of an inch in diameter.

(1) *End product.* End product must be sized, inspected and certified, and the size must be noted on the inspection certificate. The end product quality must be equal to or better than the minimum requirements of U.S. Commercial grade as defined in the United States Standards for Shelled Walnuts (Juglans regia).

(2) *Walnut meal.* Walnut meal that is accumulated during the cutting or dicing of shelled walnuts to create end product must be presented with the smallest end product from that manufacturing run that is inspected and certified. If the end product meets the applicable U.S. Commercial grade requirements, the walnut meal accumulated during the manufacture of that end product shall be identified and referenced on a separate meal certificate as "meal derived from walnut pieces that meet U.S. Commercial grade requirements." The certificate number of the smallest end product will be referenced on the meal certificate.

(3) *Failed lots.* If the end product fails to meet applicable U.S. Commercial grade requirements, the end product may be reconditioned, re-sampled, inspected again, and certified. However, the walnut meal accumulated during the manufacture of that end product shall be rejected and disposed of pursuant to the requirements of §984.64.

[41 FR 54476, Dec. 14, 1976, as amended at 75 FR 51929, Aug. 24, 2010]

§984.451 Inspection and certification of inshell and shelled walnuts.

(a) The inspection service shall be the DFA of California.

(b) Each handler shall make each container of each lot of walnuts accessible for sampling and sealing or stamping in connection with the inspection and certification of any lot of inshell or shelled walnuts.

(c) Inshell and shelled walnuts for export pursuant to §984.56(b) shall have been inspected and certified not more than 60 days prior to shipment from the handler's plant.

[41 FR 54476, Dec. 14, 1976]

§984.452 Certification of shelled walnuts for processing.

Each certificate issued for shelled walnuts for processing pursuant to §984.52(b) shall bear the notation "Certified for Processing Only." Shelled walnuts so certified for processing may not be processed by anyone other than the handler obtaining such certificate.

[28 FR 1863, Feb. 28, 1963]

§984.456 Disposition of reserve walnuts and walnuts used for reserve disposition credit.

(a) Beginning September 1 of any marketing year, a handler may become an agent of the Board to dispose of reserve walnuts of such marketing year. The agency shall be established upon execution of an "Agency Agreement for Reserve Walnuts" setting forth the terms and conditions specified by the Board for the sale of reserve walnuts in authorized outlets.

(b) Any handler who desires to transfer disposition credit in excess of his/her reserve obligation to another handler shall submit a request to the Board for such transfer on CWB Form No. 17 signed by both handlers and the Board shall credit such transfer.

(c) Any reserve walnuts that a handler has not disposed of by the end of the marketing year shall be delivered to the Board for pooling on demand. The Board may delay such demand with respect to reserve walnuts for which the handler has agreed to undertake disposition pursuant to the Board's authority. Each lot of reserve or substandard walnuts delivered to the Board for pooling and disposition shall be separately weighed at the handler's expense by a public weighmaster either upon removal from the handler's premises or in transit to Board storage facilities or diversion point. A copy of each weighmaster's certificate showing the net weight of the walnuts shall be forwarded to the Board by the handler. Walnuts delivered to the Board shall be delivered F.O.B. handler's warehouse or point of storage.

[41 FR 54476, Dec. 14, 1976, as amended at 74 FR 56696, Nov. 3, 2009]

§ 984.459 Reports of interhandler transfers.

(a) Any handler who transfers walnuts to another handler within the State of California shall submit to the Board, not later than 10 calendar days following such transfer, a report showing the following:

(1) The date of transfer;

(2) The net weight, in pounds, of the walnuts transferred;

(3) Whether such walnuts were certified by the inspection service;

(4) Whether such walnuts were inshell or shelled;

(5) The name and address of the transferring handler; and

(6) The name and address of the receiving handler.

(b) The transferring handler shall send two copies of the report to the receiving handler at the time the report is submitted to the Board. The receiving handler shall certify, on one copy of the report, to the receipt of such walnuts and submit it to the Board within 10 calendar days after the walnuts, or copies of such report, have been received, whichever is later.

[65 FR 39286, June 26, 2000]

§ 984.464 Disposition of substandard walnuts.

(a) Whenever free and reserve percentages are in effect during a marketing year substandard walnuts meeting the requirements of § 984.450 may be delivered by a handler to the Board for pooling at any time during the year and the disposition credited to the handler's reserve obligation.

(b) The Board shall maintain a list of approved crushers, livestock feed manufacturers and livestock feeders, and of the locations of the facilities within the area of production to which substandard walnuts may be shipped. The Board may deny approval to any applicant or remove any approved crusher, feed manufacturer, or feeder from the list when such denial or removal is deemed necessary to insure control of substandard walnut disposition or the Board determines that substandard walnuts are not shipped to such facilities. Substandard walnuts disposed of to an approved crusher, livestock feed manufacturer, or livestock feeder, may only be shipped directly to an approved location where the crushing, feed manufacture, or feeding is to take place. Applications for approval to crush, manufacture livestock feed, or feed substandard walnuts shall be submitted to the Board on a form prescribed by the Board and which includes the location and a description of the disposal facilities to be used and a certification to the Board and the Secretary of Agriculture that the applicant will:

(1) Crush, manufacture feed, or feed such walnuts at the location;

(2) Use such walnuts for no other purpose than for crushing into oil, manufacturing into livestock feed, or livestock feeding;

(3) Permit such inspection of his premises and of walnuts received and held by him, and such examination of his books and records covering walnut transactions as the Board may require;

(4) Keep a record of his receipts, holdings, and use of substandard walnuts available for examination by authorized representatives of the Board and

the U.S. Department of Agriculture for a period of two years after the end of the marketing year in which the recorded transactions are completed; and

(5) Make such reports, certified to the Board and the Secretary as to their correctness, as the Board may require.

(c) Each handler who disposes of substandard walnuts to an approved crusher, livestock feed manufacturer or livestock feeder shall upon shipment report to the Board on CWB Form No. 20, the quantities disposed of or shipped.

[41 FR 54476, Dec. 14, 1976, as amended at 74 FR 56696, Nov. 3, 2009]

REPORTS

§ 984.471 Reports of handler inventory.

Reports of handler inventory as of September 1, January 1, and April 1 of each marketing year shall be submitted to the Board on CWB Form No. 4 for inshell walnuts and on CWB Form No. 5 for shelled walnuts, on or before September 15, January 15, and April 15 respectively, of that marketing year.

[74 FR 56696, Nov. 3, 2009]

§ 984.472 Reports of merchantable walnuts shipped.

(a) Reports of merchantable walnuts shipped during a month shall be submitted to the Board on CWB Form No. 6 not later than the 5th day of the following month. Such reports shall include all shipments during the preceding month and shall show for inshell and shelled walnuts: the quantity shipped; whether they were shipped into domestic or export channels; and for exports, the quantity by country of destination. If a handler makes no shipments during any month he/she shall submit a report marked "None." If a handler has completed his/her shipments for the season, he/she shall mark the report "Completed," and he/she shall not be required to submit any additional CWB Form No. 6 reports during the remainder of that marketing year.

(b) Reports of walnuts purchased directly from growers by handlers who are manufacturers or retailers shall be submitted to the Board on CWB Form No. 6, not later than the 5th day of the month following the month in which the walnuts were purchased. Such reports shall show the quantity of walnuts purchased and the quantity inspected and certified as merchantable walnuts.

[74 FR 56696, Nov. 3, 2009]

§ 984.473 Report of walnut receipts.

Each handler shall file a report of his walnut receipts from growers on or before January 15 of each marketing year on forms supplied by the Board.

[40 FR 22267, May 22, 1975]

§ 984.476 Report of walnut receipts from outside of the United States.

Each handler who receives walnuts from outside of the United States shall file with the Board, on CWB Form No. 7, a report of the receipt of such walnuts. The report shall be filed as follows: On or before December 5 for such walnuts received during the period September 1 to November 30; on or before March 5 for such walnuts received during the period December 1 to February 28 (February 29 in a leap year); on or before June 5 for such walnuts received during the period March 1 to May 31; and on or before September 5 for such walnuts received during the period June 1 to August 31. The report shall include the quantity of such walnuts received, the country of origin for such walnuts, and whether such walnuts are inshell or shelled. With each report, the handler shall submit a copy of a product tag issued by a DFA of California inspector for each receipt of such walnuts that includes the name of the person from whom such walnuts were received, the date such walnuts were received by the handler, the number of containers and the U.S. Custom's Service entry number, whether such walnuts are inshell or shelled, the quantity of such walnuts received, the country of origin for such walnuts, the name of the DFA of California inspector who issued the product tag, and the date such tag was issued.

[74 FR 56696, Nov. 3, 2009]

§ 984.480 Books and other records.

Each handler shall maintain true and complete records of all inshell and shelled walnuts and walnut material,

509

by categories, received, held, or disposed of by him. The records shall be maintained in such form as to permit verification of all transactions involved and shall be made available during normal business hours to authorized representatives of the Board or the Secretary of Agriculture. These records shall include the following:

(a) The names and addresses of the persons from whom received, and the quantities received from each such person;

(b) The names and addresses of the persons to whom disposal is made, and the quantities disposed of to each such person;

(c) The quantities used by the handler for such purposes as manufacturing, production of oil, and livestock feeding; and

(d) The quantities held on September 1, January 1, and April 1 of each marketing year.

[40 FR 22268, May 22, 1975, as amended at 74 FR 56697, Nov. 3, 2009]

PART 985—MARKETING ORDER REGULATING THE HANDLING OF SPEARMINT OIL PRODUCED IN THE FAR WEST

Subpart A—Order Regulating Handling

Subpart B—Administrative Requirements

AUTHORITY: 7 U.S.C. 601–674.

SOURCE: 45 FR 25040, Apr. 14, 1980, unless otherwise noted.

Subpart A—Order Regulating Handling

DEFINITIONS

§ 985.1 Secretary.

Secretary means the Secretary of Agriculture of the United States, or any other officer or employee of the U.S. Department of Agriculture who is, or who may be, authorized to perform the duties of the Secretary of Agriculture of the United States.

§ 985.2 Act.

Act means Public Act No. 10, 73d Congress, as amended, and reenacted and amended by the Agricultural Marketing Agreement Act of 1937, as amended (sections 1–19, Stat. 31, as amended; 7 U.S.C. 601–674).

§ 985.3 Person.

Person means an individual, partnership, corporation, association, or any other business unit.

§ 985.4 Spearmint oil.

Spearmint oil, hereinafter referred to as *oil*, means essential oil extracted by distillation from plants, grown in the production area, of the genus Mentha, species Cardiaca (commonly referred to as Scotch Spearmint), Spicata (commonly referred to as Native Spearmint), or such other species, grown in the production area, that produce a spearmint flavored oil. Oil shall be segregated into the following classes:

Class 1: Oil extracted from the first cutting of Scotch Spearmint.
Class 2: Oil extracted from the second cutting of Scotch Spearmint.
Class 3: Oil extracted from Native Spearmint.
Class 4: Oil which has a spearmint flavor, extracted from plants other than Scotch or Native Spearmint.

The Committee, with the approval of the Secretary, may change these classes to recognize new, or delete obsolete, classes.

§ 985.5 Production area.

Production area means all the area within the States of Washington, Idaho, Oregon, and that portion of Nevada north of the 37th parallel and that portion of Utah west of the 111th meridian. The area shall be divided into the following districts:

(a) District 1. State of Washington
(b) District 2. The State of Idaho and that portion of the States of Nevada and Utah included in the production area.
(c) District 3. The State of Oregon.

[61 FR 32924, June 26, 1996]

§ 985.6 Producer.

Producer is synonymous with *grower* and means any person engaged in a proprietary capacity in the commercial production of oil or who causes it to be produced.

§ 985.7 Handler.

Handler means any person who handles oil.

§ 985.8 Handle.

Handle means to prepare oil for market, acquire oil from a producer, use oil commercially of own production, or sell, transport, or ship (except as a common or contract carrier of oil owned by another), or otherwise place oil into the current of commerce within the production area or from the area to points outside thereof: *Provided, That:*

(a) The preparation for market of salable oil by producers who are not dealers or users,

(b) The sale or transportation of salable oil by a producer to a handler of record within the production area, or

(c) The transfer of excess oil by the producer to another producer to enable that producer to fill a deficiency in an annual allotment, or

(d) The delivery of excess oil by the producer to the Committee or its designees, shall not be construed as handling.

§ 985.9 Marketing year.

Marketing year means the 12 months from June 1 to the following May 31, inclusive, or such other period as the Committee, with the approval of the Secretary, may establish.

§ 985.10 Crop.

Crop means that oil produced by a producer during the marketing year.

511

§ 985.11　Salable oil.

Salable oil means that oil which is free to be handled.

§ 985.12　Salable quantity.

Salable quantity means the total quantity of each class of oil which handlers may purchase from, or handle on behalf of, producers during a marketing year.

§ 985.13　Annual allotment.

Annual allotment means that portion of the salable quantity prorated to a producer.

§ 985.14　Part and subpart.

Part means the order regulating the handling of oil grown in the production area, and all rules and regulations issued thereunder. The order shall be a *subpart* of such part.

ADMINISTRATIVE COMMITTEE

§ 985.20　Establishment and membership.

A Spearmint Oil Administrative Committee is hereby established (hereinafter referred to as *Committee*) and shall consist of eight members, each of whom shall have an alternate, to administer the terms and provisions of this part. Four of the members and alternates shall be producers in District 1; two members and alternates shall be producers in District 2; and one member and alternate shall be a producer in District 3. One member and alternate shall represent the public.

§ 985.21　Eligibility.

Each member and alternate member of the Committee shall be, at the time of selection and during the term of office, a producer, or an officer or employee of a producer, in the district for which selected: *Provided,* That these requirements should not apply to the public member and alternate member.

§ 985.22　Term of office.

The term of office of each member and alternate member of the Committee shall be for two calendar years: *Provided,* That one-half of the initial members and alternates shall serve for terms ending December 31, 1980, and one-half of the initial members and alternates shall serve for terms ending December 31, 1981. Members and alternates shall serve in such capacity for the term of office for which they are selected and have qualified and until their respective successors are selected and have qualified. No member shall serve more than two consecutive terms as member and no alternate shall serve more than two consecutive terms as alternate.

§ 985.23　Nominations.

(a) *Procedure.* (1) Nominations for producer members of the Committee and their alternates shall be made at nomination meetings of producers in each District. Such meetings shall be held at such times (on or before November 1 of each year) and places as the Committee shall designate. One nominee shall be elected for each position to be filled. The names and addresses of each nominee shall be submitted to the Secretary not later than December 1 of each year.

(2) Only producers, including duly authorized officers or employees of producers present and eligible to serve as producer members of the Committee, shall participate in the nomination. If a producer produces oil in more than one district, the producer shall select the district in which that producer will participate and notify the Committee of the choice.

(3) Should the Committee find it impractical to hold nomination meetings, nominations may be submitted to the Secretary based on the results of balloting by mail. Ballots to be used may contain the names of candidates and a blank space for write-in candidates for each position, together with voting instructions. The eligible person receiving the highest number of votes for a member or alternate position shall be the nominee for that position.

(4) The producer members of the Committee shall nominate the public member and alternate and member at the first meeting following the selection of members for a new term of office.

(b) *Initial members.* As soon as practicable following the effective date of this subpart, the Secretary shall hold, or cause to be held, nomination meetings of producers in each district to nominate the initial members of the Committee.

(c) The Committee with the approval of the Secretary shall issue rules and regulations necessary to carry out the provisions of this section or to change the procedures in this section in the event they are no longer practical.

§985.24 Selection.

Committee members shall be selected by the Secretary from nominees submitted by the Committee or from among other eligible persons. Each person so selected shall qualify by filing a written acceptance with the Secretary prior to assuming the duties of the position.

§985.25 Alternate members.

An alternate for a member shall act in the place of such member (a) in the member's absence, (b) in the event of the member's death, removal, resignation, or disqualification, until a successor for the member's unexpired term has been selected and has qualified, or (c) when requested and designated by the member.

§985.26 Vacancies.

To fill any vacancy occasioned by the failure of any person appointed as a member or as an alternate member of the Committee to qualify, or in the event of the death, removal, resignation, or disqualification of any member or alternate member of the Committee, a successor to fill the unexpired term shall be nominated and appointed in the manner specified in §§985.23 and 985.24. If the names of the nominees to fill any such vacancy are not made available to the Secretary within 30 days after such vacancy occurs, the Secretary may fill such vacancy without regard to nominations, which appointment shall be made on the basis of representation provided for in §985.20.

§985.27 Powers.

The Committee shall have the following powers:

(a) To administer this subpart in accordance with its terms and provisions;

(b) To make rules and regulations to effectuate the terms and provisions of this subpart;

(c) To receive, investigate, and report to the Secretary complaints of violations of this part; and

(d) To recommend to the Secretary amendments to this subpart.

§985.28 Duties.

The Committee shall have, among others, the following duties;

(a) To select from among its membership such officers and adopt such rules or by-laws for the conduct of its meetings as it deems necessary;

(b) To appoint such employees as it may deem necessary, and to determine the compensation and to define the duties of each employee;

(c) To appoint such subcommittees and consultants as it may deem necessary;

(d) To keep minutes, books, and records which will reflect all of the acts and transactions of the Committee and which shall be subject to examination by the Secretary;

(e) To prepare periodic statements of the financial operations of the Committee and to make copies of each such statement available to producers and handlers for examination at the office of the Committee;

(f) To cause the books of the Committee to be audited by a certified public accountant at such times as the Committee may deem necessary, or as the Secretary may request, to submit copies of each audit report to the Secretary, and to make available a copy which does not contain confidential data for inspection at the offices of the committee by producers and handlers;

(g) To act as intermediary between the Secretary and any producer or handler;

(h) To investigate and assemble data on the growing, handling, and marketing conditions with respect to oil;

(i) To submit to the Secretary such available information as may be requested or that the Committee may deem desirable and pertinent;

(j) To notify producers and handlers of all meetings of the Committee to consider recommendations for regulations and of all regulatory actions taken affecting producers and handlers;

(k) To give the Secretary the same notice of meetings of the Committee and its subcommittees as is given to its members;

(l) To investigate compliance and use means available to prevent violations of the provisions of this part;

(m) With the approval of the Secretary, to redefine the districts into which the production area is divided and to reapportion the representation of any district on the Committee: *Provided,* That such changes shall reflect insofar as practical, shifts in oil production within the production area and numbers of producers; and

(n) To establish with the approval of the Secretary such rules and regulations as are necessary or incidental to administration of this subpart, as are consistent with its provisions, and as would tend to accomplish the purposes of this subpart and the act.

§ 985.29 Procedure.

(a) At an assembled meeting, all votes shall be cast in person and seven members of the Committee shall constitute a quorum. Decisions of the Committee shall require the concurring vote of at least six members. If both a Committee member and appropriate alternate are unable to attend a Committee meeting, the Committee may designate any other alternate from the same district who is present at the meeting to serve in the member's place.

(b) The Committee may vote by mail, telephone, telegraph, or other means of communication: *Provided,* That each proposition is explained accurately, fully, and identically to each member. All votes shall be confirmed promptly in writing. Seven concurring votes and no dissenting votes shall be required for approval of a Committee action by such method.

§ 985.30 Expenses and compensation.

Members of the Committee, their alternates, subcommittees including any special subcommittees, shall serve without compensation but shall receive such allowances for necessary expenses, incurred in performing their duties, as may be approved by the Committee.

RESEARCH

§ 985.31 Research and development projects.

The Committee, with the approval of the Secretary, may establish or provide for the establishment of production research, marketing research and development projects designed to assist, improve, or promote the marketing, distribution and consumption or efficient production of oil. The Committee shall consider ongoing research, by industry and grower organizations, in making its recommendations. The expense of such projects shall be paid from funds collected pursuant to § 985.41.

EXPENSES AND ASSESSMENTS

§ 985.40 Expenses.

The Committee is authorized to incur such expenses as the Secretary finds are reasonable and likely to be incurred by it for such purposes as the Secretary may, pursuant to this subpart, determine to be appropriate, and for the maintenance and functioning of the Committee during each marketing year. The Committee shall submit to the Secretary a budget for each marketing year, including an explanation of the items appearing therein, and a recommendation as to the rate of assessment for such year.

§ 985.41 Assessments.

(a) *Requirements for payment.* Each person who first handles salable oil shall pay to the Committee, upon demand, that handler's pro rata share of the expenses authorized by the Secretary for each marketing year. Each handler's pro rata share shall be the rate of assessment fixed by the Secretary times the quantity of oil which the handler handles as the first handler thereof. The payment of assessments for the maintenance and functioning of the Committee and for such purposes as the Secretary may, pursuant to this subpart, determine to be appropriate, may be required under this part throughout the period it is in effect, irrespective of whether particular provisions thereof are suspended or become inoperative.

(b) *Rate of assessment.* The Secretary shall fix the rate of assessment to be paid by each handler. At any time during or after the marketing year, the Secretary may increase the rate of assessment as necessary to cover authorized expenses. Such increase shall be applied to all oil handled during the applicable marketing year. In order to provide funds for the administration of this part before sufficient operating income is available from assessments, the Committee may accept advance assessments and may also borrow money for such purpose. Advance assessments received from a handler shall be credited toward assessments levied against the handler during the marketing year.

§985.42 Accounting.

(a) *Excess funds.* At the end of a marketing year, funds in excess of the year's expenses may be placed in an operating reserve not to exceed approximately one marketing year's operational expenses or such lower limits as the Committee, with the approval of the Secretary, may establish. Funds in such reserve shall be available for use by the Committee for expenses authorized pursuant to §985.40. Funds in excess of those placed in the operating reserve shall be refunded to handlers: *Provided,* That any sum paid by a first handler in excess of that handler's pro rata share of the expenses during any marketing year may be applied by the Committee at the end of such marketing year to any outstanding obligations due the Committee from such person. Each handler's share of such excess funds shall be the amount of assessments paid in excess of that handler's pro rata share.

(b) *Disposition of funds upon termination of order.* Upon termination of this part, any funds not required to defray the necessary expenses of liquidation shall be disposed of in such manner as the Secretary may determine to be appropriate: *Provided,* That to the extent practicable, such funds will be returned pro rata to the first handler from whom such funds were collected.

VOLUME LIMITATIONS

§985.50 Marketing policy.

(a) The Committee shall meet on or before January 15 of each year to adopt a marketing policy for the ensuing marketing year or years. As soon as is practical following the meeting or meetings, the Committee shall submit to the Secretary recommendations for volume regulations deemed necessary to meet market requirements and establish orderly marketing conditions. Additional reports shall be submitted to the Secretary of the Committee subsequently adopts a new or revised policy because of changes in the demand and supply situation with respect to the various classes of oil.

(b) In determining such marketing policy, Committee consideration shall include but not be limited to:

(1) The estimated quantity of salable oil of each class held by producers and handlers;

(2) The estimated demand for each class of oil;

(3) Prospective production of each class of oil;

(4) Total of allotment bases of each class of oil for the current marketing year and the estimated total of allotment bases of each class for the ensuing marketing year;

(5) The quantity of reserve oil, by class, in storage;

(6) Producer prices of oil, including prices for each class of oil;

(7) General market conditions for each class of oil, including whether the estimated season average price to producers is likely to exceed parity.

(c) Notice of the marketing policy recommendations for a marketing year and any later changes shall be announced publicly by the Committee, and be submitted promptly to the Secretary and all producers and handlers. The Committee shall publicly announce its marketing policy or revision thereof and notice and contents thereof shall be submitted to producers and handlers by bulletins or through appropriate media.

(d) As soon as practicable following the effective date of this subpart and the organization of the Committee, the Committee may adopt a marketing policy for the 1980–81 marketing year.

§985.51 Recommendations for volume regulation.

(a) If the Committee's marketing policy considerations indicate a need for limiting the quantity of oil of each class marketed, the Committee shall recommend to,the Secretary a salable quantity and allotment percentage for the ensuing marketing year. Such recommendations shall be made prior to February 15, or such other date as the Committee, with the approval of the Secretary, may establish.

(b) At any time during the marketing year for which the Secretary, pursuant to § 985.52(a), has established a salable quantity and an allotment percentage for each class of oil, the Committee may recommend to the Secretary that such quantity be increased with an appropriate increase in the allotment percentage. Each such recommendation, together with the Committee's reason for such recommendation, shall be submitted promptly to the Secretary.

(c) As soon as practical following the effective date of this subpart and the organization of the Committee, the Committee may recommend a salable quantity for the 1980-81 marketing year.

§ 985.52   Issuance of volume regulation.

(a) Whenever the Secretary finds, on the basis of the Committee's recommendation or other information, that limiting the total quantity of a class of oil of any crop that handlers may purchase from or handle on behalf of producers during a marketing year, would tend to effectuate the declared policy of the act, the Secretary shall establish the salable quantity for that oil.

The salable quantity shall be prorated among producers by applying an allotment percentage to each producer's allotment base for that class of oil. The allotment percentage shall be established for each class of oil by dividing the salable quantity by the total of all producers' allotment bases for the same class of oil.

(b) When an allotment percentage for a class of oil is established for any marketing year, no handler shall purchase from or handle on behalf of producers any oil of that class during such year unless:

(1) It is, at the time of handling, within the unused portion of a producer's annual allotment, and

(2) Such handler notifies the Committee of the handling in such manner as it may prescribe.

§ 985.53   Allotment base.

(a) *Initial issuance.* Each producer desiring an allotment base for one or more classes of oil shall register with the Committee and furnish to it, on forms provided by the Committee, a report of the number of pounds of each class of oil sold during each of the marketing years of 1977, of 1978, and of 1979, which is the representative base period, and the number of pounds of each class of oil currently available for sale and the location of such oil, the name and address of each handler, the quantity of oil by class sold to each handler, the acreage and location of each year's production of spearmint, and any additional information requested by the Committee. A producer who has changed or changes identity from an individual producer to a partnership or corporate producer, or from a partnership to a corporate or individual producer, or from a corporate to a partnership or individual producer, may for the purpose of establishing the initial and subsequent allotment base, register with the Committee as one and the same person.

(b)(1) Initially, the allotment base for each class of oil shall be established by the Committee for each registered producer, at the option of such producer, as follows:

(i) The average annual number of pounds of oil of that class sold during any two marketing years of the representative base period; or

(ii) The average annual number of pounds of that class of oil sold during the representative period plus 33⅓ percent of oil of that class currently available for sale; or

(iii) The quantity of that class of oil sold during the 1979 marketing year, plus the quantity of that class of oil currently available for sale.

(2) If a producer has spearmint planted by February 27, 1979, but has no sales history during the representative period, the producer's allotment base shall be established by multiplying its acreage to be harvested for spearmint oil by the average amount of oil per acre sold in the allotment base of other

producers in the state or area, whichever is more representative, in which the acreage is located: *Provided,* That, the Committee shall review and adjust these allotment bases in accordance with paragraph (c) of this section on the basis of the producer's sales of spearmint oil.

(c) Periodically, but at least once every five years, the Committee shall review and adjust each producer's allotment base to recognize changes and trends in production and demand. Any such adjustment shall be made in accordance with a formula prescribed by the Committee with the approval of the Secretary.

(d)(1) Beginning with the 1982–83 marketing year, the Committee annually shall make additional allotment bases available for each class of oil in the amount of no more than 1 percent of the total allotment base for that class of oil. Fifty percent of these additional allotment bases shall be made available for new producers and 50 percent made available for existing producers.

(2) Any person may apply for an additional allotment base for any class of oil by filing an application with the Committee on or before December 1 of the marketing year preceding the marketing year for which the additional allotment bases will be made available.

(3) The Committee shall, with the approval of the Secretary, establish rules and regulations to be used for determining the distribution of additional allotment bases. In establishing such rules, the Committee shall take into account, among other things, the minimum economic enterprise requirements for oil production, the applicant's ability to produce oil, the area where the oil will be produced and other economic and marketing factors.

(e) The right to each producer receiving an allotment base, or any legal successor in interest, to retain all or part of an allotment base, shall be dependent on continuance to make a bona fide effort to produce the annual allotment referable thereto and failing to do so, such allotment base shall be reduced by an amount equivalent to such unproduced portions.

EFFECTIVE DATE NOTE: At 50 FR 41480, Oct. 11, 1985, in §985.53, paragraph (d)(2) was suspended indefinitely.

§985.54 Issuance of annual allotments.

(a) Whenever the Secretary establishes a salable quantity and allotment percentage for a class of oil that may be freely marketed during a marketing year, the Committee shall issue an annual allotment to each producer holding an allotment base for that class of oil. Each producer's annual allotment for a class of oil shall be determined by multiplying the producer's allotment base for that class of oil by the applicable allotment percentage.

(b) On or before December 1, the Committee shall furnish each registered holder of an allotment base a form for the producer to apply for an annual allotment for the ensuing marketing year. The Committee, with the approval of the Secretary, shall establish rules and regulations prescribing the information to be submitted on this form. The Committee shall notify each producer of the producer's annual allotment for each class of oil within 10 days after the Secretary establishes the salable quantity and allotment percentage.

(c) Through 1981, a handler may acquire oil of a producer's own production to fulfill a written contract entered into by these two persons prior to February 27, 1979. The terms of this contract shall require the producer to deliver to that handler a specified quantity of a class of oil from that producer's production at a specific price from a specified acreage and produced prior to 1982. The quantity of oil acquired by the handler pursuant to that contract during the 1980–81 or 1981–82 marketing year may exceed the producer's annual allotment for the applicable marketing year, but shall be charged against the producer's annual allotment for that year.

§985.55 Identification.

(a) Each producer shall, under supervision of the Committee, identify each class of oil within 15 days following its production, or such other period of time as is recommended by the Committee with the approval of the Secretary. Identification of oil shall be accomplished before its delivery either to a handler for handling as salable oil, or to the Committee or its designees for storage as excess oil.

(b) Identification shall indicate whether the oil is salable or excess oil and include the name of the producer, the class of oil, the net weight, the container number and such other information as may be required by the Committee.

(c) Identification shall be accomplished in accordance with rules and regulations established by the Committee with the approval of the Secretary.

(d) No handler shall handle as salable oil, and the Committee shall not receive as excess oil, any oil that has not been identified as provided in this section, and no person shall alter or remove any identification except when incidental to final disposition.

§ 985.56 Excess oil.

Oil of any class in excess of a producer's applicable annual allotment shall be identified as excess oil and shall be disposed of as follows:

(a) Before October 15, or such date as the Committee, with the approval of the Secretary, may establish, a producer, following notification of the Committee, may transfer excess oil to another producer to enable that producer to fill a deficiency in that producer's annual allotment, or

(b) Before November 1, or such other date as the Committee, with the approval of the Secretary, may establish, excess oil, not used to fill another producer's deficiency, shall be delivered to the Committee or its designees for storage. Such oil shall be stored for the account of the producer. All costs of storage including identification and insurance shall be paid by the producer of excess oil. No handler shall handle excess oil and no producer shall deliver excess oil to other than the Committee or its designees.

(c) The Committee, with the approval of the Secretary, may establish such rules and regulations as it deems necessary for the transfer or storage of excess oil.

§ 985.57 Reserve pool requirements.

(a) On November 1, or such other date as the Committee, with the approval of the Secretary may establish, the Committee shall pool identified excess oil as reserve oil in such manner as to ac-

curately account for its receipt, storage, and disposition. The Committee shall store reserve oil for the account of the producer and maintain the identity of the reserve oil by producer's name, the year produced, the class of oil, and such other identification as may be used in normal commercial trade practices. The Committee shall designate a Committee employee as reserve pool manager.

(b) *Disposition.* (1) When, in any marketing year, a producer has produced less than the annual allotment of a class of oil, the producer may, upon notification of the Committee, fill the deficiency with the same class of reserve oil from the producer's prior production.

(2) Prior to March 15 of any year, or such other date as recommended by the Committee and approved by the Secretary, a producer may notify the Committee of a possible deficiency in the producer's ensuing year's production of oil and wishes to use reserve oil from own production to fill the ensuing year's annual allotment. The Committee shall approve the producer's request if the oil is still available at the time of the request.

(3) Under supervision of the Committee, a producer may exchange salable oil for the same class and quantity of reserve oil from own production so long as the oil is properly identified.

(4) When the Committee finds that additional oil is needed to fill the normal market demand, it shall offer all or a portion of the reserve oil for sale to handlers. Offers to sell, extension of offers and withdrawal of offers shall be subject to disapproval by the Secretary. The Committee may establish rules and regulations governing the offers and sale to handlers.

(5) The Committee may use reserve oil for market development projects approved by the Secretary. Such projects may be conducted by the Committee or in conjunction with or through handlers.

(c) *Pool expenses and proceeds.* Expenses incurred by the Committee in handling and storing reserve oil shall be paid by the equity holders. The proceeds from the disposition of reserve oil shall be distributed, after deduction of any expenses incurred by the Committee in receiving, handling, storing, and disposing thereof, to the equity holders or their successors in interest, on the basis of the number of pounds, class of oil and quality credited to each equity holder's account in the pool. A full accounting to each equity holder, or successor in interest, in each reserve pool shall be made by the Committee annually.

§985.58 Exempt oil.

Oil held by a producer or handler on the effective date of this subpart shall not be regulated under this subpart if reported and identified to the Committee not later than 60 days after that date. Any such oil not reported and identified to the Committee shall be subject to all regulation under this subpart.

§985.59 Transfers.

(a) Nothing contained in this part shall prevent a producer from transferring the location where that producer's annual allotment is produced to another location except that the producer shall report the transfer to the Committee within 30 days after the transfer.

(b) A producer may transfer all or part of an allotment base to another producer under rules and regulations established by the Committee, with the approval of the Secretary: *Provided,* That the allotment base obtained by transfer from another producer or issued pursuant to §985.53(d)(1) shall not be transferred for at least 2 years following transfer or issuance, and that the person receiving the allotment base submit to the Committee, evidence of an ability to produce and sell oil from such allotment base in the first marketing year following the transfer or issuance of the allotment base.

REPORTS AND RECORDS

§985.60 Reports.

(a) *Inventory.* Each handler shall file with the Committee a certified report showing such information as the Committee may specify with respect to any oil which was held by the handler at such times as the Committee may designate.

(b) *Receipts.* Each handler shall, upon request of the Committee, file with the Committee a certified report showing for each lot of oil received, the identifying marks, class of oil, weight, place of production, and the producer's name and address at such times as the Committee may designate.

(c) *Other reports.* Upon the request of the Committee, each handler shall furnish such other information as may be necessary to enable the Committee to exercise its powers and perform its duties under this part.

§985.61 Records.

Each handler shall maintain such records pertaining to all oil handled as will substantiate the required reports. All such records shall be maintained for not less than 2 years after the termination of the marketing year to which such records relate.

§985.62 Verification of reports and records.

For the purpose of assuring compliance with record keeping requirements and verifying reports filed by producers and handlers, the Secretary and the Committee, through its duly authorized employees, shall have access to any premises where applicable records are maintained, where oil is received or held, and at any time during reasonable business hours, shall be permitted to inspect such handlers' premises, and any and all records of such handlers with respect to matters within the purview of this part.

§985.63 Confidential information.

All reports and records furnished or submitted by handlers to, or obtained by the employees of the Committee, which contain data or information constituting a trade secret or disclosing the trade position, financial condition, or business operations of the particular handler from whom received, shall be treated as confidential and the reports and all information obtained from records shall, at all times, be kept in the custody and under the control of one or more employees of the Committee who shall disclose such information to no person other than the Secretary.

MISCELLANEOUS PROVISIONS

§ 985.64 Compliance.

No person shall handle oil except in conformity with the provisions of this part.

§ 985.65 Rights of the Secretary.

Members of the Committee and subcommittees, and any agents, employees or representatives thereof, shall be subject to removal or suspension by the Secretary at any time. Each and every decision, determination, and other act of the Committee shall be subject to the continuing right of disapproval by the Secretary at any time. Upon such disapproval, the disapproved action of the Committee shall be deemed null and void, except as to acts done in reliance thereon or in accordance therewith prior to such disapproval by the Secretary.

§ 985.66 Derogation.

Nothing contained in this part is, or shall be construed to be, in derogation or in modification of the rights of the Secretary or of the United States (a) to exercise any powers granted by the act or otherwise, or (b) in accordance with such powers, to act in the premises whenever such action is deemed advisable.

§ 985.67 Agents.

The Secretary may, by designation in writing, name any officer or employee of the United States or name any agency or division in the U.S. Department of Agriculture, to act as the Secretary's agent or representative in connection with any of the provisions of this part.

§ 985.68 Personal liability.

No member or alternative member of the Committee and no employee or agent of the Committee shall be held personally responsible, either individually or jointly with others, in any way whatsoever, to any person for errors in judgment, mistakes, or other acts, either of commission or omission, as such member, alternate, employee, or agent, except for acts of dishonesty, willful misconduct, or gross negligence.

§ 985.69 Duration of immunities.

The benefits, privileges, and immunities conferred upon any person by virtue of this part shall cease upon its termination, except with respect to acts done under and during the existence of this part.

§ 985.70 Separability.

If any provision of this part is declared invalid or the applicability thereof to any person, circumstance or thing is held invalid, the validity of the remainder of this part or the applicability thereof to any other person, circumstance, or thing shall not be affected thereby.

§ 985.71 Effective time.

The provisions of this subpart, and of any amendment thereto, shall become effective at such time as the Secretary may declare and shall continue in force until terminated or suspended in one of the ways specified in § 985.72.

§ 985.72 Termination.

(a) *Failure to effectuate.* The Secretary shall terminate or suspend the operation of any or all of the provisions of this part upon a finding that such provisions obstruct or do not tend to effectuate the declared policy of the act.

(b) *Referendum.* The Secretary shall terminate the provisions of this subpart at the end of any marketing year upon a finding that such termination is favored by a majority of the producers who, during the preceding marketing year, produced for market more than 50 percent of the volume of oil so produced: *Provided,* That termination shall be effective only if announced before May 31 of the then current marketing year.

(c) *Termination of act.* The provisions of this subpart shall, in any event, terminate whenever the provisions of the act authorizing them cease to be in effect.

§ 985.73 Proceedings after termination.

Upon termination of the provisions of this part, the Committee shall, for the purpose of liquidating the affairs of the Committee, continue as trustees of all the funds and property then in its possession or under its control, including claims for any funds unpaid or property not delivered at the time of such termination. The said trustees shall (a) continue in such capacity until discharged by the Secretary; (b) from time to time account for all receipts and disbursements and deliver all property on hand, together with all books and records of the Committee and of the trustees, to such persons as the Secretary may direct; and (c) upon the request of the Secretary execute such assignments or other instruments necessary or appropriate to vest in such person full title and right to all of the funds, property, and claims vested in the Committee or the trustees pursuant thereto. Any person to whom funds, property, or claims have been transferred or delivered, pursuant to this section, shall be subject to the same obligation imposed upon the Committee and upon trustees.

§985.74 Effect of termination or amendment.

Unless otherwise expressly provided by the Secretary, the termination of this subpart or of any regulation issued pursuant to this subpart, or the issuance of any amendment to either thereof, shall not (a) affect or waive any right, duty, obligation, or liability which shall have risen or which may thereafter arise in connection with any provision of this subpart or any regulation issued hereunder, or (b) release or extinguish any violation of this subpart or any regulation issued hereunder, or (c) affect or impair any rights or remedies of the Secretary or any other person with respect to any such violation.

Subpart B—Administrative Requirements

§985.104 Changed classes of spearmint oil.

Pursuant to §985.4, the classes of spearmint oil contained in that section are changed by deleting the term and definition *Class 2* Oil and changing the definition of *Class 1* Oil. The changed classes are as follows:

Class 1: Oil extracted from Scotch Spearmint.

Class 3: Oil extracted from Native Spearmint.
Class 4: Oil which has a spearmint flavor, extracted from plants other than Scotch or Native Spearmint.

[48 FR 53400, Nov. 28, 1983]

§985.141 Assessment rate.

On and after June 1, 2019, an assessment rate of $0.10 per pound is established for Far West spearmint oil. Unexpended funds may be carried over as a reserve.

[84 FR 41885, Aug. 16, 2019]

§985.152 Handling report.

Whenever an allotment percentage has been established for a class of oil, each handler shall furnish to the Committee at least the following information for each lot of that class of oil acquired by the handler from a producer: (a) Name of producer; (b) name of handler; (c) class of oil acquired; (d) date of acquisition; (e) date when oil was produced; (f) net weight of oil in the lot; (g) quantity of that class of oil in the producer's annual allotment available for handler before this acquisition; and (h) quantity of oil remaining in the producer's annual allotment after this acquisition. This information shall be furnished in such manner as the Committee may prescribe. Upon acquisition the handler or the handler's agent also shall include the applicable information on the back of the producer's Annual Allotment Certificate, showing that the acquired oil was within the unused portion of the producer's annual allotment.

[46 FR 43130, Aug. 27, 1981]

§985.153 Issuance of additional allotment base to new and existing producers.

(a) *Definitions.* (1) *New producer* means any person who never was issued an allotment base by the Committee for a class of oil in any capacity either as an individual, or as a member of a partnership, corporation, or any other business unit.

(2) *Existing producer* means any person who was issued an allotment based by the Committee for a class of oil in any capacity either as an individual, or as a member of a partnership, corporation, or any other business unit. Any

person who was initially issued an allotment base for a class of oil and changed identity of operation, as set forth in § 985.53(a), since April 14, 1980, and requests additional allotment base for that class of oil pursuant to this section, shall be deemed to be an existing producer.

(b) *Requests.* Any new or existing producer desiring additional allotment base for any class of oil made available by the Committee pursuant to § 985.53(d)(1) shall request such base by a date specified by the Committee prior to the marketing year for which such base will be made available.

(c) *Issuance*—(1) *New producers*—(i) *Regions:* For the purpose of issuing additional allotment base to new producers, the production area is divided into the following regions:

(A) *Region A.* The State of Washington.

(B) *Region B.* All areas of the production area outside the State of Washington.

(ii) The Committee shall review all requests from new producers for additional allotment base made available pursuant to § 985.53(d)(1).

(iii) Each year, the Committee shall determine the size of the minimum economic enterprise required to produce each class of oil. The Committee shall thereafter calculate the number of new producers who will receive allotment base under this section for each class of oil. The Committee shall include that information in its announcements to new producers in each region informing them when to submit requests for allotment base. The Committee shall determine whether the new producers requesting additional base have the ability to produce spearmint oil. The names of all eligible new producers from each region shall be placed in separate lots per class of oil. For each class of oil, separate drawings shall be held from a list of all applicants from Region A and from a list of all applicants from Region B. If, in any marketing year, there are no requests for additional base in a class of oil from eligible new producers in a region, such unallocated additional allotment base shall be issued to an eligible new producer whose name is selected by drawing from a list containing the

names of all remaining eligible new producers from the other region for that class of oil. The Committee shall immediately notify each new producer whose name was drawn and issue that producer an allotment base in the appropriate amount. Allotment base issued to new producers under this section shall not be transferred for at least five years following issuance.

(2) *Existing producers.* (i) The Committee shall review all requests from existing producers for additional allotment base.

(ii) *Class 1 base.* With respect to the issuance of additional Class 1 allotment base to existing producers for the 2014–2015 through the 2016–2017 marketing years, existing producers with less than 5,121 pounds of allotment base as of October 17, 2012, who request additional allotment base and who have the ability to produce additional quantities of Class 1 spearmint oil, shall be issued additional allotment base sufficient to bring them up to a level not to exceed 5,121 pounds: *Provided,* That such additional Class 1 allotment base shall be allocated to eligible producers on a pro-rata basis from available additional Class 1 allotment base: *Provided further,* That additional allotment base shall not be issued to any person if such additional allotment base would replace all or part of an allotment base that such person has previously transferred to another producer. Additional allotment base in excess of the amount needed to bring eligible producers up to 5,121 pounds of Class 1 allotment base shall be distributed on a prorated basis among all existing producers who apply and who have the ability to produce additional quantities of spearmint oil.

(iii) *Class 3 base.* With respect to the issuance of additional Class 3 allotment base for existing producers for the 2014–2015 through the 2017–2018 marketing years, existing producers with less than 5,812 pounds of allotment base as of October 17, 2012, who request additional allotment base and who have the ability to produce additional quantities of Class 3 spearmint oil, shall be issued additional allotment base sufficient to bring them up to a level not to exceed 5,812 pounds: *Provided,* That such additional Class 3 allotment base shall be allocated to eligible producers on a pro-rata basis from available additional Class 3 allotment base: *Provided further,* That additional allotment base shall not be issued to any person if such additional allotment base would replace all or part of an allotment base that such person has previously transferred to another producer. Additional allotment base in excess of the amount needed to bring eligible producers up to 5,812 pounds of Class 3 allotment base shall be distributed on a prorated basis among all existing producers who apply and who have the ability to produce additional quantities of spearmint oil.

(iv) For each marketing year after 2016–2017 for Class 1 oil and 2017–2018 for Class 3 oil, each existing producer of a class of spearmint oil who requests additional allotment base, and who has the ability to produce additional quantities of that class of spearmint oil, shall be eligible to receive a share of the additional allotment base issued for that class of oil. Additional allotment base issued by the Committee for a class of oil shall be distributed on a prorated basis among the eligible producers for that class of oil. The Committee shall immediately notify each producer who is to receive additional allotment base by issuing that producer an allotment base in the appropriate amount. Allotment base issued to existing producers under this section shall not be transferred for at least two years following issuance, except that additional allotment base allocated pursuant to paragraph (c)(2)(ii) and (c)(2)(iii) of this section shall not be transferred for at least five years following issuance.

(d) The person receiving additional allotment base pursuant to this section shall submit to the Committee evidence of an ability to produce and sell oil from such allotment base in the first marketing year following issuance of such base.

[47 FR 41332, Sept. 20, 1982, as amended at 50 FR 41480, Oct. 11, 1985; 51 FR 45450, Dec. 19, 1986; 56 FR 51829, Oct. 16, 1991; 57 FR 28595, June 26, 1992; 62 FR 43465, Aug. 14, 1997; 65 FR 30344, May 11, 2000; 68 FR 25486, May 13, 2003; 79 FR 37936, July 3, 2014]

§985.154 Issuance of annual allotments.

(a) Each producer who is a registered holder of an allotment base, and desiring an annual allotment for the ensuing marketing year, shall apply to the Committee for that allotment. The registered holder shall furnish at least the following information:

(1) The number of acres of each species (Scotch or Native) of spearmint planted, or intended to be planted for harvest in the ensuing marketing year;

(2) Whether the spearmint to be harvested in the ensuing marketing year is baby mint (first year harvest) or mature mint (second year or older harvest); and

(3) Any changes in location or production as reported for the preceding year.

(b) In order to enable the Committee to insure compliance and verify furnished information, each producer requesting an annual allotment shall permit the Committee or its representatives, whenever necessary, to measure the producer's spearmint acreage.

[46 FR 43130, Aug. 27, 1981]

§985.155 Identification of oil by producer.

Following the distillation of oil and prior to delivery either to a handler for handling as salable oil or to the Committee or its designee for storage as excess oil, each producer shall furnish the following information to the Committee: *Provided,* That any class of oil retained by a producer shall be reported to the Committee within 15 days following the completion of its distillation.

(a) Producer's name and address;

(b) Date the oil was put into the drum;

(c) Class of oil in the drum;

(d) Drum identification number;

(e) Approximate net weight of the oil;

(f) Handler's pickup receipt number, when applicable;

(g) Destination of oil for storage;

(h) Name of the firm where the oil was distilled; and

(i) Name of the person submitting the information.

[46 FR 43130, Aug. 27, 1981]

§ 985.156 Transfer of excess oil by producers.

(a) Pursuant to § 985.56(a), before December 1 of each marketing year, a producer, following notification of the Committee, may transfer excess oil to another producer to enable that producer to fill a deficiency in that producer's annual allotment.

(b) Pursuant to § 985.56(b), before December 1 of each marketing year, excess oil not used to fill another producer's deficiency shall be delivered to the Committee or its designees for storage.

[78 FR 9577, Feb. 11, 2013]

§ 985.157 Reserve pool requirements.

Pursuant to § 985.57(a), on December 1, the Committee shall pool identified excess oil as reserve oil in such manner as to accurately account for its receipt, storage, and disposition.

[78 FR 9577, Feb. 11, 2013]

§ 985.233 Salable quantities and allotment percentages.

The salable quantity and allotment percentage for each class of spearmint oil during the marketing year beginning on June 1, 2018, shall be as follows:

(a) Class 1 (Scotch) oil—a salable quantity of 760,660 pounds and an allotment percentage of 35 percent.

(b) Class 3 (Native) oil—a salable quantity of 1,431,350 pounds and an allotment percentage of 58 percent.

[83 FR 34940, Aug. 23, 2018, as amended at 84 FR 17059, Apr. 24, 2019]

§ 985.234 Salable quantities and allotment percentages—2019–2020 marketing year.

The salable quantity and allotment percentage for each class of spearmint oil during the marketing year beginning on June 1, 2019, shall be as follows:

(a) Class 1 (Scotch) oil—a salable quantity of 832,081 pounds and an allotment percentage of 38 percent.

(b) Class 3 (Native) oil—a salable quantity of 1,395,813 pounds and an allotment percentage of 56 percent.

[84 FR 19709, May 6, 2019]

PART 986—PECANS GROWN IN THE STATES OF ALABAMA, ARKANSAS, ARIZONA, CALIFORNIA, FLORIDA, GEORGIA, KANSAS, LOUISIANA, MISSOURI, MISSISSIPPI, NORTH CAROLINA, NEW MEXICO, OKLAHOMA, SOUTH CAROLINA, AND TEXAS

Subpart A—Order Regulating Handling of Pecans

DEFINITIONS

AUTHORITY: 7 U.S.C. 601–674.

SOURCE: 81 FR 51302, Aug. 4, 2016, unless otherwise noted.

Subpart A—Order Regulating Handling of Pecans

DEFINITIONS

§986.1 Accumulator.

Accumulator means a person who compiles inshell pecans from other persons for the purpose of resale or transfer.

§986.2 Act.

Act means Public Act No. 10, 73d Congress, as amended and as reenacted and amended by the Agricultural Marketing Agreement Act of 1937, as amended (7 U.S.C. 601 *et seq.*).

§986.3 Affiliation.

Affiliation. This term normally appears as "affiliate of" or "affiliated with," and means a person such as a grower or sheller who is: A grower or handler that directly, or indirectly through one or more intermediaries, owns or controls, or is controlled by, or is under common control with the grower or handler specified; or a grower or handler that directly, or indirectly through one or more intermediaries, is connected in a proprietary capacity, or shares the ownership or control of the specified grower or handler with one or more other growers or handlers. As used in this part, the term "control" (including the terms "controlling," "controlled by," and "under the common control with") means the possession, direct or indirect, of the power to direct or cause the direction of the management and policies of a handler or a grower, whether through voting securities, membership in a cooperative, by contract or otherwise.

§986.4 Blowouts.

Blowouts mean lightweight or underdeveloped inshell pecan nuts that are considered of lesser quality and market value.

§ 986.5   To certify.

To certify means the issuance of a certification of inspection of pecans by the inspection service.

§ 986.6   Confidential data or information.

Confidential data or information submitted to the Council consists of data or information constituting a trade secret or disclosure of the trade position, financial condition, or business operations of a particular entity or its customers.

§ 986.7   Container.

Container means a box, bag, crate, carton, package (including retail packaging), or any other type of receptacle used in the packaging or handling of pecans.

§ 986.8   Council.

Council means the American Pecan Council established pursuant to § 986.45, American Pecan Council.

§ 986.9   Crack.

Crack means to break, crack, or otherwise compromise the outer shell of a pecan so as to expose the kernel inside to air outside the shell.

§ 986.10   Cracks.

Cracks refer to an accumulated group or container of pecans that have been cracked in harvesting or handling.

§ 986.11   Custom harvester.

Custom harvester means a person who harvests inshell pecans for a fee.

§ 986.12   Department or USDA.

Department or *USDA* means the United States Department of Agriculture.

§ 986.13   Disappearance.

Disappearance means the difference between the sum of grower-cleaned production and handler-cleaned production (whether from improved orchards or native and seedling groves) and the sum of inshell and shelled merchantable pecans reported on an inshell weight basis.

§ 986.14   Farm Service Agency.

Farm Service Agency or *FSA* means that agency of the U.S. Department of Agriculture.

§ 986.15   Fiscal year.

Fiscal year means the twelve months from October 1 to September 30, both inclusive, or any other such period deemed appropriate by the Council and approved by the Secretary.

§ 986.16   Grade and size.

Grade and size means any of the officially established grades of pecans and any of the officially established sizes of pecans as set forth in the United States standards for inshell and shelled pecans or amendments thereto, or modifications thereof, or other variations of grade and size based thereon recommended by the Council and approved by the Secretary.

§ 986.17   Grower.

(a) *Grower* is synonymous with producer and means any person engaged within the production area in a proprietary capacity in the production of pecans if such person:

(1) Owns an orchard and harvests its pecans for sale (even if a custom harvester is used); or

(2) Is a lessee of a pecan orchard and has the right to sell the harvest (even if the lessee must remit a percentage of the crop or rent to a lessor).

(b) The term ''grower'' shall only include those who produce a minimum of 50,000 pounds of inshell pecans during a representative period (average of four years) or who own a minimum of 30 pecan acres according to the FSA, including acres calculated by the FSA based on pecan tree density. In the absence of any FSA delineation of pecan acreage, the regular definition of an acre will apply. The Council may recommend changes to this definition subject to the approval of the Secretary.

§ 986.18   Grower-cleaned production.

Grower-cleaned production means production harvested and processed through a cleaning plant to determine volumes of improved pecans, native and seedling pecans, and substandard pecans to transfer to a handler for sale.

526

§986.19 Handler.

Handler means any person who handles inshell or shelled pecans in any manner described in §986.20.

§986.20 To handle.

To handle means to receive, shell, crack, accumulate, warehouse, roast, pack, sell, consign, transport, export, or ship (except as a common or contract carrier of pecans owned by another person), or in any other way to put inshell or shelled pecans into any and all markets in the stream of commerce either within the area of production or from such area to any point outside thereof. The term "to handle" shall not include: sales and deliveries within the area of production by growers to handlers; grower warehousing; custom handling (except for selling, consigning or exporting) or other similar activities paid for on a fee-for-service basis by a grower who retains the ownership of the pecans; or transfers between handlers.

§986.21 Handler inventory.

Handler inventory means all pecans, shelled or inshell, as of any date and wherever located within the production area, then held by a handler for their account.

§986.22 Handler-cleaned production.

Handler-cleaned production is production that is received, purchased or consigned from the grower by a handler prior to processing through a cleaning plant, and then subsequently processed through a cleaning plant so as to determine volumes of improved pecans, native and seedling pecans, and substandard pecans.

§986.23 Hican.

Hican means a tree resulting from a cross between a pecan and some other type of hickory (members of the genus *Carya*) or the nut from such a hybrid tree.

§986.24 Inshell pecans.

Inshell pecans are nuts whose kernel is maintained inside the shell.

§986.25 Inspection Service.

Inspection service means the Federal-State Inspection Service or any other inspection service authorized by the Secretary.

§986.26 Inter-handler transfer.

Inter-handler transfer means the movement of inshell pecans from one handler to another inside the production area for the purposes of additional handling. Any assessments or requirements under this part with respect to inshell pecans so transferred may be assumed by the receiving handler.

§986.27 Merchantable pecans.

(a) *Inshell.* Merchantable inshell pecans mean all inshell pecans meeting the minimum grade regulations that may be effective pursuant to §986.69, Authorities regulating handling.

(b) *Shelled.* Merchantable shelled pecans means all shelled pecans meeting the minimum grade regulations that may be effective pursuant to §986.69, Authorities regulating handling.

§986.28 Pack.

Pack means to clean, grade, or otherwise prepare pecans for market as inshell or shelled pecans.

§986.29 Pecans.

(a) *Pecans* means and includes any and all varieties or subvarieties of Genus: *Carya,* Species: *illinoensis,* expressed also as *Carya illinoinensis (syn. C. illinoenses)* including all varieties thereof, excluding hicans, that are produced in the production area and are classified as:

(1) *Native or seedling* pecans harvested from non-grafted or naturally propagated tree varieties;

(2) *Improved pecans* harvested from grafted tree varieties bred or selected for superior traits of nut size, ease of shelling, production characteristics, and resistance to certain insects and diseases, including but not limited to: Desirable, Elliot, Forkert, Sumner, Creek, Excel, Gracross, Gratex, Gloria Grande, Kiowa, Moreland, Sioux, Mahan, Mandan, Moneymaker, Morrill, Cunard, Zinner, Byrd, McMillan, Stuart, Pawnee, Eastern and Western Schley, Wichita, Success, Cape Fear,

Choctaw, Cheyenne, Lakota, Kanza, Caddo, and Oconee; and

(3) *Substandard pecans* that are blow-outs, cracks, stick-tights, and other inferior quality pecans, whether native or improved, that, with further handling, can be cleaned and eventually sold into the stream of commerce.

(b) The Council, with the approval of the Secretary, may recognize new or delete obsolete varieties or sub-varieties for each category.

§ 986.30   Person.

Person means an individual, partnership, corporation, association, or any other business unit.

§ 986.31   Production area.

Production area means the following fifteen pecan-producing states within the United States: Alabama, Arkansas, Arizona, California, Florida, Georgia, Kansas, Louisiana, Mississippi, Missouri, North Carolina, New Mexico, Oklahoma, South Carolina, and Texas.

§ 986.32   Proprietary capacity.

Proprietary capacity means the capacity or interest of a grower or handler that, either directly or through one or more intermediaries or affiliates, is a property owner together with all the appurtenant rights of an owner, including the right to vote the interest in that capacity as an individual, a shareholder, member of a cooperative, partner, trustee or in any other capacity with respect to any other business unit.

§ 986.33   Regions.

(a) *Regions* within the production area shall consist of the following:

(1) *Eastern Region,* consisting of: Alabama, Florida, Georgia, North Carolina, South Carolina

(2) *Central Region,* consisting of: Arkansas, Kansas, Louisiana, Mississippi, Missouri, Oklahoma, Texas

(3) *Western Region,* consisting of: Arizona, California, New Mexico

(b) With the approval of the Secretary, the boundaries of any region may be changed pursuant to § 986.58, Reapportionment and reestablishment of regions.

§ 986.34   Representative period.

Representative period is the previous four fiscal years for which a grower's annual average production is calculated, or any other period recommended by the Council and approved by the Secretary.

§ 986.35   Secretary.

Secretary means the Secretary of Agriculture of the United States, or any other officer or employee of the United States Department of Agriculture who is, or who may be, authorized to perform the duties of the Secretary of Agriculture of the United States.

§ 986.36   Sheller.

Sheller refers to any person who converts inshell pecans to shelled pecans and sells the output in any and all markets in the stream of commerce, both within and outside of the production area; *Provided,* That the term "sheller" shall only include those who shell more than 1 million pounds of inshell pecans in a fiscal year. The Council may recommend changes to this definition subject to the approval of the Secretary.

§ 986.37   Shelled pecans.

Shelled pecans are pecans whose shells have been removed leaving only edible kernels, kernel pieces or pecan meal. *Shelled pecans* are synonymous with *pecan meats.*

§ 986.38   Stick-tights.

Stick-tights means pecans whose outer shuck has adhered to the shell causing their value to decrease or be discounted.

§ 986.39   Trade supply.

Trade supply means the quantity of merchantable inshell or shelled pecans that growers will supply to handlers during a fiscal year for sale in the United States and abroad or, in the absence of handler regulations § 986.69 setting forth minimum grade regulations for merchantable pecans, the sum of handler-cleaned and grower-cleaned production.

§986.40 Unassessed inventory.

Unassessed inventory means inshell pecans held by growers or handlers for which no assessment has been paid to the Council.

§986.41 Varieties.

Varieties mean and include all cultivars, classifications, or subdivisions of pecans.

§986.42 Warehousing.

Warehousing means to hold assessed or unassessed inventory.

§986.43 Weight.

Weight means pounds of inshell pecans, received by handler within each fiscal year; *Provided,* That for shelled pecans the actual weight shall be multiplied by two to obtain an inshell weight.

ADMINISTRATIVE BODY

§986.45 American Pecan Council.

The American Pecan Council is hereby established consisting of 17 members selected by the Secretary, each of whom shall have an alternate member nominated with the same qualifications as the member. The 17 members shall include nine (9) grower seats, six (6) sheller seats, and two (2) at-large seats allocated to one accumulator and one public member. The grower and sheller nominees and their alternates shall be growers and shellers at the time of their nomination and for the duration of their tenure. Grower and sheller members and their alternates shall be selected by the Secretary from nominees submitted by the Council. The two at-large seats shall be nominated by the Council and appointed by the Secretary.

(a) Each region shall be allocated the following member seats:

(1) *Eastern Region:* Three (3) growers and two (2) shellers;

(2) *Central Region:* Three (3) growers and two (2) shellers;

(3) *Western Region:* Three (3) growers and two (2) shellers.

(b) Within each region, the grower and sheller seats shall be defined as follows:

(1) *Grower seats:* Each region shall have a grower Seat 1 and Seat 2 allocated to growers whose acreage is equal to or exceeds 176 pecan acres. Each region shall also have a grower Seat 3 allocated to a grower whose acreage is less than 176 pecan acres.

(2) *Sheller seats:* Each region shall have a sheller Seat 1 allocated to a sheller who handles more than 12.5 million pounds of inshell pecans in the fiscal year preceding nomination, and a sheller Seat 2 allocated to a sheller who handles less than or equal to 12.5 million pounds of inshell pecans in the fiscal year preceding nomination.

(c) The Council may recommend, subject to the approval of the Secretary, revisions to the above requirements for grower and sheller seats to accommodate changes within the industry.

§986.46 Council nominations and voting.

Nomination of Council members and alternate members shall follow the procedure set forth in this section, or as may be changed as recommended by the Council and approved by the Secretary. All nominees must meet the requirements set forth in §§986.45, American Pecan Council, and 986.48, Eligibility, or as otherwise identified by the Secretary, to serve on the Council.

(a) *Initial members.* Nominations for initial Council members and alternate members shall be conducted by the Secretary by either holding meetings of shellers and growers, by mail, or by email, and shall be submitted on approved nomination forms. Eligibility to cast votes on nomination ballots, accounting of nomination ballot results, and identification of member and alternate nominees shall follow the procedures set forth in this section, or by any other criteria deemed necessary by the Secretary. The Secretary shall select and appoint the initial members and alternate members of the Council.

(b) *Successor members.* Subsequent nominations of Council members and alternate members shall be conducted as follows:

(1) *Call for nominations.* (i) Nominations for the grower member seats for each region shall be received from growers in that region on approved forms containing the information stipulated in this section.

(ii) If a grower is engaged in producing pecans in more than one region, such grower shall nominate in the region in which they grow the largest volume of their production.

(iii) Nominations for the sheller member seats for each region shall be received from shellers in that region on approved forms containing the information stipulated in this section.

(iv) If a sheller is engaged in handling in more than one region, such sheller shall nominate in the region in which they shelled the largest volume in the preceding fiscal year.

(2) *Voting for nominees.* (i) Only growers, through duly authorized officers or employees of growers, if applicable, may participate in the nomination of grower member nominees and their alternates. Each grower shall be entitled to cast only one nomination ballot for each of the three grower seats in their region.

(ii) If a grower is engaged in producing pecans in more than one region, such grower shall cast their nomination ballot in the region in which they grow the largest volume of their production. Notwithstanding this stipulation, such grower may vote their volume produced in any or all of the three regions.

(iii) Only shellers, through duly authorized officers or employees of shellers, if applicable, may participate in the nomination of the sheller member nominees and their alternates. Each sheller shall be entitled to cast only one nomination ballot for each of the two sheller seats in their region.

(iv) If a sheller is engaged in handling in more than one region, such sheller shall cast their nomination ballot in the region in which they shelled the largest volume in the preceding fiscal year. Notwithstanding this stipulation, such sheller may vote their volume handled in all three regions.

(v) If a person is both a grower and a sheller of pecans, such person may not participate in both grower and sheller nominations. Such person must elect to participate either as a grower or a sheller.

(3) *Nomination procedure for grower seats.* (i) The Council shall mail to all growers who are on record with the Council within the respective regions a grower nomination ballot indicating the nominees for each of the three grower member seats, along with voting instructions. Growers may cast ballots on the proper ballot form either at meetings of growers, by mail, or by email as designated by the Council. For ballots to be considered, they must be submitted on the proper forms with all required information, including signatures.

(ii) On the ballot, growers shall indicate their vote for the grower nominee candidates for the grower seats and also indicate their average annual volume of inshell pecan production for the preceding four fiscal years.

(iii) *Seat 1* (growers with equal to or more than 176 acres of pecans). The nominee for this seat in each region shall be the grower receiving the highest volume of production (pounds of inshell pecans) votes from the respective region, and the grower receiving the second highest volume of production votes shall be the alternate member nominee for this seat. In case of a tie vote, the nominee shall be selected by a drawing.

(iv) *Seat 2* (growers with equal to or more than 176 acres of pecans). The nominee for this seat in each region shall be the grower receiving the highest number of votes from their respective region, and the grower receiving the second highest number of votes shall be the alternate member nominee for this seat. In case of a tie vote, the nominee shall be selected by a drawing.

(v) *Seat 3* (grower with less than 176 acres of pecans). The nominee for this seat in each region shall be the grower receiving the highest number of votes from the respective region, and the grower receiving the second highest number of votes shall be the alternate member nominee for this seat. In case of a tie vote, the nominee shall be selected by a drawing.

(4) *Nomination procedure for sheller seats.* (i) The Council shall mail to all shellers who are on record with the Council within the respective regions the sheller ballot indicating the nominees for each of the two sheller member seats in their respective regions, along with voting instructions. Shellers may cast ballots on approved ballot forms either at meetings of shellers, by

mail, or by email as designated by the Council. For ballots to be considered, they must be submitted on the approved forms with all required information, including signatures.

(ii) *Seat 1* (shellers handling more than 12.5 million lbs. of inshell pecans in the preceding fiscal year). The nominee for this seat in each region shall be assigned to the sheller receiving the highest number of votes from the respective region, and the sheller receiving the second highest number of votes shall be the alternate member nominee for this seat. In case of a tie vote, the nominee shall be selected by a drawing.

(iii) *Seat 2* (shellers handling equal to or less than 12.5 million lbs. of inshell pecans in the preceding fiscal year). The nominee for this seat in each region shall be assigned to the sheller receiving the highest number of votes from the respective region, and the sheller receiving the second highest number of votes shall be the alternate member nominee for this seat. In case of a tie vote, the nominee shall be selected by a drawing.

(5) *Reports to the Secretary.* Nominations in the foregoing manner received by the Council shall be reported to the Secretary on or before 15 of each July of any year in which nominations are held, together with a certified summary of the results of the nominations and other information deemed by the Council to be pertinent or requested by the Secretary. From those nominations, the Secretary shall select the fifteen grower and sheller members of the Council and an alternate for each member, unless the Secretary rejects any nomination submitted. In the event the Secretary rejects a nomination, a second nomination process may be conducted to identify other nominee candidates, the resulting nominee information may be reported to the Secretary after July 15 and before September 15. If the Council fails to report nominations to the Secretary in the manner herein specified, the Secretary may select the members without nomination. If nominations for the public and accumulator at-large members are not submitted by September 15 of any year in which their nomination is due, the Secretary may select such members without nomination.

(6) *At-large members.* The grower and sheller members of the Council shall select one public member and one accumulator member and respective alternates for consideration, selection and appointment by the Secretary. The public member and alternate public member may not have any financial interest, individually or corporately, or affiliation with persons vested in the pecan industry. The accumulator member and alternate accumulator member must meet the criteria set forth in §986.1, Accumulator, and may reside or maintain a place of business in any region.

(7) *Nomination forms.* The Council may distribute nomination forms at meetings, by mail, by email, or by any other form of distribution recommended by the Council and approved by the Secretary.

(i) *Grower nomination forms.* Each nomination form submitted by a grower shall include the following information:

(A) The name of the nominated grower;

(B) The name and signature of the nominating grower;

(C) Two additional names and respective signatures of growers in support of the nomination;

(D) Any other such information recommended by the Council and approved by the Secretary.

(ii) *Sheller nomination forms.* Each nomination form submitted by a sheller shall include the following:

(A) The name of the nominated sheller;

(B) The name and signature of the nominating sheller;

(C) One additional name and signature of a sheller in support of the nomination;

(D) Any other such information recommended by the Council and approved by the Secretary.

(8) *Changes to the nomination and voting procedures.* The Council may recommend, subject to the approval of the Secretary, a change to these procedures should the Council determine that a revision is necessary.

§ 986.47 Alternate members.

(a) Each member of the Council shall have an alternate member to be nominated in the same manner as the member.

(b) An alternate for a member of the Council shall act in the place and stead of such member in their absence or in the event of their death, removal, resignation, or disqualification, until the next nomination and elections take place for the Council or the vacancy has been filled pursuant to § 986.48, Eligibility.

(c) In the event any member of the Council and their alternate are both unable to attend a meeting of the Council, any alternate for any other member representing the same group as the absent member may serve in the place of the absent member.

§ 986.48 Eligibility.

(a) Each grower member and alternate shall be, at the time of selection and during the term of office, a grower or an officer, or employee, of a grower in the region and in the classification for which nominated.

(b) Each sheller member and alternate shall be, at the time of selection and during the term of office, a sheller or an officer or employee of a sheller in the region and in the classification for which nominated.

(c) A grower can be a nominee for only one grower member seat. If a grower is nominated for two grower member seats, he or she shall select the seat in which he or she desires to run, and the grower ballot shall reflect that selection.

(d) Any member or alternate member who at the time of selection was employed by or affiliated with the person who is nominated shall, upon termination of that relationship, become disqualified to serve further as a member and that position shall be deemed vacant.

(e) No person nominated to serve as a public member or alternate public member shall have a financial interest in any pecan grower or handling operation.

§ 986.49 Acceptance.

Each person to be selected by the Secretary as a member or as an alternate member of the Council shall, prior to such selection, qualify by advising the Secretary that if selected, such person agrees to serve in the position for which that nomination has been made.

§ 986.50 Term of office.

(a) Selected members and alternate members of the Council shall serve for terms of four years: *Provided*, That at the end of the first four (4) year term and in the nomination and selection of the second Council only, four of the grower member and alternate seats and three of the sheller member and alternate seats shall be seated for terms of two years so that approximately half of the memberships' and alternates' terms expire every two years thereafter. Member and alternate seats assigned two-year terms for the seating of the second Council only shall be as follows:

(1) Grower member Seat 2 in all regions shall be assigned a two-year term;

(2) Grower member Seat 3 in all regions shall, by drawing, identify one member seat to be assigned a two-year term; and,

(3) Sheller Seat 2 in all regions shall be assigned a two-year term.

(b) Council members and alternates may serve up to two consecutive, four-year terms of office. Subject to paragraph (c) of this section, in no event shall any member or alternate serve more than eight consecutive years on the Council as either a member or an alternate. However, if selected, an alternate having served up to two consecutive terms may immediately serve as a member for two consecutive terms without any interruption in service. The same is true for a member who, after serving for up to two consecutive terms, may serve as an alternate if nominated without any interruption in service. A person having served the maximum number of terms as set forth above may not serve again as a member or an alternate for at least twelve consecutive months. For purposes of determining when a member or alternate has served two consecutive terms, the accrual of terms shall begin following any period of at least twelve consecutive months out of office.

(c) Each member and alternate member shall continue to serve until a successor is selected and has qualified.

(d) A term of office shall begin as set forth in the by-laws or as directed by the Secretary each year for all members.

(e) The Council may recommend, subject to approval of the Secretary, revisions to the start day for the term of office, the number of years in a term, and the number of terms a member or an alternate can serve.

§986.51 Vacancy.

Any vacancy on the Council occurring by the failure of any person selected to the Council to qualify as a member or alternate member due to a change in status making the member ineligible to serve, or due to death, removal, or resignation, shall be filled, by a majority vote of the Council for the unexpired portion of the term. However, that person shall fulfill all the qualifications set forth in this part as required for the member whose office that person is to fill. The qualifications of any person to fill a vacancy on the Council shall be certified in writing to the Secretary. The Secretary shall notify the Council if the Secretary determines that any such person is not qualified.

§986.52 Council expenses.

The members and their alternates of the Council shall serve without compensation, but shall be reimbursed for the reasonable and necessary expenses incurred by them in the performance of their duties under this part.

§986.53 Powers.

The Council shall have the following powers:

(a) To administer the provisions of this part in accordance with its terms;

(b) To make bylaws, rules and regulations to effectuate the terms and provisions of this part;

(c) To receive, investigate, and report to the Secretary complaints of violations of this part; and

(d) To recommend to the Secretary amendments to this part.

§986.54 Duties.

The duties of the Council shall be as follows:

(a) To act as intermediary between the Secretary and any handler or grower;

(b) To keep minute books and records which will clearly reflect all of its acts and transactions, and such minute books and records shall at any time be subject to the examination of the Secretary;

(c) To furnish to the Secretary a complete report of all meetings and such other available information as he or she may request;

(d) To appoint such employees as it may deem necessary and to determine the salaries, define the duties, and fix the bonds of such employees;

(e) To cause the books of the Council to be audited by one or more certified public accountants at least once for each fiscal year and at such other times as the Council deems necessary or as the Secretary may request, and to file with the Secretary three copies of all audit reports made;

(f) To investigate the growing, shipping and marketing conditions with respect to pecans and to assemble data in connection therewith;

(g) To investigate compliance with the provisions of this part; and,

(h) To recommend by-laws, rules and regulations for the purpose of administering this part.

§986.55 Procedure.

(a) The members of the Council shall select a chairman from their membership, and shall select such other officers and adopt such rules for the conduct of Council business as they deem advisable.

(b) The Council may provide for meetings by telephone, or other means of communication, and any vote cast at such a meeting shall be confirmed promptly in writing. The Council shall give the Secretary the same notice of its meetings as is given to members of the Council.

(c) *Quorum.* A quorum of the Council shall be any twelve voting Council members. The vote of a majority of members present at a meeting at which there is a quorum shall constitute the act of the Council; *Provided,* That:

(1) Actions of the Council with respect to the following issues shall require a two-thirds (12 members) concurring vote of the Council:

(i) Establishment of or changes to by-laws;

(ii) Appointment or administrative issues relating to the program's manager or chief executive officer;

(iii) Budget;

(iv) Assessments;

(v) Compliance and audits;

(vi) Reestablishment of regions and reapportionment or reallocation of Council membership;

(vii) Modifying definitions of grower and sheller;

(viii) Research or promotion activities under § 986.68;

(ix) Grade, quality and size regulation under § 986.69(a)(1) and (2);

(x) Pack and container regulation under § 986.69(a)(3); and,

(2) Actions of the Council with respect to the securing of commercial bank loans for the purpose of financing start-up costs of the Council and its activities or securing financial assistance in emergency situations shall require a unanimous vote of all members present at an in-person meeting; *Provided,* That in the event of an emergency that warrants immediate attention sooner than a face-to-face meeting is possible, a vote for financing may be taken. In such event, the Council's first preference is a videoconference and second preference is phone conference, both followed by written confirmation of the members attending the meeting.

§ 986.56　Right of the Secretary.

The members and alternates for members and any agent or employee appointed or employed by the Council shall be subject to removal or suspension by the Secretary at any time. Each and every regulation, decision, determination, or other act shall be subject to the continuing right of the Secretary to disapprove of the same at any time, and, upon such disapproval, shall be deemed null and void, except as to acts done in reliance thereon or in compliance therewith prior to such disapproval by the Secretary.

§ 986.57　Funds and other property.

(a) All funds received pursuant to any of the provisions of this part shall be used solely for the purposes specified in this part, and the Secretary may require the Council and its members to account for all receipts and disbursements.

(b) Upon the death, resignation, removal, disqualification, or expiration of the term of office of any member or employee, all books, records, funds, and other property in their possession belonging to the Council shall be delivered to their successor in office or to the Council, and such assignments and other instruments shall be executed as may be necessary to vest in such successor or in the Council full title to all the books, records, funds, and other property in the possession or under the control of such member or employee pursuant to this subpart.

§ 986.58　Reapportionment and reestablishment of regions.

The Council may recommend, subject to approval of the Secretary, reestablishment of regions, reapportionment of members among regions, and may revise the groups eligible for representation on the Council. In recommending any such changes, the following shall be considered:

(a) Shifts in acreage within regions and within the production area during recent years;

(b) The importance of new production in its relation to existing regions;

(c) The equitable relationship between Council apportionment and regions;

(d) Changes in industry structure and/or the percentage of crop represented by various industry entities; and

(e) Other relevant factors.

EXPENSES, ASSESSMENTS, AND
MARKETING POLICY

§ 986.60　Budget.

As soon as practicable before the beginning of each fiscal year, and as may be necessary thereafter, the Council shall prepare a budget of income and expenditures necessary for the administration of this part. The Council may recommend a rate of assessment calculated to provide adequate funds to defray its proposed expenditures. The Council shall present such budget to the Secretary with an accompanying report showing the basis for its calculations, and all shall be subject to Secretary approval.

§986.61 Assessments.

(a) Each handler who first handles inshell pecans shall pay assessments to the Council. Assessments collected each fiscal year shall defray expenses which the Secretary finds reasonable and likely to be incurred by the Council during that fiscal year. Each handler's share of assessments paid to the Council shall be equal to the ratio between the total quantity of inshell pecans handled by them as the first handler thereof during the applicable fiscal year, and the total quantity of inshell pecans handled by all regulated handlers in the production area during the same fiscal year. The payment of assessments for the maintenance and functioning of the Council may be required under this part throughout the period it is in effect irrespective of whether particular provisions thereof are suspended or become inoperative. Handlers may avail themselves of an inter-handler transfer, as provided for in §986.62, Inter-handler transfers.

(b) Based upon a recommendation of the Council or other available data, the Secretary shall fix three base rates of assessment for inshell pecans handled during each fiscal year. Such base rates shall include one rate of assessment for any or all varieties of pecans classified as native and seedling; one rate of assessment for any or all varieties of pecans classified as improved; and one rate of assessment for any pecans classified as substandard.

(c) Upon implementation of this part and subject to the approval of the Secretary, initial assessment rates per classification shall be set within the following prescribed ranges: Native and seedling classified pecans shall be assessed at one-cent to two-cents per pound; improved classified pecans shall be assessed at two-cents to three-cents per pound; and, substandard classified pecans shall be assessed at one-cent to two-cents per pound. These assessment ranges shall be in effect for the initial four years of the order.

(d) Subsequent assessment rates shall not exceed two percent of the aggregate of all prices in each classification across the production area based on Council data, or the average of USDA reported average price received by growers for each classification, in the preceding fiscal year as recommended by the Council and approved by the Secretary. After four years from the implementation of this part, the Council may recommend, subject to the approval of the Secretary, revisions to this calculation or assessment ranges.

(e) The Council, with the approval of the Secretary, may revise the assessment rates if it determines, based on information including crop size and value, that the action is necessary, and if the revision does not exceed the assessment limitation specified in this section and is made prior to the final billing of the assessment.

(f) In order to provide funds for the administration of the provisions of this part during the first part of a fiscal year, before sufficient operating income is available from assessments, the Council may accept the payment of assessments in advance and may also borrow money for such purposes; *Provided*, That no loan may amount to more than 50 percent of projected assessment revenue projected for the year in which the loan is secured, and the loan must be repaid within five years.

(g) If a handler does not pay assessments within the time prescribed by the Council, the assessment may be increased by a late payment charge and/or an interest rate charge at amounts prescribed by the Council with approval of the Secretary.

(h) On August 31 of each year, every handler warehousing inshell pecans shall be identified as the first handler of those pecans and shall be required to pay the assessed rate on the category of pecans in their possession on that date. The terms of this paragraph may be revised subject to the recommendation of the Council and approval by the Secretary.

(i) On August 31 of each year, all inventories warehoused by growers from the current fiscal year shall cease to be

535

eligible for inter-handler transfer treatment. Instead, such inventory will require the first handler that handles such inventory to pay the assessment thereon in accordance with the prevailing assessment rates at the time of transfer from the grower to the said handler. The terms of this paragraph may be revised subject to the recommendation of the Council and approval by the Secretary.

§ 986.62 Inter-handler transfers.

Any handler inside the production area, except as provided for in § 986.61(h) and (i), Assessments, may transfer inshell pecans to another handler inside the production area for additional handling, and any assessments or other marketing order requirements with respect to pecans so transferred may be assumed by the receiving handler. The Council, with the approval of the Secretary, may establish methods and procedures, including necessary reports, to maintain accurate records for such transfers. All inter-handler transfers will be documented by forms or electronic transfer receipts approved by the Council, and all forms or electronic transfer receipts used for inter-handler transfers shall require that copies be sent to the selling party, the receiving party, and the Council. Such forms must state which handler has the assessment responsibilities.

§ 986.63 Contributions.

The Council may accept voluntary contributions. Such contributions may only be accepted if they are free from any encumbrances or restrictions on their use and the Council shall retain complete control of their use. The Council may receive contributions from both within and outside of the production area.

§ 986.64 Accounting.

(a) Assessments collected in excess of expenses incurred shall be accounted for in accordance with one of the following:

(1) Excess funds not retained in a reserve, as provided in paragraph (a)(2) of this section shall be refunded proportionately to the persons from whom they were collected; or

(2) The Council, with the approval of the Secretary, may carry over excess funds into subsequent fiscal periods as reserves: *Provided,* That funds already in reserves do not equal approximately three fiscal years' expenses. Such reserve funds may be used:

(i) To defray expenses during any fiscal period prior to the time assessment income is sufficient to cover such expenses;

(ii) To cover deficits incurred during any fiscal period when assessment income is less than expenses;

(iii) To defray expenses incurred during any period when any or all provisions of this part are suspended or are inoperative; and

(iv) To cover necessary expenses of liquidation in the event of termination of this part.

(b) Upon such termination, any funds not required to defray the necessary expenses of liquidation shall be disposed of in such manner as the Secretary may determine to be appropriate. To the extent practical, such funds shall be returned pro rata to the persons from whom such funds were collected.

(c) All funds received by the Council pursuant to the provisions of this part shall be used solely for the purposes specified in this part and shall be accounted for in the manner provided for in this part. The Secretary may at any time require the Council and its members to account for all receipts and disbursements.

(d) Upon the removal or expiration of the term of office of any member of the Council, such member shall account for all receipts and disbursements and deliver all property and funds in their possession to the Council, and shall execute such assignments and other instruments as may be necessary or appropriate to vest in the Council full title to all of the property, funds, and claims vested in such member pursuant to this part.

(e) The Council may make recommendations to the Secretary for one or more of the members thereof, or any other person, to act as a trustee for holding records, funds, or any other Council property during periods of suspension of this subpart, or during any period or periods when regulations are not in effect and if the Secretary determines such action appropriate, he or she may direct that such person or persons shall act as trustee or trustees for the Council.

§986.65 Marketing policy.

By the end of each fiscal year, the Council shall make a report and recommendation to the Secretary on the Council's proposed marketing policy for the next fiscal year. Each year such report and recommendation shall be adopted by the affirmative vote of at least two-thirds (⅔) of the members of the Council and shall include the following and, where applicable, on an inshell basis:

(a) Estimate of the grower-cleaned production and handler-cleaned production in the area of production for the fiscal year;

(b) Estimate of disappearance;

(c) Estimate of the improved, native, and substandard pecans;

(d) Estimate of the handler inventory on August 31, of inshell and shelled pecans;

(e) Estimate of unassessed inventory;

(f) Estimate of the trade supply, taking into consideration imports, and other factors;

(g) Preferable handler inventory of inshell and shelled pecans on August 31 of the following year;

(h) Projected prices in the new fiscal year;

(i) Competing nut supplies; and

(j) Any other relevant factors.

AUTHORITIES RELATING TO RESEARCH, PROMOTION, DATA GATHERING, PACKAGING, GRADING, COMPLIANCE, AND REPORTING

§986.67 Recommendations for regulations.

Upon complying with §986.65, Marketing policy, the Council may propose regulations to the Secretary whenever it finds that such proposed regulations may assist in effectuating the declared policy of the Act.

§986.68 Authority for research and promotion activities.

The Council, with the approval of the Secretary, may establish or provide for the establishment of production research, marketing research and development projects, and marketing promotion, including paid generic advertising, designed to assist, improve, or promote the marketing, distribution, and consumption or efficient production of pecans including product development, nutritional research, and container development. The expenses of such projects shall be paid from funds collected pursuant to this part.

§986.69 Authorities regulating handling.

(a) The Council may recommend, subject to the approval of the Secretary, regulations that:

(1) Establish handling requirements or minimum tolerances for particular grades, sizes, or qualities, or any combination thereof, of any or all varieties or classifications of pecans during any period;

(2) Establish different handling requirements or minimum tolerances for particular grades, sizes, or qualities, or any combination thereof for different varieties or classifications, for different containers, for different portions of the production area, or any combination of the foregoing, during any period;

(3) Fix the size, capacity, weight, dimensions, or pack of the container or containers, which may be used in the packaging, transportation, sale, preparation for market, shipment, or other handling of pecans; and

(4) Establish inspection and certification requirements for the purposes of (a)(1) through (3) of this section.

(b) Regulations issued hereunder may be amended, modified, suspended, or terminated whenever it is determined:

(1) That such action is warranted upon recommendation of the Council and approval by the Secretary, or other available information; or

(2) That regulations issued hereunder no longer tend to effectuate the declared policy of the Act.

(c) The authority to regulate as put forward in this subsection shall not in any way constitute authority for the Council to recommend volume regulation, such as reserve pools, producer allotments, or handler withholding requirements which limit the flow of product to market for the purpose of reducing market supply.

(d) The Council may recommend, subject to the approval of the Secretary, rules and regulations to effectuate this subpart.

§ 986.70 Handling for special purposes.

Regulations in effect pursuant to § 986.69, Authorities regulating handling, may be modified, suspended, or terminated to facilitate handling of pecans for:

(a) Relief or charity;

(b) Experimental purposes; and

(c) Other purposes which may be recommended by the Council and approved by the Secretary.

§ 986.71 Safeguards.

The Council, with the approval of the Secretary, may establish through rules such requirements as may be necessary to establish that shipments made pursuant to § 986.70, Handling for special purposes, were handled and used for the purpose stated.

§ 986.72 Notification of regulation.

The Secretary shall promptly notify the Council of regulations issued or of any modification, suspension, or termination thereof. The Council shall give reasonable notice thereof to industry participants.

REPORTS, BOOKS, AND OTHER RECORDS

§ 986.75 Reports of handler inventory.

Each handler shall submit to the Council in such form and on such dates as the Council may prescribe, reports showing their inventory of inshell and shelled pecans.

§ 986.76 Reports of merchantable pecans handled.

Each handler who handles merchantable pecans at any time during a fiscal year shall submit to the Council in such form and at such intervals as the Council may prescribe, reports showing the quantity so handled and such other information pertinent thereto as the Council may specify.

§ 986.77 Reports of pecans received by handlers.

Each handler shall file such reports of their pecan receipts from growers, handlers, or others in such form and at such times as may be required by the Council with the approval of the Secretary.

§ 986.78 Other handler reports.

Upon request of the Council made with the approval of the Secretary each handler shall furnish such other reports and information as are needed to enable the Council to perform its duties and exercise its powers under this part.

§ 986.79 Verification of reports.

For the purpose of verifying and checking reports filed by handlers on their operations, the Secretary and the Council, through their duly authorized representatives, shall have access to any premises where pecans and pecan records are held. Such access shall be available at any time during reasonable business hours. Authorized representatives of the Council or the Secretary shall be permitted to inspect any pecans held and any and all records of the handler with respect to matters within the purview of this part. Each handler shall maintain complete records on the receiving, holding, and disposition of all pecans. Each handler shall furnish all labor necessary to facilitate such inspections at no expense to the Council or the Secretary. Each handler shall store all pecans held by him in such manner as to facilitate inspection and shall maintain adequate storage records which will permit accurate identification with respect to inspection certificates of respective lots and of all such pecans held or disposed of theretofore. The Council, with the approval of the Secretary, may establish any methods and procedures needed to verify reports.

§ 986.80 Certification of reports.

All reports submitted to the Council as required in this part shall be certified to the Secretary and the Council as to the completeness and correctness of the information contained therein.

§986.81 Confidential information.

All reports and records submitted by handlers to the Council, which include data or information constituting a trade secret or disclosing the trade position, or financial condition or business operations of the handler shall be kept in the custody of one or more employees of the Council and shall be disclosed to no person except the Secretary.

§986.82 Books and other records.

Each handler shall maintain such records of pecans received, held and disposed of by them as may be prescribed by the Council for the purpose of performing its duties under this part. Such books and records shall be retained and be available for examination by authorized representatives of the Council and the Secretary for the current fiscal year and the preceding three (3) fiscal years.

ADDITIONAL PROVISIONS

§986.86 Exemptions.

(a) Any handler may handle inshell pecans within the production area free of the requirements of this part if such pecans are handled in quantities not exceeding 1,000 inshell pounds during any fiscal year.

(b) Any handler may handle shelled pecans within the production area free of the requirements of this part if such pecans are handled in quantities not exceeding 500 shelled pounds during any fiscal year.

(c) Mail order sales are not exempt sales under this part.

(d) The Council, with the approval of the Secretary, may establish such rules, regulations, and safeguards, and require such reports, certifications, and other conditions, as are necessary to ensure compliance with this part.

§986.87 Compliance.

Except as provided in this subpart, no handler shall handle pecans, the handling of which has been prohibited by the Secretary in accordance with provisions of this part, or the rules and regulations thereunder.

§986.88 Duration of immunities.

The benefits, privileges, and immunities conferred by virtue of this part shall cease upon termination hereof, except with respect to acts done under and during the existence of this part.

§986.89 Separability.

If any provision of this part is declared invalid, or the applicability thereof to any person, circumstance, or thing is held invalid, the validity of the remaining provisions and the applicability thereof to any other person, circumstance, or thing shall not be affected thereby.

§986.90 Derogation.

Nothing contained in this part is or shall be construed to be in derogation of, or in modification of, the rights of the Secretary or of the United States to exercise any powers granted by the Act or otherwise, or, in accordance with such powers, to act in the premises whenever such action is deemed advisable.

§986.91 Liability.

No member or alternate of the Council nor any employee or agent thereof, shall be held personally responsible, either individually or jointly with others, in any way whatsoever, to any party under this part or to any other person for errors in judgment, mistakes, or other acts, either of commission or omission, as such member, alternate, agent or employee, except for acts of dishonesty, willful misconduct, or gross negligence. The Council may purchase liability insurance for its members and officers.

§986.92 Agents.

The Secretary may name, by designation in writing, any person, including any officer or employee of the USDA or the United States to act as their agent or representative in connection with any of the provisions of this part.

§986.93 Effective time.

The provisions of this part and of any amendment thereto shall become effective at such time as the Secretary may declare, and shall continue in force until terminated in one of the ways specified in § 986.94.

§ 986.94   **Termination.**

(a) The Secretary may at any time terminate this part.

(b) The Secretary shall terminate or suspend the operation of any or all of the provisions of this part whenever he or she finds that such operation obstructs or does not tend to effectuate the declared policy of the Act.

(c) The Secretary shall terminate the provisions of this part applicable to pecans for market or pecans for handling at the end of any fiscal year whenever the Secretary finds, by referendum or otherwise, that such termination is favored by a majority of growers; *Provided*, That such majority of growers has produced more than 50 percent of the volume of pecans in the production area during such fiscal year. Such termination shall be effective only if announced on or before the last day of the then current fiscal year.

(d) The Secretary shall conduct a referendum within every five-year period beginning from the implementation of this part, to ascertain whether continuance of the provisions of this part applicable to pecans are favored by two-thirds by number or volume of growers voting in the referendum. The Secretary may terminate the provisions of this part at the end of any fiscal year in which the Secretary has found that continuance of this part is not favored by growers who, during an appropriate period of time determined by the Secretary, have been engaged in the production of pecans in the production area: *Provided*, That termination of this part shall be effective only if announced on or before the last day of the then current fiscal year.

(e) The provisions of this part shall, in any event, terminate whenever the provisions of the Act authorizing them cease to be in effect.

§ 986.95   **Proceedings after termination.**

(a) Upon the termination of this part, the Council members serving shall continue as joint trustees for the purpose of liquidating all funds and property then in the possession or under the control of the Council, including claims for any funds unpaid or prop-erty not delivered at the time of such termination.

(b) The joint trustees shall continue in such capacity until discharged by the Secretary; from time to time accounting for all receipts and disbursements; delivering all funds and property on hand, together with all books and records of the Council and of the joint trustees to such person as the Secretary shall direct; and, upon the request of the Secretary, executing such assignments or other instruments necessary and appropriate to vest in such person full title and right to all of the funds, property, or claims vested in the Council or in said joint trustees.

(c) Any funds collected pursuant to this part and held by such joint trustees or such person over and above the amounts necessary to meet outstanding obligations and the expenses necessarily incurred by the joint trustees or such other person in the performance of their duties under this subpart, as soon as practicable after the termination hereof, shall be returned to the handlers pro rata in proportion to their contributions thereto.

(d) Any person to whom funds, property, or claims have been transferred or delivered by the Council, upon direction of the Secretary, as provided in this part, shall be subject to the same obligations and duties with respect to said funds, property, or claims as are imposed upon said joint trustees.

§ 986.96   **Amendments.**

Amendments to this part may be proposed from time to time by the Council or by the Secretary.

§ 986.97   **Counterparts.**

Handlers may sign an agreement with the Secretary indicating their support for this marketing order. This agreement may be executed in multiple counterparts by each handler. If more than fifty percent of the handlers, weighted by the volume of pecans handled during an appropriate period of time determined by the Secretary, enter into such an agreement, then a marketing agreement shall exist for the pecans marketing order. This marketing agreement shall not alter the terms of this part. Upon the termination of this part, the marketing agreement has no further force or effect.

§986.98 Additional parties.

After this part becomes effective, any handler may become a party to the marketing agreement if a counterpart is executed by the handler and delivered to the Secretary.

§986.99 Order with marketing agreement.

Each signatory handler hereby requests the Secretary to issue, pursuant to the Act, an order for regulating the handling of pecans in the same manner as is provided for in this agreement.

Subpart B—Administrative Provisions

§986.161 Assessment rates.

On and after October 1, 2016, assessment rates of $0.03 per pound for pecans classified as improved, $0.02 per pound for pecans classified as native and seedling, and $0.02 per pound for pecans classified as substandard pecans are established.

[82 FR 43669, Sept. 19, 2017]

§986.162 Inter-handler transfers.

(a) Inter-handler transfers of inshell pecans, pursuant to §986.62, shall be reported to the Council on APC Form 4. Handlers shall file reports by the tenth day of the month following the month of transfer. Should the tenth day of the month fall on a weekend or holiday, reports are due by the first business day following the tenth day of the month; Provided, that for the 2016–17 fiscal year, all inter-handler transfer forms shall be submitted by December 28, 2017. The report shall contain the following information:

(1) Month of transfer;

(2) The type and weight of pecans transferred;

(3) The amount of assessments owed on the pecans transferred;

(4) The names and signatures for both the transferring and receiving handlers; and

(5) Handler assuming the reporting and assessment obligations on the pecans transferred.

[82 FR 56155, Nov. 28, 2017]

§986.175 Handler inventory.

(a) Handlers shall submit to the Council a year-end inventory report following August 31 each fiscal year. Handlers shall file such reports by September 10. Should September 10 fall on a weekend, reports are due by the first business day following September 10. Such reports shall be reported to the Council on APC Form 7. For the purposes of this form, "crop year" is the same as the "fiscal year." The report shall include:

(1) The name and address of the handler;

(2) The total weight and type of inshell pecans in inventory, regardless of country of origin;

(3) The total weight and type of shelled pecans in inventory, regardless of country of origin;

(4) The total weight and type of inshell pecans committed, not shipped, for export and domestic shipments, and any uncommitted inventory, regardless of country of origin;

(5) The total weight and type of shelled pecans committed, not shipped, for export and domestic shipments, and any uncommitted inventory, regardless of country of origin;

(6) The combined total inventory for inshell and shelled pecans calculated on an inshell basis, and combined weight committed, not shipped, for exports and domestic shipments, and any uncommitted inventory;

(7) Total weight and type of domestic pecans handled for the fiscal year;

(8) Total assessments owed, assessments paid to date, and remaining assessments due to be paid by the due date of the year-end inventory report for the fiscal year;

(9) The average price paid for all inshell pecans purchased during the fiscal year regardless of how the pecans are handled, including pecans from outside the production area; and

(10) The average yield of shelled pecans per pound of inshell pecans shelled during the fiscal year.

(b) [Reserved]

[82 FR 56155, Nov. 28, 2017, as amended at 84 FR 8411, Mar. 8, 2019]

§ 986.177 Reports of pecans received by handlers.

(a) *Summary report U.S. pecans received for your own account.* Handlers shall submit to the Council, by the tenth day of the month, a summary report of inshell domestic pecans received during the preceding month. Should the tenth day of the month fall on a weekend or holiday, reports are due by the first business day following the tenth day of the month. The report shall be submitted to the Council on APC Form 1 and contain the following information:

(1) The name and address of the handler;

(2) The month covered by the report;

(3) The total weight and type of inshell pecans received, and the weight by variety for improved pecans received during the reporting period;

(4) The total weight and type of inshell pecans received, and the weight by variety for improved pecans received year to date; and,

(5) Assessments due on pecans received during the reporting period to be paid by the due date of the report.

(b) *Pecans purchased outside the United States.* Handlers shall submit to the Council, by the tenth day of the month, a summary report of shelled and inshell pecans imported during the preceding month. Should the tenth day of the month fall on a weekend or holiday, reports are due by the first business day following the tenth day of the month. The report shall be submitted to the Council on APC Form 6 and contain the following information:

(1) The name and address of the handler;

(2) The month covered by the report;

(3) The date the pecans were imported;

(4) The country of origin; and,

(5) The total weight of shelled and inshell pecans received, and the weight by variety for improved pecans received.

[83 FR 7361, Feb. 21, 2018]

§ 986.178 Other reports.

(a) *Report of shipments and inventory on hand.* Handlers shall submit to the Council, by the tenth day of the month following the month of activity, a report of all shipments, inventory, and committed inventory for pecans. Should the tenth day of the month fall on a weekend or holiday, reports are due by the first business day following the tenth day of the month. The report shall be submitted to the Council on APC Form 2 and contain the following information:

(1) The name and address of the handler;

(2) The month covered by the report;

(3) The weight of all shipments of pecans, inshell and shelled, and inter-handler transfers shipped and received during the reporting period;

(4) The weight of all shipments of pecans, inshell and shelled, and inter-handler transfers shipped and received in the previous month and year to date;

(5) Total inventory held by handler;

(6) All the inventory committed (pecans not shipped, but sold or otherwise obligated) whether for domestic sale or export; and,

(7) The weight of all shelled or inshell pecans under contract for purchase from other handlers.

(b) *Exports by country of destination.* Handlers shall submit to the Council, by the tenth day of the month following the month of shipment, a report of exports. Should the tenth day of the month fall on a weekend or holiday, reports are due by the first business day following the tenth day of the month. The report shall be reported to the Council on APC Form 3 and contain the following information:

(1) The name and address of the handler;

(2) The month covered by the report;

(3) The total weight of pecans shipped for export, whether inshell, shelled, or substandard during the reporting period;

(4) The total weight of pecans shipped for export, whether inshell, shelled, or substandard during the previous period and year to date; and,

(5) The destination(s) of such exports.

(c) *Inshell pecans exported to Mexico for shelling and returned to the United States as shelled meats.* Handlers shall submit to the Council, by the tenth day of the month following the month of shipment, a report of all inshell pecans exported to Mexico for shelling and returned to the United States as shelled pecans. Should the tenth day of the month fall on a weekend or holiday, reports are due by the first business day following the tenth day of the month. The report shall be submitted to the Council on APC Form 5 and contain the following information:

(1) The name and address of the handler;

(2) The month covered by the report;

(3) The date of inshell shipment(s);

(4) The weight of pecans exported for shelling;

(5) The date shelled pecans returned to the United States after shelling;

(6) The weight of shelled pecans returned to the United States after shelling; and

(7) The total weight of inshell pecans exported to Mexico for shelling, and shelled pecans returned from Mexico, year to date.

[83 FR 7361, Feb. 21, 2018]

PART 987—DOMESTIC DATES PRODUCED OR PACKED IN RIVERSIDE COUNTY, CALIFORNIA

Subpart A—Order Regulating Handling

Subpart A—Order Regulating Handling

SOURCE: 27 FR 6818, July 19, 1962, unless otherwise noted.

DEFINITIONS

§ 987.1   Secretary.

Secretary means the Secretary of Agriculture of the United States, or any officer or employee of the Department to whom authority has heretofore been delegated, or to whom authority may hereafter be delegated, to act in his stead.

§ 987.2   Act.

Act means Public Act No. 10, 73d Congress, as amended and as reenacted and amended by the Agricultural Marketing Agreement Act of 1937, as amended (sections 1–19, 48 Stat. 31, as amended; 7 U.S.C. 601 *et seq.*).

§ 987.3   Person.

Person means an individual, partnership, corporation, association, or any other business unit.

§ 987.4   Area of production.

Area of production means Riverside County, Calif.

[36 FR 15037, Aug. 12, 1971]

§ 987.5   Dates.

Dates means the Deglet Noor, Zahidi, Halawy, and Khadrawy varieties of domestic dates produced or packed in the area of production.

§ 987.6   Crop year.

Crop year means the 12-month period beginning October 1 of each year and ending September 30 of the following year.

[43 FR 4250, Feb. 1, 1978]

§ 987.7   Producer.

Producer is synonymous with grower and means any person engaged in a proprietary capacity in the production of dates for sale.

§ 987.8   Handler.

Handler means any person handling dates which have not been inspected and certified for handling in the hands of a previous holder and any repacker: *Provided,* That for the purposes of §§ 987.21 and 987.24 a person shall qualify as a handler only if he has acquired the dates directly from producers.

[32 FR 12595, Aug. 31, 1967, as amended at 43 FR 4250, Feb. 1, 1978]

§987.8a Repacker.

Repacker means any wholesaler or jobber who receives packed dates certified for handling pursuant to §987.41(a), repackages them in containers other than those in which received, and handles such repackaged dates.

[32 FR 12595, Aug. 31, 1967]

§987.9 Handle.

Handle means to sell, consign, transport, or ship (except as a common or contract carrier of dates owned by another person) or in any way to put dates into the current of commerce including the shipment or delivery of utility dates or cull dates into nonhuman consumption outlets, except that sales or deliveries, by producers, of other than cull dates, to a handler within the area of production, or the movement of dates by a handler to storage for his account within the area of production, or counties adjoining the area of production, shall not be considered handling. The Committee, with the approval of the Secretary, may establish monitoring procedures for storage of dates in Orange, San Diego, and Yuma Counties.

[36 FR 15037, Aug. 12, 1971, as amended at 43 FR 4250, 4251, Feb. 1, 1978]

§987.10 Handler carry-over.

Handler carry-over means, as of any date, all marketable dates then held by a handler or for his account (whether or not sold), plus the estimated quantity of marketable dates in ungraded or unprocessed lots then held by said handler.

§987.11 Trade demand.

Trade demand means those quantities of marketable dates which the Committee finds are required to satisfy the need for dates in specific outlets in which marketable dates are handled.

[43 FR 4250, Feb. 1, 1978]

§987.12 Marketable dates.

Marketable dates means those dates which are certified as equal to or higher than the applicable minimum grade and size requirements in effect pursuant to §987.39, and any additional applicable requirements in effect pursuant to §987.40. Marketable dates shall include but not be limited to the following:

(a) *DAC dates.* DAC dates are marketable whole or pitted dates that are inspected and certified as meeting the grade, size, container, and identification requirements established by the Committee, with the approval of the Secretary, for a specific variety for handling in the United States and Canada.

(b) *Dates for further processing.* Dates for further processing (FP) are marketable whole dates acquired by one handler from another handler that are certified as meeting the same grade and size requirements for DAC dates, with the exception of moisture requirements, and such identification requirements applicable to FP dates that are established by the Committee, with the approval of the Secretary, for any specific variety.

(c) *Export dates.* Export dates are marketable whole or pitted dates that are inspected and certified as meeting the grade, size, container, and identification requirements established by the Committee, with the approval of the Secretary, for a specific variety, to be handled in export to any country or group of countries with the exeption of Canada. The Committee may establish different requirements for different countries.

(d) *Product dates.* Product dates are marketable dates that are inspected and certified as meeting the applicable grade and size requirements for dates to be handled in such forms as rings, chunks, pieces, butter, macerate, paste, or any other forms which the Committee deems appropriate and which will result in dates moving into consumption in a form other than that of whole or pitted dates.

[43 FR 4250, Feb. 1, 1978]

§ 987.13 Free dates.

Free dates means dates of any variety that are at the time of certification destined for consumption in whole or pitted form in the United States and Canada (and such other countries as the Committee determines are likely to acquire them at prices reasonably comparable with prices received domestically) and which are free to be handled pursuant to any free percentage established by the Secretary in accordance with § 987.44.

[43 FR 4251, Feb. 1, 1978]

§ 987.14 Restricted dates.

Restricted dates means those dates which must be withheld by handlers pursuant to any restricted percentage established by the Secretary in accordance with § 987.44.

§ 987.15 Substandard dates.

Utility dates means those dates which fail to meet the requirements for marketable dates but are not cull dates.

[27 FR 6818, July 19, 1962, as amended at 43 FR 4251, Feb. 1, 1978]

§ 987.16 Cull dates.

Cull dates means dates which fail to meet the requirements (with respect to freedom from defects) prescribed in title 3, group 4, article 24, section 24, section 1434 of the Food and Agricultural Code of California for dates for use in products or by-products other than alcohol, brandy, and products not intended for human consumption and any dates residual from field or packinghouse grading operations.

[27 FR 6818, July 19, 1962, as amended at 43 FR 4251, Feb. 1, 1978]

§ 987.17 Graded dates.

Graded dates means those dates which are eligible for certification as marketable dates.

§ 987.18 Committee.

Committee means the California Date Administrative Committee established pursuant to § 987.21.

[36 FR 15037, Aug. 12, 1971]

§ 987.20 Part and subpart.

Part means the order regulating the handling of domestic dates produced or packed in Riverside County, Calif., and all rules, regulations, and supplementary orders issued thereunder. The aforesaid order shall be a *subpart* of such part.

[36 FR 15037, Aug. 12, 1971]

CALIFORNIA DATE ADMINISTRATIVE COMMITTEE

§ 987.21 Establishment and membership.

A California Date Administrative Committee consisting of nine members is hereby established to administer the terms and conditions of this part. For each member there shall be an alternate member, and the provisions of this part applicable to the number, nomination, qualification and selection of members shall apply in like manner to alternate members. Three of the members, referred to in this part as "producer members", shall be producers or officers or employees of producers, and shall not be handlers, or directors, officers, or employees exercising a supervisory or managerial function of a handler. The six remaining members, referred to in this part as "producer-handlers", shall be selected from (a) handlers, or directors, officers or employees of a handler, or (b) producers who are also handlers or directors, officers or employees exercising a supervisory or managerial function of a handler. The Committee, with the approval of the Secretary, may issue rules and regulations covering matters of eligibility for producer members, or revising the composition of the Committee prescribed in this section if it no longer is representative following a substantial change in the industry.

[43 FR 4251, Feb. 1, 1978]

§ 987.23 Term of office.

The term of office for members and alternate members shall be three years beginning August 1, except that such term may be shorter if the Committee composition is changed in the interim pursuant to § 987.21. *Provided,* That the terms of office of all members and alternates currently serving at the time of the amendment will end on July 31, 2014. Each member and alternate member shall, unless otherwise ordered by the Secretary, continue to serve until his or her successor has been selected and has qualified.

[77 FR 37765, June 25, 2012]

§987.24 Nomination and selection.

(a) Nomination for members and alternate members of the Committee shall be made not later than June 15 of every third year.

(b) Opportunity shall be provided producers and handlers to nominate individuals to serve on the Committee by establishing a day for polling and also for casting absentee ballots. Persons will only be able to vote in nominations for the group in which they would be qualified to serve on the Committee, and shall nominate the applicable number of individuals for the positions prescribed pursuant to §987.21. Each producer, regardless of the number and locations of his date gardens, voting in the nominations for producer members and producer alternate members, shall be entitled to one vote for each member and alternate member position to be filled. The individual receiving the highest number of votes for a position shall be the nominee. Each person voting in the nominations for producer-handler members and producer-handler alternate members, shall be given the opportunity to vote for one member and one alternate member position. His ballot shall be weighted by the pounds of dates he had certified as marketable dates, from the beginning of the then current crop year through April which he produced in his own gardens or acquired from other producers. The individual receiving the highest weighted vote for a producer-handler position shall be the nominee. The Committee, with the approval of the Secretary, may issue rules and regulations on the manner in which nominees for a position may be obtained, polling, balloting, absentee ballots, and the weighting of votes for producer-handler positions when the Committee is restructured during a term of office.

[43 FR 4251, Feb. 1, 1978, as amended at 77 FR 37765, June 25, 2012]

§987.25 Qualification.

Each person selected as a member or alternate member of the Committee shall, prior to serving on the Committee, qualify by filing with the Secretary a written acceptance after receiving notice of his selection. Any member or alternate who, at the time of his selection, was a member of or employed by a member of the group which nominated him shall, upon ceasing to be such member or employee, become disqualified to serve further and his position on the Committee shall be deemed vacant.

§987.26 Vacancies.

In the event of any vacancy occasioned by the failure to qualify, declination to serve, removal, resignation, disqualification, or death of any person nominated to serve on the Committee, or any member or alternate member selected by the Secretary, the Committee shall promptly submit its recommendation to the Secretary of a nominee eligible to serve in accordance with the requirements specified for the group in §987.21. If the vacancy is for a member position, the Committee shall recommend appointment of the alternate member if that person is willing to serve in that position. If the Committee's recommendation is not submitted within 30 calendar days after such vacancy occurs, the Secretary may fill such vacancy without regard to nominations, and the selection shall be made on the basis of representation provided in §987.21.

[43 FR 4251, Feb. 1, 1978]

§987.27 Alternates.

An alternate for a member of the Committee shall act in the place and stead of such member during his absence or in the event of his removal, resignation, disqualification, or death, until a successor for such member's unexpired term has been selected and has qualified. In the event a member and his alternate are unable to attend a meeting of the Committee, such member or alternate, in that order, may designate an alternate from the group he represents to act in his place. If neither a member nor his alternate has designated an alternate as his replacement, or such designated alternate is unable to serve as the replacement, the chairman may, with the concurrence of a majority of the members including alternates acting as members, representing such group, designate an alternate from such group who is present at the meeting and is not acting as a member to act in the place and stead of the absent member.

[27 FR 6818, July 19, 1962, as amended at 36 FR 15038, Aug. 12, 1971; 43 FR 4251, Feb. 1, 1978]

§ 987.28 Expenses.

The members of the Committee shall serve without compensation but shall be allowed their necessary expenses.

§ 987.29 Powers.

The Committee shall have the following powers:

(a) To administer the terms and provisions of this subpart.

(b) To make rules and regulations to effectuate the terms and provisions of this subpart.

(c) To receive, investigate, and report to the Secretary complaints of violations of this subpart, and

(d) To recommend to the Secretary amendments to this subpart.

§ 987.30 Duties.

The Committee shall have, among other things, the following duties:

(a) To act as intermediary between the Secretary and any producer or handler.

(b) To keep minutes, books, and records which will clearly reflect all of its transactions and such minutes, books, and other records shall be subject to examination by the Secretary at any time.

(c) To investigate the growing, handling, and marketing conditions with respect to dates, to assemble data in connection therewith.

(d) To furnish to the Secretary such available information as may be deemed pertinent to the administration of this subpart or as he may request and to give to the Secretary the same notice of meetings of the Committee as is given to the members of the Committee.

(e) To appoint such employees as it may deem necessary and to determine the salaries, define the duties and where desirable fix the bonds of such employees.

(f) To cause the books of the Committee to be audited by a certified public accountant at least once each crop year and at such other times as the Committee may deem necessary or the Secretary may request. The report of each such audit shall show among other things the receipt and expenditure of funds pursuant hereto. Two copies of such audit shall be submitted to the Secretary.

(g) To investigate compliance and to use means available to the Committee to prevent violations of this part.

(h) To furnish the Committee viewpoints of the consumer, the Committee may utilize a consumer consultant. The consumer consultant shall have no financial interest in the date industry and shall receive no compensation, however, such person shall be reimbursed for necessary expenses attendant to those assignments that the Committee has given prior support and approval.

[27 FR 6818, July 19, 1962, as amended at 43 FR 4252, Feb. 1, 1978]

§ 987.31 Procedure.

(a) A majority of the Committee shall constitute a quorum.

(b) The Committee shall, from among its members, select a chairman and such other officers and adopt such rules for the conduct of its business as it may deem advisable.

(c) For any decision of the Committee to be valid, a concurring vote of at least five members is required, except as follows:

(1) In matters relating to restructuring Committee composition pursuant to § 987.21, concurrence by at least eight members is required;

(2) In matters relating to establishment, modification and application of free and restricted percentages pursuant to §§ 987.44 and 987.46, concurrence by at least seven members is required; and

(3) In matters relating to recommendation of any program of paid

advertising or major program of market promotion pursuant to §987.33, concurrence by at least six members is required.

(d) At the discretion of the chairperson, Committee meetings may be assembled or conducted by means of teleconference, video conference, or other means of communication that may be developed. Assembled meetings may also allow for participation by means of teleconference or video conference or other communication methods, at the discretion of the chair. Members participating in meetings via any of these alternative means retain the same voting privileges that they would otherwise have.

(e) The Committee may vote upon any proposition by mail, or by telephone when confirmed in writing within two weeks, upon due notice and full and identical explanation to all members, including alternates acting as members, but any such action shall not be considered valid unless unanimously approved.

(f) If the total number of members of the Committee is changed pursuant to §987.21, the minimum voting requirements shall be in the same ratio to the revised total number of members, as nearly as practicable, as the minimum voting requirements prescribed in paragraph (c) of this section are to nine.

[36 FR 15038, Aug. 12, 1971, as amended at 43 FR 4252, Feb. 1, 1978; 77 FR 37765, June 25, 2012]

MARKET DEVELOPMENT

§987.33 Research and promotion.

(a) The Committee, with the approval of the Secretary, may establish or provide for the establishment of marketing research and development projects, including marketing promotion and paid advertising, designed to assist, improve, or promote the marketing, distribution, and consumption of dates. The expenses of such projects shall be paid from funds collected pursuant to §987.72. Upon conclusion of each program, but at least annually, the Committee shall summarize and report on the program status and accomplishments, to its members and the Secretary. A similar report to the Committee shall be required of any contracting party on any paid advertising or major program. Also, for each advertising or major program the contracting party shall be required to maintain records of money received and expenditures and such shall be available to the Committee and the Secretary. The Committee shall, with the approval of the Secretary, establish criteria which will determine such major program.

(b) [Reserved]

[32 FR 12595, Aug. 31, 1967, as amended at 36 FR 15038, Aug. 12, 1971; 43 FR 4252, Feb. 1, 1978]

MARKETING POLICY

§987.34 Development.

As early as practicable, but no later than October 31, the Committee shall prepare and submit to the Secretary, a report setting forth its marketing policy, including data on which it is based, by variety, for regulation of dates in the crop year.

(a) The committee shall consider such factors as:

(1) The estimated production of dates during the crop year;

(2) The estimated production of DAC dates, export dates, and product dates;

(3) The handler carryin on October 1 of dates of those qualities;

(4) The estimated trade demand in each outlet during the crop year; and

(5) The desirable carryout, by outlet.

(b) If dates to be handled as free dates are not synonymous with those to be handled in DAC outlets, the Committee shall consider such additional factors as:

(1) The supply of marketable dates that will be available from the estimated production, and from the October 1 carryin, that could be used as free dates, and

(2) The estimated trade demand for free dates during the current crop year, and the desirable carryout for free dates.

(c) The Committee shall submit its recommendation as to grade, size, and container regulations and its recommendation whether free and restricted percentages should be established and if so, the free and restricted percentages and the appropriate withholding factor.

[43 FR 4252, Feb. 1, 1978]

§ 987.35 Modifications.

In the event the Committee subsequently determines that the marketing policy should be modified due to changing supply or demand conditions, it shall formulate and submit to the Secretary its modified marketing policy along with the data which it considered in connection with such modification.

§ 987.36 Notice.

The Committee shall give notice through newspapers having general circulation in the area of production or by other means of communication to producers and handlers of the contents of each marketing policy report submitted to the Secretary and of each report modifying such marketing policy. Copies of all such reports shall be maintained in the office of the Committee where they shall be available for examination by producers and handlers.

§ 987.38 Handlers of record.

Each crop year but no later than October 10 for continuing handlers and prior to handling dates in the case of new handlers, any person desiring to handle dates shall submit a report to the Committee on a form prescribed by it containing the following information with respect to all dates which such person expects to handle:

(a) The name and address of each producer;

(b) The location of each date garden; and

(c) The acreage and estimated current season's production thereon.

Those reports required to be filed by October 10 shall reflect producers who are signed up with the handlers as of October 1 of the then current crop year. The Committee, with the approval of the Secretary, may issue rules and regulations to carry out the provisions of this section.

[43 FR 4252, Feb. 1, 1978]

GRADE REGULATION

§ 987.39 The establishment of minimum standards.

In order to effectuate the declared policy of the act, all dates handled as marketable dates shall meet the requirements of U.S. Grade C, or if for further processing, U.S. Grade C (Dry) of the effective U.S. Standards for Grades of Dates, 7 CFR 52.1001: *Provided,* That the Secretary, may upon recommendation of the Committee, prescribe other minimum standards of grades and sizes for marketable dates of any variety to be handled in any designated outlet. To aid the Secretary in prescribing such other minimum standards, the Committee shall furnish to the Secretary the data upon which it acted in recommending such standards. The provisions hereof relating to minimum standards of grades and sizes for marketable dates and inspection requirements, within the meaning of section 2(3) of the act, and any other provisions relating to the administration and enforcement thereof shall continue in effect irrespective of whether the season average price to producers for dates is or is not in excess of the parity level specified in section 2(1) of the act. Notice of the minimum standard regulation shall be sent by the Committee to all handlers of record. On and after the effective date of such regulations no handler shall handle dates except in accordance with such minimum standard.

[32 FR 12596, Aug. 31, 1967, as amended at 43 FR 4252, Feb. 1, 1978]

§ 987.40 Additional grade or size regulations.

Whenever the Committee deems it advisable to establish grade or size requirements for any variety of dates, in addition to the minimum standard provided pursuant to § 987.39, to govern dates of such variety to be handled in any designated outlet or to be withheld to meet withholding obligation, or both, it shall recommend to the Secretary requirements as to grade based on the effective United States Standards for Grades of Dates or any modification thereof, and such size requirements as it may deem appropriate. If the Secretary finds, upon the basis of such recommendation or other information available to him, that such additional grade or size regulation, or both such regulations, will tend to effectuate the declared policy of the act, he shall establish such regulations. Notice thereof, showing the effective date, shall be sent by the Committee to all handlers of record. On and after the effective date no handler shall handle dates of such variety in any designated outlet or withhold such dates to meet withholding obligation except in accordance with such regulations.

[27 FR 6818, July 19, 1962, as amended at 43 FR 4252, Feb. 1, 1978]

§987.41 Inspection.

(a) *Packed dates.* Prior to handling any dates packed for handling each handler shall, at his own expense, cause:

(1) An inspection to be made of such dates in order to ascertain if such dates meet the applicable grade and size regulations prescribed or provided for in this part; and

(2) A certification for handling to be made of all such dates as meet such grade and size regulations.

(b) *Dates for further processing.* Prior to handling any dates for further processing each handler shall, at his own expense, cause: (1) An inspection to be made to ascertain if such dates meet the applicable grade and size requirements effective pursuant to §987.39 or §987.40, except for character associated with moisture; and (2) a certification for further processing to be made of all such dates as meet such grade and size requirements: *Provided,* That such inspection and certification requirements shall not apply to inter-handler transfers within the area of production of field-run dates or graded dates.

(c) *Identification and service.* All dates handled shall be identified by seals, stamps, or other means prescribed by the Committee and affixed to the containers by the handlers under the supervision of the Committee or the designated inspectors. Inspection shall be performed by inspectors of the United States Department of Agriculture's Processed Products Standardization and Inspection Branch or such other inspection service as may be recommended by the committee and approved by the Secretary. Handlers shall cause a copy of each inspection certificate to be furnished to the Committee.

VOLUME REGULATION

§987.43 Outlets and specifications for marketable dates.

Marketable dates shall not be handled or otherwise disposed of except as provided in this subpart. This shall not preclude dates of better grades or sizes being handled or otherwise disposed of in any outlet established for dates of lesser grades or sizes. The Committee, with the approval of the Secretary, may modify the designations specified in §987.12 to reflect new major outlets and regulatory requirements needed because of changes in marketing conditions. Marketable dates shall include but not be limited to the following: DAC dates, Dates for further processing, Export dates, and Product dates.

[43 FR 4252, Feb. 1, 1978]

§987.44 Free and restricted percentages.

(a) Whenever the Committee finds that the available supply of marketable dates of applicable grade and size available to supply the trade demand for free dates of any variety is likely to be excessive, and that limiting the volume of marketable dates to be handled as free dates through establishment of free and restricted percentages applicable to such variety of such dates would tend to effectuate the declared policy of the act, it shall recommend such percentages to the Secretary. If the Secretary finds, upon the basis of the Committee's recommendation and supporting data or other information available to him, that the establishment of such percentages would tend to effectuate the declared policy of the act, he shall establish such percentages. The sum of the free and restricted percentages for any crop year shall equal 100 percent.

(b) The dates handled by any handler in accordance with the provisions hereof shall be determined to be that handler's quota fixed by the Secretary within the meaning of section 8a(5) of the act.

[27 FR 6818, July 19, 1962, as amended at 43 FR 4252, Feb. 1, 1978]

§ 987.45 Withholding restricted dates.

(a) Whenever free and restricted percentages for any variety of dates have been established for a crop year by the Secretary in accordance with § 987.44, each handler shall, at the time of having dates of such variety certified for handling as free dates (including those for further processing that are to be handled as free dates), withhold from handling a quantity of marketable dates of such variety having a weight equal to the restricted percentage for such variety referrable to the dates so far certified. The withholding requirement shall not apply to dates certified for delivery directly to an excess supply removal program of the Secretary. The weight required to be withheld shall be determined by dividing the restricted percentage by the free percentage and applying the resultant withholding factor, rounded to the nearest one-tenth of one percent, to the weight of dates so certified. The withholding factor, computed as aforesaid, shall be established by the Secretary. When pitted dates are certified, the weight to be withheld shall be determined by dividing the weight of the pitted dates certified for handling or further processing by a divisor established by the Committee with the approval of the Secretary and applying the withholding factor.

(b) Compliance by any handler with the withholding of restricted dates may be deferred to any date not later than January 31 of any crop year, upon request to the Committee and when accompanied by a written undertaking that on or prior to such date, he will have fully satisfied his withholding obligation. Such undertaking shall be secured by a bond or bonds to be filed with, and acceptable to, the Committee and with a surety or sureties acceptable to the Committee, running in favor of the Committee and the Secretary in an amount conditioned upon full compliance with such undertaking. The amount shall be determined by multiplying the poundage of the deferred restricted obligation by a bonding rate per pound which would provide funds estimated to be sufficient for the Committee to purchase on the open market a volume of dates equivalent to the deferred obligation. Such bonding rate shall be established annually, and modified as necessary, by the Committee. Any sums collected through default by a handler on his bond shall be used by the Committee to purchase dates to meet the violated restricted obligation, reimburse the Committee for expenses relative to the default, and any excess money remaining shall be refunded to the defaulting handler. The dates so purchased by the Committee shall be turned over to the defaulting handler for disposition as restricted dates. In the event the Committee is unable to purchase a poundage of dates equal to the defaulted volume, the sums collected shall, after reimbursement of Committee expenses in connection with the default, be distributed among all handlers other than the defaulting handler in proportion to the volume of certified dates handled as free dates (including those for further processing that were handled as free dates), during the crop year in which the default occurred.

(c) At any time during the crop year free dates may be inspected and certified for handling or for further processing as provided in § 987.41. Dates so certified shall, at the time of certification, be identified by appropriate seals, stamps, or tags to be furnished by the Committee and to be affixed to the containers by the handler under the direction and supervision of the Committee or its designated inspectors. The assessment requirements in § 987.72 as well as the withholding obligation prescribed in paragraph (a) of this section shall be met at the time of certification. However, a handler who has had more free dates certified for handling or further processing than he subsequently shipped or otherwise handled may, upon request to the Committee and with its approval, have any of such excess quantity of the certified dates suspended from certification of record or, if damaged or the outlet changed, removed from certification, and his withholding and assessment obligations adjusted accordingly. A handler, who has had dates certified for handling or further processing and has not had them so suspended from certification of record or removed from certification, may carry such certified free dates over into the new crop year and need not pay the assessment nor meet the requirements of any withholding percentages established for such year.

(d) Dates withheld to meet the withholding obligation shall be stored at the expense of the handler, in storage of his own choosing and disposed of in accordance with §987.55. All such dates shall be inspected and identified by appropriate seals, stamps, or tags to be furnished by the Committee and to be affixed to the containers by the handler under the direction and supervision of the Committee or its designated inspectors. All withholding and movement of restricted dates, shall be subject to the supervision and accounting control of the Committee and reports shall be filed as required by this part. Any handler who during a crop year disposes in restricted outlets of a quantity of marketable dates in excess of his withholding obligation of such year may: (1) On written request delivered to the Committee not later than September 30 of such crop year have a part or all of such excess transferred, by the Committee, to such other handler or handlers as he may name, for crediting such other handlers' withholding obligations incurred in that crop year; and in addition (2) have a part or all of the remainder of such excess credited to his restricted obligation of the subsequent crop year: *Provided*, That the amount of any such credit shall not exceed that established by the Committee, with the approval of the Secretary, as the percentage of such withholding obligation.

(e) On request to the Committee and with its approval, a handler may, in accordance with the provisions of this paragraph and any applicable rules and regulations which the Committee may prescribe with the approval of the Secretary, defer until any date not later than September 30 of the crop year the meeting of any portion of his obligation to withhold restricted dates by setting aside such amount of graded dates as will assure a quantity of marketable dates equal at least to the

quantity needed to be withheld to meet his withholding obligation. With respect to any such dates the handler may set aside in connection with such a deferment, the Committee may require, if it deems it necessary, the handler to have made, at his own expense, such inspection as may be necessary for a determination as to whether such dates conform to the applicable requirements for dates that may be set aside under this paragraph. As a condition to the Committee approving the deferment, the handler shall agree in writing that:

(1) He will adequately mark and identify the set-aside graded dates as such and hold them separate and apart from other dates;

(2) The graded dates will not be removed from the stacks in which so set aside without the prior written permission of the Committee;

(3) Inspection of the dates by the Committee will be permitted at any reasonable time; and

(4) If the quantity, quality, or size of the set-aside dates is found by the Committee at any time to be deficient, the handler will promptly set aside such additional or substitute quantity of graded dates as is necessary to correct the deficiency.

(f) Upon the Committee prescribing, with the approval of the Secretary, minimum standards for inspection of field-run dates and appropriate administrative rules and regulations, a handler may, in accordance therewith and the provisions of this paragraph, satisfy all or any part of his obligation to withhold restricted dates by setting aside field-run dates or by disposing of field-run dates in outlets prescribed in, or pursuant to, §987.56. The field-run dates shall be of such quality or size as shall be prescribed in such rules and regulations. The setting aside, direct disposal, and disposal of any field-run dates set aside shall occur prior to September 30 of the crop year in which the withholding obligation occurs. Prior to the disposal or setting aside of the field run dates, the handler shall have had them inspected to determine the weight of dates eligible to satisfy withholding obligation. Upon such disposal or setting aside of the field-run dates, the handler shall be credited with satisfaction of his restricted obligation to the extent of the eligible weight of dates. In permitting the handler to so satisfy his withholding obligation the Committee shall require the handler to agree in writing that:

(1) Any field-run dates set aside will be held separate and apart from other dates and appropriately marked;

(2) Such dates will not be removed from the stacks in which so set aside for substitution of other dates, disposition, or for any other reason without prior written permission of the Committee; and

(3) Inspection of said dates by the Committee will be permitted at any reasonable time. In order to satisfy a withholding obligation by direct disposal of field-run dates into cull outlets, the disposal shall be under the supervision of the Committee and through persons on a Committee approved list of feeders and manufacturers. The handler may, upon giving prior notice to the Committee of any of the following proposed actions with respect to field-run dates withheld and obtaining its approval, (i) dispose of any such set-aside, field-run dates in the same manner as provided for direct disposal (ii) grade such dates and have the graded dates certified as marketable dates and withhold or dispose of such marketable dates as restricted dates, or (iii) substitute for the set-aside, field-run dates an equivalent quantity of marketable dates which he shall withhold or dispose of as restricted dates.

[27 FR 6818, July 19, 1962, as amended at 29 FR 9707, July 18, 1964; 36 FR 15039, Aug. 12, 1971; 43 FR 4253, Feb. 1, 1978]

§ 987.46   Revisions of percentages.

The Secretary may, on recommendation of the Committee submitted prior to January 31 of the crop year, or on the basis of other information available to him, increase the free percentage to conform with such new relation as may be found to exist between trade demand for free dates and available supply of marketable dates of applicable grade and size. Upon any revision in the free and restricted percentages the control obligation of each handler with respect to free dates handled or cer-

tified for handling or for further processing by him for the entire crop year shall be recomputed in accordance with such revised control percentages. The handler shall be permitted to select, insofar as practicable, under the supervision and direction of the Committee, the particular dates to be removed from any dates withheld.

[27 FR 6818, July 19, 1962, as amended at 43 FR 4253, Feb. 1, 1978]

§ 987.47   Surplus.

All cull dates and all substandard dates, including such dates blended with varieties within the generic term *dates* not regulated by this part, except any utility dates released to human consumption outlets pursuant to § 987.56, are surplus dates of any crop year. No handler shall ship or deliver such surplus dates to other than the Committee or its designee(s) for disposition in eligible outlets for such dates, except that any producer or handler may dispose of any such surplus dates of his own production within his own livestock feeding operations. Surplus dates delivered to the Committee shall be disposed of by it, in those outlets specified in § 987.56, at the best prices attainable and the proceeds returned pro rata, after deduction of Committee costs, to equity holders. The Committee may assist handlers with the cleaning, storage, or delivery of surplus dates and may, with the approval of the Secretary, establish rules and regulations necessary and incidental to administration of this regulation.

[27 FR 6818, July 19, 1962, as amended at 43 FR 4251, Feb. 1, 1978]

CONTAINER REGULATION

§ 987.48   Container regulation.

Whenever the Committee deems it advisable to establish a container regulation for any variety of dates, it shall recommend to the Secretary the size, capacity, weight, or pack of the container, or containers, which may be used in the handling or packaging of dates, or both. If the Secretary finds upon the basis of such recommendation or other information available to him that such container regulation would tend to effectuate the declared policy of the act he shall establish such regulation and notice thereof showing the effective date shall be sent by the Committee to all handlers of record. After the effective date of such regulation, no handler shall handle dates of such variety except in accordance with such regulation and all other applicable requirements in effect pursuant to this part.

QUALIFICATIONS TO REGULATION

§ 987.50 Application after end of crop year.

Unless otherwise specified the regulations and the bonding rates established for any crop year shall continue in effect with respect to all free dates for which control obligations have not been previously met, until regulations and bonding rates are established for the new crop year. Thereupon the withholding obligations for all free dates handled or certified for handling or for further processing during such crop year shall be adjusted to the newly established percentages and a similar adjustment shall be made in any bond or bonds already given for that crop year.

[27 FR 6818, July 19, 1962, as amended at 43 FR 4253, Feb. 1, 1978]

§ 987.51 Interhandler transfers.

Transfers of dates may be made from one handler to another, and each handler who so transfers any such dates shall immediately upon the completion of the particular transfer notify the Committee of the transfer, specifying the date of the transfer, the quantity and variety of dates involved, and the name of the receiving handler. If such transfer is wholly within the area of production, the assessment and withholding obligations shall be placed on the handler agreeing to assume them: *Provided,* That in the absence of the Committee receiving notice of a specific agreement on such obligations, the buying handler shall be held accountable. If such transfer is from within the area of production to any point outside thereof, the assessment and withholding obligations shall be met by the handler within the area of production. Except for packed dates inspected and certified for handling prior to transfer and which are not repacked, any receiving handler (other than a repacker not otherwise a handler, who shall comply with § 987.53) shall comply with the requirements of § 987.41 on all dates, but this shall apply to repacked dates previously inspected and certified for handling only if the handler also packs dates received as field-run dates.

[32 FR 12596, Aug. 31, 1967]

§ 987.52 Exemption.

(a) The Committee may exempt from regulation, upon written request of any producer or handler, the dates he sells to consumers through roadside stands, local date shops, mail order or specialty outlets, if it determines that the particular request is not likely to materially interfere with the objectives of this part. All dates handled pursuant to exemptions under this section shall be reported to the Committee in such manner and in such form as the Committee may prescribe. The Committee shall issue, with the approval of the Secretary, appropriate rules and regulations establishing the bases on which exemptions may be granted.

(b) The Committee may, with the approval of the Secretary, recommend that the handling of any date variety be exempted from regulations established pursuant to §§ 987.39 through 987.51 and §§ 987.61 through 987.72.

[27 FR 6818, July 19, 1962, as amended at 77 FR 37765, June 25, 2012]

§ 987.53 Application of regulations to repackers.

Repackers shall be exempt from those requirements of this part, including reporting requirements, with respect to packed dates which had been certified for handling, pursuant to § 987.41(a), prior to receipt, except that: (a) A repacker who processes such dates by machine pitting shall comply with the grade, size, inspection, certification, and identification requirements, and (b) a repacker who repackages such dates in containers other than those in which received, shall comply with the then effective container regulations established pursuant to § 987.48.

[32 FR 12596, Aug. 31, 1967]

DISPOSITION OF OTHER THAN FREE DATES

§ 987.55　Outlets for restricted dates.

Restricted dates may be disposed of only through exportation to such countries as the Committee may approve or by diversion in product outlets described in § 987.43 which the Committee concludes to be appropriate and which will result in dates moving into consumption in a form other than that of whole or pitted dates. To facilitate sales and promote orderly marketing of any variety of restricted dates handled in export, the Committee may participate in or negotiate for handlers, the sale of such dates to meet all or a substantial part of the needs of the particular country, and, in connection with each such sale, the Committee shall extend to all handlers an opportunity to participate therein and shall distribute the returns therefrom to participating handlers according to their respective contributions of dates. The Committee, with the approval of the Secretary, may prescribe rules and regulations governing the opportunity to participate in such sales. The provisions of this section shall not preclude restricted dates being disposed of in outlets for utility and cull dates prescribed in § 987.56.

[43 FR 4253, Feb. 1, 1978]

§ 987.56　Outlets for utility and cull dates.

Subject to the provisions of § 987.47, utility dates and cull dates may be disposed of without inspection, but only in feed, non-table syrup, alcohol, or brandy outlets, or in such other outlets for non-human food products as the Committee with the approval of the Secretary, may specify: *Provided*, That whenever the Committee concludes and the Secretary finds that the use of utility dates of any variety in certain products for human consumption would tend to effectuate the declared policy of the act, the Secretary shall specify such products, and dates of such variety that are inspected and certified as utility dates may be disposed of for use, or used, in such products: *And provided further*, That whenever the Committee concludes and the Secretary finds that the disposition of utility dates of any variety through any export outlet would tend to effectuate the declared policy of the act, the Secretary shall specify such export outlet, and dates of such variety that are inspected and certified as meeting such grade, size, container, and identification requirements as may be prescribed by the Committee with the approval of the Secretary for such outlet may be so exported.

[29 FR 9707, July 18, 1964, as amended at 43 FR 4251, 4253, Feb. 1, 1978]

§ 987.57　Approved manufacturers or feeders.

(a) Diversion of dates, pursuant to § 987.55 or § 987.56, shall be accomplished only by such persons (which may include handlers) as are approved manufacturers or feeders. Any person may become an approved manufacturer or feeder if he (1) submits an application to the Committee in which he agrees, as a condition to approval of his application, to furnish to the Committee such information as it may require and to comply with the requirements and restrictions relative to the use and disposition of such dates, as set forth in this part, and (2) receives from the Committee written approval of his application. The application and approval shall be in accordance with such rules, regulations and safeguards as may be prescribed pursuant to § 987.59.

(b) [Reserved]

[27 FR 6818, July 9, 1962, as amended at 43 FR 4251, 4253, Feb. 1, 1978]

§ 987.58　Terminal date.

Dates covered by §§ 987.55 and 987.56 shall, by September 30 of the subsequent crop year (a) in accordance with the applicable requirements of such sections, be disposed of, or be converted from their whole or pitted form; or (b) be set aside and marked for disposition pursuant to the applicable requirements of such sections. The Committee may prescribe, with the approval of the Secretary, such rules, regulations and safeguards, pursuant to § 987.59, as may be necessary to prevent dates covered by §§ 987.55 and 987.56 from interfering with the objectives of this part.

§987.59 Safeguards.

The Committee may prescribe, with the approval of the Secretary, such rules, regulations and safeguards as are necessary to prevent dates covered by §§987.55 and 987.56 from interfering with the objectives of this part.

REPORTS AND RECORDS

§987.61 Reports of handler carryover.

Each handler shall file each year with the Committee written reports of his carryover of dates as of March 1, October 1, and at such other times as the Committee may prescribe: *Provided*, That during those seasons when volume regulations are established by the Secretary, the handler shall file an additional report on his January 1 carryover. Such reports shall be filed within 10 days of the date of the carryover. These reporting dates specified may be changed, upon recommendation of the Committee, together with substantiation of the need therefore, with the approval of the Secretary.

[43 FR 4253, Feb. 1, 1978]

§987.62 Reports of dates shipped.

Each handler who ships dates during a crop year shall submit to the Committee, in such form and at such intervals as the Committee may prescribe, reports showing the net weight of dates shipped by him and such other information pertinent thereto as the Committee may specify.

§987.63 Reports on restricted dates withheld.

Each handler from time to time, on demand of the Committee, shall file with it a report of the restricted dates withheld by him in satisfaction of his withholding obligation. Such reports shall show such information as the Committee may require and may be in such form as the Committee may prescribe.

§987.64 Reports on disposition of restricted, other marketable, utility, and cull dates.

Each handler disposing of any dates pursuant to §§987.55 and 987.56 shall promptly thereafter report such disposition to the Committee in such form as the Committee may prescribe.

[43 FR 4253, Feb. 1, 1978]

§987.65 Other reports.

Upon request of the Committee each handler shall furnish to it in such manner and at such times as it prescribes, such other information as will enable the Committee to perform its duties and exercise its powers hereunder.

§987.66 Certification of reports.

All reports submitted to the Committee as required in this part shall be certified to the United States Department of Agriculture and to the Committee as to the completeness and correctness of the information therein.

§987.67 Confidential information.

All data or other information constituting a trade secret or disclosing a trade position or business condition shall be received by, and kept in the custody of, one or more designated employees of the Committee and information which would reveal the circumstances of a single handler shall be disclosed to no person other than the Secretary.

§987.68 Verification of reports and records.

For the purpose of checking compliance with record keeping requirements and verifying reports filed by handlers, the Secretary and the Committee, through its duly authorized employees, shall have access to any premises where dates are held and, at any time during reasonable business hours, shall be permitted to examine any dates held and any and all records with respect to matters within the purview of this part. Handlers shall furnish labor necessary to facilitate such examinations at no expense to the Committee. All handlers shall maintain complete records establish and which accurately show the quantity of dates handled, disposed of, and withheld. The Committee, with the approval of the Secretary, may establish the type of records to be maintained. Such records shall be retained by handlers for not less than two years subsequent to the termination of each crop year.

[27 FR 6818, July 19, 1962, as amended at 36 FR 15039, Aug. 12, 1971; 43 FR 4253, Feb. 1, 1978]

EXPENSES AND ASSESSMENTS

§ 987.71 Expenses.

The Committee is authorized to incur such expenses, including maintenance of an operating reserve fund, as the Secretary may find are reasonable and are likely to be incurred by it during each crop year for the maintenance and functioning of the Committee and for such other purposes as he determines to be appropriate. The recommendation of the Committee as to total expenses and allocation thereof for each crop year, together with all data supporting such recommendation, shall be submitted to the Secretary within a reasonable time after the marketing policy for each crop year is recommended.

§ 987.72 Assessments.

(a) *Requirement for payment.* Each handler shall pay to the Committee upon demand, on all dates he has certified as meeting the requirements for marketable dates and utility dates utilized in product outlets including the eligible portion of any field-run dates certified and set aside or disposed of pursuant to § 987.45(f), his pro rata share of all expenses which the Secretary finds are reasonable and likely to be incurred by the Committee during each crop year. Should the condition arise wherein the utility portion of dates handled in certain other outlets should not be, in the opinion of the Committee, subject to the payment of assessments on that portion, the Committee may recommend and the Secretary approve by rulemaking, such exclusion. Each handler's pro rata share shall be the rate of assessment per hundredweight fixed by the Secretary. At any time during or after a crop year the Secretary may increase such assessment rate to secure sufficient funds to cover unanticipated expenses or a deficit in assessable poundage. Any such increase shall apply to all assessable poundage of the crop year. The Committee may accept payments of assessments in advance and may borrow money in any amount not to exceed 10 percent of the estimated expenses set forth in its budget for the then crop year. The assessment weight of pitted dates shall be determined by dividing the weight of such dates by a divisor established by the Committee with the approval of the Secretary.

(b) *Delinquent payments.* Any assessment not paid by a handler within a period of time prescribed by the Committee may be subject to an interest or late payment charge, or both. The period of time, rate of interest, and late payment charge shall be as recommended by the Committee and approved by the Secretary.

(c) *Surplus expenses.* The Committee is authorized to use temporarily funds derived from assessments collected pursuant to paragraph (a) of this section to defray expenses incurred in disposing of surplus dates. All such expenses shall be deducted from the proceeds obtained by the Committee from such disposal.

(d) *Operating reserve.* The Committee, with the approval of the Secretary, may establish and maintain during one or more crop years an operating monetary reserve in an amount not to exceed the average of one year's expenses incurred during the most recent five preceding crop years, except that an established reserve need not be reduced to conform to any recomputed average. Funds in reserve shall be available for use by the Committee for expenses authorized pursuant to § 987.71.

(e) *Refunds.* Funds held by the Committee at the conclusion of the crop year in excess of the crop year's expenses, including reserve requirements, may be used to defray expenses for no more than the ensuing four months, and thereafter within a reasonable time the Committee shall credit, or upon demand, refund the aforesaid excess to handlers who contributed to such excess: *Provided,* That the excess due any handler may be applied, in whole or in part, by the Committee to any outstanding obligation due the Committee from such handler. A handler's share of the excess funds shall be the amount of assessments he paid in excess of his actual pro rata share of the expenses, including reserve requirements, of the Committee for the preceding crop year. Upon termination of this subpart any money in possession of the Committee shall be distributed in such manner as the Secretary may direct: *Provided,* That, to the extent practicable, such funds shall be returned pro rata to the persons from whom such funds were collected.

[27 FR 6818, July 19, 1962, as amended at 29 FR 9707, July 18, 1964; 43 FR 4253, Feb. 1, 1978; 77 FR 37765, June 25, 2012]

MISCELLANEOUS PROVISIONS

§987.76 Compliance.

No handler shall handle any dates (including dates for further processing) except in conformity with, and as authorized by or pursuant to, the applicable provisions of this part, including but not being limited to the regulations relating to grade, size, and volume; and no handler shall use or otherwise dispose of restricted dates or any other dates which have not been certified for handling or for further processing except in conformity with, and as authorized by or pursuant to, the applicable provisions of this part.

§987.77 Personal liability.

No member or alternate member of the committee, or any employee or agent thereof, shall be held personally responsible, either individually or jointly with others, in any way whatsoever, to any handler or any other person for errors in judgment, mistakes, or other acts either of commission or omission, as such member, alternate member, agent, or employee, except for acts of dishonesty, willful misconduct or gross negligence.

§987.78 Separability.

If any provision of this part is declared invalid, or the applicability thereof to any person, circumstance, or thing is held invalid, the validity of the remainder of this part or the applicability of this part to any other person, circumstance, or thing shall not be affected thereby.

§987.79 Derogation.

Nothing contained in this part is, or shall be construed to be, in derogation or in modification of the rights of the Secretary or of the United States to exercise any powers granted by the act or otherwise, or, in accordance with such powers, to act in the premises whenever such action is deemed advisable.

§987.80 Duration of immunities.

The benefits, privileges, and immunities conferred upon any person by virtue of this subpart shall cease upon its termination except with respect to acts done under and during its existence.

§987.81 Agents.

The Secretary may, by a designation in writing, name any person, including any officer or employee of the United States Government, or name any bureau or division of the United States Department of Agriculture, to act as his agent or representative in connection with any of the provisions of this part.

§987.82 Effective time, suspension, or termination.

(a) *Effective time.* The provisions of this part, as well as any amendments hereto, shall become effective at such time as the Secretary may declare, and shall continue in force until terminated or suspended in one of the ways hereinafter specified in this section.

(b) *Suspension or termination*—(1) *Failure to effectuate policy of act.* The Secretary shall terminate or suspend the operation of any or all of the provisions of this part, whenever he finds that such provisions do not tend to effectuate the declared policy of the act.

(2) *When favored by growers.* The Secretary shall terminate the provisions of this part at the end of any crop year whenever he finds that such termination is favored by a majority of the growers of dates who, during that crop year, have been engaged in the production for market of dates in the area of production: *Provided,* That such majority have, during such period, produced for market more than 50 percent of the volume of such dates produced for market within said area; but such termination shall be effective only if announced on or before August 1 of the then current crop year.

(3) *If enabling legislation is terminated.* The provisions of this part shall, in any event, terminate whenever the provisions of the act authorizing them cease to be in effect.

(c) *Proceedings after termination*—(1) *Designation of trustees.* Upon the termination of the provisions hereof, the members of the Committee then functioning shall continue as joint trustees, for the purpose of liquidating the affairs of the Committee, of all funds and property then in the possession or under the control of the Committee, including claims for any funds unpaid or property not delivered at the time of such termination. Action by said trusteeship shall require the concurrence of a majority of the said trustees.

(2) *Duties of trustees.* Said trustees shall continue in such capacity until discharged by the Secretary; shall, from time to time, account for all receipts and disbursements and deliver all property on hand, together with all books and records of the Committee and the joint trustees, to such person as the Secretary may direct; and shall, upon request of the Secretary, execute such assignments or other instruments necessary or appropriate to vest in such persons full title and right to all funds, property, and claims vested in the Committee or the joint trustees pursuant hereto.

(3) *Obligations of persons other than Committee members and trustees.* Any person to whom funds, property, or claims have been transferred or delivered by the Committee or its members, pursuant to this section, shall be subject to the same obligations imposed upon the members of the said Committee and upon the said joint trustees.

[27 FR 6818, July 19, 1962, as amended at 36 FR 15039, Aug. 12, 1971]

§ 987.83 Effect of termination or amendment.

Unless otherwise expressly provided by the Secretary, the termination hereof or of any regulation issued pursuant to this part, or the issuance of any amendment to either thereof, shall not—

(a) Affect or waive any right, duty, obligation, or liability which shall have arisen or which may thereafter arise in connection with any provision of this part or any regulation issued hereunder, or

(b) Release or extinguish any violation of this part or of any regulation issued hereunder, or

(c) Affect or impair any rights or remedies of the Secretary or of any other person, with respect to any such violation.

§ 987.84 Amendments.

Amendments hereto may be proposed, from time to time by any person or by the Committee.

Subpart B—Administrative Requirements

Source: 37 FR 23325, Nov. 2, 1972, unless otherwise noted.

Definitions

§ 987.101 Lot.

Lot means the aggregate quantity of dates of the same variety, style, type and grade in like containers with like identification either (a) packed as a continuous production segment, or (b) offered for inspection as a shipping, storage, or other unit.

§ 987.102 Lot number.

Lot number is synonymous with code and means a combination of letters or numbers, or both, acceptable to the Committee, showing at least the date of packing, the variety, and the outlet category of the dates. The combination of letters or numbers, or both, imprinted on the containers shall differ from those of any other lot coded within a 3-year period.

§ 987.104 Major marketing promotion.

A major marketing promotion program is one requiring the expenditure of more than $500 of Committee funds.

[43 FR 28435, June 30, 1978]

§ 987.105 Whole equivalent of pitted dates.

For the purposes of this part, the whole date equivalent weight of pitted dates shall be determined by dividing the weight of the pitted dates by 0.83.

[53 FR 39226, Oct. 6, 1988]

§ 987.112 Identification of dates.

(a) *General.* Prior to applying the markings required by this section, each handler shall remove or delete from each container all former identifying marks which conflict with those applicable to the dates currently in the container. Dates of each outlet category shall be held, stored, or shiped in a manner to preserve their identity. Except as provided in paragraph (d) of this section, the markings on the containers shall be not less than five-sixteenths ($\frac{5}{16}$) inch in height on containers exceeding 5 pounds net weight and not less than one-eighth ($\frac{1}{8}$) inch in height on smaller containers. All markings shall be legible.

(b) *DAC dates.* Each handler shall mark every shipping or storage container (excluding subcontainers) of DAC dates with his name or that of the distributor for whom the handler is packing, and the lot number. Under the supervision of the inspection service every container shall be marked with the date of inspection, the name or insignia of the inspection service, and the letters "DAC".

(c) *FP dates.* Each handler shall mark every shipping or storage container (excluding subcontainers) of FP dates with his name or that of the distributor for whom the handler is packing, and the lot number. Under the supervision of the inspection service every container shall be marked with the date of inspection, the name or insignia of the inspection service, and the letters "FP".

(d) *Export dates.* Each handler shall mark every shipping or storage container (excluding subcontainers) of Export dates with his name or that of the exporting firm, and the lot number. If the dates, including fieldrun dates with cull dates removed, are certified as meeting the grade and size requirements for export to approved countries other than Mexico, the containers shall be marked "Export". Dry dates for processing packed for shipment to approved countries shall be marked "Export Dry". Dates packed for export to Mexico shall be marked "Export Mexico". However, "Export Mexico" shall be in letters not less than three-fourths ($\frac{3}{4}$) inch in height on containers exceeding 5 pounds net weight, and not less than one-eighth ($\frac{1}{8}$) inch in height on smaller containers. DAC dates and FP dates, marked pursuant to paragraphs (b) and (c), respectively, of this section, may be exported without change of marking.

(e) *Product and utility dates.* Each handler shall mark every shipping or storage container (excluding subcontainers) of Product dates, or Utility dates when approved for use in products, with the lot number and, if for shipment outside the area of production, with the word *Product* or *Utility*, as applicable. Whenever a handler, or an approved date product manufacturer, utilizes a procedure that maintains the identity of the lot and assures that the dates will be used in products or exported, the Committee may waive the requirements of this paragraph for that lot.

(f) *Unidentified dates.* If a handler loses the identity of any lot of dates previously inspected and certified as marketable dates, the certification as to such lot shall be void.

[43 FR 28436, June 30, 1978]

§ 987.112a Grade, size, and container requirements for each outlet category.

(a) In lieu of the minimum standards prescribed in § 987.39, the following standards are prescribed as the minimum grades and sizes for marketable dates to be handled in the applicable outlets. These standards shall continue in effect irrespective of whether the season average price to producers for dates is or is not in excess of the parity level specified in section 2(1) of the act.

(b) *DAC dates.* (1) All varieties of DAC dates may be handled in the United States, Canada, or any other outlet established for dates of lesser grades or sizes.

(2) DAC dates of any variety shall at least meet the requirements of U.S. Grade B, except that up to 25 percent, by weight, of the dates may possess semi-dry of dry calyx ends, but not more than 5 percent, by weight, of the dates may possess dry calyx ends. Also, with respect to whole dates of the Deglet Noor variety, the individual

561

dates in the sample from the lot shall weigh at least 6.5 grams, but up to 10 percent, by weight, may weigh less than 6.5 grams, except beginning February 21, 1997, and ending October 31, 1997, the 10 percent tolerance shall be increased to 15 percent. These size requirements are in addition to, and do not supersede, the requirements as to uniformity of size prescribed in the grade standards.

(3) DAC dates of any variety, when packed in plastic containers, other than bags and master shipping containers, shall contain a net weight (i) for whole dates, of either eight ounces, twelve ounces, 1 pound 8 ounces, or two pounds or more, and (ii) for pitted dates, of either seven ounces, ten ounces, one pound, one pound eight ounces, or two pounds or more. DAC dates packed in other than plastic containers may be handled without regard to the net weight content. For the purpose of this subparagraph, *plastic container* means any containers of any shape made from plastic and in which dates are packed without the use of cardboard boats, trays, or other like stiffening material: *Provided,* That DAC dates shipped for sale in Canada in plastic containers are exempt from the net weight requirements of this subparagraph.

(4) The California Date Administrative Committee may designate with the approval of the Secretary such other types and sizes of containers for testing in connection with a research project conducted by or in cooperation with the Committee. The time period and the quantity of dates which may be marketed by handlers during that period shall be designated by the Committee for each market research project. The handling of each lot of dates in such test containers shall be subject to the prior approval, and under the supervision, of the Committee.

(c) *Dates for further processing.* (1) Except as provided in § 987.152(b)(1), all varieties of FP dates may be disposed of only (i) to persons in the United States capable of processing and packing the dates and having them certified as DAC dates, or (ii) exported to the countries designated in paragraph (d)(2) of this section.

(2) FP dates of any variety shall at least meet the requirements of U.S. Grade B (dry). Also, with respect to whole dates of the Deglet Noor variety, the individual dates in the sample from the lot shall weigh at least 6.5 grams, but up to 10 percent, by weight, may weigh less than 6.5 grams, except beginning February 21, 1997, and ending October 31, 1997, the 10 percent tolerance shall be increased to 15 percent. These size requirements are in addition to, and do not supersede, the requirements as to uniformity of size prescribed in the grade standards.

(d) *Export dates.* (1) Dates of any variety identified as "Export" dates and inspected and certified as meeting the requirements of this subparagraph may only be exported to any country except Canada. Such dates shall at least meet the requirements of U.S. Grade C: *Provided,* That Deglet Noor dates shall score not less than 31 points for character and 24 points for absence of defects but up to 40 percent, by weight, of the dates may be damaged by broken skin.

(2) *Export of dry dates.* Dates of any variety identified and certified as meeting the requirements of this subparagraph only may be exported to the following designated date producing and processing countries in North Africa: Morocco, Algeria, Tunisia, Libya,[1] Egypt, and Sudan; the following date processing and consuming countries north of the Mediterranean Sea: Spain, France, Belgium, West Germany, Italy, France, Greece, and the Netherlands; and the following date processing and consuming country in Asia: Japan. Such dates shall at least meet U.S. Grade C (dry) except for defects removable by washing: *Provided,* That Deglet Noor dates shall score not less than 31

[1] Executive Order 12543 of January 7, 1986 (51 FR 875), prohibits trade and certain transactions involving Libya, and is applicable to exports of dates under this marketing order as long as the executive order is in effect. That order, among other things, prohibits the exports to Libya of any goods, technology (including technical data or other information) or services from the United States, except publications and donations of articles intended to relieve human suffering, such as food, clothing, medicine and medical supplies intended strictly for medical purposes.

points for character and 24 points for absence of defects but up to 40 percent, by weight, of the dates may be damaged by broken skin.

(3) Dates of any variety identified as "Export—Mexico" and inspected and certified as at least meeting the requirements of U.S. Grade C may be exported only to Mexico. No dates shall be exported to Mexico unless the handler certifies to the Committee and the U.S. Department of Agriculture, on CDAC Form No. 11(a), which shall be submitted to the Committee, that the importing buyer has agreed that such dates will not reenter the United States or be shipped to Canada. The form shall show the identity of the handler, the trucker, the importer, the destination of the dates, the location of the border-crossing station, and such other information as the Committee deems appropriate to perform its duties and excercise its powers under this part.

(4) Whenever field-run dates of any variety are authorized for export to any country, each lot shall consist of at least 85 percent, by weight, of sound dates. *Sound dates* means individual dates which are at least U.S. Grade C in character and are free of the defects—other than those removable by washing—scored to determine the point requirement applicable to their intended destination.

(5) Dates meeting the grade and size requirements of this paragraph may be disposed of in outlet categories established for dates of lesser grades and sizes.

(e) *Product dates.* (1) Dates of any variety identified as "Product" dates and inspected and certified as at least meeting the requirements of this paragraph may be disposed of by handlers for use or used by them in the production of table syrup, rings, chunks, pieces, butter, paste, and macerated dates or other products approved by the Committee. If the handler does not use the dates in products, he may sell them to: (i) Other handlers within the area of production for conversion into products, or (ii) to date product manufacturers approved by the Committee regardless of their location. Once the dates have been converted from their whole or pitted form, they may be shipped to any market in the United States, Canada, or foreign country.

(2) Product dates of any variety and identified as "Product" shall meet the requirements of U.S. Grade C, except that mashing and mechanical injury not affecting eating quality shall not be considered in determining the defect factor.

(f) *Change of outlet.* A handler may change the outlet category for any lot of dates: *Provided,* That prior to such change, the handler files a completed CDAC Form No. 1(a) and a new inspection certificate with the Committee. If the grade and size requirements of the new outlet category are the same as or less than the requirements of the outlet category previously intended, only a condition inspection is required. If the grade and size requirements of the new outlet category are greater, a complete inspection is required. The handler shall change the marking on the containers to conform with the identification requirements prescribed in §987.112 for the new outlet.

(g) *Deteriorated dates.* Any marketable dates which deteriorated in quality so that they are either utility or cull dates may be disposed of only in the applicable outlets for such dates or they may be reconditioned and upon reconditioning, the modified lot may be reinspected and recertified, as applicable.

[43 FR 28436, June 30, 1978, as amended at 47 FR 4489, Feb. 1, 1982; 47 FR 23417, May 28, 1982; 48 FR 176, Jan. 11, 1983; 51 FR 4478, Feb. 5, 1986; 52 FR 35530, Sept. 22, 1987; 53 FR 35994, Sept. 16, 1988; 56 FR 778, Jan. 9, 1991; 57 FR 61779, Dec. 29, 1992; 62 FR 7663, Feb. 20, 1997]

NOMINATIONS

§987.124   Nomination and polling.

(a) Date producers and producer-handlers shall be provided an opportunity to nominate and vote for individuals to serve on the Committee. For this purpose, the Committee shall, no later than June 15 of every third year, provide date producers and producer-handlers nomination and balloting material by mail or equivalent electronic means, upon which producers and producer-handlers may nominate candidates and cast their votes for members and alternate members of the

Committee in accordance with the requirements in paragraphs (b)(1) and (b)(2) of this section, respectively. All ballots are subject to verification. Balloting material should be provided to voters at least two weeks before the due date and should contain, at least, the following information:

(1) The names of incumbents who are willing and eligible to continue to serve on the Committee;

(2) The names of other persons willing and eligible to serve;

(3) Instructions on how voters may add write-in candidates;

(4) The date on which the ballot is due to the Committee or its agent; and

(5) How and where to return ballots.

(b)(1) *Producers.* Each producer may vote for three producer members and three producer alternate members. No producer may vote more than once for any one person. The three individuals receiving the highest number of votes for the producer member positions shall be the producer member nominees. Individuals nominated for producer member and failing to receive enough votes to become a producer member nominee shall have their names listed with those nominated for producer alternate members and the votes cast for them as member shall be counted with any votes they received for producer alternate member. The three individuals receiving the highest number of votes for the producer alternate member positions shall be the producer alternate member nominees.

(2) *Producer-handlers.* Each producer-handler may vote for one producer-handler member and one producer-handler alternate member, and these votes shall be weighted as provided in § 987.24. No producer-handler may vote more than once for any one person. The six individuals receiving the highest weighted votes for the producer-handler member positions shall be the producer-handler member nominees. Individuals nominated for producer-handler member and failing to receive enough votes to become a producer-handler member nominee shall have their names listed with those nominated for producer-handler alternate members and the votes cast for them as member shall be counted with any votes they received for producer-handler alternate member. The six individuals receiving the highest weighted vote for producer-handler alternate member positions shall be the alternate member nominees.

[43 FR 28437, June 30, 1978, as amended at 74 FR 61267, Nov. 24, 2009; 77 FR 37766, June 25, 2012]

§ 987.138 **Handlers of record.**

Prior to handling dates, each person shall file CDAC Form No. 18 with the Committee at the times, and containing the information, prescribed in § 987.38.

[43 FR 28437, June 30, 1978]

INSPECTION

§ 987.141 **Inspection and certification.**

Each handler shall furnish, or cause the inspection service to furnish, to the Committee a copy of the inspection certificate issued to him on each lot of dates, and such certificate shall contain at least the following information: (a) The date of inspection; (b) the name of the handler; (c) the lot number and the applicable outlet category set forth in § 987.112a; (d) the variety of dates and weight of the lot; (e) the number and type of containers in the lot; and (f) if the dates (1) are other than field-run dates, a certification as to the grade of the dates and whether or not they meet the applicable grade, size, container, and identification requirements, or (2) are field-run dates, a certification showing the percentage by weight, of sound dates in the lot, and whether or not they meet the identification requirements for such dates.

[43 FR 28437, June 30, 1978]

VOLUME REGULATION

§ 987.145 **Withholding obligation.**

(a) *Satisfying the withholding obligation.* Any handler may satisfy all or part of his withholding obligation for any variety of dates for which free and restricted percentages have been established by having an adequate quantity of that variety inspected and certified as meeting the applicable grade, size, and container requirements prescribed by the Committee for any approved restricted date outlet.

(b) *Credit for excess disposition in restricted outlets.* Disposition of marketable dates in restricted outlets in excess of a handler's withholding obligation may be: (1) Transferred pursuant to §987.45 upon such handler filing a completed CDAC Form No. 14 with the Committee, or (2) credited to the handler's withholding obligation of the following crop year so long as the excess disposition exceds 199 pounds. However, the quantity so credited shall never exceed 40 percent of the handler's withholding obligation of the crop year in which the excess disposition occurred and 100 percent of the withholding obligation incurred by him during October through December of the crop year following the crop year in which such excess disposition occurred. All such crediting or accumulation shall be contingent upon the Committee receiving, in due course, confirmation that the dates were disposed of in eligible restricted outlets. With respect to exports, the withholding credit shall be granted upon the Committee receiving notification from the inspection service, and in due course a copy of the on board bill of lading or other documentary evidence satisfactory to the Committee.

(c) *FP dates.* Withholding obligations on FP dates shall be based on the weight of such dates when they are inspected and certified. However, if such dates are subsequently processed and packed within the area of production, the withholding obligation shall be adjusted to reflect any increase in weight.

(d) *Dates for deferment of withholding.* Any handler may defer his certification and withholding or disposition of restricted dates by pledging a comparable volume of graded or field-run dates as a surety that he will meet his withholding obligation at a later date. Such deferment shall not be effective until: (1) The handler files with the Committee a CDAC Form No. 12 to set aside graded dates or CDAC Form No. 13 to set aside field-run dates; and (2) the pledged dates are set aside as a lot and identified by the handler as "Restricted" and as "Graded" or "Field-Run", as appropriate, and as to the number of containers, the date of set-aside and whether or not the dates

have been inspected. If the handler sets aside field-run dates or disposes of field-run dates in outlets prescribed in or pursuant to §987.56 to obtain withholding credit for the sound date portion in the lot, the field-run dates shall meet the requirements prescribed in paragraph (f) of this section for eligible field-run dates, as determined by the inspection service.

(e) *Identification of restricted dates.* Any lot of restricted dates not immediately disposed of through exportation to countries approved by the Committee or directed to approved product outlets shall be stored as a lot separate from all other dates and in a specified location with a USDA inspection service tag marked "Restricted".

(f) *Field-run dates.* Field-run dates set aside for the purpose of deferring or meeting any part or all of a withholding obligation shall consist of at least 70 percent, by weight, of sound dates but may contain 10 percent, by weight, of cull dates of which not more than 5 percent may be hidden culls—i.e., dates with internal defects including souring, mold, fermentation, insect infestation, or foreign material.

(g) *Substitution.* Any handler may, under the direction and supervision of the Committee or the inspection service, substitute for any quantity of restricted dates held by him a like quantity of dates of the same variety and of the same or more recent year's production which have been certified and identified as meeting the requirements for restricted dates.

[43 FR 28437, June 30, 1978]

§987.147 Surplus.

(a) *General.* Surplus dates delivered to the Committee pursuant to §987.47 shall be pooled for sale to livestock feeders, distillers, or manufacturers of inedible products: *Provided,* That if any portion of the deliveries differs sufficiently to require separate handling, and earn a different average return, such portion shall be handled as a separate pool. The income from sale of surplus, after deduction of committee expenses, shall be paid to the respective equity holders in the pool or pools, or to their assignees, on the basis of the weight of dates each delivered.

(b) *Delivery.* The Committee may refuse delivery of any surplus dates which it determines are excessively soured, fermented, or adulterated by palm debris, rocks, paper, wood, plastic liners, or other foreign material. If the Committee refuses delivery, the deliverer shall be permitted to clean such dates sufficiently to make them acceptable to the Committee. The weight of each accepted delivery shall be that determined by a public weightmaster or, in the absence of such weight, that determined by the Committee on the basis of the number and size of the containers used in the delivery. Upon delivery of surplus dates to the Committee, the deliverer, or a designee of the Committee shall execute CDAC Form SP–1, Delivery Manifest, showing:

(1) The person to receive payment of the net proceeds for the surplus,

(2) The date and place of loading,

(3) If field surplus, the location and owner of the garden,

(4) The type and number of containers loaded or dumped,

(5) The net weight of the load, and

(6) If the delivery is directly to a buyer's truck, the driver, truck and buyer.

QUALIFICATION TO REGULATION

§ 987.151 Interhandler transfers.

When any handler transfers dates, other than product dates, to another handler, the selling handler shall promptly notify the Committee by filing with it a completed CDAC Form No. 1 and shall show the name and address of the transferring or selling handler and of the receiving or buying handler, the variety and processed category or classification of the dates, the lot number and inspection certificate number on any lot of packed and certified dates, the number and type of containers, the net weight of the transferred dates, and if applicable, the transferring handler's statement on assuming the withholding and assessment obligation. A transfer of products dates between handlers shall be reported as a disposition by the selling

handler filing with the Committee a completed CDAC Form No. 8.

[37 FR 23325, Nov. 2, 1972, as amended at 43 FR 28438, June 30, 1978]

§ 987.152 Exemption from regulations.

(a) *Producer exemption.* The Committee may permit any producer to sell dates from such producer's own production free of the requirements of §§ 987.41, 987.45, 987.48, and 987.72 when sold directly to consumers through a roadside stand or date shop owned or operated by the producer within 25 miles of the city limits of Indio, California, through shipments by parcel post or express, or by certified producers at certified farmers' markets, as these terms are defined by the State of California. Permission to so sell dates shall be granted only upon the producer filing with the Committee a completed CDAC Form No. 9 wherein the producer describes how the producer plans to sell such dates and agrees to sell only dates of DAC date quality of the producer's own production in direct sales; and to report such sales to the Committee. If the producer fails to comply with this agreement, the Committee may revoke any or all exemptions granted the producer.

(b) *Handler exemptions*—(1) *Specialty sales.* The Committee may permit any handler to sell to health food stores or health food outlets, dates which at least meet the requirements for FP dates. It may permit any handler to sell to a candy manufacturer hand-pitted dates which meet the grade requirements for DAC dates except for size, or damage due to cutting and pitting. Also, it may permit any handler to sell hand-layered dates in tin, wood, plastic, or other type of container exempt from §§ 987.41(a) and 987.48, or to make shipments by common carrier of up to 150 pounds to any one purchaser in any one day exempt from the provisions of § 987.41(a): *Provided,* That the hand-layered dates or the shipment to a single purchaser in any 1 day have been packed from dates certified as meeting the grade requirements for DAC dates and have not been commingled with other dates. Permission to use these exemptions shall be granted only upon the handler filing with the Committee its CDAC Form No. 10

wherein he describes how he plans to sell, and agrees to sell only specific dates and to report such sales.

(2) *Donations.* The Committee may permit any handler to donate marketable dates other than DAC dates to needy persons, prisoners, or Indians on reservations. Before such donation is made, such handler shall file a request for donation with the Committee detailing the quantity and grade of dates involved and the name and address of the intended donee. The donation may be subject to Committee surveillance, verification by written documentation of receipt by the donee, and any other safeguards necessary to assure consumption in these outlets.

[37 FR 23325, Nov. 2, 1972, as amended at 43 FR 28438, June 30, 1978; 53 FR 35994, Sept. 16, 1988; 57 FR 39112, Aug. 28, 1992]

§987.157 Approved date product manufacturers.

Any person, including date handlers, with facilities for converting dates into products may apply to the committee, by filing CDAC Form No. 3, for listing as an approved date product manufacturer.

(a) The applicant shall indicate on such form: The products he/she intends to make; the quantity of dates he/she may use; the location of his/her facilities; and agree that all dates obtained for manufacturing into products shall be used for that purpose, none shall be resold or disposed of as whole or pitted dates.

(b) As a condition to become an approved date product manufacturer: Each applicant is subject to an inspection of his/her manufacturing plant to verify that proper equipment to convert dates into products is in place and that the plant meets appropriate sanitation requirements; the applicant also shall agree to file a report of the disposition of each lot of dates on the Committee's CDAC Form No. 8 within 24 hours of the transaction, and to file an annual usage and inventory report on CDAC Form No. 4 by October 10 of each year; and an applicant who is also a handler under the order shall be in compliance with the order, including the assessment payment and reporting requirements.

(c) The committee shall approve each such application on the basis of information furnished or its own investigation, and may revoke any approval for cause. The name and address of all approved manufacturers shall be placed on a list and made available to each date handler in Riverside County.

(d) If an application is disapproved, the committee shall notify the applicant in writing of the reasons for disapproval, and allow the applicant an opportunity to respond to the disapproval. When the applicant has complied with all the qualification requirements to become an approved manufacturer, the committee shall notify the applicant in writing of such approval. The applicant's name shall be added to the list of approved manufacturers, which shall be made available to each date handler in Riverside County.

(e) Each approved manufacturer of date products is required to renew their approved manufacturer status with the committee by submitting an updated CDAC Form No. 3 at the end of a crop year, but no later than October 10 of the new crop year. In addition, the approved manufacturer must continue to meet the other approved manufacturer qualification requirements.

(f) In the event an approved date product manufacturer who is also a regulated date handler within the area of production does not remain in compliance with the order, or fails or refuses to submit reports or to pay assessments required by the committee, such date product manufacturer shall become ineligible to continue as an approved date product manufacturer. Prior to making a determination to remove a date product manufacturer from the approved date product manufacturer list, the committee shall notify such manufacturer in writing of its intention and the reasons for removal. The committee shall allow the date product manufacturer an opportunity to respond. In the event that a date product manufacturer's name has been removed from the list of approved date product manufacturers, a new application must be submitted to the committee and the applicant must await approval.

[70 FR 11119, Mar. 8, 2005]

REPORTS AND RECORDS

§ 987.161 Handler carryover.

Each handler shall file with the Committee, a report of his carryover of dates as of March 1 and October 1 and, when volume regulation is established, as of January 1. This report shall be on CDAC Form No. 5 and shall show, by variety, at least:

(a) The quantity of DAC dates held within and outside the area,

(b) The quantity of FP dates held within the area,

(c) The quantity of export dates, and

(d) The quantity of dates held graded but not certified, and as field-run, segregated as to outlet category.

[43 FR 28438, June 30, 1978, as amended at 53 FR 35995, Sept. 16, 1988]

§ 987.162 Handler acquisition and disposition.

(a) Handlers shall file CDAC Form No. 6 with the committee by the 16th of each month or such other date as the committee may prescribe, reporting at least the following for the preceding month:

(1) Their acquisitions of field run dates;

(2) Their shipments of marketable dates in each outlet category;

(3) Their shipments of free dates and disposition of restricted dates, whenever applicable; and

(4) Their purchases from other handlers of DAC, export, product, graded, and field run dates.

(b) In addition, this report shall include the names and addresses of any producers not previously identified pursuant to § 987.38, the quantity of dates acquired from each producer, the location of such producer's date garden, the acreage of that garden, and the estimated current season's production from that garden.

[74 FR 61267, Nov. 24, 2009]

§ 987.164 Shipments of product dates and disposition of restricted dates in approved product outlets.

Each handler shall file with the Committee a completed CDAC Form No. 8 showing the shipment of each lot of product dates or the disposition of restricted dates in approved product outlets. This report shall be filed promptly after shipment or disposition of those dates and shall identify the lot, the outlet, the number of containers, and the net weight of the dates. If such dates are sold to an approved date product manufacturer, a copy of the completed form shall be signed and dated by the manufacturer and returned to the Committee. If the lot was certified as product dates and is exported to Mexico, the handler shall submit completed CDAC Form No. 8 together with completed CDAC Form No. 11(a) to the Committee.

[43 FR 28439, June 30, 1978, as amended at 53 FR 35995, Sept. 16, 1988]

§ 987.165 Other reports.

(a) *Exempt sales.* Each handler shall file with the Committee, a completed CDAC Form No. 2 showing the quantity and variety of dates sold under exemption during the crop year. The report shall be filed upon the completion of such sales or promptly after the end of the crop year.

(b) *Products.* Each approved date product manufacturer shall file with the Committee a completed CDAC Form No. 4 showing his beginning and ending inventories of product dates, the quantity received during the crop year, the quantity used, the type and quantity of products manufactured, and his year-end inventory of products. This report shall be filed promptly after the end of each crop year.

[37 FR 23325, Nov. 2, 1972, as amended at 43 FR 28439, June 30, 1978]

§ 987.168 Handler records.

Each handler shall establish complete records which accurately show the quantity of dates handled, disposed of, and withheld. These records shall be maintained for at least 2 years after the end of the crop year of record. Records shall show:

(a) For grower deliveries of dates, the name of each grower, the varieties delivered and the net weight of each variety;

(b) For shipments of dates, the variety, type of pack, net weight and destination or name and address of the person to whom each shipment was sent;

(c) If different from shipments, the variety, type of pack, net weight and purchaser of each quantity of dates sold; and

(d) Manifests, invoices, weight certificates, inventory tabulations, or any other documents necessary to prepare, file, or substantiate the reports required to be filed with the Committee.

[37 FR 23325, Nov. 2, 1972, as amended at 43 FR 28439, June 30, 1978]

§ 987.172 Adjustment of assessment obligation, and late payment and interest charges.

(a) In accordance with §§ 987.45 and 987.72, the assessment obligation of FP dates shall be based on the weight of the dates at the time of inspection and certification. However, if such dates are subsequently processed and packed within the area of production, the assessment obligation shall be adjusted to reflect any increase in weight and the obligation shall be placed on the handler agreeing to assume it.

(b) Pursuant to § 987.72, the committee shall impose an interest charge on any handler whose assessment payment has not been received in the committee's office, or the envelope containing the payment legibly postmarked by the U.S. Postal Service, within 60 days of the invoice date shown on the handler's statement. The interest charge shall be a rate of one and one half percent per month, and shall be applied to the unpaid assessment balance for the number of days all or any part of the unpaid balance is delinquent beyond the 60-day payment period.

(c) In addition to the interest charge specified in paragraph (b) of this section, the committee shall impose a late payment charge on any handler whose payment has not been received in the committee's office, or the envelope containing the payment legibly postmarked by the U.S. Postal Service, within 60 days of the invoice date. The late payment charge shall be 10 percent of the unpaid balance.

[43 FR 28439, June 30, 1978, as amended at 79 FR 41417, July 16, 2014]

Subpart C—Assessment Rates

§ 987.339 Assessment rate.

On and after October 1, 2018, an assessment rate of $0.15 per hundredweight is established for dates produced or packed in Riverside County, California.

[84 FR 9695, Mar. 18, 2019]

PART 989—RAISINS PRODUCED FROM GRAPES GROWN IN CALIFORNIA

Subpart A—Order Regulating Handling

DEFINITIONS

AUTHORITY: 7 U.S.C. 601–674.

Subpart A—Order Regulating Handling

SOURCE: 25 FR 12813, Dec. 14, 1960; 27 FR 2506, Mar. 16, 1962, unless otherwise noted.

EDITORIAL NOTE: Nomenclature changes to part 989 appear at 83 FR 53973, Oct. 26, 2018.

DEFINITIONS

§989.1 Secretary.

Secretary means the Secretary of Agriculture of the United States or any officer or employee of the United States Department of Agriculture to whom authority has heretofore been delegated or to whom authority may hereafter be delegated, to act in his stead.

§989.2 Act.

Act means Public Act No. 10, 73d Congress, as amended, and as re-enacted and amended by the Agricultural Marketing Agreement Act of 1937, as amended (sections 1–19, 48 Stat. 31, as amended; 7 U.S.C. 601–674).

[42 FR 37201, July 20, 1977]

§989.3 Person.

Person means an individual, partnership, corporation, association, or any other business unit.

§989.4 Area.

Area means the State of California.

§989.5 Raisins.

Raisins means grapes of any variety grown in the area, from which a significant part of the natural moisture has been removed by sun-drying or artificial dehydration, either prior to or after such grapes have been removed from the vines. Removal of a significant part of the natural moisture means removal which has progressed to the point where the grape skin develops wrinkles characteristic of wrinkles in fully formed raisins.

[37 FR 19622, Sept. 21, 1972]

§989.7 Golden Seedless raisins.

Golden Seedless raisins means raisins, the production of which includes soda dipping, sulfuring, and artificial dehydration.

§989.8 Natural condition raisins.

Natural condition raisins means raisins the production of which includes sun-drying or artificial dehydration but which have not been further processed to a point where they meet any of the conditions for "packed raisins", as defined in §989.9.

[25 FR 12813, Dec. 14, 1960, as amended at 42 FR 37201, July 20, 1977]

§989.9 Packed raisins.

Packed raisins means raisins which have been stemmed, graded, sorted, cleaned, or seeded, and placed in any container customarily used in the marketing of raisins or in any container suitable or usable for such marketing. Raisins in the process of being packed or raisins which are partially packed shall be subject to the same requirements as packed raisins.

§989.10 Varietal types.

Varietal types means raisins generally recognized as possessing characteristics differing from other raisins in a degree sufficient to make necessary or desirable separate identification and classification. Varietal types are the following: Natural (sun-dried) Seedless, Dipped Seedless, Golden Seedless, Muscats (including other raisins with seeds), Sultana, Zante Currant, Monukka, and Oleate and Related Seedless: *Provided,* That the Committee may, subject to approval of the Secretary, change this list of varietal types.

[48 FR 32974, July 20, 1983]

§989.11 Producer.

Producer means any person engaged in a proprietary capacity in the production of grapes which are sun-dried or dehydrated by artificial means until they become raisins.

[83 FR 53968, Oct. 26, 2018]

§989.12 Dehydrator.

Dehydrator means any person who produces raisins by dehydrating grapes by artificial means.

§989.12a Cooperative bargaining association.

Cooperative bargaining association means a nonprofit cooperative association of raisin producers engaged within the area in bargaining with handlers as to price and otherwise arranging for the sale of natural condition raisin of its members.

[32 FR 12158, Aug. 24, 1967]

§ 989.13   Processor.

Processor means any person who receives or acquires natural condition raisins, off-grade raisins, other failing raisins or raisin residual material and uses them or it within the area, with or without other ingredients, in the production of a product other than raisins, for market or distribution.

[32 FR 12158, Aug. 24, 1967; 33 FR 2983, Feb. 15, 1968, as amended at 42 FR 37201, July 20, 1977]

§ 989.14   Packer.

Packer means any person who, within the area, stems, sorts, cleans, or seeds raisins, grades stemmed raisins, or packages raisins for market as raisins: *Provided,* That:

(a) No producer with respect to the raisins produced by him, and no group of producers with respect to raisins produced by the producers comprising the group, and not otherwise a packer, shall be deemed a packer if he or it sorts or cleans (with or without water) such raisins in their unstemmed form;

(b) Any dehydrator shall be deemed to be a packer, with respect to raisins dehydrated by him, only if he stems, cleans with water subsequent to such dehydration, seeds or packages them for market as raisin;

(c) The Committee may, with the approval of the Secretary restrict the exceptions as to permitted cleaning if necessary to cause delivery of sound raisins; and

(d) No person shall be deemed a packer by reason of the fact he repackages for market (with or without additional preparation) packed raisins which, in the hands of a previous holder, have been inspected and certified as meeting the applicable minimum grade standards for packed raisins.

[32 FR 12158, Aug. 24, 1967]

§ 989.15   Handler.

Handler means: (a) Any processor or packer; (b) any person who places, ships, or continues natural condition raisins in the current of commerce from within the area to any point outside thereof; (c) any person who delivers off-grade raisins, other failing raisins or raisin residual material to other than a packer or other than into any eligible non-normal outlet; or (d) any person who blends raisins: *Provided,* That blending shall not cause a person not otherwise a handler to be a handler on account of such blending if he is either: (1) A producer who, in his capacity as a producer, blends raisins entirely of his own production in the course of his usual and customary practices of preparing raisins for delivery to processors, packers, or dehydrators; (2) a person who blends raisins after they have been placed in trade channels by a packer with other such raisins in trade channels; or (3) a dehydrator who, in his capacity as a dehydrator, blends raisins entirely of his own manufacture.

[37 FR 19622, Sept. 21, 1972]

§ 989.16   Blend.

Blend means to mix or commingle raisins.

§ 989.17   Acquire.

Acquire means to have or obtain physical possession of raisins by a handler at his packing or processing plant or at any other established receiving station operated by him: *Provided,* That a handler shall not be deemed to acquire any raisins (including raisins produced or dehydrated by him) while:

(a) He stores them for another person or as handler-produced tonnage in compliance with the provisions of §§ 989.58 and 989.70;

(b) He reconditions them, or;

(c) He has them in his possession for the purpose of inspection; and *Provided further,* That the term shall apply only to the handler who first acquires the raisins.

§ 989.18   Committee.

Committee means the Raisin Administrative Committee established under § 989.26.

[41 FR 32412, Aug. 3, 1976]

§ 989.20   Ton.

Ton means a short ton of 2,000 pounds.

§989.21 Crop year.

Crop year means the 12-month period beginning with August 1 of any year and ending with July 31 of the following year.

[41 FR 32412, Aug. 3, 1976]

§989.22 District.

District means any one of the geographical areas referred to in §989.26, and designated in the rules and regulations.

[48 FR 32974, July 20, 1983]

§989.23 File.

File means transmit or deliver to the Secretary or Committee, as the case may be, and such act shall be deemed to have been accomplished at the time:

(a) Of actual receipt by the Secretary or Committee in the event of personal delivery;

(b) Of receipt at the office of the telegraph company, in case submission is by telegram; or

(c) Shown by the postmark, in case submission is by mail.

§989.24 Standard raisins, off-grade raisins, other failing raisins, and raisin residual material.

(a) *Standard raisins* means raisins which meet the then effective minimum grade and condition standards for natural condition raisins.

(b) *Off-grade raisins* means raisins which do not meet the then effective minimum grade and condition standards for natural condition raisins: *Provided*, That raisins which are certified as off-grade raisins shall continue to be such until successfully reconditioned or become "other failing raisins."

(c) *Other failing raisins* means any raisins received or acquired by a handler, either as standard raisins or off-grade raisins, which are processed to a point where they qualify as packed raisins but fail to meet the applicable minimum grade standards for packed raisins.

(d) *Raisin residual material* means defective raisins, stemmer waste, sweepings, and other residue accumulated by a handler from reconditioning raisins or from processing standard raisins and other failing raisins.

[25 FR 12813, Dec. 14, 1960, as amended at 42 FR 37201, July 20, 1977]

§989.24a Non-normal outlets.

Non-normal outlets means outlets other than those customarily used for commercial disposition of raisins meeting the then applicable minimum standards for natural condition raisins or packed raisins.

[29 FR 9483, July 11, 1964]

§989.25 Part and subpart.

Part means the order regulating the handling of raisins produced from grapes grown in California, and all rules, regulations, and supplementary orders issued thereunder. This order regulating the handling of raisins produced from grapes grown in California shall be a *subpart* of such part.

RAISIN ADMINISTRATIVE COMMITTEE

§989.26 Establishment and membership.

A Raisin Administrative Committee is hereby established consisting of 47 members of whom 35 shall represent producers, 10 shall represent handlers, 1 shall represent the cooperative bargaining association(s) and 1 shall be a public member. The producer members shall be selected as follows:

(a) Producer members representing the cooperative marketing association(s) shall be members of such association(s) engaged in the handling of raisins, each of which acquired not less than 10 percent of the total raisin acquisitions during the preceding crop year, and those members shall be equal to the product, rounded to the nearest whole number, obtained by multiplying 35 by the ratio the cooperative marketing association(s) raisin acquisitions are to the acquisitions of all handlers during the preceding crop year.

(b) Producer members representing cooperative bargaining association(s) shall be members of such associations, and the number of those members shall be equal to the product, rounded to the nearest whole number, obtained by multiplying 35 by the ratio the raisins acquired by handlers from bargaining association members are to the total acquisitions of all handlers during the preceding crop year.

(c) All other producer members who shall not be members of a cooperative bargaining association(s), cooperative marketing association(s) engaged in the handling of raisins which acquired 10 percent or more of the total acquisitions during the preceding crop year, nor sold for cash to cooperative marketing association(s), shall represent all producers not defined in paragraph (a) or (b) of this section and shall be selected in the number and, when appropriate, for the districts as designated in the rules and regulations.

(d) The handler members shall be divided into two groups and include the following:

(1) Handler members shall be selected from and represent cooperative marketing association(s) engaged in the handling of raisins each of which acquired not less than 10 percent of the total raisin acquisitions during the preceding crop year, and the number of those members shall be equal to the product, rounded to the nearest whole number, obtained by multiplying 10 by the ratio of the cooperative marketing association(s) raisin acquisitions are to the total acquisitions of all handlers during the preceding crop year.

(2) The remaining handler members shall be selected from and represent all other handlers, which would include all independent handlers and small cooperative marketing association(s) who acquired less than 10 percent of the total raisin acquisitions during the preceding crop year. Handler nominees for this group shall be nominated by all handlers in the group in a manner determined by the Committee, with the approval of the Secretary, and specified in the rules and regulations.

(e) The "cooperative" bargaining association'(s) member shall be selected from the cooperative bargaining association(s). The public member shall be nominated by the Committee and selected by the Secretary as public member.

(f) For each member of the Committee there shall be an alternate member who shall have the same qualifications as the member for whom he is an alternate.

[48 FR 32974, July 20, 1983]

§ 989.27 Eligibility.

No person shall be selected or continue to serve as a member or alternate member of the Committee who is not actively engaged in the business of the group which he represents either in his own behalf, or as an officer, agent, or employee of a business unit engaged in such business: *Provided,* That only producers, as defined in § 989.11, engaged as such with respect to the most recent grape crop, are eligible to serve on the Committee. Only handlers who packed or processed raisins during the then current crop year shall be eligible to represent handlers on the Committee. Any handler eligible to represent a particular group shall continue to represent handlers for the entire term for which he was selected.

[48 FR 32974, July 20, 1983]

§ 989.28 Term of office.

The term of office of all representatives serving on the Committee shall be for two years and shall end on April 30 of even numbered calendar years, but each such member and alternate member shall continue to serve until their successor is selected and has qualified.

[48 FR 32975, July 20, 1983]

§ 989.29 Initial members and nomination of successor members.

(a) *Initial members.* Members and alternate members of the Committee serving immediately prior to the effective date of this amended subpart shall, if thereafter they are eligible, serve on the Committee until April 30, 1984, and until their respective successors have been selected and qualified.

(b) *Nominations for successor members.* Nominations for successor members and alternate members of the Committee shall be made as follows:

(1) The Committee shall notify the cooperative marketing association(s) engaged in handling not less than 10 percent of the total raisin acquisitions during the preceding crop year, and cooperative bargaining association(s), of

the date by which nominations to fill member and alternate member positions shall be made. The Committee shall give reasonable publicity of a meeting or meetings of producers who are not members of cooperative bargaining association(s), or cooperative marketing association(s) which handled 10 percent or more of the total raisin acquisitions during the preceding crop year, and of independent handlers and cooperative marketing association(s) who handled less than 10 percent of the total raisin acquisitions during the preceding crop year, for the purpose of making nominations to fill the member and alternate member positions prescribed in §989.26 (c) and (d): *Provided*, That member and alternate member nominations by independent handlers and cooperative marketing association(s) who acquired less than 10 percent of the total raisin acquisitions during the preceding crop year may be made to the Committee by mail in lieu of meetings.

(2)(i) Any producer representing independent producer and producers who are affiliated with cooperative marketing association(s) handling less than 10 percent of the total raisin acquisitions during the preceding crop year must have produced grapes which were made into raisins in the particular district for which they are nominated to represent said district as a producer member or alternate producer member on the Committee. In the event any such nominee is engaged as a producer in more than one district, such producer may be a nominee for only one district. One or more producers may be nominated for each such producer member or alternate member position.

(ii) Each such producer whose name is offered in nomination for producer member positions to represent on the Committee independent producers or producers who are affiliated with cooperative marketing association(s) handling less than 10 percent of the total raisin acquisitions during the preceding crop year shall be given the opportunity to provide the Committee a short statement outlining qualifications and desire to serve if selected. Similarly, each such producer whose name is offered in nomination for pro-

ducer alternate member positions to represent on the Committee independent producers or producers who are affiliated with cooperative marketing association(s) handling less than 10 percent of the total raisin acquisitions during the preceding crop year shall be given the opportunity to provide the Committee a short statement outlining qualifications and desire to serve if selected. These brief statements, together with a ballot and voting instructions, shall be mailed to all independent producers and producers who are affiliated with cooperative marketing associations handling less than 10 percent of the total raisin acquisitions during the preceding crop year of record with the Committee in each district. The producer member candidate receiving the highest number of votes shall be designated as the first member nominee, the second highest shall be designated as the second member nominee until nominees for all producer member positions have been filled. Similarly, the producer alternate member candidate receiving the highest number of votes shall be designated as the first alternate member nominee, the second highest shall be designated as the second alternate member nominee until nominees for all member positions have been filled.

(iii) In the event that there are more producer member nominees than positions to be filled and not enough producer alternate member nominees to fill all positions, producer member nominees not nominated for a member seat may be nominated to fill vacant alternate member seats. Member seat nominees shall indicate, prior to the nomination vote, whether they are willing to accept nomination for an alternate seat in the event they are not nominated for a member seat and there are vacant alternate member seats. Member seat nominees that do not indicate willingness to be considered for vacant alternate member seats shall not be considered.

(iv) Each independent producer or producer affiliated with cooperative marketing association(s) handling less than 10 percent of the total raisin acquisitions during the preceding crop year shall cast only one vote with respect to each position for which nominations are to be made. Write-in candidates shall be accepted. The person receiving the most votes with respect to each position to be filled, in accordance with paragraph (b)(2)(ii) and (iii) of this section, shall be the person to be certified to the Secretary as the nominee. The Committee may, subject to the approval of the Secretary, establish rules and regulations to effectuate this section.

(3) One or more eligible handlers for each handler position to be filled may be proposed for nomination to represent independent handlers and cooperative marketing association(s) which acquired less than 10 percent of the total raisin acquisitions during the preceding crop year on the Committee. Nominations shall be made by and from handlers, or employees, representatives or agents of handlers falling within such groups. Each handler shall cast only one vote with respect to each position for which nomination is to be made. The person receiving the most votes with respect to each handler member of handler alternate member position shall be the person to be certified to the Secretary as the nominee for each such position.

(4) Each vote cast shall be on behalf of the person voting, the person's agent, subsidiaries, affiliates, and representatives. Voting at each handler meeting shall be in person. The results of each ballot at each handler meeting shall be announced at that meeting.

(5) Each nomination shall be certified by the Committee to the Secretary on or before April 5 immediately preceding the commencement of the term of office of the member or alternate member position for which the nomination is certified.

[48 FR 32975, July 20, 1983, as amended at 54 FR 34137, Aug. 18, 1989; 83 FR 53968, Oct. 26, 2018]

§ 989.30 Selection.

The Secretary shall select producer, handler, cooperative bargaining association(s), and public members and alternate members in the number specified in 989.26, as applicable, and with the qualifications specified in § 989.27. Such selections may be made from

nominations certified pursuant to § 989.29 or from other eligible producers, handlers, or cooperative bargaining association(s) officers or employees.

[48 FR 32975, July 20, 1983]

§ 989.31 Failure to nominate.

In the event nomination for a member or alternate member position on the Committee is not certified pursuant to and within the time specified in § 989.29, the Secretary may select an eligible person to fill such position without regard to nomination.

[41 FR 32412, Aug. 3, 1976]

§ 989.32 Acceptance.

Each person to be selected by the Secretary as a member or as an alternate member of the Committee shall, prior to such selection, qualify by advising the Secretary that he/she agrees to serve in the position for which nominated for selection.

[48 FR 32975, July 20, 1983]

§ 989.33 Alternate members.

The alternate for a member of the Committee shall act in the place and stead of such member (a) during his absence, and (b) in the event of his removal, resignation, disqualification, or death, until a successor for such member's unexpired term has been selected and has qualified.

[41 FR 32412, Aug. 3, 1976]

§ 989.34 Vacancies.

To fill any vacancy occasioned by the failure of any person selected as a member or as an alternate member of the Committee to qualify, or in the event of the removal, resignation, disqualification, or death of any member or alternate member, a successor for such person's unexpired term shall be nominated and selected in the manner set forth in §§ 989.29 and 989.30, insofar as such provisions are applicable. If nomination to fill any vacancy is not filed within 40 calendar days after such vacancy occurs, the Secretary may select an eligible person to fill such vacancy without regard to nomination.

[41 FR 32412, Aug. 3, 1976]

§989.35 Powers.

The Committee shall have the following powers:

(a) To administer the terms and provisions of this part;

(b) To make rules and regulations to effectuate the terms and provisions of this part;

(c) To recommend to the Secretary amendments to this part; and

(d) To receive, investigate, and report to the Secretary complaints of violations of this part.

[41 FR 32412, Aug. 3, 1976, as amended at 48 FR 32975, July 20, 1983]

§989.36 Duties.

The Committee shall have, among others, the following duties:

(a) To act as intermediary between the Secretary and any producer, packer, dehydrator, processor or cooperative bargaining association;

(b) To investigate compliance and to use means available to it to prevent violations of this part;

(c) To keep minutes, books, and other records, which shall clearly reflect all of its acts and transactions, and such minutes, books, and other records shall be subject to examination by the Secretary at any time;

(d) To investigate and assemble data on the production, handling and market conditions with respect to raisins;

(e) To submit to the Secretary such available information with respect to raisins and grapes as he may request, and such other information as the Committee may deem desirable and pertinent;

(f) To select from among its members a chairman and other officers, and to adopt such rules and regulations for the conduct of its business as it may deem advisable;

(g) To appoint or employ such other persons as it may deem necessary, and to determine the salaries and define the duties of each such person;

(h) To cause the books of the Committee to be audited by certified public accountants at least once each year, or at such other times as the Committee may deem necessary or as the Secretary may request, and the report of each such audit shall show, among other things, the receipts and expenditures of funds, and at least two copies of each such audit shall be submitted to the Secretary;

(i) To prepare quarterly statements of its financial operations and make such statements, together with the minutes of its meetings, available at the office of the Committee for inspection by producers, handlers and dehydrators;

(j) To give reasonable advance notice of the times, places, and purposes of its meetings by mail or other appropriate means to each member and alternate member and such notice shall be given as widespread publicity as is practicable;

(k) To conduct meetings for the purpose of making nominations for membership on the Committee and the certifying of nominations made for such purposes to the Secretary;

(l) To establish, with the approval of the Secretary, such rules and procedures relative to administration of this subpart as may be consistent with the provisions contained in this subpart and as may be necessary to accomplish the purposes of the act and the efficient administration of this subpart.

[41 FR 32412, Aug. 3, 1976, as amended at 48 FR 32975, July 20, 1983]

§989.37 Obligation.

Upon the removal, resignation, disqualification, or expiration of the term of office of any member or alternate member, such member or alternate member shall account for all receipts and disbursements and deliver to his successor, to the Committee, or to a designee of the Secretary all property (including, but not limited to, all books and records) in his possession or under his control as member or alternate member, and he shall execute such assignments and other instruments as may be necessary or appropriate to vest in such successor, Committee, or designee full title to such property and funds, and all claims vested in such member or alternate member. Upon the death of any member or alternate member of the Committee, full title to such property, funds, and claims vested in such member or alternate member shall be vested in his successor or, until such successor has been

577

selected and has qualified, in the Committee.

[41 FR 32412, Aug. 3, 1976]

§ 989.38   Procedure.

The Committee shall meet at the call of the chairman, or vice-chairman when acting as chairman, or at the call of any three members. All decisions of the Committee reached shall be by majority vote of the members present. All votes shall be cast in person and a quorum must be present. The presence of 25 members shall be required to constitute a quorum. The Committee shall give to the Secretary the same notice of meetings of the Committee as it gives to its members.

[48 FR 32976, July 20, 1983]

§ 989.39   Compensation and expenses.

The members and alternate members of the Committee shall serve without compensation, but shall be allowed their necessary expenses as approved by the Committee.

[54 FR 34137, Aug. 18, 1989]

RESEARCH AND DEVELOPMENT

§ 989.53   Research and development.

(a) *General.* The Committee, with the approval of the Secretary, may establish or provide for the establishment of projects involving production research, market research and development, marketing promotion including paid advertising, designed to assist, improve, or promote the production, marketing, distribution, and consumption of raisins in domestic and foreign markets. These projects may include, but need not be limited to those designed to:

(1) Improve through research the accuracy of raisin production estimates;

(2) Improve through research the preparation for market, sanitation, quality, condition, storability, processing, or packaging of raisins;

(3) Ascertain through research the factors affecting acceptance of raisins by manufacturers or consumers;

(4) Promote the marketing, distribution, or consumption of raisins in domestic and foreign markets by collecting data thereon, consulting with members of the trade, and making the information available to producers, handlers, and exporters; and

(5) Promote the marketing, distribution, or consumption of raisins in foreign markets through the use of merchandising programs.

(b) *Creditable expenditures.* The Committee, with the approval of the Secretary, may provide for crediting all or any portion of a handler's direct expenditures for marketing promotion, including paid advertising, that promotes the sale of raisins, raisin products, or their use. No handler shall receive credit for any allowable direct expenditures that would exceed the total of his assessment obligation which is attributable to that portion of his assessment designated for marketing promotion including paid advertising.

(c) *Criteria.* Before any project involving marketing promotion, including paid advertising, and the

crediting of the handler's pro rata expense assessment obligation of handlers is undertaken pursuant to this section, the Secretary after recommendation by the Committee, shall approve appropriate criteria to effectively regulate such activity.

[48 FR 32976, July 20, 1983, as amended at 83 FR 53969, Oct. 26, 2018]

MARKETING POLICY

§ 989.54   Marketing policy.

(a) *Marketing policy.* Each crop year, the Committee shall prepare and submit to the Secretary a report setting forth its recommended marketing policy, including quality regulations for the pending crop. In developing the marketing policy, the Committee may give consideration to the production, harvesting, processing, and storage conditions of that crop, as well as the following factors:

(1) The estimated tonnage held by producers and handlers at the beginning of the crop year;

(2) The expected general quality and any modifications of the minimum grade standards;

(3) The estimated tonnage of standard and off-grade raisins which will be produced;

(4) An estimated desirable carryout at the end of the crop year;

(5) The estimated market demand for raisins, considering the estimated world raisin supply and demand situation;

(6) Current prices being received and the probable general level of prices to be received for raisins by producers and handlers;

(7) The trend and level of consumer income;

(8) Any prohibition of trade practices, pursuant to §989.62 intended for the crop year; and

(9) Any other pertinent factors bearing on the marketing of raisins including the estimated supply of and demand for other varietal types and regulations applicable thereto.

(b) *Modification.* In the event the Committee subsequently deems it advisable to modify its marketing policy on any crop, because of national emergency, crop failure, or other major change in economic conditions, it shall hold a meeting for that purpose, and file a report thereof with the Secretary within 5 days (exclusive of Saturdays, Sundays, and holidays) after the holding of such meeting, which report shall show such modification and the basis therefor.

(c) *Publicity.* The Committee shall promptly give reasonable publicity to producers, dehydrators, handlers, and the cooperative bargaining association(s) of each meeting to consider a marketing policy or any modification thereof, and each such meeting shall be open to them. Similar publicity shall be given to producers, dehydrators, handlers, and the cooperative bargaining association(s) of each marketing policy report or modification thereof, filed with the Secretary and of the Secretary's action thereon. Copies of all marketing policy reports shall be maintained in the office of the Committee, where they shall be made available for examination by any producer, dehydrator, handler, or cooperative bargaining association representative. The Committee shall notify handlers, dehydrators and the cooperative bargaining association(s), and give reasonable publicity to producers of its computation.

[48 FR 32976, July 20, 1983, as amended at 50 FR 1831, Jan. 14, 1985; 54 FR 24670, June 9, 1989; 83 FR 53969, Oct. 26, 2018]

§§989.55—989.56 [Reserved]

GRADE, QUALITY, AND CONDITION STANDARDS

§989.58 Natural condition raisins.

(a) *Regulation.* No handler shall acquire or receive natural condition raisins which fail to meet such minimum grade, quality, and condition standards as the Committee may establish, with the approval of the Secretary, in applicable rules and regulations: *Provided,* That a handler may receive raisins for inspection, may receive off-grade raisins for reconditioning and may receive or acquire off-grade raisins for use in eligible non-normal outlets: *And provided further,* That a handler may acquire natural condition raisins which exceed the tolerance established for maturity under a weight dockage system established pursuant to rules and regulations recommended by the Committee and approved by the Secretary. Nothing contained in this paragraph shall apply to the acquisition or receipt of natural condition raisins of a particular varietal type for which minimum grade, quality, and condition standards are not applicable or then in effect pursuant to this part.

(b) *Changes in minimum grade, quality, and condition standards for natural condition raisins.* The Committee may recommend to the Secretary changes in the minimum grade, quality, and condition standards for natural condition raisins of any varietal type and may recommend to the Secretary that minimum grade, quality, and condition standards for any varietal type be added to or deleted. The Committee shall submit with its recommendation all data and information upon which it acted in making its recommendation, and such other information as the Secretary may request. The Secretary shall approve any such change if he finds, upon the basis of data submitted to him by the Committee or from other pertinent information available to him, that to do so would tend to effectuate the declared policy of the Act.

(c) *Publicity and notice.* The Committee shall give prompt and reasonable publicity to producer, dehydrators, and handlers of each recommendation submitted by it to the

Secretary and of each regulation issued by the Secretary. Notice of such regulation shall be given to all handlers registered or certified mail.

(d) *Inspection and certification.* (1) Each handler shall cause an inspection and certification to be made of all natural condition raisins acquired or received by him, except with respect to:

(i) An inter-plant or inter-handler transfer of off-grade raisins as described in paragraph (e)(2) of this section, unless such inspection and certification are required by rules and procedures made effective pursuant to this amended subpart;

(ii) An inter-plant or inter-handler transfer of standard raisins as described in § 989.59(e);

(iii) Raisins received from a dehydrator which have been previously inspected pursuant to paragraph (d)(2) of this section;

(iv) Any raisins for which minimum grade, quality, and condition standards are not then in effect;

(v) Raisins received from a cooperative bargaining association which have been inspected and are in compliance with requirements established pursuant to paragraph (d)(3) of this section; and

(vi) Any raisins, if permitted in accordance with such rules and procedures as the Committee may establish with the approval of the Secretary, acquired or received for disposition in eligible non-normal outlets. Except as otherwise provided in this section, prior to blending raisins, acquiring raisins, storing raisins, reconditioning raisins, or acquiring raisins which have been reconditioned, each handler shall obtain an inspection certification showing whether or not the raisins meet the applicable grade, quality, and condition standards: *Provided,* That the initial inspection for infestation shall not be required if the raisins are fumigated in accordance with such rules and procedures as the Committee shall establish with the approval of the Secretary. The handler shall submit or cause to be submitted to the Committee a copy of such certification, together with such other documents or records as the Committee may require. Such certification shall be issued by inspectors of the Specialty Crops Inspection Division of the U.S. Department of Agriculture, unless the Committee determines, and the Secretary concurs in such determination, that inspection by another agency would improve the administration of this amended subpart. The Committee may require that raisins held on memorandum receipt be re-inspected and certified as a condition for their acquisition by a handler.

(2) The Committee may, in accordance with rules and procedures established with the approval of the Secretary, authorize handlers to receive or acquire natural condition raisins which have been produced by any dehydrator by dehydrating grapes by artificial means and have been inspected and certified on his premises. In the event there shall have been compliance with Committee requirements, any handler who receives or acquires such inspected and certificated raisins shall be deemed to have satisfied the requirements contained in paragraph (d)(1) of this section with respect to inspection and certification of natural condition raisins received or acquired by him.

(3) The Committee may, in accordance with rules and the procedures established with the approval of the Secretary, authorize handlers to receive or acquire without further inspection and certification, natural condition raisins, standard or off-grade, which have been inspected, certified and held, in compliance with Committee requirements, at a receiving station of a cooperative bargaining association.

(e) *Off-grade raisins.* (1) Any natural condition raisins tendered to a handler which fail to meet the applicable minimum grade, quality, and condition standards may:

(i) Be received or acquired by the handler for disposition, without further inspection, in eligible non-normal outlets;

(ii) Be returned unstemmed to the person tendering the raisins; or

(iii) Be received by the handler for reconditioning. Off-grade raisins received by a handler under any one of the three described categories may be changed to any other of the categories under such rules and procedures as the Committee, with the approval of the Secretary, shall establish. No handler shall ship or otherwise dispose of off-grade raisins which he does not return to the tenderer, transfer to another handler as provided in paragraph (e)(2) of this section, or recondition so that they at least meet the minimum standards prescribed in or pursuant to this amended subpart, except into eligible non-normal outlets.

(2) Off-grade raisins may be transferred from the plant of the handler where received to another plant of his or to that of another handler within the State of California under such rules and procedures as the Committee, with the approval of the Secretary, shall establish to safeguard the objectives of this part.

(3) Each handler shall, while holding any off-grade raisins, store them separate and apart from other raisins and the off-grade raisins shall be stored in accordance with disposition and reconditioning categories. The Committee with the approval of the Secretary may prescribe rules and procedures for the storage of the raisins.

(4) If the handler is to acquire the raisins after they are reconditioned, his obligation with respect to such raisins shall be based on the weight of the raisins (if stemmed, adjusted to natural condition weight) after they have been reconditioned.

(5) The Committee shall establish, with the approval of the Secretary, such additional rules and procedures as may be necessary to insure adequate control of off-grade raisins, including, but not limited to, the reconditioning of off-grade raisins, the disposition and use of unsuccessfully reconditioned raisins, and the disposition and use of residual matter from reconditioning operations.

(f) *Blending.* No handler shall blend raisins except: (1) Incidental to reconditioning raisins as permitted under rules and procedures established by the Committee, with the approval of the Secretary; (2) blending standard raisins with standard raisins; or (3) blending raisins which meet the minimum grade standards for packed raisins with other raisins which meet such standards.

[25 FR 12813, Dec. 14, 1960, as amended at 29 FR 9483, July 11, 1964; 32 FR 12161, Aug. 24, 1967; 32 FR 18086, Dec. 19, 1967; 42 FR 37201, July 20, 1977; 83 FR 53969, 53973, Oct. 26, 2018]

§989.59 Regulation of the handling of raisins subsequent to their acquisition by handlers.

(a) *Regulation.* Unless otherwise provided in this part, no handler shall:

(1) Ship or otherwise make final disposition of natural condition raisins unless they at least meet the effective and applicable minimum grade, quality, and condition standards for natural condition raisins; or

(2) Ship or otherwise make final disposition of packed raisins unless they at least meet such minimum grade quality, and condition standards established by the Committee, with the approval of the Secretary, in applicable rules and regulations or as later changed or prescribed pursuant to the provisions of paragraph (b) of this section: *Provided,* That nothing contained in this paragraph shall prohibit the shipment or final disposition of any raisins of a particular varietal type for which minimum standards are not applicable or then in effect pursuant to this part. *And provided further,* That a handler may grind raisins, which do not meet the minimum grade, quality, and condition standards for packed raisins because of mechanical damage or sugaring, into a raisin paste. The Committee may establish, with approval of the Secretary, different grade, quality, and condition regulations for different markets.

(b) *Changes to minimum grade, quality, or condition standards.* The Committee may recommend changes in the minimum grade, quality, or condition standards for packed raisins of any varietal type and may recommend to the Secretary that minimum grade, quality, or condition standards for any varietal type be added or deleted. The Committee shall submit with its recommendation all data and information upon which it acted in making its recommendation, and such other information as the Secretary may request. The Secretary shall approve any such change if he finds, upon the basis of data submitted to him by the Committee or from other pertinent information available to him, that to do so would tend to effectuate the declared policy of the Act.

(c) *Publicity and notice.* The Committee shall give prompt and reasonable notice to producers, dehydrators, handlers, and the cooperative bargaining association(s) of each recommendation submitted by it to the Secretary and of each regulation issued by the Secretary. Notice of such regulation shall be given to all handlers of record by registered or certified mail.

(d) *Inspection and certification.* Unless otherwise provided in this section, each handler shall, at his own expense, before shipping or otherwise making final disposition of raisins, cause an inspection to be made of such raisins to determine whether they meet the then applicable minimum grade, quality, and condition standards for natural condition raisins or the then applicable minimum standards for packed raisins. Such handler shall obtain a certificate that such raisins meet the aforementioned applicable minimum standards and shall submit or cause to be submitted to the Committee a copy of such certificate together with such other documents or records as the Committee may require. The certificate shall be issued by the Specialty Crops Inspection Division of the United States Department of Agriculture, unless the Committee determines, and the Secretary concurs in such determination, that inspection by another agency will improve the administration of this amended subpart. Any certificate issued pursuant to this paragraph shall be valid only for such period of time as the Committee may specify, with the approval of the Secretary, in appropriate rules and regulations.

(e) *Inter-plant and inter-handler transfers.* Any handler may transfer from his plant to his own or another handler's plant within the State of California any raisins without having had such raisins inspected as provided in paragraph (d) of this section. The transferring handler shall transmit promptly to the Committee a report of such transfer, except that transfers between plants owned or operated by the same handler need not be reported. Before shipping or otherwise making final disposition of such raisins, the receiving handler shall comply with the requirements of this section.

(f) *Disposition of off-grade raisins, other failing raisins, and raisin residual material in eligible non-normal outlets.* Any off-grade raisins, except those returned unstemmed to the tenderer or successfully reconditioned, and any raisin residual material which may be received or acquired by a handler or accumulated by a handler from reconditioning raisins or from processing standard raisins and other failing raisins, shall be disposed of or marketed by the handler, without further inspection, in eligible non-normal outlets: *Provided,* That no packer shall be precluded from recovering raisins from such accumulations or acquisitions: *Provided further,* That whenever the Secretary concludes, on the basis of a recommendation of the Committee, that to specify one or more non-normal outlets as ineligible for any class of such receipts, acquisitions, or accumulations will tend to effectuate the declared policy of the act, he shall specify such ineligible outlets and prohibit the shipment thereto or final disposition therein of such class by handlers as well as the receipt and use thereof by processors: *And provided further,* That no processor who is a distiller shall be precluded from receiving or using for distillation (1) the standard raisins which subsequently fail to meet the said applicable standards, (2) the raisin residual material accumulated from processing standard raisins, or (3) the raisin residual material referable to the standard raisin equivalent recovered in reconditioning; and any handler may ship such raisins and raisin residual material to such processor. The Committee shall establish, with the approval of the Secretary, such rules and procedures as may be necessary to insure adequate control over the off-grade raisins, other failing raisins, and raisin residual material subject to this paragraph. Such rules may include a requirement that the disposition and use of all or any class of off-grade raisins, other failing raisins, or raisin residual material be confined to the area. The provisions of this paragraph are not intended to excuse any failure to comply with all applicable food and sanitary rules and regulations of city, county, State, Federal, or other agencies having jurisdiction.

(g) *Exemption of experimental and specialty packs.* The Committee may establish, with the approval of the Secretary, rules and procedures providing for the exemption of raisins in experimental and specialty packs from one or more of the requirements of the minimum grade, quality, or condition standards of this section, together with the inspection and certification requirements if applicable.

[25 FR 12813, Dec. 14, 1960, as amended at 32 FR 12161, Aug. 24, 1967; 37 FR 19622, Sept. 21, 1972; 42 FR 37202, July 20, 1977; 83 FR 53970, 53973, Oct. 26, 2018]

§ 989.60 Exemption.

(a) Notwithstanding any other provisions of this amended subpart, the Committee may establish, with the approval of the Secretary, such rules and procedures as may be necessary to permit the acquisition and disposition of any off-grade raisins, free from any or all regulations, for uses in non-normal outlets.

(b) The Committee may establish, with the approval of the Secretary, such rules and procedures as may be necessary to exempt from any or all regulations raisins produced in southern California (i.e., the counties of Riverside, Imperial, San Bernardino, Ventura, Orange, Los Angeles, and San Diego) and disposed of for distillation, livestock feed, or by export in natural condition to Mexico.

(c) The Committee may designate such raisins as it deems appropriate for production, processing, and marketing research and development. The period of such designation shall be for not more than five years unless extended by the Committee. The volume which may be acquired by all handlers shall not exceed 500 natural condition tons annually for each designated project, unless increased by the Secretary upon a recommendation of the Committee. Such designated raisins may be acquired and disposed of free from those regulations specified by the Committee. In any crop year, when the total industry acquisitions of the designated raisins exceed 500 natural condition tons or a larger quantity approved by the Secretary upon a rec-

ommendation of the Committee, the exemption shall not apply.

[29 FR 9484, July 11, 1964, as amended at 32 FR 18086, Dec. 19, 1967; 37 FR 19623, Sept. 21, 1972; 42 FR 37202, July 20, 1977; 83 FR 53971, Oct. 26, 2018]

§ 989.61 Above parity situations.

The provisions of this part relating to minimum grade, quality, and condition standards and inspection requirements, within the meaning of section 2(3) of the Act, and any other provisions pertaining to the administration and enforcement of the Order, shall continue in effect irrespective of whether the estimated season average price to producers for raisins is in excess of the parity level specified in section 2(1) of the Act.

[83 FR 53971, Oct. 26, 2018]

TRADE PRACTICES

§ 989.62 Authorization for prohibition of trade practices.

Whenever the Secretary finds, upon recommendation of the Committee or other information, that continuance of certain practices in trade channels would tend to interfere with the achieving of the objectives of this part, he may prohibit handlers from using such practices, for any crop year or portion thereof, in selling raisins in containers exceeding four pounds net weight. The prohibited practices may include:

(a) Any provision within or added to a sales contract, or action or agreement outside such contract, whereby the handler is obligated to reflect declines in market prices of raisins by charging the buyer a subsequent market price in lieu of the sales price specified in the contract.

(b)(1) Any agreement in an undertaking to hold raisins in reserve for possible future delivery to a buyer, or action or agreement outside such undertaking, whereby the handler is obligated to not reflect increases in market prices by charging the buyer a price specified in the agreement.

(2) Prior to any such practices being prohibited in any crop year, the Committee shall recommend, for the approval of the Secretary, such rules and

procedures and such record keeping requirements as are necessary to administer these prohibitions and obtain compliance therewith.

[25 FR 12813, Dec. 14, 1960; 27 FR 2506, Mar. 16, 1962, as amended at 84 FR 30863, June 28, 2019]

§§ 989.65—989.72 [Reserved]

REPORTS AND RECORDS

§ 989.73 Reports.

(a) *Inventory reports.* Each handler shall, upon request of the Committee, file promptly with the Committee a certified report, showing such information as the Committee shall specify with respect to any raisins which were held by him on a date designated by the Committee, which information as specified may include, but not be limited to:

(1) The quantity of any raisins so held, segregated as to varietal type, natural condition, packed, standard quality or off-grade quality; and

(2) The locations of the raisins.

(b) *Acquisition reports.* Each handler shall submit to the Committee in accordance with such rules and procedures as are prescribed by the Committee, with the approval of the Secretary, certified reports, for such periods as the Committee may require, with respect to his acquisitions of each varietal type of raisins during the particular period covered by such report, which report shall include, but not be limited to:

(1) The total quantity of standard raisins acquired;

(2) The total quantity of off-grade raisins acquired pursuant to § 989.58(e)(1)(i); and

(3) Cumulative totals of such acquisitions from the beginning of the then current crop year to and including the end of the period for which the report is made. Upon written application made to the Committee, a handler may be relieved of submitting such reports after completing his packing operations for the season. Upon request of the Committee, each handler shall furnish to the Committee, in such manner and at such times as it may require, the name and address of each person from whom he acquired raisins and the quantity of each varietal type of raisins acquired from each such person.

(c) Each handler shall file such reports of creditable promotion including paid advertising as recommended by the Committee and approved by the Secretary.

(d) *Other reports.* Upon the request of the Committee, with the approval of the Secretary, each handler shall furnish to the Committee such other information as may be necessary to enable it to exercise its powers and perform its duties under this amended part.

[25 FR 12813, Dec. 14, 1960, as amended at 32 FR 12164, Aug. 24, 1967; 32 FR 18086, Dec. 19, 1967; 48 FR 32978, July 20, 1983; 83 FR 53971, Oct. 26, 2018]

§ 989.75 Confidential information.

All reports and records furnished or submitted by a handler to the Committee shall be received by, and at all times kept under the custody or control of, one or more employees of the Committee, who shall disclose to no person, except the Secretary upon request therefor, data or information obtained or extracted therefrom which would constitute a trade secret or the disclosure of which might affect the trade position, financial condition, or business operations of the particular handler from whom received: *Provided,* That the Committee may require such an employee to disclose to it, or to any person designated by it or by the Secretary, information and data of a general nature, compilations of data affecting handlers as a group, and any data affecting one or more handlers, so long as the identity of the individual handlers involved is not disclosed.

§ 989.76 Records.

Each handler shall maintain such records of all raisins received, and of all raisins acquired, by him as prescribed by the Committee. Such records shall include, but not be limited to, the quantity of raisins of each varietal type acquired from each person and the name and address of each such person, total acquisitions, total sales, and total other disposition of each varietal type which he handles, and each handler shall maintain such records for at least two years after the termination of the crop year in which the transactions occurred. The Committee, with the approval of the Secretary, may prescribe rules and regulations to include under this section handler records that detail promotion and advertising activities which the Committee may need to perform its functions under §989.53.

[25 FR 12813, Dec. 14, 1960, as amended at 48 FR 32978, July 20, 1983]

§989.77 Verification of reports and records.

For the purpose of checking and verifying reports filed by handlers and records prescribed in or pursuant to this amended subpart, the Committee, through its duly authorized representatives, shall have access to any handler's premises during regular business hours and shall be permitted at any such times to inspect such premises and any raisins held by such handler, and any and all records of the handler with respect to the holding or disposition of raisins by him and promotion and advertising activities conducted by handlers under §989.53. Each handler shall furnish all labor and equipment necessary to make such inspections. Each handler shall store raisins in a manner which will facilitate inspection, and shall maintain storage records which will permit accurate identification of raisins held by him or theretofore disposed of. Insofar as is practicable and consistent with the carrying out of the provisions of this amended subpart, all data and information obtained or received through checking and verification of reports and records shall be treated as confidential information.

[25 FR 12813, Dec. 14, 1960, as amended at 48 FR 32978, July 20, 1983]

EXPENSES AND ASSESSMENTS

§989.79 Expenses.

The Committee is authorized to incur such expenses as the Secretary finds are reasonable and likely to be incurred by it during each crop year, for the maintenance and functioning of the Committee and for such purposes as he may, pursuant to this subpart, determine to be appropriate. The funds to cover such expenses shall be obtained levying assessments as provided in §989.80. The Committee shall file with the Secretary for each crop year a proposed budget of these expenses and a proposal as to the assessment rate to be fixed pursuant to §989.80, together with a report thereon. Such filing shall be not later than October 5 of the crop year, but this date may be extended by the Committee not more than 5 days if warranted by a late crop.

[83 FR 53971, Oct. 26, 2018]

§989.80 Assessments.

(a) Each handler shall pay to the Committee, upon demand, his pro rata share of the expenses which the Secretary finds will be incurred, as aforesaid, by the Committee during each crop year less any amounts credited pursuant to §989.53. Such handler's pro rata share of such expenses shall be equal to the ratio between the total raisin tonnage acquired by such handler during the applicable crop year and the total raisin tonnage acquired by all handlers during the same crop year.

(b) Each handler who reconditions off-grade raisins but does not acquire the standard raisins recovered therefrom shall, with respect to his assessable portion of all such standard raisins, pay to the Committee, upon demand, his pro rata share of the expenses which the Secretary finds will be incurred by the Committee each crop year. Such handler's pro rata share of such expenses shall be equal to the ratio between the handler's assessable portion (which shall be a quantity equal to such handler's standard raisins which are acquired by some other handler or handlers) during the applicable crop year and the total raisin tonnage acquired by all handlers.

(c) The Secretary shall fix the rate of assessment to be paid by all handlers on the basis of a specified rate per ton. At any time during or after a crop year, the Secretary may increase the rate of assessment to obtain sufficient funds to cover any later finding by the Secretary relative to the expenses of the Committee. Each handler shall pay such additional assessment to the Committee upon demand. In order to provide funds to carry out the functions of the Committee, the Committee may accept advance payments from any handler to be credited toward such assessments as may be levied pursuant to this section against such handler during the crop year. The payment of assessments for the maintenance and functioning of the Committee, and for such purposes as the Secretary may pursuant to this subpart determine to be appropriate, may be required under this part throughout the period it is in effect, irrespective of whether particular provisions thereof are suspended or become inoperative.

(d) Each handler shall, with respect to administrative assessments not paid within 30 calendar days of the date of the Committee's invoice, pay to the Committee interest on the unpaid assessment at the rate of the prime rate established by the bank in which the Committee has its administrative assessment funds deposited, on the day that the administrative assessment becomes delinquent plus 2 percent; and further, that such rate of interest be added to the bill monthly until the delinquent handler's assessment plus applicable interest has been paid: *Provided,* That the Committee may, with the approval of the Secretary, modify the interest rate applicable to delinquent handler's assessment through the establishment of applicable rules and regulations.

[29 FR 9484, July 11, 1964, as amended at 32 FR 12165, Aug. 24, 1967; 42 FR 37202, July 20, 1977; 48 FR 32978, July 20, 1983; 81 FR 44764, July 11, 2016; 83 FR 53971, Oct. 26, 2018]

§ 989.81 Accounting.

(a) If, at the end of the crop year, the assessments collected are in excess of expenses incurred, such excess shall be accounted for in accordance with one of the following:

(1) If such excess is not retained in a reserve, as provided in paragraph (a)(2) of this section, it shall be refunded proportionately to the persons from whom collected in accordance with § 989.80; *Provided,* That any sum paid by a person in excess of his or her pro rata share of expenses during any crop year may be applied by the Committee at the end of such crop year as credit for such person, toward the Committee's administrative operations for the following crop year; *Provided further,* That

the Committee may credit the excess to any outstanding obligations due the Committee from such person.

(2) The Committee may carry over such excess funds into subsequent crop years as a reserve; *Provided,* That funds already in the reserve do not exceed one crop year's budgeted expenses as averaged over the past six years. In the event that funds exceed one crop year's expenses, funds in excess of one crop year's budgeted expenses shall be distributed in accordance with paragraph (a)(1) of this section. Such funds may be used:

(i) To defray essential administrative expenses (*i.e.,* staff wages/salaries and related benefits, office rent, utilities, postage, insurance, legal expenses, audit costs, consulting, Web site operation and maintenance, office supplies, repairs and maintenance, equipment leases, domestic staff travel and Committee mileage reimbursement, international Committee travel, international staff travel, bank charges, computer software and programming, costs of compliance activities, and other similar essential administrative expenses) exclusive of promotional expenses during any crop year, prior to the time assessment income is sufficient to cover such expenses;

(ii) To cover deficits incurred during any period when assessment income is less than expenses;

(iii) To defray expenses incurred during any period when any or all provisions of this part are suspended;

(iv) To meet any other such expenses recommended by the Committee and approved by the Secretary; and

(v) To cover the necessary expenses of liquidation in the event of termination of this part. Upon such termination, any funds not required to defray the necessary expenses of liquidation shall be disposed of in such manner as the Secretary may determine to be appropriate; Provided, That to the extent practicable, such funds shall be returned pro rata to the persons from whom such funds were collected.

(b) The Committee may, with the approval of the Secretary, maintain in its own name or in the name of its members, a suit against any handler for the collection of such handler's pro rata share of the expenses.

[25 FR 12813, Dec. 14, 1960; 27 FR 2506, Mar. 16, 1962, as amended at 81 FR 44764, July 11, 2016]

§989.82 [Reserved]

§989.83 Funds.

All funds received by the Committee pursuant to the provisions of this part, shall be used solely for the purposes authorized, and shall be accounted for in the manner provided, in this part. The Secretary may, at any time, require the Committee and its members and alternate members to account for all receipts and disbursements.

MISCELLANEOUS PROVISIONS

§989.84 Disposition limitation.

No handler shall dispose of standard raisins, off-grade raisins, or other failing raisins, except in accordance with the provisions of this subpart or pursuant to regulations issued by the Committee.

[83 FR 53972, Oct. 26, 2018]

§989.85 Personal liability.

No member or alternate member of the Committee or any employee or agent thereof shall be held personally responsible, either individually or jointly with others, in any way whatsoever, to any handler or any person, for errors in judgment, mistakes, or other acts either of commission or omission, as such member, alternate member, employee, or agent, except for acts of dishonesty.

[41 FR 32417, Aug. 3, 1976, as amended at 48 FR 32978, July 20, 1983]

§989.86 Separability.

If any provision of this amended subpart is declared invalid, or the applicability thereof to any person, circumstance, or thing is held invalid, the validity of the remainder of this amended subpart or the applicability thereof to any other person, circumstance, or thing shall not be affected thereby.

§989.87 Derogation.

Nothing contained in this amended subpart is, or shall be construed to be, in derogation or in modification of the rights of the Secretary or of the United States to exercise any powers granted by the act or otherwise, or, in accordance with such powers, to act in the premises whenever such action is deemed advisable.

§989.88 Duration of immunities.

The benefits, privileges, and immunities conferred upon any person by virtue of this amended subpart shall cease upon the termination of this amended subpart, except with respect to acts done under and during the existence of this subpart.

§989.89 Agents.

The Secretary may, by a designation in writing, name any person, including any officer or employee of the United States Government, or name any bureau or division in the United States Department of Agriculture, to act as his agent or representative in connection with any of the provisions of this amended subpart.

§989.90 Effective time.

The provisions of this amended subpart, as well as any amendments to this amended subpart shall become effective at such time as the Secretary may declare, and shall continue in force until terminated, or during suspension, in one of the ways specified in §989.91.

§989.91 Suspension or termination.

(a) The Secretary may, at any time, terminate the provisions of this amended subpart by giving at least one day's notice by means of a press release or in any other manner which he may determine.

(b) The Secretary shall terminate or suspend the operation of any or all of the provisions of this amended subpart, whenever he finds that such provisions do not tend to effectuate the declared policy of the act.

(c) No less than five crop years and no later than six crop years after the effective date of this amendment, the Secretary shall conduct a referendum to ascertain whether continuance of this part is favored by producers. Subsequent referenda to ascertain continuance shall be conducted every six crop years thereafter. The Secretary may terminate the provisions of this part at the end of any crop year in which the Secretary has found that continuance of this part is not favored by a two-thirds majority of voting producers, or a two-thirds majority of volume represented thereby, who, during a representative period determined by the Secretary, have been engaged in the production for market of grapes used in the production of raisins in the State of California. Such termination shall be announced on or before the end of the crop year.

(d) The Secretary shall terminate the provisions of this amended subpart at the end of any crop year whenever he finds that such termination is favored by a majority of the producers who, during a representative period determined by the Secretary, have been engaged in the production for market of grapes used in the production of raisins in the State of California: *Provided,* That such majority have, during such representative period, produced for market more than 50 percent of the volume of such grapes produced for market within said State; but such termination shall be effective only if announced before July 31 of the then current crop year.

(e) The provisions of this amended subpart shall, in any event, terminate whenever the provisions of the act authorizing them cease to be in effect.

[25 FR 12813, Dec. 14, 1960, as amended at 41 FR 32417, Aug. 3, 1976; 83 FR 53972, Oct. 26, 2018]

§ 989.92 Proceedings after termination.

(a) Upon the termination of the provisions of this amended subpart, the members of the Committee then functioning shall continue as joint trustees for the purpose of liquidating the affairs of the Committee, of all funds and property then in the possession or under the control of the Committee, including claims for any funds unpaid or property not delivered at the time of such termination. Action by said trusteeship shall require the concurrence of a majority of the said trustees.

(b) Said trustees shall continue in such capacity until discharged by the Secretary; shall, from time to time, account for all receipts and disburse-ments and deliver all property on hand, together with all books and records of the Committee and the joint trustees, to such person as the Secretary may direct; and shall, upon the request of the Secretary, execute such assignments or other instruments necessary or appropriate to vest in such person full title and right to all of the funds, property, and claims vested in the Committee or the joint trustees pursuant to this subpart.

(c) Any person to whom funds, property or claims have been transferred or delivered by the Committee or its members, pursuant to this section, shall be subject to the same obligations imposed upon the members of the said Committee and upon said joint trustees.

§ 989.93 Effect of termination or amendment.

Unless otherwise expressly provided by the Secretary, the termination of this amended subpart or any regulation issued pursuant to this amended subpart, or the issuance of any amendment to either thereof, shall not (a) affect or waive any right, duty, obligation, or liability which shall have arisen or which may thereafter arise in connection with any provision of this amended subpart or any regulation issued under this amended subpart, (b) release or extinguish any violation of this amended subpart, or of any regulation issued under this amended subpart, or (c) affect or impair any rights or remedies of the Secretary or of any other person, with respect to any such violation.

§ 989.94 Amendments.

Amendments to this amended subpart may be proposed from time to time, by any person or by the Committee.

§ 989.95 Right of Secretary.

The members of the Committee (including alternates and successors) and any agent or employee appointed or employed by the Committee, shall be subject to removal or suspension by the Secretary, in his discretion, at any time. Every decision, determination, or other act of the Committee shall be subject to the continuing right of the

Secretary to disapprove of the same at any time. Upon such disapproval, the disapproved action of the Committee shall be deemed null and void.

[41 FR 32417, Aug. 3, 1976, as amended at 48 FR 32978, July 20, 1983]

§ 989.96 Storage of raisins held on memorandum receipt and of packer-owned tonnage.

All raisins stored by a handler for another person on memorandum or warehouse receipt, or raisins produced and stored by a handler, shall be stored separate and apart from other raisins and shall be clearly marked or tagged as raisins stored on memorandum or warehouse receipt or as raisins produced by the handler but not acquired by him in his capacity as a handler.

[25 FR 12813, Dec. 14, 1960. Redesignated at 83 FR 53971, Oct. 26, 2018]

Subpart B—Administrative Requirements

SOURCE: 27 FR 3112, Mar. 31, 1962, unless otherwise noted.

DEFINITIONS

§ 989.102 Inspection service.

Inspection service means the Specialty Crops Inspection Division, Agricultural Marketing Service of the United States Department of Agriculture.

[49 FR 18730, May 2, 1984, as amended at 83 FR 53973, Oct. 26, 2018]

§ 989.104 Lot.

(a) *Natural condition raisins—(1) Basic definition.* For the purpose of incoming and outgoing inspection of natural condition raisins, *lot* means, except as otherwise provided in this paragraph, the quantity of such raisins of the same varietal type or of differing varietal types when commingled within their containers (including sweat and picking boxes and bins), which does not exceed a car, truck, or truck-trailer load, and which is submitted for inspection at one time and in the same place.

(2) *Separation of large units.* If a quantity of raisins in excess of a car, truck, or truck-trailer load is submitted for inspection, the total quantity may, at the discretion of the inspector, be separated into such readily identifiable portions, either prior to or in the course of inspection, as can be conveniently and properly inspected, and each such portion shall constitute a lot.

(3) *Resubmission after reconditioning.* Raisins which are submitted for inspection after reconditioning (such as sorting or drying) and whose original lot identity is no longer applicable, shall be a new lot.

(4) *Meeting and failing portions.* Where a portion of a quantity of raisins submitted for inspection meets the minimum grade and condition standards and has been separated from the remainder of the raisins failing to meet such standards:

(i) The meeting portion shall be one lot; and

(ii) The remainder shall be one or more lots as necessary to cause each lot to contain either (a) a single defect in excess of tolerance or (b) two or more of the same defects in excess of tolerance occurring together within each of the individual containers.

(5) *Entire quantity failing.* Where the entire quantity of raisins submitted for inspection fails to meet such standards, then, whether such quantity shall be one or more lots shall be determined in the same manner as for the failing remainder referred to in paragraph (a)(4) of this section.

(6) *Special condition.* Notwithstanding other provisions of this section, any quantity of raisins failing to meet such standards and which are not to be reconditioned may be a single lot.

(b) *Packed raisins.* For the purpose of outgoing inspection of packed raisins, lot means: (1) For in-line inspection (i.e., where samples are drawn from a flow of raisins prior to packaging), the aggregate quantity of raisins of the same varietal type, subtype, or size (or in their mixed form), processed in any continuous production of one calendar day and packaged in one size and style of package but excluding those rejected by inspection; and (2) for floor inspection (i.e., where samples are drawn from containers of raisins), the aggregate quantity of such raisins in like containers but not necessarily processed in one continuous production or during one calendar day, identifiable and offered for inspection as a lot.

§ 989.105 Inspection point.

Inspection point means any plant or receiving station of a handler, or any other place where raisins are received by a handler, and which is so designated by the Committee. The inspection point(s) of the handler shall include any area(s) in which he receives grapes or raisins for dehydration unless he keeps his raisin dehydration business separate, physically and by records, from his business of handling raisins.

[31 FR 16305, Dec. 21, 1966]

§ 989.106 Ship.

Ship means the physical movement of raisins other than to storage for the handler's account within the general locality of the packing plant.

§ 989.107 Inspection certificate.

Inspection certificate means any written certification, finding, or attestation as to the quality or condition of any lot or lots issued by an authorized member of the inspection service.

§ 989.110 Varietal types.

Pursuant to § 989.10, specific definitions for each varietal type of raisins contained in that section are as follows:

(a) Natural (sun-dried) Seedless includes all sun-dried seedless raisins possessing similar identifiable characteristics as raisins produced from Thompson Seedless grapes or similar grape varieties, whether dried on trays or on the vine, with or without the application of a drying agent that is a food-grade additive such as, soda, oil, Ethyl Oleate, or Methyl Oleate prior to, during, or after the drying process.

(b) Dipped Seedless includes all raisins produced by artificial dehydration of seedless grapes that possess the characteristics similar to Thompson Seedless grapes which, in order to expedite drying, have been dipped in or sprayed with water only after such grapes have been removed from the .vine.

(c) Golden Seedless includes all seedless raisins whose color generally varies from golden yellow to dark amber.

(d) Muscats (including other raisins with seeds) include all raisins which usually contain seeds and possess characteristics similar to Muscat raisins.

(e) Sultana includes all raisins which usually contain an undeveloped (vestigial) seed and possess characteristics similar to Sultana raisins.

(f) Zante Currant includes all raisins that possess characteristics similar to those produced from Black Corinth or White Corinth grapes.

(g) Monukka includes all raisins produced from Monukka grapes.

(h) Other Seedless includes all raisins produced from Ruby Seedless, Kings Ruby Seedless, Flame Seedless and other seedless grapes not included in any of the varietal categories for Seedless raisins defined in paragraphs (a), (b), (c), (d) or (h) above.

(i) Other Seedless-Sulfured includes all raisins produced from Ruby Seedless, Kings Ruby Seedless, Flame Seedless and other seedless grapes not included in any of the varietal categories for Seedless raisins defined in paragraphs (a), (b), (c), (d), (h), or (i) of this section which have been artificially dehydrated and sulfured.

[49 FR 18730, May 2, 1984, as amended at 53 FR 34714, Sept. 8, 1988; 55 FR 32598, Aug. 10, 1990; 67 FR 36792, May 28, 2002; 68 FR 42947, July 21, 2003]

§ 989.111 Independent producer and small cooperative producer.

(a) *Independent producer* means any producer who is not a member of a cooperative bargaining association or a cooperative marketing association, nor has sold for cash to a cooperative marketing association.

(b) *Small cooperative producer* means any producer who is a member of a cooperative marketing association which acquired less than 10 percent of total raisin acquisitions during the crop year preceding the year in which nominations are held.

[49 FR 18730, May 2, 1984]

§ 989.115 Independent handler, major cooperative marketing association handler, and small cooperative marketing association handler.

(a) *Independent handler* means any handler who is not a cooperative marketing association of producers.

(b) *Major cooperative marketing association handler* means any handler who is a cooperative marketing association of producers which acquired not less than 10 percent of the total raisin acquisitions during the crop year preceding nominations.

(c) *Small cooperative marketing association handler* means any handler who is a cooperative marketing association of producers which acquired less than 10 percent of the total raisin acquisitions during the crop year preceding nominations.

[49 FR 18730, May 2, 1984]

RAISIN ADMINISTRATIVE COMMITTEE

§ 989.122   Districts for independent and small cooperative producer representation on the Committee.

For the purposes of § 989.26(c) and commencing with the term of office beginning May 1, 1984, independent and small cooperative producer districts are as follows:

(a) *District No. 1.* All of the counties north of Fresno County.

(b) *District No. 2.* All of the counties south of Fresno County.

(c) *District No. 3* All of Fresno County.

[49 FR 18730, May 2, 1984]

§ 989.126   Representation of the Committee.

(a) To provide independent and small cooperative producers equitable representation throughout the production area commencing with the term of office beginning May 1, 1984, representation shall be apportioned among the three districts specified in § 989.122. Districts 1 and 2 shall each have one producer member, and District 3 shall have the remaining producer members to which independent and small cooperative producers are entitled pursuant to § 989.26(c).

(b) Pursuant to section 989.26(d) and commencing with the term of office beginning May 1, 1994, apportionment of the independent and small cooperative marketing association handlers shall be:

(1) Two members selected from and representing the four handler(s) other than major cooperative marketing association handler(s) who acquired the largest percentage of the total raisin acquisitions during the preceding crop year;

(2) Three members selected from and representing the six handlers other than major cooperative marketing association handler(s) who acquired the next largest percentage of the total raisin acquisitions during the preceding crop year; and

(3) The remaining member(s) selected from and representing all other handlers, including small cooperative marketing association handler(s) and all processors.

[49 FR 18730, May 2, 1984, as amended at 59 FR 27226, May 26, 1994]

§ 989.129   Voting at nomination meetings.

Any person (defined in § 989.3 as an individual, partnership, corporation, association, or any other business unit) who is engaged, in a proprietary capacity, in the production of grapes which are sun-dried or dehydrated by artificial means to produce raisins and who qualifies under the provisions of § 989.29(b)(2) shall be eligible to cast one ballot for a nominee for each producer member position and one ballot for a nominee for each producer alternate member position on the committee which is to be filled for his district. Such person must be the one who or which: Owns and farms land resulting in his or its ownership of such grapes produced thereon; rents and farms land, resulting in his or its ownership of all or a portion of such grapes produced thereon; or owns land which he or it does not farm and, as rental for such land, obtains the ownership of a portion of such grapes or the raisins. In this connection, a partnership shall be deemed to include two or more persons (including a husband and wife) with respect to land the title to which, or leasehold interest in which, is vested in them as tenants in common, joint tenants, or under community property laws, as community property. In a landlord-tenant relationship, wherein each of the parties is a producer, each such producer shall be entitled to one vote for a nominee for each producer member position and one vote for each producer alternate member position. Hence, where two persons operate land as landlord and tenant on a share-crop basis, each person is entitled to one vote for each such position to be filled.

Where land is leased on a cash rental basis, only the person who is the tenant or cash renter (producer) is entitled to vote. A partnership or corporation, when eligible, is entitled to cast only one vote for a nominee for each producer position to be filled in its district.

[83 FR 53972, Oct. 26, 2018]

§ 989.139 Compensation for attendance of alternates at Committee meetings.

Whenever a member of the Raisin Administrative Committee has reason to believe that he will be unable to attend a Committee meeting and has so notified his alternate or the Committee manager, such notification or a request from the manager shall be held to be a request for the alternate to attend and he shall be reimbursed for reasonable expenses subject to the limitations contained in § 989.39.

[42 FR 52376, Sept. 30, 1977, as amended at 49 FR 18731, May 2, 1984]

§§ 989.154—989.156 [Reserved]

QUALITY CONTROL

§ 989.157 Raisins produced from grapes grown outside of California.

(a) Any raisins produced from grapes grown outside the State of California that are received by a handler shall be observed and marked for identification by an inspector. As provided in § 989.173(b)(7), the inspection service may request information needed to properly mark such raisins for identification; it shall be the handler's responsibility to arrange for such identification and furnish required documentation promptly.

(b) In the absence of an inspector to observe and mark such raisins for identification, the handler shall not permit the unloading to occur unless the handler has a written statement from the inspection service that an inspector cannot be furnished within a reasonable time: *Provided*, That raisins so unloaded shall be observed and marked properly upon an inspector being available.

(c) The handler shall notify the inspection service in writing at least one business day in advance of the time

such handler plans to begin receiving raisins produced from grapes grown outside the State of California, unless a shorter period is acceptable to the inspection service.

(d) Raisins produced from grapes grown outside of the State of California and received by a handler shall be marked for identification by the inspector affixing to one container on each pallet or to each bin in each lot a prenumbered RAC control card (to be furnished by the Committee) which shall remain affixed until the raisins are processed and disposed of or disposed of as natural condition raisins. The cards shall be removed only by an inspector of the inspection service or authorized Committee personnel.

(e) Each handler shall store raisins produced from grapes grown outside the State of California separate and apart from all other raisins held by such handler to the satisfaction of the Committee. Storage of such raisins shall be deemed "separate and apart" if the containers are marked as raisins produced from grapes grown outside the State of California and placed so as to be readily and clearly identified.

(f) Any raisins received by a handler produced from grapes grown outside the State of California shall be processed and/or disposed of under the surveillance of the inspection service. The handler shall notify the inspection service in writing at least one business day in advance of the time such processing and/or disposition will occur, unless a shorter period is acceptable to the inspection service.

(g) The handler receiving raisins produced from grapes grown outside of California shall pay fees assessed by the inspection service to identify and maintain surveillance of such raisins.

[55 FR 28019, July 9, 1990]

§ 989.158 Natural condition raisins.

(a) *Incoming inspection.* (1)(i) The Committee shall, upon request of a handler who complies with the requirements of this part with respect to inspection points, designate as his inspection point any place (including his plant or receiving station) where the handler receives raisins.

(ii) Each handler shall, at his expense, provide at each of his inspection points reasonably safe and adequate facilities for receiving raisins, drawing samples, and efficient inspection of natural condition raisins. At the time of inspection of any lot, the handler shall, at his expense, provide the inspector with any assistance necessary in the inspection of the raisins, including the movement of individual containers. Each handler, other than a processor, shall maintain with the Committee a current written description, defining the boundaries and other pertinent details, of each of his inspection points. In the event the Committee determines that any inspection point, or any modification thereof, does not comply with the definition or the requirements of this part, it shall notify the handler of the changes necessary for compliance. The handler shall make such changes promptly. In the event any of his inspection points is the same as that of another handler or person receiving raisins or grapes in any form, the handler shall maintain his raisins separate and apart from any other raisins.

(iii) The weight of each lot of raisins tendered for receiving, storage, reconditioning, acquisition, or disposition shall be substantiated by an official "State Certificate of Weights and Measures" issued by a public weighmaster, whether located at the inspection point or otherwise, or such other document approved by the Committee which accurately reflects the weight of each lot tendered. The net weight of such raisins for the purposes of this part, shall be determined by deducting the sand tare and box tare from the gross weight of the raisins. The sand tare shall be the weight of the sand and other foreign material removed from the raisins by passing the raisins over a screen (of a type commonly used by the industry for such purpose) having 36 square openings to the square inch, with each opening being one-eighth of an inch square.

(2) No handler, other than a processor, shall receive at points other than at an inspection point, natural condition raisins from a tenderer, either for acquisition, storage, reconditioning, inspection, or for disposition in eligible non-normal outlets: *Provided*, That this requirement shall not preclude a handler from dehydrating, free from the provisions of this part, at separate dehydrating facilities recognized in §989.105 and located in California, raisins not delivered to an inspection point. Any handler who accepts raisins at an inspection point for drying or other reconditioning shall be deemed to have received the raisins for reconditioning and shall be subject to the provisions of this part with respect to such raisins.

(3) For each lot of natural condition raisins received by a handler for acquisition, reconditioning, storage, inspection, or for disposition in eligible non-normal outlets, the handler shall, immediately upon physical receipt and tentative acceptance thereof, issue a prenumbered (numbered serially in advance) door receipt or weight certificate showing the name and address of the tenderer, the weight of the lot, the number and type of containers in the lot, and any other information necessary to identify the lot. For the purposes of identifying incoming lots of raisins, other than dehydrated raisins covered by paragraph (e) of this section, a handler, if it is impracticable for him to issue immediately a door receipt or weight certificate, may issue for temporary use only a prenumbered "Request for USDA Inspection" on a form furnished by the Committee. Any such raisins so received by a handler shall, prior to their acceptance, be inspected at an inspection point during the unloading process, and if certified as standard raisins shall be, unless returned to the tenderer, either promptly acquired by the handler or received for storage on memorandum receipt: *Provided*, That in the absence of an inspector to perform inspection during unloading, the handler shall not permit unloading to occur unless such absence is during normal business hours and the handler has a written statement from the inspection service to the effect that inspection cannot be furnished within a reasonable time: *And provided further*, That the raisins so unloaded shall be inspected promptly upon an inspector being available. It shall be the handler's responsibility in any case to arrange for the inspection, other than with respect to dehydrated raisins covered by paragraph (e) of this

section, and furnish weight certificates promptly. Any raisins received by a handler as off-grade for disposition in eligible non-normal outlets or for reconditioning may be accepted uninspected: *Provided,* That an application for receiving such uninspected raisins shall be submitted by the handler, on a form furnished by the Committee, to the Inspection Service prior to, or upon physical receipt of, such off-grade raisins. Such form shall provide for at least the name and address of the tenderer (equity holder), date, number, and type of containers, net weight of the raisins, and the particular defect(s) the handler indicates would cause the raisins to be off-grade. Handlers shall complete and sign the form. The application for such uninspected raisins shall not be acceptable unless signed by the tenderer. The uninspected raisins shall be subject to surveillance by the Inspection Service. Each lot of raisins accepted by a handler for reconditioning shall be reconditioned separately from any other lot.

(4) If any lot of natural condition raisins tendered to a handler is separated into two or more lots because a portion of the original lot failed to meet minimum grade and condition standards, or because the entire lot failed due to more than one defect, the handler shall issue a prenumbered weight certificate for each such new lot not returned to the tenderer, showing the name and address of the tenderer, the weight of the lot, and the number and type of containers in the lot. The weight of any meeting lot shall be determined by weighing it, or by weighing the failing portion of the incoming lot and deducting the weight thereof from the weight of the incoming lot. The weight of each failing lot shall be determined by weighing it, or by deriving such weight by applying the original average container weight to the number of containers.

(5) Any financially interested party may, upon the payment of any fees assessed by the inspection service, obtain an appeal inspection. An appeal inspection shall be applicable only to raisins which have not been removed from their containers, with pallet control cards still affixed, are readily identifiable, and have not been removed from

the original inspection point: *Provided,* That when the condition of a lot of such raisins may have changed subsequent to the original inspection, an additional inspection, rather than an appeal inspection, may be obtained.

(6) Raisins produced by a handler shall be subject to the requirements of paragraph (a) (3) and (4) of this section upon delivery to an inspection point. Raisins produced by a handler by dehydration within an inspection point shall be subject to the requirements of paragraph (a) (3) and (4) of this section immediately upon completion of said dehydration.

(7) The inspection certificate for a mixed lot of natural condition raisins (raisins of different varietal types commingled within their containers) shall show the percentage which the raisins of each varietal type is of the total raisins contained in the lot.

(8) With respect to any lot of natural condition raisins being received and inspected at a handler's inspection point pursuant to paragraph (a)(3) of this section, the handler shall notify the inspection service if he elects to have the raisins inspected for infestation. If the handler elects not to have the raisins inspected for infestation, he shall: (i) Fumigate promptly all raisins he receives; (ii) notify the inspection service in advance of the time he plans to fumigate such raisins; (iii) permit the inspection service to monitor the fumigation; and (iv) permit the inspection service to make periodic incubation checks of his packed raisins. The inspection service shall certify the raisins received as standard raisins if they meet all other grade and condition standards. If the handler elects to have the raisins inspected for infestation, the inspector shall afford such handler the opportunity to fumigate such raisins during the inspection and certification process. Such raisins shall remain under the supervision of the inspector during the fumigation. The inspection certificate shall not be issued until the fumigation is completed: *Provided,* That the inspection certificate shall be issued, whether or not the fumigation is completed, not later than five business days after the date the inspection and certification process is

suspended by the inspector to permit fumigation.

(9) With respect to any lot (as defined in § 989.104(a)(1)) of natural condition raisins being received and inspected at a handler's inspection point pursuant to paragraph (a)(3) of this section and notwithstanding separation of the meeting portion of the original lot from the failing portions thereof for the purposes on § 989.104 and paragraph (a)(4) of this section, any tenderer may, when permitted by the handler and when notified by the inspector of defects during the inspection and certification process, and in accordance with the provisions of this subparagraph, perform any one or more of the following on an individual box basis: (i) Mix raisins within boxes containing raisins that are wet, or of high moisture content in some areas of the box; (ii) dump raisins from wet boxes into dry boxes; (iii) remove wet raisins; or (iv) remove foreign material such as sandburs, puncture vine seed, Eucalyptus pods or leaves, rocks, and sticks. This authorization to the tenderer shall not extend to raisins in containers larger than sweat boxes; and the number of boxes in the original lot on which the aforesaid actions may be performed during such process shall not exceed ten, or five percent of the total number of containers in the lot, whichever is less. Where the percentage computation results in a fraction of a box and is less than ten boxes, it shall be rounded upward to the next number. The entire lot of raisins shall remain under surveillance of the inspector during such process. The actions of the tenderer shall be done without delay, take place at the unloading dock in the inspection point, or in the immediate area thereof, and be under observation of the inspector.

(b) *Submission of inspection certificates to the Committee.* A copy of each inspection certificate which a handler is required to submit to the Committee pursuant to § 989.58(d) shall be submitted not later than Wednesday of the week following the week for which such certificate was issued. This may be accomplished by authorizing the inspection service to submit a copy of each such inspection certificate directly to the Committee. A copy of such authorization shall be furnished to the Committee.

(c) *Off-grade raisins*—(1) *Holding and identification.* The inspection certificates covering any lot of off-grade raisins shall state whether or not such off-grade raisins are storable. Any raisins which do not meet the applicable grade and condition standards shall be classified in one of the three categories specified in § 989.58(e)(1) within 5 business days (excluding Saturdays, Sundays, and holidays) after inspection or 3 such business days after issuance of the inspection certificate, whichever is later: *Provided,* That these time limits may be extended by the Committee under such conditions as it may deem necessary in the circumstances. The handler shall report to the Committee the information as required and specified in § 989.173(b)(5). Any such lot of off-grade raisins shall be identified immediately following inspection by fixing to a container on each pallet a prenumbered RAC control card (to be furnished by the Committee), and kept separate and apart from any other raisins in the handler's possession. In the event the handler does not normally use pallets in his operation the RAC control card shall be affixed to one or more of the containers in each lot. The RAC control cards shall remain fixed to the containers until the raisins are (i) disposed of by the handler in eligible non-normal outlets, (ii) returned unstemmed to the tenderer, or (iii) submitted for reconditioning. The cards shall be removed only by an inspector of the inspection service or authorized Committee personnel, except control cards designating lots held only for fumigation may be removed by the handler after the completion of fumigation to the satisfaction of the inspection service. Each lot of off-grade raisins not returned to the tenderer shall be stored by the handler separate and apart by varietal types from all other raisins and by disposition and conditioning categories which preserve the lot identity and, if for reconditioning, the defect identity. Off-grade raisins shall be stored in such a manner as to be accessible to the Committee.

(2) *Change in off-grade categories.* After raisins have been classified as to the categories in § 989.58(e)(1), any lot

of natural condition off-grade raisins held by a handler under paragraph (i) or (iii) of § 989.58(e)(1), may be changed to the other category, or to paragraph (ii). Prior to making such change, the handler shall notify the inspection service at least one business day in advance of the time such handler plans to begin such change. Such notification shall be provided verbally or by other means of communication, including e-mail. Any off-grade lot under paragraph (ii) of § 989.58(e)(1) which has not been removed from the handler premises and is identifiable with the original inspection, may be tendered to the handler for the purposes of paragraph (i) or (iii) of § 989.58(e)(1) and, if accepted, the handler shall so report to the Committee. It shall be the responsibility of the handler to establish and maintain the identity of the raisins in the changed categories in accordance with the applicable provisions of paragraph (c)(1) of this section. Where the tenderer has a financial interest in the raisins the handler shall, before making any change in category, submit to the Committee evidence of the tenderer's permission to make any such change, except for changes from paragraph (i) or paragraph (iii) to paragraph (ii) of § 989.58(e)(1).

(3) *Inter-plant and inter-packer transfer of off-grade raisins.* Any packer may, pursuant to § 989.58(e)(2) and under the surveillance of the inspection service, transfer to or from another packer's plant in California, any off-grade raisins for reconditioning. Such transfer may be for the packer's convenience or that of a financially interested person. Where a tenderer or other person has a financial interest in the raisins, the handler shall first obtain the tenderer's or other interested person's written agreement to the transfer. The handler shall notify the inspection service in advance of the time such handler plans to transfer each lot. Such notification shall be provided verbally or by other means of communication, including e-mail. The notification shall be at least 1 business day in advance of the transfer unless a shorter period is acceptable to the inspection service. In the same manner except for the tenderer's or other person's written agreement, any packer may transfer off-grade raisins

from one of his plants or inspection points to another of his plants in California. In both cases such raisins may be removed directly to the premises of the receiving packer or another plant of the packer under the surveillance of the inspection service. Upon completion of the transfer all applicable provisions of this part shall apply with respect to such raisins and the packer receiving them.

(4) *Reconditioning off-grade raisins—reconditioning requirements.* (i) The handler shall notify the inspection service at least one business day in advance of the time such handler plans to begin reconditioning each lot of raisins, unless a shorter period is acceptable to the inspection service. Such notification shall be provided verbally or by other means of communication, including email. Natural condition raisins which have been reconditioned shall continue to be considered natural condition raisins for purposes of reinspection (inspection pursuant to § 989.58(d)) after such reconditioning has been completed, if no water or moisture has been added; otherwise, such raisins shall be considered as packed raisins. The weight of the raisins reconditioned successfully shall be determined by re-weighing, except where a lot, before reconditioning, failed due to excess moisture only. The weight of such raisins resulting from reconditioning a lot failing account excess moisture may be determined by deducting 1.2 percent of the weight for each percent of moisture in excess of the allowable tolerance. When necessary due to the presence of sand, as determined by the inspection service, the requirement for deducting sand tare and the manner of its determination, as prescribed in paragraph (a)(1) of this section, shall apply in computing the net weight of any such successfully reconditioned natural condition raisins. The weight of the reconditioned raisins acquired as packed raisins shall be adjusted to natural condition weight by the use of factors applicable to the various degrees of processing accomplished. The applicable factor shall be that selected by the inspector of the reconditioned raisins from among factors established by the Committee with the approval of the Secretary.

(ii) In reconditioning off-grade raisins, a handler shall use methods designed to remove the defects whereby the lot fails to qualify as standard raisins. Lots with identical defects may be reconditioned simultaneously (commingled basis) but lots with differing defects shall be reconditioned as separate lots.

(5) *General.* Reconditioning of off-grade raisins by a handler shall be done in accordance with such procedure as will enable the inspector to observe the off-grade raisins at any time and to make a proper inspection. A packer may recover raisins from residual raisin material obtained from his reconditioning operations in conformity with the applicable provisions of § 989.159(g)(1).

(6) *Off-grade raisins which are not reconditioned successfully.* (i) Except as provided in paragraph (c)(6)(ii) of this section, no handler shall return to the tenderer any off-grade raisins received for reconditioning which, after his reconditioning of them is complete, have been stemmed and which then fail to meet the applicable minimum grade standards. Any raisins which fail to meet the applicable minimum grade and condition standards or minimum grade standards after reconditioning and all residual material from reconditioning, held by the handler, shall be identified promptly by affixing to one or more containers in each lot, or to a container in each pallet if pallets are used, a prenumbered RAC control card as prescribed in paragraph (c)(1) of this section: *Provided,* That such failing raisins and residual material which are placed directly into trucks or trailers for immediate disposition need not be identified by affixing thereto a RAC control card. The handler shall hold the failing raisins and the residual material separate and apart from all other raisins. The control cards shall be removed from the containers only by an inspector of the inspection service or authorized Committee personnel. The handler shall physically dispose of the residual material, and any failing raisins which he does not return unstemmed to the tenderer, only in eligible non-normal outlets as provided in § 989.159(g)(2).

(ii) Any packer may arrange for or permit the tenderer to remove the stemmed raisins (described in paragraph (c)(6)(i) of this section), but not the residual, directly to the premises, within California, of another packer for further reconditioning of the raisins at the latter's premises. Such removal and transfer shall be made under the surveillance of the inspection service. The packer shall notify the inspection service as required in paragraph (c)(3) of this section. Such raisins may be received by the other packer without inspection. On and after such receipt of the raisins for further reconditioning, all applicable provisions of this part shall apply with respect to such raisins and the packer so receiving them.

(7) *Return of off-grade raisins to tenderer.* Any off-grade raisins which are to be returned unstemmed to the tenderer pursuant to § 989.58(e)(1)(ii), shall be physically returned within five business days after the issuance of the inspection certificate: *Provided,* That such time limit may be extended by the Committee as it may deem justified by extenuating circumstances. The handler shall file with the Committee a report of the returned raisins as required in § 989.173(b)(4).

(i) *Unstemmed and stemmed raisins.* For the purpose of determining whether or not off-grade raisins may be returned to the person tendering such raisins, "unstemmed" raisins shall be defined as lots of raisins that contain 150 or more capstems per pound. "Stemmed" raisins means lots of raisins that contain less than 150 capstems per pound.

(d) *Reinspection of raisins held more than one hundred and twenty days on memorandum receipt.* No handler shall acquire raisins held on memorandum receipt for a period longer than one hundred and twenty (120) days unless such raisins have been reinspected and certified immediately prior to acquisition as meeting the minimum requirements for standard raisins: *Provided,* That the Committee at any other time may require such reinspection and certification of raisins held on memorandum receipt as a prerequisite to acquisition if it has reason to believe that the raisins do not then meet such requirements.

(e) *Inspection of raisins on dehydrator's premises*—(1) *Application and agreement.* (i) Any dehydrator may submit to the Committee for approval, and the Committee may approve, in accordance with the provisions of this paragraph an application and agreement, on a form furnished by the Committee, providing for dehydrator on-premise inspection of natural condition raisins produced by the dehydrator by subjecting grapes to artificial heat. Raisins so produced are referred to in paragraph (a)(3) of this section and in this paragraph as "dehydrated raisins."

(ii) The provisions of such application and agreement shall include at least the following:

(a) The dehydrator shall request the inspection service to inspect all dehydrated raisins which the dehydrator produces and to issue a related memorandum report of inspection at the time of loading any quantity of such raisins for delivery to a packer's inspection point;

(b) The dehydrator will arrange with the inspection service for the necessary inspection service to be performed by the service, and the dehydrator will submit to the Committee a statement from the inspection service that the dehydrator has adequate facilities for the inspection and that such arrangements have been made;

(c) All necessary reconditioning of dehydrated raisins, identification and segregation of raisins, and movement of inspected dehydrated raisins on or from the dehydrator's premises shall be done in such manner and under such conditions as the inspection service may require;

(d) The dehydrator shall, at the time of the packer's receipt of such raisins, furnish to the packer to whose inspection point the inspected raisins are delivered the original and one copy of the memorandum report of inspection covering such raisins;

(e) The dehydrator shall maintain such records and furnish such reports and permit access to such records and the dehydrator's premises as required in the application and agreement or as the Committee may subsequently request; and

(f) The application and agreement may be suspended or terminated as provided therein.

(iii) The Committee will notify raisin packers of each dehydrator whose application and agreement has been approved by the Committee (such dehydrator is referred to in this subpart as "authorized dehydrator"); similarly, the Committee will notify packers of each suspension or termination of a previously approved application and agreement.

(2) *Delivery of inspected dehydrated raisins.* Any dehydrated raisins which (i) are inspected on an authorized dehydrator's premises where produced; (ii) are moved promptly and directly to a packer's inspection point from the premises of the authorized dehydrator; (iii) are accompanied by an applicable memorandum report of inspection to be furnished to the packer; and (iv) are otherwise in compliance with the provisions of such approved application and agreement and this paragraph may be received by the packer without the inspection at time of receipt required by § 989.58(d). With respect to such dehydrated raisins, the packer shall comply with all applicable requirements and procedures of this part, including, but not limited to, inspection after any necessary reconditioning and the inspection prescribed in § 989.59.

(3) *Packer's obligations.* Immediately upon a packer's receiving any such already inspected dehydrated raisins accompanied by the applicable memorandum report of inspection, the packer shall give to the inspector at the packer's inspection point where the dehydrated raisins were received, the original and one copy of such memorandum report so that the inspector may enter the net weight and scale ticket number on such memorandum report of inspection and copy thereof. Whenever a packer receives off-grade raisins from an authorized dehydrator he shall so advise the inspector at the packer's inspection point at the time of such receipt; and such raisins shall not be unloaded except in the presence of the inspector or in accordance with such prior arrangements as may have been made between the packer and the inspection service.

(f) *Inspection of raisins at cooperative bargaining association's receiving station*—(1) *Application and agreement.* (i) In accordance with the provisions of this paragraph, any cooperative bargaining association may submit to the Committee for approval, and the Committee may approve, an application and agreement, on a form furnished by the Committee, providing that where the association receives from individual producers lots of natural condition raisins at any of its receiving points and the raisins are inspected and stored consistent with such application and agreement, such lots shall be eligible for delivery to handlers, pursuant to paragraph (f)(3) of this section, without reinspection. Any raisins which upon inspection by the inspection service do not meet the applicable grade and condition standards shall be identified immediately following inspection and kept separate and apart from any other raisins in the association's possession.

(2) *Terms and conditions.* The provisions of such application and agreement shall include at least the following terms and conditions:

(i) That the association shall, prior to delivery of any raisins to handlers, arrange for inspection services at the association's receiving station(s), and cause to be submitted to the Committee a statement by the inspection service of such arrangement and of the association's having adequate laboratory and other facilities for such services available at the association's receiving station(s).

(ii) That the association shall maintain such facilities satisfactory to the inspection service.

(iii) That the association shall request inspection of each lot of raisins immediately upon physical arrival thereof at the association's receiving station(s), and shall provide the inspector with any assistance necessary in the inspection of such raisins, including the movement of individual containers.

(iv) That the association shall fumigate all raisins received at the association's receiving station(s) as necessary to assure that the raisins are free from active infestation and maintain them as such while on such premises, and that fumigation shall be performed to the satisfaction of the inspection service.

(v) That the association shall, with respect to all raisins entering its premises which are not returned to the producer as provided in paragraph (f)(2)(vi) of this section, promptly affix to one or more containers in each lot, or to a container in each pallet if pallets are used, a Committee control card showing thereon such information as the Committee requires to maintain the producer identity of each lot and prevent commingling with any other lot. The association shall not move all or any portion of a lot of raisins on the premises of the association's receiving station(s) or load any such raisins for shipment, except in the presence of an inspector of the inspection service.

(vi) That the association shall store any standard raisins and any off-grade raisins which are held by it after receipt and inspection on the premises of its receiving station(s) under conditions which protect the raisins from rain, infestation and contamination, and which can be expected to maintain their respective conditions except for normal and natural deterioration and shrinkage. Any raisins which after receipt and inspection are not accepted and held by the association shall be returned to the producer within 5 business days after the issuance of the inspection certificate.

(vii) That the association shall furnish the inspection service with a completed Committee form requesting issuance, at the time of loading any lot of inspected raisins for delivery to any handler's inspection point, of a memorandum report of inspection covering such lot.

(viii) That the association shall deliver to the handler at the time of receipt of any such lot of eligible raisins at the handler's inspection point the original and one copy of the inspection service's related memorandum report of inspection; and such original and copy shall accompany the shipment of such lot from the premises of the association's receiving station(s) to the handler's inspection point.

(ix) That the association shall maintain complete records of the receipt, holding and disposition of each lot of raisins and retain such records for at least 2 years after the crop year in which such transactions occurred.

(x) That the association shall file promptly with the Committee certified reports showing such information as the Committee may request relative to the association's receipts, holdings, and dispositions of raisins.

(xi) That the association shall permit the Committee, the inspection service, and the Secretary of Agriculture, through their duly authorized representatives, to have access to the premises of the association's receiving station(s) to inspect such premises and any raisins thereon and any and all records with respect to the association's receipts, holdings and dispositions of raisins.

(xii) That upon approval of the application and agreement the Committee will notify handlers of such approval and that eligible lots of inspected raisins will not require incoming inspection at handler inspection points; will notify the interested handlers of any suspension or revocation, for good cause, of the eligibility of a particular lot of raisins; and will notify handlers of any suspension or termination of the application and agreement.

(xiii) That the Committee will request the inspection service to establish a fee to the association for the services to be rendered at the same rate as is charged handlers.

(xiv) That the application and agreement may be suspended or terminated as provided therein.

(3) *Waiver of requirement for incoming inspection at handler inspection point.* Any lot of raisins which (i) is inspected on the premises of the association's receiving station(s) pursuant to an approved application and agreement, (ii) is in compliance with the provisions of such application and agreement and this paragraph, (iii) is moved under the surveillance of the inspection service to a handler's inspection point from the association's receiving station(s) after issuance of the related memorandum report of inspection, and (iv) is accompanied by such memorandum report to be furnished to the handler may be received by the handler without the inspection as required by § 989.58(d) at time of receipt.

(4) *Handler's obligations.* With respect to such raisins received by the handler, the handler shall comply with all applicable requirements and procedures of this part, including, but not limited to, the inspection prescribed in § 989.59 and that required, as prescribed in § 989.58(d), prior to the handler acquiring reconditioned raisins. Immediately upon a handler receiving any such raisins accompanied by the applicable memorandum report of inspection, the handler shall give to the inspector at the handler's inspection point where such raisins are received, the original and one copy of the memorandum report so that the inspector may enter the net weight and scale ticket number on such memorandum report of inspection and copy thereof.

[27 FR 3112, Mar. 31, 1962, as amended at 27 FR 10249, Oct. 19, 1962; 27 FR 10409, Oct. 25, 1962; 28 FR 13544, Dec. 14, 1963; 31 FR 16305, Dec. 21, 1966; 32 FR 15915, Nov. 21, 1967; 32 FR 17467, Dec. 6, 1967; 35 FR 16037, Oct. 13, 1970; 38 FR 20237, July 30, 1973; 42 FR 52377, Sept. 30, 1977; 49 FR 18731, May 2, 1984; 49 FR 33994, Aug. 28, 1984; 55 FR 2226, Jan. 23, 1990; 55 FR 36608, Sept. 6, 1990; 73 FR 42259, July 21, 2008; 83 FR 53972, Oct. 26, 2018; 84 FR 30863, June 28, 2019]

§ 989.159   **Regulation of the handling of raisins subsequent to their acquisition.**

(a) *Inspection facilities.* At each of the premises where packed raisins are to be inspected each handler shall, at his expense provide reasonably safe and adequate space and other facilities necessary for the proper and efficient inspection of such raisins.

(b) *Identification of inspected raisins.* (1) Each handler shall mark each shipping container with legible code or other identification, satisfactory to the Committee and the inspection service which shall indicate with respect to packed raisins, the date that the raisins in such shipping containers were packed, and with respect to shipments of natural condition raisins, the date on which such raisins were inspected.

(2) Each handler shall furnish promptly to the Committee, through the inspection service, a certified report on a form furnished by the Committee showing the handler's count and weight of the raisins of each pack and varietal type packed each day.

(c) *Outgoing inspection.* (1) Outgoing inspection and certification of raisins as required by §989.59(d) shall be made as set forth in this paragraph.

(2) Such inspection of natural condition raisins (which is subject to exceptions and exemptions provided in this part) shall be made of each individual lot and in each case not more than five days before the date of shipment or other final disposition of the lot. The certificate that the raisins meet the applicable minimum grade and condition standards for natural condition raisins, which the handler is required to obtain and submit to the Committee pursuant to §989.59(d), shall be on Form FV 146 labeled "Certificate of Quality and Condition (Processed Foods)." If shipment involves exportation to a foreign country, the handler shall surrender to the United States Customs Service at the port of exit two copies of such inspection certificate. Such an inspection and certification (on Form FV 146) may, if requested by the handler, be made at the time of his receipt or acquisition of the raisins. In such an event, no additional inspection shall be required if the lot remains intact and identifiable, and shipment or other final disposition takes place within five days after the date of the inspection.

(3) Such inspection of packed raisins shall be made prior to shipment or other final disposition, and unless made during the final processing or packing operations so as to facilitate proper sampling, the inspector shall perform the inspection on the basis of representative samples drawn from shipping containers of the packed raisins.

(4) Except as otherwise provided in this part, where there is presented for inspection a lot of packed raisins consisting of raisins of different varietal types or sub-types commingled within their containers, each such type and sub-type shall be inspected separately, except that inspection for moisture shall be performed on the lot as a whole. The inspection certificate shall show the respective percentages which the raisins of the various types and sub-types are of the lot and whether each meets the applicable minimum grade standards. In the event the rai-

sins of any such varietal type or sub-type contained in the lot fail to meet the applicable requirements, other than for moisture, none of the lot shall be certified as meeting minimum grade standards unless it is found to be practicable to separate the raisins into two new lots, one which meets and the other which fails to meet all of the applicable minimum standards, respectively. Any lot of mixed types or sub-types of packed raisins for which minimum grade standards are prescribed for each type or sub-type pursuant to §989.59 (a) and (b) but which in their commingled form cannot be inspected against the standards for the respective varietal types or sub-types and hence are excluded from the category of "Mixed types" as defined in the then effective United States Standards for Grades of Processed Raisins, or any raisins which as a mixed lot contain moisture in excess of 18 percent, shall not be certified as meeting the minimum grade standard for packed raisins.

(d) *Submission of inspection certificates to the Committee.* A copy of each inspection certificate which a handler is required to submit to the Committee pursuant to §989.59(d) shall be submitted not later than Wednesday of the week following the week in which the certificate was issued. This may be accomplished by authorizing the inspection service in writing to submit a copy of each such inspection certificate directly to the Committee. A copy of such authorization shall be furnished to the Committee.

(e) *Term of inspection certificate.* Any handler who:

(1) Fails to ship or make other final disposition for human consumption of any lot of packed raisins within 90 calendar days, or of any lot of natural condition raisins within 5 calendar days, after the date of the last inspection of the lot; or

(2) Has any shipment or portion of a shipment returned to his inspection point or storage premises within the area, shall, before any such shipment or final disposition, or before blending with other raisins, have such raisins inspected for condition and shall furnish promptly to the Committee (which may be through the inspection

service as provided in § 989.158(b)) a copy of the inspection certificate showing that the raisins meet the respective requirements of this part for shipment, final disposition or blending.

(f) *Exemption of experimental and specialty packs*—(1) *Shipment under exemption.* Upon obtaining approval of the Committee as provided in this paragraph, any handler may ship or dispose of raisins in experimental or specialty packs without regard to one or more of the requirements of the minimum grade standards for packed raisins and inspection and certification requirements, prescribed pursuant to § 989.59. For the purpose of this exemption, experimental and specialty packs means raisins processed using methods, materials, or techniques that are not normally employed in packing raisins.

(2) *Application for exemption.* Each application for exemption shall be filed with the Committee in triplicate. The application shall at least contain information as to:

(i) The name and address of the handler;

(ii) The estimated quantity of each varietal type of raisins for which the exemption is requested;

(iii) The specific requirements in the minimum grade standards from which exemption is requested;

(iv) The special processing involved;

(v) The net weight of each type of container;

(vi) Whether disposition will be made direct to consumers, wholesalers, retailers, persons, or organizations, and any special uses to be made of such raisins; and

(vii) The general quality, style, and condition of the raisins for which the exemption is requested.

(3) *Committee action on application.* The Committee in its discretion shall approve each application for exemption of raisins, if it concludes that such exemption shall not jeopardize the objectives of the marketing order program. The Committee shall notify the handler promptly in writing of its approval or disapproval of his application and, if the application is approved, the maximum quantity for which approval is granted. If the application is disapproved, the Committee shall inform the handler of the reasons therefor.

(4) *Reports.* The handler shall report shipments or other dispositions under an approved exemption as required pursuant to § 989.173(e).

(g) *Off-grade raisins, other failing raisins, and raisin residual material*—(1) *Recovery of raisins.* (i) For the purposes of §§ 989.59(f) and 989.158(c)(4), a packer may recover raisins from:

(A) Residual raisins from his or her processing of standard raisins;

(B) Any raisins acquired as standard raisins which fail to meet the applicable outgoing grade and condition standards;

(C) Any raisins rejected on a condition inspection; and

(D) Residual raisins from reconditioning of off-grade raisins.

(ii) *Provided,* That such recovery under paragraphs (g)(1)(i)(B) and (C) of this section must occur without blending, if the failure to meet the minimum grade standards for packed raisins is due to a defect or defects affecting the wholesomeness of the raisins: *And provided further,* That such recovery under paragraph (g)(1)(i)(D) of this section must occur without blending, except as permitted in § 989.158(c)(4)(ii), and the weight of standard raisins in residual from off-grade raisins shall be credited equitably to the same lot or lots from which the residual was obtained. The provisions of this paragraph (g)(1) are not intended to excuse any failure to comply with all applicable food and sanitary rules and regulations of city, county, state, federal, or other agencies having jurisdiction.

(2) *Disposition.* (i) Except as authorized in this part, no handler shall ship or otherwise dispose of any off-grade raisins, other failing raisins, or raisin residual material. Any handler may ship, transfer, or otherwise dispose of off-grade raisins, other failing raisins, and raisin residual material to or at points within the continental United States (other than Alaska) for use in eligible non-normal outlets only after filing with the Committee a written application to make such shipment, transfer, or other disposition and receiving its written approval thereof. However, the requirements of prior filing and approval of any such application shall not apply to:

(A) The transfer of any such raisins or residual material by a handler from one of his plants to another of his plants in the State of California, except any transfer of raisins which are for reconditioning shall be in accordance with §989.158(c)(3);

(B) Any inter-packer transfer or removal of off-grade raisins made in accordance with §989.158(c)(3) and of unsuccessfully reconditioned off-grade raisins which have been stemmed (other failing raisins) made in accordance with §989.158(c)(6)(ii);

(C) Any return by a handler of unstemmed off-grade raisins to the tenderer in accordance with §989.158(c)(7);

(D) Any shipment or transfer of off-grade raisins, other failing raisins, or raisin residual material by any handler to a processor within the State of California for use, within the State, in eligible non-normal outlets;

(E) Any shipment or transfer of off-grade raisins, other failing raisins, or raisin residual material by a handler to any person with an effective agreement with the Committee, in which he agrees to use such raisins and raisin residual material only in eligible non-normal outlets, if not so used, to pay to the Committee liquidated damages in the amount and under the conditions specified in paragraph (g)(2)(iii) of this section, and to maintain complete, accurate, and current records regarding his dealings in raisins and raisin residual material, retain the records for at least 2 years, and permit representatives of the Committee and Secretary of Agriculture to examine all of his books and records relating to raisins and residual material; and

(F) Any direct use by the handler of such raisins or material in eligible non-normal outlets within the State of California.

(ii) Each such application shall, in addition to the agreement specified in paragraph (g)(2)(iii) of this section, include as a minimum:

(A) The names and addresses of the handler, the buyer, the consignee, and the user;

(B) The quantity of off-grade and other failing raisins and the quantity of raisins residual material to be shipped or otherwise disposed of;

(C) A description of such off-grade raisins and other failing raisins and raisin residual material, as to type or origin;

(D) The present location of such raisins and raisin residual material;

(E) The particular use to be made of the raisins; and

(F) A copy of the sales contract, which may be on a form furnished by the Committee, wherein the buyer agrees:

(1) Not to ship such raisins or raisin residual material to points outside the continental United States or to Alaska;

(2) To dispose of the raisins or raisin residual material only for uses in eligible non-normal outlet(s); and

(3) To maintain complete, accurate, and current records regarding his or her dealings in raisins, retain the records for at least 2 years, and permit representatives of the Committee and of the Secretary of Agriculture to examine all of his or her books and records relating to raisins and residual material.

(iii) Each such application shall also include a provision for liquidated damages wherein the handler, in consideration of the Committee approving his application, agrees that in the event any raisins or raisin residual material covered by the approved application should be shipped to points outside of the continental United States or to Alaska, or disposed of in other than eligible non-normal outlets, by any person, it will cause serious and substantial damage to the Committee, to producers, and to handlers of raisins and will be difficult, if not impossible, to prove the extent of such damage. Therefore, the handler shall pay to the Committee a sum equal to the established field price as liquidated damages for each ton so shipped or disposed of, such sum being a fair measure of damages and not a penalty.

(iv) The Committee shall notify the applicant in writing of its approval action. In acting on an application, the Committee may disapprove the application when: The application is incomplete, or any required information has not been submitted; the Committee has cause to believe that the raisins or raisin residual material covered by the application will not be shipped or disposed of in accordance with the application; or the handler, or any of the parties involved in the proposed shipment or disposition, had shipped or made disposition or use of raisins or raisin residual material covered by a previously approved application inconsistent with that application. When the use or the name and address of the user or consignee are not known to the handler, the Committee shall not approve the application until it has been informed as to such use and user and consignee of the raisins or residual material.

(v) The Committee may, for cause, revoke any previously approved application of a handler if the handler, buyer, consignee or user covered by the application has shipped or made disposition inconsistent with any approved application. The Committee shall notify the handler in writing of such revocation.

(vi) The handler shall furnish the Committee with a copy of the shipping document or other documentary evidence of the disposition as may be satisfactory to the Committee and at such times as the Committee may direct.

(h) *Appeal inspection.* An appeal inspection on an original inspection may be obtained from the inspection service upon the request of any financially interested party and upon the payment of any fees assessed by the inspection service for such appeal inspection.

[27 FR 3112, Mar. 31, 1962, as amended at 30 FR 6906, May 21, 1965; 31 FR 16306, Dec. 21, 1966; 36 FR 13980, July 29, 1971; 38 FR 13012, May 18, 1973; 38 FR 20237, July 30, 1973; 42 FR 52377, Sept. 30, 1977; 49 FR 18731, May 2, 1984; 84 FR 30863, June 28, 2019]

§ 989.160 Exemptions.

(a) Any processor may receive or acquire any raisins for use in eligible non-normal outlets, and dispose of them for such use, without having them inspected and certified. Processors receiving or acquiring raisins under such exemption, or otherwise receiving or acquiring raisins which do not meet the applicable minimum grade and condition standards, shall not ship or otherwise dispose of any such raisins except in conformity with the provisions of § 989.159(g)(2). Proc-

essors shall report receipts and acquisitions and make such other reports as are or may be required pursuant to §§ 989.73 and 989.173.

(b) *Disposition of raisins produced in Southern California.* Raisins produced from grapes dried on the vine in the counties of Riverside, Imperial, San Bernardino, Ventura, Orange, Los Angeles, and San Diego, which are disposed of for use in distillation or livestock feed, shall be exempt from the provisions of this part.

[30 FR 6906, May 21, 1965, as amended at 38 FR 13013, May 18, 1973; 59 FR 44031, Aug. 26, 1994]

§§ 989.166—989.167 [Reserved]

REPORTS AND RECORDS

§ 989.173 Reports.

(a) *Inventory reports.* Each handler shall submit to the Committee as of the close of business on July 31 of each crop year, and not later than the following August 6, an inventory report which shall show, with respect to each varietal type of raisins held by such handler, the quantity of off-grade raisins segregated as to those for reconditioning and those for disposition as such. *Provided,* That, for the Other Seedless varietal type, handlers shall report the information required in this paragraph separately for the different types of Other Seedless raisins. Upon request by the Committee, each handler shall file at other times, and as of other dates, any of the said information which may reasonably be necessary and which the Committee shall specify in its request.

(b) *Reports of raisins received or acquired*—(1) *General.* (i) Except as otherwise provided in paragraph (i) of this section, each handler shall submit to the Committee (on forms furnished by it) for each week (Sunday through Saturday or such other 7-day period for which the handler has submitted a proposal to and received approval from the Committee) and not later than the following Wednesday, the reports specified in paragraphs (b)(2), (3), (4), and (5) of this section.

(ii) For each report required to be submitted pursuant to this paragraph, the required information shall be shown separately for each varietal type: *Provided*, That, for the Other Seedless varietal type, the required information shall be shown separately for the different types of Other Seedless raisins. With each report, other than that specified in paragraph (b)(4) of this section, the handler shall submit a copy of the door receipt, weight certificate or such other document approved by the Committee that accurately reflects the weight of each lot tendered, for each lot of raisins received or acquired by him during the reporting period and for each lot of raisins stored on memorandum or warehouse receipt which was returned to the tenderer during such period, which shall show the information to be contained on such receipts or weight certificates as specified in §989.158(a)(3). At the time he submits the reports specified in paragraphs (b) (2) and (3) of this section to the Committee, each handler shall submit a copy of each such report to the Inspection Service.

(2) *Acquisition of standard raisins.* Each handler shall report:

(i) The total net weight of the standard raisins acquired during the reporting period; and

(ii) The cumulative totals of such acquisitions from the beginning of the then current crop year.

(3) *Standard raisins received for memorandum storage.* Each handler shall, with respect to all standard raisins held for memorandum receipt, storage, bailment, or warehousing (raisins received other than by acquisition or inter-handler transfer), report:

(i) The net weight of such standard raisins held at the start of the reporting period;

(ii) The net weight of such standard raisins received during the reporting period;

(iii) The net weight of such standard raisins acquired during such period and included with the acquisitions required to be reported pursuant to paragraph (b)(2) of this section;

(iv) The net weight of such raisins returned during such period to the persons from whom they were received; and

(v) The net weight(s) and location(s) of such raisins held at the end of such period.

(4) *Off-grade raisins returned to tenderers.* Each handler shall report with respect to each lot of off-grade raisins which the handler returned during the reporting period to the tenderer pursuant to paragraph (1) of §989.58(e):

(i) The inspection certificate number;

(ii) The net weight;

(iii) The name of the tenderer; and

(iv) The date the lot was returned to the tenderer.

(5) *Off-grade raisins received for reconditioning or disposition in eligible nonfood channels.* Each handler who is not a processor shall, with respect to all off-grade raisins received by the handler and retained by him for reconditioning or for disposition or use in eligible non-normal outlets, report for each category received or reconditioned during the reporting period:

(i) The name of each tenderer;

(ii) The net weight of such raisins;

(iii) The locations where received;

(iv) The inspection certificate number covering each receipt;

(v) The name and address of each person to whom residual or off-grade lots were delivered for disposition, and the respective net weight delivered; and

(vi)(A) The total net weight (according to location) of each category of off-grade raisins held by him at the end of the reporting period.

(B) Each non-acquiring handler shall report also the weight of standard raisins recovered from reconditioning, their inspection certificate number(s) and the handler or other person to whom the standard raisins were delivered.

(6) *Monthly report of raisins received or acquired by processors.* Each processor who receives or acquires off-grade raisins, or who avail himself of the exemptions from the grade and inspection requirements provided in §§989.58, 989.59(f), and 989.160 and receives or acquires raisins or raisin residual material, shall submit to the Committee on or before the 7th day of each month a report of such raisins, raisin residual material, and off-grade raisins received or acquired during the preceding month. Each report shall show for each varietal type:

(i) The name and address of each handler, producer, or other person from

605

whom such raisins or raisin residual material was received or acquired; and

(ii) The net weight of such raisins and raisin residual material.

(7) *Receipt of raisins produced from grapes grown outside the State of California.* Each handler who receives raisins produced from grapes grown outside the State of California shall submit to the Committee, on an appropriate form provided by the Committee so that it is received by the Committee not later than the eighth day of each month, a report of the receipt of such raisins. This report shall include: The varietal type of raisins received; the net weight (pounds) of raisins received for the current month as well as a cumulative quantity from August 1; and the state or country where the raisins were produced. With each report, the handler shall submit a copy of the door receipt, weight certificate, or such other document as required by the Committee that includes, but is not limited to, the name of the tenderer (equity holder) from whom such raisins were received, the varietal type(s) of raisins, the net fruit weight, the number and type of containers in the lot, the date of delivery, and the address including State or country where such raisins were produced.

(c) *Reports of disposition*—(1) Each month each handler who is not a processor shall furnish to the Committee, on an appropriate form provided by the Committee and so that it is received by the Committee not later than the seventh day of the month, a report showing the aggregate quantity of each varietal type of packed raisins and standard natural condition raisins which were shipped or otherwise disposed of by such handler during the preceding month (exclusive of transfers within the State of California between plants of any such handler and from such handler to other handlers): *Provided,* That, for the Other Seedless varietal type, handlers shall report such information for the different types of Other Seedless raisins. Such required information shall be segregated as to:

(i) Domestic outlets (exclusive of Federal Government purchases) according to the quantity shipped in consumer cartons, the quantity shipped in bags having a net weight content of

four pounds or less, and the quantity shipped in bulk packs (including, but not limited to those in bags having a net weight content of more than four pounds);

(ii) Federal Government purchases;

(iii) The varietal type of raisin, with organically-produced raisins as specified in paragraph (g) of this section separated out, net weight, and condition of the raisins transferred: *Provided,* That, for the Other Seedless varietal type, handlers shall report such information for the different types of Other Seedless raisins; and

(iv) Export outlets, by countries of destination; and

(v) Each of any other outlets in which the handler has made disposition of such raisins other than by any transfer which is excluded by the preceding sentence.

(2) *Disposition by handlers (other than processors) of off-grade raisins, other failing raisins, and raisin residual material.* Each handler who is not a processor shall submit to the Committee on or before the seventh day of each month a report of all shipments and other dispositions made during the preceding month of off-grade raisins, other failing raisins, and raisin residual material. Such report shall be submitted on a form furnished by the Committee and shall include the following information:

(i) Date of each shipment and other disposition;

(ii) Name and address of each buyer and receiver; and

(iii) Description and net weight of the raisins and raisin residual material in each shipment or other disposition.

(3) *Disposition by handlers of raisins produced from grapes grown outside the State of California.* Each handler who receives raisins produced from grapes grown outside the State of California shall submit to the Committee, on or before the eighth day of each month, a report, on the appropriate form provided by the Committee, of all shipments of such raisins made during the preceding month. This report shall include:

(i) The varietal type(s) of raisins shipped;

(ii) The net weight (pounds) of raisins shipped;

(iii) The destination (domestic, export, and other disposition such as distilleries, livestock feeders, or concentrate) of such shipments; and

(iv) The area of origin (state or country) of the raisins shipped.

(4) *Disposition reports by processors.* Each processor shall submit to the Committee, upon its request, such of the following information and for such period as the Committee shall specify;

(i) The quantity of raisins and raisin material sold or otherwise disposed of by processing operations, segregated as to the processing outlets and the kinds of raisins or raisin material which the Committee shall specify; and

(ii) The quantity of raisins or raisin material sold or otherwise disposed of by the processor, segregated as to specified outlets and kinds of raisins or raisin material.

(d) *Reports of inter-handler transfers.* (1) Any handler who transfers raisins to another handler within the State of California shall submit to the Committee not later than five calendar days following such transfer a report showing:

(i) The date of transfer;

(ii) The name(s) and address(es) of the handler or handlers and the locations of the plants;

(iii) The varietal type of raisin, with organically-produced raisins as specified in paragraph (g) of this section separated out, net weight, and condition of the raisins transferred: Provided, That, for the Other Seedless varietal type, handlers shall report such information for the different types of Other Seedless raisins;

(iv) If packed, the inspection certificate number in the event such raisins have been inspected prior to such transfer and a certificate issued. Two copies of such report shall be forwarded to the receiving handler at the time the report is submitted to the Committee, on one of which the receiving handler shall certify to the receipt of such raisins and submit it to the Committee within five calendar days after the raisins or the copies of such report have been received by him, whichever is later; and

(v) If packed, the transferring handler shall certify that such handler is transferring only acquired raisins that meet all applicable marketing order requirements, including reporting, incoming inspection, and assessments.

(2) [Reserved]

(e) *Report of shipments of experimental or specialty packs under exemption.* Each handler who obtains an exemption pursuant to §989.59(g) for the shipment of experimental or specialty packs of raisins shall submit to the Committee on a copy of the approved application for exemption a report showing the quantity of raisins shipped or disposed of under such exemption. The handler shall submit the report promptly after the end of the crop year or after completion by him of all shipments of such exempted raisins, whichever is earlier.

(f) *Organically-produced raisins.* For purposes of this section, organically-produced raisins means raisins that have been certified by an organic certification organization currently registered with the California Department of Food and Agriculture or such certifying organization accredited under the National Organic Program. Handlers of such raisins shall submit the following reports to the Committee by varietal type: *Provided:* That, for the Other Seedless varietal type, handlers shall report such information for the different types of Other Seedless raisins.

(1) *Inventory report of organically-produced raisins.* Each handler shall submit to the Committee by the close of business on July 31 of each crop year, and not later than the following August 6, on an appropriate form provided by the Committee, a report showing, with respect to the organically-produced raisins held by such handler:

(i) The quantity of raisins, segregated as to locations where they are stored and whether they are natural condition or packed;

(ii) The quantity of off-grade raisins segregated as to those for reconditioning and those for disposition as such.

(2) *Acquisition report of organically-produced standard raisins.* Each handler shall submit to the Committee for each week (Sunday through Saturday or such other 7-day period for which the handler has submitted a proposal to and received approval from the Committee) and not later than the following Wednesday, on an appropriate

form provided by the Committee, a report showing the following:

(i) The total net weight of the standard raisins acquired during the reporting period; and

(ii) The location of the reserve tonnage; and

(iii) The cumulative totals of such acquisitions (as so segregated) from the beginning of the current crop year.

(iv) Upon request of the Committee, each handler shall provide copies of the organic certificate(s) applicable to the quantity of raisins reported as acquired.

(3) *Disposition report of organically-produced raisins.* No later than the seventh day of each month, handlers who are not processors shall submit to the Committee, on an appropriate form provided by the Committee, a report showing the aggregate quantity of packed raisins and standard natural condition raisins which were shipped or otherwise disposed of by such handler during the preceding month (exclusive of transfer within the State of California between the plants of any such handler and from such handler to other handlers). Such information shall include:

(i) Domestic outlets (exclusive of Federal government purchases) according to the quantity shipped in consumer cartons, the quantity of bags having a net weight content of 4 pounds or less, and the quantity shipped in bulk packs (including, but not limited to those in bags having a net weight content of more than 4 pounds);

(ii) Federal government purchases;

(iii) Export outlets according to quantity shipped in consumer cartons, the quantity shipped in bags having a net weight of 4 pounds or less, and the quantity shipped in bulk packs (including, but not limited to those in bags having a net weight content of more than 4 pounds);

(iv) Export outlets, by countries of destination; and

(v) Each of any other outlets in which the handler disposed of such raisins other than by any transfer which is excluded by the preceding sentence.

(g) [Reserved]

(h) *Certification of report.* All reports submitted to the Committee pursuant to this part shall be dated, and certified to the United States Department of Agriculture and to the Raisin Administrative Committee as to the truthfulness, accuracy and completeness of the information shown thereon.

(i) *Reporting by non-profit cooperative associations.* Non-profit cooperative associations need not submit door tags, door receipts, weight certificates or other similar documents with its report as to raisins received or acquired from its members.

(j) *Exemption from filing report.* A handler may be relieved by the Committee of submitting any of the reports required pursuant to paragraph (b) of this section which he shall specify in a written application therefor to the Committee stating that no transactions subject to such reports are contemplated for the balance of the crop year: *Provided,* That any such exemption shall remain in effect only so long as said handler has no such transactions subject to such reports.

[27 FR 3112, Mar. 31, 1962]

EDITORIAL NOTE: For FEDERAL REGISTER citations affecting § 989.173, see the List of CFR Sections Affected, which appears in the Finding Aids section of the printed volume and at *www.govinfo.gov.*

§ 989.176 Records.

Each handler shall maintain complete, accurate, and current records of all of his business affairs concerning which he is required to submit reports with the Committee, and shall maintain such records for at least two years after the termination of the crop year in which the transactions occurred.

Subpart C—Supplementary Requirements

§ 989.210 Handling of varietal types of raisins acquired pursuant to a weight dockage system.

(a) *General.* A handler may acquire as standard raisins lots of Natural (sundried) Seedless, Golden Seedless, Dipped Seedless, Monukka, Other Seedless, Sultana, Zante Currant, Muscat (including other raisins with seeds), and Other Seedless-Sulfured raisins under the weight dockage provisions described in §§ 989.212 and 989.213. The creditable weight of each lot of raisins acquired in this manner shall be that obtained by multiplying the net weight of the raisins in the lot by the applicable factor(s) from the appropriate dockage table(s) included in those sections.

(b) *Assessments.* Assessments on any lot of raisins of the varietal types specified in paragraph (a) of this section acquired by a handler pursuant to a weight dockage system shall be applicable to the creditable weight of such lot.

(c) *Identification.* Any lot of raisins of the varietal types specified in paragraph (a) of this section acquired pursuant to a weight dockage system shall be so identified by the inspection service affixing to one container on each pallet, or to each bin, in such lot, a prenumbered RAC control card (to be furnished by the Committee) which shall remain affixed to the container or bin until the raisins are processed or disposed of as natural condition raisins. The control card shall only be removed by, or under the supervision of an inspector of, the inspection service, or authorized Committee personnel.

(d) *Application of dockage factors.* A lot of raisins acquired which may be subject to both a substandard and maturity dockage factor shall have only the highest of the two dockage factors applied to determine the creditable weight.

[53 FR 49296, Dec. 7, 1988, as amended at 67 FR 36792, May 28, 2002; 68 FR 42947, July 21, 2003; 83 FR 53973, Oct. 26, 2018]

§989.212 Substandard dockage.

(a) *General.* Subject to prior agreement between handler and tenderer, Natural (sun-dried) Seedless, Golden Seedless, Dipped Seedless, Monukka, Other Seedless, and Other Seedless-Sulfured raisins containing from 5.1 through 17.0 percent, by weight, of substandard raisins may be acquired by a handler under a weight dockage system. A handler may also, subject to prior agreement, acquire as standard raisins any lot of Muscat (including other raisins with seeds), Sultana, and Zante Currant raisins containing from 12.1 through 20.0 percent, by weight, of substandard raisins under a weight dockage system. The creditable weight of each lot of raisins acquired under

the substandard dockage system shall be obtained by multiplying the net weight of the lot of raisins by the applicable dockage factor from the appropriate dockage table prescribed in paragraph (b) or (c) of this section.

(b) *Substandard dockage table applicable to Natural (sun-dried) Seedless, Golden Seedless, Dipped Seedless, Monukka, Other Seedless, and Other Seedless-Sulfured raisins.*

Percent substandard	Dockage factor
5.0 or less	(¹)
5.1	.999
5.2	.998
5.3	.997
5.4	.996
5.5	.995

¹ No dockage.

NOTE TO PARAGRAPH (b): Percentages in excess of the last percentage shown in the table shall be expressed in the same increment as the foregoing, and the dockage factor for each such increment shall be .001 less than the dockage factor for the preceding increment. Deliveries in excess of 17.0 percent would be off-grade; therefore, the dockage factor does not apply.

(c) *Substandard dockage table applicable to Muscat (including other raisins with seeds), Sultana and Zante Currant raisins.*

Percent substandard	Dockage factor
12.0 or less	(¹)
12.1	.999
12.2	.998
12.3	.997
12.4	.996
12.5	.995

¹ No dockage.

NOTE TO PARAGRAPH (c): Percentages in excess of the last percentage shown in the table shall be expressed in the same increments as the foregoing, and the dockage factor for each increment shall be .001 less than the dockage factor for the preceding increment. Deliveries in excess of 20.0 percent would be off-grade; therefore, the dockage factor does not apply.

[57 FR 28597, June 26, 1992, as amended at 63 FR 56785, Oct. 23, 1998; 67 FR 36792, May 28, 2002; 68 FR 42947, July 21, 2003]

§989.213 Maturity dockage.

(a) *General.* Subject to prior agreement between handler and tenderer, Natural (sun-dried) Seedless, Golden Seedless, Dipped Seedless, Monukka,

Other Seedless, and Other Seedless-Sulfured raisins containing from 35.0 percent through 49.9 percent, by weight, of well-matured or reasonably well-matured raisins may be acquired by a handler under a weight dockage system. The creditable weight of each lot of raisins acquired under the maturity dockage system shall be obtained by multiplying the net weight of the lot of raisins by the applicable dockage factor from the dockage table prescribed in paragraphs (b), (c), and (d) of this section.

(b) Maturity dockage table applicable to lots of Natural (sun-dried) Seedless, Golden Seedless, Dipped Seedless, Monukka, Other Seedless, and Other Seedless-Sulfured raisins which contain 45.0 percent through 49.9 percent well-matured or reasonably well-matured raisins:

Percent well-matured or reasonably well-matured:	Dockage factor
50.0 or more	(¹)
49.9	0.9995
49.8	.9990
49.7	.9985
49.6	.9980
49.5	.9975

¹ No dockage.

NOTE: Percentages less than the last percentage shown in the table, down to 45.0 percent, shall be expressed in the same increments as the foregoing, and the dockage for each such increment shall be .0005 less than the dockage factor for the preceding increment.

(c) Maturity dockage table applicable to lots of Natural (sun-dried) Seedless, Golden Seedless, Dipped Seedless, Monukka, Other Seedless, and Other Seedless-Sulfured raisins which contain 40.0 percent through 44.9 percent well-matured or reasonably well-matured raisins:

Percent well-matured or reasonably well-matured:	Dockage factor
44.9	0.974
44.8	.973
44.7	.972
44.6	.971
44.5	.970
44.4	.969

NOTE: Percentages less than the last percentage shown in the table, down to 40.0 percent, shall be expressed in the same increments as the foregoing, and the dockage factor for each such increment shall be .001 less

than the dockage factor for the preceding increment.

(d) Maturity dockage table applicable to lots of Natural (sun-dried) Seedless, Golden Seedless, Dipped Seedless, Monukka, Other Seedless, and Other Seedless-Sulfured raisins which contain 35.0 percent through 39.9 percent well-matured or reasonably well-matured raisins:

Percent well-matured or reasonably well-matured:	Dockage factor
39.9	0.9235
39.8	.9220
39.7	.9205
39.6	.9190
39.5	.9175
39.4	.9160

NOTE TO PARAGRAPH (d): Percentages less than the last percentage shown in the table shall be expressed in the same increments as the foregoing, and the dockage factor for each such increment shall be .0015 less than the dockage factor for the preceding increment.

[25 FR 12813, Dec. 14, 1960; 27 FR 2506, Mar. 16, 1962, as amended at 52 FR 32776, Aug. 31, 1987; 53 FR 34715, Sept. 8, 1988; 53 FR 49296, Dec. 7, 1988; 54 FR 43041, Oct. 20, 1989; 63 FR 56785, Oct. 23, 1998; 67 FR 36792, May 28, 2002; 68 FR 42947, July 21, 2003]

§§ 989.221—989.257　[Reserved

Subpart D—Assessment Rates

§ 989.347　Assessment rate.

On and after August 1, 2018, an assessment rate of $22.00 per ton is established for assessable raisins produced from grapes grown in California.

[84 FR 2051, Feb. 6, 2019]

Subpart E—Conversion Factors

§ 989.601　Conversion factors for raisin weight.

The following factors for the named varietal types of raisins shall be used to convert the net weight of reconditioned raisins acquired by handlers as packed raisins to natural condition weight. The net weight of the raisins after the completion of processing shall be divided by the applicable factor to obtain the natural condition weight: *Provided,* That the adjusted weight does not exceed the original weight of the raisins prior to reconditioning; and *Provided further,* That, if the adjusted weight exceeds the original weight, the original weight will be used.

Varietal type	Conversion factor
Natural (sun-dried) Seedless	0.92
Golden Seedless, Dipped Seedless, Other Seedless, and Other Seedless-Sulfured	0.95
Muscats (including raisins with seeds):	
Seeded	0.80
Unseeded	0.92
Sultana	0.92
Zante Currant	0.91

[54 FR 41587, Oct. 11, 1989, as amended at 67 FR 36792, May 28, 2002; 68 FR 42947, July 21, 2003]

Subpart F—Quality Control

§ 989.701 Minimum grade and condition standards for natural condition raisins.

Effective pursuant to § 989.58, raisins meeting the varietal standards hereinafter set forth shall be considered as standard raisins and those failing to meet such standards shall be considered as off-grade raisins. Where the raisins in any lot consist of two or more varietal types commingled within their containers, the lot shall be considered as a mixed lot and as standard raisins if they meet for each defect the most restrictive requirements for the varietal types of raisins comprising the lot. In the event layered Muscats (including other raisins with seeds) or Cluster Seedless raisins are commingled within their containers with loose Muscats (including other raisins with seeds) or loose Cluster Seedless raisins respectively, the entire lot shall be considered as loose Muscats (including other raisins with seeds) or Natural (sun-dried) Seedless raisins. The raisins shall be considered as standard raisins if the lot as a whole meets the minimum standards for loose Muscats (including other raisins with seeds) or Natural (sun-dried) Seedless raisins: *Provided,* That with respect to the requirements peculiar to a varietal type such as possessing characteristic color, flavor, or odor, the raisins shall be considered as meeting such requirements if they have been properly prepared as raisins. In each category, only those raisins which have been properly dried and cured in original natural condition, are free from active infestation, and are in such condition that they are capable of being received, stored, and packed without undue deterioration or spoilage, shall be considered as storable raisins.

(a) *Natural (sun-dried) Seedless, Monukka and Other Seedless raisins.* Natural condition Natural (sun-dried) Seedless, Monukka and Other Seedless raisins shall have been prepared from sound, wholesome, matured grapes properly dried and cured, and shall meet the following additional requirements: (1) Shall be fairly free from damage by sugaring, mechanical injury, sunburn, or other similar injury; (2) shall have a normal characteristic color, flavor, and odor of properly prepared raisins; (3) shall contain not more than 5 percent, by weight, of substandard raisins (raisins that show development less than that characteristic of raisins prepared from fairly well-matured grapes), and shall also contain at least 50 percent well-matured or reasonably well-matured raisins; (4) shall not exceed 16 percent moisture as determined by the dried fruit moisture tester method, except that there shall be no maximum moisture content for Cluster Seedless raisins; and (5) shall be of such quality and condition as can be expected to withstand storage as provided in the order and that when processed in accordance with good commercial practice will meet the minimum standards for processed raisins established by the Committee, and that with respect to Cluster Seedless raisins, in addition to the above requirements the raisins shall be fairly free from shattered (or loose end) berries, and be uniformly cured; shall contain 30 percent or more "2 Crown" or larger size berries; and shall be of such quality and condition that when processed in accordance with good commercial practice will, except for moisture content, meet the minimum standards for processed raisins established by the Committee.

(b) *Dipped Seedless, Oleate and Related Seedless, and Other Seedless-Sulfured raisins.* Natural condition Dipped Seedless, and Other Seedless-Sulfured raisins shall have been prepared from sound, wholesome, matured grapes properly dried and cured, and shall meet the following additional requirements:

(1) Shall be fairly free from damage by sugaring, mechanical injury, sunburn, or other similar injury;

(2) Shall have a normal characteristic flavor and odor of properly prepared raisins;

(3) Shall contain not more than 5 percent, by weight, of substandard raisins (raisins that show development less than that characteristic of raisins prepared from fairly well-matured grapes), and for the 1985–86 and subsequent years also contain at least 50 percent well-matured or reasonably well-matured raisins;

(4) Shall not exceed 14 percent moisture as determined by the dried fruit moisture tester method,

(5) Shall be of such quality and condition as can be expected to withstand storage as provided in the order and that when processed in accordance with good commercial practice will meet the minimum standards for processed raisins established by the Committee.

(c) *Golden Seedless.* Natural condition Golden Seedless raisins shall have been prepared from sound, wholesome, matured grapes properly dried and cured, and shall meet the following additional requirements: (1) Shall be fairly free from damage by sugaring, mechanical injury, sunburn, or other similar injury; (2) shall have a normal characteristic flavor and odor of properly prepared raisins; (3) shall contain not more than 5 percent, by weight, of substandard raisins (raisins that show development less than that characteristic of raisins prepared from fairly well-matured grapes), and for the 1985–86 and subsequent crop years also contain at least 50 percent well-matured or reasonably well-matured raisins; (4) shall not exceed 14 percent moisture as determined by the dried fruit moisture tester method, (5) shall be of such quality and condition as can be expected to withstand storage as provided in the

order and that when processed in accordance with good commercial practice will meet the minimum standards for processed raisins established by the Committee; and (6) shall possess a color varying from yellowish green to dark amber or dark greenish amber with not more than 15 percent, by weight, of all the raisins being definitely dark berries. *Definitely dark berries* means raisins which are definitely darker than dark amber and characteristic of "naturally" raisined grapes.

(d) *Muscats (including other raisins with seeds).* Natural condition Muscat raisins (including other raisins with seeds) shall have been prepared from sound, wholesome, matured grapes properly dried and cured, and shall meet the following additional requirements:

(1) Shall be fairly free from damage by sugar, mechanical injury, sunburn or other similar injury;

(2) Shall have a normal characteristic color, flavor, and odor of properly prepared raisins and shall contain not more than 12 percent, by weight, of substandard raisins (raisins that show development less than that characteristic of raisins prepared from fairly well-matured grapes);

(3) Shall not exceed 16 percent moisture as determined by the dried fruit moisture tester method, except that water dipped, vine sprayed or similarly treated Muscats (including other raisins with seeds) shall not exceed 14 percent moisture, and that there shall be no maximum moisture content for layered Muscats (including other raisins with seeds);

(4) The raisins shall be of such quality and condition as can be expected to withstand storage as provided in the marketing agreement and order, and that when processed in accordance with good commercial practice will meet the minimum standards for processed raisins established by the Committee, and that with respect to layered Muscats (including other raisins with seeds), in addition to the above requirements the raisins shall be fairly free from shattered (or loose end) berries; uniformly cured; 30 percent or more "3 Crown" or larger size; of such

quality and condition that when processed in accordance with good commercial practice will, except for moisture content, meet the minimum standards for processed raisins established by the Committee.

(e) *Sultana Raisins.* Natural condition Sultana raisins shall have been prepared from sound, wholesome, matured grapes properly dried and cured, and shall meet the following additional requirements:

(1) Shall be fairly free from damage by sugaring, mechanical injury, sunburn, or other similar injury;

(2) Shall have a normal characteristic color, flavor, and odor of properly prepared raisins and shall contain not more than 12 percent, by weight, of substandard raisins (raisins that show development less than that characteristic of raisins prepared from fairly well-matured grapes);

(3) Shall not exceed 16 percent moisture as determined by the dried fruit moisture tester method; and

(4) The raisins shall be of such quality and condition as can be expected to withstand storage as provided in the marketing agreement and order, and that when processed in accordance with good commercial practice will meet the minimum standards for processed raisins established by the Committee.

(f) *Zante Currant Raisins.* Natural condition Zante Currant raisins shall have been prepared from sound, wholesome, matured grapes properly dried and cured, and shall meet the following additional requirements; (1) Shall be fairly free from damage by sugaring, mechanical injury, sunburn, or other similar injury; (2) shall have a normal characteristic color, flavor, and odor of properly prepared raisins and shall contain not more than 12 percent, by weight, of substandard raisins (raisins that show development less than that characteristic of raisins prepared from fairly well-matured grapes); (3) shall not exceed 16 percent moisture as determined by the dried fruit moisture tester method; and (4) the raisins shall be of such quality and condition as can be expected to withstand storage as provided in the marketing agreement and order, and that when processed in accordance with good commercial prac-

tice will meet the minimum standards for processed raisins established by the Committee.

[42 FR 52378, Sept. 30, 1977, as amended at 46 FR 39121, July 31, 1981; 48 FR 49215, Oct. 25, 1983; 49 FR 1669, Jan. 13, 1984; 49 FR 33994, Aug. 28, 1984; 50 FR 35771, Sept. 4, 1985; 53 FR 34715, Sept. 8, 1988; 67 FR 36793, May 28, 2002; 68 FR 42947, July 21, 2003]

§989.702 Minimum grade standards for packed raisins.

Effective pursuant to §989.59, the minimum grade standards for packed raisins shall be as follows:

(a) *Natural (sun-dried) Seedless, Dipped Seedless, and Other Seedless-Sulfured raisins.* Packed Natural (sun-dried) Seedless, Dipped Seedless, and Other Seedless-Sulfured raisins shall meet the requirements of U.S. Grade C as defined in the effective United States Standards for Grades of Processed Raisins (§§ 52.1841 through 52.1858 of this title): *Provided,* That at least 70.0 percent, by weight, of the raisins shall be well-matured or reasonably well-matured. With respect to select-sized and mixed-sized raisin lots, the raisins shall at least meet the U.S. Grade B tolerances for pieces of stem, and underdeveloped and substandard raisins, and small sized raisins shall meet the U.S. Grade C tolerances for those factors.

(b) *Golden Seedless Raisins.* Packed Golden Seedless raisins shall at least meet the requirements prescribed in paragraph (a) of this section, and the color requirements for "colored" as defined in said standards.

(c) *Monukka and Other Seedless Raisins.* Packed Monukka and Other Seedless raisins shall at least meet the requirements prescribed in paragraph (a) of this section, except that the tolerance for moisture shall be 19 percent rather than 18 percent.

(d) *Muscat (including other raisins with seeds) Raisins.* Packed Muscat (including other raisins with seeds) raisins shall at least meet the requirements of U.S. Grade C of the said standards. Layer Muscat (including other raisins with seeds) raisins shall at least meet U.S. Grade B as defined for "Layer or Cluster Raisins With Seeds" in said standards, except for the provisions therein relating to moisture content.

(e) *Sultana Raisins.* Packed Sultana raisins shall at least meet the requirements of U.S. Grade C as defined in said standards.

(f) *Zante Currant Raisins.* Packed Zante Currant raisins shall at least meet the requirements of U.S. Grade B as defined in said standards.

(g) *Cluster Seedless Raisins*—(1) *Description.* Raisins referred to as *Cluster Seedless raisins* means the raisins have not been detached from the main bunch. Cluster Seedless raisins shall at least meet the requirements of Marketing Order Grade B prescribed in this paragraph. The processed raisins are prepared from clean, sound, dried grapes; are stored or cleaned, or both, and are washed with water to assure a wholesome product.

(2) *Grades.* (i) Marketing Order Grade A is a quality of Cluster Seedless raisins that have similar varietal characteristics; have a good typical color; have a good characteristic flavor; are uniformly cured and show development characteristics of raisins prepared from well-matured grapes; contain not more than 23 percent, by weight, of moisture; that not less than 30 percent, by weight, of the raisins, are "2 Crown" size or larger and meet the additional requirements as outlined in the table in paragraph (2)(iv) of this paragraph.

(ii) Marketing Order Grade B is the quality of the Cluster Seedless raisins that have similar varietal characteristics; have a reasonably good typical color; have a good characteristic flavor; are uniformly cured and show characteristics of raisins prepared from reasonably well-matured grapes; contain not more than 23 percent, by weight, of moisture; that not less than 30 percent, by weight, of raisins, exclusive of stems and branches, are "2 Crown" size or larger and meet the additional requirements as outlined in the table in paragraph (2)(iv) of this paragraph.

(iii) Substandard is the quality of Cluster Seedless raisins that fail to meet the requirements of Marketing Order Grade B.

(iv) Allowances for defects in Cluster Seedless raisins:

Defects	Marketing order grade A	Marketing order grade B
	Maximum (percent by weight)	
Sugared	5	10
Discolored, damaged, or moldy.	5	7
Provided these limits are not exceed:		
Damaged	3	4
Moldy	2	3
Substandard Development and Undeveloped.	2	5
Shattered (or loose) individual berries and small clusters of 2 or 3 berries each.	Practically free	Reasonably free.
	Appearance or edibility of product	
Slightly discolored or damaged by fermentation or any other defect not described above.	May not be affected	May not be more than slightly affected.
Grit, sand, or silt	None of any consequence may be present that affects the appearance or edibility of the product.	

Defects	Marketing order grade A	Marketing order grade B
	Maximum (percent by weight)	
Sugared	5	10
Discolored, damaged, or moldy.	5	7
Provided these limits are not exceed:		
Damaged	3	4
Moldy	2	3
Substandard Development and Undeveloped.	2	5
Shattered (or loose) individual berries and small clusters of 2 or 3 berries each.	Practically free	Reasonably free.
	Appearance or edibility of product	
Slightly discolored or damaged by fermentation or any other defect not described above.	May not be affected.	May not be more than slightly affected.
Grit, sand, or silt	None of any consequence may be present that affects the appearance or edibility of the product.	

(h) A handler may grind raisins which do not meet the minimum grade standards prescribed in paragraphs (a) through (g) of this section because of mechanical damage or sugaring, into a raisin paste.

[49 FR 33994, Aug. 28, 1984, as amended at 50 FR 35772, Sept. 4, 1985; 53 FR 34715, Sept. 8, 1988; 67 FR 36793, May 28, 2002; 68 FR 42947, July 21, 2003; 81 FR 84403, Nov. 23, 2016]

Subpart G—Antitrust Immunity and Liability

§ 989.801　Restrictions applicable to Committee personnel.

Members and employees of the Raisin Administrative Committee are immune from prosecution under the United States antitrust laws only insofar as their conduct in administering the Raisin Marketing Order is authorized by the Agricultural Marketing Agreement Act of 1937, 7 U.S.C. 601 *et seq.*, or the provisions of the order. Under the antitrust laws. Committee members and employees may not engage in any unauthorized agreement or concerted action that unreasonably restrains United States domestic or foreign commerce. For example, Committee members and employees have no authority to participate, either directly or indirectly, whether on an informal or formal, written or oral basis, in any bilateral or international undertaking or agreement with any competing foreign producer or seller or with any foreign government, agency, or instrumentality acting on behalf of competing foreign producers or sellers to (a) raise, fix, stabilize, or set a floor for raisin, sultana, or currant prices, or (b) limit the quantity or quality of raisins, sultanas, or currants imported into or exported from the United States. Participation in any such unauthorized agreement or joint undertaking could result in prosecution under the antitrust laws by the United States Department of Justice and/or suit by injured private persons seeking treble damages, and could also result in expulsion of members from the Committee or termination of employment with the Committee.

[46 FR 39984, Aug. 6, 1981]

PART 990—DOMESTIC HEMP PRODUCTION PROGRAM

Subpart A—Definitions

Sec.
990.1　Meaning of terms.

Subpart B—State and Tribal Hemp Production Plans

990.2　State and Tribal plans; General authority.
990.3　State and Tribal plans; Plan requirements.
990.4　USDA approval of State and Tribal plans.
990.5　Audit of State or Tribal plan compliance.
990.6　Violations of State and Tribal plans.
990.7　Establishing records with USDA Farm Service Agency.
990.8　Production under Federal law.

Subpart C—USDA Hemp Production Plan

990.20　USDA requirements for the production of hemp.
990.21　USDA hemp producer license.
990.22　USDA hemp producer license approval.
990.23　Reporting hemp crop acreage with USDA Farm Service Agency.
990.24　Responsibility of a USDA licensed producer prior to harvest.
990.25　Standards of performance for detecting delta-9 tetrahydrocannabinol (THC) concentration levels.
990.26　Responsibility of a USDA producer after laboratory testing is performed.
990.27　Non-compliant cannabis plants.
990.28　Compliance.
990.29　Violations.
990.30　USDA producers; License suspension.
990.31　USDA licensees; Revocation.
990.32　Recordkeeping requirements.

Subpart D—Appeals

990.40　General adverse action appeal process.
990.41　Appeals under the USDA hemp production plan.
990.42　Appeals under a State or Tribal hemp production plan.

Subpart E—Administrative Provisions

990.60　Agents.
990.61　Severability.
990.62　Expiration of this part.
990.63　Interstate transportation of hemp.

Subpart F—Reporting Requirements

990.70　State and Tribal hemp reporting requirements.

990.71 USDA plan reporting requirements.

AUTHORITY: 7 U.S.C. 1639o note, 1639p, 16939q, and 1639r.

SOURCE: 84 FR 58554, Oct. 31, 2019, unless otherwise noted.

EFFECTIVE DATE NOTE: At 84 FR 58554, Oct. 31, 2019, part 990 was added, effective Oct. 31, 2019 through Nov. 1, 2021.

Subpart A—Definitions

§ 990.1 Meaning of terms.

Words used in this subpart in the singular form shall be deemed to impart the plural, and vice versa, as the case may demand. For the purposes of provisions and regulations of this part, unless the context otherwise requires, the following terms shall be construed, respectively, to mean:

Acceptable hemp THC level. When a laboratory tests a sample, it must report the delta-9 tetrahydrocannabinol content concentration level on a dry weight basis and the measurement of uncertainty. The acceptable hemp THC level for the purpose of compliance with the requirements of State, Tribal, or USDA hemp plans is when the application of the measurement of uncertainty to the reported delta-9 tetrahydrocannabinol content concentration level on a dry weight basis produces a distribution or range that includes 0.3% or less. For example, if the reported delta-9 tetrahydrocannabinol content concentration level on a dry weight basis is 0.35% and the measurement of uncertainty is ±0.06%, the measured delta-9 tetrahydrocannabinol content concentration level on a dry weight basis for this sample ranges from 0.29% to 0.41%. Because 0.3% is within the distribution or range, the sample is within the acceptable hemp THC level for the purpose of plan compliance. This definition of "acceptable hemp THC level" affects neither the statutory definition of hemp, 7 U.S.C. 1639o(1), in the 2018 Farm Bill nor the definition of "marihuana," 21 U.S.C. 802(16), in the CSA.

Act. Agricultural Marketing Act of 1946.

Agricultural Marketing Service or *AMS.* The Agricultural Marketing Service of the U.S. Department of Agriculture.

Applicant. An applicant is:

(1) A State or Indian Tribe that has submitted a State or Tribal hemp production plan to USDA for approval under this part; or

(2) A producer in a State or territory of an Indian Tribe who is not subject to a State or Tribal hemp production plan and who has submitted an application for a license under the USDA hemp production plan under this part.

Cannabis. A genus of flowering plants in the family Cannabaceae of which *Cannabis sativa* is a species, and *Cannabis indica* and *Cannabis ruderalis* are subspecies thereof. Cannabis refers to any form of the plant in which the delta-9 tetrahydrocannabinol concentration on a dry weight basis has not yet been determined.

Controlled Substances Act (CSA). The Controlled Substances Act as codified in 21 U.S.C. 801 *et seq.*

Conviction. Means any plea of guilty or nolo contendere, or any finding of guilt, except when the finding of guilt is subsequently overturned on appeal, pardoned, or expunged. For purposes of this part, a conviction is expunged when the conviction is removed from the individual's criminal history record and there are no legal disabilities or restrictions associated with the expunged conviction, other than the fact that the conviction may be used for sentencing purposes for subsequent convictions. In addition, where an individual is allowed to withdraw an original plea of guilty or nolo contendere and enter a plea of not guilty and the case is subsequently dismissed, the individual is no longer considered to have a conviction for purposes of this part.

Corrective action plan. A plan established by a State, Tribal government, or USDA for a licensed hemp producer to correct a negligent violation or noncompliance with a hemp production plan and this part.

Criminal History Report. Criminal history report means the Federal Bureau of Investigation's Identity History Summary.

Culpable mental state greater than negligence. To act intentionally, knowingly, willfully, or recklessly.

Decarboxylated. The completion of the chemical reaction that converts THC-acid (THC-A) into delta-9-THC, the intoxicating component of cannabis. The decarboxylated value is also calculated using a conversion formula that sums delta-9-THC and eighty-seven and seven tenths (87.7) percent of THC-acid.

Decarboxylation. The removal or elimination of carboxyl group from a molecule or organic compound.

Delta-9 tetrahydrocannabinol or *THC.* Delta-9-THC is the primary psychoactive component of cannabis. For the purposes of this part, delta-9-THC and THC are interchangeable.

Drug Enforcement Administration or DEA. The United States Drug Enforcement Administration.

Dry weight basis. The ratio of the amount of moisture in a sample to the amount of dry solid in a sample. A basis for expressing the percentage of a chemical in a substance after removing the moisture from the substance. Percentage of THC on a dry weight basis means the percentage of THC, by weight, in a cannabis item (plant, extract, or other derivative), after excluding moisture from the item.

Entity. A corporation, joint stock company, association, limited partnership, limited liability partnership, limited liability company, irrevocable trust, estate, charitable organization, or other similar organization, including any such organization participating in the hemp production as a partner in a general partnership, a participant in a joint venture, or a participant in a similar organization.

Farm Service Agency or *FSA.* An agency of the United States Department of Agriculture.

Gas chromatography or *GC.* A type of chromatography in analytical chemistry used to separate, identify, and quantify each component in a mixture. GC relies on heat for separating and analyzing compounds that can be vaporized without decomposition.

Geospatial location. For the purposes of this part, "geospatial location" means a location designated through a global system of navigational satellites used to determine the precise ground position of a place or object.

Handle. To harvest or store hemp plants or hemp plant parts prior to the delivery of such plants or plant parts for further processing. "Handle" also includes the disposal of cannabis plants that are not hemp for purposes of chemical analysis and disposal of such plants.

Hemp. The plant species *Cannabis sativa* L. and any part of that plant, including the seeds thereof and all derivatives, extracts, cannabinoids, isomers, acids, salts, and salts of isomers, whether growing or not, with a delta-9 tetrahydrocannabinol concentration of not more than 0.3 percent on a dry weight basis.

High-performance liquid chromatography or *HPLC.* A type of chromatography technique in analytical chemistry used to separate, identify, and quantify each component in a mixture. HPLC relies on pumps to pass a pressurized liquid solvent containing the sample mixture through a column filled with a solid adsorbent material to separate and analyze compounds.

Indian Tribe. As defined in section 4 of the Indian Self-Determination and Education Assistance Act (25 U.S.C. 5304).

Information sharing system. The database mandated under the Act which allows USDA to share information collected under State, Tribal, and USDA plans with Federal, State, Tribal, and local law enforcement.

Key participants. A sole proprietor, a partner in partnership, or a person with executive managerial control in a corporation. A person with executive managerial control includes persons such as a chief executive officer, chief operating officer and chief financial officer. This definition does not include non-executive managers such as farm, field, or shift managers.

Law enforcement agency. Any Federal, State, or local law enforcement agency.

Lot. A contiguous area in a field, greenhouse, or indoor growing structure containing the same variety or strain of cannabis throughout the area.

Marijuana. As defined in the CSA, "marihuana" means all parts of the plant Cannabis sativa L., whether growing or not; the seeds thereof; the resin extracted from any part of such plant; and every compound, manufacture, salt, derivative, mixture, or preparation of such plant, its seeds or resin. The term 'marihuana' does not include hemp, as defined in section 297A of the Agricultural Marketing Act of 1946, and does not include the mature stalks of such plant, fiber produced from such stalks, oil or cake made from the seeds of such plant, any other compound, manufacture, salt, derivative, mixture, or preparation of such mature stalks (except the resin extracted therefrom), fiber, oil, or cake, or the sterilized seed of such plant which is incapable of germination (7 U.S.C. 1639o). "Marihuana" means all cannabis that tests as having a concentration level of THC on a dry weight basis of higher than 0.3 percent.

Measurement of Uncertainty (MU). The parameter, associated with the result of a measurement, that characterizes the dispersion of the values that could reasonably be attributed to the particular quantity subject to measurement.

Negligence. Failure to exercise the level of care that a reasonably prudent person would exercise in complying with the regulations set forth under this part.

Phytocannabinoid. Cannabinoid chemical compounds found in the cannabis plant, two of which are Delta-9 tetrahydrocannabinol (delta-9 THC) and cannabidiol (CBD).

Plan. A set of criteria or regulations under which a State or Tribal government, or USDA, monitors and regulates the production of hemp.

Postdecarboxylation. In the context of testing methodologies for THC concentration levels in hemp, means a value determined after the process of decarboxylation that determines the total potential delta-9 tetrahydrocannabinol content derived from the sum of the THC and THC-A content and reported on a dry weight basis. The postdecarboxylation value of THC can be calculated by using a chromatograph technique using heat, gas chromatography, through which THCA is converted from its acid form to its neutral form, THC. Thus, this test calculates the total potential THC in a given sample. The postdecarboxylation value of THC can also be calculated by using a high-performance liquid chromatograph technique, which keeps the THC-A intact, and requires a conversion calculation of that THC-A to calculate total potential THC in a given

sample. See the definition for decarboxylation.

Produce. To grow hemp plants for market, or for cultivation for market, in the United States.

Producer. Producer means a producer as defined in 7 CFR 718.2 that is licensed or authorized to produce hemp under this part.

Reverse distributor. A person who is registered with the DEA in accordance with 21 CFR 1317.15 to dispose of marijuana under the Controlled Substances Act.

Secretary. The Secretary of Agriculture of the United States.

State. Any one of the fifty States of the United States of America, the District of Columbia, the Commonwealth of Puerto Rico, and any other territory or possession of the United States.

State department of agriculture. The agency, commission, or department of a State government responsible for agriculture in the State.

Territory of the Indian Tribe has the same meaning as "Indian Country" in 18 U.S.C. 1151.

Tribal government. The governing body of an Indian Tribe.

USDA licensed hemp producer or licensee. A person, partnership, or corporation authorized by USDA to produce hemp.

Subpart B—State and Tribal Hemp Production Plans

§ 990.2 State and Tribal plans; General authority.

States or Indian Tribes desiring to have primary regulatory authority over the production of hemp in the State or territory of the Indian Tribe for which it has jurisdiction shall submit to the Secretary for approval, through the State department of agriculture (in consultation with the Governor and chief law enforcement officer of the State) or the Tribal government, as applicable, a plan under which the State or Indian Tribe monitors and regulates that production.

§ 990.3 State and Tribal plans; Plan requirements.

(a) *General requirements.* A State or Tribal plan submitted to the Secretary for approval must include the practice and procedures described in this paragraph (a).

(1) A State or Tribal plan must include a practice to collect, maintain, and report to the Secretary relevant, real-time information for each producer licensed or authorized to produce hemp under the State or Tribal plan regarding:

(i) Contact information as described in § 990.70(a)(1);

(ii) A legal description of the land on which the producer will produce hemp in the State or territory of the Indian Tribe including, to the extent practicable, its geospatial location; and

(iii) The status and number of the producer's license or authorization.

(2) A State or Tribal plan must include a procedure for accurate and effective sampling of all hemp produced, to include the requirements in this paragraph (a)(2).

(i) Within 15 days prior to the anticipated harvest of cannabis plants, a Federal, State, local, or Tribal law enforcement agency or other Federal, State, or Tribal designated person shall collect samples from the flower material from such cannabis plants for delta-9 tetrahydrocannabinol concentration level testing as described in §§ 990.24 and 990.25.

(ii) The method used for sampling from the flower material of the cannabis plant must be sufficient at a confidence level of 95 percent that no more than one percent (1%) of the plants in the lot would exceed the acceptable hemp THC level. The method used for sampling must ensure that a representative sample is collected that represents a homogeneous composition of the lot.

(iii) During a scheduled sample collection, the producer or an authorized representative of the producer shall be present at the growing site.

(iv) Representatives of the sampling agency shall be provided with complete and unrestricted access during business hours to all hemp and other cannabis plants, whether growing or harvested, and all land, buildings, and other structures used for the cultivation, handling, and storage of all hemp and other cannabis plants, and all locations listed in the producer license.

(v) A producer shall not harvest the cannabis crop prior to samples being taken.

(3) A State or Tribal plan must include a procedure for testing that is able to accurately identify whether the sample contains a delta-9 tetrahydrocannabinol content concentration level that exceeds the acceptable hemp THC level. The procedure must include a validated testing methodology that uses postdecarboxylation or other similarly reliable methods. The testing methodology must consider the potential conversion of delta-9 tetrahydrocannabinolic *acid* (THC-A) in hemp into THC and the test result measures total available THC derived from the sum of the THC and THC-A content. Testing methodologies meeting the requirements of this paragraph (a)(3) include, but are not limited to, gas or liquid chromatography with detection. The total THC concentration level shall be determined and reported on a dry weight basis.

(i) Any test of a representative sample resulting in higher than the acceptable hemp THC level shall be conclusive evidence that the lot represented by the sample is not in compliance with this part. Lots tested and not certified by the DEA-registered laboratory at or below the acceptable hemp THC level may *not* be further handled, processed or enter the stream of commerce and the producer shall ensure the lot is disposed of in accordance with § 990.27.

(ii) Samples of hemp plant material from one lot shall not be commingled with hemp plant material from other lots.

(iii) Analytical testing for purposes of detecting the concentration levels of THC shall meet the following standards:

(A) Laboratory quality assurance must ensure the validity and reliability of test results;

(B) Analytical method selection, validation, and verification must ensure that the testing method used is appropriate (fit for purpose), and that the laboratory can successfully perform the testing;

(C) The demonstration of testing validity must ensure consistent, accurate analytical performance;

(D) Method performance specifications must ensure analytical tests are sufficiently sensitive for the purposes of the detectability requirements of this part; and

(E) An effective disposal procedure for hemp plants that are produced that do not meet the requirements of this part. The procedure must be in accordance with DEA reverse distributor regulations found at 21 CFR 1317.15.

(F) Measurement of uncertainty (MU) must be estimated and reported with test results. Laboratories shall use appropriate, validated methods and procedures for all testing activities and evaluate measurement of uncertainty.

(4) A State or Indian Tribe shall promptly notify the Administrator by certified mail or electronically of any occurrence of cannabis plants or plant material that do not meet the definition of hemp in this part and attach the records demonstrating the appropriate disposal of all of those plants and materials in the lot from which the representative samples were taken.

(5) A State or Tribal plan must include a procedure to comply with the enforcement procedures in § 990.6.

(6) A State or Tribal plan must include a procedure for conducting annual inspections of, at a minimum, a random sample of producers to verify that hemp is not produced in violation of this part. These procedures must enforce the terms of violations as stated in the Act and defined under § 990.6.

(7) A State or Tribal plan must include a procedure for submitting the information described in § 990.70 to the Secretary not more than 30 days after the date on which the information is received. All such information must be submitted to the USDA in a format that is compatible with USDA's information sharing system.

(8) The State or Tribal government must certify that the State or Indian Tribe has the resources and personnel to carry out the practices and procedures described in paragraphs (a)(1) through (7) of this section.

(9) The State or Tribal plan must include a procedure to share information with USDA to support the information sharing requirements in 7 U.S.C. 1639q(d). The procedure must include the requirements described in this paragraph (a)(9).

(i) The State or Tribal plan shall require producers to report their hemp crop acreage to the FSA, consistent with the requirement in § 990.7.

(ii) The State or Tribal government shall assign each producer with a license or authorization identifier in a format prescribed by USDA.

(iii) The State or Tribal government shall require producers to report the total acreage of hemp planted, harvested, and, if applicable, disposed. The State or Tribal government shall collect this information and report it to AMS.

(b) *Relation to State and Tribal law.* A State or Tribal plan may include any other practice or procedure established by a State or Indian Tribe, as applicable; *Provided,* That the practice or procedure is consistent with this part and Subtitle G of the Act.

(1) *No preemption.* Nothing in this part preempts or limits any law of a State or Indian Tribe that:

(i) Regulates the production of hemp; and

(ii) Is more stringent than this part or Subtitle G of the Act.

(2) *References in plans.* A State or Tribal plan may include a reference to a law of the State or Indian Tribe regulating the production of hemp, to the extent that the law is consistent with this part.

§ 990.4 USDA approval of State and Tribal plans.

(a) *General authority.* No plans will be accepted by USDA prior to October 31, 2019. No later than 60 calendar days after the receipt of a State or Tribal plan for a State or Tribal Nation in which production of hemp is legal, the Secretary shall:

(1) Approve the State or Tribal plan only if the State or Tribal plan complies with this part; or

(2) Disapprove the State or Tribal plan if the State or Tribal plan does not comply with this part. USDA shall provide written notification to the State or Tribe of the disapproval and the cause for the disapproval.

(b) *Amended plans.* A State or Tribal government, as applicable, must submit to the Secretary an amended plan if:

(1) The Secretary disapproves a State or Tribal plan if the State or Tribe wishes to have primary jurisdiction over hemp production within its State or territory of the Indian Tribe; or

(2) The State or Tribe makes substantive revisions to its plan or its laws which alter the way the plan meets the requirements of this part. If this occurs, the State or Tribal government must re-submit the plan with any modifications based on laws and regulation changes for USDA approval. Such re-submissions should be provided to USDA within 365 days from the date that the State or Tribal laws and regulations are effective. Producers shall continue to comply with the requirements of the existing plan while such modifications are under consideration by USDA. If State or Tribal government laws or regulations in effect under the USDA-approved plan change but the State or Tribal government does not re-submit a modified plan within one year from the effective date of the new law or regulation, the existing plan is revoked.

(3) USDA approval of State or Tribal government plans shall remain in effect unless an amended plan must be submitted to USDA because of a substantive revision to a State's or Tribe's plan, a relevant change in State or Tribal laws or regulations, or approval of the plan is revoked by USDA.

(c) *Technical assistance.* The Secretary may provide technical assistance to help a State or Indian Tribe develop or amend a plan. This may include the review of draft plans or other informal consultation as necessary.

(d) *Approved State or Tribal plans.* If the Secretary approves a State or Tribal plan, the Secretary shall notify the State or Tribe by letter or email.

(1) In addition to the approval letter, the State or Tribe shall receive their plan approval certificate either as an attachment or assessable via website link.

(2) The USDA shall post information regarding approved plans on its website.

(3) USDA approval of State or Tribal government plans shall remain in effect unless:

(i) The State or Tribal government laws and regulations in effect under the USDA-approved plan change, thus requiring such plan to be re-submitted for USDA approval.

(ii) A State or Tribal plan must be amended in order to comply with amendments to Subtitle G the Act and this part.

(e) *Producer rights upon revocation of State or Tribal plan.* If USDA revokes approval of the State or Tribal plan due to noncompliance as defined in §990.5, producers licensed or authorized to produce hemp under the revoked State or Tribal plan may continue to produce for the remainder of the calendar year in which the revocation became effective. Producers may then apply to be licensed under the USDA plan for 90 days after the notification even if the time period does note coincide with the annual application window.

§990.5 Audit of State or Tribal plan compliance.

The Secretary may conduct an audit of the compliance of a State or Indian Tribe with an approved plan.

(a) *Frequency of audits.* Compliance audits may be scheduled, at minimum, once every three years and may include an onsite-visit, a desk-audit, or both. The USDA may adjust the frequency of audits if deemed appropriate based on program performance, compliance issues, or other relevant factors identified and provided to the State or Tribal governments by USDA.

(b) *Scope of audit review.* The audit may include, but is not limited to, a review of the following:

(1) The resources and personnel employed to administer and oversee its approved plan;

(2) The process for licensing and systematic compliance review of hemp producers;

(3) Sampling methods and laboratory testing requirements and components;

(4) Disposal of non-compliant hemp plants or hemp plant material practices, to ensure that correct reporting to the USDA has occurred;

(5) Results of and methodology used for the annual inspections of producers; and

(6) Information collection procedures and information accuracy (*i.e.*, geospatial location, contact information reported to the USDA, legal description of land).

(c) *Audit reports.* (1) Audit reports will be issued to the State or Tribal government within 60 days after the audit concluded. If the audit reveals that the State or Tribal government is not in compliance with its USDA approved plan, USDA will advise the State or Indian Tribe of non-compliances and the corrective measures that must be completed to come into compliance with the regulations in this part. The USDA will require the State or Tribe to develop a corrective action plan, which will be reviewed and approved by the USDA, and the State or Tribe will be able to demonstrate its compliance with the regulations in this part through a second audit by USDA. If the State or Tribe requests USDA assistance to develop a corrective action plan in the case of a first instance of noncompliance, the State or Tribe must request this assistance not later than 30 days after the issuance of the audit report. The USDA will approve or deny the corrective action plan within 60 days of its receipt.

(2) If the USDA determines that the State or Indian Tribe is not in compliance after the second audit, the USDA may revoke its approval of the State or Tribal plan for a period not to exceed one year. USDA will not approve a State or Indian Tribe's plan until the State or Indian Tribe demonstrates upon inspection that it is in compliance with all regulations in this part.

§ 990.6 Violations of State and Tribal plans.

(a) *Producer violations.* Producer violations of USDA-approved State and Tribal hemp production plans shall be subject to enforcement in accordance with the terms of this section.

(b) *Negligent violations.* Each USDA-approved State or Tribal plan shall contain provisions relating to negligent producer violations as defined under this part. Negligent violations shall include, but not be limited to:

(1) Failure to provide a legal description of land on which the producer produces hemp;

(2) Failure to obtain a license or other required authorization from the State department of agriculture or Tribal government, as applicable; or

(3) Production of cannabis with a delta-9 tetrahydrocannabinol concentration exceeding the acceptable hemp THC level. Hemp producers do not commit a negligent violation under this paragraph (b)(3) if they make reasonable efforts to grow hemp and the cannabis (marijuana) does not have a delta-9 tetrahydrocannabinol concentration of more than 0.5 percent on a dry weight basis.

(c) *Corrective action for negligent violations.* Each USDA-approved State or Tribal plan shall contain rules and regulations providing for the correction of negligent violations. Each correction action plan shall include, at minimum, the following terms:

(1) A reasonable date by which the producer shall correct the negligent violation.

(2) A requirement that the producer shall periodically report to the State department of agriculture or Tribal government, as applicable, on its compliance with the State or Tribal plan for a period of not less than the next 2 years from the date of the negligent violation.

(3) A producer that negligently violates a State or Tribal plan approved under this part shall not as a result of that violation be subject to any criminal enforcement action by the Federal, State, Tribal, or local government.

(4) A producer that negligently violates a USDA-approved State or Tribal plan three times in a 5-year period shall be ineligible to produce hemp for a period of 5 years beginning on the date of the third violation.

(5) The State or Tribe shall conduct an inspection to determine if the corrective action plan has been implemented as submitted.

(d) *Culpable violations.* Each USDA-approved State or Tribal plan shall contain provisions relating to producer violations made with a culpable mental state greater than negligence, including that:

(1) If the State department of agriculture or Tribal government with an approved plan determines that a producer has violated the plan with a culpable mental state greater than negligence, the State department of agriculture or Tribal government, as applicable, shall immediately report the producer to:

(i) The U.S. Attorney General; and

(ii) The chief law enforcement officer of the State or Indian Tribe, as applicable.

(2) Paragraphs (b) and (c) of this section shall not apply to culpable violations.

(e) *Felonies.* Each USDA-approved State or Tribal plan shall contain provisions relating to felonies. Such provisions shall state that:

(1) A person with a State or Federal felony conviction relating to a controlled substance is subject to a 10-year ineligibility restriction on participating in the plan and producing hemp under the State or Tribal plan from the date of the conviction. An exception applies to a person who was lawfully growing hemp under the 2014 Farm Bill before December 20, 2018, and whose conviction also occurred before that date.

(2) Any producer growing hemp lawfully with a license, registration, or authorization under a pilot program authorized by section 7606 of the Agricultural Act of 2014 (7 U.S.C. 5940) before October 31, 2019 shall be exempted from paragraph (e)(1) of this section.

(3) For producers that are entities, the State or Tribal plan shall determine which employee(s) of a producer shall be considered to be participating in the plan and subject to the felony conviction restriction for purposes of paragraph (e)(1) of this section.

(f) *False statement.* Each USDA-approved State or Tribal plan shall state that any person who materially falsifies any information contained in an application to participate in such program shall be ineligible to participate in that program.

(g) *Appeals.* For States and Tribes who wish to appeal an adverse action, subpart D of this part will apply.

§ 990.7 **Establishing records with USDA Farm Service Agency.**

All producers licensed to produce hemp under an USDA-approved State or Tribal plan shall report hemp crop acreage with FSA and shall provide, at minimum, the following information:

(a) Street address and, to the extent practicable, geospatial location for each lot or greenhouse where hemp will be produced. If an applicant operates in more than one location, that information shall be provided for all production sites.

(b) If an applicant has production sites licensed under a USDA-approved State or Tribal plan, those sites will be covered under the respective plan and will not need to be included under the producer's application to become licensed under the USDA plan.

(c) Acreage dedicated to the production of hemp, or greenhouse or indoor square footage dedicated to the production of hemp.

(d) License or authorization identifier.

§ 990.8 **Production under Federal law.**

Nothing in this subpart prohibits the production of hemp in a State or the territory of an Indian Tribe for which a State or Tribal plan is not approved under this subpart if the production of that hemp is in accordance with subpart C of this part, and if the production of hemp is not otherwise prohibited by the State or Indian Tribe.

Subpart C—USDA Hemp Production Plan

§ 990.20 **USDA requirements for the production of hemp.**

(a) *General hemp production requirements.* The production of hemp in a State or territory of an Indian Tribe where there is no USDA approved State or Tribal plan must be produced in accordance with this subpart provided that the production of hemp is not prohibited by the State or territory of an Indian Tribe where production will occur.

(b) *Convicted felon ban.* A person with a State or Federal felony conviction relating to a controlled substance is subject to a 10-year ineligibility restriction on participating in the plan and producing hemp under the USDA plan from the date of the conviction. An exception applies to a person who was lawfully growing hemp under the 2014 Farm Bill before December 20, 2018, and whose conviction also occurred before December 20, 2018.

(c) *Falsifying material information on application.* Any person who materially falsifies any information contained in an application to for a license under the USDA plan shall be ineligible to participate in the USDA plan.

§ 990.21 USDA hemp producer license.

(a) *General application requirements—*
(1) *Requirements and license application.* Any person producing or intending to produce hemp must have a valid license prior to producing, cultivating, or storing hemp. A valid license means the license is unexpired, unsuspended, and unrevoked.

(2) *Application window.* Applicants may submit an application for a new license to USDA between December 2, 2019 and November 2, 2020. In subsequent years, applicants may submit an application for a new license or renewal of an existing license to USDA from August 1 through October 31 of each year.

(3) *Required information on application.* The applicant shall provide the information requested on the application form, including:

(i) *Contact information.* Full name, residential address, telephone number and email address. If the applicant is a business entity, the full name of the business, the principal business location address, full name and title of the key participants, title, email address (if available) and employer identification number (EIN) of the business; and

(ii) *Criminal history report.* A current criminal history report for all key participants dated within 60 days prior to the application submission date. A license application will not be considered complete without all required criminal history reports.

(4) *Submission of completed application forms.* Completed application forms shall be submitted to USDA.

(5) *Incomplete application procedures.* Applications missing required informa-tion shall be returned to the applicant as incomplete. The applicant may resubmit a completed application.

(6) *License expiration.* USDA-issued hemp producer licenses shall be valid until December 31 of the year three years after the year in which license was issued.

(b) *License renewals.* USDA hemp producer licenses must be renewed prior to license expiration. Licenses are not automatically renewed. Applications for renewal shall be subject to the same terms, information collection requirements, and approval criteria as provided in this subpart for initial applications unless there has been an amendment to the regulations in this part or the law since approval of the initial or last application.

(c) *License modification.* A license modification is required if there is any change to the information submitted in the application including, but not limited to, sale of a business, the production, handling, or storage of hemp in a new location, or a change in the key participants producing under a license.

§ 990.22 USDA Hemp producer license approval.

(a) A license shall not be issued unless:

(1) The application submitted for USDA review and approval is complete and accurate.

(2) The criminal history report(s) submitted with the license application confirms that all key participants to be covered by the license have not been convicted of a felony, under State or Federal law, relating to a controlled substance within the past ten (10) years unless the exception in § 990.20(b) applies.

(3) The applicant has submitted all reports required as a participant in the hemp production program by this part.

(4) The application contains no materially false statements or misrepresentations and the applicant has not previously submitted an application with any materially false statements or misrepresentations.

(5) The applicant's license is not currently suspended.

(6) The applicant is not applying for a license as a stand-in for someone

whose license has been suspended, revoked, or is otherwise ineligible to participate.

(7) The State or territory of Indian Tribe where the person produces or intends to produce hemp does not have a USDA-approved plan or has not submitted a plan to USDA for approval and is awaiting USDA's decision. For the first year, USDA will not accept request for licenses under the USDA plan until December 2, 2019 to allow States and Tribes to submit their plans.

(8) The State or territory of Indian Tribe where the person produces or intends to produce hemp does not prohibit the production of hemp.

(b) USDA shall provide written notification to applicants whether the application has been approved or denied unless the applicant is from a State or territory of an Indian Tribe that has a plan submitted to USDA and is awaiting USDA approval.

(1) If an application is approved, a license will be issued. Information regarding approved licenses will be available on the AMS website.

(2) Licenses will be valid until December 31 of the year three after the year in which the license was issued.

(3) Licenses may not be sold, assigned, transferred, pledged, or otherwise disposed of, alienated or encumbered.

(4) If a license application is denied, the notification from USDA will explain the cause for denial. Applicants may appeal the denial in accordance with subpart D of this part.

(c) If the applicant is producing in more than one location, the applicant may have more than one license to grow hemp. If the applicant has operations in a location covered under a State or Tribal plan, that operation must be licensed under the State or Tribal plan, not a USDA plan.

§990.23 Reporting hemp crop acreage with USDA Farm Service Agency.

All USDA plan producers shall report hemp crop acreage with FSA and shall provide, at minimum, the following information:

(a) Street address and, to the extent practicable, geospatial location of the lot, greenhouse, building, or site where hemp will be produced. All locations where hemp is produced must be reported to FSA.

(b) Acreage dedicated to the production of hemp, or greenhouse or indoor square footage dedicated to the production of hemp.

(c) The license number.

§990.24 Responsibility of a USDA licensed producer prior to harvest.

(a) Within 15 days prior to the anticipated harvest of cannabis plants, a producer shall have an approved Federal, State, local law enforcement agency or other USDA designated person collect samples from the flower material of such cannabis material for delta-9 tetrahydrocannabinol concentration level testing.

(b) The method used for sampling from the flower material of the cannabis plant must be sufficient at a confidence level of 95 percent that no more than one percent (1%) of the plants in the lot would exceed the acceptable hemp THC level. The method used for sampling must ensure that a representative sample is collected that represents a homogeneous composition of the lot.

(c) During a scheduled sample collection, the producer or an authorized representative of the producer shall be present at the growing site.

(d) Representatives of the sampling agency shall be provided with complete and unrestricted access during business hours to all hemp and other cannabis plants, whether growing or harvested, and all land, buildings, and other structures used for the cultivation, handling, and storage of all hemp and other cannabis plants, and all locations listed in the producer license.

(e) A producer shall not harvest the cannabis crop prior to samples being taken.

§990.25 Standards of performance for detecting delta-9 tetrahydrocannabinol (THC) concentration levels.

(a) Analytical testing for purposes of detecting the concentration levels of delta-9 tetrahydrocannabinol (THC) in the flower material of the cannabis plant shall meet the following standard:

(1) Laboratory quality assurance must ensure the validity and reliability of test results;

(2) Analytical method selection, validation, and verification must ensure that the testing method used is appropriate (fit for purpose) and that the laboratory can successfully perform the testing;

(3) The demonstration of testing validity must ensure consistent, accurate analytical performance; and

(4) Method performance specifications must ensure analytical tests are sufficiently sensitive for the purposes of the detectability requirements of this part.

(b) At a minimum, analytical testing of samples for delta-9 tetrahydrocannabinol concentration levels must use post-decarboxylation or other similarly reliable methods approved by the Secretary. The testing methodology must consider the potential conversion of delta-9 tetrahydrocannabinolic acid (THCA) in hemp into delta-9 tetrahydrocannabinol (THC) and the test result reflect the total available THC derived from the sum of the THC and THC-A content. Testing methodologies meeting the requirements of this paragraph (b) include, but are not limited to, gas or liquid chromatography with detection.

(c) The total delta-9 tetrahydrocannabinol concentration level shall be determined and reported on a dry weight basis. Additionally, measurement of uncertainty (MU) must be estimated and reported with test results. Laboratories shall use appropriate, validated methods and procedures for all testing activities and evaluate measurement of uncertainty.

(d) Any sample test result exceeding the acceptable hemp THC level shall be conclusive evidence that the lot represented by the sample is not in compliance with this part.

§ 990.26 Responsibility of a USDA producer after laboratory testing is performed.

(a) The producer shall harvest the crop not more than fifteen (15) days following the date of sample collection.

(b) If the producer fails to complete harvest within fifteen (15) days of sample collection, a secondary pre-harvested sample of the lot shall be required to be submitted for testing.

(c) Harvested lots of hemp plants shall not be commingled with other harvested lots or other material without prior written permission from USDA.

(d) Lots that meet the acceptable hemp THC level may enter the stream of commerce.

(e) Lots tested and *not* certified by the DEA-registered laboratory not exceeding the acceptable hemp THC level may *not* be further handled, processed, or enter the stream of commerce and the licensee shall ensure the lot is disposed of in accordance with § 990.27.

(f) Any producer may request additional testing if it is believed that the original delta-9 tetrahydrocannabinol concentration level test results were in error.

§ 990.27 Non-compliant cannabis plants.

(a) Cannabis plants exceeding the acceptable hemp THC level constitute marijuana, a schedule I controlled substance under the Controlled Substances Act (CSA), 21 U.S.C. 801 *et seq.*, and must be disposed of in accordance with the CSA and DEA regulations found at 21 CFR 1317.15.

(b) Producers must notify USDA of their intent to dispose of non-conforming plants and verify disposal by submitting required documentation.

§ 990.28 Compliance.

(a) *Audits.* Producers may be audited by the USDA. The audit may include a review of records and documentation, and may include site visits to farms, fields, greenhouses, storage facilities, or other locations affiliated with the producer's hemp operation. The inspection may include the current crop year, as well as any previous crop year(s). The audit may be performed remotely or in person.

(b) *Frequency of audit verifications.* Audit verifications may be performed once every three (3) years unless otherwise determined by USDA. If the results of the audit find negligent violations, a corrective action plan may be established.

(c) *Assessment of producer's hemp operations for conformance.* The producer's operational procedures, documentation, and recordkeeping, and other practices may be verified during the onsite audit verification. The auditor may also visit the production, cultivation, or storage areas for hemp listed on the producer's license.

(1) *Records and documentation.* The auditor shall assess whether required reports, records, and documentation are properly maintained for accuracy and completeness.

(2) [Reserved]

(d) *Audit reports.* Audit reports will be issued to the licensee within 60 days after the audit is concluded. If USDA determines under an audit that the producer is not compliant with this part, USDA shall require a corrective action plan. The producer's implementation of a corrective action plan may be reviewed by USDA during a future site visit or audit.

§ 990.29 Violations.

Violations of this part shall be subject to enforcement in accordance with the terms of this section.

(a) *Negligent violations.* A hemp producer shall be subject to enforcement for negligently:

(1) Failing to provide an accurate legal description of land where hemp is produced;

(2) Producing hemp without a license; and

(3) Producing cannabis (marijuana) exceeding the acceptable hemp THC level. Hemp producers do not commit a negligent violation under this paragraph (a) if they make reasonable efforts to grow hemp and the cannabis (marijuana) does not have a delta-9 tetrahydrocannabinol concentration of more than 0.5 percent on a dry weight basis.

(b) *Corrective action for negligent violations.* For each negligent violation, USDA will issue a Notice of Violation and require a corrective action plan for the producer. The producer shall comply with the corrective action plan to cure the negligent violation. Corrective action plans will be in place for a minimum of two (2) years from the date of their approval. Corrective action plans will, at a minimum, include:

(1) The date by which the producer shall correct each negligent violation;

(2) Steps to correct each negligent violation; and

(3) A description of the procedures to demonstrate compliance must be submitted to USDA.

(c) *Negligent violations and criminal enforcement.* A producer that negligently violates this part shall not, as a result of that violation be subject to any criminal enforcement action by any Federal, State, Tribal, or local government.

(d) *Subsequent negligent violations.* If a subsequent violation occurs while a corrective action plan is in place, a new corrective action plan must be submitted with a heightened level of quality control, staff training, and quantifiable action measures.

(e) *Negligent violations and license revocation.* A producer that negligently violates the license 3 times in a 5-year period shall have their license revoked and be ineligible to produce hemp for a period of 5 years beginning on the date of the third violation.

(f) *Culpable mental state greater than negligence.* If USDA determines that a licensee has violated the terms of the license or of this part with a culpable mental state greater than negligence:

(1) USDA shall immediately report the licensee to:

(i) The U.S. Attorney General; and

(ii) The chief law enforcement officer of the State or Indian territory, as applicable, where the production is located; and

(2) Paragraphs (a) and (b) of this section shall not apply to culpable violations.

§ 990.30 USDA producers; License suspension.

(a) USDA may issue a notice of suspension to a producer if USDA or its representative receives some credible evidence establishing that a producer has:

(1) Engaged in conduct violating a provision of this part; or

(2) Failed to comply with a written order from the USDA–AMS Administrator related to negligence as defined in this part.

(b) Any producer whose license has been suspended shall not handle or remove hemp or cannabis from the location where hemp or cannabis was located at the time when USDA issued its notice of suspension, without prior written authorization from USDA.

(c) Any person whose license has been suspended shall not produce hemp during the period of suspension.

(d) A producer whose license has been suspended may appeal that decision in accordance with subpart D of this part.

(e) A producer whose license has been suspended and not restored on appeal may have their license restored after a waiting period of one year from the date of the suspension.

(f) A producer whose license has been suspended may be required to complete a corrective action plan to fully restore the license.

§ 990.31 USDA licensees; Revocation.

USDA shall immediately revoke the license of a USDA producer if such producer:

(a) Pleads guilty to, or is convicted of, any felony related to a controlled substance; or

(b) Made any materially false statement with regard to this part to USDA or its representatives with a culpable mental state greater than negligence; or

(c) Is found to be growing cannabis exceeding the acceptable hemp THC level with a culpable mental state greater than negligence or negligently violated this part three times in five years.

§ 990.32 Recordkeeping requirements.

(a) USDA producers shall maintain records of all hemp plants acquired, produced, handled, or disposed of as will substantiate the required reports.

(b) All records and reports shall be maintained for at least three years.

(c) All records shall be made available for inspection by USDA inspectors, auditors, or their representatives during reasonable business hours. The following records must be made available:

(1) Records regarding acquisition of hemp plants;

(2) Records regarding production and handling of hemp plants;

(3) Records regarding storage of hemp plants; and

(4) Records regarding disposal of all cannabis plants that do not meet the definition of hemp.

(d) USDA inspectors, auditors, or their representatives shall have access to any premises where hemp plants may be held during reasonable business hours.

(e) All reports and records required to be submitted to USDA as part of participation in the program in this part which include confidential data or business information, including but not limited to information constituting a trade secret or disclosing a trade position, financial condition, or business operations of the particular licensee or their customers, shall be received by, and at all times kept in the custody and control of, one or more employees of USDA or their representatives. Confidential data or business information may be shared with applicable Federal, State, Tribal, or local law enforcement or their designee in compliance with the Act.

Subpart D—Appeals

§ 990.40 General adverse action appeal process.

(a) Persons who believe they are adversely affected by the denial of a license application under the USDA hemp production program may appeal such decision to the AMS Administrator.

(b) Persons who believe they are adversely affected by the denial of a license renewal under the USDA hemp production program may appeal such decision to the AMS Administrator.

(c) Persons who believe they are adversely affected by the termination or suspension of a USDA hemp production license may appeal such decision to the AMS Administrator.

(d) States and territories of Indian Tribes that believe they are adversely affected by the denial of a proposed State or Tribal hemp plan may appeal such decision to the AMS Administrator.

§990.41 Appeals under the USDA hemp production plan.

(a) *Appealing a denied USDA-plan license application.* A license applicant may appeal the denial of a license application.

(1) If the AMS Administrator sustains an applicant's appeal of a licensing denial, the applicant will be issued a USDA hemp production license.

(2) If the AMS Administrator denies an appeal, the applicant's license application will be denied. The applicant may request a formal adjudicatory proceeding within 30 days to review the decision. Such proceeding shall be conducted pursuant to the U.S. Department of Agriculture's Rules of Practice Governing Adjudicatory Proceedings, 7 CFR part 1, subpart H.

(b) *Appealing a denied USDA-plan license renewal.* A producer may appeal the denial of a license renewal.

(1) If the AMS Administrator sustains a producer's appeal of a licensing renewal decision, the applicant's USDA hemp production license will be renewed.

(2) If the AMS Administrator denies the appeal, the applicant's license will not be renewed. The denied producer may request a formal adjudicatory proceeding within 30 days to review the decision. Such proceeding shall be conducted pursuant to the U.S. Department of Agriculture's Rules of Practice Governing Formal Adjudicatory Proceedings, 7 CFR part 1, subpart H.

(c) *Appealing a USDA-plan license termination or suspension.* A USDA hemp plan producer may appeal the termination or suspension of a license.

(1) If the AMS Administrator sustains the appeal of a license termination or suspension, the producer will retain their license.

(2) If the AMS Administrator denies the appeal, the producer's license will be terminated or suspended. The producer may request a formal adjudicatory proceeding within 30 days to review the decision. Such proceeding shall be conducted pursuant to the U.S. Department of Agriculture's Rules of Practice Governing Formal Adjudicatory Proceedings, 7 CFR part 1, subpart H.

(d) *Filing period.* The appeal of a denied license application, denied license renewal, suspension, or termination must be filed within the time-period provided in the letter of notification or within 30 business days from receipt of the notification, whichever occurs later. The appeal will be considered "filed" on the date received by the AMS Administrator. The decision to deny a license application or renewal, or suspend or terminate a license, is final unless a formal adjudicatory proceeding is requested within 30 days to review the decision. Such proceeding shall be conducted pursuant to the U.S. Department of Agriculture's Rules of Practice Governing Adjudicatory Proceedings, 7 CFR part 1, subpart H.

(e) *Where to file.* Appeals to the Administrator must be filed in the manner as determined by AMS.

(f) *What to include.* All appeals must include a copy of the adverse decision and a statement of the appellant's reasons for believing that the decision was not proper or made in accordance with applicable program regulations in this part, policies, or procedures.

§990.42 Appeals under a State or Tribal hemp production plan.

(a) *Appealing a State or Tribal hemp production plan application.* A State or Tribe may appeal the denial of a proposed State or Tribal hemp production plan by the USDA.

(1) If the AMS Administrator sustains a State or Tribe's appeal of a denied hemp plan application, the proposed State or Tribal hemp production plan shall be established as proposed.

(2) If the AMS Administrator denies an appeal, the proposed State or Tribal hemp production plan shall not be approved. Prospective producers located in the State or territory of the Indian Tribe may apply for hemp licenses under the terms of the USDA plan. The State or Tribe may request a formal adjudicatory proceeding be initiated within 30 days to review the decision. Such proceeding shall be conducted pursuant to the U.S. Department of Agriculture's Rules of Practice Governing Adjudicatory Proceedings, 7 CFR part 1, subpart H.

(b) *Appealing the suspension or termination of a State or Tribal hemp production plan.* A State or Tribe may appeal the revocation by USDA of an existing State or Tribal hemp production plan.

(1) If the AMS Administrator sustains a State or Tribe's appeal of a State or Tribal hemp production plan suspension or revocation, the associated hemp production plan may continue.

(2) If the AMS Administrator denies an appeal, the State or Tribal hemp production plan will be suspended or revoked as applicable. Producers located in that State or territory of the Indian Tribe may continue to produce hemp under their State or Tribal license until the end the calendar year in which the State or Tribal plan's disapproval was effective or when the State or Tribal license expires, whichever is earlier. Producers may apply for a USDA license under subpart C of this part unless hemp production is otherwise prohibited by the State or Indian Tribe. The State or Indian Tribe may request a formal adjudicatory proceeding be initiated to review the decision. Such proceeding shall be conducted pursuant to the U.S. Department of Agriculture's Rules of Practice Governing Formal Adjudicatory Proceedings, 7 CFR part 1, subpart H.

(c) *Filing period.* The appeal of a State or Tribal hemp production plan suspension or revocation must be filed within the time-period provided in the letter of notification or within 30 business days from receipt of the notification, whichever occurs later. The appeal will be considered "filed" on the date received by the AMS Administrator. The decision to deny a State or Tribal plan application or suspend or revoke approval of a plan, is final unless the decision is appealed in a timely manner.

(d) *Where to file.* Appeals to the Administrator must be filed in the manner as determined by AMS.

(e) *What to include in appeal.* All appeals must include a copy of the adverse decision and a statement of the appellant's reasons for believing that the decision was not proper or made in accordance with applicable program regulations in this part, policies, or procedures.

Subpart E—Administrative Provisions

§ 990.60 Agents.

As provided under 7 CFR part 2, the Secretary may name any officer or employee of the United States or name any agency or division in the United States Department of Agriculture, to act as their agent or representative in connection with any of the provisions of this part.

§ 990.61 Severability.

If any provision of this part is declared invalid or the applicability thereof to any person or circumstances is held invalid, the validity of the remainder of this part or the applicability thereof to other persons or circumstances shall not be affected thereby.

§ 990.62 Expiration of this part.

This part expires on November 1, 2021 unless extended by notification in the FEDERAL REGISTER. State and Tribal plans approved under subpart B of this part remain in effect after November 1, 2021 unless USDA disapproves the plan. USDA hemp producer licenses issued under subpart C of this part remain in effect until they expire unless USDA revokes or suspends the license.

§ 990.63 Interstate transportation of hemp.

No State or Indian Tribe may prohibit the transportation or shipment of hemp or hemp products lawfully produced under a State or Tribal plan approved under subpart B of this part, under a license issued under subpart C of this part, or under 7 U.S.C. 5940 through the State or territory of the Indian Tribe, as applicable.

Subpart F—Reporting Requirements

§ 990.70 State and Tribal hemp reporting requirements.

(a) *State and Tribal hemp producer report.* Each State and Tribes with a plan approved under this part shall submit to USDA, by the first of each month, a report providing the contact information and the status of the license or other authorization issued for each producer covered under the individual State and Tribal plans. If the first of the month falls on a weekend or holiday, the report is due by the first business day following the due date. The report shall be submitted using a digital format compatible with USDA's information sharing systems, whenever possible. The report shall contain the information described in this paragraph (a).

(1)(i) For each new producer who is an individual and is licensed or authorized under the State or Tribal plan, the report shall include full name of the individual, license or authorization identifier, business address, telephone number, and email address (if available).

(ii) For each new producer that is an entity and is licensed or authorized under the State or Tribal plan, the report shall include full name of the entity, the principal business location address, license or authorization identifier, and the full name, title, and email address (if available) of each employee for whom the entity is required to submit a criminal history record report.

(iii) For each producer that was included in a previous report and whose reported information has changed, the report shall include the previously reported information and the new information.

(2) The status of each producer's license or authorization.

(3) The period covered by the report.

(4) Indication that there were no changes during the current reporting cycle, if applicable.

(b) *State and Tribal hemp disposal report.* If a producer has produced cannabis exceeding the acceptable hemp THC level, the cannabis must be disposed of in accordance with the Controlled Substances Act and DEA regulations found at 21 CFR 1317.15. States and Tribes with plans approved under this part shall submit to USDA, by the first of each month, a report notifying USDA of any occurrence of non-conforming plants or plant material and providing a disposal record of those plants and materials. This report would include information regarding name and contact information for each producer subject to a disposal during the reporting period, and date disposal

was completed. If the first of the month fall on a weekend or holiday, reports are due by the first business day following the due date. The report shall contain the information described in this paragraph (b).

(1) Name and address of the producer.

(2) Producer license or authorization identifier.

(3) Location information, such as lot number, location type, and geospatial location or other location descriptor for the production area subject to disposal.

(4) Information on the agent handling the disposal.

(5) Disposal completion date.

(6) Total acreage.

(c) *Annual report.* Each State or Tribe with a plan approved under this part shall submit an annual report to USDA. The report form shall be submitted by December 15 of each year and contain the information described in this paragraph (c).

(1) Total planted acreage.

(2) Total harvested acreage.

(3) Total acreage disposed.

(d) *Test results report.* Each producer must ensure that the DEA-registered laboratory that conducts the test of the sample(s) from its lots reports the test results for all samples tested to USDA. The test results report shall contain the information described in this paragraph (d) for each sample tested.

(1) Producer's license or authorization identifier.

(2) Name of producer.

(3) Business address of producer.

(4) Lot identification number for the sample.

(5) Name and DEA registration number of laboratory.

(6) Date of test and report.

(7) Identification of a retest.

(8) Test result.

§990.71 USDA plan reporting requirements.

(a) *USDA hemp plan producer licensing application.* USDA will accept applications from December 2, 2019 through November 2, 2020. Thereafter applicants, may submit a USDA Hemp Licensing Application to USDA from August 1 through October 31 of each year. Licenses will be valid until December

31 of the year three years after the license is issued. The license application will be used for both new applicants and for producers seeking renewal of their license. The application shall include the information described in this paragraph (a).

(1) *Contact information.* (i) For an applicant who is an individual, the application shall include full name of the individual, business address, telephone number, and email address (if available).

(ii) For an applicant that is an entity, the application shall include full name of the entity, the principal business location address, and the full name, title, and email address (if available) of each key participant of the entity.

(2) *Criminal history report.* As part of a complete application, each applicant shall provide a current Federal Bureau of Investigation's Identity History Summary. If the applicant is a business entity, a criminal history report shall be provided for each key participant.

(i) The applicant shall ensure the criminal history report accompanies the application.

(ii) The criminal history report must be dated within 60 days of submission of the application submittal.

(3) *Consent to comply with program requirements.* All applicants submitting a completed license application, in doing so, consent to comply with the requirements of this part.

(b) *USDA hemp plan producer disposal form.* If a producer has produced cannabis exceeding the acceptable hemp THC level, the cannabis must be disposed of in accordance with the Controlled Substances Act and DEA regulations found at 21 CFR 1317.15. Forms shall be submitted to USDA no later than 30 days after the date of completion of disposal. The report shall contain the information described in this paragraph (b).

(1) Name and address of the producer.

(2) Producer's license number.

(3) Geospatial location, or other valid land descriptor, for the production area subject to disposal.

(4) Information on the agent handling the disposal.

(5) Date of completion of disposal.

(6) Signature of the producer.

(7) Disposal agent certification of the completion of the disposal.

(c) *USDA hemp plan producer annual report.* Each producer shall submit an annual report to USDA. The report form shall be submitted by December 15 of each year and contain the information described in this paragraph (c).

(1) Producer's license number.

(2) Producer's name.

(3) Producer's address.

(4) Lot, location type, geospatial location, total planted acreage, total acreage disposed, and total harvested acreage.

(d) *Test results report.* Each producer must ensure that the DEA-registered laboratory that conducts the test of the sample(s) from its lots reports the test results for all samples tested to USDA. The test results report shall contain the information described in this paragraph (d) for each sample tested.

(1) Producer's license number.

(2) Name of producer.

(3) Business address of producer.

(4) Lot identification number for the sample.

(5) Name and DEA registration number of laboratory.

(6) Date of test and report.

(7) Identification of a retest.

(8) Test result.

PART 993—DRIED PRUNES PRODUCED IN CALIFORNIA

Subpart A—Order Regulating Handling

DEFINITIONS

Sec.
993.1 Secretary.
993.2 Act.
993.3 Person.
993.4 Area.
993.5 Prunes.
993.6 Non-French prunes.
993.7 French prunes.
993.8 Natural condition prunes.
993.9 Processed prunes.
993.10 Standard prunes.
993.11 Standard processed prunes.
993.12 Substandard prunes.
993.13 Handle.
993.14 Handler.
993.15 Dehydrator.
993.16 Producer.
993.17 Ton.
993.18 Grade.
993.19a Size.

Subpart B—Administrative Requirements

Subpart C—Assessment Rates

AUTHORITY: 7 U.S.C. 601–674.

Subpart A—Order Regulating Handling

SOURCE: 26 FR 476, Jan. 19, 1961, unless otherwise noted.

DEFINITIONS

§ 993.1　Secretary.

Secretary means the Secretary of Agriculture of the United States, or any other officer or employee of the United States Department of Agriculture who is or who may hereafter be, authorized to exercise the powers and to perform the duties of the Secretary under the Act.

§ 993.2　Act.

Act means Public Act No. 10, 73d Congress, as amended and reenacted and amended by the Agricultural Marketing Agreement Act of 1937, as amended (7 U.S.C. 601 *et seq.*).

§ 993.3　Person.

Person means an individual, partnership, corporation, association, or any other business unit.

§ 993.4　Area.

Area means the State of California.

§ 993.5　Prunes.

Prunes means and includes all sun-dried or artificially dehydrated plums, of any type or variety, produced from plums grown in the area, except: (a) Sulfur-bleached prunes which are produced from yellow varieties of plums and are commonly known as silver prunes; and (b) plums which have not been dried or dehydrated to a point where they are capable of being stored prior to packaging, without material deterioration or spoilage unless refrigeration or other artificial means of preservation are used, and so long as they are treated by a process which is in conformity with, or generally similar to, the processes for treatment of plums of that type which have been developed or recommended by the Food Technology Division, College of Agriculture, University of California, for the specialty pack known as "high moisture content prunes," but this exception shall not apply if and when such plums are dried to the point where they are capable of being stored without material deterioration or spoilage, refrigeration or other artificial means of preservation.

§ 993.6　Non-French prunes.

Non-French prunes means prunes commonly known as Imperial, Sugar, Robe de Sargent, Burton, Standard, Jefferson, Fellenberg, Italian, President, Giant, and Hungarian (Gross), produced from such varieties of plums. This definition may be modified by the committee with the approval of the Secretary.

§ 993.7　French prunes.

French prunes means: (a) Prunes produced from plums of the following varieties of plums: French (Prune d'Agen, Petite Prune d'Agen), Coates (Cox, Double X, Saratoga); and (b) any other prunes which possess taste, flesh texture, and other characteristics similar to those of the prunes named in this section.

§993.8 Natural condition prunes.

Natural condition prunes means prunes which have not been processed.

§993.9 Processed prunes.

Processed prunes means prunes which have been cleaned, or treated with water or steam, by a handler.

§993.10 Standard prunes.

Standard prunes means any lot of natural condition prunes meeting the applicable grade and size standards prescribed pursuant to §993.49 other than pursuant to §993.49(c).

[26 FR 476, Jan. 19, 1961, as amended at 37 FR 861, Jan. 20, 1972]

§993.11 Standard processed prunes.

Standard processed prunes means any lot of processed prunes meeting the applicable grade and size standards prescribed pursuant to §993.50.

§993.12 Substandard prunes.

Substandard prunes means any lot of processed or natural condition prunes failing to meet the applicable grade and size standards prescribed pursuant to §§993.49 and 993.50 other than pursuant to §993.49(c).

[26 FR 476, Jan. 19, 1961, as amended at 37 FR 861, Jan. 20, 1972]

§993.13 Handle.

Handle means to receive, package, sell, consign, transport, or ship (except as a carrier of prunes owned by another person), or in any other way to place prunes in the current of the commerce within the area or from such area to any point outside thereof: *Provided,* That this term shall not include: (a) Sales or deliveries of prunes by a producer or dehydrator to a producer, dehydrator, or handler within the area; (b) the receiving of prunes by a producer or dehydrator from a producer or dehydrator; and (c) receipts, sales, or shipments of prunes already handled by another person other than pursuant to §993.50(f).

§993.14 Handler.

Handler means any person who handles prunes.

§993.15 Dehydrator.

Dehydrator means any person who produces prunes by drying or dehydrating plums by means of sun-drying or artificial heat.

§993.16 Producer.

Producer means any person who is engaged, in a proprietary capacity, in growing plums for drying or dehydrating into prunes.

§993.17 Ton.

Ton means a short ton of 2,000 pounds.

§993.18 Grade.

Grade means the classification of prunes for quality and condition according to the grading specifications established pursuant to the provisions of this subpart.

§993.19a Size.

Size means either (a) the number of prunes contained in a pound and may be referred to in terms of size ranges, or (b) the diameter of a round opening, expressed in multiples of one thirty-second of an inch, through which prunes pass freely.

[37 FR 861, Jan. 20, 1972]

§993.19b Undersized prunes.

Undersized prunes means prunes which pass freely through a round opening of a specified diameter.

[37 FR 861, Jan. 20, 1972]

§993.20 Crop year.

Crop year means the 12-month period beginning August 1 of any year and ending July 31 of the following year.

§993.21 Domestic.

Domestic means the United States, Canal Zone, Puerto Rico, Virgin Islands, and Canada.

§993.21a Proper storage.

Proper storage means storage of such character as will maintain prunes in the same condition as when received by a handler, except for normal and natural deterioration and shrinkage.

[30 FR 9798, Aug. 6, 1965]

§ 993.21b   Trade demand.

(a) *Domestic trade demand.* The quantity of prunes which the commercial trade will acquire from - all handlers during a crop year for distribution in domestic markets for human consumption as prunes and prune products.

(b) *Foreign trade demand.* The quantity of prunes which the commercial trade will acquire from all handlers during a crop year for distribution in other than domestic markets for human consumption as prunes and prune products.

[30 FR 9798, Aug. 6, 1965]

§ 993.21c   Salable prunes.

Salable prunes means those prunes which are free to be handled pursuant to any salable percentage established by the Secretary pursuant to § 993.54, or, if no reserve percentage is in effect for a crop year, all prunes, excluding the quantity of undersized prunes determined pursuant to § 993.49(c), received by handlers from producers and dehydrators during that year.

[46 FR 61637, Dec. 18, 1981]

§ 993.21d   Reserve prunes.

Reserve prunes means those prunes which must be withheld in satisfaction of a reserve obligation arising from application of a reserve percentage established by the Secretary pursuant to § 993.54.

[30 FR 9798, Aug. 6, 1965]

EFFECTIVE DATE NOTE: At 70 FR 30613, May 27, 2005, § 993.21d was suspended indefinitely.

§ 993.22   Consumer package.

Consumer package means: (a) Any container of prunes holding less than 10 pounds of standard processed prunes or standard prunes; or (b) any container holding less than 10 pounds of prunes and other dried fruit if more than 60 percent of the net weight of mixed dried fruit in the lot consists of standard processed prunes or standard prunes.

§ 993.23   Part and subpart.

Part means the order regulating the handling of dried prunes produced in California, and all rules, regulations, and supplementary orders issued thereunder. This order regulating the handling of dried prunes produced in California shall be a *subpart* of such part.

PRUNE MARKETING COMMITTEE

§ 993.24   Establishment and membership.

A Prune Marketing Committee (herein referred to as the "Committee"), consisting of 22 members with an alternate member for each such member, is hereby established to administer the terms and provisions of this part, of whom with their respective alternates, 14 shall represent producers, 7 shall represent handlers, and 1 shall represent the public. Committee membership shall be allocated in accordance with the following grouping with the alternate positions identically allocated:

(a) Three handler members to represent handlers who are cooperative marketing associations of producers (referred to in this part as "cooperative handlers");

(b) Three handler members to represent handlers other than cooperative handlers (referred to in this part as "independent handlers");

(c) One handler member to represent handlers who are cooperative handlers or independent handlers, whichever of such handlers handled as first handlers more than 50 percent of the prunes handled by all handlers during the crop year preceding the year in which nominations are made;

(d) Fourteen producer members to be selected from and to represent producers who are members of cooperative marketing associations (referred to in this part as "cooperative producers") and producers other than "cooperative producers" (referred to in this part as "independent producers"); the number of the producer members for the cooperative producer group or the independent producer group, as the case may be, shall be in the same proportion, as near as practicable, to the total of 14, as the tonnage of prunes handled by the respective group of co-operative handlers or independent handlers as first handlers during the crop year preceding the year in which nominations are made is to the total tonnage of prunes handled by all handlers as first handlers.

(e) The public member and alternate shall have no financial interest in the prune industry.

[26 FR 476, Jan. 19, 1961, as amended at 46 FR 61636, Dec. 18, 1981]

§ 993.25 Term of office.

The term of office of members, and their respective alternates, shall be two years, ending on May 31 of even numbered years, and any later date which may be necessary for the selection and qualification of their respective successors.

§ 993.26 Selection.

Selection of members of the committee, and their respective alternates, shall be made in the appropriate number specified in § 993.24, by the Secretary from nominees nominated pursuant to this part or, in the discretion of the Secretary, from other eligible persons.

§ 993.27 Eligibility.

Producer members of the Committee shall be at the time of their selection, and during their term of office, producers in the group, for which selected and if to represent a district also producers in the district for which selected, and, except for producer members representing cooperative producers, shall not be engaged in the handling of prunes either in a proprietary capacity or as a director, officer, or employee. Handler members of the Committee shall be handlers in the group they represent or directors, officers, or employees of such handlers. These eligibility requirements shall not apply to the public member and alternate member.

[46 FR 61636, Dec. 30, 1981]

§ 993.28 Nominees.

(a) For the purpose of obtaining nominations for producer members to represent independent producers, the Committee shall, with the approval of the Secretary, divide the area into districts giving, insofar as practicable, equal representation to numbers of independent producers and production of prune tonnage by such producers. The number of districts shall be equal to the number of such producer members or seven, whichever is the lesser. Candidates for nomination by independent producers from the various districts shall be obtained at meetings convened by the committee. Following such meetings, the committee shall prepare a separate ballot for each of the districts, or a joint ballot for two or more districts, containing (1) the names of the candidates for each district involved and (2) provision for write-in candidates. The ballot shall be mailed to each independent producer of record in each district. The voting procedure (including the casting of the ballot by mail addressed to the committee), and tabulation of votes shall be in accordance with rules and regulations prescribed by the committee, with the approval of the Secretary. Each voter shall be entitled to cast only one vote for a member nominee and only one vote for an alternate member nominee in a district in which he is a producer, and no voter shall vote for candidates in more than one district. In case he is a producer in more than one district, he shall elect in which of such districts he will vote and notify the committee as to his choice. Whenever the number of producer members to represent independent producers during the ensuing term of office is to exceed seven, one nominee shall be nominated by independent producers in each of the seven districts and an additional nominee for each member in excess of the seven members shall be nominated, without reference to districts, by such seven nominees. The committee shall recommend the establishment of districts, or any changes therein, to the Secretary prior to January 31 of each year in which nominations are made.

(b) Before April 16 of each even-numbered year nominations of producer members to represent cooperative producers and handler members to represent cooperative handlers shall be submitted to the Secretary by cooperative marketing associations engaged in the handling of prunes. The number of cooperative producer members and handler members to be nominated by each cooperative marketing association shall bear, as near as practicable, the same percentage as each cooperative marketing association's tonnage of prunes handled as first handler thereof is to the total tonnage handled by all cooperative marketing associations during the preceding crop year.

(c) In any year in which nominations are made following a crop year during which the tonnage of prunes handled by independent handlers as first handlers exceeded the tonnage of prunes handled by cooperative handlers as first handlers, nominees for member positions to represent independent handlers shall be nominated as follows:

(1) Each of the two independent handlers who handled during such preceding crop year, the two largest percentages of the prune tonnage handled by all independent handlers shall nominate from their respective organizations, one nominee for a handler member and one for an alternate member;

(2) Three independent handlers who handled during such preceding crop year the next three largest percentages of the prune tonnage handled by all independent handlers shall nominate from among their organizations, one nominee for a handler member and one for an alternate member;

(3) All other independent handlers who handled the remaining percentage of such prune tonnage shall nominate from their organizations, one nominee for a handler member and one for an alternate member.

In any year in which nominations are made following a crop year during which the tonnage of prunes handled by cooperative handlers as first handlers exceeded the tonnage of prunes handled by independent handlers as first handlers, nominees for two member and alternate positions to represent the independent handlers referred to in paragraph (c)(1) of this section shall be nominated in accordance with said paragraph (c)(1), and one nominee for the member and one for the alternate position to represent all other independent handlers shall be nominated by the handlers referred to in paragraph (c) (2) and (3) of this sec-

tion and the votes of such handlers shall be weighted by the tonnage of prunes handled during the preceding crop year by the respective handlers.

(d) The committee shall establish with the approval of the Secretary, the procedures by which such nominations, other than by cooperative marketing associations engaged in the handling of prunes, shall be obtained and shall submit such nominations to the Secretary before April 16 of the year in which nominations are made. In the event the committee determines that any nominating procedure specified in this section does not result in equitable representation, it may establish, with the prior approval of the Secretary, such modifications as will tend to assure such representation.

(e) The producer and handler members of the Committee selected for a new term of office shall nominate a public member and alternate member at the first meeting following their selection.

[31 FR 9713, July 19, 1966, as amended at 46 FR 61636, Dec. 18, 1981]

§ 993.29 Alternates.

An alternate for a member of the committee shall act in the place and stead of such member (a) during his absence, and (b) in the event of his removal, resignation, disqualification, or death, until a successor for such member's unexpired term has been selected and has qualified. Except as otherwise specifically provided in this subpart the provisions of this part applicable to members also apply to alternate members.

§ 993.30 Failure to nominate.

If a nomination for any position on the committee is not received by the Secretary by May 1, the Secretary may select an eligible individual without regard to nominations.

§ 993.31 Acceptance.

Each person selected as a member or alternate member of the committee shall, prior to serving on the committee, qualify by filing with the Secretary a written acceptance within 15 days after receiving notice of his selection.

§993.32 Vacancies.

In the event of any committee vacancy occasioned by the removal, resignation, disqualification, or death of any member, or in the event of the failure of any person selected as a member or alternate member to qualify, a successor for the unexpired term shall be nominated within 60 calendar days thereof. Such nominations shall be made in the manner provided for in this subpart, insofar as applicable, except that nominations of nominees for a producer member position to represent independent producers may, at the discretion of the committee, be made to the committee by the incumbent producer members of the committee who represent independent producers.

§993.33 Voting procedure.

Decisions of the Committee shall be by majority vote of the members present and voting and a quorum must be present: *Provided,* That decisions on marketing policy, grade or size regulations, pack specifications, salable and reserve percentages, and on any matters pertaining to the control or disposition of reserve prunes or to prune plum diversion pursuant to §993.62, including any delegation of authority for action on such matters and any recommendation of rules and procedures with respect to such matters, including any such decision arrived at by mail or telegram, shall require at least 14 affirmative votes. A quorum shall consist of at least 13 members of whom at least 8 must be producer members and at least 4 must be handler members. Except in case of emergency, a minimum of 5 days notice must be given with respect to any meeting of the Committee. In case of an emergency, to be determined within the discretion of the chairman of the Committee, as much notice of a meeting as is practicable in the circumstances shall be given. The Committee may vote by mail or telegram upon due notice to all members, but any proposition to be so voted upon first shall be explained accurately, fully, and identically by mail or telegram to all members. When any proposition is submitted to be voted on by

such method, one dissenting vote shall prevent its adoption.

[46 FR 61637, Dec. 18, 1981]

EFFECTIVE DATE NOTE: At 70 FR 30613, May 27, 2005, in §993.33, the words "salable and reserve percentages, and on any matters pertaining to the control or disposition of reserve prunes or to prune plum diversion pursuant to §993.62," were suspended indefinitely.

§993.34 Expenses.

The members of the committee, and alternates when acting as members, or when alternates' expenses are authorized by the committee, shall serve without compensation but shall be allowed their expenses.

[30 FR 9798, Aug. 6, 1965]

§993.35 Powers.

The committee shall have the following powers:

(a) To administer the terms and provisions of this subpart;

(b) To make rules and regulations to effectuate the terms and provisions of this subpart;

(c) To receive, investigate, and report to the Secretary complaints of violations of this subpart; and

(d) To recommend to the Secretary amendments to this subpart.

§993.36 Duties.

The committee shall have, among others, the following duties:

(a) To act as intermediary between the Secretary and any producer, dehydrator, or handler;

(b) To keep minutes, books, and other records which shall clearly reflect all of its acts and transactions, and such minutes, books, and other records shall be subject to examination by the Secretary at any time;

(c) To make, subject to the prior approval of the Secretary, scientific and other studies, and assemble data on the producing, handling, shipping, and marketing conditions relative to prunes, which are necessary in connection with the performance of its official duties;

(d) To select, from among its members, a chairman and other appropriate officers, and to adopt such rules and regulations for the conduct of the business of the committee as it may deem advisable;

(e) To appoint or employ such other persons as it may deem necessary, and to determine the salaries and define the duties of such persons;

(f) To submit to the Secretary not later than the fourth Tuesday of July of each year, a budget of its anticipated expenditures and the recommended rate of assessment for the ensuing crop year, and the supporting data therefor;

(g) To submit to the Secretary such available information with respect to prunes as the committee may deem appropriate, or as the Secretary may request;

(h) To prepare and submit to the Secretary quarterly statements of the financial operations of the committee, exclusive of reserve prune operations, and to make such statements, together with the minutes of the meetings of said committee, available for inspection at the offices of the committee by producers, dehydrators, and handlers;

(i) To prepare and submit to the Secretary annually, as soon as practicable after the end of each crop year and at such other times as the committee may deem appropriate or the Secretary may request, a statement of the committee's financial operations with respect to reserve prunes for such crop year and to make such statement available at the offices of the committee for inspection by producers, dehydrators, and handlers;

(j) To cause the books of the committee to be audited by a certified public accountant at least once each crop year, and at such other times as the committee may deem necessary or as the Secretary may request, and two copies of each such audit report shall be submitted to the Secretary and a copy which does not contain confidential data shall be available for inspection at the offices of the committee, by producers, dehydrators, and handlers;

(k) To give the Secretary the same notice of meetings of the committee as is given to the members of the committee;

(l) To give producers, dehydrators, and handlers reasonable advance notice of meetings of the committee, and to maintain all such meetings open to such persons;

(m) To investigate compliance with the provisions of this subpart and with any rules and regulations established pursuant to such provisions; and

(n) To establish, with the approval of the Secretary, such rules and procedures relative to administration of this subpart as may be consistent with the provisions contained in this subpart and as may be necessary to accomplish the purposes of the act and the efficient administration of this subpart.

[30 FR 9798, Aug. 6, 1965, as amended at 37 FR 861, Jan. 20, 1972]

EFFECTIVE DATE NOTE: At 70 FR 30613, May 27, 2005, in § 993.33, paragraph (i) was suspended indefinitely.

§ 993.37 Research and development.

The committee, with the approval of the Secretary, may establish or provide for the establishment of marketing research and development projects designed to assist, improve, or promote the marketing, distribution, and consumption of prunes. The expense of such projects shall be paid from funds collected pursuant to § 993.81.

MARKETING POLICY

§ 993.41 Marketing policy.

(a) On or before the first Tuesday of each July, the committee shall prepare and submit to the Secretary a report setting forth its recommended marketing policy for the ensuing crop year. If it becomes advisable to modify such policy, because of changed demand, supply, or other conditions, the committee shall formulate a new policy and shall submit a report thereon to the Secretary. Notice of the committee's marketing policy, and of any modifications thereof, shall be given promptly by reasonable publicity to producers, dehydrators, and handlers.

(b) In formulating its marketing policy for the ensuing crop year, the committee shall consider and shall include in its report to the Secretary, the following estimates (natural condition basis) and recommendations:

(1) The carryover of salable prunes as of August 1;

(2) The carryover of reserve prunes as of August 1;

(3) The grade and size composition of the salable and reserve carryovers;

(4) The quantity of prunes to be produced without regard to possible diversions of prune plums by producers;

(5) The probable quality and prune sizes in the crop;

(6) The domestic trade demand by uses of prunes;

(7) The foreign trade demand by countries or groups of countries;

(8) The desirable carryout of salable prunes at the end of the ensuing crop year;

(9) The quantity of undersized prunes in the crop, itemized as to French prunes and non-French prunes;

(10) The quantity of prunes to be withheld as reserve prunes so as to protect against errors of estimation and permit orderly marketing of the supply;

(11) The recommended salable and reserve percentages;

(12) The quantity of prune plums, dried weight basis, deemed desirable to be diverted pursuant to §993.62;

(13) Any recommended change in regulations pursuant to §§933.49 to 993.53, inclusive;

(14) The probable assessable tonnage for the purposes of §993.81; and

(15) The current prices for prunes, the trend and level of consumer income, whether producer prices are likely to exceed parity, and such other factors as may have a bearing on the marketing of prunes or the administration of this part.

[30 FR 9798, Aug. 6, 1965, as amended at 37 FR 862, Jan. 20, 1972]

EFFECTIVE DATE NOTE: At 70 FR 30613, May 27, 2005, §993.41 was suspended indefinitely.

PROHIBITION ON HANDLING

§993.48   Regulation.

No handler shall handle prunes except in accordance with the provisions of this part.

[30 FR 9799, Aug. 6, 1965]

EFFECTIVE DATE NOTE: At 70 FR 30613, May 27, 2005, §993.48 was suspended indefinitely.

GRADE AND SIZE REGULATIONS

§993.49   Incoming regulation.

(a) No handler shall receive prunes from producers or dehydrators, other than substandard prunes and undersized prunes, unless such prunes meet the minimum standards for natural condition prunes set forth in §993.97 (Exhibit A), or as such standards may be modified, or the more restrictive grade regulation established pursuant to this section, and then in effect: *Provided,* That no handler shall receive any prunes (including substandard prunes and undersized prunes) from producers or dehydrators unless such prunes have been properly dried and cured in original natural condition, without the addition of water, and are free from active insect infestation, so that they are capable of being received, stored, and packed without material deterioration or spoilage. Any "high moisture content prunes," as described in the exception in §993.5(b), in the possession of a handler, shall be held separate and apart from any prunes held by him. If such "high moisture content prunes" are dried or dehydrated to a point where they are capable of being stored, without material deterioration or spoilage, unrefrigerated or not otherwise artificially preserved, they shall be deemed, at that time, to have been received by such handler as prunes, and shall be subject to all of the conditions and restrictions of this subpart.

(b) The Secretary, on the basis of a recommendation of the committee or other information, may establish size regulations or more restrictive grade regulations with resepct to prunes that may be received by a handler from producers and dehydrators whenever he finds that such action would tend to effectuate the declared policy of the act.

(c) In no crop year shall a handler receive from producers or dehydrators prunes, other than as undersized prunes, which pass freely through a round opening with a diameter as follows: For French prunes 23/32 of an inch, and for non-French prunes 28/32 of an inch: *Provided,* That the Secretary upon a recommendation of the Committee, may establish larger openings whenever it is determined that supply conditions for a crop year warrant such regulation. The quantity of undersized prunes in any lot received by a handler from a producer or dehydrator shall be determined by the inspection service and entered on the applicable inspection certificate.

[26 FR 476, Jan. 19, 1961, as amended at 37 FR 862, Jan. 20, 1972; 46 FR 61637, Dec. 18, 1981]

EFFECTIVE DATE NOTE: At 70 FR 30613, May 27, 2005, § 993.49 was suspended indefinitely.

§ 993.50 Outgoing regulation.

(a) Except as otherwise specifically provided, no handler shall ship or otherwise make final disposition of prunes which fail to meet the applicable minimum standards set forth in § 993.97 (Exhibit A), or as such standards may be modified, for standard prunes or standard processed prunes.

(b) The Secretary, on the basis of a recommendation of the committee or other information, may establish size regulations, pack specifications, or more restrictive grade regulations with respect to prunes that may be shipped or otherwise disposed of by a handler if such action would tend to effectuate the declared policy of the act. If a more restrictive grade regulation is established in connection with § 993.97 (Exhibit A) it shall insofar as practicable apply comparably to both natural condition prunes and processed prunes. When pack specifications are in effect, no handler shall ship prunes in consumer packages, unless such prunes are identified by an appropriate label, seal, stamp, or tag affixed to such container by the handler showing the size of prunes in the lot from which the container was packed. In order to effectuate such orderly marketing of prunes as will be in the public interest, whether prices are above or below parity, no handler shall use descriptive terms in a manner inconsistent with that set forth in this subpart or in any pack specifications or other regulation issued by the Secretary pursuant to this subpart.

(c) Non-French prunes: No handler shall ship or otherwise make final disposition of any lot of standard prunes or standard processed prunes of the non-French varieties or any lot which includes non-French prunes in excess of a tolerance to be prescribed by the Secretary on recommendation of the Committee, unless the average count of such non-French prunes contained in any such lot is 40 or less per pound. However, under safeguards to be established by the Committee, any lot containing non-French prunes with an average size count of more than 40 prunes per pound may be shipped to or disposed of in prune product outlets in which they lose their form and character as prunes by conversion prior to consumption. A tolerance as to the permitted deviation of sizes about the average count shall be prescribed by the Secretary, upon recommendation of the Committee.

(d) French prunes: No handler shall ship or otherwise make final disposition of any lot of French prunes for human consumption as prunes, or any lot of mixed dried fruit containing French prunes for human consumption as mixed dried fruit, unless the average count of French prunes contained in any such lot is 100 or less per pound. However, under safeguards to be established by the Committee, any lot containing French prunes with an average size count of more than 100 prunes per pound may be shipped to or disposed of in prune product outlets in which they lose their form and character as prunes by conversion prior to consumption. In determining whether any such lot conforms to this minimum size requirement, the following tolerance shall apply: In a sample of 100 ounces, the count per pound of 10 ounces of the smallest prunes shall not vary from the count per pound of 10 ounces of the largest prunes by more than 45 points. The Secretary may, upon the basis of the recommendation and information submitted by the Committee and other available information, modify this tolerance for uniformity of size.

(e) No handler shall ship or otherwise make final disposition of any lot of substandard prunes except for use as prune products in which the prunes lose their form and character as prunes by conversion prior to consumption, or for use in non-human consumption outlets: *Provided*, That any such prunes which are shipped or otherwise disposed of for human consumption shall meet the minimum standards prescribed in II C (1), (2), and (3) of § 993.97 or as such standards as may pursuant to §993.52 be modified. The committee shall issue any such rules and regulations as may be necessary to insure such uses.

(f) Notwithstanding the restrictions contained in this section, any handler may transfer prunes from one plant owned by him to another plant owned by him within the area without having an inspection made as provided for in §993.51. Any handler may ship prunes from his plant to another handler's plant within the area without having an inspection made as provided for in §993.51, but a report of such inter-handler transfer shall be made promptly by the transferring handler to the committee. The receiving handler shall, before shipping or otherwise making final disposition of such prunes, comply with the requirements of this section and of §993.51.

(g) No handler shall ship or otherwise dispose of, for human consumption, the quantity of prunes determined by the inspection service pursuant to §993.49(c) to be undersized prunes. However, such handler may, at the direction and under the supervision of the Committee, dispose of such quantity of prunes in nonhuman consumption outlets. Prunes so disposed of shall be of the same variety as, and reasonably comparable in size, to such undersized prunes. The handler shall cause the inspection service to make a determination whether the prunes disposed of by the handler in nonhuman consumption outlets meet such requirements. In making the determination with respect to comparability in size, the inspection service shall apply a tolerance permitting a deviation from the size of the applicable opening established pursuant to §993.49(c). Any such tolerance, together with any rules and regulations to insure proper disposition of the prunes and that such prunes are reasonably comparable to the undersized prunes so received, shall be established by the Committee with the approval of the Secretary. The quantity of prunes determined pursuant to §993.49(c) shall not be deemed to be within the handler's quota for salable prunes fixed by the Secretary within the meaning of section 8a(5) of the Act.

[26 FR 476, Jan. 19, 1961, as amended at 37 FR 862, Jan. 20, 1972]

EFFECTIVE DATE NOTE: At 70 FR 30613, May 27, 2005, §993.50 was suspended indefinitely.

§993.51 Inspection and certification.

Each handler shall at his own expense, before or upon the receiving, and before the shipping or disposing of prunes, cause an inspection to be made of such prunes to determine whether they meet the applicable grade and size requirements or the pack specifications, including labeling, effective pursuant to this part. Such handler shall obtain a certificate that such prunes meet the aforementioned applicable requirements and shall submit such certificate, or cause it to be submitted, to the committee. Acceptable certificates shall be those issued by inspectors of the Dried Fruit Association of California. The Secretary may designate another inspection service in the event the services of the Association prove unsatisfactory.

EFFECTIVE DATE NOTE: At 70 FR 30613, May 27, 2005, §993.51 was suspended indefinitely.

§993.52 Modification.

Minimum standards, pack specifications or size regulations, including the openings prescribed in §993.49(c), may be modified by the Secretary, on the basis of a recommendation of the committee or other information, whenever he finds that such modification would tend to effectuate the declared policy of the act.

[26 FR 476, Jan. 19, 1961, as amended at 37 FR 862, Jan. 20, 1972]

EFFECTIVE DATE NOTE: At 70 FR 30613, May 27, 2005, §993.52 was suspended indefinitely.

§993.53 Above parity situations.

The minimum standards, the minimum sizes, including the minimum undersized regulation in §933.49(c), and the provisions of this part relating to administration shall continue in effect irrespective of whether the estimated season average price for prunes is in excess of the parity level specified in section 2(1) of the act.

[46 FR 61637, Dec. 18, 1981]

EFFECTIVE DATE NOTE: At 70 FR 30613, May 27, 2005, §993.53 was suspended indefinitely.

RESERVE CONTROL

SOURCE: 30 FR 9799, Aug. 8, 1965, unless otherwise noted.

§ 993.54 Establishment of salable and reserve percentages.

Whenever the Secretary finds, from the recommendations and supporting information supplied by the committee, or from any other available information, that to establish the percentages of prunes for any crop year which shall be salable prunes and reserve prunes, respectively, or to modify the previously established percentages, would tend to effectuate the declared policy of the act, he shall establish or modify such percentages. The salable and reserve percentages when applied to the natural condition weight of prunes, excluding the quantity of undersized prunes determined pursuant to § 993.49(c), received during the crop year by a handler from producers and dehydrators, plus that diverted tonnage (dried weight natural condition prune basis) on diversion certificates issued pursuant to § 993.62 and credited to or held by him, shall determine the weight of each handler's receipts which are salable prunes and reserve prunes. The total of the salable and reserve percentages shall equal 100 percent. A cooperative marketing association may concentrate the prunes of its producer members before applying the salable and reserve percentages.

[30 FR 9799, Aug. 6, 1965, as amended at 37 FR 862, Jan. 20, 1972]

EFFECTIVE DATE NOTE: At 70 FR 30613, May 27, 2005, § 993.54 was suspended indefinitely.

§ 993.55 Application of salable and reserve percentages after end of crop year.

The salable and reserve percentages established for any crop year shall remain in effect after that crop year until salable and reserve percentages are established for another crop year. After such percentages are established, all reserve obligations shall be adjusted to the newly established percentages.

[46 FR 61637, Dec. 18, 1981]

EFFECTIVE DATE NOTE: At 70 FR 30613, May 27, 2005, § 993.55 was suspended indefinitely.

§ 993.56 Reserve obligation.

Whenever salable and reserve percentages are in effect for any crop year, the reserve obligation of a handler shall approximate the average marketable content of the handler's receipts and shall be a weight of natural condition prunes equal to the reserve percentage applied to the natural condition weight of prunes, excluding the quantity of undersized prunes determined pursuant to § 993.49(c), such handler receives during the crop year from producers and dehydrators plus that diverted tonnage (dried weight natural condition prune basis) on diversion certificates credited to or held by him which were issued pursuant to § 993.62. However, if the committee determines the requirement as to setaside reflecting average marketable content of receipts is not essential to achieve program objectives for the crop of a particular season, it may be eliminated for that season by the committee, with the approval of the Secretary. As a prerequisite for making this determination, the committee must find that the resultant setaside procedures assure that the trade demand for manufacturing prunes, as well as prunes for consumption as prunes, will be met. The salable prunes permitted to be disposed of by any handler in accordance with the provisions of this part shall be deemed to be that handler's quota fixed by the Secretary within the meaning of section 8a(5) of the act.

[30 FR 9799, Aug. 8, 1965, as amended at 37 FR 862, Jan. 20, 1972]

EFFECTIVE DATE NOTE: At 70 FR 30613, May 27, 2005, § 993.56 was suspended indefinitely.

§ 993.57 Holding requirement and delivery.

Each handler shall at all times, hold, in his possession or under his control, in proper storage for the account of the committee, free and clear of all liens, the quantity of prunes necessary to meet his reserve obligation, less any quantity: (a) For which he has a temporary deferment pursuant to § 993.58(a); (b) of prune plums (dried weight natural condition basis) diverted pursuant to § 993.62 as shown on diversion certificates held by him, or credited by the committee against his reserve obligation; (c) disposed of by him under a sales contract of the committee; (d) delivered by him to the committee, or to a person designated by it, pursuant to its instructions; and (e) for which he is otherwise relieved by the committee of such responsibility to so hold prunes. No handler may transfer a reserve obligation but any handler may, upon notification to the committee arrange to hold reserve prunes on the premises of another handler or in approved commercial storage, under conditions of proper storage. The committee may, after giving reasonable notice, require a handler to deliver to it, or to a person designated by it, f.o.b. handler's warehouse or point of storage, reserve prunes held by him. The committee may require that such delivery consist of natural condition prunes or it may arrange for such delivery to consist of processed prunes.

EFFECTIVE DATE NOTE: At 70 FR 30613, May 27, 2005, § 993.57 was suspended indefinitely.

§ 993.58 Deferment of time for withholding.

(a) Compliance by any handler with the requirement of § 993.57 for withholding reserve prunes may be temporarily deferred to any date desired by the handler, but not later than November 15 of the crop year, upon the execution and delivery by such handler to the committee of a written undertaking that on or prior to the desired date he will have fully satisfied his holding requirement. Such undertaking shall be secured by a bond or bonds to be filed with and acceptable to the committee in the amount or amounts specified, conditioned upon full compliance with such undertaking.

(b)(1) Each bond shall be provided by and at the handler's expense, with a surety or sureties acceptable to the committee, and shall be in an amount computed by multiplying the pounds of natural condition prunes for which deferment is desired by the bonding rate. Such bonding rate shall be established by the committee at a level sufficient to achieve the objectives of this part.

(2) In case a handler defaults in meeting his deferred withholding requirement, any funds collected by the committee from the bonding company

through such default shall be used by the committee to purchase from handlers a quantity of natural condition prunes, up to but not exceeding the quantity on which default occurred. Purchases shall be made from prunes with respect to which the reserve obligation has been met, and shall be of grades, varieties, or sizes and in such containers as the committee specifies in consideration of available reserve prune outlets. Purchases shall be at prices determined to be appropriate by the committee and if more prunes are offered than required by the committee, it shall make the purchases from various handlers as nearly as practicable in proportion to the quantity of their respective offerings at the same price. The committee shall dispose of the prunes acquired as soon as practicable in the most favorable reserve prune outlets and shall deposit the proceeds from such sales, less committee expenses in connection with such transaction, with reserve pool funds for distribution to equity holders.

(3) If for any reason the committee is unable to purchase a quantity of prunes as large as the quantity of reserve prunes in default by the handler, any remaining balance of funds received because of the default less expenses of the committee, shall be deposited with reserve pool funds for distribution to equity holders.

(c) A handler who has defaulted on his bond shall be credited on his reserve obligation with, and his holding requirement reduced by, that quantity of prunes represented by the sums collected but not more than the extent of his default.

EFFECTIVE DATE NOTE: At 70 FR 30613, May 27, 2005, § 993.58 was suspended indefinitely.

§ 993.59 Payment to handlers for services.

The committee shall pay handlers for necessary services rendered by them in connection with reserve prunes including, but not limited to, inspection, receiving, storing, grading, and fumigation, in accordance with a schedule of payments and conditions established by the Secretary after recommendation by the committee.

EFFECTIVE DATE NOTE: At 70 FR 30613, May 27, 2005, § 993.59 was suspended indefinitely.

PRODUCER DIVERSION

§ 993.62   Diversion privileges.

(a) *Prune plums.* The words *prune plums* as used in this section mean plums of a variety used in the production of prunes.

(b) *Voluntary principle.* No producer shall be required to divert all or any portion of the prune plums produced by him.

(c) *Authorization.* If, on the basis of a committee recommendation for diversion operations, the availability of governing rules and procedures established by the Secretary after recommendation of the committee, and other information, the Secretary concurs that diversion operations should be permitted, he shall authorize such operations.

(d) *Diversion certificates.* After diversion operations are authorized, and subject to the applicable rules and procedures, any producer may divert prune plums of his own production for eligible purposes and receive from the committee a diversion certificate therefor: *Provided,* That diversion certificates for prune plums diverted by producer members of a cooperative marketing association shall be issued by the committee to the association if it so requests. To the extent permitted by the rules and procedures, the certificate may be submitted to any handler in lieu of reserve prunes and to the same extent the certificate shall entitle the handler to satisfy his reserve obligation. Only to the extent permitted by the rules and procedures, diversion certificates may be transferable among producers and handlers.

(e) *Eligible diversions.* Within such restrictions as may be prescribed in rules and procedures, diversion may be authorized for such dispositions as are not competitive with the normal marketing of prunes and prune products. Such eligible diversions may include: (1) Disposal of prune plums for nonhuman use; (2) leaving prune plums unharvested; and (3) such other methods of diversion as may be authorized. No diversion certificate shall be issued by the committee for prune plums which would not, under normal producer practices, be dried and delivered to a handler.

(f) *Nonparticipation in pool proceeds.* Any prune plums diverted pursuant to this section shall not be included in any reserve pool.

(g) *Payment of costs.* Prior to the issuance of a diversion certificate to a producer or a cooperative marketing association, the producer or association shall pay to the committee fees established to cover costs pertaining to the diversion.

[30 FR 9800, Aug. 6, 1965]

EFFECTIVE DATE NOTE: At 70 FR 30613, May 27, 2005, § 993.62 was suspended indefinitely.

DISPOSITION OF RESERVE PRUNES

§ 993.65   Disposition of reserve prunes.

(a) *Committee's right of disposition.* The committee shall have the power and authority to sell or dispose of any and all reserve prunes (1) to meet demand either (i) as domestic trade demand, or (ii) as foreign trade demand, or (2) for use in any outlet, defined in rules and procedures, established by the Secretary after recommendation of the committee, noncompetitive with normal outlets for salable prunes.

(b) *Methods of disposition.* The committee may, for any of the purposes of § 993.65(a), offer to sell and sell reserve prunes to handlers for disposition or sale by them in specified outlets. Sale of reserve prunes by the committee to any handler for resale in such outlets or for resale to other persons for sale in such outlets shall be governed by the provisions of a sales agreement, executed by the handler with the committee. The committee may refuse to sell reserve prunes to any handler if the handler violates the terms and conditions of the agreement or other provisions of this part. The committee may sell reserve prunes into any outlet in which direct selling is determined to be more appropriate.

(c) *Offers to sell reserve prunes.* No offer to sell reserve prunes either to handlers or to other persons shall be made by the committee until 5 days (exclusive of Saturdays, Sundays, and holidays) have elapsed from the time it files with the Secretary complete information as to the terms and conditions of the proposed offer including the basis for determining the handlers' shares: *Provided,* That at any time prior to the expiration of the 5-day period the offer may be made upon the committee receiving from the Secretary notice that he does not disapprove it.

(d) *Transfer of shares.* No handler may transfer a reserve obligation. However, any handler who is authorized by the committee to dispose of reserve prunes may arrange with another handler to dispose of his share of reserve prunes through such other handler. In that event, credit for the reserve disposition shall go to the handler whose reserve prunes are used.

(e) *Distribution of proceeds.* Expenses incurred by the committee for the receiving, handling, holding, or disposing of any quantity of reserve prunes shall be charged against the proceeds of sales of such prunes. Net proceeds from the disposition of reserve prunes shall be distributed by the committee either directly, or through handlers as agents of the committee, under safeguards to be established by the committee, to persons in proportion to their contributions thereto, or to their successors in interest, with appropriate grade and size differentials as established by the committee. Progress payments may be made by the committee as sufficient funds accumulate. Distribution of the proceeds in connection with the reserve prunes contributed by a cooperative marketing association shall be made to such association, if it so requests.

[30 FR 9800, Aug. 6, 1965]

EFFECTIVE DATE NOTE: At 70 FR 30613, May 27, 2005, §993.65 was suspended indefinitely.

REPORTS AND BOOKS AND OTHER RECORDS

§993.71 Confidential information.

All reports and records furnished or submitted by handlers to the committee which include data or information constituting a trade secret or disclosing of the trade position, financial condition, or business operations of the particular handler from whom received shall be received by, and at all times kept in the custody and under the control of one or more employees of the committee, who shall disclose such information to no person except the Sec-

retary. Notwithstanding the above provisions of this section, information may be disclosed to the committee when reasonably necessary to enable the committee to carry out its functions under this subpart.

§993.72 Reports of acquisitions, sales, uses, and shipments.

Each handler shall file such reports of his acquisitions, sales, uses, and shipments of prunes, as may be requested by the committee.

EFFECTIVE DATE NOTE: At 70 FR 30613, May 27, 2005, §993.72 was suspended indefinitely.

§993.73 Other reports.

Upon the request of the committee, each handler shall furnish such other reports and information as are needed to enable the committee to perform its functions under this subpart.

EFFECTIVE DATE NOTE: At 70 FR 30613, May 27, 2005, §993.73 was suspended indefinitely.

§993.74 Records.

Each handler shall maintain such records of prunes received, held and disposed of by him, as are prescribed by the committee and needed by it to perform its functions under this subpart. Such records shall be retained for at least two years beyond the crop year of their applicability.

EFFECTIVE DATE NOTE: At 70 FR 30613, May 27, 2005, §993.74 was suspended indefinitely.

§993.75 Verification of reports.

For the purpose of checking and verifying reports filed by handlers or the operation of handlers under the provisions of this subpart, the Secretary, and the Committee through its duly authorized agents, shall have access to any premises where prunes may be held by any handler and at any time during reasonable business hours, shall be permitted to inspect any prunes so held by such handler and any and all records of such handler with respect to the holding or disposition of all prunes which may be held or which may have been disposed of by him.

[37 FR 862, Jan. 20, 1972]

EFFECTIVE DATE NOTE: At 70 FR 30613, May 27, 2005, §993.75 was suspended indefinitely.

EXPENSES AND ASSESSMENTS

§ 993.80 Expenses.

The committee is authorized to incur such expenses as the Secretary finds are reasonable and likely to be incurred by it during each crop year for the maintenance and functioning of the committee and for such other purposes as the Secretary may, pursuant to the provisions of this subpart, determine to be appropriate.

§ 993.81 Assessments.

(a) Each handler shall pay to the committee, upon demand, with respect to all salable prunes handled by him as the first handler thereof, his pro rata share of all expenses which the Secretary finds are reasonable and likely to be incurred by the committee during each crop year. Each handler's pro rata share shall be the rate of assessment per ton fixed by the Secretary. At any time during or after a crop year the Secretary may increase the rate of assessment to cover unanticipated expenses of the committee or a deficit in assessable tonnage.

(b) In order to provide funds to carry out the functions of the committee, the committee may accept advance payments from any handler to be credited toward such assessments as may be levied pursuant to this section against the respective handler.

(c) Any money collected as assessments during any crop year and not expended in connection with the committee's operations may be used by the committee for a period of five months subsequent to such crop year. At the end of such period the committee shall, from funds on hand, refund or credit to handler accounts the aforesaid excess. Each handler's share of such excess funds shall be the amount of assessments he has paid in excess of his pro rata share of the actual net expenses of the committee for the preceding crop year. Any money collected from assessments hereunder and remaining unexpended in the possession of the committee at the termination of this part, shall be distributed in such manner as the Secretary may direct: *Provided*, That to the extent practical, such funds shall be returned pro rata to the persons from whom such funds were collected.

[26 FR 476, Jan. 19, 1961, as amended at 30 FR 9800, Aug. 6, 1965]

§ 993.82 Funds.

All funds received by the committee pursuant to the provisions of this part shall be used solely for authorized purposes. The Secretary may, at any time, require the committee or its members and alternate members to account for all receipts and disbursements.

MISCELLANEOUS PROVISIONS

§ 993.83 Rights of the Secretary.

The members of the committee (including successors or alternates) and any agent or employee appointed or employed by the committee, shall be subject to the removal or suspension by the Secretary, in his discretion, at any time. Each and every decision, determination, or other acts of the committee shall be subject to the continuing right of the Secretary to disapprove of the same at any time, and upon such disapproval, shall be deemed null and void.

§ 993.84 Personal liability.

No member or alternate member of the committee, or any employee, representative, or agent thereof shall be held personally responsible, either individually of jointly with others, in any way whatsoever, to any person, for errors in judgment, mistakes, or other acts, either of commission or omission, as such member, alternate member, employee, representative, or agent, except for acts of dishonesty.

§ 993.85 Separability.

If any provision of this subpart is declared invalid or the applicability thereof to any person, circumstance, or thing is held invalid, the validity of the remainder of this subpart or the applicability thereof to any other person, circumstance, or thing shall not be affected thereby.

§ 993.86 Derogation.

Nothing contained in this subpart is, or shall be construed to be, in derogation or in modification of the rights of the Secretary or of the United States to exercise any powers granted by the act or otherwise, or, in accordance with such powers, to act in the premises whenever such action is deemed advisable.

§993.87 Duration of immunities.

The benefits, privileges, and immunities conferred upon any person by virtue of this subpart shall cease upon the termination of this subpart, except with respect to acts done under and during the existence of this subpart.

§993.88 Agents.

(a) *Authorization by Secretary.* The Secretary may, by a designation in writing, name any person, including any officer or employee of the United States Government, or name any bureau or division in the United States Department of Agriculture, to act as his agent or representative in connection with any of the provisions of this subpart.

(b) *Authorization by committee.* The committee may authorize any person or persons or agency to act as its agent or representative in connection with the provisions of this subpart.

§993.89 Effective time.

The provisions of this subpart, as well as any amendments to this subpart, shall become effective at such time as the Secretary may declare, and shall continue in force until terminated, or during suspension, in one of the ways specified in §993.90.

§993.90 Termination or suspension.

(a) *Failure to effectuate policy of act.* The Secretary may, at any time, terminate the provisions of this subpart, by giving at least one day's notice by means of a press release or in any other manner which he may determine. The Secretary shall terminate or suspend the operation of any or all of the provisions of this subpart, whenever he finds that such provisions do not tend to effectuate the declared policy of the act.

(b) *Referendum.* The Secretary shall terminate the provisions of this subpart on or before the fifteenth day of July of any crop year, to be effective at the end of such crop year, whenever he is required to do so by the provisions of section 8c(16)(B) of the act. The Secretary may, at any time he deems it desirable, hold a referendum of producers to determine whether they favor termination of this subpart. However, beginning with 1951, if the Secretary receives a recommendation, adopted by at least a majority vote of the producer members of the committee, requesting the holding of such a referendum, the Secretary shall hold such a referendum: *Provided,* That the Secretary shall not be required to hold such a referendum upon the basis of such a request more than once every two years.

(c) *Termination of act.* The provisions of this subpart shall terminate, in any event, upon the termination of the act.

§993.91 Procedure upon termination.

Upon the termination of this subpart, the members of the committee then functioning shall continue as joint trustees, for the purpose of liquidating the affairs of the committee. Action by such trustee shall require the concurrence of a majority of the said trustees. Such trustees shall continue in such capacity until discharged by the Secretary, and shall, from time to time, account for all receipts and disbursements and deliver all property on hand, together with all books and records of the committee and the joint trustees, to such person as the Secretary may direct; and shall, upon the request of the Secretary, execute such assignments or other instruments necessary or appropriate to vest in such person full title and right to all the funds, properties, and claims vested in the committee or the joint trustees, pursuant to this subpart. Any person to whom funds, property, or claims have been transferred or delivered by the committee or the joint trustees, pursuant to this section, shall be subject to the same obligations imposed upon the members of the said committee and upon said joint trustees.

§993.92 Effect of termination or amendment.

Unless otherwise expressly provided by the Secretary, the termination of this subpart or of any regulation issued pursuant to this subpart, or the issuance of any amendment to either thereof, shall not (a) affect or waive any right, duty, obligation, or liability which shall have arisen or which may thereafter arise in connection with any provision of this subpart or any regulation issued under this subpart, or (b) release or extinguish any violation of this subpart or any regulation issued under this subpart, or (c) affect or impair any rights or remedies of the Secretary, or of any other person, with respect to such violation.

§ 993.93 Amendments.

Amendments to this subpart may be proposed from time to time, by any person or by the committee, and may be made a part of this subpart by the procedures provided under the act.

§ 993.97 Exhibit A; minimum standards.

I. Minimum standards for natural condition prunes:

A. *Defects.* Defects are: (1) Off-color; (2) inferior meat condition; (3) end cracks; (4) fermentation; (5) skin or flesh damage; (6) scab; (7) burned; (8) mold; (9) imbedded dirt; (10) insect infestation; (11) decay.

B. *Explanation of terms.* (1) *Off-color* means a dull color or skin differing noticeably in appearance from that which is characteristic of mature, properly handled fruit of a given variety or type.

(2) *Inferior meat condition* means flesh which is fibrous, woody or otherwise inferior due to immaturity to the extent that the characteristic texture of the meat is substantially affected.

(3) *End cracks* means callous growth cracks, at the blossom end of prunes, aggregating more than three-eighths of one inch (⅜″) but not more than one-half of one inch (½″) in length.

(4) *Fermentation* means damage to the flesh by fermentation to the extent that the characteristic appearance or flavor is substantially affected.

(5) *Skin or flesh damage* means growth cracks, splits, breaks in skin or flesh of the following descriptions:

(a) Callous growth cracks, except end cracks as defined in this section, aggregating more than three-eighths of one inch (⅜″) in length;

(b) Splits or skin breaks exposing flesh and affecting materially the normal appearance of the prunes;

(c) Any cracks, splits or breaks open to the pit;

(d) Healed or unhealed surface or flesh blemishes caused by insect injury and which materially affect appearance, edibility or keeping quality;

(e) Skin damage caused by rain or overdipping to the extent that the prunes cannot be

processed normally without material sloughing of the skin.

(6) *Scab* means tough or thick scab exceeding in the aggregate the area of a circle three-eighths of one inch (⅜″) in diameter or by unsightly scab of another character exceeding in the aggregate the area of a circle three-fourths of one inch (¾″) in diameter.

(7) *Burned* means injury by sunburn or excessive heat in dehydration to the extent that the characteristic appearance, flavor or edibility of the fruit is noticeably affected.

(8) *Mold* means a characteristic fungus growth and is self-explanatory.

(9) *Imbedded dirt* means the presence of dirt or other extraneous material so imbedded in, or adhering to, the prune that it cannot be removed in normal processing.

(10) *Insect infestation* means the presence of insects, insect fragments or insect remains.

C. *Maximum tolerances.* Tolerance allowances shall be on a weight basis and shall not exceed the following:

(1) The tolerance allowance for decay shall not exceed one percent (1%).

(2) The combined tolerance allowance for mold, imbedded dirt, insect infestation, and decay shall not exceed five percent (5%).

(3) The combined tolerance allowance for fermentation, skin or flesh damage, scab-burned, mold, imbedded dirt, insect infestation, and decay shall not exceed eight percent (8%).

(4) The combined tolerance allowance for end cracks, fermentation, skin or flesh damage, scab, burned, mold, imbedded dirt, insect infestation, and decay shall not exceed ten percent (10%), except that the first eight percent (8%) of end cracks shall be given one-half value and any additional percentage of end cracks shall be given full value.

(5) The combined tolerance allowance for off-color, inferior meat condition, end cracks, fermentation, skin or flesh damage, scab-burned, mold, imbedded dirt, insect infestation, and decay shall not exceed twenty percent (20%), except that the first eight percent (8%) of end cracks shall be given one-half value and any additional percentage of end cracks shall be given full value.

(6) Prunes showing obvious live insect infestation shall be fumigated prior to acceptance.

D. Natural condition prunes must be properly dried and cured in original natural condition, without the addition of water, and free from active infestation, so that they are capable of being received, stored and packed without deterioration or spoilage.

II. Minimum standards for processed prunes:

A. *Defects.* Defects are: (1) Off-color; (2) inferior meat condition; (3) end cracks; (4) fermentation; (5) skin or flesh damage; (6) scab; (7) burned; (8) mold; (9) imbedded dirt; (10) insect infestation; (11) decay.

B. *Explanation of terms.* (1) *Off-color* means a dull color or skin differing noticeably in appearance from that which is characteristic of mature, properly handled fruit of a given variety or type.

(2) *Inferior meat condition* means flesh which is fibrous, woody or otherwise inferior due to immaturity to the extent that the characteristic texture of the meat is substantially affected.

(3) *End cracks* means callous growth, cracks, at the blossom end of prunes, aggregating more than three-eighths of one inch (⅜″) but not more than one-half of one inch (½″) in length.

(4) *Fermentation* means damage to the flesh by fermentation to the extent that the characteristic appearance or flavor is substantially affected.

(5) *Skin or flesh damage* means growth cracks, splits, breaks in skin or flesh of the following descriptions:

(a) Callous growth cracks, except end cracks as defined in this section, aggregating more than three-eighths of one inch (⅜″) in length;

(b) Splits or skin breaks exposing flesh and materially affecting the normal appearance of French prunes; or markedly affecting the normal appearance of varieties other than the French variety;

(c) Any cracks, splits or breaks open to the pit;

(d) Healed or unhealed surface or flesh blemishes caused by insect injury and which materially affect appearance, edibility or keeping quality.

(6) *Scab* means tough or thick scab exceeding in the aggregate the area of a circle three-eighths of one inch (⅜″) in diameter or by unsightly scab of another character exceeding in the aggregate the area of a circle three-fourths of one inch (¾″) in diameter.

(7) *Burned* means injury by sunburn or excessive heat in dehydration to the extent that the characteristic appearance, flavor or edibility of the fruit is noticeably affected.

(8) *Mold* means a characteristic fungus growth and is self-explanatory.

(9) *Imbedded dirt* means the presence of dirt or other extraneous material so imbedded in, or adhering to, the prune that it cannot be readily removed in washing the fruit.

(10) *Insect infestation* means the presence of insects, insect fragments or insect remains.

C. *Maximum tolerances.* Tolerance allowances shall be on a weight basis and shall not exceed the following:

(1) There shall be no tolerance allowance for live insect infestation.

(2) The tolerance allowance for decay shall not exceed one percent (1%).

(3) The combined tolerance allowance for mold, imbedded dirt, insect infestation, and decay shall not exceed five percent (5%).

(4) The combined tolerance allowance for fermentation, skin or flesh damage, scab, burned, mold, imbedded dirt, insect infestation, and decay shall not exceed eight percent (8%).

(5) The combined tolerance allowance for end cracks, fermentation, skin or flesh damage, scab, burned, mold, imbedded dirt, insect infestation, and decay shall not exceed ten percent (10%), except that the first eight percent (8%) of end cracks shall be given one-half value and any additional percentage of end cracks shall be given full value.

(6) The combined tolerance allowance for off-color, inferior meat condition, end cracks, fermentation, skin or flesh damage, scab, burned, mold, imbedded dirt, insect infestation, and decay shall not exceed twenty percent (20%), except that the first eight percent (8%) of end cracks shall be given one-half value and any additional percentage of end cracks shall be given full value.

EFFECTIVE DATE NOTE: At 70 FR 30613, May 27, 2005, §993.97 was suspended indefinitely.

Subpart B—Administrative Requirements

DEFINITIONS

SOURCE: 26 FR 8278, Sept. 2, 1961, unless otherwise noted.

§ 993.101   Order.

Order means Marketing Agreement No. 110, as amended, and Order No. 993, as amended (§§ 993.1 through 993.97), regulating the handling of dried prunes produced in California, or as they may be further amended hereafter.

§ 993.102   Committee.

Committee means the Prune Marketing Committee established pursuant to § 993.24.

[26 FR 8278, Sept. 2, 1961, as amended at 48 FR 57261, Dec. 29, 1983]

§ 993.103   Terms in the order.

Terms defined in the order shall have the same meaning when used in this subpart.

§ 993.104   Lot.

(a) *Lot* for the purposes of §§ 993.49 and 993.149 means any quantity of prunes delivered by one producer or one dehydrator to a handler on which inspection is requested: *Provided,* That a lot shall be limited to (1) the prunes contained in not more than 30 "ton box'' containers or (2), if in other containers, not more than 60,000 pounds of prunes. If the prunes in any containers are markedly inferior in quality and condition to other prunes in the proffered lot, the containers shall be segregated into lots of reasonable uniform quality.

(b) *Lot* for the purposes of §§ 993.50 and 993.150 means:

(1) With respect to in-line inspection either (i) the aggregate quantity of prunes of the same size, other than those rejected by inspection, processed in any continuous production of one calendar day and packed during such day in one size and style of container or (ii) the aggregate quantity of prunes of the same size, other than those rejected by inspection, so processed and held in packing containers for later packaging.

(2) With respect to floor inspection either (i) prunes not previously inspected in-line, of the same size, in like containers, bearing the same identification (e.g., brand) if in consumer packages, and offered for inspection as a lot; or (ii) prunes previously inspected in-line but rejected as failing to meet requirements, of the same size, in like containers, processed in any continuous production of one calendar day, and offered for inspection as a new lot.

EFFECTIVE DATE NOTE: At 70 FR 30613, May 27, 2005, § 993.104 was suspended indefinitely.

§ 993.105   Size count.

Size count means the count or number of prunes per pound.

[26 FR 8278, Sept. 2, 1961. Redesignated at 35 FR 11380, July 16, 1970, and further redesignated at 37 FR 15980, Aug. 9, 1972]

EFFECTIVE DATE NOTE: At 70 FR 30613, May 27, 2005, § 993.105 was suspended indefinitely.

§ 993.106   In-line inspection.

In-line inspection means inspection of prunes where samples are drawn from a flow of prunes prior to packaging.

EFFECTIVE DATE NOTE: At 70 FR 30613, May 27, 2005, § 993.106 was suspended indefinitely.

§ 993.107   Floor inspection.

Floor inspection means inspection of prunes where samples are drawn from packaged prunes or from unpackaged prunes that are held in packing containers.

EFFECTIVE DATE NOTE: At 70 FR 30613, May 27, 2005, § 993.107 was suspended indefinitely.

§ 993.108   Non-human consumption outlet.

Non-human consumption outlet means any livestock feeder or manufacturer of inedible syrup, industrial alcohol, animal feed, or other product for non-human use, who has established, to the satisfaction of the committee, that any prunes or prune waste received for a non-human use will be used only within such outlet.

[26 FR 8278, Sept. 2, 1961; 26 FR 8483, Sept. 9, 1961]

EFFECTIVE DATE NOTE: At 70 FR 30613, May 27, 2005, § 993.108 was suspended indefinitely.

§ 993.109   Modified definition of non-French prunes.

The definition of non-French prunes set forth in § 993.6 is modified to read as follows: *Non-French Prunes* means prunes commonly known as Imperial, Sugar, Robe de Sargent, Burton, Standard, Jefferson, Fellenberg, Italian, President, Giant, Hungarian (Gross), and Moyer, produced from such varieties of plums.

[38 FR 22887, Aug. 27, 1973]

PRUNE ADMINISTRATIVE COMMITTEE

§ 993.128   Nominations for membership.

(a) *Districts.* In accordance with the provisions of § 993.28, the districts referred to therein are described as follows:

District No. 1. The counties of Colusa, Glenn, Solano and Yolo.

District No. 2. That portion of Sutter County north of a line extending along Franklin Road easterly to the Yuba County line and westerly to the Colusa County line.

District No. 3. That portion of Sutter County south of a line extending along Franklin Road easterly to the Yuba County line and westerly to the Colusa County line.

District No. 4. The counties of Alpine, Amador, Del Norte, El Dorado, Humboldt, Lake, Lassen, Mendocino, Modoc, Napa, Nevada, Placer, Plumas, Sacramento, Shasta, Sierra, Siskiyou, Sonoma, Tehama and Trinity.

District No. 5. All of Butte County.

District No. 6. All of Yuba County.

District No. 7. The counties of Fresno, Kern, Kings, Madera Merced, San Benito, San Joaquin, Santa Clara, Tulare and all other counties not included in Districts 1, 2, 3, 4, 5 and 6.

(b) *Voting procedures*—(1) *Independent producers.* Prior to March 8 of each election year, the Committee shall cause a meeting to be held, in each of the election districts established pursuant to §993.28(a) for the purpose of obtaining names of proposed candidates for nomination to the Secretary for selection as members and alternate members for the respective districts. Each such candidate must be a producer in the district for which he is proposed. Prior to March 15 of that election year, the Committee shall prepare for each district and mail to each independent producer of record in such district a ballot as prescribed in §993.28(a). Each voter shall be entitled to cast only one vote for a member nominee and only one vote for an alternate member nominee in a district in which he is a producer, and no voter shall vote for candidates in more than one district. In case he is a producer in more than one district he shall elect in which of such districts he will vote and notify the Committee as to his choice. In order to be counted, such a mail ballot must be executed and returned to the Committee postmarked not later than the following March 31. One nominee for member and one nominee for alternate member for each district shall be submitted to the Secretary by the Committee on the basis of those receiving the plurality of the mail ballots cast for the respective positions in the particular district. Returns shall be considered in light of the voting by each district separately.

(2) *Independent handler nominees.* (i) Prior to March 15 of each election year, the Committee shall notify each independent handler of record of the group of independent handlers in which he has been classified pursuant to the provisions of §993.28(c) and of the number of independent handler positions on the Committee for the ensuing term of office pursuant to the provisions of §993.24 (b) and (c). Prior to April 1 of each election year, each of the two independent handlers classified in the group specified in §993.28(c)(1) shall notify the Committee in writing of his nominee for member and nominee for alternate member as prescribed therein.

(ii) In any election year in which four member positions and four alternate member positions are assigned to independent handlers for the ensuing term of office, the Committee shall, prior to April 1 of such year, cause to be held a meeting of the three independent handlers classified in the group specified in §993.28(c)(2) and a separate meeting of all other independent handlers classified in the group specified in §993.28(c)(3). Each group at its meeting shall, from among that group, elect one member nominee and one alternate member nominee by plurality vote. Each handler present at the meeting of his group shall be entitled to one vote for a candidate for each position assigned to that group.

(iii) In any election year in which only three member positions and three alternate member positions are assigned to independent handlers for the ensuing term of office, the Committee shall, prior to April 1 thereof, cause to be held a meeting of all independent handlers except those classified in the group specified in §993.28(c)(1). At such meeting one member nominee and one alternate member nominee shall be elected in accordance with the applicable provisions of §993.28(c).

[47 FR 7389, Feb. 19, 1982, as amended at 55 FR 5571, Feb. 16, 1990; 59 FR 8518, Feb. 23, 1994; 64 FR 72912, Dec. 29, 1999]

GRADE AND SIZE REGULATIONS

§993.149 Receiving of prunes by handlers.

(a) *Receiving stations*—(1) *General.* Prunes shall be received by a handler at any receiving station so designated by the Committee. *Receiving station* shall mean any plant of a handler or a dehydrator's premises; this term shall also mean any other place where prunes are normally and usually received by a handler in any considerable volume as ranch deliveries, and at which there are adequate facilities to enable the inspection service to determine whether the prunes meet the applicable grade, size, and condition requirements.

(2) *Receiving at dehydrator.* Any handler may arrange with the committee and the inspection service for the incoming inspection and certification to be based on samples of prunes drawn as prune plums and dehydrated in the same manner as the prunes to which they are referable. Where such arrangement is acceptable to the Committee as permitting the inspection and certification of the prunes to be comparable to an inspection and certification when based on samples drawn as prunes, such certification shall be acceptable for the purposes of this section if the inspector further certifies that the dehydration process of the prunes being certified resulted in prunes eligible to be received under the terms and conditions of this part.

(b) *Inspection stations.* Prunes shall be inspected only at inspection stations established by the inspection service with the concurrence of the Committee. *Inspection station* shall mean a centralized station and any receiving station other than a handler's plant or a dehydrator's premises.

(c) *Incoming inspection*—(1) *General.* Upon any producer or dehydrator delivering prunes to a handler, the handler shall issue to the inspection service an identification tag showing the name and address of such producer or dehydrator, the date of delivery, the county of production, the number and type of containers, the approximate net weight of the prunes, the place where the prunes are to be inspected, and any other information necessary to identify such prunes to the satisfaction of the inspector and the Committee. For each such delivery, the handler shall issue to the producer or dehydrator a door receipt or weight certificate showing the name and address of the producer or dehydrator, the weight of the delivery, and any other information necessary to identify the delivery. Such information shall be available to the inspector and the Committee. Each lot shall be sampled separately and as soon as practicable following delivery. The handler shall supply any necessary information together with any assistance needed by the inspector in drawing samples including the dumping of containers.

(2) *Certification.* Following inspection of a lot not returned to the producer or dehydrator, the handler shall require the inspection service to issue, in quintuplicate, a certificate containing at least the following information: (i) The place where samples were drawn and the date and place of inspection; (ii) the name and address of the producer or dehydrator, the handler, and the inspection service; (iii) the variety of the prunes, the county in which such prunes were produced, the number and type of the containers thereof, the net weight of the prunes as shown on the applicable door receipt or weight certificate, together with the number of such receipt or certificate, and the contract or account number under which the prunes were delivered; (iv) whenever applicable, the percentage by weight of undersized prunes in the lot; (v) with respect to the balance of the lot, the inspector's computation of the percentage, by screen size of prunes and in the aggregate, of each group or combination of groups of defects for which a maximum tolerance is in effect; (vi) whether the prunes in the lot, exclusive of any undersigned prunes, are standard or substandard; (vii) the inspector's computation of the percentage of weight of each screen size and in the aggregate, of offgrade prunes (those defective pursuant to § 993.97) necessary to be removed therefrom in order for the remainder in each screen size and in the aggregate to be standard prunes, and (viii) the average size count of prunes of each screen size and of the aggregate: *Provided,* That whenever an undersized prune regulation is in effect for the crop year, the average size count shall be of all prunes except undersized prunes in the lot, by screen size and in the aggregate. The handlers shall require the inspection service to furnish promptly the producer or dehydrator with one copy of the certificate and the handler with two copies.

(d) *Conditional provisions*—(1) *Wet or slack-dry prunes.* Any prunes delivered to a handler by a producer or dehydrator which an inspector determines have not been properly dried and cured in original natural condition, or which show evidence of the addition thereto of water, may be held by the handler for the account of the producer or dehydrator for conditioning by further drying or dehydration: *Provided,* That such prunes shall be identified and kept separate and apart from any other prunes in the handler's possession until resubmitted for inspection and certificated as properly dried and cured, or returned to the producer or dehydrator. The certificate shall show, in addition to other inspection requirements, that the conditioning was performed and indicate the net weight after conditioning.

(2) *Prunes with active insect infestation.* Any prunes delivered to a handler which an inspector determines are not free from active insect infestation, may be returned to the producer or dehydrator or may be held by the handler for the account of the producer or dehydrator for conditioning by fumigation: *Provided,* That such prunes shall be identified and kept separate and apart from any other prunes in the handler's possession until resubmitted for inspection and certificated to show, in addition to other inspection requirements, performance of fumigation and freedom from active infestation.

(3) *High moisture content prunes.* The delivery of any high moisture content prunes to a handler by a producer or dehydrator shall be reported promptly by the handler to the inspection service. The inspection service shall be requested to submit a report to the committee of each such delivery which shall contain the following information: (i) The date and place of the delivery; (ii) the name and address of the producer or dehydrator, the handler, and the inspection service; and (iii) the variety of the high moisture content prunes, the county in which they were produced, and their net weight as shown on the door receipt or weight certificate, together with the number of such receipt or certificate. Any handler who, subsequent to delivery to him of high moisture content prunes, elects to dry or dehydrate them or any portion thereof to a point where they are capable of being received by such handler shall, prior to proceeding with such drying or dehydration, notify an inspector of the inspection service of his election, and the same procedure shall apply as set forth in paragraph

(d)(1) of this section. For each day on which a handler processes and packages high moisture content prunes, he shall furnish promptly to the inspector a signed statement and one copy showing the handler's name and address and the net weight of the total tonnage of high moisture content prunes processed and packaged by him on that day. The handler shall furnish promptly to the inspector two copies of the shipping or disposition order or other documents which shall show the date of each shipment or disposition, the applicable reference number thereof, and an adequate description of the shipment or disposition. One copy of each document so furnished shall be required to be forwarded to the committee. Upon request of the committee a handler shall, within ten days thereafter, file with the committee a signed report on Form PMC 3.1 "Report of High Moisture Content Prunes" which shall contain the following information: (i) The date and the name and address of the handler; (ii) the total tonnage of high moisture content prunes delivered to the handler during the crop year to the date of the report; (iii) the total tonnage of high moisture content prunes shipped or otherwise disposed of by the handler during such period; (iv) the total tonnage of high moisture content prunes delivered to the handler during such period which were dried or dehydrated and received as prunes by the handler; and (v) the total tonnage of high moisture content prunes in the handler's possession on the date of the report.

(4) *Return of prunes to producers and dehydrators.* Any lot of prunes delivered to a handler by a producer or dehydrator may be returned to the producer or dehydrator prior to an inspection thereof. Any lot of prunes so delivered whose identity has been maintained may be so returned following an inspection thereof, except prunes which have been size graded or sorted by the handler, resulting in a segregation of defects. Prunes which have been sorted for the producer or dehydrator, the identity of which have been maintained to the satisfaction of the inspector and the Committee, may be resubmitted for inspection in not more than three

655

new lots, equal in weight to the original lot, and the applicable inspections shall supersede the original inspection.

[26 FR 8278, Sept. 2, 1961, as amended at 33 FR 11812, Aug. 21, 1968; 33 FR 12033, Aug. 24, 1968; 33 FR 14172, Sept. 19, 1968; 35 FR 11380, July 16, 1970; 37 FR 15980, Aug. 9, 1972; 39 FR 30343, Aug. 22, 1974; 43 FR 40199, Sept. 11, 1978; 48 FR 57261, Dec. 29, 1983]

EFFECTIVE DATE NOTE: At 70 FR 30613, May 27, 2005, § 993.149 was suspended indefinitely.

§ 993.150 Disposition of prunes by handlers.

(a) *Inspection stations.* An inspection station shall be any plant of a handler, and any other place where he handles prunes.

(b) *Outgoing inspection.* Except as otherwise specifically provided, no handler shall ship or otherwise make final disposition of natural condition prunes or of processed prunes unless he has, prior to such shipment or final disposition, had them inspected and obtained a certificate showing that such prunes meet the effective minimum standards. Such inspection shall be made during that portion of the final preparation of the prunes for shipment or other final disposition as will permit proper sampling, whether in-line or floor inspection, and no handler shall perform such final preparation unless an inspector is present. The handler shall furnish promptly to the inspector a copy of the shipping or disposition order or other documents, which shall show the date of each shipment or disposition, the applicable reference number thereof, and an adequate description of the shipment or disposition. For the prunes inspected each day which meet the applicable minimum grade and size requirements for standard prunes, or standard processed prunes, the handler shall cause the inspector to issue in triplicate a signed certificate containing the following information:

(1) The date and place of inspection;

(2) The name and address of the handler and of the inspection service;

(3) The number and size of packages or the net weight of prunes;

(4) The number of the worksheet or worksheets on which the inspector's computations and results of tests are recorded; and

(5) A statement that the prunes meet the effective minimum standards for standard prunes, or standard processed prunes, as the case may be.

(c) *Interhandler transfers.* With the exception of those prunes held by a handler pending their disposition pursuant to § 993.49(c) and those prunes held by him for the account of the Committee pursuant to § 993.57, a handler may transfer prunes to another handler within the area. Any such interhandler transfer may be without the transferring handler having an inspection made as provided for in § 993.51: *Provided,* That before each such transfer the transferring handler shall: (1) Give written notice of the transfer to the inspection service including the proposed date of the transfer, the names of the handlers and, by plant designation, the present location and the destination of the prunes, the number of containers, variety, size designation, and total net weight of the prunes, and the manifest or billing number; and (2) receive from the inspection service a DFA Form P–5 "Shipping Inspection Report and Certificate" marked "Interhandler Transfer Report" on which the inspection service recorded the information furnished by the transferring handler. The transferring handler shall sign the "Interhandler Transfer Report" including all copies thereof that were received from the inspection service, and forward the signed original and one copy to the receiving handler at the time of the interhandler transfer. Upon receipt of the transferred prunes, the receiving handler shall enter on both the original and the copy the date he received the prunes, sign the original, and immediately forward it to the inspection service. The transferring handler shall cause the inspection service to promptly report the transfer to the Committee. As provided in § 993.50(f), the receiving handler shall, before shipping or otherwise making final disposition of such prunes, comply with the requirements of §§ 993.50 and 993.51.

(d) *Tolerances for non-French prunes.* Any lot of standard prunes or standard processed prunes containing more than 2 percent by weight of non-French prunes shall be disposed of only in prune product outlets as prescribed in § 993.50(c) unless the non-French prunes therein have an average count of 40 or less per pound and unless in a 100-ounce sample of the lot, the count per pound of 10 ounces of the smallest prunes in the sample does not vary from the count per pound of 10 ounces of the largest prunes in the sample by more than 35 points. A lot shall be deemed to exceed the 2 percent tolerance for non-French prunes whenever an inspection shows such prunes exceed 2 percent in any four consecutive sampling units of two tons or less or, if less than four such units are sampled, in such lesser number of units.

(e) *Prunes which fail to meet minimum standards*—(1) *Committee's approval of disposition*—(i) *General.* Those defective prunes accumulated by a handler by removing them from standard or substandard prunes, and those prunes received or held by a handler which fail to meet the applicable minimum standards and are held for disposition without removal of defective prunes in excess of maximum tolerances, may only be used, if within the tolerances prescribed in §993.97 II. C. (1), (2), and (3), for prune products, or if any such tolerances are exceeded and any live infestation corrected by fumigation, for non-human consumption or be destroyed. In order to insure that all such prunes are shipped or otherwise disposed of in accordance with §993.50(e), no handler shall during any crop year ship or otherwise make final disposition of any such prunes, other than prune waste subject to daily non-human disposition for sanitation purposes, unless prior thereto he had obtained during that crop year (except as otherwise provided in paragraph (e)(1)(iii) of this section) the Committee's approval of his application to do so.

(ii) *Application for approval.* The handler's application to ship or otherwise make final disposition of any such prunes shall be submitted on Form PMC 2.2 "Application for Permission to Dispose of Substandard Prunes". If the prunes are for shipment, the application shall set forth: (a) The name and address of the handler's vendee and the name and address of the consignee whether the same as or different from the vendee; (b) the particular use to be made of the prunes; (c) if such use is to

be by a person other than the handler's vendee or the consignee, the name and address of such user; and (d) the crop year or the period within, or the portion of, the crop year during which shipments are to be made. When the use or the name and address of the consignee or user are not known by the handler, the handler shall arrange for the submission of such information to the Committee. If use is to be by the handler, the application shall so indicate and shall set forth all applicable information. Each application for shipment shall be limited to the handler's vendee and the consignee if different from the vendee, and to a specific user and use, and may be open as to quantity: *Provided,* That, when the use or name and address of the user are not known by the handler, the application shall include the quantity of prunes to be shipped and be limited to that quantity. Each application for final disposition for a particular use by the handler shall be limited to such handler and use.

(iii) *Approval of applications.* The Committee's approval of a handler's application shall be transmitted to the handler on Form PMC 2.3 "Permission to Dispose of Substandard Prunes". In approving an application, the Committee shall specify the crop year, or the period within or the portion of the crop year, for which the approval is granted: *Provided,* That, the Committee may approve in July any such application that is submitted during that month by the handler for shipment or other final disposition of the prunes covered thereby in the succeeding crop year. When the use or the name and address of the user or consignee are not known to the handler, the Committee shall not approve the application until it has been informed as to such use and user and consignee of the prunes.

(iv) *Disapproval of applications; or revocation of approved applications.* In acting on an application, the Committee may disapprove the application when: (a) The application does not conform with the requirements of paragraph (e)(1)(ii) of this section; (b) the Committee has cause to believe that the prunes covered by the application will not be shipped or disposed of in accordance with the application; or (c) the handler, or any of the parties involved in the proposed shipment or disposition, had shipped or made other disposition of prunes covered by a previously approved application inconsistent with that application. The Committee may for cause revoke a handler's previously approved application if he ships or makes other disposition inconsistent with such application. Whenever a user uses prunes inconsistent with an approved application, the Committee may for cause revoke such application, and such other approved applications applicable to such user as the Committee deems necessary to assure that the prunes covered by such applications will not be used in a manner inconsistent with those applications or the order. The Committee shall notify the handler in writing of each disapproval and each revocation.

(v) *Evidence of non-human disposition.* Whenever defective or substandard prunes or prune waste are shipped to or otherwise disposed of in non-human consumption outlets, or destroyed, the handler shall furnish the Committee with a copy of the shipping document or other documentary evidence of the disposition as may be satisfactory to the Committee and at such times as the Committee may direct.

(vi) *Books and records.* Each handler who ships or otherwise disposes of defective or substandard prunes or prune waste shall make available for examination by the Committee, at his business office at any reasonable time during business hours, copies of all applicable purchase orders, sales contracts, or disposition documents, together with any further information which the Committee may deem necessary or desirable to enable it to determine whether such prunes or prune waste have been or will likely be utilized as authorized.

(2) *Out of the area shipments.* Whenever substandard prunes for human consumption are packed in closed containers, and if for shipment outside the area they shall be so packed, each such container shall be clearly marked "For Manufacturing Purposes Only". Whenever substandard prunes restricted to non-human usage are shipped in closed containers, each such container shall

be clearly marked "For Non-Human Usage". In each instance, the letters shall be of reasonable prominence and in a conspicuous place on the container.

(3) *Inspection of substandard prunes.* Each handler shall cause substandard prunes, for use in prune products, to be inspected (prior to disposition or shipment by a handler) by an inspector, and that such inspector issue, in triplicate, a signed clearance certificate (for the preparation of which the handler shall make available to the inspector the necessary data) containing the following information:

(i) The date and place of inspection and clearance;

(ii) The name and address of the inspection service and of the handler;

(iii) The number and kind of packages, the net weight, and the adequacy of the marking;

(iv) The lot number or shipping or disposition order number;

(v) The committee's approval number;

(vi) The destination; and

(vii) The actual percentage of offgrade prunes of each group, or combination of groups, of defects in excess of the then current tolerances for standard prunes or standard processed prunes.

(f) *Pitted prunes*—(1) *For human consumption as such.* (i) No handler shall ship or otherwise make final disposition of any lot of pitted prunes for human consumption as pitted prunes unless the lot, before pitting, met (A) the applicable minimum standard set forth in § 993.97 (Exhibit A), or as such standards may be modified, for standard prunes or standard processed prunes, and (B) the requirements specified in § 993.50 (c) and (d).

(ii) No handler shall ship or otherwise make final disposition of any lot of pitted prunes for human consumption as pitted prunes unless these prunes do not exceed an average of 0.5 percent by count of prunes with whole pits and/or pit fragments 2 mm or longer; and four of ten subsamples examined have no more than 0.5 percent by count of prunes with whole pits and/or pit fragments 2 mm or longer. For the purposes of this paragraph (f)(1)(ii), pitted prunes means prunes with the pit removed that are characterized by a uniform depression and minimal skin break where the pit has been removed.

(iii) No handler shall ship or otherwise make final disposition of any lot of macerated prunes for human consumption as pitted prunes unless these prunes do not exceed an average of 2 percent by count of prunes with whole pits and/or pit fragments 2 mm or longer; and four of ten subsamples examined have no more than 2 percent by count with whole pits and/or pit fragments 2 mm or longer. For the purposes of this paragraph (f)(1)(iii), macerated prunes means prunes with the pit removed that are characterized by a flattened appearance with slightly more skin breaks where the pit has been removed than with pitted prunes.

(2) *For use in prune products.* Any lot of substandard prunes, whether natural condition or processed, if within the applicable tolerances prescribed in §993.97 II C (1), (2), and (3), may be pitted and shipped or disposed of for use and used in prune products for human consumption: *Provided,* That prior to shipment or other final disposition by handler, such prunes have lost their form and character as prunes to the satisfaction of the inspector and the committee. An inspection certificate on such lot shall not be issued until the inspector has determined that the prunes therein have lost their form and character as prunes. Disposition of pitted prunes by handlers for use in prune products shall be in accordance with the applicable provisions of paragraph (e) of this section.

(g) *Disposition of undersized prunes—* (1) *Application for and approval of disposition.* Undersized prunes accumulated by a handler pursuant to section 993.49(c) shall be disposed of in non-human consumption outlets during the crop year in which the prunes establishing such obligations were received from producers or dehydrators, or such later date that a handler may request in a notice, filed with the Committee at least 30 days prior to July 31 of the year of accumulation: *Provided,* That, such handler has made a bona fide effort to dispose of its undersized prunes as demonstrated by the shipment of at least 65 percent of its undersized obligation by May 31; such handler has a

sufficient quantity of undersized prunes held in storage to meet its remaining obligation; and the extension of time requested is not later than 60 days beyond the end of the crop year. Prior to making any such disposition, the handler shall obtain the Committee's approval of his application to do so. The handler's application to ship or otherwise make final disposition of any such undersized prunes shall be submitted on Form PMC 2.21 "Application for Permission to Dispose of Undersized Prunes" which shall set forth: (i) The name and address of the handler's vendee and the name and address of the consignee whether the same as or different from the vendee; (ii) the particular use to be made of the prunes; (iii) if such use is to be by a person other than the handler's vendee or the consignee, the name and address of such user; and (iv) the crop year or the period within, or portion of, the crop year during which shipment or other disposition is to be made. When the use or the name and address of the consignee or user are not known by the handler, the handler shall arrange for the submission of such information to the Committee. If use is to be by the handler, the application shall so indicate and shall set forth all applicable information. Each application for shipment shall be limited to the handler's vendee and the consignee, if different from the vendee, and to a specific user and use. Each application for final disposition for a particular use by the handler shall be limited to such handler and use. The Committee's approval of a handler's application shall be transmitted to the handler on Form PMC 2.31 "Permission to Dispose of Undersized Prunes." In approving an application, the Committee shall specify the crop year or the period within, or the portion of, the crop year for which the approval is granted. When the use or name and address of the user or consignee are not known to the handler, the Committee shall not approve the application until it has been informed as to such use and user and consignee of the prunes. The requirements of §993.150(e)(1)(iv) (except item (a) thereof), (v), and (vi) with regard to disapproval of applications or revocation of approved applications, evidence of nonhuman disposition, and the maintenance of books and records, applicable to prunes which fail to meet minimum standards, shall also apply to undersized prunes.

(2) *Documentation of disposition of undersized prunes*—(i) *Documentation of shipment or other disposition.* For each quantity of undersized prunes so shipped or otherwise disposed of, the handler shall promptly forward to the Committee one copy of the applicable bill of lading, truck receipt, or related documentation of disposition which shall show: (*a*) The name of the consignee; (*b*) the destination by name and address of the person designated to receive the prunes; (*c*) the date of shipment or other disposition; (*d*) the net weight of the prunes; and (*e*) identification of the prunes as undersized prunes.

(3) *Tolerances permitting a deviation in prune sizes from applicable undersized openings*—(i) *Undersized French prunes.* Whenever an undersized regulation specifies an opening for French prunes, any quantity of any size of French prunes disposed of by a handler in compliance with § 993.50(g) shall satisfy a handler's undersized disposition.

(ii) *Undersized non-French prunes.* Whenever an undersized regulation specifies an opening for non-French prunes, any quantity of any size of non-French prunes disposed of by a handler in compliance with § 993.50(g) shall satisfy a handler's undersized disposition.

[26 FR 8280, Sept. 2, 1961, as amended at 27 FR 458, Jan. 17, 1962; 29 FR 2331, Feb. 11, 1964; 33 FR 14172, Sept. 19, 1968; 35 FR 5108, Mar. 24, 1970; 35 FR 11381, July 16, 1970; 37 FR 15980, Aug. 9, 1972; 40 FR 52838, Nov. 13, 1975; 48 FR 57261, Dec. 29, 1983; 57 FR 56243, Nov. 27, 1992; 58 FR 13698, Mar. 15, 1993; 59 FR 10228, Mar. 3, 1994]

EFFECTIVE DATE NOTE: At 70 FR 30613, May 27, 2005, § 993.150 was suspended indefinitely.

RESERVE CONTROL

§ 993.156 Application of reserve percentage.

The reserve obligation of each handler shall be determined by applying the reserve percentage to the weight of prunes in each lot, after deducting the weight of prunes in such lot shown as a percentage on the applicable inspection certificate as necessary to be removed therefrom pursuant to § 993.49(c), in such manner as may be prescribed in such reserve control regulation established for the crop year in which such lot is received by a handler from a producer or dehydrator.

[30 FR 13311, Oct. 20, 1965]

EFFECTIVE DATE NOTE: At 70 FR 30613, May 27, 2005, § 993.156 was suspended indefinitely.

§ 993.157 Holding and delivery of reserve prunes.

(a) *Sales and deliveries.* Committee sales and deliveries of reserve prunes from the holdings of any handler shall not exceed the quantity of reserve prunes required to be held by him. The reserve prune holding requirement of the handler shall be reduced by the tonnage so sold or delivered.

(b) *Assistance to handlers.* As assistance to handlers, the committee shall furnish each handler a monthly tabulation, beginning as soon as possible after the start of the crop year, showing his reserve obligation and holding requirement based on records on file with the committee.

(c) *Failure to hold and deliver reserve prunes in accordance with reserve obligation.* In the event a handler fails to hold for the committee and deliver his total reserve prune obligation in any category and is unable to rectify such a deficiency with salable prunes, he shall compensate the committee in an amount computed by multiplying the pounds of natural condition prunes so deficient by the applicable values established by the committee: *Provided,* That the remedies prescribed herein shall be in addition to, and not exclusive of, any of the remedies or penalties prescribed in the act with respect to noncompliance. The determination of any such deficiency shall include application of any tolerance allowance for shrinkage in weight, increase in the number of prunes per pound, and normal and natural deterioration and spoilage which may then be in effect.

(d) *Excess delivery of prunes to the committee.* In the event a handler delivers to the committee as reserve prunes a quantity of prunes in excess of his holding requirement for reserve prunes, the committee shall make such practical adjustments as are consistent with this part and this may include compensating the handler for such excess (nonreserve prunes) by paying to him the proceeds received by the committee for such excess.

(e) *Holding reserve prunes on other than a handler's premises.* No handler shall hold reserve prunes on the premises of another handler, or in approved commercial storage other than on his own premises, unless prior thereto he notifies the committee in a certified report on Form PMC 5.1 "Notice of Proposed Intent to Store Reserve Prunes" which shall contain at least the following information: (1) The date and the name and address of the handler; (2) the name and address of the person on whose premises the reserve prunes will be stored for the handler; (3) the approximate quantity to be so stored and the exact location and description of the storage facilities; and (4) the proposed date that such storage will begin. The report shall be accompanied by a signed statement by the persons on whose premises the reserve prunes are to be stored agreeing to hold such prunes under conditions of proper storage and further agreeing to permit access to such premises by the committee at any time during business hours for the purpose of examining or taking delivery of such prunes in accordance with the provisions of this part. No handler shall be permitted to hold reserve prunes on any premises outside the area.

(f) *Exchange of salable prunes for reserve prunes.* No handler shall exchange salable prunes for reserve prunes unless he has entered into a sales agreement authorized pursuant to §993.65(b) whereby the value of any such exchange, and payment therefor to the committee, shall be determined.

(g) *Delivery by nonsignatory handlers.* Any handler not signing the sales agreement authorized pursuant to §993.65(b), shall deliver to the Committee, upon demand, the total weight of his reserve obligation by such variety, grade, and size categories, and at the count per pound for each size category as is required by the reserve control regulation of the applicable crop year. Such deliveries of prunes may be either graded prunes or any lot of ungraded prunes, or portion thereof, identifiable to the satisfaction of the committee as being in the same form as when received: *Provided,* That the percent of standard prunes in each lot shall be taken into account but with respect to any lot of graded prunes, no credit shall be given to the standard obligation of the handler if in a sample of 100 ounces, the count per pound of 10 ounces of the smallest prunes exceeds the count per pound of 10 ounces of the largest prunes by more than 45 prunes per pound.

[33 FR 19162, Dec. 24, 1968, as amended at 48 FR 57261, Dec. 29, 1983]

EFFECTIVE DATE NOTE: At 70 FR 30613, May 27, 2005, §993.157 was suspended indefinitely.

§993.158 Deferment of reserve withholding.

Any handler who desires to defer withholding pursuant to the provisions of §993.58 shall notify the committee on Form PMC 9.1, "Notification of Desire for Deferment of Reserve Withholding", containing at least the following information: (a) The date and the name and address of the handler; (b) the total salable prunes acquired or under contract with producers and dehydrators; (c) the period for which deferment is requested; and (d) the tonnage of reserve prunes, by categories, on which deferment is requested. The notification shall be accompanied by the undertaking and bond or bonds required by §993.58. No handler shall defer withholding of reserve prunes until he has filed the required undertaking and bond or bonds with the committee and has received its acceptance.

[30 FR 13311, Oct. 20, 1965, as amended at 48 FR 57261, Dec. 29, 1983]

EFFECTIVE DATE NOTE: At 70 FR 30613, May 27, 2005, §993.158 was suspended indefinitely.

§993.159 Payments for services performed with respect to reserve tonnage prunes.

(a) *Payment for crop year of acquisition.* Each handler shall, with respect to reserve prunes held by the handler for the account of the Committee pursuant to §993.59, be paid at a rate computed by the Committee (natural condition rate) for necessary services rendered by the handler in connection with such prunes so held during all or any part of the crop year in which the prunes were physically received from producers or dehydrators. Each handler holding reserve prunes shall perform such services to assure that the prunes are maintained in good condition. No payment will be made for prunes released by handler acceptance of diversion certificates if the handler has not stored the released prunes. The rate of payment shall be established by the Committee and must be approved by the Secretary. Following such approval, it shall be publicized as required in paragraph (e) of this section.

(1) On or before July 20 of each crop year when the Committee recommends a reserve pool (except the Committee may extend this date by not more than ten business days if warranted by a late crop), the Committee shall hold a meeting to review the costs for necessary services rendered by handlers in connection with reserve prunes.

(2) Such amount shall, together with the additional payments, as provided in this section, be in full payment for the costs incurred in connection with but not be limited to the following services: Inspection, receiving, storing, grading, fumigation, and handling. The costs include, but are not limited to:

(i) Acquisition costs, which include those for salaries, commission, or brokerage fees, transportation and handling between plants and receiving stations, inspection, and other costs, including container expenses, incidental to acquisition or storage;

(ii) Direct labor costs, which include those for weighing, receiving and stacking, grading, preliminary sorting and storing (including that performed by the handler at the receiving station), and loading for shipment or other delivery to the Committee or its designee;

(iii) Plant overhead costs, which include those for supervision, indirect labor, fuel, power and water, taxes and insurance on facilities, depreciation and rent, repairs and maintenance (clean-up, etc.), factory supplies and expense, and employee benefits (payroll taxes, compensation insurance, health insurance, pension plan contributions, vacation pay, holiday and other paid days off, and other such costs).

(3) The Committee shall survey all handlers to obtain their costs for services performed with respect to reserve tonnage prunes. The Committee will compute the average industry cost for holding reserve pool prunes by adding each handlers' cost data, and dividing the composite figure by the number of handlers participating in the survey. In the event that any handler's cost data is too low or too high, the Committee may choose to exclude the high and low data in computing an industry average. The industry average costs may be rounded to the nearest $0.25. The industry average costs computed by the Committee shall be publicized by the Committee pursuant to paragraph (e) of this section.

(b) *Reimbursement for required insurance costs.* Each handler holding reserve prunes for the account of the Committee shall maintain proper insurance thereon, including fire and extended coverage, in valuations (according to grade and/or size) established by, or acceptable to, the Committee for the particular crop year. The Committee shall reimburse the handler for the actual costs of such insurance. Prior to the receipt of reserve prunes at the beginning of each crop year, the handler shall certify to the Committee and the Secretary of Agriculture, on Form PMC 4.5, that such handler has a fire and extended coverage policy fully insuring all reserve prunes received by the handler during such crop year. Such certification shall contain the following information:

(1) The name and address of the handler;

(2) The location(s) where reserve prunes will be held for the account of the Committee and the premium rate per $100 value per annum at each location;

(3) The value per ton at which the reserve prunes are insured; and

(4) The name and address of the insurance underwriter.

(c) *Certain additional payments in connection with the holding of reserve prunes for the account of the Committee.* (1) Whenever a handler is directed by the Committee to move and dump containers or reserve prunes held by the handler for the account of the Committee for the purpose of causing an inspection to be made of the prunes as provided in §993.75, but without taking delivery of the prunes at that time, the handler shall be paid for such services at a rate per ton (natural condition weight) determined by the Committee and approved by the Secretary of Agriculture. Such reimbursement rate shall be computed as described in paragraph (a)(3) of this section and publicized as required in paragraph (e) of this section.

(2) Additional payment for reserve tonnage prunes held beyond the crop year of acquisition shall be made in accordance with this paragraph. Each handler holding reserve prunes shall complete such services so that the Committee is assured that the prunes are maintained in good condition.

(i) For storage and necessary fumigation, each handler shall be compensated at a per ton rate announced by the Committee in accordance with paragraph (a)(3) of this section:

(A) For all or any part of the first 3 months of the succeeding crop year, the rate per ton shall be 10 percent of the yearly rate established for the crop year of acquisition;

(B) For all or any part of the second 3 months of the succeeding crop year, the rate per ton shall be 50 percent of the rate established for the first 3 months of the succeeding crop year;

(C) For all or any part of the third 3 months of the succeeding crop year, the rate per ton shall be 25 percent of the rate established for the first 3 months of the succeeding crop year;

(D) For all or any part of the fourth 3 months of the succeeding crop year, the rate per ton shall be 25 percent of the rate established for the first 3 months of the succeeding crop year;

(ii) For all or part of the succeeding crop year, the Committee shall determine the per ton rate for bin rental within the industry and announce bin rental rate to the industry pursuant to paragraph (e) of this section.

(iii) For insurance as prescribed in paragraph (b) of this section.

(d) *Certain additional payments in connection with the delivery of reserve prunes to the Committee or its designee.* (1) Whenever a handler is directed by

the Committee to deliver to it or its designee reserve prunes in natural condition, the Committee shall furnish the handler with the containers in which to deliver the prunes, or reimburse the handler, at cost, for any containers which the handler furnishes pursuant to an agreement with the Committee.

(2) Whenever the Committee arranges with a handler for the reserve prunes delivered to it or its designee to be in processed and packaged condition, the Committee shall reimburse the handler at the agreed rate, determined by the Committee to be reasonable, for the processing, container, and packaging costs.

(e) The Committee shall give reasonable publicity to producer and handler members and alternates who serve on the Committee, commercial dehydrators, handlers, and the cooperative bargaining association(s) of each meeting to consider handler payment rates or any modification thereof, and each such meeting shall be open to them. Similar publicity shall be given to producer and handler members and alternates who serve on the Committee, commercial dehydrators, handlers, and the cooperative bargaining association(s) of each payment rate modification submitted to USDA for review and approval. The Committee shall notify producer and handler members and alternates who serve on the Committee, commercial dehydrators, handlers, and cooperative bargaining association(s) of USDA's action on payment rates and conditions for payment by first class mail and/or by electronic communications.

[68 FR 17543, Apr. 10, 2003]

EFFECTIVE DATE NOTE: At 70 FR 30613, May 27, 2005, §993.159 was suspended indefinitely.

VOLUNTARY DIVERSION

§993.162 Voluntary prune plum diversion.

(a) *Quantity to be diverted.* The Committee shall indicate the quantity of prune plums that producers may divert pursuant to §993.62 whenever it recommends to the Secretary that diversion operations for a crop year be permitted. Whenever diversion operation for a crop year have been authorized by the Secretary, the Committee shall notify producers, commercial dehydrators, and handlers, known to it of such authorization and diversion program procedures. The Committee shall compute the dried weight equivalent of prune plums so diverted on a dryaway basis as follows:

(1) For prune plums of the French variety, the Committee shall survey at least eight commercial prune dehydrators that are geographically dispersed within the production area to obtain their annual dryaway ratios for each of the preceding five crop years, and compute a five-year average dryaway ratio for each dehydrator. The Committee shall then add together the participating commercial dehydrators' five-year average dryaway ratios for each producing region within the production area, and divide the total by the number of participating commercial dehydrators in that region to compute the dryaway ratio by producing region. In the event any of the annual dryaway ratios for any of the crop years is abnormally high or low in any year, the Committee may replace the abnormal year's data with that of an earlier year. The prune producing regions for which dryaway ratios shall be computed for prune plums of the French variety are as follows:

(i) North Sacramento Valley, which includes the counties of Butte, Glenn, Shasta, and Tehama;

(ii) South Sacramento, Napa, Sonoma, and Santa Clara Valleys, which includes the counties of Amador, Colusa, Lake, Placer, Solano, Sutter, Yolo, Yuba, Napa, Sonoma, San Benito, and Santa Clara; and

(iii) San Joaquin Valley, which includes the counties of Fresno, Kern, Kings, Madera, Merced, San Joaquin, Stanislaus, and Tulare.

(A) *New producing counties within the area.* If there were new producing counties within the State of California, the Committee will, with the approval of the Secretary, assign the new prune producing county or counties, as the case may be, to one of the prune producing regions based on geographic proximity and/or production/dehydration characteristics. The addition of a county or counties, as the case may be,

to one of the producing regions will be announced to the industry.

(B) *Removal of a county from a production area.* When prune acreage ceases to exist in a county, the Committee will, with the approval of the Secretary, remove that county from the existing region. Removal of a county from a production region also will be announced to the industry.

(2) For prune plums of the non-French variety, the dryaway ratio shall be 1 pound for each 3.50 pounds of prune plums diverted. The prune-producing region for prune plums of non-French varieties is the State of California.

(b) *Eligible diversions.* Eligible diversions shall preclude prune plums from becoming prunes and may include the following methods:

(1) Disposing of harvested prune plums under Committee supervision for nonhuman use at a location and in a manner satisfactory to the Committee;

(2) Leaving unharvested the entire production of prune plums from a solid block of bearing trees designated by the producer applying for the diversion of removing prune plum trees prior to harvest; and/or

(3) Such other diversions as may be authorized by he Committee and approved by the Secretary.

(4) In accordance with § 993.62(c), eligible diversion shall not apply to prune plums, which would not, under normal producer practices, be dried and delivered to a handler. On or before July 20 of each crop year when the Committee recommends a reserve pool and diversion program (except the Committee with the approval of the Secretary may extend this date by not more than 10 business days if warranted by a late crop), the Committee shall identify, with the approval of the Secretary, the acceptable method(s) of voluntary prune plum diversion through reasonable publicity to producers, commercial dehydrators, handlers, and the cooperative bargaining association(s). For the purposes of this section, cooperative bargaining association means a nonprofit cooperative association of dried prune producers engaged within the production area in bargaining with

handlers as to price and otherwise arranging for the sale of natural condition dried prunes of its members.

(c) *Applications for diversion*—(1) *By producers.* Each producer desiring to divert prune plums of his own production shall, prior to diversion, file with the Committee a certified application on Form PMC 10.1 "Application for Prune Plum Diversion" containing at least the following information:

(i) The name and address of the producer; whether the producer is an owner-operator, share-landlord, share-tenant, or cash tenant; and the name and address of any other person or persons sharing a proprietary interest in such prune plums;

(ii) The proposed method of diversion and the location where the diversion is to take place;

(iii) The quantity and variety of prune plums proposed to be diverted; and

(iv) The approximate period of diversion.

(v) A deposit fee shall accompany each producer's application to cover costs associated with processing the application and administering the diversion program. The Committee shall compute, with the approval of the Secretary, and announce to the industry, the deposit fee. The deposit fee announced shall be a set dollar amount or a per ton cost based on the tonnage to be diverted. The fee paid by the applicant shall be the greater of these amounts.

(2) *By dehydrator as agent.* Any producer, or group of producers, may authorize a dehydrator to act as an agent to divert harvested prune plums. Prior to diversion such dehydrator shall submit to the Committee an application on Form PMC 10.1 "Application for Prune Plum Diversion" for each producer or group of producers under contract with the dehydrator. A deposit fee shall accompany each such application to cover the costs associated with processing the application and administration of the program. With respect to any group of four or more producers under contract with a dehydrator, the deposit fee for the group shall be the greater of either double the single deposit fee, pursuant to paragraph (c)(1) of this section, or the amount obtained

by multiplying the total tonnage of prune plums to be diverted by the group of producers covered in the dehydrator's application times the per ton deposit rate announced by the Committee pursuant to (c)(1) of this section.

(3) *Receipt of applications.* The Committee shall establish, and give prompt notice to the industry, a final date for receipt of applications for diversion: *Provided,* That the Committee may extend such deadline if the total tonnage represented in all applications is substantially less than the total tonnage established by the Committee pursuant to paragraph (a) of this section.

(d) *Approval of applications.* No certificate of diversion shall be issued by the Committee unless it has approved the application covering such diversion.

(1) The Committee's approval of an application shall be in writing, and include at least the following:

(i) The details as to the method of diversion to be followed;

(ii) The method of appraisal to be used by the Committee to determine the quantity of prune plums diverted;

(iii) The lesser of either the quantity specified in the application to be diverted, or modification of that quantity as a result of any Committee action to prorate the total quantity to be diverted by all producers; and

(iv) Such other information as may be necessary to assist the applicant in meeting the requirements of this section, including the conditions for proof of diversion.

(2) If the Committee determines that it cannot approve an application it shall notify the applicant promptly. The Committee shall state the reason(s) for failing to approve the application, and request the applicant to submit, if practicable, an amended application correcting the deficiencies in the original application.

(3) The Committee shall establish, and give prompt notice to the industry of a final date by which a producer or dehydrator may modify an approved application, including changing the method of diversion or the quantity of prune plums to be diverted: *Provided,* That any such change shall include information on the location or quantity

of such diversion and shall be accompanied by a payment of a second deposit fee, calculated pursuant to paragraph (c)(1) or (c)(2), as applicable, of this section, plus a $2 per ton service charge for any increase in tonnage to be diverted.

(4) If an applicant cancels an approved diversion application prior to diversion, no part of the deposit fee shall be refunded, except upon approval by the Committee following review of all circumstances in the matter.

(e) *Report of diversion.* (1) When diversion of prune plums has been completed, the diverter (whether producer or dehydrator as agent of a producer) shall submit the required proof of such diversion to the Committee. When the Committee concludes that diversion has been completed pursuant to the requirements of this section, it shall furnish the producer whose prune plums were diverted with a listing of the total quantity of prune plums concluded to be so diverted: *Provided,* That a producer shall be given credit for any quantity of his prune plums diverted in excess of the quantity approved by the Committee pursuant to paragraph (d) of this section but not in excess of 120 percent of such approved quantity and then only to the extent that such creditable excess is already covered by his applicable deposit fee or such fee is increased by an additional deposit to cover such excess.

(2) Upon completion of the computation of dryaway pursuant to paragraph (a) of this section applicable to the diverter's diversion of prune plums, the Committee shall issue a report of diversion to the producer whose prune plums were diverted for the total quantity, dried weight equivalent, credited for diversion setting forth the computations by which such total quantity was derived.

(f) *Transferable certificate of diversion*—(1) *General.* As hereinafter set forth, transferable certificates of diversion shall be issued by the Committee. Any transferable certificate of diversion issued to a handler that is a cooperative marketing association, or submitted to a handler and accepted by him, shall be returned to the Committee by the handler for credit against the handler's reserve obliga-

tion of the crop year in accordance with § 993.57. Such credit shall be based on the amount shown on the certificate, and shall be applied to reduce the handler's holding requirement for such crop year. With respect to such creditable certificate of a handler with a holding requirement prior to issuance or acceptance, as applicable, of the transferable certificate of diversion, such credit shall result in an adjustment downward in the handler's then applicable holding requirement in an amount equal to that computed by applying the applicable salable percentage to the total quantity on such certificate. Any adjustment in a handler's holding requirement shall not affect his obligation, if any, to continue to hold reserve prunes that are undersized prunes. The term *undersized prunes* shall have the same meaning as prescribed by the Secretary for the then current crop year. If the Committee determines that effective administration of diversion operations requires establishment of a final date for submission of transferable certificates of diversion by producers to handlers, or a final date for return of such certificates by handlers to the Committee for crediting against their reserve obligations, or both, it shall establish such dates.

(2) *Issuance to producers.* Except as provided in paragraph (f)(3) of this section, the Committee shall issue transferable certificates of diversion to each producer diverting prune plums and to whom a report of diversion was issued. Prior to issuance of any such transferable certificate of diversion, the producer shall advise the Committee, in writing: (i) Of the name of the handler to whom the transferable certificate of diversion is to be submitted and who is holding reserve prunes referable to prunes received from such producer; and (ii) how much of the quantity shown on his report of diversion he desires to use in lieu of reserve prunes but not in excess of the quantity of reserve prunes referable to prunes received by the handler from such producer. The Committee shall enter on the transferable certificate of diversion the name of the handler and the quantity covered by the certificate. The transferable certificate of diversion shall be endorsed by the producer and

the handler prior to its return to the Committee in order to be credited by the Committee against such handler's reserve obligation. If any portion of the quantity shown on the producer's report of diversion remains unused and he desires to transfer a transferable certificate of diversion covering all or any part of such unused portion to another producer, he shall advise the Committee, in writing, of the name and address of such producer, together with the applicable quantity desired to be covered by the transfer, and, if known, the name of the handler to whom such a transferable certificate is to be submitted. However, the quantity to be covered by the transfer shall not exceed the quantity of reserve prunes referable to prunes received by the handler from the transferee-producer. The Committee shall enter on the transferable certificate of diversion the names of the transferee-producer and the handler, and the quantity covered by the certificate. Prior to submission of any such transferred diversion certificate to a handler, the transferee-producer shall advise the Committee, in writing, of the name and address of the handler to whom the transferable certificate is to be submitted and who is holding reserve prunes referable to prunes received from such producer. Such transferred diversion certificate shall be endorsed by both producers and the handler in order to be credited by the Committee against such handler's reserve obligation.

(3) *Issuance to a cooperative marketing association.* In connection with prune plums diverted by producers who are members of a cooperative marketing association, the Committee shall, when so requested by the association, issue the applicable transferable certificates of diversion to it. The quantity entered on the report of diversion of a cooperative producer shall be entered on or annexed to the applicable transferable certificate of diversion issued to the association. Such transferable certificates of diversion shall be returned to the Committee by the association endorsed by an authorized officer of the association in order to be credited by the Committee against the association's reserve obligation.

(4) *Applicability of certain payments.* The provisions of §§993.59 and 993.159 governing payments to a handler for necessary services rendered by the handler in connection with reserve prunes shall not be applicable to prunes no longer required to be held as reserve prunes due to a downward adjustment by the Committee in the handler's holding requirement on the basis of applicable transferable certificates of diversion returned to the Committee.

(g) *Costs.* Pursuant to §993.62(g), the costs pertaining to diversion are to be defrayed by payment of fees by the producer or cooperative marketing association to whom a diversion certificate is issued. After authorized diversion operations for a crop year are completed, the Committee shall ascertain its costs of diversion operations during such crop year. If the total amount represented by the deposit fees which accompanied the applications for diversion exceeds such costs, each producer, and each cooperative marketing association, entitled thereto shall receive a proportionate refund of the net amount. Such refund shall be calculated in the same proportion as the quantity of prune plums diverted by each such producer, and each such cooperative marketing association, is to the total quantity of prune plums diverted: *Provided,* That the Committee may prescribe a minimum charge to cover costs of processing each application for diversion submitted to it.

[35 FR 12323, Aug. 1, 1970, as amended at 36 FR 15039, Aug. 12, 1971; 48 FR 57261, Dec. 29, 1983; 68 FR 17270, Apr. 9, 2003]

EFFECTIVE DATE NOTE: At 70 FR 30613, May 27, 2005, §993.162 was suspended indefinitely.

DISPOSITION OF RESERVE PRUNES

§993.165 Disposition of reserve prunes.

(a) *General.* For purposes of §993.65(a)(2), normal outlets for salable prunes (herein referred to as "normal outlets") and outlets noncompetitive with normal outlets for salable prunes (herein referred to as "noncompetitive outlets") are defined in paragraphs (b) and (c) of this section.

(b) *Normal outlets. Normal outlets* means all outlets not specifically set forth in paragraph (c) of this section as noncompetitive outlets.

(c) *Noncompetitive outlets. Noncompetitive outlets* means (1) the U.S. Government or any agency thereof and any State or local government, except when such outlets are normally serviced through regular commercial trade channels, (2) any foreign government or any agency thereof, except any which normally is serviced through regular commercial trade channels, (3) any foreign country with an average of annual commercial imports of California prunes of less than 5 tons, based on imports during the most recent 5 years, (4) diced prunes for use as an ingredient in, or the manufacture of, food products for human consumption, other than for use in the manufacture of prune juice, prune concentrate, baby food, puree, butter, jam, and low moisture nuggets, granules, and powder, (5) charities, (6) research or educational activities, and (7) animal feed, distillation, and other salvage use.

[31 FR 5751, Apr. 14, 1966, as amended at 37 FR 5600, Mar. 17, 1972]

EFFECTIVE DATE NOTE: At 70 FR 30613, May 27, 2005, § 993.165 was suspended indefinitely.

REPORTS AND BOOKS AND OTHER RECORDS

§ 993.172 Reports of holdings, receipts, uses, and shipments.

(a) *Holdings as of March 31.* Each handler shall, on or before the 15th day of April, file with the committee a signed report of holdings of prunes which have not been inspected or received by him as a handler as of March 31. The report shall show for such prunes the name and address of the producer or dehydrator, the date of each identification tag assigned to such prunes, the numbers and dates of door receipts or weight certificates or any other identifying documents assigned to such prunes, the net weight shown on each, the total net weight of all prunes so held, and the name and address of the handler making the report.

(b) *Receipts by handlers.* Each handler shall file with the committee, for each month, not later than the 5th working day of the next succeeding month, a signed report on Form PMC 11.1, "New Crop Supply and Inbound Prune Re-

port", containing at least the following information: (1) The date, the name and address of the handler, and the period covered by the report; and (2) the total tonnage received during the month from each of (i) producers and dehydrators, (ii) other handlers, including interhandler transfers, and (iii) sources other than producers, dehydrators and other handlers.

(c) [Reserved]

(d) *Shipments by handlers.* Each handler shall file with the Committee for each quarter, not later than the 5th working day of the months of November, February, May and August, signed reports on Form PMC 12.1, "Reports of Shipments," and Form 12.1A, "Cumulative Prune Export Shipments" reporting shipments of prunes during the crop year through the last day of the immediately preceding quarter. Such reports shall contain at least the following information:

(1) The date, the name, and address of the handler, and the period covered by the report;

(2) The pounds of prunes shipped or otherwise disposed of, other than shipments to or for the account of other handlers as follows: Domestic outlets segregated by uses (including Federal Government agencies); export markets segregated by regions; both domestic and export totals segregated by type of pack (bulk and consumer pack); and pitted prunes (pitted weight) segregated as to total to domestic outlets and total to export markets segregated by regions;

(3) The total pounds shipped to or for the account of other handlers, including interhandler transfers; and

(4) The total pounds of prunes not covered by, or excluded from, the definition of the term "prunes" (§ 993.5) shipped.

(e) *Holding of reserve prunes.* Upon request of the committee, a handler shall file with the committee, within 10 calendar days thereafter, a certified report on Form PMC 4.1, "Reserve Prunes Held by Handler", containing the following information as of the date specified by the committee in its request: (1) The date and name and address of the handler; (2) the effective date of the report; and (3) the tonnages of reserve prunes physically held by or for the handler, itemized by plants, together with the location of the plants and itemized by the tonnages and average size count by category held at each such plant.

[26 FR 8281, Sept. 2, 1961, as amended at 30 FR 13311, Oct. 20, 1965; 31 FR 14988, Nov. 29, 1966; 33 FR 19162, Dec. 24, 1968; 48 FR 57261, Dec. 29, 1983; 49 FR 1469, Jan. 12, 1984; 68 FR 37393, June 24, 2003]

EFFECTIVE DATE NOTE: At 70 FR 30613, May 27, 2005, §993.172 was suspended indefinitely.

§993.173 Reports of accounting.

(a) *Independent handler's reports of accounting.* Within 10 days (exclusive of Saturdays, Sundays, and legal holidays) after a handler, other than a non-profit cooperative agricultural marketing association, makes an accounting or settlement with a producer or dehydrator for prunes delivered to him, he shall submit to the committee a copy of the accounting or settlement record, which shall contain the following information:

(1) The names and addresses of the producer or dehydrator, any other person having a financial interest in the prunes, and the handler;

(2) The date of the accounting or settlement;

(3) The contract or account number;

(4) An itemized statement listing each lot of prunes in the delivery, showing the date received, receiving point, weight certificate, or door receipt number, inspection certificate number, variety, crop year of production, and the net weight, if any, of prunes shown by the applicable incoming inspection certificate to be disposed of for nonhuman consumption in accordance with §993.150(g);

(5) The total net weight of prunes to be set aside for nonhuman consumption, and the total net weight received; and

(6) The total net weight of each lot, itemized as to salable and reserve prunes by category as developed from inspection certificates.

(b) *Cooperative marketing associations' reports of accounting.* Upon written notice by the committee, non-profit cooperative agricultural marketing associations which are handlers shall file with the committee within 10 days (exclusive of Saturdays, Sundays, and legal

holidays) thereafter a signed cumulative report of the prunes received from its members and any other producers or dehydrators for whom it performs handling services, which shall contain the following information:

(1) The name and address of the association and the date of the report;

(2) The aggregate net weight of prunes, as shown by the applicable incoming inspection certificates, required to be disposed of for nonhuman consumption in accordance with §993.150(g); and

(3) The total net weight of prunes received, itemized by crop years of production, and itemized as to salable and reserve prunes by category as developed from inspection certificates.

(c) *Carryover and marketing policy information.* Upon request of the committee, a handler shall within 10 days (exclusive of Saturdays, Sundays, and legal holidays) thereafter, file with the committee a signed report on Form PMC 14.1 "Report of Carryover and Marketing Policy Information," containing such of the following items of information as may be requested by the committee: (1) The tonnage of prunes held by the handler by size and grade, as of the date specified in the committee's request and the tonnage of reserve prunes by size in each category; and (2) the handler's estimate of the tonnage of prunes held by producers and dehydrators from whom the handler received prunes during the current or preceding crop year, of the tonnage and quality and size of prunes expected to be produced by such producers and dehydrators during the current or following crop year, of current prices being received by producers, dehydrators, and handlers, and of the probable trade demand.

[30 FR 13311, Oct. 20, 1965, as amended at 35 FR 11381, July 16, 1970; 37 FR 15981, Aug. 9, 1972; 48 FR 57261, Dec. 29, 1983]

EFFECTIVE DATE NOTE: At 70 FR 30613, May 27, 2005, §993.173 was suspended indefinitely.

§993.174 Records.

Each handler shall maintain such records as are necessary to furnish the reports required to be submitted to the Committee by him under this subpart including, but not limited to, records of all transactions on prunes received,

669

held and disposed of by him, and he shall retain such records for at least two years after the end of the crop year in which the applicable transaction occurred.

[26 FR 8281, Sept. 2, 1961; 26 FR 8483, Sept. 9, 1961. Redesignated at 30 FR 13312, Oct. 20, 1965]

EDITORIAL NOTES At 70 FR 30613, May 27, 2005, § 993.174 was suspended indefinitely.

Subpart C—Assessment Rates

§ 993.347   Assessment rate.

On and after August 1, 2019, an assessment rate of $0.25 per ton of salable dried prunes is established for California dried prunes.

[84 FR 64971, Nov. 26, 2019]

Subpart D—Undersized Prune Requirements

§ 993.400   Modifications.

Pursuant to the authority in § 993.52, the provisions in § 993.49(c) prescribing size openings for undersized prune regulations are hereby modified to permit larger size openings. For French prunes, any undersized regulation may prescribe an opening of $23/32$ of an inch or $24/32$ of an inch; for non-French prunes, any undersized regulation may prescribe an opening of $28/32$ of an inch or $30/32$ of an inch.

[40 FR 42531, Sept. 15, 1975]

EFFECTIVE DATE NOTE: At 70 FR 30613, May 27, 2005, § 993.400 was suspended indefinitely.

§ 993.409   Undersized prune regulation for the 2002–03 crop year.

Pursuant to §§ 993.49(c) and 993.52, an undersized prune regulation for the 2002–03 crop year is hereby established. Undersized prunes are prunes which pass through openings as follows: for French prunes, $24/32$ of an inch in diameter; for non-French prunes, $30/32$ of an inch in diameter.

[67 FR 31722, May 10, 2002]

EFFECTIVE DATE NOTE: At 70 FR 30613, May 27, 2005, § 993.409 was suspended indefinitely.

Subpart E—Pack Specification as to Size

SOURCE: 26 FR 8281, Sept. 2, 1961, unless otherwise noted.

DEFINITIONS

§ 993.501   Consumer package of prunes.

Consumer package of prunes means *consumer package* as defined in § 993.22.

EFFECTIVE DATE NOTE: At 70 FR 30613, May 27, 2005, § 993.501 was suspended indefinitely.

§ 993.502   Size count.

Size count means the count or number of prunes per pound.

§ 993.503   Size category.

Size category means each of the size categories listed in § 993.515 and fixes the range or the limits of the various size counts.

EFFECTIVE DATE NOTE: At 70 FR 30613, May 27, 2005, § 993.503 was suspended indefinitely.

§ 993.504   In-line inspection.

In-line inspection means inspection of prunes where samples are drawn from a flow of prunes prior to packaging.

EFFECTIVE DATE NOTE: At 70 FR 30613, May 27, 2005, § 993.504 was suspended indefinitely.

§ 993.505   Floor inspection.

Floor inspection means inspection of prunes where samples are drawn from packaged prunes or from unpackaged prunes that are held in packing containers for later packaging.

EFFECTIVE DATE NOTE: At 70 FR 30613, May 27, 2005, § 993.505 was suspended indefinitely.

§ 993.506   Lot.

Lot for the purposes of this subpart shall have the same meaning as defined in § 993.104(b) of the Subpart—Administrative Rules and Regulations.

EFFECTIVE DATE NOTE: At 70 FR 30613, May 27, 2005, § 993.506 was suspended indefinitely.

SPECIFICATIONS AS TO SIZE

§ 993.515   Size categories.

For the purpose of this part, the pack specifications prescribed for the packing of prunes in consumer packages shall, subject to the limitation prescribed in § 993.516, be according to those commercially recognized size categories as are listed in paragraph (a) of this section by numerical designation or in paragraph (b) of this section by nomenclature designation.

(a) *Numerical designations.* Each of the following is a numerical size category described by the range of the size counts of prunes per pounds included in the respective size categories expressed as follows or in an applicable equivalent range expressed in the metric system per 500 grams: 15/20, 15/22, 18/24, 20/30, 25/35, 30/40, 35/45, 40/50, 50/60, 60/70, 70/80, 75/85, 80/90, and 90/100.

(b) *Nomenclature designations.* Each of the following is a nonmenclature size category:

(1) Extra large;

(2) Large;

(3) Medium; and

(4) Small, breakfast, petite, or economy.

(c) *Nomenclature designations defined.* As used in paragraph (b) of this section:

(1) *Extra large* means any size count which falls within the range of 25 to 40 prunes, inclusive, per pound;

(2) *Large* means any size count which falls within the range of 40 to 60 prunes, inclusive, per pound;

(3) *Medium* means any size count which falls within the range of 60 to 85 prunes, inclusive, per pound; and

(4) *Small, breakfast, petite,* or *economy* means any size count which falls within the range of 85 to 100 prunes, inclusive, per pound.

[26 FR 8281, Sept. 2, 1961, as amended at 49 FR 35930, Sept. 13, 1984]

EFFECTIVE DATE NOTE: At 70 FR 30613, May 27, 2005, § 993.515 was suspended indefinitely.

§ 993.516 Tolerances and limitations.

With respect to in-line inspections and floor inspections, prunes in a particular lot shall, subject to the other applicable requirements of this section, be considered as being according to a particular size category prescribed in § 993.515 if the average size count of the prunes in such lot falls within the range of the size counts specified for such size category, and the count per pound of 10 ounces of the smallest prunes in a sample of 100 ounces varies

from the count per pound of 10 ounces of the largest prunes in such sample by no more than 45 points.

EFFECTIVE DATE NOTE: At 70 FR 30613, May 27, 2005, § 993.516 was suspended indefinitely.

LABELING

§ 993.517 Identification.

The size category of the prunes in any lot shall be clearly marked by the handler on each consumer package of such prunes, on the parts or panels of the package or label which are normally presented in retail display, in terms of the applicable numerical or nomenclature designation prescribed in § 993.515, which designation shall not be lacking in prominence and conspicuousness. Any handler may, at his option, clearly mark on such consumer package additional information describing in numerical terms the average size count, or particular range of size counts, of the prunes in such lot so long as such numerical terms fall within the range of the size counts of the applicable numerical or nomenclature designation and do not tend to be deceptive as to the actual average size count, or range of the size counts, of the prunes in such lot. Descriptive terms other than synonyms of the prescribed nomenclature designation or words of like connotation, describing the style of pack, variety of prune, or other item of commercial significance may also be marked on the consumer package. Prunes in any lot of which the maximum size count is less than 25 shall be clearly marked by the handler in terms of the applicable numerical designation prescribed in § 993.515(a); and the handler may use nomenclature terms descriptive of size other than the nomenclature designations prescribed in § 993.515(b).

[26 FR 8281, Sept. 2, 1961, as amended at 46 FR 38070, July 24, 1981]

EFFECTIVE DATE NOTE: At 70 FR 30613, May 27, 2005, § 993.517 was suspended indefinitely.

COMPLIANCE

§ 993.518 Compliance.

Whenever the season average price to producers for prunes is below the parity level specified in section 2(1) of the act, no handler shall ship consumer packages of prunes unless such prunes are packed and labeled in accordance with the specifications prescribed in this subpart; and whether prices are above or below parity, no handler shall use the nomenclature designations in § 993.515(b) to describe size categories other than those prescribed pursuant to § 993.515(c).

EFFECTIVE DATE NOTE: At 70 FR 30613, May 27, 2005, § 993.518 was suspended indefinitely.

Subpart F—Grade Requirements

§ 993.601 More restrictive grade regulation.

(a) *Incoming and outgoing regulation.* Whenever the estimated season average price to producers for prunes does not exceed the parity level specified in section 2(1) of the act, the minimum standards which handlers' receipts of natural condition prunes are required to meet pursuant to § 993.49(a) shall be the standards specified in §§ 993.49(a) and 993.97(I), and the minimum standards which handlers' shipments or other final dispositions of prunes are required to meet pursuant to § 993.50(a) shall be the applicable standards set forth in § 993.97 *Exhibit A*; *minimum standards*, except that the following revised tolerance allowances shall apply in lieu of the tolerance allowances prescribed in paragraphs I C(2), II C(3), I C(5) and II C(6) of § 993.97 as follows:

(1) The combined tolerance allowance for off-color, inferior meat condition, end cracks, fermentation, skin or flesh damage, scab, burned, mold, imbedded dirt, insect infestation, and decay shall not exceed fifteen percent (15%), except that the first eight percent (8%) of end cracks shall be given one-half value and any additional percentage of end cracks shall be given full value.

(2) The combined tolerance allowance for mold, brown rot, imbedded dirt, insect infestation, and decay shall not exceed five percent (5%), and, within such tolerance, brown rot shall not exceed three percent (3%).

(b) *Above parity situations.* Whenever the estimated season average price to producers for prunes exceeds the parity level specified in section 2(1) of the act,

the minimum standards set forth in § 993.97 shall apply in their entirety.

[27 FR 7540, Aug. 1, 1962, as amended at 59 FR 38113, July 27, 1994]

EFFECTIVE DATE NOTE: At 70 FR 30613, May 27, 2005, § 993.601 was suspended indefinitely.

§ 993.602 Maximum tolerances.

In lieu of the provision prescribed in I C of § 993.97 that the tolerance allowances prescribed therein shall be on a weight basis, the tolerance allowance percentage for each defect or group of defects in I C of § 993.97 shall be derived by dividing the number of prunes in the applicable sample affected with the applicable defect(s) by the total number of prunes in such sample.

[39 FR 30344, Aug. 22, 1974]

EFFECTIVE DATE NOTE: At 70 FR 30613, May 27, 2005, § 993.602 was suspended indefinitely.

PART 996—MINIMUM QUALITY AND HANDLING STANDARDS FOR DOMESTIC AND IMPORTED PEANUTS MARKETED IN THE UNITED STATES

DEFINITIONS

996.60 Safeguard procedures for imported peanuts.

AUTHORITY: 7 U.S.C. 7958.

SOURCE: 67 FR 57140, Sept. 9, 2002, unless otherwise noted.

DEFINITIONS

§ 996.1 Act and scope.

Act means Public Law 107–171, or the Farm Security and Rural Investment Act of 2002, enacted May 13, 2002. None of the definitions or provisions of this part shall apply to any other part or program (including, but not limited to, any program providing for payments or loans to peanut producers or other persons interested in peanuts or peanut quotas) unless explicitly adopted in such other part or program.

§ 996.2 Conditional release.

Conditional release means release from U.S. Customs Service custody to the importer for purposes of handling and USDA required sampling, inspection and chemical analysis.

[68 FR 1157, Jan. 9, 2003]

§ 996.3 Crop year.

Crop year means the calendar year in which the peanuts were planted as documented by the applicant for inspection.

[81 FR 50287, Aug. 1, 2016]

§ 996.4 Handle.

Handle means to engage in the receiving or acquiring, cleaning and shelling, cleaning inshell, or crushing of domestic or imported peanuts and in the shipment (except as a common or contract carrier of peanuts owned by another) or sale of cleaned-inshell or shelled peanuts or other activity causing peanuts to enter into human consumption channels of commerce: *Provided,* That this term does not include sales or deliveries of peanuts by a producer to a handler or to an intermediary person engaged in delivering peanuts to handler(s): *And provided further,* That this term does not include sales or deliveries of peanuts by such intermediary person(s) to a handler.

§ 996.5 Handler.

Handler means any person who handles peanuts, in a capacity other than that of a custom cleaner or dryer, an assembler, a warehouseman or other intermediary between the producer and the person handling peanuts.

§ 996.6 Importation.

Importation means the arrival of foreign produced peanuts at a port-of-entry with the intent to enter the peanuts into channels of commerce of the United States.

§ 996.7 Importer.

Importer means a person who engages in the importation of foreign produced peanuts into the United States.

[68 FR 1157, Jan. 9, 2003]

§ 996.8 Incoming inspection.

Incoming inspection means the sampling, inspection, and certification of farmers stock peanuts to determine segregation and grade quality.

§ 996.9 Inshell peanuts.

Inshell peanuts means peanuts, the kernel or edible portions of which are contained in the shell in their raw or natural state which are milled but unshelled.

[81 FR 50288, Aug. 1, 2016]

§ 996.10 Inspection Service.

Inspection Service means the Federal Inspection Service, Specialty Crops Program, Agricultural Marketing Service, USDA, or the Federal-State Inspection Service.

[81 FR 50288, Aug. 1, 2016]

§ 996.11 Negative aflatoxin content.

Negative aflatoxin content means 15 parts per billion (ppb) or less for peanuts that have been certified as meeting edible quality grade standards.

§ 996.12 Outgoing inspection.

Outgoing inspection means the sampling, inspection, and certification of either: shelled peanuts which have been cleaned, sorted, sized, and otherwise prepared for further processing; or inshell peanuts which have been cleaned, sorted, and otherwise prepared for further processing.

[81 FR 50288, Aug. 1, 2016]

§ 996.13　Peanuts.

Peanuts means the seeds of the legume *Arachis hypogaea* and includes both inshell and shelled peanuts produced in the United States or imported from foreign countries and intended for further processing prior to consumption by humans or animals, other than those intended for wildlife or those in green form for consumption as boiled peanuts.

(a) *Farmers Stock.* "Farmers stock peanuts" means picked and threshed peanuts which have not been shelled, crushed, cleaned or otherwise changed (except for removal of foreign material, loose shelled kernels, and excess moisture) from the form in which customarily marketed by producers.

(b) *Segregation 1.* "Segregation 1 peanuts" means farmers stock peanuts with not more than 3.49 percent damaged kernels nor more than 1.00 percent concealed damage caused by rancidity, mold, or decay and which are free from visible *Aspergillus flavus.*

(c) *Segregation 2.* "Segregation 2 peanuts" means farmers stock peanuts with more than 3.49 percent damaged kernels or more than 1.00 percent concealed damage caused by rancidity, mold, or decay and which are free from visible *Aspergillus flavus.*

(d) *Segregation 3.* "Segregation 3 peanuts" means farmers stock peanuts with visible *Aspergillus flavus.*

[67 FR 57140, Sept. 9, 2002, as amended at 81 FR 50288, Aug. 1, 2016; 82 FR 48758, Oct. 20, 2017]

§ 996.14　Person.

Person means an individual, partnership, corporation, association, any other business unit or legal entity.

§ 996.15　Positive lot identification.

Positive lot identification is a means of identifying those peanuts meeting outgoing quality regulations as defined in § 996.31 and relating the inspection certificate issued by the Inspection Service, as defined in § 996.10, to the lot covered so that there is no doubt that the peanuts in the lot are the same peanuts described on the inspection certificate.

[81 FR 50288, Aug. 1, 2016]

§ 996.16　Producer.

Producer means any person in the United States engaged in a proprietary capacity in the production of peanuts for market.

§ 996.17　[Reserved]

§ 996.18　Secretary.

Secretary means the Secretary of Agriculture of the United States or any officer, employee, or agent of the United States Department of Agriculture who is, or who may hereafter be authorized to act in the Secretary's stead.

§ 996.19　Shelled peanuts.

Shelled peanuts means the kernels or portions of kernels of peanuts in their raw or natural state after the shells are removed.

[81 FR 50288, Aug. 1, 2016]

§ 996.20　USDA.

USDA means the United States Department of Agriculture, including any officer, employee, service, program, or branch of the Department of Agriculture, or any other person acting as the Secretary's agent or representative in connection with any provisions of this part.

§ 996.21　USDA laboratory.

USDA laboratory means laboratories of the Science and Technology Programs, Agricultural Marketing Service, USDA, which chemically analyze peanuts for aflatoxin content.

§ 996.22　USDA-approved laboratory.

USDA-approved laboratory means laboratories approved by the Science and Technology Programs, Agricultural Marketing Service, USDA, that chemically analyze peanuts for aflatoxin content.

QUALITY AND HANDLING STANDARDS

§996.30 Incoming quality standards.

(a) All farmers stock peanuts received or acquired by a handler shall be officially inspected by the Inspection Service, and certified as to segregation, moisture content, and foreign material.

(b) *Moisture.* Domestic and imported peanuts shall be dried to 18 percent or less prior to inspection and to 10.49 percent or less prior to storing or milling: *Provided,* That Virginia-type peanuts used for seed shall be dried to 18 percent or less prior to inspection and to 11.49 percent or less prior to storing or milling.

[67 FR 57140, Sept. 9, 2002, as amended at 68 FR 1157, Jan. 9, 2003; 70 FR 44046, Aug. 1, 2005; 81 FR 50288, Aug. 1, 2016]

§996.31 Outgoing quality standards.

(a) *Shelled peanuts:* No handler or importer shall ship or otherwise dispose of shelled peanuts for human consumption unless such peanuts are positive lot identified, chemically analyzed by a USDA laboratory or USDA-approved laboratory and certified "negative" as to aflatoxin, and certified by the Inspection Service as meeting the following quality standards:

MINIMUM QUALITY STANDARDS—PEANUTS FOR HUMAN CONSUMPTION
[Whole kernels and splits: Maximum limitations]

Type and grade category	Unshelled peanuts and damaged kernels and minor defects (percent)	Total fall through sound whole kernels and/or sound split and broken kernels	Foreign materials (percent)	Moisture (percent)
Excluding Lots of "splits"				
Runner	3.50	6.00%; 17/64 inch round screen.	.20	9.00
Virginia (except No. 2)	3.50	6.00%; 17/64 inch round screen.	.20	9.00
Spanish and Valencia	3.50	6.00%; 16/64 inch round screen.	.20	9.00
No. 2 Virginia	3.50	6.00%; 17/64 inch round screen.	.20	9.00
Runner with splits (not more than 15% sound splits).	3.50	6.00%; 17/64 inch round screen.	.20	9.00
Virginia with splits (not more than 15% sound splits).	3.50	6.00%; 17/64 inch round screen.	.20	9.00
Spanish and Valencia with splits (not more than 15% sound splits).	3.50	6.00%; 16/64 inch round screen.	.20	9.00
Lots of "splits"				
Runner (not less than 90% splits)	3.50	6.00%; 17/64 inch round screen.	.20	9.00
Virginia (not less than 90% splits)	3.50	6.00%; 17/64 inch round screen.	.20	9.00
Spanish and Valencia (not less than 90% splits)	3.50	6.00%; 16/64 inch round screen.	.20	9.00

(b) *Cleaned-inshell peanuts:* No handler or importer shall ship or otherwise dispose of cleaned-inshell peanuts for human consumption unless such peanuts are Positive lot identified and are determined by the Inspection Service to contain:

(1) Not more than 1.00 percent kernels with mold unless a sample of such peanuts, drawn by an inspector of the Inspection Service, is analyzed chemically by a USDA laboratory or a USDA-approved laboratory and certified "negative" as to aflatoxin;

(2) Not more than 3.50 percent peanuts with damaged or defective kernels;

(3) Not more than 10.00 percent moisture; or

(4) Not more than 0.50 percent foreign material.

[67 FR 57140, Sept. 9, 2002; 67 FR 63503, Oct. 11, 2002, as amended at 68 FR 46924, Aug. 7, 2003; 68 FR 53490, Sept. 11, 2003; 81 FR 50288, Aug. 1, 2016]

§ 996.40　Handling standards.

(a) *Identification:* Each lot of shelled or cleaned- inshell peanuts intended for human consumption shall be identified by positive lot identification prior to being shipped or otherwise disposed of. Positive lot identification (PLI) methods are tailored to the size and containerization of the lot, by warehouse storage or space requirements, or by necessary further movement of the lot prior to certification. Positive lot identification is established by the Inspection Service and includes the following methods of identification. For domestic lots and repackaged import lots, PLI includes PLI stickers, tags or seals applied to each individual package or container in such a manner that is acceptable to the Inspection Service and maintains the identity of the lot. For imported lots, PLI tape may be used to wrap bags or boxes on pallets, PLI stickers may be used to cover the shrink-wrap overlap, doors may be sealed to isolate the lot, bags or boxes may be stenciled with a lot number, or any other means that is acceptable to the Inspection Service. The crop year means the calendar year in which the peanuts were planted as documented by the applicant. All lots of shelled and cleaned-inshell peanuts shall be shipped under positive lot identification procedures. However, peanut lots failing to meet quality requirements may be moved from a handler's facility to another facility owned by the same handler or another handler without PLI so long as such handler maintains a satisfactory records system for traceability purposes as defined in § 996.73.

(b) *Sampling and testing shelled peanuts for outgoing Inspection:* Prior to shipment, the following sampling and inspection procedures shall be conducted on each lot of shelled peanuts intended for human consumption. The lot size of shelled or cleaned-inshell peanuts presented for outgoing inspec-

tion in bags or bulk shall not exceed 200,000 pounds.

(1) Each handler or importer shall cause appropriate samples, based on a sampling plan approved by the Inspection Service, of each lot of shelled peanuts intended for human consumption to be drawn by the Inspection Service. The gross amount of peanuts drawn shall be large enough to provide for a grade analysis, for a grading check-sample, and for three 48-pound samples for aflatoxin chemical analysis. The three 48-pound samples shall be designated by the Inspection Service as "Sample 1," "Sample 2," and "Sample 3" and each sample shall be placed in a suitable container and positive lot identified by means acceptable to the Inspection Service. Sample 1 may be prepared for immediate testing or Sample 1, Sample 2, and Sample 3 may be returned to the handler or importer for testing at a later date. Imported peanuts shall be labeled "Sample 1IMP," "Sample 2IMP," and "Sample 3IMP" and handled accordingly.

(2) Before shipment of a lot of shelled peanuts to a buyer, the handler or importer shall cause Sample 1 to be ground by the Inspection Service, a USDA laboratory or a USDA-approved laboratory, in a "subsampling mill." The resultant ground subsample from Sample 1 shall be of a size specified by the Inspection Service and shall be designated as "Subsample 1–AB" and at the handler's, importer's or buyer's option, a second subsample may also be extracted from Sample 1. It shall be designated as "Subsample 1–CD." Subsample 1–CD may be sent as requested by the handler or buyer, for aflatoxin assay, to a USDA laboratory or USDA-approved laboratory that can provide analyses results on such samples in 36 hours. The cost of sampling and testing Subsample 1–CD shall be for the account of the applicant. Subsample 1–AB shall be analyzed only in a USDA laboratory or USDA-approved laboratory. Both Subsamples 1–AB and 1–CD shall be accompanied by a notice of sampling or grade certificate, signed by the inspector, containing, at least, identifying information as to the handler or importer, and the positive lot identification of the shelled peanuts.

(3) The samples designated as Sample 2 and Sample 3 shall be held as aflatoxin check-samples by the Inspection Service or the handler or importer and shall not be included in the shipment to the buyer until the analyses results from Sample 1 are known.

(4) Upon call from the laboratory, the handler or importer shall cause Sample 2 to be ground by the Inspection Service, USDA or USDA-approved laboratory in a "subsampling mill." The resultant ground subsample from Sample 2 shall be of a size specified by the Inspection Service and it shall be designated as "Subsample 2–AB." Upon call from the laboratory, the handler shall cause Sample 3 to be ground by the Inspection Service, USDA or USDA-approved laboratory in a "subsampling mill." The resultant ground subsample from Sample 3 shall be of a size specified by Inspection Service and shall be designated as "Subsample 3–AB." "Subsamples 2–AB and 3–AB" shall be analyzed only in a USDA laboratory or a USDA-approved laboratory and each shall be accompanied by a notice of sampling. The results of each assay shall be reported by the laboratory to the handler and to USDA.

(5) Handlers and importers may make arrangements for required inspection and certification by contacting the Inspection Service office closest to where the peanuts will be made available for sampling. For questions regarding inspection services, a list of Federal or Federal-State Inspection Service offices, or for further assistance, handlers and importers may contact: Specialty Crops Inspection Division, Specialty Crops Program, AMS, USDA, 1400 Independence Avenue SW., Room 1536–S, (STOP 0240), Washington, DC, 20250–0240; Telephone: (202) 720–5870; Fax: (202) 720–0393.

(6) Handlers and importers may make arrangements for required chemical analysis for aflatoxin content at the nearest USDA or USDA-approved laboratory. For further information concerning chemical analysis and a list of laboratories authorized to conduct such analysis contact: Science and Technology Program, AMS, USDA, 1400 Independence Avenue SW., STOP 0270, Washington, DC 20250–0270; Telephone (202) 690–0621; Fax (202) 720–4631.

(c) *Appeal inspections.* Any "holder of the title" to any lot of peanuts may request an appeal inspection if it is believed that the original aflatoxin test results were in error. Appeal inspections would be conducted in accordance with Federal or Federal-State inspection procedures for milled peanuts. The aflatoxin appeal sample would be drawn by Federal or Federal-State Inspection Service officials and the appeal analysis would be conducted by USDA or USDA-approved laboratories. Any financially interested person may request an appeal inspection if it is believed that the original quality inspection is in error. Quality appeals would be conducted by Federal or Federal-State Inspection Service inspectors in accordance with the Federal or Federal-State inspection procedures for milled peanuts. The person requesting the appeal inspection would pay the cost of such appeals. The appeal inspection results shall be issued to the person requesting the appeal inspection and a copy shall be mailed to USDA or its agent.

[67 FR 57140, Sept. 9, 2002, as amended at 68 FR 1157, Jan. 9, 2003; 68 FR 46924, Aug. 7, 2003; 81 FR 50289, Aug. 1, 2016]

§ 996.50 Reconditioning failing quality peanuts.

(a) Lots of peanuts which have not been certified as meeting the requirements for disposition to human consumption outlets may be disposed for non-human consumption uses: *Provided,* That each such lot is positive lot identified using red tags, identified using a traceability system as defined in § 996.73, or other methods acceptable to the Inspection Service, and certified as to aflatoxin content (actual numerical count), unless they are designated for crushing. However, on the shipping papers covering the disposition of each such lot, the handler or importer shall cause the following statement to be shown: "The peanuts covered by this bill of lading (or invoice, etc.) are not to be used for human consumption."

(b)(1) Sheller oil stock residuals shall be positive lot identified using red tags, identified using a traceability system as defined in § 996.73, or other methods acceptable to the Inspection

Service, and may be disposed of domestically or to the export market in bulk or bags or other suitable containers. Disposition to crushing may be to approved crushers. However, sheller oil stock residuals may be moved from a handler's facility to another facility owned by the same handler or another handler without PLI so long as such handler maintains a satisfactory records system for traceability purposes as defined in § 996.73.

(2) If such peanuts are not tested and certified as to aflatoxin content, pursuant to paragraph (a) of this section, the handler or importer shall cause the following statement to be shown on the shipping papers: "The peanuts covered by this bill of lading (or invoice, etc.) are limited to crushing only and may contain aflatoxin."

(c) *Remilling.* Handlers and importers may remill, or cause to have remilled, lots of shelled or cleaned-inshell peanuts failing to meet the applicable outgoing quality standards in the table in § 996.31(a). If, after remilling, such peanut lot meets the applicable quality standards in § 996.31, the lot may be moved for human consumption under positive lot identification procedures and accompanied by applicable grade and aflatoxin certificates.

(d) *Blanching.* Handlers and importers may blanch, or cause to have blanched, shelled peanuts failing to meet the outgoing quality standards specified in the table in § 996.31(a). If after blanching, such peanut lot meets the quality standards in § 996.31(a), the lot may be moved for human consumption under positive lot identification procedures and accompanied by applicable grade and aflatoxin certificates. Peanut lots certified as meeting the fall through standard or the damaged kernels and minor defects standard as specified in § 996.31(a), prior to blanching shall be exempt from fall through, damaged kernels and minor defects standards after blanching.

(e) Lots of shelled peanuts moved for remilling or blanching shall be positive lot identified and accompanied by valid grade inspection certificate, *Except* That, a handler's shelled peanuts may be moved without PLI and grade inspection to the handler's blanching facility that blanches only the handler's

peanuts. Lots of shelled peanuts may be moved for remilling or blanching to another handler without PLI if the handler uses a traceability system as defined in § 996.73, *Except* That, any grade inspection certificates associated with these lots would no longer be valid. The title of such peanuts shall be retained by the handler or importer until the peanuts have been certified by the Inspection Service as meeting the outgoing quality standards specified in the table in § 996.31(a). Remilling or blanching under the provisions of this paragraph shall be performed only by those remillers and blanchers approved by USDA. Such approved entities must agree to comply with the handling standards in this part and to report dispositions of all failing peanuts and residual peanuts to USDA, unless they are designated for crushing.

(f) Residual peanuts resulting from remilling or blanching of peanuts shall be red tagged, identified using a traceability system as defined in § 996.73, or identified by other means acceptable to the Inspection Service, and returned directly to the handler for further disposition or, in the alternative, such residual peanuts shall be positive lot identified by the Inspection Service and shall be disposed of to handlers who are crushers, or to approved crushers, *Except* That, a handler may move the residual peanuts without PLI to a facility for crushing owned by the handler. Handlers who are crushers and crushers approved by USDA must agree to comply with the terms and conditions of this part.

(g) *Re-inspection.* Whenever USDA has reason to believe that domestic or imported peanuts may have been damaged or deteriorated while in storage, USDA may reject the then effective inspection certificate and may require the owner of the peanuts to have a re-inspection to establish whether or not such peanuts may be disposed of for human consumption.

(h) The cost of transportation, sampling, inspection, certification, chemical analysis, and identification, as well as remilling and blanching, and further inspection of remilled and blanched lots, and disposition of failing peanuts, shall be borne by the applicant. Whenever peanuts are presented for inspection, the handler or importer shall furnish any labor and pay any costs incurred in moving, opening containers, and shipping samples as may be necessary for proper sampling and inspection. The Inspection Service shall bill the applicant or other responsible entity separately for applicable fees covering sampling and inspection, delivering aflatoxin samples to laboratories, positive lot identification measures, and other certifications as may be necessary to certify edible quality or non-edible disposition. The USDA and USDA-approved laboratories shall bill the applicant or other responsible entity separately for applicable fees for aflatoxin assays.

[67 FR 57140, Sept. 9, 2002, as amended at 68 FR 46924, Aug. 7, 2003]

§996.60 Safeguard procedures for imported peanuts.

(a) Prior to arrival of a foreign-produced peanut lot at a port-of-entry, the importer, or customs broker acting on behalf of the importer, shall submit information electronically to the United States Customs and Border Protection, which includes the following: The Customs Service entry number; the container number(s) or other identification of the lot(s); the volume of the peanuts in each lot being entered; the inland shipment destination where the lot will be made available for inspection; and a contact name or telephone number at the destination.

(b) *Additional standards.* (1) Nothing contained in this section shall preclude any importer from milling or reconditioning, prior to importation, any shipment of peanuts for the purpose of making such lot eligible for importation into the United States. However, all peanuts entered for human consumption use must be certified as meeting the quality standards specified in §996.31(a) prior to such disposition. Failure to fully comply with quality and handling standards as required under this section, will result enforcement action by USDA.

(2) Imported peanut lots sampled and inspected at the port-of-entry, or at other locations, shall meet the quality standards of this part in effect on the date of inspection.

(3) A foreign-produced peanut lot entered for consumption or for warehouse may be transferred or sold to another person: *Provided,* That the original importer shall be the importer of record unless the new owner applies for bond and files Customs Service documents pursuant to 19 CFR 141.113 and 141.20: *And provided further,* That such peanuts must be certified and reported to USDA pursuant to §996.71 of this part.

(4) The provisions of this section do not supersede any restrictions or prohibitions on peanuts under the Federal Plant Quarantine Act of 1912, the Federal Food, Drug and Cosmetic Act, any other applicable laws, or regulations of other Federal agencies, including import regulations and procedures of the Customs Service.

[67 FR 57140, Sept. 9, 2002, as amended at 68 FR 1158, Jan. 9, 2003; 81 FR 50289, Aug. 1, 2016]

REPORTS AND RECORDS

§996.71 Reports and recordkeeping.

(a) Each handler and importer shall maintain a satisfactory records system for traceability purposes as defined in §996.73.

(b) USDA shall maintain copies of grade and aflatoxin certificates on all peanut lots inspected and chemically tested. USDA and USDA-approved laboratories shall file copies of all aflatoxin certificates completed by such laboratories with the Southeast Marketing Field Office, Marketing Order and Agreement Division, Specialty Crops Program, AMS, USDA, 1124 1st Street South, Winter Haven, Florida 33880; Telephone (863) 324–3375, Fax: (863) 291–8614, or other address as determined by USDA.

[67 FR 57140, Sept. 9, 2002, as amended at 81 FR 50289, Aug. 1, 2016]

§996.72 Confidential information.

All reports and records furnished or submitted by handlers and importers to USDA which include data or information constituting a trade secret or disclosing a trade position, financial condition, or business operations of the particular handlers or their customers shall be received by, and at all times kept in the custody and control of one or more employees of USDA, and, except as provided in §996.74 or otherwise provided by law, such information shall not be disclosed to any person outside USDA.

§ 996.73　Verification of reports.

(a) For the purpose of checking and verifying reports kept by handlers and importers and the operation of handlers and importers under the provisions of this Part, the officers, employees or duly authorized agents of USDA shall have access to any premises where peanuts may be held at any time during reasonable business hours and shall be permitted to inspect any peanuts that meet outgoing quality regulations, so held by such handler or importer and any and all records of such handler with respect to the acquisition, holding, or disposition of all peanuts meeting outgoing quality regulations, which may be held or which may have been disposed by handler.

(b) Reports shall be maintained by the handler for nonconforming products to assure traceability throughout the supply chain. The traceability system must include documented records, which enable a full product history to be produced in a timely manner and must ensure product can be traced forward (raw material to distribution) and backwards from distribution to the warehouse feeding the shelling plant, and ensure that all associated tests and all relevant records have been completed. The traceability system shall include identification of all raw materials, process parameters (for specific lot), packaging and final disposition. The handler shall be able to identify the warehouse in which the peanuts were stored immediately prior to shelling. Traceability must be maintained throughout production runs with specific lot codes, and there shall be complete linkage from raw material receipt through final disposition.

[81 FR 50290, Aug. 1, 2016]

§ 996.74　Compliance.

(a) A handler or importer shall be subject to withdrawal of inspection services, for a period of time to be determined by USDA, if the handler or importer:

(1) Fails to obtain outgoing inspection on shelled or cleaned-inshell peanuts, pursuant to § 996.31, and ships

such peanuts for human consumption use;

(2) Ships failing quality peanuts, pursuant to § 996.31, for human consumption use;

(3) Commingles failing quality peanuts with certified edible quality peanuts and ships the commingled lot for human consumption use without meeting outgoing quality regulations;

(4) Fails to maintain positive lot identification, pursuant to § 996.40(a), on peanut lots certified for human consumption use;

(5) Fails to maintain and provide access to records, pursuant to § 996.71, and the standards for traceability and nonconforming product disposition pursuant to § 996.73, on the reconditioning or disposition of peanuts acquired by such handler or importer; and on lots that meet outgoing quality standards; or

(6) Otherwise violates any provision of section 1308 of the Act or any provision of this part.

(b) Any peanut lot shipped which fails to meet the outgoing quality standards specified in § 996.31, and is not reconditioned to meet such standards, or is not disposed to non-human consumption outlets as specified in § 996.50, shall be reported by USDA to the Food and Drug Administration and listed on an Agricultural Marketing Service Web site.

[67 FR 57140, Sept. 9, 2002, as amended at 81 FR 50290, Aug. 1, 2016]

§ 996.75　Effective time.

The provisions of this part, as well as any amendments, shall apply to current crop year peanuts, subsequent crop year peanuts, and prior crop year peanuts not yet inspected, or failing peanut lots that have not met disposition standards, and shall continue in force and effect until modified, suspended, or terminated.

[81 FR 50290, Aug. 1, 2016]

PART 999—SPECIALTY CROPS; IMPORT REGULATIONS

AUTHORITY: 7 U.S.C. 601–674.

§999.1 Regulation governing the importation of dates.

(a) *Definitions.* (1) *Dates in retail packages* means whole or pitted dates, other than dates prepared or preserved, wrapped or packaged for sale at retail.

(2) *Dates for packaging* means whole or pitted dates in bulk containers which are to be repacked, in whole or part, in the United States as dates in retail packages.

(3) *Bulk container* means any container of dates which, together with the dates therein, weighs more than ten pounds.

(4) *Dates for processing* means any dates for use in a bakery, confectionery, or other product and includes dates coated with a substance materially altering their color.

(5) *Dates prepared or preserved* means dates processed into a confection or other product, dates coated with a substance materially altering their color, or dates prepared for incorporation into a product by chopping, slicing, or other processing which materially alters their form.

(6) *Person* means any individual, partnership, corporation, association, or other business unit.

(7) *USDA inspector* means an inspector of the Specialty Crops Inspection Division, Fruit and Vegetable Program, or any other duly authorized employee of the USDA.

(8) *Inspection certificate* means a written statement or memorandum report issued by a USDA inspector setting forth in addition to appropriate descriptive information the quality and condition of the product inspected, and in the case of imported dates, a statement of meeting or failing, as applicable, the U.S. import requirements

under section 8e of the AMA Act of 1937.

(9) *Importation* means release from custody of United States Customs and Border Protection.

(b) *Grade requirements.* (1) Except as provided in paragraph (d) of this section, the importation into the United States of any lot of dates for packaging or dates in retail packages is prohibited unless the dates are wholesome and unadulterated and meet the following grade requirements which are determined to be comparable to those imposed upon domestic dates handled pursuant to Order No. 987, as amended (part 987 of this chapter: The whole or pitted dates in the lot are of one variety, and are of such quality and condition that upon inspection on the basis of a representative sample thereof, with hydration (of the sample) in accordance with good commercial practice or without any hydration, the dates possess a reasonably good color, are reasonably uniform in size, are reasonably free from defects, possess a reasonably good character, and score not less than 80 points when scored in accordance with the scoring system applicable to U.S. Grade B dates, as prescribed in the U.S. Standards for Grades of Dates (§§ 52.1001 through 52.1011 of this chapter): *Provided,* That not more than 25 percent, by weight, of the dates may possess semidry or dry calyx ends except that not more than 5 percent, by weight, of the dates may possess dry calyx ends: *And provided further,* That in determining the grade for pitted dates, the pitted dates shall not be scored as damaged because of the longitudinal slit caused by removing the pit or the mashing resulting therefrom unless the flesh is seriously torn or mangled.

(2) Compliance with the grade requirements shall be determined on the basis of an inspection and certification by a USDA inspector.

(c) *Inspection and certification requirements*—(1) *Inspection.* Inspection shall be performed by USDA inspectors in accordance with the Regulations Governing the Inspection and Certification of Processed Fruits and Vegetables and Related Products (part 52 of this title). The cost of each such inspection and related certification shall be borne by the applicant. Applicants shall provide USDA inspectors with the entry number and such other identifying information for each lot as the inspector may request.

(2) *Certification.* Each lot of dates inspected in accordance with paragraph (c)(1) of this section shall be covered by an inspection certificate. Each such certificate shall set forth, among other things, the following:

(i) The date and place of inspection.

(ii) The name of the applicant.

(iii) The Customs entry number pertaining to the lot or shipment covered by the certificate;

(iv) The variety, quantity, and identifying marks of the lot inspected.

(v) The statement, if applicable: "Meets U.S. import requirements under section 8e of the AMA Act of 1937".

(vi) If the lot fails to meet the import requirements, a statement to that effect and the reasons therefor.

(d) *Exemptions.* (1) Notwithstanding any other provisions of this section, any lot of dates for importation which in the aggregate does not exceed 70 pounds and any dates that are so denatured as to render them unfit for human consumption may be imported exempt from the provisions of this section.

(2) The grade, size, quality, and maturity requirements of this section shall not apply to dates which are donated to needy persons, prisoners, or Native Americans on reservations; dates for processing; or dates prepared or preserved, but all such dates shall be subject to the safeguard provisions contained in § 999.500.

(3) Dates for packaging or dates in retail packages that fail to meet the grade, size, quality, and maturity requirements of this section may be reclassified as dates for processing for importation, but such dates shall be subject to the safeguard provisions contained in § 999.500.

(e) *Reconditioning.* Nothing contained in this section shall preclude the reconditioning of failing lots of dates, prior to importation, so that such dates may be made eligible to meet the grade requirements prescribed in paragraph (b) of this section.

(f) *Books and records.* Each person subject to this section shall maintain true and complete records of his transactions with respect to imported dates. Such records and copies of executed forms shall be retained for not less than two years subsequent to the calendar year of acquisition. The Secretary, through his duly authorized representatives, shall have access to any such person's premises during regular business hours and shall be permitted at any such times to inspect such records and any dates held by such person.

(g) *Other restrictions.* The provisions of this section do not supersede any restrictions or prohibitions on the importation of dates under the Plant Quarantine Act of 1912, the Federal Food, Drug, and Cosmetic Act, or any other applicable laws or regulations or the need to comply with applicable food and sanitary regulations of city, county, State, or Federal agencies.

(h) *Compliance.* Any person who violates any provision of this section shall be subject to a forfeiture in the amount prescribed in section 8a(5) of the Agricultural Marketing Agreement Act of 1937, as amended (sections 1–19, 48 Stat. 31, as amended; 7 U.S.C. 601–674), or, upon conviction, a penalty in the amount prescribed in section 8c(14) of said act, or to both such forfeiture and penalty. False representations to an agency of the United States on any matter within its jurisdiction, knowing it to be false, is a violation of 18 U.S.C. 1001 which provides for a fine or imprisonment or both.

[28 FR 3469, Apr. 10, 1963, as amended at 31 FR 960, Jan. 25, 1966; 33 FR 15986, Oct. 31, 1968; 36 FR 6736, Apr. 8, 1971; 58 FR 69190, Dec. 30, 1993; 74 FR 2808, Jan. 16, 2009; 80 FR 15678, Mar. 25, 2015; 81 FR 87412, Dec. 5, 2016]

§ 999.100 Regulation governing imports of walnuts.

(a) *Definitions.* (1) *Walnuts* means all walnuts commonly known as English or Persian walnuts (Juglans regia).

(2) *Inshell walnuts* means walnuts, the kernels or edible portions of which are contained in the shell.

(3) *Shelled walnuts* means the kernels of walnuts after the shells are removed.

(4) *Person* means any individual, partnership, corporation, association, or other business unit.

(5) *USDA Inspector* means any Federal or Federal-State inspector of the Fresh Products Standardization and Inspection Branch of the Fruit and Vegetable Division, Consumer and Marketing Service, United States Department of Agriculture.

(6) *Importation of walnuts* means the release of walnuts from the custody of the United States Customs Service.

(b) *Grade and size regulations.* No person may import walnuts (Juglans regia) into the United States unless such walnuts have been inspected and certified by a USDA inspector as meeting the following requirements:

(1) *Inshell walnuts.* All inshell walnuts shall be of a quality equal to or better than the requirements of U.S. No. 2 and "baby" size as prescribed in the United States Standards for Walnuts (Juglans regia) in the Shell (§§ 51.2945 through 51.2966 of this title); or

(2) *Shelled walnuts.* All shelled walnuts shall be of a quality equal to or better than the requirements for U.S. Commercial Grade as prescribed in the United States Standards for Shelled Walnuts (Juglans regia) (§§ 51.2275 through 51.2294 of this title excluding §§ 51.2278(b), 51.2284 and 51.2285) effective January 25, 1959, except that the minimum size shall be pieces not more than five percent of which will pass through a round opening $^6/_{64}$ inch in diameter and no other size requirements shall apply.

(c) *Inspection and certification.* (1) All inspections and certifications required by paragraph (b) of this section shall be made by USDA inspectors in accordance with the regulations governing the inspection and certification of fresh fruits, vegetables, and other products (Part 51 of this title). The cost of inspection and certification shall be borne by the applicant.

(2) Each inspection certificate shall set forth among other things the following:

(i) The date and place of inspection;

(ii) The name of the applicant;

(iii) The name of the importer;

(iv) The Customs entry number pertaining to the lot or shipment covered by the certificate;

(v) The quantity and identifying marks of the container; and

(vi) The statement, if applicable, "Meets U.S. import requirements under section 8e of the Agricultural Marketing Agreement Act of 1937".

(3) Whenever walnuts are offered for inspection, the applicant shall furnish any labor and pay any costs incurred in moving and opening containers as may be necessary for proper sampling and inspection. The applicant shall also furnish the USDA inspector the entry number and such other identifying information for each lot as he may request.

(4) Inspection must be completed prior to the importation of walnuts. To avoid delay the applicant should make advance arrangements with the USDA inspection office.

(d) *Reconditioning prior to importation.* Nothing contained in this section shall be deemed to preclude reconditioning walnuts prior to importation, in order that such walnuts may be made eligible to meet the grade and size regulations prescribed in paragraph (b) of this section.

(e)(1) *Minimum quantity.* Notwithstanding any other provision of this section, the importation of any lot of walnuts which does not exceed, in net weight, 60 pounds of shelled walnuts or 115 pounds of inshell walnuts shall be exempt from the requirements of this section.

(2) *Exemptions.* The grade, size, quality and maturity requirements of this section shall not apply to walnuts which are: green walnuts (so immature that they cannot be used for drying and sale as dried walnuts); walnuts used in non-competitive outlets such as use by charitable institutions, relief agencies, governmental agencies for school lunch programs, and diversion to animal feed or oil manufacture, but such walnuts shall be subject to the safeguard provisions contained in § 999.500.

(f) *Other import requirements.* The provisions of this section do not supersede any restrictions or prohibitions on walnuts under the Federal Plant Quarantine Act of 1912, or any other applicable laws or regulations of city, county, State, or Federal Agencies including the Federal Food, Drug and Cosmetic Act.

(g) *Compliance.* Any person violating any of the provisions of this regulation is subject to a forfeiture in the amount prescribed in section 608a(5) of the Agricultural Marketing Agreement Act of 1937, as amended (7 U.S.C. 601–674), or, upon conviction, a penalty in the amount prescribed in section 608c(14) of said act, or to both such forfeiture and penalty. False representations in any matter within the jurisdiction of any agency of the United States, knowing it to be false, is a violation of 18 U.S.C. 1001 which provides for a fine or imprisonment or both.

[29 FR 230, Jan. 9, 1964, as amended at 40 FR 29263, July 11, 1975; 41 FR 2075, Jan. 14, 1976; 42 FR 35146, July 8, 1977; 58 FR 69190, Dec. 30, 1993; 74 FR 2809, Jan. 16, 2009]

§ 999.200 Regulation governing the importation of prunes.

(a) *Definitions.* (1) *Prunes* means and includes all sun-dried or artificially dehydrated plums, of any type of variety, produced from plums, except: Sulfur-bleached prunes which are produced from yellow varieties of plums and are commonly known as silver plums; (ii) plums which have not been dried or dehydrated to a point where they are capable of being stored prior to packing, without material deterioration or spoilage unless refrigeration or other artificial means of preservation are used, and so long as they are treated by a process which is in conformity with, or generally similar to, the processes for treatment of plums of that type which have been developed or recommended by the Food Technology Division, College of Agriculture, University of California, for the specialty pack known as "high moisture content prunes", but this exception shall not apply if and when such plums are dried to the point where they are capable of being stored without material deterioration or spoilage, refrigeration or other artificial means of preservation; and (iii) brine dried prunes that have been impregnated with brine or salt

during the dehydration process to the extent that they have lost their form and character as prunes, and cannot be reconstituted to permit economic use of the individual fruits as prunes, and are imported under International Harmonized Tariff Schedule No. 0813.20.1000.

(2) *Pitted prunes* means prunes with the pit removed that are characterized by a uniform depression and minimal skin break where the pit has been removed.

(3) *Macerated prunes* means dried prunes with the pit removed that are characterized by a flattened appearance with slightly more skin break where the pit has been removed than with pitted prunes.

(4) *Standard prunes* means any lot of prunes meeting the grade and size requirements prescribed in paragraph (b)(1) of this section.

(5) *Standard pitted prunes* means any lot of pitted prunes meeting the grade requirements prescribed in paragraphs (b)(2) and (b)(3) of this section.

(6) *Standard pitted macerated prunes* means any lot of pitted macerated prunes meeting the grade requirements in paragraphs (b)(2) and (b)(4) of this section.

(7) *Manufacturing grade substandard prunes* means any lot of prunes which meets the grade requirements prescribed in paragraph (b)(5) of this section but fails to meet the requirements for standard prunes, standard pitted prunes and standard pitted macerated prunes.

(8) *Size* means the number of prunes contained in a pound.

(9) *Person* means any individual, partnership, corporation, association, or other business unit.

(10) *Fruit and Vegetable Division* means the Fruit and Vegetable Division of the Agricultural Marketing Service, U.S. Department of Agriculture, Washington, DC 20250.

(11) *USDA inspector* means an inspector of the Processed Products Standardization and Inspection Branch, Fruit and Vegetable Division, or any other duly authorized employee of the USDA.

(12) *Importation* means release from custody of the U.S. Bureau of Customs.

(13) *Undersized prunes* means those prunes that pass freely through a round opening 23/32 of an inch in diameter.

(b) *Grade and size requirements.* (1) Except as provided in paragraph (b)(5) or paragraph (d) of this section, no person may import any lot of prunes into the United States unless the prunes are inspected and an inspection certificate issued with respect thereto, and the lot meets the applicable grade requirements specified in exhibit A of this section and the average count (i.e., number) of the prunes in such lot is 100 or less per pound. In determining whether any lot conforms to the size requirement, the following tolerance shall apply: In a sample of 100 ounces, the count per pound of 10 ounces of smallest prunes may not vary from the count per pound of 10 ounces of the largest prunes by more than 45 points.

(2) No person may import any lot of pitted prunes or pitted macerated prunes for human consumption as pitted or pitted macerated prunes unless the lot meets the applicable minimum grade requirements set forth in §999.200 (exhibit A), except that skin or flesh damage shall not be scored as a defect in determining whether the prunes meet the grade requirements. Pitted and pitted macerated prunes shall not be subject to size and undersized requirements.

(3) No person may import any lot of pitted prunes for human consumption as pitted prunes unless the lot does not exceed an average of 0.5 percent by count of prunes with whole pits and/or pit fragments 2 mm or longer and four of ten subsamples examined have no more than 0.5 percent by count of prunes with whole pits and/or pit fragments 2 mm or longer.

(4) No person may import any lot of pitted macerated prunes for human consumption as pitted macerated prunes unless the lot does not exceed an average of 2 percent by count of prunes with whole pits and/or pit fragments 2 mm or longer; and four of ten subsamples examined have no more than 2 percent by count with whole pits and/or pit fragments 2 mm or longer.

(5) Any person may import any lot of prunes, except any lot containing undersized prunes, pitted prunes or pitted macerated prunes, into the United States for use in human consumption outlets as prune products in which the prunes lose their form and character as prunes by conversion prior to consumption if the prunes are inspected and an inspection certificate issued with respect thereto, and each lot meets the grade requirements set forth in paragraphs (1), (2), and (3) of exhibit A of this section, and the importer first files as a condition of such importation an executed Form FV-6—'Importer's Exempt Commodity Form.'.

(c) *Inspection and certification requirements*—(1) *Inspection.* Inspection shall be performed by a USDA inspector in accordance with the regulations governing inspection and certification of processed fruits and vegetables, processed products thereof, and certain other processed food products (part 52 of this title). The cost of each such inspection and related certification shall be borne by the applicant.

(2) *Certification.* Each lot of prunes inspected in accordance with paragraph (c)(1) of this section shall be covered by an inspection certificate. Each such certificate shall set forth, among other things, the following:

(i) The date and place of inspection.

(ii) The name of the applicant.

(iii) The Customs entry number pertaining to the lot or shipment covered by the certificate;

(iv) The quantity and identifying marks of the lot inspected.

(v) The statement, as applicable: "Meets U.S. import requirements for standard prunes, standard pitted and standard pitted macerated prunes under section 8e of the AMA Act of 1937"; "Meets U.S. import requirements for manufacturing grade substandard prunes under section 8e of the AMA Act of 1937"; or "Fails to meet U.S. import requirements for prunes under section 8e of the AMA Act of 1937".

(vi) If the lot fails to meet the import requirements, a statement of the reason therefor.

(d) *Exemptions.* Notwithstanding any other provisions of this section, the importation of any lot of prunes which in the aggregate does not exceed 150 pounds, net weight, and any prunes that are so denatured as to render them unfit for human consumption shall be exempt from the requirements of this section.

(e) *Additional requirements*—(1) *General.* Prior to importation of any prunes, the person importing such prunes shall file an inspection certificate with the Collector of Customs at the port at which the customs entry is filed. In addition, if such prunes are manufacturing grade substandard prunes, such person shall also file with the Collector of Customs an executed Form FV–6—'Importer's Exempt Commodity Form.' Promptly after such filing, such person shall transmit a copy of this form to the Fruit and Vegetable Division. No person may import, sell, or use any manufacturing grade substandard prunes other than for use as set forth in paragraph (b)(5) of this section. Each person importing manufacturing grade substandard prunes shall obtain from each purchaser, no later than the time of delivery to such purchaser, and file with the Fruit and Vegetable Division not later than the 5th day of the month following the month in which the prunes were delivered, an executed Form FV–6—'Importer's Exempt Commodity Form.' One copy of this executed form shall be retained by the importer and one copy shall be retained by the purchaser.

(2) *Manufacturing Grade Substandard Prune—sale by other than importer.* Each wholesaler or other reseller of manufacturing grade substandard prunes should, for his or her protection, obtain from each purchaser and hold in his or her files an executed Form FV–6—'Importer's Exempt Commodity Form' covering each sale during the calendar year.

(f) *Reconditioning.* Nothing contained in this section shall preclude the reconditioning of failing lots of prunes, prior to importation, so that such prunes may be made eligible to meet the requirements prescribed pursuant to paragraphs (b)(1) through (5), as applicable, of this section.

(g) *Books and records.* Each person subject to this section shall maintain true and complete records of his transactions with respect to imported prunes. Such records and copies of executed forms shall be retained for not less than 2 years subsequent to the calendar year of acquisition. The Sec-

retary, through his duly authorized representatives, shall have access to any such person's premises during regular business hours and shall be permitted at any such times to inspect such records and any prunes held by such person.

(h) *Other restrictions.* The provisions of this section do not supersede any restrictions or prohibitions on the importation of prunes under the Plant Quarantine Act of 1912, the Federal Food, Drug and Cosmetic Act, or any other applicable laws or regulations or the need to comply with applicable food and sanitary regulations of city, county, State, or Federal agencies.

(i) *Compliance.* Any person who violates any provision of this section shall be subject to a forfeiture in the amount prescribed in section 8a(5) of the Agricultural Marketing Agreement Act of 1937, as amended (sections 1–19, 48 Stat. 31, as amended; 7 U.S.C. 601–674), or, upon conviction, a penalty in the amount prescribed in section 8c(14) of said act, or to both such forfeiture and penalty. False representations to an agency of the United States on any matter within its jurisdiction, knowing it to be false, is a violation of 18 U.S.C. 1001 which provides for a fine or imprisonment or both.

EXHIBIT A

GRADE REQUIREMENTS

A. *Defects.* Defects are: (1) Off-color; (2) inferior meat condition; (3) end cracks; (4) fermentation; (5) skin or flesh damage; (6) scab; (7) burned; (8) mold; (9) imbedded dirt; (10) insect infestation; (11) decay.

B. *Explanation of terms.* (1) *Off-color* means a dull color or skin differing noticeably in appearance from that which is characteristic of mature, properly handled fruit of a given variety or type.

(2) *Inferior meat condition* means flesh which is fibrous, woody, or otherwise inferior due to immaturity to the extent that the characteristic texture of the meat is substantially affected.

(3) *End cracks* means callous growth cracks, at the blossom end of prunes, aggregating more than three-eighths of one inch (⅜″) but not more than one-half of one inch (½″) in length.

(4) *Fermentation* means damage to the flesh by fermentation to the extent that the characteristic appearance or flavor is substantially affected.

(5) *Skin or flesh damage* means growth cracks, splits, breaks in skin or flesh of the following descriptions:

(a) Callous growth cracks, except end cracks as defined in this section, aggregating more than three-eighths of one inch (⅜″) in length;

(b) Splits or skin breaks exposing flesh and materially affecting the normal appearance of the prunes;

(c) Any cracks, splits, or breaks open to the pit;

(d) Healed or unhealed surface or flesh blemishes caused by insect injury and which materially affect appearance, edibility or keeping quality.

(6) *Scab* means tough or thick scab exceeding in the aggregate the area of a circle three-eighths of one inch (⅜″) in diameter or by unsightly scab of another character exceeding in the aggregate the area of a circle three-fourths of one inch (¾″) in diameter.

(7) *Burned* means injury by sunburn or excessive heat in dehydration to the extent that the characteristic appearance, flavor or edibility of the fruit is noticeably affected.

(8) *Mold* means a characteristic fungus growth and is self-explanatory.

(9) *Imbedded dirt* means the presence of dirt or other extraneous material so imbedded in, or adhering to, the prune that it cannot readily be removed in washing the fruit.

(10) *Insect infestation* means the presence of insects, insect fragments or insect remains.

C. *Maximum tolerances.* Tolerance allowances shall be on a weight basis and shall not exceed the following:

(1) There shall be no tolerance allowance for live insect infestation.

(2) The tolerance allowances for decay shall not exceed one percent (1%).

(3) The combined tolerance allowance for mold, brown rot, imbedded dirt, insect infestation, and decay shall not exceed five percent (5%), and, within such tolerance, brown rot shall not exceed three percent (3%).

(4) The combined tolerance allowance for fermentation, skin or flesh damage, scab, burned, mold, imbedded dirt, in-

sect infestation, and decay shall not exceed eight percent (8%).

(5) The combined tolerance allowance for end cracks, fermentation, skin or flesh damage, scab, burned, mold, imbedded dirt, insect infestation, and decay shall not exceed ten percent (10%), except that the first eight percent (8%) of end cracks shall be given one-half value and any additional percentage of end cracks shall be given full value.

(6) The combined tolerance allowance for off-color, inferior meat condition, end cracks, fermentation, skin or flesh damage, scab, burned, mold, imbedded dirt, insect infestation, and decay shall not exceed fifteen percent (15%), except that the first eight percent (8%) of end cracks shall be given one-half value and any additional percentage of end cracks shall be given full value.

[36 FR 18782, Sept. 22, 1971, as amended at 47 FR 47230, Oct. 25, 1982; 57 FR 56245, Nov. 27, 1992; 59 FR 38113, July 27, 1994; 60 FR 57910, Nov. 24, 1995; 74 FR 2809, Jan. 16, 2009]

EFFECTIVE DATE NOTE: At 74 FR 2809, Jan. 16, 2009, §999.200 was suspended indefinitely.

§999.300 Regulation governing importation of raisins.

(a) *Definitions.* For purposes of this section:

(1) *Raisins* means grapes from which a part of the natural moisture has been removed.

(2) *Varietal type* means the applicable one of the following: Thompson Seedless raisins, Muscat raisins, Layer Muscat raisins, Currant raisins, Monukka raisins, Other Seedless raisins, Golden Seedless raisins, and Other Seedless-Sulfured raisins.

(3) *Thompson Seedless Raisins* includes those raisins commonly referred to in international trade as Sultana raisins and means raisins made from Thompson Seedless (Sultana) grapes and from grapes with characteristics similar to Thompson Seedless (Sultanina) grapes.

(4) *Person* means any individual, partnership, corporation, association, or other business unit.

(5) *USDA inspector* means an inspector of the Specialty Crops Inspection Division, Fruit and Vegetable Program, Agricultural Marketing Service, U.S. Department of Agriculture, or any

other duly authorized employee of the U.S. Department of Agriculture.

(6) *Importation of raisins* means the release of raisins from custody of the U.S. Customs and Border Protection.

(b) *Grade and size requirements.* The importation of raisins into the United States is prohibited unless the raisins are inspected and certified as provided in this section. Except as provided in paragraph (e)(2) of this section, no person may import raisins into the United States unless such raisins have been inspected and certified by a USDA inspector as to whether or not the raisins are of a varietal type, and if a varietal type, as at least meeting the following applicable grade and size requirements, which requirements are the same as those imposed upon domestic raisins handled pursuant to Order No. 989, as amended (part 989 of this chapter):

(1) With respect to Thompson Seedless and Other Seedless-Sulfured raisins—the requirements of U.S. Grade C as defined in the effective United States Standards of Grades of Processed Raisins (§§ 52.1841 through 52.1858 of this title): *Provided,* That, at least 70 percent, by weight, of the raisins shall be well-matured or reasonably well-matured. With respect to select-sized and mixed-sized lots, the raisins shall at least meet the U.S. Grade B tolerances for pieces of stem and undeveloped and substandard raisins, and small sized raisins shall meet the U.S. Grade C tolerances for those factors;

(2) With respect to Muscat raisins—the requirements of U.S. Grade C as defined in said standards;

(3) With respect to Layer Muscat raisins—the requirements of U.S. Grade B as defined for "Layer or Cluster Raisins with Seeds" in said standards, except for the provisions therein relating to moisture content;

(4) With respect to Currant raisins—the requirements of U.S. Grade B as defined in said standards;

(5) With respect to Monukka and Other Seedless raisins—the requirements for Thompson Seedless Raisins prescribed in paragraph (b)(1) of this section, except that the tolerance for moisture shall be 19 percent rather than 18 percent;

(6) With respect to Golden Seedless raisins—the requirements prescribed in

paragraph (b)(1) of this section for Thompson Seedless raisins and the color requirements for "colored" as defined in said standards.

(c) *Inspection and certification requirements.* (1) All inspections and certifications required by paragraph (b) of this section shall be made by USDA inspectors in accordance with the regulations governing inspection and certification of processed fruits and vegetables, processed products thereof, and certain other processed food products (part 52 of this title). The cost of each such inspection and certification shall be borne by the applicant.

(2) Each lot of raisins inspected in accordance with paragraph (c)(1) of this section shall be covered by an inspection certificate. Each such certificate shall set forth, among other things, the following:

(i) The date and place of inspection;

(ii) The name of the applicant;

(iii) The name of the importer;

(iv) The Customs entry number pertaining to the lot or shipment covered by the certificate;

(v) The quantity and identifying marks of the lot inspected;

(vi) The statement, as applicable, "Meets U.S. import requirements under section 8e of the AMA Act of 1937" or "Fails to meet U.S. import requirements under section 8e of the AMA Act of 1937"; and

(vii) If the lot fails to meet the import requirements, a statement of the reasons therefor.

(3) Whenever raisins are offered for inspection, the applicant shall furnish any labor and pay any costs incurred in moving and opening containers as may be necessary for proper sampling and inspection. The applicant shall also furnish the USDA inspector the entry number and such other identifying information for each lot as he may request. To avoid delay in scheduling the inspection the applicant should make advance arrangements with the USDA inspection office.

(d) *Reconditioning.* Nothing contained in this section shall preclude the reconditioning of failing lots of raisins prior to importation of raisins in order that such raisins may be made eligible to meet the applicable grade and size requirements in paragraph (b) of this section.

(e) *Exemptions.* (1) Notwithstanding any other provision of this section, any lot of raisins which in the aggregate does not exceed 100 pounds, net weight, may be imported without regard to the restrictions of this section.

(2) Any lot of raisins which does not meet the applicable grade and size requirements of paragraph (b) of this section may be imported for use in the production of alcohol, syrup for industrial use, or any lot of raisins which does not meet such requirements with respect to mechanical damage or sugaring may be imported for use in the production of raisin paste, but all such raisins shall be subject to the safeguard provisions contained in §999.500.

(f) *Books and records.* Each person subject to this section shall maintain true and complete records of his transactions with respect to imported raisins. Such records shall be retained for not less than 2 years subsequent to the calendar year of importation. The Secretary, through his duly authorized representatives, shall have access to any such person's premises during regular business hours and shall be permitted at any such time to inspect such records and any imported raisins held by such person.

(g) *Other restriction.* The provisions of this section do not supersede any restrictions or prohibitions on the importation of raisins under the Federal Plant Quarantine Act of 1912, the Federal Food, Drug and Cosmetic Act, or any other applicable laws or regulations, or the need to comply with applicable food and sanitary regulations of city, county, State, or Federal agencies.

(h) *Compliance.* Any person violating any of the provisions of this regulation is subject to a forfeiture in the amount prescribed in section 8a(5) of the Agricultural Marketing Agreement Act of 1937, as amended (sections 1–19, 48 Stat. 31, as amended; 7 U.S.C. 601–674), or, upon conviction, a penalty in the amount prescribed in section 8c(14) of said act, or to both such forfeiture and penalty. False representation to an agency of the United States in any matter within its jurisdiction, knowing it to be false, is a violation of 18 U.S.C. 1001 which provides for a fine or imprisonment or both.

[37 FR 5282, Mar. 14, 1972, as amended at 37 FR 13635, July 12, 1972; 37 FR 23820, Nov. 9, 1972; 41 FR 52646, Dec. 1, 1976; 43 FR 47972, Oct. 18, 1978; 43 FR 57863, Dec. 11, 1978; 45 FR 65513, Oct. 3, 1980; 47 FR 51731, Nov. 17, 1982; 50 FR 45808, Nov. 4, 1985; 53 FR 34715, Sept. 8, 1988; 67 FR 57505, Sept. 11, 2002; 74 FR 2809, Jan. 16, 2009; 80 FR 15678, Mar. 25, 2015; 81 FR 84403, Nov. 23, 2016]

§999.400 Regulation governing the importation of filberts.

(a) *Definitions.* (1) *Filberts* means filberts or hazelnuts.

(2) *Inshell filberts* means filberts, the kernels or edible portions of which are contained in the shell.

(3) *Shelled filberts* means the kernels of filberts after the shells are removed.

(4) *Person* means any individual, partnership, corporation, association, or other business unit.

(5) *USDA inspector* means a Federal or Federal-State inspector, Food Safety and Quality Service, United States Department of Agriculture, or any other duly authorized employee of the USDA.

(6) *Importation* means release from custody of the United States Bureau of Customs.

(b) *Grade and size requirements.* Except as provided in paragraph (d) of this section, no person shall import into the United States any lot of filberts unless the filberts meet the following requirements, which are identical to those for filberts grown in Oregon and Washington and handled pursuant to Order No. 982, as amended (7 CFR part 982):

(1) *Inshell filberts.* All inshell filberts shall be of a quality equal to or better than the requirements of U.S. No. 1 grade and medium size as defined in the U.S. Standards for Filberts in the Shell (7 CFR part 51), except that the tolerance for insect injury shall be two percent. With this modification, the U.S. No. 1 grade, medium size is identical to the Oregon No. 1 grade, medium size (as defined in the Oregon Grade Standards Filberts in Shell) and prescribed for inshell filberts under Order No. 982, as amended.

(2) *Shelled filberts.* All shelled filberts shall be of a quality equal to or better than the requirements prescribed in exhibit A of this section.

(c) *Inspection and certification requirements*—(1) *General.* Compliance with the grade and size requirements of paragraph (b) of this section shall be determined on the basis of an inspection and certification by a USDA inspector.

(2) *Inspection.* Inspection shall be performed by USDA inspectors in accordance with the Regulations Governing the Inspection and Certification of Fresh Fruits and Vegetables and Related Products (7 CFR part 51). The cost of each such inspection and related certification shall be borne by the applicant. Whenever filberts are offered for inspection, the applicant shall furnish any labor and pay any costs incurred in moving and opening containers as may be necessary for proper sampling and inspection. The applicant shall also furnish the USDA inspector the entry number and such other identifying information for each lot as he may request. Inspection must be completed prior to the importation of filberts. The applicant should make advance arrangements with the USDA inspection office to avoid delay in scheduling the inspection.

(3) *Certification.* Each lot of filberts inspected in accordance with paragraph (c)(1) of this section shall be covered by an inspection certificate. Each such certificate shall set forth, among other things, the following:

(i) The date and place of inspection.

(ii) The name of the applicant.

(iii) The name of the importer.

(iv) The Customs entry number pertaining to the lot or shipment covered by the certificate;

(v) The quantity, and identifying marks of the lot inspected.

(vi) The statement, if applicable: "Meets U.S. import requirements under section 8e of the AMA Act of 1937".

(vii) If the lot fails to meet the import requirements, a statement to that effect and the reasons therefor.

(d) *Exemptions.* Notwithstanding any other provisions of this section, the importation of any lot of filberts which does not exceed 115 pounds in net weight shall be exempt from the requirements of this section.

(e) *Reconditioning prior to importation.* Nothing contained in this section shall be deemed to preclude reconditioning

filberts prior to importation, in order that such filberts may be made eligible to meet the applicable grade and size regulations prescribed in paragraph (b) of this section.

(f) *Other restrictions.* The provisions of this section do not supersede the Federal Plant Quarantine Act of 1912, the Federal Food, Drug, and Cosmetic Act, or any other applicable laws or regulations or the need to comply with applicable food and sanitary regulations of city, county, State or Federal agencies.

(g) *Compliance.* Any person who violates any provision of this section shall be subject to a forfeiture in the amount prescribed in section 8a(5) of the Agricultural Marketing Agreement Act of 1937, as amended (sections 1–19, 48 Stat. 31, as amended; 7 U.S.C. 601–674), or, upon conviction, a penalty in the amount prescribed in section 8c(14) of said act, or to both such forfeiture and penalty. False representations to any agency of the United States on any matter within its jurisdiction, knowing it to be false, is a violation of 18 U.S.C. 1001 which provides for a fine or imprisonment or both.

EXHIBIT A

GRADE REQUIREMENTS FOR SHELLED FILBERTS

Filbert kernels or portions of filbert kernels shall meet the following requirements:

(1) Well dried and clean;

(2) Free from foreign material, mold, rancidity, decay or insect injury; and

(3) Free from serious damage caused by serious shriveling, or other means.

TOLERANCES

In order to allow for variations incident to proper grading and handling the following tolerances, by weight, are permitted as specified:

(1) For Foreign Material: 0.02 of one percent, for foreign material.

(2) For Defects: Five percent for kernels or portions of kernels which are below the requirements of this grade, including not more than the following: Two percent for mold, rancidity, decay or insect injury: *Provided,* That not more than one percent shall be for mold, rancidity, or insect injury.

DEFINITIONS

(1) *Well dried* means that the kernels are firm and crisp, not containing more than 6 percent moisture.

(2) *Clean* means practically free from plainly visible adhering dirt or other foreign material.

(3) *Foreign material* means any substance other than the filbert kernels, or portions of kernels. (Loose skins, pellicles or corky tissue which have become separated from the kernels shall not be considered as foreign material, provided that this material does not exceed .02 of one percent by weight.)

(4) *Serious damage* means any specific defect described in this section, or any equally objectionable variation of any one of these defects, or any other defects, or any combination of defects, which seriously detracts from the appearance or the edible or marketing quality of the individual portion of the kernel or of the lot as a whole. The following defects shall be considered as serious damage.

(i) *Serious shriveling* means when the kernel is seriously shrunken, wrinkled and tough.

(ii) *Mold* means that there is a visible growth of mold either on the outside or inside of the kernel.

(iii) *Rancidity* means that the kernel is noticeably rancid to the taste. An oily appearance of the flesh does not necessarily indicate a rancid condition.

(iv) *Decay* means that any portion of the kernel is decomposed.

(v) *Insect injury* means that the insect, frass or web is present, or the kernel or portion of kernel show definite evidence of insect feeding.

[42 FR 64899, Dec. 29, 1977, as amended at 45 FR 63482, Sept. 25, 1980; 47 FR 12612, Mar. 24, 1982; 48 FR 34015, July 27, 1983; 74 FR 2809, Jan. 16, 2009]

§ **999.500 Safeguard procedures for walnuts, dates, pistachios, and raisins exempt from grade, size, quality, and maturity requirements.**

(a) Each person who imports or receives any of the commodities listed in paragraphs (a)(1) through (4) of this section shall file (electronically or paper) an "Importer's Exempt Commodity Form" (FV–6) with the Marketing Order and Agreement Division, Fruit and Vegetable Program, AMS,

USDA. A "person who imports" may include a customs broker, acting as an importer's representative (hereinafter referred to as "importer"). A copy of the form (electronic or paper) shall be provided to the U.S. Customs and Border Protection. If a paper form is used, a copy of the form shall accompany the lot to the exempt outlet specified on the form. Any lot of any commodity offered for inspection or aflatoxin testing and, all or a portion thereof, subsequently imported as exempt under this provision shall also be reported on an FV–6. Such form (electronic or paper) shall be provided to the Marketing Order and Agreement Division in accordance with paragraph (d) of this section. The applicable commodities are:

(1) Dates which are donated to needy persons, prisoners or Native Americans on reservations; dates for processing; dates prepared or preserved; or dates for packaging or dates in retail packages that fail to meet grade, size, quality, and maturity requirements and are reclassified as dates for processing;

(2) Walnuts which are: Green walnuts (so immature that they cannot be used for drying and sale as dried walnuts); walnuts used in non-competitive outlets such as use by charitable institutions, relief agencies, governmental agencies for school lunch programs, and diversion to animal feed or oil manufacture;

(3) Substandard pistachios which are for non-human consumption purposes; or

(4) Raisins which do not meet grade and size requirements and are used in the production of alcohol, or syrup for industrial use, or which do not meet grade requirements with respect to mechanical damage or sugaring and are used in the production of raisin paste.

(b) *Certification of exempt use.* (1) Each importer of an exempt commodity as specified in paragraph (a) of this section shall certify on the FV–6 form (electronic or paper) as to the intended exempt outlet (*e.g.*, processing, charity, livestock feed). If certification is made using a paper FV–6 form, the importer shall provide a handwritten signature on the form.

(2) Each receiver of an exempt commodity as specified in paragraph (a) of this section shall also receive a copy of

the associated FV–6 form (electronic or paper) filed by the importer. Within two days of receipt of the exempt lot, the receiver shall certify on the form (electronic or paper) that such lot has been received and will be utilized in the exempt outlet as certified by the importer. If certification is made using a paper FV–6 form, the receiver shall provide a handwritten signature on the form.

(c) It is the responsibility of the importer to notify the Marketing Order and Agreement Division of any lot of exempt commodity rejected by a receiver, shipped to an alternative exempt receiver, exported, or otherwise disposed of. In such cases, a second FV–6 form must be filed by the importer, providing sufficient information to determine ultimate disposition of the exempt lot, and such disposition shall be so certified by the final receiver.

(d) All FV–6 forms and other correspondence regarding entry of exempt commodities must be submitted electronically, by mail, or by fax to the Marketing Order and Agreement Division, Fruit and Vegetable Program, AMS, USDA, 1400 Independence Avenue SW, STOP 0237, Washington, DC 20250–0237; telephone (202) 720–2491; email *ComplianceInfo@ams.usda.gov;* or fax (202) 720–5698.

[80 FR 15678, Mar. 25, 2015]

§ 999.600 **Regulation governing the importation of pistachios.**

(a) *Definitions.* As used in this part:

(1) *Aflatoxin* is one of a group of mycotoxins produced by the molds *Aspergillus flavus* and *Aspergillus parasiticus.* Aflatoxins are naturally occurring compounds produced by molds, which can be spread in improperly processed and stored nuts, dried fruits, and grains.

(2) *Aflatoxin inspection certificate* means a certificate issued by a USDA or USDA-accredited laboratory.

(3) *Certified lots of pistachios* are those for which aflatoxin inspection certificates have been issued.

(4) *Customs* means the U.S. Customs and Border Protection.

(5) *Importation of pistachios* means the release of pistachios from the custody of U.S. Customs and Border Protection.

(6) *Importer* means a person who engages in the importation of pistachios into the United States.

(7) *Inshell pistachios* means pistachios that have shells that have not been removed.

(8) *Inspection Service* means the Federal Inspection Service, Fruit and Vegetable Programs, Agricultural Marketing Service, USDA, or the Federal-State Inspection Programs.

(9) *Inspector* means any inspector authorized by USDA to draw and prepare pistachio samples.

(10) *Lot* means any quantity of pistachios that is submitted for testing purposes under this part.

(11) *Person* means an individual, partnership, limited-liability corporation, corporation, trust, association, or any other business unit.

(12) *Pistachio* means the nut of the pistachio tree, *Pistachia vera,* whether inshell or shelled.

(13) *Secretary* means the Secretary of Agriculture of the United States or any officer or employee of the United States Department of Agriculture who is, or who may hereafter be, authorized to act in his/her stead.

(14) *Shelled pistachios* means pistachio kernels, or portions of kernels, after the pistachio shells have been removed.

(15) *Substandard pistachios* means pistachios, inshell or shelled, that do not comply with the aflatoxin regulations of this section.

(16) *USDA* means the United States Department of Agriculture, including any officer, employee, service, program, or branch of the Department of Agriculture, or any other person acting as the Secretary's agent or representative in connection with any provisions of this section.

(17) *USDA laboratory* means laboratories of the Science and Technology Programs, Agricultural Marketing Service, USDA, that perform chemical analyses of pistachios for aflatoxin content.

(18) *USDA-accredited laboratory* means a laboratory that has been approved or accredited by the U.S. Department of Agriculture to perform chemical analyses of pistachios for aflatoxin content.

(b) *Importation Requirements.* The importation of any lot of pistachios for human consumption is prohibited unless it meets the requirements contained in this section, which are determined to be the same as or comparable to those imposed upon domestic pistachios handled pursuant to Order No. 983, as amended (part 983 of this chapter).

(c) *Maximum aflatoxin tolerance.* No importer shall ship for domestic human consumption lots of pistachios that exceed an aflatoxin level of 15 ppb. Compliance with the aflatoxin requirements of this section shall be determined upon the basis of sampling by a USDA-authorized inspector and testing by a USDA or USDA-accredited laboratory. All shipments must be covered by an aflatoxin inspection certificate issued by the laboratory. Testing and certification must be completed prior to the importation of pistachios.

(d) *Sampling.* (1) All sampling for aflatoxin testing shall be performed by USDA-authorized inspectors in accordance with USDA rules and regulations governing the inspection and certification of fresh fruits, vegetables, and other products (7 CFR part 51). The cost of each such sampling and related certification shall be borne by the importer. Whenever pistachios are offered for sampling and testing, the importer shall furnish any labor and pay any costs incurred for storing, moving, and opening containers as may be necessary for proper sampling and testing. The importer shall furnish the USDA inspector with the customs entry number and such other identifying information for each lot as he or she may request. Importers may make arrangements for required sampling by contacting the Inspection Service office closest to where the pistachios will be made available for sampling. For questions regarding sampling, a list of Federal or Federal-State Inspection Program offices, or for further assistance, importers may contact: Specialty Crops Inspection Division, Specialty Crops Program, AMS, USDA, 1400 Independence Avenue SW., Room 1536–S, Washington, DC 20250; Telephone: (202) 720–5870; Fax: (202) 720–0393.

(2) Lot samples shall be drawn from each lot of pistachios designated for aflatoxin testing, and individual test samples shall be prepared by, or under the supervision of, an inspector. Each sample shall be drawn and prepared in accordance with the sample size requirements outlined in Tables 1 and 2 below. The gross weight of the inshell lot and test samples for aflatoxin testing and the minimum number of incremental samples required are shown in Table 1. The gross weight of the kernel lot and test samples for aflatoxin testing and the minimum number of incremental samples required is shown in Table 2. If more than one test sample is necessary, the test samples shall be designated by the inspector as Test Sample #1 and Test Sample #2. Each sample shall be placed in a suitable container, with the lot number clearly identified, and the importer shall submit it, along with a copy of the customs entry documentation, to a USDA or USDA-accredited laboratory. The importer shall assume all costs for shipping samples to the laboratory.

TABLE 1—INSHELL PISTACHIO LOT SAMPLING INCREMENTS FOR AFLATOXIN CERTIFICATION

Lot weight (lbs.)	Minimum number of incremental samples for the lot sample	Total weight of lot sample (kilograms)	Weight of test sample (kilograms)
220 or less	10	2.0	2.0
221–440	15	3.0	3.0
441–1,100	20	4.0	4.0
1,101–2,200	30	6.0	6.0
2,201–4,400	40	8.0	8.0
4,401–11,000	60	12.0	6.0
11,001–22,000	80	16.0	8.0
22,001–150,000	100	20.0	10.0

TABLE 2—SHELLED PISTACHIO KERNEL LOT SAMPLING INCREMENTS FOR AFLATOXIN CERTIFICATION

Lot weight (lbs.)	Minimum number of incremental samples for the lot sample	Total weight of lot sample (kilograms)	Weight of test sample (kilograms)
220 or less	10	1.0	1.0
221–440	15	1.5	1.5
441–1,100	20	2.0	2.0
1,101–2,200	30	3.0	3.0
2,201–4,400	40	4.0	4.0
4,401–11,000	60	6.0	3.0
11,001–22,000	80	8.0	4.0
22,001–150,000	100	10.0	5.0

(e) *Aflatoxin testing.* Importers may make arrangements for required chemical analysis for aflatoxin content at the nearest USDA or USDA-accredited laboratory. For further information concerning chemical analysis and a list of laboratories authorized to conduct such analysis contact: Science and Technology Programs, AMS, USDA, 1400 Independence Avenue SW., STOP 0270, Washington, DC 20250–0270; Telephone: (202) 720–5231; Fax: (202) 720–6496.

(1) Aflatoxin test samples shall be received and logged by a USDA or USDA-accredited laboratory, and each test sample shall be prepared and analyzed using High Pressure Liquid Chromatography (HPLC) or the AOAC-approved immunoaffinity column with direct fluorometry method. The aflatoxin level shall be calculated on a kernel weight basis.

(2) Lots that require a single test sample will be certified as "negative" on the aflatoxin inspection certificate if the sample has an aflatoxin level at or below 15 ppb. If the aflatoxin level is above 15 ppb, the lot fails and the laboratory shall fill out an *Imported Pistachios—Failed Lot Notification* report (Form FV–249) as described in paragraph (h)(1) of this section.

(3) Lots that require two test samples will be certified as "negative" on the aflatoxin inspection certificate if Test Sample #1 has an aflatoxin level at or below 10 ppb. If the aflatoxin level of Test Sample #1 is above 20 ppb, the lot fails and the laboratory shall fill out an *Imported Pistachios—Failed Lot Notification* report (Form FV–249). If the aflatoxin level of Test Sample #1 is above 10 ppb and at or below 20 ppb, the laboratory may, at the importer's discretion, analyze Test Sample #2 and

average the test results of Test Samples #1 and #2. Alternately, the importer may elect to withdraw the lot from testing, rework the lot, and resubmit it for testing after reworking. If the importer directs the laboratory to proceed with the analysis of Test Sample #2, a lot will be certified as negative to aflatoxin and the laboratory shall issue an aflatoxin inspection certificate if the averaged result of Test Samples #1 and #2 is at or below 15 ppb. If the average aflatoxin level of Test Samples #1 and #2 is above 15 ppb, the lot fails and the laboratory shall fill out an *Imported Pistachios—Failed Lot Notification* report (Form FV–249).

(4) If an importer does not elect to use Test Sample #2 for certification purposes, the importer may request that the laboratory return the sample to the importer.

(f) *Certification.* Each lot of pistachios sampled and tested in accordance with paragraphs (d) and (e) of this section shall be covered by an aflatoxin inspection certificate completed by the laboratory. The certification expires for the lot or remainder of the lot after 12 months. Each such certificate shall set forth the following:

(1) The date and place of sampling and testing.

(2) The name of the applicant.

(3) The Customs entry number pertaining to the lot or shipment covered by the certificate.

(4) The quantity and identifying marks of the lot tested.

(5) The aflatoxin level of the lot, stated on a kernel weight basis.

(6) The statement, if applicable: "Meets U.S. import requirements under section 8e of the AMA Act of 1937."

(7) If the lot fails to meet the import requirements, a statement to that effect and the reasons therefore.

(g) *Failed lots/rework procedure.* Any lot or portion thereof that fails to meet the import requirements prior to or after reconditioning may be exported, sold for non-human consumption, or disposed of under the supervision of Customs and, if necessary for verification purposes, the Federal or Federal-State Inspection Programs, with the costs of certifying the disposal of such lot paid by the importer.

(1) *Inshell rework procedure for aflatoxin.* If inshell rework is selected as a remedy to meet the aflatoxin requirements of this part, then 100 percent of the product within that lot shall be removed from the bulk and/or retail packaging containers and reworked to remove the portion of the lot that caused the failure. Reworking shall consist of mechanical, electronic, or manual procedures normally used in the handling of pistachios. After the rework procedure has been completed, the total weight of the accepted product and the total weight of the rejected product shall be reported by the importer to Customs and USDA on an *Imported Pistachios—Rework and Failed Lot Disposition* report (Form FV–251) as described in paragraph (h)(2) of this section. The reworked lot shall be sampled and tested for aflatoxin as specified in paragraphs (d) and (e) of this section, except that the lot sample size and the test sample size shall be doubled. If, after the lot has been reworked and tested, it fails the aflatoxin test for a second time, the lot may be shelled and the kernels reworked, sampled, and tested in the manner specified for an original lot of kernels, or the failed lot may be exported, used for non-human consumption, or otherwise disposed of.

(2) *Kernel rework procedure for aflatoxin.* If pistachio kernel rework is selected as a remedy to meet the aflatoxin requirements of this part, then 100 percent of the product within that lot shall be removed from the bulk and/or retail packaging containers and reworked to remove the portion of the lot that caused the failure. Reworking shall consist of mechanical, electronic, or manual procedures normally used in the handling of pistachios. After the

rework procedure has been completed the total weight of the accepted product and the total weight of the rejected product shall be reported to Customs and USDA on an *Imported Pistachios—Rework and Failed Lot Disposition* report (Form FV–251). The reworked lot shall be sampled and tested for aflatoxin as specified in paragraphs (d) and (e) of this section.

(3) *Failed lot reporting.* If a lot fails to meet the aflatoxin requirements of this part, the testing laboratory shall complete an *Imported Pistachios—Failed Lot Notification* report (Form FV–249) as described in paragraph (h)(1) of this section, and shall submit it to Customs, the importer, and USDA within 10 working days of the test failure. This form must be completed and submitted each time a lot fails aflatoxin testing.

(h) *Reports and Recordkeeping.*

(1) *Form FV–249 Imported Pistachios— Failed Lot Notification.* Each USDA or USDA-accredited laboratory shall notify the importer; Customs; and the Marketing Order and Agreement Division, Fruit and Vegetable Programs, AMS, USDA; of all lots that fail to meet the maximum aflatoxin requirements by completing this form and submitting it within 10 days of failed aflatoxin testing.

(2) *Form FV–251 Imported Pistachios— Rework and Failed Lot Disposition.* Each importer who reworks a failing lot of pistachios shall complete this report and shall forward it to Customs and the Marketing Order and Agreement Division, Fruit and Vegetable Programs, AMS, USDA, no later than 10 days after the rework is completed. If rework is not selected as a remedy, the importer shall complete and submit this form within 10 days of alternate disposition of the lot.

(i) *Exemptions.* Any importer may import pistachios free of the requirements of this section if such importer imports a quantity not exceeding a total of 5,000 dried pounds between September 1 and August 31 of each year. Substandard pistachios imported for use in non-human consumption outlets shall be subject to the safeguard provisions contained in § 999.500.

(j) *Reconditioning prior to importation.* Nothing contained in this section shall be deemed to preclude reconditioning pistachios prior to importation, in order that such pistachios may be made eligible to meet the applicable aflatoxin regulations prescribed in paragraphs (c) through (f) of this section.

(k) *Comingling.* Certified lots of pistachios may be comingled with other certified lots, but the comingling of certified lots and uncertified lots shall cause the loss of certification for the comingled lots.

(1) *Retesting.* Whenever USDA has reason to believe that imported pistachios may have been damaged or deteriorated while in storage, USDA may reject the then effective inspection certificate and may require the owner of the pistachios to have them retested to establish whether or not such pistachios may be shipped for human consumption.

(m) *Compliance.* Any person who violates any provision of this section shall be subject to a forfeiture in the amount prescribed in section 8a(5) of the Agricultural Marketing Agreement Act of 1937, as amended (7 U.S.C. 601–674), or, upon conviction, a penalty in the amount prescribed in section 8c(14) of the said Act, or to both such forfeiture and penalty. False representation to any agency of the United States on any matter within its jurisdiction, knowing it to be false, is a violation of 18 U.S.C. 1001, which provides for a fine or imprisonments or both.

(n) *Other import requirements.* The provisions of this section do not supersede any restrictions or prohibitions on pistachios under the Federal Plant Quarantine Act of 1912, or any other applicable laws or regulations of city, county, State, or Federal Agencies including the Federal Food, Drug and Cosmetic Act.

[77 FR 51691, Aug. 27, 2012, as amended at 81 FR 87412, Dec. 5, 2016]

FINDING AIDS

A list of CFR titles, subtitles, chapters, subchapters and parts and an alphabetical list of agencies publishing in the CFR are included in the CFR Index and Finding Aids volume to the Code of Federal Regulations which is published separately and revised annually.

Table of CFR Titles and Chapters

(Revised as of January 1, 2020)

Title 1—General Provisions

Title 2—Grants and Agreements

Title 2—Grants and Agreements—Continued

Title 3—The President

Title 4—Accounts

Title 5—Administrative Personnel

Title 5—Administrative Personnel—Continued

701

Title 8—Aliens and Nationality

Title 9—Animals and Animal Products

Title 10—Energy

Title 11—Federal Elections

Title 12—Banks and Banking

Title 20—Employees' Benefits—Continued

Title 21—Food and Drugs

Title 22—Foreign Relations

Title 23—Highways

Title 25—Indians

Title 26—Internal Revenue

Title 27—Alcohol, Tobacco Products and Firearms

Title 28—Judicial Administration

Title 29—Labor

Title 29—Labor—Continued

Title 30—Mineral Resources

Title 31—Money and Finance: Treasury

Title 32—National Defense

Title 33—Navigation and Navigable Waters

Title 34—Education

Title 34—Education—Continued

Title 35 [Reserved]

Title 36—Parks, Forests, and Public Property

Title 37—Patents, Trademarks, and Copyrights

712

Title 38—Pensions, Bonuses, and Veterans' Relief

Title 39—Postal Service

Title 40—Protection of Environment

Title 41—Public Contracts and Property Management

Title 41—Public Contracts and Property Management—Continued

Title 42—Public Health

Title 43—Public Lands: Interior

Title 44—Emergency Management and Assistance

Title 45—Public Welfare

Title 45—Public Welfare—Continued

Title 46—Shipping

Title 47—Telecommunication

Title 48—Federal Acquisition Regulations System

Title 49—Transportation

Title 50—Wildlife and Fisheries

Alphabetical List of Agencies Appearing in the CFR

(Revised as of January 1, 2020)

724

List of CFR Sections Affected

All changes in this volume of the Code of Federal Regulations (CFR) that were made by documents published in the FEDERAL REGISTER since January 1, 2015 are enumerated in the following list. Entries indicate the nature of the changes effected. Page numbers refer to FEDERAL REGISTER pages. The user should consult the entries for chapters, parts and subparts as well as sections for revisions.

For changes to this volume of the CFR prior to this listing, consult the annual edition of the monthly List of CFR Sections Affected (LSA). The LSA is available at *www.govinfo.gov*. For changes to this volume of the CFR prior to 2001, see the "List of CFR Sections Affected, 1949–1963, 1964–1972, 1973–1985, and 1986–2000" published in 11 separate volumes. The "List of CFR Sections Affected 1986–2000" is available at *www.govinfo.gov*.

2015

7 CFR—Continued

7 CFR—Continued

○